Intermediate Algebra

Mark D. Turner
Charles P. McKeague

xyz textbooks

Intermediate Algebra

Mark D. Turner
Charles P. McKeague

Publisher: XYZ Textbooks

Project Managers: Jennifer Thomas
Katherine Heistand Shields

Composition: Katherine Heistand Shields,
Matthew Hoy, Jennifer Thomas

Sales: Amy Jacobs, Richard Jones, Bruce Spears

Cover Design: Kyle Schoenberger

© 2016 by McKeague Textbooks, LLC

ISBN-13: 978-1-63098-050-4 / ISBN-10: 1-63098-050-1

For product information and technology assistance, contact us at
XYZ Textbooks, 1-877-745-3499

For permission to use material from this text or product,
e-mail: **info@mathtv.com**

XYZ Textbooks
1339 Marsh Street
San Luis Obispo, CA 93401
USA

For your course and learning solutions, visit www.xyztextbooks.com

Brief Contents

Contents

10 Sequences and Series   743

11 Conic Sections   793

Preface: A Note to Instructors

Intermediate Algebra
by Mark D. Turner and Charles P. McKeague

Description

Intermediate Algebra fits the traditional, one-semester, intermediate algebra course. The prerequisite is Elementary Algebra in college, or Algebra I in high school.

The normal sequence of topics is enriched with modeling applications, study skills, and other built-in features such as the "Spotlight on Success" (as detailed below). Encourage your students to make regular use of these features— you'll find that they will gradually build a foundation of successful studying practices that will benefit them in all their future courses.

This textbook is more than just the book itself. We built this book to be the hub of a math "toolbox" of sorts. While most students still prefer to use the printed book in their studies, the eBook extends the reach of the book, giving students and instructors access to a wide array of supporting tools:

- The eBook includes free access to the Elementary Algebra eBook, plus over 20 eBooks covering 8 math courses—great for remediation.
- It also includes free access to 10,000 MathTV videos, with 3-4 tutorials for every single example.
- Plus all of the accompanying worksheets and digital supplements for the book.
- Additionally, QR code technology connects the printed textbook directly to the digital resources through the students' smartphones and tablets.
- The associated Matched Problems Worksheets gives you the opportunity to flip your classroom, without the need to create new materials yourself.

Textbook Features

Every section has been written so that it can be discussed in a single class session. The clean layout and conversational style used by the authors make it easy for students to read, and important information, such as definitions and properties, are highlighted so that they can easily be located and referenced by students.

In addition, the following features provide both instructors and students a vast array of resources which can be used to enrich the learning environment and promote student success.

Chapter Introductions   We begin each chapter with a brief application that involves some of the concepts to be presented later in the chapter. These "glimpses" of what is to come help motivate students by showing them an interesting, real-world scenario that requires the knowledge and skills they will be attaining. Acting as a mini-theme for the chapter, students will encounter examples and problems in the following sections that directly relate to this application.

Learning Objectives   Every section now begins with a short set of learning objectives, listing the specific, measurable knowledge and skills students will be acquiring in that section. Objectives help students identify and focus on the important concepts in each section and increase the likelihood of their success by having established and clear goals. For instructors, the objectives can help in organizing class lessons and learning activities, and in creating student assessments.

Getting Ready for Class   It's always a challenge to get students to read, especially in math. To encourage them, we place four key questions at the end of every section, under the heading "Getting Ready for Class." If students have read the section, they should be able to answer each of these with ease. Even a minimal attempt to answer these questions can enhance the students' in-class experience.

We've heard some innovative strategies from early adopters of the textbook. Instructors in one department use these questions as a student's "ticket" to class. They collect the answers as students come in the door. A simple action that can start the class off right!

Getting Ready for the Next Section   When students finish a section, they can feel a great sense of accomplishment. We want to help maintain their momentum as much as possible, which is why we offer a brief preview of the next stop in their mathematical journey. A small set of "Getting Ready for the Next Section" problems appear near the end of every problem set. These are the exact same problems that students will see when they read through the next section of the text. Students who consistently work these problems will be much better prepared for class.

Learning Objectives Assessments   Every problem set concludes with a short section of multiple-choice questions that can be used by students or instructors to measure the extent to which students have met the learning objectives for that section. These questions can easily be adapted for use with clicker technology, or used as a quick exit quiz at the close of each class session, to verify that students understand the main points from the lesson.

Paying Attention to Instructions   Students tend to get stuck making the same mistakes over and over again, developing bad habits. The more time you spend teaching, the more you can start to anticipate these problems. For example, we know that students don't always pay attention to instructions when they are doing their homework, and it can get them into trouble on tests and quizzes.

The only way to fix this is to address it head-on. Our strategy was to build problems into the text itself, under the heading "Paying Attention to Instructions," that challenge students to carefully read the instructions for each problem. Small nuances make a big difference in mathematics, and we hope that this feature will help you demonstrate to your students just how important this is.

"How To" Segments   Many sections include a "How To" segment that outlines the steps in the method or process used to solve certain types of problems. These summaries help students internalize the particular problem solving strategy introduced in that section.

Spotlight on Success   Each student has a unique approach to learning. A one-size-fits-all strategy doesn't work, which is why our MathTV videos have always used multiple tutors for each example. We want to give every student the chance to succeed!

The "Spotlight on Success" feature is designed in the same spirit. Scattered throughout the text, this feature offers students a variety of strategies from many different sources. Some are from peers, others are from instructors. Many of the spotlights feature the same peer tutors that students will see on the MathTV videos that accompany the text.

Real-Data Application Problems   Students do better in math when they see its application to real-world problems. That's why we've included as many applied problems as we can. Many times the charts and graphics in the text look like the types of charts and graphics students see in the media.

Not only does this help them make the connection, but it can give these a greater mathematical "sense," and even make the concepts a little easier to understand.

Facts from Geometry   Anything we can do to help solidify abstract concepts is a good thing. Geometry can give students another view of the problem, and potentially, another avenue for understanding algebraic concepts. Students who are visual learners will love this feature.

Chapter Summaries and Tests   Every chapter concludes with a Chapter Summary and Chapter Test. The chapter summary lists the main properties and definitions found in the chapter, with examples given in the margin. The chapter test provides a representative sample of the various types of problems students have encountered in the chapter. These features are valuable assets to students in preparing for exams or refreshing their skills with previously learned concepts.

Connecting Print and Digital   We want students to get the most out of their course materials, which is why we think of the textbook as "toolbox" for students. QR codes are integrated throughout the textbook, to connect the printed version to the digital assets. As they read their printed book, support material is quickly and conveniently available via one scan of the accompanying QR code. No hunting and searching and scrolling—a direct link to additional support.

Supplements for the Instructor

Please contact your sales representative—or see xyztextbooks.com/instructors for more info.

MathTV.com Every example in every XYZ Textbook is worked on video by multiple student instructors. Students benefit from seeing multiple approaches, and gain confidence in learning from their peers. The MathTV library contains over 10,000 videos, from basic math through calculus. It's great for learning the material at hand, and for remediation too.

Complete Solutions Manual Available online, this manual contains complete solutions to all the exercises in the problem sets.

Printable Test Items Choose from a bank of pre-created tests for most of our textbooks.

Supplements for the Student

MathTV.com Students have access to math instruction 24 hours a day, seven days a week. Assistance with any problem or subject is never more than a few clicks away.

XYZ eBook This textbook is available online for both instructors and students. Tightly integrated with MathTV.com, students can read the book and watch videos of the author and peer instructors explaining each example. Access to the online book is available free with the purchase of a new book.

QR Codes QR Codes connect print to digital! Scan the QR codes located inside printed textbooks* with your mobile device, and it will take you directly to the accompanying MathTV video.

*QR codes are not available in every textbook.

Student Solutions Manual Contains complete solutions to all the odd-numbered exercises in the text. Available for purchase separately.

A Note from Charles P. McKeague

I am extremely pleased to have Mark Turner as my co-author on this book. We have worked together for years—Mark is the co-author of the last four editions of my trigonometry book. He is an award-winning teacher and, as you will see, an excellent writer. He is an innovative instructor, bringing many new ideas into the classroom—and he gives regular presentations to share his knowledge and strategies with other instructors.

There are many things I like about working with Mark. First, and maybe most important, is that the integration of Mark's writing style with mine is seamless. When I read over our material, I often cannot tell which one of us wrote it. Mark also has an attention to detail and an eye for consistency that very few authors have. I know this is a better book because of Mark's contributions to it.

Acknowledgements

Many thanks to the following instructors who participated as reviewers during the production of this book. Your suggestions were invaluable to us!

David Adams, *San Diego Mesa College*
Lill Birdsall, *American River College*
Yvette Butterworth, *Foothill & Canada Colleges*
James A Edson, *Palo Alto College*
Patricia Elliott-Traficante, *Holyoke Community College*
Beth Hentges, *Century College*
Calvin Gatson, *Alabama State University*
Jeffrey M. Gervasi, *Cuesta College*
Barbara Granger, *Holyoke Community College*
Michael Headley, *Metropolitan Community College Kansas City*
Richard Kaiser, *Solano Community College*
Dr. Joe Kemble, *Lamar University*
Honey Kirk, *Palo Alto College*
Alexander Kolesnik, *Ventura College*
Chaitanya Mistry, *SUNY Ulster County Community College*
John Petru, *Northwest Vista College*
Rick Ponticelli, *North Shore Community College and Salem State University*
Kathy Renfro, *Lakeland Community College*
Elgin Schilhab, *Austin Community College, Highland Campus*

Preface:
A Note to Students

We want you to succeed.

Welcome to the XYZ Textbooks/MathTV community. We want you to succeed in this course. As you will see as you progress through this book, and access the other tools we have for you, we are different that other publishers. Here's how:

Our Authors are Real Teachers The best textbooks are written by the best teachers, period. We select our authors based on one factor only: Can they teach? All of our authors have won awards for teaching. The result is the best instruction you can get in written form, produced by award-winning, experienced instructors.

Innovative Products The foundation of our products is the textbook, which is the hub for all the resources you will need to do well in your math course. These include eBooks, videos, worksheets, and a variety of ways to access these resources, from QR codes built into our books, to our MathTV Mobile site.

Peer Tutors Math can be difficult. Sometimes your class time and textbook are not enough. We understand that, which is why we created MathTV, providing you with a set of instructional videos by students just like you, who have found a way to master the material you are studying. You'll get to see how your peers solve each problem, sometimes offering a different view from how an instructor solves the problem, and other times, solving the problem in the same way your instructor does, giving you confidence that that is the way the problem should be solved. It also means that you can get help anytime (not just during office hours).

Fair Prices We're small, independently owned and independently run. Why does that matter to you? Because we do not have the overhead and expenses that the larger publishers have. Yes, we want to be a profitable business, but we believe that we can keep our prices reasonable and still give you everything you need to be successful. Also, we want you to use this book, and the best way to make sure that happens is to make it affordable.

Unlimited Access When you purchase one of our products, we give you access to all of our products. Why? Because everything you need to know about math is not contained in one book. Suppose you need to review a topic from a math course you completed previously? No problem. Suppose you want to see an alternate approach? It's all yours. As a member of our XYZ Textbooks/MathTV community, you have access to everything we produce, including all our eBooks.

We know you can do it.

We Believe in You We have seen students with all varieties of backgrounds and levels in mathematics do well in the courses we supply books and materials for. In fact, we have never run across a student that could not be successful in algebra. And that carries over to you: We believe in you; we believe you can be successful in whatever math class you are taking. Our job is to supply you with the tools you can use to attain success, you supply the drive and ambition.

We Know College can be Difficult It is not always the material you are studying that makes college difficult. We know that many of you are working, some part time, some full time. We know many of you have families to support, or look after. We understand that your time can be limited. We take all this to heart when we create the materials we think you will need. For example, we make our videos available on your smart phone, tablet, and on the Internet. That way, no matter where you are, you will have access to help when you get stuck on a problem.

We Believe in what We Do We know you will see the value in the things we have created. That's why the first chapter in every one of our eBooks is free, and so are all the resources that come with it. We want you to try us out for free. See what you think. We wouldn't do that if we didn't believe in what we do here.

Here are some strategies to help.

I often find my students getting frustrated and asking themselves the question, "Why can't I understand this stuff the first time through?" The answer is, "You're not expected to."

Learning a topic in mathematics isn't always accomplished the first time through the material. If you don't understand a topic the first time you see it, that's perfectly normal.

Stick with it. Understanding mathematics takes time. You may find that you need to read over new material a number of times before you can begin to work problems. The process of understanding requires reading the book, studying the examples, working problems, and getting your questions answered.

How to Be Successful in Mathematics

1. **If you are in a lecture class, be sure to attend all class sessions on time.** You simply will not know exactly what went on in class unless you were there. Missing class and then expecting to find out what went on from someone else is not a good strategy. Make the time to be there—and to be attentive.

2. **Read the book.** It is best to read the section that will be covered in class beforehand. It's OK if you don't fully understand everything you read! Reading in advance at least gives you a sense of what will be discussed, which puts you in a good position when you get to class.

3. **Work problems every day and check your answers.** The secret to success in mathematics is working problems. The more problems you work, the better you will perform. It's really that simple. The answers to the odd-numbered problems are given in the back of the book. When you have finished an assignment, be sure to compare your answers with those in the book. If you have made a mistake, find out what it is, and try to correct it.

4. **Do it on your own.** Having someone else show you how to work a problem is not the same as working the problem yourself. It is absolutely OK to get help when you are stuck. As a matter of fact, it is a good idea. Just be sure you do the work yourself. After all, when it's test time, it's all you! Get confident in every problem type, and you will do well.

5. **Review every day.** After you have finished the problems your instructor has assigned, take another 15 minutes and review a section you have already completed. This simple trick works wonders. The more you review, the longer you will retain the material you have learned. Since math topics build upon one another, this will help you throughout the term.

6. **Don't expect to understand every new topic the first time you see it.** Sometimes it will come easy and sometimes it won't. Don't beat yourself up over it—that's just the way things are in mathematics. It's perfectly normal. Expecting to understand each new topic the first time you see it can lead to disappointment and frustration. The process of understanding takes time and practice. It requires that you read the book, work problems, and get your questions answered.

7. **Spend as much time as it takes for you to master the material.** What's the exact amount of time you need to spend on mathematics to master it? There's no way to know except to do it. You will find out as you go what is or isn't enough time for you. Some sections may take less time, and some may take more. If you end up spending 2 or more hours on each section, OK. Then that's how much time it takes; trying to get by with less will not work.

8. **Relax.** It's probably not as difficult as you think. You might get stuck at points. That's OK, everyone does. Take a break if you need to. Seek some outside help. Watch a MathTV video of the problem. There is a solution, and you *will* find it—even if it takes a while.

Real Numbers and Algebraic Expressions

1

iStockphoto.com © dolgachov

We use mathematics to get a better understanding of the world we live in. One topic that concerns college students is the amount of money they pay for textbooks and other course materials. The information in the chart below was gathered by the National Association of College Stores, using a survey of undergraduate students at 23 universities. The table and chart below give a more detailed account of the data from the survey.

School Year	Average Spending on Course Materials
2007/08	$701
2009/10	$667
2011/12	$655
2012/13	$662
2013/14	$638
2014/15	$563

Average Spending on Course Materials —
$701 $667 $655 $662 $638 $563
2007/08 2009/10 2011/12 2012/13 2013/14 2014/15
Year

Source: National Association of College Stores

Reading tables and charts is a skill you need, if you want to understand some of the things happening around you. Studying mathematics helps give you experience with tables and charts, and will help you develop critical thinking skills that assist you in understanding the data in the tables and charts. For example, did you notice that two of the school years are missing from the chart? Can you think of a reason why those years are missing?

Later in this course, we will translate items in tables and charts into points on a coordinate plane. Once we have done that, we will be able to associate equations with the data in tables and charts. For example, an equation that would summarize some of the information in the chart above could look like this:

$$y = -19.7x + 701$$

where x is the number of school years since the 2007/08 school year. We could use this equation to predict the average spending of course materials for the 2017/18 school year by letting $x = 10$, and solving for y.

Success Skills

Some of the students enrolled in my college algebra classes develop difficulties early in the course. Their difficulties are not associated with their ability to learn mathematics; they all have the potential to pass the course. Students who get off to a poor start do so because they have not developed the study skills necessary to be successful in algebra. Here is a list of things you can do to begin to develop effective study skills.

1. **Put Yourself on a Schedule** The general rule is that you spend 2 hours on homework for every hour you are in class. Make a schedule for yourself in which you set aside 2 hours each day to work on algebra. Once you make the schedule, stick to it. Don't just complete your assignments and stop. Use all the time you have set aside. If you complete an assignment and have time left over, read the next section in the book, and then work more problems.

2. **Find Your Mistakes and Correct Them** There is more to studying algebra than just working problems. You must always check your answers with the answers in the back of the book. When you have made a mistake, find out what it is and correct it. Making mistakes is part of the process of learning mathematics. In the prologue to *The Book of Squares*, Leonardo Fibonacci (ca. 1170–ca. 1250) had this to say about the content of his book:

> I have come to request indulgence if in any place it contains something more or less than right or necessary; for to remember everything and be mistaken in nothing is divine rather than human . . .

Fibonacci knew, as you know, that human beings make mistakes. You cannot learn algebra without making mistakes.

3. **Gather Information on Available Resources** You need to anticipate that you will need extra help sometime during the course. One resource is your instructor; you need to know your instructor's office hours and where the office is located. Another resource is the math lab or study center, if they are available at your school. It also helps to have the phone numbers of other students in the class, in case you miss class. You want to anticipate that you will need these resources, so now is the time to gather them together.

The Real Numbers

Learning Objectives

In this section we will learn how to:

1. Locate and label points on the number line.

2. Change a fraction to an equivalent fraction with a new denominator.

3. Identify the opposite of a number.

4. Find the absolute value of a number.

5. Factor a number into a product of primes.

6. Identify the reciprocal of a number.

7. Find the perimeter and area of squares, rectangles, and triangles.

The Number Line

In this section, we start our work with negative numbers. To represent negative numbers in algebra, we use what is called the **real number line.** Here is how we construct a real number line: First, we draw a straight line and label a convenient point on the line with 0. Then we mark off equally spaced distances in both directions from 0. We label the points to the right of 0 with the numbers 1, 2, 3, . . . (the dots mean "and so on"). The points to the left of 0 we label in order, -1, -2, -3, ... Here is what it looks like:

Note: If there is no sign (+ or −) in front of a number, the number is assumed to be positive (+).

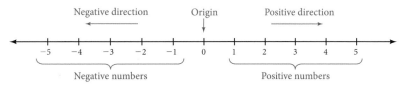

The numbers increase in value going from left to right. If we "move" to the right, we are moving in the positive direction. If we move to the left, we are moving in the negative direction. When we compare two numbers on the number line, the number on the left is always less than the number on the right. For instance, $-3 < -1$ because it is to the left of -1 on the number line.

NOTATION

We use the following inequality notation when comparing two numbers:

$a < b$ Indicates a lies to the left of b on the number line
In words: a is less than b.

$a > b$ Indicates a lies to the right of b on the number line
In words: a is greater than b.

We can extend our inequality notation to $a \leq b$, which we read *a is less than, or equal to, b.* And $a \geq b$ which we read as *a is greater than, or equal to, b.*

Note: There are other numbers on the number line that you may not be as familiar with. They are irrational numbers such as π, $\sqrt{2}$, $\sqrt{3}$. You will see these later in the book.

EXAMPLE 1   Locate and label the points on the real number line associated with the numbers -3.5, $-1\frac{1}{4}$, $\frac{1}{2}$, $\frac{3}{4}$, and 2.5.

SOLUTION   We draw a real number line from -4 to 4 and label the points in question.

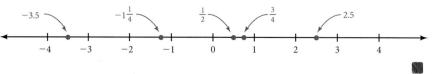

DEFINITION   *coordinate*

The number associated with a point on the real number line is called the **coordinate** of that point.

In the preceding example, the numbers $\frac{1}{2}$, $\frac{3}{4}$, 2.5, -3.5, and $-1\frac{1}{4}$ are the coordinates of the points they represent.

DEFINITION   *real numbers*

The numbers that can be represented with points on the real number line are called **real numbers**.

Real numbers include whole numbers, fractions, decimals, and other numbers that are not as familiar to us as these.

Equivalent Fractions on the Number Line

As we proceed through the chapter, from time to time we will review some of the major concepts associated with fractions. To begin, here is the formal definition of a fraction:

DEFINITION   *fraction*

If a and b are real numbers with $b \neq 0$, then the expression

$$\frac{a}{b}$$

is called a **fraction**. The top number a is called the **numerator**, and the bottom number b is called the **denominator**. The restriction $b \neq 0$ keeps us from writing an expression that is undefined. (As you will see later in this chapter, division by zero is not allowed.)

The number line can be used to visualize fractions. Recall that for the fraction $\frac{a}{b}$, a is called the numerator and b is called the denominator. The denominator indicates the number of equal parts in the interval from 0 to 1 on the number line. The numerator indicates how many of those parts we have. If we take that part of the number line from 0 to 1 and divide it into **three equal parts**, we say that we have divided it into ***thirds*** (Figure 1). Each of the three segments is $\frac{1}{3}$ (one third) of the whole segment from 0 to 1.

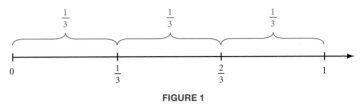

FIGURE 1

Two of these smaller segments together are $\frac{2}{3}$ (two thirds) of the whole segment. Three of them would be $\frac{3}{3}$ (three thirds), or the whole segment.

Let's do the same thing again with six equal divisions of the segment from 0 to 1 (Figure 2). In this case, we say each of the smaller segments has a length of $\frac{1}{6}$ (one sixth).

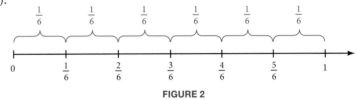

FIGURE 2

The same point we labeled with $\frac{1}{3}$ in Figure 1 is now labeled with $\frac{2}{6}$. Likewise, the point we labeled earlier with $\frac{2}{3}$ is now labeled $\frac{4}{6}$. It must be true then that

$$\frac{2}{6} = \frac{1}{3} \qquad \text{and} \qquad \frac{4}{6} = \frac{2}{3}$$

Actually, there are many fractions that name the same point as $\frac{1}{3}$. If we were to divide the segment between 0 and 1 into 12 equal parts, 4 of these 12 equal parts $\left(\frac{4}{12}\right)$ would be the same as $\frac{2}{6}$ or $\frac{1}{3}$; that is,

$$\frac{4}{12} = \frac{2}{6} = \frac{1}{3}$$

Even though these three fractions look different, each names the same point on the number line, as shown in Figure 3. All three fractions have the same value because they all represent the same number.

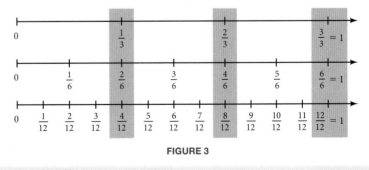

FIGURE 3

(def **DEFINITION**   *equivalent fractions*

Fractions that represent the same relationship between the numerator and denominator are said to be *equivalent*. Equivalent fractions may look different, but they must have the same value.

It is apparent that every fraction has many different representations, each of which is equivalent to the original fraction. The next two properties give us a way of changing the terms of a fraction without changing its value.

> **⟨Δ≠Σ⟩ PROPERTY**   *Multiplication and Equivalent Fractions*
>
> Multiplying the numerator and denominator of a fraction by the same nonzero number results in an equivalent fraction.

Because division can always be expressed in terms of multiplication (as we will see in Section 1.3), dividing the numerator and denominator of a fraction by the same nonzero number also results in an equivalent fraction.

■ EXAMPLE 2   Write $\frac{3}{4}$ as an equivalent fraction with denominator 20.

SOLUTION   The denominator of the original fraction is 4. The fraction we are trying to find must have a denominator of 20. We know that if we multiply 4 by 5, we get 20. The preceding property indicates that we are free to multiply the denominator by 5 as long as we do the same to the numerator.

$$\frac{3}{4} = \frac{3 \cdot 5}{4 \cdot 5} = \frac{15}{20}$$

The fraction $\frac{15}{20}$ is equivalent to the fraction $\frac{3}{4}$.   ■

Opposites

An important concept associated with numbers on the number line is that of opposites. Here is the definition:

> **⟨def⟩ DEFINITION**   *opposite*
>
> Any two real nonzero numbers the same distance from zero, but in opposite directions from zero on the number line, are called ***opposites,*** or *additive inverses.*

> **⟨Δ≠Σ⟩ NOTATION**
>
> If x is any real number, then we denote the opposite of x by $-x$.

■ EXAMPLE 3   Give the opposite of each number.

a. 5   **b.** -3 **c.** $\frac{1}{4}$ **d.** -2.3

SOLUTION

	Number	Opposite	
a.	5	-5	The opposite of 5 is -5
b.	-3	3	The opposite of -3 is $-(-3) = 3$
c.	$\frac{1}{4}$	$-\frac{1}{4}$	The opposite of $\frac{1}{4}$ is $-\frac{1}{4}$
d.	-2.3	2.3	The opposite of -2.3 is $-(-2.3) = 2.3$ ■

Each negative number is the opposite of some positive number, and each positive number is the opposite of some negative number.

The negative sign in front of a number can be read in a few different ways. It can be read as "negative" or "the opposite of." We say -4 is the opposite of 4, or negative 4. The one we use will depend on the situation. For instance, the expression $-(-3)$ is best read "the opposite of negative 3." Because the opposite of -3 is 3, we have $-(-3) = 3$. In general, if a is any positive real number, then

$$-(-a) = a \quad \text{(The opposite of a negative is a positive.)}$$

Opposites are also called additive inverses, because when you add any two opposites, the result is always zero.

$$a + (-a) = 0$$

Absolute Value

Representing numbers on the number line lets us give each number two important properties: a direction from zero and a distance from zero. The direction from zero is represented by the sign in front of the number. (A number without a sign is understood to be positive.) The distance from zero is called the absolute value of the number, as the following definition indicates:

> (dĕf **DEFINITION** *absolute value (geometric definition)*
>
> The **absolute value** of a real number is its distance from zero on the number line. If x represents a real number, then the absolute value of x is written $|x|$.

This definition of absolute value is geometric in form because it defines absolute value in terms of the number line. Here is an alternative definition of absolute value that is algebraic in form because it involves only symbols.

Note It is important to recognize that if x is a real number, $-x$ is not necessarily negative. For example, if x is 5, then $-x$ is -5. On the other hand, if x were -5, then $-x$ would be $-(-5)$, which is 5.

> (dĕf **DEFINITION** *absolute value (algebraic definition)*
>
> If x represents a real number, then the **absolute value** of x is written $|x|$, and is given by
> $$|x| = \begin{cases} x & \text{if } x \geq 0 \\ -x & \text{if } x < 0 \end{cases}$$

If the original number is positive or 0, then its absolute value is the number itself. If the number is negative, its absolute value is its opposite.

EXAMPLE 4 Write each expression without absolute value symbols.

a. $|5|$ **b.** $|-5|$ **c.** $\left|-\dfrac{1}{2}\right|$ **d.** $-\left|-\dfrac{1}{2}\right|$

SOLUTION

a. $|5| = 5$ The number 5 is 5 units from zero

b. $|-5| = -(-5) = 5$ The number -5 is 5 units from zero

c. $\left|-\dfrac{1}{2}\right| = -\left(-\dfrac{1}{2}\right) = \dfrac{1}{2}$ The number $-\dfrac{1}{2}$ is $\dfrac{1}{2}$ unit from zero

d. $-\left|-\dfrac{1}{2}\right| = -\left(\dfrac{1}{2}\right) = -\dfrac{1}{2}$ The opposite of $\left|-\dfrac{1}{2}\right|$ is $-\dfrac{1}{2}$

The absolute value of a number is never negative. It is the distance the number is from zero without regard to which direction it is from zero.

Prime Numbers and Factoring

The following diagram shows the relationship between multiplication and factoring:

$$\text{Multiplication}$$

$$\text{Factors} \longrightarrow 3 \cdot 4 = 12 \longleftarrow \text{Product}$$

$$\text{Factoring}$$

When we read the problem from left to right, we say the product of 3 and 4 is 12. We can also say we multiply 3 and 4 to get 12. When we read the problem in the other direction, from right to left, we say we have *factored* 12 into the product 3 times 4, or 3 and 4 are *factors* of 12.

> (dĕf) **DEFINITION**   *factor*
>
> If a and b represent integers, then a is said to be a *factor* (or divisor) of b if a divides b evenly — that is, if a divides b with no remainder.

Note Recall from your previous math classes that integers include whole numbers (0, 1, 2…) and their opposites (−1, −2, −3…).

The number 12 can be factored still further:

$$12 = 4 \cdot 3$$
$$= 2 \cdot 2 \cdot 3$$
$$= 2^2 \cdot 3$$

The numbers 2 and 3 are called *prime* factors of 12 because neither can be factored any further.

Note Recall that we can use exponents to rewrite $2 \cdot 2 \cdot 3$ as $2^2 \cdot 3$. Exponents indicate repeated multiplication.

> (dĕf) **DEFINITION**   *prime, composite*
>
> A *prime* number is any positive integer larger than 1 whose only positive factors (divisors) are itself and 1. An integer greater than 1 that is not prime is said to be *composite*.

Here is a list of the first few prime numbers:

$$\text{Prime numbers} = \{2, 3, 5, 7, 11, 13, 17, 19, 23, 29, 31, 37, 41,\ldots\}$$

When a number is not prime, we can factor it into the product of prime numbers. To factor a number into the product of primes, we simply factor it until it cannot be factored further.

 **EXAMPLE 5**   Factor 525 into the product of primes.

SOLUTION   Because 525 ends in 5, it is divisible by 5.

$$525 = 5 \cdot 105$$
$$= 5 \cdot 5 \cdot 21$$
$$= 5 \cdot 5 \cdot 3 \cdot 7$$
$$= 3 \cdot 5^2 \cdot 7$$

EXAMPLE 6   Reduce $\dfrac{210}{231}$ to lowest terms.

SOLUTION   First we factor 210 and 231 into the product of prime factors. Then we reduce to lowest terms by dividing the numerator and denominator by any factors they have in common.

> **Note**   The multiplication property for equivalent fractions allows us to obtain an equivalent fraction by dividing out common factors from the numerator and denominator.

$$\frac{210}{231} = \frac{2 \cdot 3 \cdot 5 \cdot 7}{3 \cdot 7 \cdot 11}$$ Factor the numerator and denominator completely

$$= \frac{2 \cdot 3 \cdot 5 \cdot 7}{3 \cdot 7 \cdot 11}$$ Divide the numerator and denominator by $3 \cdot 7$

$$= \frac{2 \cdot 5}{11}$$

$$= \frac{10}{11}$$

Reciprocals

Before we go further with our study of the number line, we need to review multiplication with fractions. Recall that for the fraction $\frac{a}{b}$, a is called the numerator and b is called the denominator. To multiply two fractions, we simply multiply numerators and multiply denominators.

> **Note**   In past math classes, you may have written fractions like $\frac{8}{5}$ (improper fractions) as mixed numbers, such as $1\frac{3}{5}$. In algebra, it is usually better to leave them as improper fractions.

EXAMPLE 7   Multiply.

a. $\dfrac{3}{5} \cdot \dfrac{7}{8} = \dfrac{3 \cdot 7}{5 \cdot 8} = \dfrac{21}{40}$

b. $8 \cdot \dfrac{1}{5} = \dfrac{8}{1} \cdot \dfrac{1}{5} = \dfrac{8 \cdot 1}{1 \cdot 5} = \dfrac{8}{5}$

c. $\left(\dfrac{2}{3}\right)^4 = \dfrac{2}{3} \cdot \dfrac{2}{3} \cdot \dfrac{2}{3} \cdot \dfrac{2}{3} = \dfrac{16}{81}$

Multiplication of fractions is useful in understanding the concept of the reciprocal of a number. Here is a formal definition:

> **(def) DEFINITION**   *reciprocals*
>
> Any two real numbers whose product is 1 are called *reciprocals*, or *multiplicative inverses*.

EXAMPLE 8   Give the reciprocal of each number.

a. 5   **b.** $\dfrac{1}{3}$ **c.** $\dfrac{3}{4}$ **d.** x

SOLUTION

	Number	Reciprocal	
a.	5	$\dfrac{1}{5}$	Because $5\left(\dfrac{1}{5}\right) = \dfrac{5}{1}\left(\dfrac{1}{5}\right) = \dfrac{5}{5} = 1$
b.	$\dfrac{1}{3}$	3	Because $\dfrac{1}{3}(3) = \dfrac{1}{3}\left(\dfrac{3}{1}\right) = \dfrac{3}{3} = 1$
c.	$\dfrac{3}{4}$	$\dfrac{4}{3}$	Because $\dfrac{3}{4}\left(\dfrac{4}{3}\right) = \dfrac{12}{12} = 1$
d.	x	$\dfrac{1}{x}$	Because $x\left(\dfrac{1}{x}\right) = \dfrac{x}{1}\left(\dfrac{1}{x}\right) = \dfrac{x}{x} = 1, x \neq 0$

Although we will not develop multiplication with negative numbers until later in the chapter, you should know that the reciprocal of a negative number is also a negative number. For example, the reciprocal of -4 is $-\dfrac{1}{4}$.

Formulas for Area and Perimeter

Note The vertical line labeled h in the triangle is its height, or altitude. It extends from the top of the triangle down to the base, meeting the base at an angle of 90°. The altitude of a triangle is always perpendicular to the base. The small square shown where the altitude meets the base is used to indicate that the angle formed is 90°.

FACTS FROM GEOMETRY   *Formulas for Area and Perimeter*

A square, rectangle, and triangle are shown in the following figures. Note that we have labeled the dimensions of each with variables. The formulas for the perimeter and area of each object are given in terms of its dimensions.

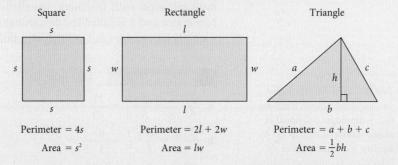

Square	Rectangle	Triangle
Perimeter = $4s$	Perimeter = $2l + 2w$	Perimeter = $a + b + c$
Area = s^2	Area = lw	Area = $\dfrac{1}{2}bh$

The formula for perimeter gives us the distance around the outside of the object along its sides, whereas the formula for area gives us a measure of the amount of surface the object has.

EXAMPLE 9   Find the perimeter and area of each figure.

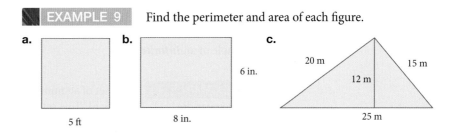

a.

5 ft

b.

6 in.

8 in.

c.

20 m   15 m

12 m

25 m

SOLUTION   We use the preceding formulas to find the perimeter and the area. In each case, the units for perimeter are linear units, whereas the units for area are square units.

a. Perimeter $= 4s = 4 \cdot 5$ feet $= 20$ feet

Area $= s^2 = (5$ feet$)^2 = 25$ square feet

b. Perimeter $= 2l + 2w = 2(8$ inches$) + 2(6$ inches$) = 28$ inches

Area $= lw = (8$ inches$)(6$ inches$) = 48$ square inches

c. Perimeter $= a + b + c = (20$ meters$) + (25$ meters$) + (15$ meters$)$

$= 60$ meters

Area $= \dfrac{1}{2}bh = \dfrac{1}{2}(25$ meters$)(12$ meters$) = 150$ square meters   ◼

Sets and Subsets

The concept of a set can be considered the starting point for all the branches of mathematics.

(dĕf **DEFINITION**   *set*

A **set** is a collection of objects or things. The objects in the set are called **elements**, or **members**, of the set.

Sets are usually denoted by capital letters, and elements of sets are denoted by lowercase letters. We use braces, { }, to enclose the elements of a set.

To show that an element is contained in a set, we use the symbol $\in$. That is,

$x \in A$ is read "x is an element (member) of set A"

For example, if A is the set $\{1, 2, 3\}$, then $2 \in A$. However, $5 \notin A$ means 5 is not an element of set A.

(dĕf **DEFINITION**   *subset*

Set **A** is a **subset** of set **B**, written $A \subset B$, if every element in A is also an element of B. That is,

$A \subset B$ if and only if   **A** is contained in **B**

EXAMPLE 10   The set of numbers used to count things is {1, 2, 3, ...}. The dots mean the set continues indefinitely in the same manner. This is an example of an *infinite set*.

EXAMPLE 11   The set of all numbers represented by the dots on the faces of a regular die is {1, 2, 3, 4, 5, 6}. This set is a subset of the set in Example 10. It is an example of a *finite set* because it has a limited number of elements.

> **(def) DEFINITION**   *empty set*
>
> The set with no members is called the *empty*, or *null set*. It is denoted by the symbol $\varnothing$. The empty set is considered a subset of every set.

Another notation we can use to describe sets is called *set-builder notation*. Suppose we want set A to contain all the real numbers that are less than 2. Here is how we would write that set

$$A = \{x \mid x < 2\}$$

The right side of this statement is read "the set of all x such that x is less than 2." As you can see, the vertical line after the first x is read "such that."

EXAMPLE 12   If $A = \{1, 2, 3, 4, 5, 6\}$, find $C = \{x \mid x \in A \text{ and } x \geq 4\}$.

SOLUTION   We are looking for all the elements of A that are also greater than or equal to 4. They are 4, 5 and 6. Using set notation we have

$$C = \{4, 5, 6\}$$

Getting Ready for Class

Each section of the book will end with some problems and questions like the ones that follow. They are for you to answer after you have read through the section, but before you go to class. All of them require that you give written responses in complete sentences. Writing about mathematics is a valuable exercise. If you write with the intention of explaining and communicating what you know to someone else, you will find that you understand the topic you are writing about even better than you did before you started writing.

After reading through the preceding section, respond in your own words and in complete sentences.

A. What is a real number?

B. What is an equivalent fraction?

C. How do you find the reciprocal of a number?

D. Explain how you find the perimeter and the area of a rectangle.

Draw a number line that extends from -5 to 5. Label the points with the following coordinates.

1. 5   **2.** -2 **3.** -4 **4.** -3

5. 1.5   **6.** -1.5 **7.** $\dfrac{9}{4}$ **8.** $\dfrac{8}{3}$

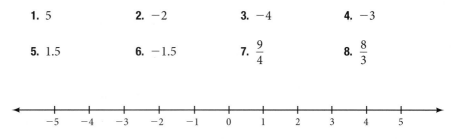

Write each of the following fractions as an equivalent fraction with denominator 24.

9. $\dfrac{3}{4}$ **10.** $\dfrac{5}{6}$ **11.** $\dfrac{1}{2}$ **12.** $\dfrac{1}{8}$ **13.** $\dfrac{5}{8}$ **14.** $\dfrac{7}{12}$

Write each fraction as an equivalent fraction with denominator 60.

15. $\dfrac{3}{5}$ **16.** $\dfrac{5}{12}$ **17.** $\dfrac{11}{30}$ **18.** $\dfrac{9}{10}$ **19.** $-\dfrac{5}{6}$ **20.** $-\dfrac{7}{12}$

For each of the following numbers, give the opposite, the reciprocal, and the absolute value. (Assume all variables are nonzero.)

21. 10   **22.** 8   **23.** $\dfrac{3}{4}$ **24.** $\dfrac{5}{7}$

25. $\dfrac{11}{2}$ **26.** $\dfrac{16}{3}$ **27.** -3 **28.** -5

29. $-\dfrac{2}{5}$ **30.** $-\dfrac{3}{8}$ **31.** x **32.** a

Simplify each absolute value expression.

33. $|-2|$ **34.** $|-7|$ **35.** $\left|-\dfrac{3}{4}\right|$ **36.** $\left|\dfrac{5}{6}\right|$

37. $|\pi|$ **38.** $|-\sqrt{2}|$ **39.** $-|4|$ **40.** $-|5|$

41. $-|-2|$ **42.** $-|-10|$ **43.** $-\left|-\dfrac{3}{4}\right|$ **44.** $-\left|\dfrac{7}{8}\right|$

Place one of the symbols $<$ or $>$ between each of the following to make the resulting statement true.

45. $-5 \quad -3$ **46.** $-8 \quad -1$ **47.** $-3 \quad -7$ **48.** $-6 \quad 5$

49. $|-4| \quad -|-4|$ **50.** $3 \quad -|-3|$ **51.** $7 \quad -|-7|$ **52.** $-7 \quad |-7|$

53. $-\dfrac{3}{4} \quad -\dfrac{1}{4}$ **54.** $-\dfrac{2}{3} \quad -\dfrac{1}{3}$ **55.** $-\dfrac{3}{2} \quad -\dfrac{3}{4}$ **56.** $-\dfrac{8}{3} \quad -\dfrac{17}{3}$

Factor each number into the product of prime factors.

57. 266 **58.** 385 **59.** 111 **60.** 735 **61.** 369 **62.** 1,155

Reduce each fraction to lowest terms.

63. $\dfrac{165}{385}$ **64.** $\dfrac{550}{735}$ **65.** $\dfrac{385}{735}$ **66.** $\dfrac{266}{285}$

67. $\dfrac{111}{185}$ **68.** $\dfrac{279}{310}$ **69.** $\dfrac{75}{135}$ **70.** $\dfrac{38}{30}$

71. $\dfrac{6}{8}$ **72.** $\dfrac{10}{25}$ **73.** $\dfrac{200}{5}$ **74.** $\dfrac{240}{6}$

Find the perimeter and area of each figure.

75. 1 in. 1 in.

76. 15 mm 15 mm

77. 0.75 in. 1.5 in.

78. 1.5 cm 4.5 cm

79. 2.75 cm 3.5 cm 2.5 cm 4 cm

80. 1.8 in. 1 in. 1.2 in. 2 in.

Applying the Concepts

Wind Chill The table below shows wind temperatures. The first column gives the air temperature, and the first row is wind speed in miles per hour. The numbers within the table indicate how cold the weather will feel. For example, if the thermometer reads 30°F and the wind is blowing at 15 miles per hour, the wind chill temperature is 9°F. Use Table 1 to answer Problems 81 and 82.

Wind Chill Temperatures					
Air Temperature (°F)	Wind Speed (mph)				
	10	15	20	25	30
30°	16°	9°	4°	1°	−2°
25°	10°	2°	−3°	−7°	−10°
20°	3°	−5°	−10°	−15°	−18°
15°	−3°	−11°	−17°	−22°	−25°
10°	−9°	−18°	−24°	−29°	−33°
5°	−15°	−25°	−31°	−36°	−41°
0°	−22°	−31°	−39°	−44°	−49°
−5°	−27°	−38°	−46°	−51°	−56°

TABLE 1

81. Reading Tables Find the wind chill temperature if the thermometer reads 20°F and the wind is blowing at 25 miles per hour.

82. Reading Tables Which will feel colder: a day with an air temperature of 10°F with a 25-mile-per-hour wind, or a day with an air temperature of 25°F and a 10-mile-per-hour wind?

83. iPad Apps The chart shows the top categories for iPad applications. Use the chart to answer the following questions.

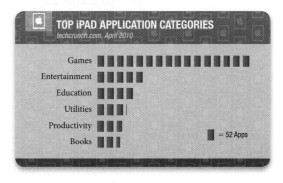

a. How many entertainment applications are there?

b. True or false? There are more than 150 book applications.

c. True or false? There are fewer than 880 game applications.

84. eBook Memory The chart shows the memory capacity for various eBook readers. Write a mathematical statement using one of the symbols $=$, $<$, $>$ to compare the following capacities:

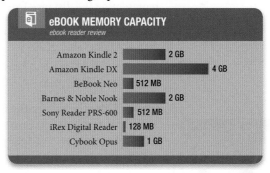

a. Amazon Kindle 2 to the Amazon Kindle DX

b. Barnes and Noble Nook to the Cybook Opus

c. BeBook Neo to the Sony Reader PRS-600

85. Geometry Find the area and perimeter of an $8\frac{1}{2}$-by-11-inch piece of notebook paper.

86. Geometry Find the area and perimeter of an $8\frac{1}{2}$-by-$5\frac{1}{2}$-inch piece of paper.

Calories and Exercise The table here gives the amount of energy expended per hour for various activities for a person weighing 120, 150, or 180 pounds. Use the table to answer questions 87–90.

87. Suppose you weigh 120 pounds. How many calories will you burn if you play handball for 2 hours and then ride your bicycle for an hour?

Energy Expended from Exercising			
	Calories per Hour		
Activity	120 lb	150 lb	180 lb
Bicycling	299	374	449
Bowling	212	265	318
Handball	544	680	816
Horseback trotting	278	347	416
Jazzercise	272	340	408
Jogging	544	680	816
Skiing (downhill)	435	544	653

88. How many calories are burned by a person weighing 150 pounds who jogs for $\frac{1}{2}$ hour and then goes bicycling for 2 hours?

89. Two people go skiing. One weighs 180 pounds and the other weighs 120 pounds. If they ski for 3 hours, how many more calories are burned by the person weighing 180 pounds?

90. Two people spend 3 hours bowling. If one weighs 120 pounds and the other weighs 150 pounds, how many more calories are burned during the evening by the person weighing 150 pounds?

Distance and Absolute Value These problems are based on absolute value $|x|$ representing the distance from 0 on a number line.

91. Find two numbers with an absolute value of 20.

92. Find two numbers x such that $|x| = 13$.

93. Is it possible for x to satisfy $|x| = -5$? Explain.

94. Is it possible for x to satisfy $|x| < 0$? Explain.

Freeway Exits The exit for the Griffith Observatory is exit #141 from Interstate 5. The exit numbers increase by 1 for every mile as you travel north. Use this information for the next four problems.

95. If you are 16 miles north of the exit for the observatory, at what exit number are you located?

96. If you are 17 miles south of the exit for the observatory, at what exit number are you located?

97. If you are 20 miles away from the exit for the observatory, at what exit numbers could you be located? (Hint: there are two answers.)

98. If you are 35 miles away from the exit for the observatory, at what exit numbers could you be located? (Hint: there are two answers.)

If $A = \{1, 3, 5, 7\}$ and $B = \{1, 4, 9, 16\}$, mark each of the following as True or False.

99. $1 \in A$

100. $A \subset B$

101. $1 \in A$ and $1 \in B$

102. $9 \in A$

Use set-builder notation to write each statement as a set.

103. All real numbers greater than 5.

104. All real numbers less than -2.

105. All real numbers greater than or equal to 5.

106. All real numbers less than or equal to -2.

107. If $A = \{1, 2, 3, 4\}$, find

    a. $C = \{x \mid x \in A \text{ and } x < 3\}$

    b. $D = \{x \mid x \in A \text{ and } x \geq 3\}$

108. If $B = \{1, 3, 5, 7, \ldots\}$, find

    a. $C = \{x \mid x \in B \text{ and } x < 11\}$

    b. $D = \{x \mid x \in B \text{ and } x \leq 3\}$

Learning Objectives Assessment

The following problems can be used to help assess if you have successfully met the learning objectives for this section.

109. Which of the following could be the coordinate of the point shown in the figure below?

    a. $2\frac{2}{3}$
    b. -1.5
    c. $-\frac{1}{3}$
    d. $-\frac{4}{3}$

110. Write $\frac{3}{4}$ as an equivalent fraction with denominator 36.

    a. $\frac{27}{36}$
    b. $\frac{24}{36}$
    c. $\frac{12}{36}$
    d. $\frac{3}{36}$

111. Find the opposite of $-\frac{2}{3}$.

    a. $-\frac{3}{2}$
    b. $\frac{3}{2}$
    c. $-\frac{2}{3}$
    d. $\frac{2}{3}$

112. Simplify $|-8|$.

    a. 8
    b. -8
    c. 0
    d. $-\frac{1}{8}$

113. Factor 84 into a product of prime factors.

    a. $4 \cdot 21$
    b. $2^2 \cdot 3 \cdot 7$
    c. $2 \cdot 3^2 \cdot 7$
    d. $2 \cdot 6 \cdot 7$

114. Find the reciprocal of $-\frac{2}{3}$.

    a. $\frac{3}{2}$
    b. $-\frac{3}{2}$
    c. $\frac{2}{3}$
    d. $-\frac{2}{3}$

115. Find the area of a square with sides of length 6 centimeters.

    a. 24 cm
    b. 12 cm
    c. 36 cm^2
    d. 12 cm^2

SPOTLIGHT ON SUCCESS *Student Instructor Cynthia*

Each time we face our fear, we gain strength, courage, and confidence in the doing.
—Unknown

I must admit, when it comes to math, it takes me longer to learn the material compared to other students. Because of that, I was afraid to ask questions, especially when it seemed like everyone else understood what was going on. Because I wasn't getting my questions answered, my quiz and exam scores were only getting worse. I realized that I was already paying a lot to go to college and that I couldn't afford to keep doing poorly on my exams. I learned how to overcome my fear of asking questions by studying the material before class, and working on extra problem sets until I was confident enough that at least I understood the main concepts. By preparing myself beforehand, I would often end up answering the question myself. Even when that wasn't the case, the professor knew that I tried to answer the question on my own. If you want to be successful, but you are afraid to ask a question, try putting in a little extra time working on problems before you ask your instructor for help. I think you will find, like I did, that it's not as bad as you imagined it, and you will have overcome an obstacle that was in the way of your success.

Learning Objectives

In this section we will learn how to:

1. Recognize and apply the properties of real numbers.

2. Add fractions.

3. Simplify algebraic expressions.

Properties of Real Numbers

We know that adding 3 and 7 gives the same answer as adding 7 and 3. The order of two numbers in an addition problem can be changed without changing the result. This fact about numbers and addition is called the ***commutative property of addition***.

For all the properties listed in this section, *a*, *b*, and *c* represent real numbers.

[Δ≠Σ] PROPERTY   *Commutative Property of Addition*

In symbols: $a + b = b + a$
In words: The *order* of the numbers in a sum does not affect the result.

[Δ≠Σ] PROPERTY   *Commutative Property of Multiplication*

In symbols: $a \cdot b = b \cdot a$
In words: The *order* of the numbers in a product does not affect the result.

EXAMPLE 1

a. The statement $3 + 7 = 7 + 3$ is an example of the commutative property of addition.

b. The statement $3 \cdot x = x \cdot 3$ is an example of the commutative property of multiplication.

VIDEO EXAMPLES

SECTION 1.2

The other two basic operations (subtraction and division) are not commutative. If we change the order in which we are subtracting or dividing two numbers, we change the result.

Another property of numbers you have used many times has to do with grouping. When adding $3 + 5 + 7$, we can add the 3 and 5 first and then the 7, or we can add the 5 and 7 first and then the 3. Mathematically, it looks like this:

$$(3 + 5) + 7 = 3 + (5 + 7)$$

Operations that behave in this manner are called ***associative*** operations.

> **[Δ≠Σ] PROPERTY**   *Associative Property of Addition*
>
> *In symbols:* $a + (b + c) = (a + b) + c$
> *In words:*   The *grouping* of the numbers in a sum does not affect the result.

Note The associative properties also do not apply to subtraction or division.

> **[Δ≠Σ] PROPERTY**   *Associative Property of Multiplication*
>
> *In symbols:* $a(bc) = (ab)c$
> *In words:*   The *grouping* of the numbers in a product does not affect the result.

The following examples illustrate how the associative properties can be used to simplify expressions that involve both numbers and variables.

EXAMPLE 2   Simplify by using the associative property.

a. $2 + (3 + y) = (2 + 3) + y$ Associative property
$$= 5 + y$$ Add

b. $5(4x) = (5 \cdot 4)x$ Associative property
$$= 20x$$ Multiply

c. $\frac{1}{4}(4a) = \left(\frac{1}{4} \cdot 4\right)a$ Associative property
$$= 1a$$ Multiply
$$= a$$

d. $2\left(\frac{1}{2}x\right) = \left(2 \cdot \frac{1}{2}\right)x$ Associative property
$$= 1x$$ Multiply
$$= x$$

e. $6\left(\frac{1}{3}x\right) = \left(6 \cdot \frac{1}{3}\right)x$ Associative property
$$= 2x$$ Multiply

Our next property involves both addition and multiplication. It is called the *distributive property* and is stated as follows:

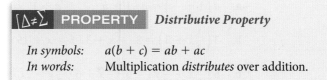

PROPERTY   *Distributive Property*

In symbols: $a(b + c) = ab + ac$
In words: Multiplication *distributes* over addition.

You will see as we progress through the book that the distributive property is used very frequently in algebra. To see that the distributive property works, compare the following:

$$3(4 + 5) \qquad\qquad 3(4) + 3(5)$$
$$= 3(9) \qquad\qquad = 12 + 15$$
$$= 27 \qquad\qquad = 27$$

In both cases the result is 27. Because the results are the same, the original two expressions must be equal, or $3(4 + 5) = 3(4) + 3(5)$.

> *Note*   Although the properties we are listing are stated for only two or three real numbers, they hold for as many numbers as needed. For example, the distributive property holds for expressions like $3(x + y + z + 2)$. That is,
>
> $3(x + y + z + 2) = 3x + 3y + 3z + 6$

EXAMPLE 3   Apply the distributive property to each expression and then simplify the result.

a. $5(4x + 3) = 5(4x) + 5(3)$ Distributive property
$\qquad\qquad = 20x + 15$ Multiply

b. $6(3x + 2y) = 6(3x) + 6(2y)$ Distributive property
$\qquad\qquad\quad = 18x + 12y$ Multiply

c. $\dfrac{1}{2}(3x + 6) = \dfrac{1}{2}(3x) + \dfrac{1}{2}(6)$ Distributive property

$\qquad\qquad\quad = \dfrac{3}{2}x + 3$ Multiply

d. $2(3y + 4) + 2 = 2(3y) + 2(4) + 2$ Distributive property
$\qquad\qquad\qquad = 6y + 8 + 2$ Multiply
$\qquad\qquad\qquad = 6y + 10$ Add

We can combine our knowledge of the distributive property with multiplication of fractions to manipulate expressions involving fractions. Here are some examples that show how we do this:

 **EXAMPLE 4**   Apply the distributive property to each expression and then simplify the result.

a. $a\left(1 + \dfrac{1}{a}\right) = a \cdot 1 + a \cdot \dfrac{1}{a} = a + 1$

b. $3\left(\dfrac{1}{3}x + 5\right) = 3 \cdot \dfrac{1}{3}x + 3 \cdot 5 = x + 15$

c. $6\left(\dfrac{1}{3}x + \dfrac{1}{2}y\right) = 6 \cdot \dfrac{1}{3}x + 6 \cdot \dfrac{1}{2}y = 2x + 3y$

Combining Similar Terms

The distributive property can also be used to combine similar terms. (For now, a **term** is the product of a number with one or more variables.) **Similar terms** are terms with the same variable part. The terms $3x$ and $5x$ are similar, as are $2y$, $7y$, and $-3y$, because the variable parts are the same.

EXAMPLE 5   Use the distributive property to combine similar terms.

a. $3x + 5x = (3 + 5)x$ Distributive property

          $= 8x$ Add

b. $3y + y = (3 + 1)y$ Distributive property

          $= 4y$ Add

Review of Addition with Fractions

To add fractions, each fraction must have the same denominator.

> **(déf) DEFINITION**   *least common denominator*
>
> The **least common denominator** (LCD) for a set of denominators is the smallest number divisible by *all* the denominators.

The first step in adding fractions is to find the least common denominator for all the denominators. We then rewrite each fraction (if necessary) as an equivalent fraction with the common denominator. Finally, we add the numerators and reduce to lowest terms if necessary.

EXAMPLE 6   Add: $\dfrac{5}{12} + \dfrac{7}{18}$.

SOLUTION   The least common denominator for the denominators 12 and 18 must be the smallest number divisible by both 12 and 18. We can factor 12 and 18 completely and then build the LCD from these factors.

Note   The LCD will contain every factor the *greatest* number of times it appears in any one denominator.

$$\left.\begin{array}{l} 12 = 2 \cdot 2 \cdot 3 \\ 18 = 2 \cdot 3 \cdot 3 \end{array}\right\} \quad \text{LCD} = 2 \cdot 2 \cdot 3 \cdot 3 = 36$$

12 divides the LCD

18 divides the LCD

Next, we rewrite our original fractions as equivalent fractions with denominators of 36. To do so, we multiply each original fraction by an appropriate form of the number 1.

$$\frac{5}{12} + \frac{7}{18} = \frac{5}{12} \cdot \frac{3}{3} + \frac{7}{18} \cdot \frac{2}{2}$$

$$= \frac{15}{36} + \frac{14}{36}$$

Finally, we add numerators and place the result over the common denominator, 36. (Remember, this is an application of the distributive property.)

$$\frac{15}{36} + \frac{14}{36} = \frac{15 + 14}{36} = \frac{29}{36}$$

Simplifying Expressions

We can use the commutative, associative, and distributive properties together to simplify algebraic expressions.

EXAMPLE 7   Simplify: $7x + 4 + 6x + 3$.

SOLUTION   We begin by applying the commutative and associative properties to group similar terms.

$$7x + 4 + 6x + 3 = (7x + 6x) + (4 + 3) \qquad \text{Commutative and associative properties}$$

$$= (7 + 6)x + (4 + 3) \qquad \text{Distributive property}$$

$$= 13x + 7 \qquad \text{Add}$$

EXAMPLE 8   Simplify: $4 + 3(2y + 5) + 8y$.

SOLUTION   Because $2y$ and 5 are not similar terms, we cannot perform the addition in the grouping symbol. We must distribute the 3 across $2y + 5$ first.

$$4 + 3(2y + 5) + 8y = 4 + 6y + 15 + 8y \qquad \text{Distributive property}$$

$$= (6y + 8y) + (4 + 15) \qquad \text{Commutative and associative properties}$$

$$= (6 + 8)y + (4 + 15) \qquad \text{Distributive property}$$

$$= 14y + 19 \qquad \text{Add}$$

The numbers 0 and 1 are called the *additive identity* and *multiplicative identity*, respectively. Adding 0 to a number does not change the value of that number. Likewise, multiplying a number by 1 does not alter the value of that number. We see below that 0 is to addition what 1 is to multiplication.

[Δ≠Σ]   PROPERTY   *Additive Identity Property*

There exists a unique number 0 such that for any real number a
In symbols: $a + 0 = a$ and $0 + a = a$

> **⟨Δ≠Σ⟩ PROPERTY**   *Multiplicative Identity Property*
>
> There exists a unique number 1 such that for any real number a
> *In symbols:* $a(1) = a$ and $1(a) = a$

The opposite of a real number is formally called its ***additive inverse***. We can add the additive inverse of a number in order to obtain the additive identity, which is 0.

> **⟨Δ≠Σ⟩ PROPERTY**   *Additive Inverse Property*
>
> For each real number a, there exists a unique real number $-a$ such that
> *In symbols:* $a + (-a) = 0$
> *In words:*   Opposites add to 0.

Likewise, the reciprocal of a number is called its ***multiplicative inverse***. When we multiply a number by its multiplicative inverse, the result is the multiplicative identity, which is 1.

> **⟨Δ≠Σ⟩ PROPERTY**   *Multiplicative Inverse Property*
>
> For every real number a, except 0, there exists a unique real number $\frac{1}{a}$ such that
> *In symbols:* $a\left(\dfrac{1}{a}\right) = 1$
> *In words:*   Reciprocals multiply to 1.

EXAMPLE 9

a. $7\left(\dfrac{1}{7}x\right) = \left(7 \cdot \dfrac{1}{7}\right)x$ Associative property

$\qquad\qquad = 1 \cdot x$ Multiplicative inverse property

$\qquad\qquad = x$ Multiplicative identity property

b. $x + 5 + (-5) = x + (5 + (-5))$ Associative property

$\qquad\qquad\quad = x + 0$ Additive inverse property

$\qquad\qquad\quad = x$ Additive identity property

Subsets of the Real Numbers

Next, we consider some of the more important subsets of the real numbers. Each set listed here is a subset of the real numbers:

Counting (or natural) numbers = $\{1, 2, 3, \ldots\}$

Whole numbers = $\{0, 1, 2, 3, \ldots\}$

Integers = $\{\ldots, -3, -2, -1, 0, 1, 2, 3, \ldots\}$

Rational numbers = $\left\{ \dfrac{a}{b} \,\middle|\, a \text{ and } b \text{ are integers}, b \neq 0 \right\}$

Remember, the notation used to write the rational numbers is read "the set of numbers $\frac{a}{b}$ such that a and b are integers and b is not equal to zero." Any number that can be written in the form

$$\frac{\text{integer}}{\text{integer}}$$

is a rational number. Rational numbers are numbers that can be written as the ratio of two integers.

There are still other numbers on the number line that are not members of the subsets we have listed so far. They are real numbers, but they cannot be written as the ratio of two integers. That is, they are not rational numbers. For that reason, we call them irrational numbers.

Irrational numbers = $\{x \mid x \text{ is real, but not rational}\}$

The following are irrational numbers:

$$\sqrt{2} \qquad -\sqrt{3} \qquad 4 + 2\sqrt{3} \qquad \pi \qquad \pi + 5\sqrt{6}$$

> **Note** Each of the following is a rational number:
>
> | | |
> |---|---|
> | $\frac{3}{4}$ | the ratio of 3 to 4 |
> | -8 | the ratio of -8 to 1 |
> | 0.75 | the ratio of 75 to 100 |
> | $0.333\ldots$ | the ratio of 1 to 3 |

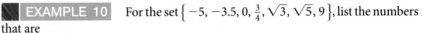 **EXAMPLE 10** For the set $\left\{ -5, -3.5, 0, \frac{3}{4}, \sqrt{3}, \sqrt{5}, 9 \right\}$, list the numbers that are

a. whole numbers **b.** integers **c.** rational numbers

d. irrational numbers **e.** real numbers

SOLUTION

a. whole numbers: $0, 9$

b. integers: $-5, 0, 9$

c. rational numbers: $-5, -3.5, 0, \frac{3}{4}, 9$

d. irrational numbers: $\sqrt{3}, \sqrt{5}$

e. They are all real numbers.

Getting Ready for Class

After reading through the preceding section, respond in your own words and in complete sentences.

A. Describe the commutative property of multiplication.

B. Explain why subtraction and division are not commutative operations.

C. What is the distributive property?

D. Explain the additive inverse property.

Problem Set 1.2

Identify the property of real numbers that justifies each of the following.

1. $3 + 2 = 2 + 3$

2. $3(ab) = (3a)b$

3. $5x = x5$

4. $2 + 0 = 2$

5. $4 + (-4) = 0$

6. $1(6) = 6$

7. $x + (y + 2) = (y + 2) + x$

8. $(a + 3) + 4 = a + (3 + 4)$

9. $4(5 \cdot 7) = 5(4 \cdot 7)$

10. $6(xy) = (xy)6$

11. $4 + (x + y) = (4 + y) + x$

12. $(r + 7) + s = (r + s) + 7$

13. $3(4x + 2) = 12x + 6$

14. $5\left(\dfrac{1}{5}\right) = 1$

15. $-\dfrac{2}{3}\left(-\dfrac{3}{2}\right) = 1$

16. $3(2x - 6) = 6x - 18$

Use the associative property to rewrite each of the following expressions and then simplify the result.

17. $4 + (2 + x)$

18. $6 + (5 + 3x)$

19. $(a + 3) + 5$

20. $(4a + 5) + 7$

21. $5(3y)$

22. $7(4y)$

23. $\dfrac{1}{3}(3x)$

24. $\dfrac{1}{5}(5x)$

25. $4\left(\dfrac{1}{4}a\right)$

26. $7\left(\dfrac{1}{7}a\right)$

27. $\dfrac{2}{3}\left(\dfrac{3}{2}x\right)$

28. $\dfrac{4}{3}\left(\dfrac{3}{4}x\right)$

Apply the distributive property to each expression. Simplify when possible.

29. $3(x + 6)$

30. $5(x + 9)$

31. $2(6x + 4)$

32. $3(7x + 8)$

33. $5(3a + 2b)$

34. $7(2a + 3b)$

35. $\dfrac{1}{3}(4x + 6)$

36. $\dfrac{1}{2}(3x + 8)$

37. $\dfrac{1}{5}(10 + 5y)$

38. $\dfrac{1}{6}(12 + 6y)$

39. $(5t + 1)8$

40. $(3t + 2)5$

Apply the distributive property, then simplify if possible.

41. $3(3x + y - 2z)$

42. $2(2x - y + z)$

43. $10(0.3x + 0.7y)$

44. $10(0.2x + 0.5y)$

45. $100(0.06x + 0.07y)$

46. $100(0.09x + 0.08y)$

47. $3\left(x + \dfrac{1}{3}\right)$

48. $5\left(x - \dfrac{1}{5}\right)$

49. $2\left(x - \dfrac{1}{2}\right)$

50. $7\left(x + \dfrac{1}{7}\right)$

51. $x\left(1 + \dfrac{2}{x}\right)$

52. $x\left(1 - \dfrac{1}{x}\right)$

53. $a\left(1 - \dfrac{3}{a}\right)$

54. $a\left(1 + \dfrac{1}{a}\right)$

55. $8\left(\dfrac{1}{8}x + 3\right)$

56. $4\left(\dfrac{1}{4}x - 9\right)$

57. $6\left(\dfrac{1}{2}x - \dfrac{1}{3}y\right)$

58. $12\left(\dfrac{1}{4}x - \dfrac{1}{6}y\right)$

59. $12\left(\dfrac{1}{4}x + \dfrac{2}{3}y\right)$

60. $12\left(\dfrac{2}{3}x - \dfrac{1}{4}y\right)$

61. $20\left(\dfrac{2}{5}x + \dfrac{1}{4}y\right)$

62. $15\left(\dfrac{2}{3}x + \dfrac{2}{5}y\right)$

Apply the distributive property to each expression. Simplify when possible.

63. $3(5x + 2) + 4$

64. $4(3x + 2) + 5$

65. $4(2y + 6) + 8$

66. $6(2y + 3) + 2$

67. $5(1 + 3t) + 4$

68. $2(1 + 5t) + 6$

69. $3 + (2 + 7x)4$

70. $4 + (1 + 3x)5$

Add the following fractions.

71. $\dfrac{2}{5} + \dfrac{1}{15}$

72. $\dfrac{5}{8} + \dfrac{1}{4}$

73. $\dfrac{17}{30} + \dfrac{11}{42}$

74. $\dfrac{19}{42} + \dfrac{13}{70}$

75. $\dfrac{9}{48} + \dfrac{3}{54}$

76. $\dfrac{6}{28} + \dfrac{5}{42}$

77. $\dfrac{25}{84} + \dfrac{41}{90}$

78. $\dfrac{23}{70} + \dfrac{29}{84}$

Simplify each expression.

79. $210\left(\dfrac{3}{14} + \dfrac{7}{30}\right)$

80. $210\left(\dfrac{3}{10} + \dfrac{11}{42}\right)$

81. $32\left(\dfrac{3}{4}\right) - 16\left(\dfrac{3}{4}\right)^2$

82. $32\left(\dfrac{3}{2}\right) - 16\left(\dfrac{3}{2}\right)^2$

Use the commutative, associative, and distributive properties to simplify the following.

83. $5a + 7 + 8a + a$

84. $3y + y + 5 + 2y + 1$

85. $2(5x + 1) + 2x$

86. $7 + 2(4y + 2)$

87. $3 + 4(5a + 3) + 4a$

88. $5x + 2(3x + 8) + 4$

89. $5x + 3(x + 2) + 7$

90. $2a + 4(2a + 6) + 3$

91. $5(x + 2y) + 4(3x + y)$

92. $3x + 4(2x + 3y) + 7y$

93. $5b + 3(4b + a) + 6a$

94. $4 + 3(2x + 3y) + 6(x + 4)$

For $\left\{ -6, -5.2, -\sqrt{7}, -\pi, 0, 1, 2, 2.3, \dfrac{9}{2}, \sqrt{17} \right\}$, list all the elements of the set that are named in each of the following problems.

95. Counting numbers

96. Whole numbers

97. Rational numbers

98. Integers

99. Irrational numbers

100. Real numbers

101. Nonnegative integers

102. Positive integers

Applying the Concepts

103. Rhind Papyrus In approximately 1650 BC, a mathematical document was written in ancient Egypt called the Rhind Papyrus. An "exercise" in this document asked the reader to find "a quantity such that when it is added to one fourth of itself results in 15." Verify this quantity must be 12.

104. Clock Arithmetic In a normal clock with 12 hours on its face, 12 is the additive identity because adding 12 hours to any time on the clock will not change the hands of the clock. Also, if we think of the hour hand of a clock, the problem 10 + 4 can be taken to mean: The hour hand is pointing at 10; if we add 4 more hours, it will be pointing at what number? Reasoning this way, we see that in clock arithmetic 10 + 4 = 2 and 9 + 6 = 3.

Find the following in clock arithmetic.

a. 10 + 5   **b.** 10 + 6   **c.** 10 + 1   **d.** 10 + 12

Griffith Park Fires Three major fires have occurred since Griffith Park was formed. The acres burned are given in the table:

Year	Acres
1933	47
1961	814
2007	817

Source: www.nfpa.org

The total acreage of Griffith Park is 4,217 acres.

105. How many more acres burned in 1961 as compared with 1933?

106. What fraction of the total acreage of the park burned in 2007?

107. What percent of the total acres was burned in 1933?

108. What percent of the total acres was burned in 1961?

Drinking and Driving The chart shows that if the risk of getting in an accident with no alcohol in your system is 1, then the risk of an accident with a blood-alcohol level of 0.24 is about 147 times as high. Use this chart for Problems 101 and 102.

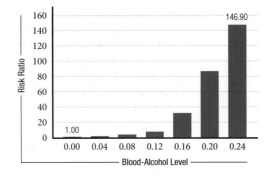

109. A person driving with a blood-alcohol level of 0.20 is how many times more likely to get in an accident than if she was driving with a blood-alcohol level of 0?

110. If the probability of getting in an accident while driving for an hour on surface streets in a certain city is 0.02%, what is the probability of getting in an accident in the same circumstances with a blood-alcohol level of 0.16?

111. Cost of Course Materials The chart below shows the average amount of money college students paid for course materials over a six-year period.

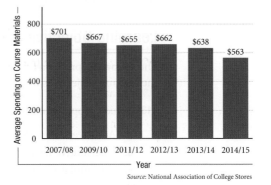

Source: National Association of College Stores

a. Notice that two school years are missing from the chart. Which ones are they?

b. Use a negative number to represent the change in spending on course materials from the 2012/13 school year to the 2014/15 school year.

c. The article also mentioned that, in the survey, most students preferred a print textbook to an eBook. Which do you prefer?

112. **The Best Selling Car Is a Truck** The numbers in the table below were reported by cars.com. They show the 2014 sales of cars and trucks for the month of September and cumulative from the first of the year through September of that year.

Rank	Model	Monthly Sales for September 2014	Monthly Change vs 2013	YTD for 2014	YTD Change vs 2013
1	Ford F-Series	59,863	−1.0%	557,037	−0.4%
2	Chevrolet Silverado	50,176	54.4%	382,153	5.9%
3	Ram Pickup	36,612	30.1%	319,868	21.7%
4	Honda Accord	32,956	30.9%	304,382	7.9%
5	Toyota Camry	28,507	−10.6%	334,978	5.0%
6	Honda CR-V	23,722	10.6%	241,015	5.2%
7	Toyota RAV4	22,724	42.7%	202,069	26.1%
8	Honda Civic	22,263	−3.1%	253,430	−0.1%
9	Ford Escape	21,718	−3.9%	230,162	0.8%
10	Ford Fusion	21,693	8.6%	240,585	6.3%

   a. How many more Ford F-Series pick up trucks were sold in the month of September than Honda Accords?

   b. How many Ford F-Series trucks were sold through the end of August?

   c. Find the difference in 2014 sales through September between the Ford Escape and the Ford Fusion.

Learning Objectives Assessment

The following problems can be used to help assess if you have successfully met the learning objectives for this section.

113. Which property of real numbers justifies $(3 + x) + 2 = 2 + (3 + x)$?

   a. Additive inverse property

   b. Distributive property

   c. Commutative property of addition

   d. Associative property of addition

114. Add: $\frac{11}{30} + \frac{5}{24}$.

   a. $\frac{23}{40}$ **b.** $\frac{8}{27}$ **c.** $\frac{2}{15}$ **d.** $\frac{23}{240}$

115. Simplify: $3(2x + 5) + 4$.

   a. $6x + 19$ **b.** $21x + 4$ **c.** $6x + 9$ **d.** $6x + 27$

Arithmetic with Real Numbers

Learning Objectives

In this section we will learn how to:

1. Add and subtract real numbers.

2. Multiply and divide real numbers.

3. Use the Order of Operations to evaluate expressions.

4. Divide fractions.

5. Find the value of an expression.

Adding Real Numbers

This section reviews the rules for arithmetic with real numbers and the justification for those rules. We can justify the rules for addition of real numbers geometrically by use of the real number line. Consider the *sum* of -5 and 3:

$$-5 + 3$$

We can interpret this expression as meaning "start at the origin and move 5 units in the negative direction and then 3 units in the positive direction." With the aid of a number line, we can visualize the process.

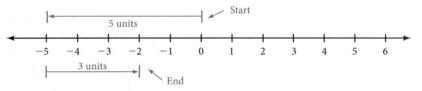

Because the process ends at -2, we say the sum of -5 and 3 is -2.

$$-5 + 3 = -2$$

We can use the real number line in this way to add any combination of positive and negative numbers.

The sum of -4 and -2, $-4 + (-2)$, can be interpreted as starting at the origin, moving 4 units in the negative direction, and then 2 more units in the negative direction.

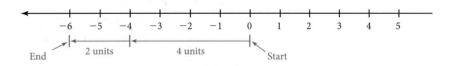

Because the process ends at -6, we say the sum of -4 and -2 is -6.

$$-4 + (-2) = -6$$

We can eliminate actually drawing a number line by simply visualizing it mentally. The following example gives the results of all possible sums of positive and negative 5 and 7.

VIDEO EXAMPLES

SECTION 1.3

 EXAMPLE 1   Add all combinations of positive and negative 5 and 7.

SOLUTION

$$5 + 7 = 12$$
$$-5 + 7 = 2$$
$$5 + (-7) = -2$$
$$-5 + (-7) = -12$$

Looking closely at the relationships in Example 1 (and trying other similar examples, if necessary), we can arrive at the following rule for adding two real numbers:

> **HOW TO**   *Add Two Real Numbers*
>
> With the *same* sign:
> **Step 1:** Add their absolute values.
> **Step 2:** Attach their common sign. If both numbers are positive, their sum is positive; if both numbers are negative, their sum is negative.
>
> With *opposite* signs:
> **Step 1:** Subtract the smaller absolute value from the larger.
> **Step 2:** Attach the sign of the number whose absolute value is larger.

Subtracting Real Numbers

To have as few rules as possible, we will not attempt to list new rules for the **difference** of two real numbers. We will instead define it in terms of addition and apply the rule for addition.

> **DEFINITION**   *difference*
>
> If a and b are any two real numbers, then the **difference** of a and b, written $a - b$, is given by
>
> $$a - b \qquad = \qquad a + (-b)$$
>
> To subtract b,   add the opposite of b.

We define the process of subtracting b from a as being equivalent to adding the opposite of b to a. In short, we say, "Subtraction is addition of the opposite."

EXAMPLE 2   Subtract.

a. $5 - 3 = 5 + (-3)$ Subtracting 3 is equivalent to adding -3
          $= 2$

b. $-7 - 6 = -7 + (-6)$ Subtracting 6 is equivalent to adding -6
          $= -13$

c. $9 - (-2) = 9 + 2$ Subtracting -2 is equivalent to adding 2
          $= 11$

d. $-6 - (-5) = -6 + 5$ Subtracting -5 is equivalent to adding 5
          $= -1$

EXAMPLE 3   Subtract -3 from -9.

SOLUTION   Because subtraction is not commutative, we must be sure to write the numbers in the correct order. Because we are subtracting -3, the problem looks like this when translated into symbols:

$$-9 - (-3) = -9 + 3 \qquad \text{Change to addition of the opposite}$$
$$= -6 \qquad \text{Add}$$

EXAMPLE 4   Add -4 to the difference of -2 and 5.

SOLUTION   The difference of -2 and 5 is written $-2 - 5$. Adding -4 to that difference gives us

$$(-2 - 5) + (-4) = -7 + (-4) \qquad \text{Simplify inside parentheses}$$
$$= -11 \qquad \text{Add}$$

Multiplying Real Numbers

Multiplication with whole numbers is simply a shorthand way of writing repeated addition. For example, $3(-2)$ can be evaluated as follows:

$$3(-2) = -2 + (-2) + (-2) = -6$$

We can evaluate the product $-3(2)$ in a similar manner if we first apply the commutative property of multiplication.

$$-3(2) = 2(-3) = -3 + (-3)$$

From these results, it seems reasonable to say that the product of a positive and a negative is a negative number.

The last case we must consider is the product of two negative numbers, such as $-3(-2)$. To evaluate this product we will look at the expression $-3[2 + (-2)]$ in two different ways. First, since $2 + (-2) = 0$, we know the expression $-3[2 + (-2)]$ is equal to 0. On the other hand, we can apply the distributive property to get

$$-3[2 + (-2)] = -3(2) + (-3)(-2) = -6 + \ ?$$

Because we know the expression is equal to 0, it must be true that our ? is 6; 6 is the only number we can add to -6 to get 0. Therefore, we have

$$-3(-2) = 6$$

Here is a summary of what we have so far:

Original Numbers Have		Answer Is
The same sign	$3(2) = 6$	Positive
Different signs	$3(-2) = -6$	Negative
Different signs	$-3(2) = -6$	Negative
The same sign	$-3(-2) = 6$	Positive

HOW TO   *Multiply Two Real Numbers*

Step 1: Multiply their absolute values.

Step 2: If the two numbers have the *same* sign, the product is positive. If the two numbers have *opposite* signs, the product is negative.

EXAMPLE 5   Multiply all combinations of positive and negative 7 and 3.

SOLUTION

$$7(3) = 21$$
$$7(-3) = -21$$
$$-7(3) = -21$$
$$-7(-3) = 21$$

Dividing Real Numbers

Just as we defined subtraction in terms of addition, we define division in terms of multiplication.

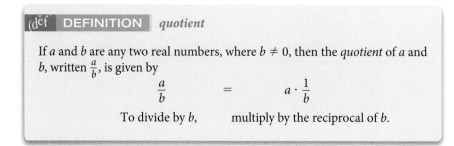

DEFINITION   *quotient*

If a and b are any two real numbers, where $b \neq 0$, then the *quotient* of a and b, written $\frac{a}{b}$, is given by

$$\frac{a}{b} \qquad = \qquad a \cdot \frac{1}{b}$$

To divide by b,   multiply by the reciprocal of b.

Dividing a by b is equivalent to multiplying a by the reciprocal of b. In short, we say, "Division is multiplication by the reciprocal."

Because division is defined in terms of multiplication, the same rules hold for assigning the correct sign to a quotient as held for assigning the correct sign to a product. That is, the quotient of two numbers with like signs is positive, while the quotient of two numbers with unlike signs is negative.

EXAMPLE 6   Divide.

a. $\dfrac{6}{3} = 6 \cdot \left(\dfrac{1}{3}\right) = 2$

b. $\dfrac{6}{-3} = 6 \cdot \left(-\dfrac{1}{3}\right) = -2$

c. $\dfrac{-6}{3} = -6 \cdot \left(\dfrac{1}{3}\right) = -2$

d. $\dfrac{-6}{-3} = -6 \cdot \left(-\dfrac{1}{3}\right) = 2$

Note The problems in Example 6 indicate that if a and b are positive real numbers, then

$$\frac{-a}{b} = \frac{a}{-b} = -\frac{a}{b}$$

and

$$\frac{-a}{-b} = \frac{a}{b}$$

The second step in the preceding examples is written only to show that each quotient can be written as a product. It is not actually necessary to show this step when working problems.

Order of Operations

It is important when evaluating arithmetic expressions in mathematics that each expression have only one answer in reduced form. Consider the expression

$$3 \cdot 7 + 2$$

If we find the product of 3 and 7 first, then add 2, the answer is 23. On the other hand, if we first combine the 7 and 2, then multiply by 3, we have 27. The problem seems to have two distinct answers depending on whether we multiply first or add first. To avoid this, we will decide that multiplication in a situation like this will always be done before addition. In this case, only the first answer, 23, is correct.

The complete set of rules for evaluating expressions follows:

[Δ≠Σ] RULE   *Order of Operations*

When evaluating a mathematical expression, we will perform the operations in the following order:
1. Begin with the expression in the innermost parentheses or brackets and work your way out.
2. Simplify all numbers with exponents, working from left to right if more than one of these expressions is present.
3. Work all multiplications and divisions left to right.
4. Perform all additions and subtractions left to right.

In the examples that follow, we find a combination of operations. In each case, we use the rule for order of operations.

EXAMPLE 7   Simplify the expression: $5 + 3(2 + 4)$.

SOLUTION $5 + 3(2 + 4) = 5 + 3(6)$ Simplify inside parentheses

$\qquad\qquad\qquad\quad = 5 + 18$ Multiply

$\qquad\qquad\qquad\quad = 23$ Add

EXAMPLE 8   Simplify each expression as much as possible.

a. $(-2 - 3)(5 - 9) = (-5)(-4)$ Simplify inside parentheses

$\qquad\qquad\qquad\qquad = 20$ Multiply

b. $2 - 5(7 - 4) - 6 = 2 - 5(3) - 6$ Simplify inside parentheses

$\qquad\qquad\qquad\qquad\quad = 2 - 15 - 6$ Multiply

$\qquad\qquad\qquad\qquad\quad = -19$ Subtract, left to right

c. $2(4 - 7)^3 + 3(-2 - 3)^2 = 2(-3)^3 + 3(-5)^2$ Simplify inside parentheses

$\qquad\qquad\qquad\qquad\qquad\quad = 2(-27) + 3(25)$ Evaluate numbers with exponents

$\qquad\qquad\qquad\qquad\qquad\quad = -54 + 75$ Multiply

$\qquad\qquad\qquad\qquad\qquad\quad = 21$ Add

d. $2|5 - 8| + 9 = 2|-3| + 9$ Simplify inside absolute value symbols

$\qquad\qquad\qquad\quad = 2(3) + 9$ Evaluate absolute value

$\qquad\qquad\qquad\quad = 6 + 9$ Multiply

$\qquad\qquad\qquad\quad = 15$ Add

> *Note*   Notice that absolute value symbols work like other grouping symbols when following the order of operations.

e. $3\left(\dfrac{t}{3} - 2\right) = 3 \cdot \dfrac{t}{3} - 3 \cdot 2$ Distributive Property

$\qquad\qquad\quad = t - 6$ Multiply

Our next examples involve more complicated fractions. The fraction bar works like parentheses to separate the numerator from the denominator. Although we don't usually write expressions this way, here is one way to think of the fraction bar:

$$\frac{-8 - 8}{-5 - 3} = (-8 - 8) \div (-5 - 3)$$

As you can see, if we apply the rule for order of operations to the expression on the right, we would work inside each set of parentheses first, then divide. Applying this to the expression on the left, we work on the numerator and denominator separately, then we divide, or reduce the resulting fraction to the lowest terms.

EXAMPLE 9

a. $\dfrac{-8 - 8}{-5 - 3} = \dfrac{-16}{-8}$ Simplify numerator and denominator separately

$= 2$ Divide

b. $\dfrac{-5(-4) + 2(-3)}{2(-1) - 5} = \dfrac{20 - 6}{-2 - 5}$ Simplify numerator and denominator separately

$= \dfrac{14}{-7}$

$= -2$ Divide

c. $\dfrac{2^3 + 3^3}{2^2 - 3^2} = \dfrac{8 + 27}{4 - 9}$ Simplify numerator and denominator separately

$= \dfrac{35}{-5}$

$= -7$ Divide

Remember, because subtraction is defined in terms of addition, we can restate the distributive property in terms of subtraction. That is, if a, b, and c are real numbers, then $a(b - c) = ab - ac$.

EXAMPLE 10   Simplify: $3(2y - 1) + y$.

SOLUTION   We begin by multiplying 3 and $2y - 1$. Then, we combine similar terms.

$3(2y - 1) + y = 6y - 3 + y$ Distributive property

$= 7y - 3$ Combine similar terms

EXAMPLE 11   Simplify: $8 - 3(4x - 2) + 5x$.

SOLUTION   First, we distribute -3 across $4x - 2$.

$8 - 3(4x - 2) + 5x = 8 - 12x + 6 + 5x$ Distributive property

$= -7x + 14$ Combine similar terms

EXAMPLE 12   Simplify: $5(2a + 3) - (6a - 4)$.

SOLUTION   We begin by applying the distributive property to remove the parentheses. The expression $-(6a - 4)$ can be thought of as $-1(6a - 4)$. Thinking of it in this way allows us to apply the distributive property.

$-1(6a - 4) = -1(6a) - (-1)(4) = -6a + 4$

Here is the complete problem:

$5(2a + 3) - (6a - 4) = 10a + 15 - 6a + 4$ Distributive property

$= 4a + 19$ Combine similar terms

Dividing Fractions

In the first section of this book, we reviewed multiplication with fractions. Now we'll review division with fractions.

EXAMPLE 13   Divide and reduce to lowest terms.

a. $\dfrac{3}{4} \div \dfrac{6}{11} = \dfrac{3}{4} \cdot \dfrac{11}{6}$ Definition of division

$\qquad\qquad = \dfrac{33}{24}$ Multiply numerators; multiply denominators

$\qquad\qquad = \dfrac{11}{8}$ Divide numerator and denominator by 3

b. $10 \div \dfrac{5}{6} = \dfrac{10}{1} \cdot \dfrac{6}{5}$ Definition of division

$\qquad\qquad = \dfrac{60}{5}$ Multiply numerators; multiply denominators

$\qquad\qquad = 12$ Divide

c. $-\dfrac{3}{8} \div 6 = -\dfrac{3}{8} \cdot \dfrac{1}{6}$ Definition of division

$\qquad\qquad = -\dfrac{3}{48}$ Multiply numerators; multiply denominators

$\qquad\qquad = -\dfrac{1}{16}$ Divide numerator and denominator by 3

Division with the Number 0

For every division problem an associated multiplication problem involving the same numbers exists. For example, the following two problems say the same thing about the numbers 2, 3, and 6:

Division	*Multiplication*
$\dfrac{6}{3} = 2$	$6 = 2(3)$

We can use this relationship between division and multiplication to clarify division involving the number 0.

First, dividing 0 by a number other than 0 is allowed and always results in 0. To see this, consider dividing 0 by 5. We know the answer is 0 because of the relationship between multiplication and division. This is how we write it:

$$\frac{0}{5} = 0 \qquad \text{because} \qquad 0 = 0(5)$$

On the other hand, dividing a nonzero number by 0 is not allowed in the real numbers. Suppose we were attempting to divide 5 by 0. We don't know whether there is an answer to this problem, but if there is, let's say the answer is a number that we can represent with the letter n. If 5 divided by 0 is a number n, then

$$\frac{5}{0} = n \quad \text{and} \quad 5 = n(0)$$

But this is impossible, because no matter what number n is, when we multiply it by 0 the answer must be 0. It can never be 5. In algebra, we say expressions like $\frac{5}{0}$ are **undefined**, because there is no answer to them. That is, division by 0 is not allowed in the real numbers.

> *Note* You may wonder why we can't divide 0 by 0, since
> $$0 = (0)(0)$$
> Dividing zero by itself is a more complicated concept, one analyzed in calculus, not algebra. So for our purposes, we will still consider $\frac{0}{0}$ as undefined.

Finding the Value of an Algebraic Expression

Recall that an algebraic expression is a combination of numbers, variables, and operation symbols. Each of the following is an algebraic expression:

$$7a \qquad x^2 - y^2 \qquad 2(3t - 4) \qquad \frac{2x - 5}{6}$$

An expression such as $2(3t - 4)$ will take on different values depending on what number we substitute for t. For example, if we substitute -8 for t then the expression $2(3t - 4)$ becomes $2[3(-8) - 4)]$ which simplifies to -56. If we apply the distributive property to $2(3t - 4)$ we have

$$2(3t - 4) = 6t - 8$$

Substituting -8 for t in the simplified expression gives us $6(-8) - 8 = -56$, which is the same result we obtained previously. As you would expect, substituting the same number into an expression, and any simplified form of that expression, will yield the same result.

 **EXAMPLE 14**   Evaluate the following expressions when a is -2, 0, and 3:
$$(a + 4)^2, a^2 + 16, \text{ and } a^2 + 8a + 16$$

SOLUTION Organizing our work with a table, we have

a	$(a + 4)^2$	$a^2 + 16$	$a^2 + 8a + 16$
-2	$(-2 + 4)^2 = 4$	$(-2)^2 + 16 = 20$	$(-2)^2 + 8(-2) + 16 = 4$
0	$(0 + 4)^2 = 16$	$0^2 + 16 = 16$	$0^2 + 8(0) + 16 = 16$
3	$(3 + 4)^2 = 49$	$3^2 + 16 = 25$	$3^2 + 8(3) + 16 = 49$

When we study polynomials later in the book, you will see that the expressions $(a + 4)^2$ and $a^2 + 8a + 16$ are equivalent, and that neither one is equivalent to $a^2 + 16$ (except when a is replaced by 0).

Getting Ready for Class

After reading through the preceding section, respond in your own words and in complete sentences.

A. Explain the steps you use to add two real numbers.

B. Use symbols to define the word difference.

C. How do we define division in terms of multiplication?

D. Find the value of the expression $3(4t - 2)$ if $t = 2$.

Find each of the following sums.

1. $6 + (-2)$ **2.** $11 - 5$ **3.** $-6 + 2$ **4.** $-11 + 5$

Find each of the following differences.

5. $-7 - 3$ **6.** $-6 - 9$ **7.** $-7 - (-3)$ **8.** $-6 - (-9)$

9. $\dfrac{3}{4} - \left(-\dfrac{5}{6}\right)$ **10.** $\dfrac{2}{3} - \left(-\dfrac{7}{5}\right)$ **11.** $\dfrac{11}{42} - \dfrac{17}{30}$ **12.** $\dfrac{13}{70} - \dfrac{19}{42}$

13. Subtract 5 from -3.   **14.** Subtract -3 from 5.

15. Find the difference of -4 and 8.   **16.** Find the difference of 8 and -4.

17. Subtract $4x$ from $-3x$.   **18.** Subtract $-5x$ from $7x$.

19. What number do you subtract from 5 to get -8?   **20.** What number do you subtract from -3 to get 9?

21. Add -7 to the difference of 2 and 9.   **22.** Add -3 to the difference of 9 and 2.

23. Subtract $3a$ from the sum of $8a$ and a.   **24.** Subtract $-3a$ from the sum of $3a$ and $5a$.

Find the following products.

25. $3(-5)$ **26.** $-3(5)$ **27.** $-3(-5)$ **28.** $4(-6)$

29. $2(-3)(4)$ **30.** $-2(3)(-4)$ **31.** $-2(5x)$ **32.** $-5(4x)$

33. $-\dfrac{1}{3}(-3x)$ **34.** $-\dfrac{1}{6}(-6x)$ **35.** $-\dfrac{2}{3}\left(-\dfrac{3}{2}y\right)$ **36.** $-\dfrac{2}{5}\left(-\dfrac{5}{2}y\right)$

Simplify each expression as much as possible following the order of operations.

37. $1(-2) - 2(-16) + 1(9)$ **38.** $6(1) - 1(-5) + 1(2)$

39. $1(1) - 3(-2) + (-2)(-2)$ **40.** $-2(-14) + 3(-4) - 1(-10)$

41. $-4(0)(-2) - (-1)(1)(1) - 1(2)(3)$

42. $1(0)(1) + 3(1)(4) + (-2)(2)(-1)$

43. $1[0 - (-1)] - 3(2 - 4) + (-2)(-2 - 0)$

44. $-3(-1 - 1) + 4(-2 + 2) - 5[2 - (-2)]$

45. $3(-2)^2 + 2(-2) - 1$ **46.** $4(-1)^2 + 3(-1) - 2$

47. $2(-2)^3 - 3(-2)^2 + 4(-2) - 8$ **48.** $5 \cdot 2^3 - 3 \cdot 2^2 + 4 \cdot 2 - 5$

49. $\dfrac{0 - 4}{0 - 2}$ **50.** $\dfrac{0 + 6}{0 - 3}$ **51.** $\dfrac{-4 - 4}{-4 - 2}$ **52.** $\dfrac{6 + 6}{6 - 3}$

53. $\dfrac{-6 + 6}{-6 - 3}$ **54.** $\dfrac{4 - 4}{4 - 2}$ **55.** $\dfrac{-2 - 4}{2 - 2}$ **56.** $\dfrac{3 + 6}{3 - 3}$

57. $\dfrac{3 - (-1)}{-3 - 3}$ **58.** $\dfrac{-1 - 3}{3 - (-3)}$ **59.** $\dfrac{-3^2 + 9}{-4 - 4}$ **60.** $\dfrac{-2^2 + 4}{-5 - 5}$

Simplify each expression following the order of operations.

61. $|8 - 2|$

62. $|1 - 6|$

63. $|5 \cdot 2^3 - 2 \cdot 3^2|$

64. $|2 \cdot 10^2 + 3 \cdot 10|$

65. $|7 - 2| - |4 - 2|$

66. $|10 - 3| - |4 - 1|$

67. $10 - |7 - 2(5 - 3)|$

68. $12 - |9 - 6(7 - 5)|$

69. $15 - |8 - 2(3 \cdot 4 - 9)| - 10$

70. $25 - |9 - 3(4 \cdot 5 - 18)| - 20$

Simplify each expression.

71. $-2(4x - 3)$

72. $-2(-5t + 6)$

73. $-\dfrac{1}{2}(6a - 8)$

74. $-\dfrac{1}{3}(6a - 9)$

75. $3(5x + 4) - x$

76. $4(7x + 3) - x$

77. $6 - 7(3 - m)$

78. $3 - 5(5 - m)$

79. $7 - 2(3x - 1) + 4x$

80. $8 - 5(2x - 3) + 4x$

81. $5(3y + 1) - (8y - 5)$

82. $4(6y + 3) - (6y - 6)$

83. $4(2 - 6x) - (3 - 4x)$

84. $7(1 - 2x) - (4 - 10x)$

85. $10 - 4(2x + 1) - (3x - 4)$

86. $7 - 2(3x + 5) - (2x - 3)$

87. $3x - 5(x - 3) - 2(1 - 3x)$

88. $4x - 7(x - 3) + 2(4 - 5x)$

Find the following quotients.

89. $\dfrac{4}{0}$

90. $\dfrac{-7}{0}$

91. $\dfrac{0}{-3}$

92. $\dfrac{0}{5}$

Use the definition of division to write each division problem as a multiplication problem, then simplify.

93. $-\dfrac{3}{4} \div \dfrac{9}{8}$

94. $-\dfrac{2}{3} \div \dfrac{4}{9}$

95. $-8 \div \left(-\dfrac{1}{4}\right)$

96. $-12 \div \left(-\dfrac{2}{3}\right)$

97. $-40 \div \left(-\dfrac{5}{8}\right)$

98. $-30 \div \left(-\dfrac{5}{6}\right)$

99. $\dfrac{4}{9} \div (-8)$

100. $\dfrac{3}{7} \div (-6)$

Simplify as much as possible following the order of operations.

101. $\dfrac{3(-1) - 4(-2)}{8 - 5}$

102. $\dfrac{6(-4) - 5(-2)}{7 - 6}$

103. $8 - (-6)\left[\dfrac{2(-3) - 5(4)}{-8(6) - 4}\right]$

104. $-9 - 5\left[\dfrac{11(-1) - 9}{4(-3) + 2(5)}\right]$

105. $6 - (-3)\left[\dfrac{2 - 4(3 - 8)}{1 - 5(1 - 3)}\right]$

106. $8 - (-7)\left[\dfrac{6 - 1(6 - 10)}{4 - 3(5 - 7)}\right]$

Complete each of the following tables.

107.

a	b	Sum $a + b$	Difference $a - b$	Product ab	Quotient $\frac{a}{b}$
3	12				
−3	12				
3	−12				
−3	−12				

108.

a	b	Sum $a + b$	Difference $a - b$	Product ab	Quotient $\frac{a}{b}$
8	2				
−8	2				
8	−2				
−8	−2				

109.

x	$3(5x - 2)$	$15x - 6$	$15x - 2$
−2			
−1			
0			
1			
2			

110.

x	$(x + 1)^2$	$x^2 + 1$	$x^2 + 2x + 1$
−2			
−1			
0			
1			
2			

111. Find the value of $-\dfrac{b}{2a}$ when

    a. $a = 3, b = -6$

    b. $a = -2, b = 6$

    c. $a = -1, b = -2$

    d. $a = -0.1, b = 27$

112. Find the value of $b^2 - 4ac$ when

    a. $a = 3, b = -2,$ and $c = 4$

    b. $a = 1, b = -3,$ and $c = -28$

    c. $a = 1, b = -6,$ and $c = 9$

    d. $a = 0.1, b = -27,$ and $c = 1700$

Use a calculator to simplify each expression. If rounding is necessary, round your answers to the nearest ten thousandth (4 places past the decimal point). You will see these types of problems later in the book.

113. $\dfrac{1.380}{0.903}$

114. $\dfrac{1.0792}{0.6690}$

115. $\dfrac{1}{2}(-0.1587)$

116. $\dfrac{1}{2}(-0.7948)$

117. $\dfrac{1}{2}\left(\dfrac{1.2}{1.4} - 1\right)$

118. $\dfrac{1}{2}\left(\dfrac{1.3}{1.1} - 1\right)$

119. $\dfrac{(6.8)(3.9)}{7.8}$

120. $\dfrac{(2.4)(1.8)}{1.2}$

121. $\dfrac{0.0005(200)}{(0.25)^2}$

122. $\dfrac{0.0006(400)}{(0.25)^2}$

123. $-500 + 27(100) - 0.1(100)^2$

124. $-500 + 27(170) - 0.1(170)^2$

125. $-0.05(130)^2 + 9.5(130) - 200$

126. $-0.04(130)^2 + 8.5(130) - 210$

Applying the Concepts

127. Football Yardage A football team gains 6 yards on one play and then loses 8 yards on the next play. To what number on the number line does a loss of 8 yards correspond? The total yards gained or lost on the two plays corresponds to what negative number?

128. Checking Account Balance Nancy has a balance of $20 in her checking account. If she writes a check for $30, what negative number can be used to represent the new balance in her checking account?

Temperature In the United States, temperature is measured on the Fahrenheit temperature scale. On this scale, water boils at 212 degrees and freezes at 32 degrees. To denote a temperature of 32 degrees on the Fahrenheit scale, we write 32°F, which is read "32 degrees Fahrenheit."

Use this information for Problems 129 and 130.

129. Temperature and Altitude Marilyn is flying from Seattle to San Francisco on a Boeing 737 jet. When the plane reaches an altitude of 35,000 feet, the temperature outside the plane is 64 degrees below zero Fahrenheit. Represent the temperature with a negative number. If the temperature outside the plane gets warmer by 10 degrees, what will the new temperature be?

130. Temperature Change At 10:00 in the morning in White Bear Lake, Minn., John notices the temperature outside is 10 degrees below zero Fahrenheit. Write the temperature as a negative number. An hour later it has warmed up by 6 degrees. What is the temperature at 11:00 that morning?

131. Scuba Diving Steve is scuba diving near his home in Maui. At one point he is 100 feet below the surface. Assuming the surface is 0, represent this number with a negative number. If he descends another 5 feet, what negative number will represent his new position?

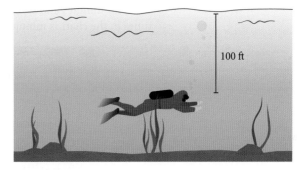

132. Oceans and Mountains The deepest ocean depth is 35,840 feet, found in the Pacific Ocean's Mariana Trench. The tallest mountain is Mount Everest, with a height of 29,028 feet. What is the difference between the highest point on Earth and the lowest point on Earth?

City and State Parks The Trust for Public Land lists the size of Griffith Park as 4,217 acres.

133. South Mountain Preserve in Phoenix, Arizona is the largest municipally–owned city park at 16,094 acres. How many times larger is this park than Griffith Park? (Round to the nearest tenth.)

134. The largest state park within a city is Chugache State Park in Anchorage, Alaska at 490,125 acres. How many times would Griffith Park fit inside Chugache State Park? (Round to the nearest tenth.)

135. Convenience Stores The chart shows the states with the most convenience stores. Given the area of the state, find the number of convenience stores per square mile.

a. Georgia: 59,424 square miles

b. Texas: 268,580 square miles

c. Which state has a higher concentration of stores, Georgia or Texas?

136. Downhill Skiing In our number system, everything is in terms of powers of 10. With minutes and seconds, we think in terms of 60's. The format for the times in the chart is minutes:seconds:hundreths.

FASTEST DOWNHILL SKIER - VANCOUVER 2010
www.vancouver2010.com

Lindsey Vonn	1:44:19
Julia Mancuso	1:44:75
Elisabeth Goergl	1:45:65
Andrea Fischbacher	1:45:68

minutes:seconds:hundredths

Find the difference between each of the following downhill ski times:

a. Lindsey Vonn and Julia Mancuso

b. Elizabeth Goergl and Andrea Fischbacher

c. Lindsey Vonn and Andrea Fischbacher

Learning Objectives Assessment

The following problems can be used to help assess if you have successfully met the learning objectives for this section.

137. Simplify: $-4 - 9 - (-5)$.

    a. 10   **b.** -18 **c.** 0   **d.** -8

138. Simplify: $\dfrac{2(-6)}{-3}$.

    a. -4 **b.** 4   **c.** $\dfrac{4}{3}$ **d.** $-\dfrac{4}{3}$

139. Divide: $\dfrac{5}{6} \div \left(-\dfrac{7}{12}\right)$.

    a. $\dfrac{10}{7}$ **b.** $-\dfrac{10}{7}$ **c.** $\dfrac{35}{2}$ **d.** $-\dfrac{35}{2}$

140. Find the value of $x^2 - 3y$ when $x = -2$ and $y = -5$.

    a. 11   **b.** -19 **c.** 19   **d.** -11

Learning Objectives

In this section we will learn how to:

1. Simplify expressions using the properties of exponents.

2. Convert numbers between scientific notation and expanded form.

3. Multiply and divide expressions written in scientific notation.

The following figure shows a square and a cube, each with a side of length 1.5 centimeters. To find the area of the square, we raise 1.5 to the second power: 1.5^2. To find the volume of the cube, we raise 1.5 to the third power: 1.5^3.

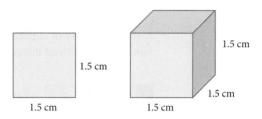

1.5 cm

1.5 cm

1.5 cm

1.5 cm

1.5 cm

1.5 cm

Because the area of the square is 1.5^2, we say second powers are *squares*; that is, x^2 is read "x squared." Likewise, because the volume of the cube is 1.5^3, we say third powers are *cubes*; that is, x^3 is read "x cubed." Exponents and the vocabulary associated with them are topics we will study in this section.

Properties of Exponents

In this section, we will be concerned with the simplification of expressions that involve exponents. We begin by making some generalizations about exponents, based on specific examples.

VIDEO EXAMPLES

SECTION 1.4

EXAMPLE 1   Write the product $x^3 \cdot x^4$ with a single exponent.

SOLUTION

$$x^3 \cdot x^4 = (x \cdot x \cdot x)(x \cdot x \cdot x \cdot x)$$
$$= (x \cdot x \cdot x \cdot x \cdot x \cdot x \cdot x)$$
$$= x^7 \qquad \text{Note: } 3 + 4 = 7$$

We can generalize this result into the first property of exponents.

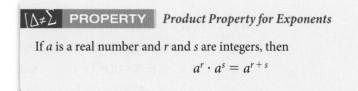

[Δ≠Σ PROPERTY   *Product Property for Exponents*

If a is a real number and r and s are integers, then
$$a^r \cdot a^s = a^{r+s}$$

EXAMPLE 2   Write $(5^3)^2$ with a single exponent.

SOLUTION

$$(5^3)^2 = 5^3 \cdot 5^3$$

$$= 5^6 \qquad \text{Note: } 3 \cdot 2 = 6$$

Generalizing this result, we have the second property of exponents.

> **$\lceil\Delta\neq\Sigma$ PROPERTY**   *Power Property for Exponents*
>
> If a is a real number and r and s are integers, then
> $$(a^r)^s = a^{r \cdot s}$$

A third property of exponents arises when we have the product of two or more numbers raised to an integer power.

EXAMPLE 3   Expand $(3x)^4$ and then multiply.

SOLUTION

$$(3x)^4 = (3x)(3x)(3x)(3x)$$

$$= (3 \cdot 3 \cdot 3 \cdot 3)(x \cdot x \cdot x \cdot x)$$

$$= 3^4 \cdot x^4 \qquad \text{Note: The exponent 4 distributes over the product } 3x$$

$$= 81x^4$$

Generalizing Example 3, we have our first distributive property for exponents.

> **$\lceil\Delta\neq\Sigma$ PROPERTY**   *Distributive Property for Exponents over Multiplication*
>
> If a and b are any two real numbers and r is an integer, then
> $$(ab)^r = a^r \cdot b^r$$

Here are some examples that use combinations of the first three properties of exponents to simplify expressions involving exponents.

EXAMPLE 4   Simplify each expression using the properties of exponents.

a. $(-3x^2)(5x^4) = (-3)(5)(x^2 \cdot x^4)$ Commutative and associative properties

$\qquad\qquad\quad = -15x^6$ Product property for exponents

b. $(-2x^2)^3(4x^5) = (-2)^3(x^2)^3(4x^5)$ Distributive property for exponents

$\qquad\qquad\quad\; = (-8)(x^6)(4x^5)$ Power property for exponents

$\qquad\qquad\quad\; = (-8 \cdot 4)(x^6 \cdot x^5)$ Commutative and associative properties

$\qquad\qquad\quad\; = -32x^{11}$ Product property for exponents

c. $(x^2)^4(x^2y^3)^2(y^4)^3 = x^8 \cdot x^4 \cdot y^6 \cdot y^{12}$ Power property and distributive property for exponents

$\qquad\qquad\qquad\quad = x^{12}y^{18}$ Product property for exponents

The next property of exponents deals with negative integer exponents.

Note   The negative exponent property for exponents is actually a definition. That is, we are defining negative integer exponents as indicating reciprocals. Doing so gives us a way to write an expression with a negative exponent as an equivalent expression with a positive exponent.

[Δ≠Σ]   **PROPERTY**   *Negative Exponent Property*

If a is any nonzero real number and r is a positive integer, then

$$a^{-r} = \frac{1}{a^r}$$

EXAMPLE 5   Write with positive exponents, and then simplify.

a. $5^{-2} = \dfrac{1}{5^2} = \dfrac{1}{25}$

b. $(-2)^{-3} = \dfrac{1}{(-2)^3} = \dfrac{1}{-8} = -\dfrac{1}{8}$

c. $\left(\dfrac{3}{4}\right)^{-2} = \dfrac{1}{\left(\dfrac{3}{4}\right)^2} = \dfrac{1}{\dfrac{9}{16}} = \dfrac{16}{9}$

If we generalize the result in Example 5c, we have the following extension of the negative exponent property,

$$\left(\frac{a}{b}\right)^{-r} = \left(\frac{b}{a}\right)^{r}$$

which indicates that raising a fraction to a negative power is equivalent to raising the reciprocal of the fraction to the positive power.

The first distributive property for exponents indicated that exponents distribute over products. Because division is defined in terms of multiplication, we can expect that exponents will distribute over quotients as well. The second distributive property for exponents is the formal statement of this fact.

[Δ≠Σ]   **PROPERTY**   *Distributive Property for Exponents over Division*

If a and b are any two real numbers with $b \neq 0$, and r is an integer, then

$$\left(\frac{a}{b}\right)^{r} = \frac{a^r}{b^r}$$

Proof of Distributive Property for Exponents over Division

$$\left(\frac{a}{b}\right)^r = \underbrace{\left(\frac{a}{b}\right)\left(\frac{a}{b}\right)\left(\frac{a}{b}\right)\cdots\left(\frac{a}{b}\right)}_{r \text{ factors}}$$

$$= \frac{a \cdot a \cdot a \cdots a}{b \cdot b \cdot b \cdots b} \quad\begin{array}{l}\leftarrow r \text{ factors}\\ \leftarrow r \text{ factors}\end{array}$$

$$= \frac{a^r}{b^r}$$

Because multiplication with the same base resulted in addition of exponents, it seems reasonable to expect division with the same base to result in subtraction of exponents.

> ### ⟨Δ≠Σ⟩ **PROPERTY**   *Quotient Property for Exponents*
>
> If a is any nonzero real number, and r and s are any two integers, then
>
> $$\frac{a^r}{a^s} = a^{r-s}$$

Notice again that we have specified r and s to be any integers. Our definition of negative exponents is such that the properties of exponents hold for all integer exponents, whether positive or negative integers. Here is a proof of the quotient property for exponents.

Proof of The Quotient Property for Exponents

Our proof is centered on the fact that division by a number is equivalent to multiplication by the reciprocal of the number.

$$\frac{a^r}{a^s} = a^r \cdot \frac{1}{a^s} \qquad \text{Dividing by } a^s \text{ is equivalent to multiplying by } \frac{1}{a^s}$$

$$= a^r a^{-s} \qquad \text{Negative exponent property}$$

$$= a^{r+(-s)} \qquad \text{Product property for exponents}$$

$$= a^{r-s} \qquad \text{Definition of subtraction}$$

▧ **EXAMPLE 6**   Apply the quotient property for exponents to each expression, and then simplify the result. All answers that contain exponents should contain positive exponents only.

a. $\dfrac{2^8}{2^3} = 2^{8-3} = 2^5 = 32$

b. $\dfrac{x^2}{x^{18}} = x^{2-18} = x^{-16} = \dfrac{1}{x^{16}}$

c. $\dfrac{a^6}{a^{-8}} = a^{6-(-8)} = a^{14}$

d. $\dfrac{m^{-5}}{m^{-7}} = m^{-5-(-7)} = m^2$ ▧

Let's complete our list of properties by looking at how the numbers 0 and 1 behave when used as exponents.

Here is the definition we use when the number 1 is used as an exponent.

$$a^1 = a$$
$$\uparrow$$
$$\text{1 factor}$$

For 0 as an exponent, consider the expression $\dfrac{3^4}{3^4}$. Because $3^4 = 81$, we have

$$\frac{3^4}{3^4} = \frac{81}{81} = 1$$

However, because we have the quotient of two expressions with the same base, we can subtract exponents.

$$\frac{3^4}{3^4} = 3^{4-4} = 3^0$$

Hence, 3^0 must be the same as 1.

Summarizing these results, we have our last two properties for exponents.

> **[Δ≠Σ] PROPERTY** *Identity Property for Exponents*
>
> If a is any real number, then
> $$a^1 = a$$

> **[Δ≠Σ] PROPERTY** *Zero Exponent Property*
>
> If a is any real number, then
> $$a^0 = 1 \quad \text{(as long as } a \neq 0)$$

EXAMPLE 7 Simplify.

a. $(2x^2y^4)^0 = 1$

b. $(2x^2y^4)^1 = 2x^2y^4$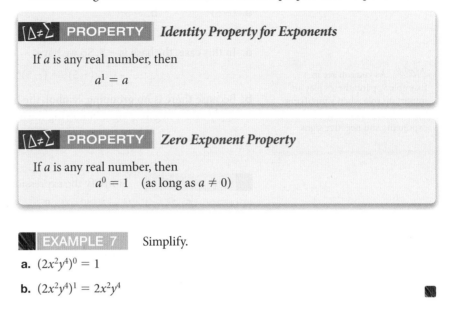

Here are some examples that use many of the properties of exponents. There are a number of ways to proceed on problems like these. You should use the method that works best for you.

EXAMPLE 8 Simplify.

a. $\dfrac{(x^3)^{-2}(x^4)^5}{(x^{-2})^7} = \dfrac{x^{-6}x^{20}}{x^{-14}}$ Power property for exponents

$\qquad = \dfrac{x^{14}}{x^{-14}}$ Product property for exponents

$\qquad = x^{28}$ Quotient property for exponents: $x^{14-(-14)} = x^{28}$

b. $\dfrac{6a^5b^{-6}}{12a^3b^{-9}} = \dfrac{6}{12} \cdot \dfrac{a^5}{a^3} \cdot \dfrac{b^{-6}}{b^{-9}}$ Write as separate fractions

$\qquad = \dfrac{1}{2}a^2b^3$ Quotient property for exponents

Note The answer to Example 8(b) can also be written as $\dfrac{a^2b^3}{2}$. Either answer is correct.

c. $\dfrac{(4x^{-5}y^3)^2}{(x^4y^{-6})^{-3}} = \dfrac{16x^{-10}y^6}{x^{-12}y^{18}}$ Power property and distributive property for exponents

$\qquad = 16x^2y^{-12}$ Quotient property for exponents

$\qquad = 16x^2 \cdot \dfrac{1}{y^{12}}$ Negative exponent property

$\qquad = \dfrac{16x^2}{y^{12}}$ Multiply

With more complex expressions involving exponents, we follow the order of operations that we introduced in the previous section.

EXAMPLE 9   Simplify.

a. $(-5)^2$ **b.** -5^2

SOLUTION

a. In this case, the base is -5. So we have

$$(-5)^2 = (-5)(-5) = 25$$

b. Because there is no grouping symbol, the base is 5, and the exponent does not apply to the negative sign. Following the order of operations, we have

$$-5^2 = -(5^2) = -25$$

EXAMPLE 10   Simplify the expression $5 \cdot 2^3 - 4 \cdot 3^2$.

SOLUTION $5 \cdot 2^3 - 4 \cdot 3^2 = 5 \cdot 8 - 4 \cdot 9$ Simplify exponents left to right

$$= 40 - 36$$ Multiply left to right

$$= 4$$ Subtract

EXAMPLE 11   Simplify the expression $20 - (2 \cdot 5^2 - 30)$.

SOLUTION $20 - (2 \cdot 5^2 - 30) = 20 - (2 \cdot 25 - 30)$ ⎱ Simplify inside parentheses,
$$= 20 - (50 - 30)$$ evaluating exponents first,
$$= 20 - (20)$$ then multiply, and finally
$$= 0$$ subtract

The absolute value symbol acts as a grouping symbol. When working with the absolute value of sums and differences, we must simplify the expression inside the absolute value symbols first and then find the absolute value of the simplified expression.

EXAMPLE 12   Simplify each expression.

a. $|8 - 3|$ **b.** $|3 \cdot 2^3 + 2 \cdot 3^2|$ **c.** $|9 - 2| - |8 - 6|$

SOLUTION

a. $|8 - 3| = |5|$ Simplify inside the absolute value

$$= 5$$ Find the absolute value of 5

b. $|3 \cdot 2^3 + 2 \cdot 3^2| = |3 \cdot 8 + 2 \cdot 9|$ Simplify exponents

$$= |24 + 18|$$ Multiply

$$= |42|$$ Add

$$= 42$$ Find the absolute value of 42

c. $|9 - 2| - |8 - 6| = |7| - |2|$ Simplify inside each absolute value

$$= 7 - 2$$ Evaluate each absolute value

$$= 5$$ Subtract

Scientific Notation

Scientific notation is a method for writing very large or very small numbers in a more manageable form. Here is the definition:

> **(dëf DEFINITION** *scientific notation*
>
> A number is written in *scientific notation* if it is written as the product of a number between 1 and 10 and an integer power of 10. A number written in scientific notation has the form
>
> $$n \times 10^r$$
>
> where $1 \leq n < 10$ and $r =$ an integer.

EXAMPLE 13 Write 376,000 in scientific notation.

SOLUTION We must rewrite 376,000 as the product of a number between 1 and 10 and a power of 10. To do so, we move the decimal point five places to the left so that it appears between the 3 and the 7. Then we multiply this number by 10^5. The number that results has the same value as our original number and is written in scientific notation.

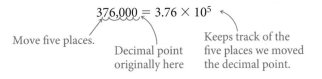

$$376,000 = 3.76 \times 10^5$$

Move five places. Decimal point originally here Keeps track of the five places we moved the decimal point.

If a number written in expanded form is greater than or equal to 10, then when the number is written in scientific notation the exponent on 10 will be positive. A number that is less than one and greater than zero will have a negative exponent when written in scientific notation.

EXAMPLE 14 Write 4.52×10^3 in expanded form.

SOLUTION Because 10^3 is 1,000, we can think of this as simply a multiplication problem. That is,

$$4.52 \times 10^3 = 4.52 \times 1,000 = 4,520$$

On the other hand, we can think of the exponent 3 as indicating the number of places we need to move the decimal point to write our number in expanded form. Because our exponent is positive 3, we move the decimal point three places to the right.

$$4.52 \times 10^3 = 4,520$$

The following table lists some additional examples of numbers written in expanded form and in scientific notation. In each case, note the relationship between the number of places the decimal point is moved and the exponent on 10.

Number Written in Expanded Form		Number Written in Scientific Notation
376,000	=	3.76×10^5
49,500	=	4.95×10^4
3,200	=	$3.2 \ \times 10^3$
591	=	5.91×10^2
46	=	$4.6 \ \times 10^1$
8	=	$8 \ \ \times 10^0$
0.47	=	$4.7 \ \times 10^{-1}$
0.093	=	$9.3 \ \times 10^{-2}$
0.00688	=	6.88×10^{-3}
0.0002	=	$2 \ \ \times 10^{-4}$
0.000098	=	$9.8 \ \times 10^{-5}$

USING TECHNOLOGY   *Scientific Notation*

Some scientific calculators have a key that allows you to enter numbers in scientific notation. The key is labeled

$$\boxed{\text{EXP}} \quad \text{or} \quad \boxed{\text{EE}} \quad \text{or} \quad \boxed{\text{SCI}}$$

To enter the number 3.45×10^6, you would enter the decimal number, press the scientific notation key, and then enter the exponent.

$$3.45 \ \boxed{\text{EXP}} \ 6$$

To enter 6.2×10^{-27}, you would use the following sequence:

$$6.2 \ \boxed{\text{EXP}} \ 27 \ \boxed{+/-}$$

Simplifying Expressions with Scientific Notation

We can use the properties of exponents to do arithmetic with numbers written in scientific notation. Here are some examples:

EXAMPLE 15   Simplify each expression and write all answers in scientific notation.

a. $(2 \times 10^8)(3 \times 10^{-3}) = (2)(3) \times (10^8)(10^{-3})$

$$= 6 \times 10^5$$

b. $\dfrac{4.8 \times 10^9}{2.4 \times 10^{-3}} = \dfrac{4.8}{2.4} \times \dfrac{10^9}{10^{-3}}$

$$= 2 \times 10^{9-(-3)}$$

$$= 2 \times 10^{12}$$

c. $\dfrac{(6.8 \times 10^5)(3.9 \times 10^{-7})}{7.8 \times 10^{-4}} = \dfrac{(6.8)(3.9)}{7.8} \times \dfrac{(10^5)(10^{-7})}{10^{-4}}$

$$= 3.4 \times 10^2$$

Getting Ready for Class

After reading through the preceding section, respond in your own words and in complete sentences.

A. Explain the difference between -2^4 and $(-2)^4$.

B. If a positive base is raised to a negative exponent, can the result be a negative number?

C. State the product property for exponents in your own words.

D. What is scientific notation?

Problem Set 1.4

Evaluate each of the following.

1. 4^2
2. $(-4)^2$
3. -4^2
4. $-(-4)^2$

5. -0.3^3
6. $(-0.3)^3$
7. 2^5
8. 2^4

9. $\left(\dfrac{1}{2}\right)^3$
10. $\left(\dfrac{3}{4}\right)^2$
11. $\left(\dfrac{5}{6}\right)^2$
12. $\left(\dfrac{2}{3}\right)^3$

13. $\left(\dfrac{1}{10}\right)^4$
14. $\left(\dfrac{1}{10}\right)^5$
15. $\left(-\dfrac{5}{6}\right)^2$

16. $\left(-\dfrac{7}{8}\right)^2$
17. $\left(-\dfrac{3}{7}\right)^2$
18. $\left(-\dfrac{4}{5}\right)^3$

Use the properties of exponents to simplify each of the following as much as possible.

19. $x^5 \cdot x^4$
20. $x^6 \cdot x^3$
21. $(2^3)^2$
22. $(3^2)^2$

23. $-3a^2(2a^4)$
24. $5a^7(-4a^6)$
25. $(4x^2)^2$
26. $(-3y^2)^2$

Write each of the following with positive exponents. Then simplify as much as possible.

27. 3^{-2}
28. $(-5)^{-2}$
29. $(-2)^{-5}$
30. 2^{-5}

31. $\left(\dfrac{3}{4}\right)^{-2}$
32. $\left(\dfrac{3}{5}\right)^{-2}$
33. $\left(\dfrac{1}{3}\right)^{-2} + \left(\dfrac{1}{2}\right)^{-3}$
34. $\left(\dfrac{1}{2}\right)^{-2} + \left(\dfrac{1}{3}\right)^{-3}$

Multiply.

35. $8x^3 \cdot 10y^6$
36. $5y^2 \cdot 4x^2$
37. $8x^3 \cdot 9y^3$
38. $4y^3 \cdot 3x^2$

39. $3x \cdot 5y$
40. $3xy \cdot 5z$
41. $4x^6y^6 \cdot 3x$
42. $16x^4y^4 \cdot 3y$

43. $27a^6c^3 \cdot 2b^2c$
44. $8a^3b^3 \cdot 5a^2b$
45. $12x^3y^4 \cdot 3xy^2$
46. $-8x^2y \cdot 2xy^4$

Divide. (Assume all variables are nonzero.)

47. $\dfrac{10x^5}{5x^2}$
48. $\dfrac{-15x^4}{5x^2}$
49. $\dfrac{20x^3}{5x^2}$
50. $\dfrac{25x^7}{-5x^2}$

51. $\dfrac{8x^3y^5}{-2x^2y}$
52. $\dfrac{-16x^2y^2}{-2x^2y}$
53. $\dfrac{4x^4y^3}{-2x^2y}$
54. $\dfrac{10a^4b^2}{4a^2b^2}$

Use the properties of exponents to simplify each expression. Write all answers with positive exponents only. (Assume all variables are nonzero.)

55. $\dfrac{x^{-1}}{x^9}$
56. $\dfrac{x^{-3}}{x^5}$
57. $\dfrac{a^4}{a^{-6}}$
58. $\dfrac{a^5}{a^{-2}}$

59. $\dfrac{t^{-10}}{t^{-4}}$
60. $\dfrac{t^{-8}}{t^{-5}}$
61. $\left(\dfrac{x^5}{x^3}\right)^6$
62. $\left(\dfrac{x^7}{x^4}\right)^5$

63. $\dfrac{(x^5)^6}{(x^3)^4}$
64. $\dfrac{(x^7)^3}{(x^4)^5}$
65. $\dfrac{(x^{-2})^3(x^3)^{-2}}{x^{10}}$
66. $\dfrac{(x^{-4})^3(x^3)^{-4}}{x^{10}}$

67. $\dfrac{5a^8b^3}{20a^5b^{-4}}$ **68.** $\dfrac{7a^6b^{-2}}{21a^2b^{-5}}$ **69.** $\dfrac{(3x^{-2}y^8)^4}{(9x^4y^{-3})^2}$ **70.** $\dfrac{(6x^{-3}y^{-5})^2}{(3x^{-4}y^{-3})^4}$

71. $\left(\dfrac{8x^2y}{4x^4y^{-3}}\right)^4$ **72.** $\left(\dfrac{5x^4y^5}{10xy^{-2}}\right)^3$ **73.** $\left(\dfrac{x^{-5}y^2}{x^{-3}y^5}\right)^{-2}$ **74.** $\left(\dfrac{x^{-8}y^{-3}}{x^{-5}y^6}\right)^{-1}$

75. $\left(\dfrac{ab^{-3}c^{-2}}{a^{-3}b^0c^{-5}}\right)^0$ **76.** $\left(\dfrac{a^3b^2c^1}{a^{-1}b^{-2}c^{-3}}\right)^0$ **77.** $\left(\dfrac{x^2}{x^{-3}}\right)^0$ **78.** $\left(\dfrac{2x^2y}{xy^5}\right)^0$

Simplify each expression.

79. a. $3 \cdot 5 + 4$ **80. a.** $3 \cdot 7 - 6$
  b. $3(5 + 4)$ **b.** $3(7 - 6)$
  c. $3 \cdot 5 + 3 \cdot 4$ **c.** $3 \cdot 7 - 3 \cdot 6$

81. a. $6 + 3 \cdot 4 - 2$ **82. a.** $8 + 2 \cdot 7 - 3$
  b. $6 + 3(4 - 2)$ **b.** $8 + 2(7 - 3)$
  c. $(6 + 3)(4 - 2)$ **c.** $(8 + 2)(7 - 3)$

83. a. $(5 + 7)^2$ **84. a.** $(8 - 3)^2$
  b. $5^2 + 7^2$ **b.** $8^2 - 3^2$
  c. $5^2 + 2 \cdot 5 \cdot 7 + 7^2$ **c.** $8^2 - 2 \cdot 8 \cdot 3 + 3^2$

85. a. $2 + 3 \cdot 2^2 + 3^2$ **86. a.** $3 + 4 \cdot 4^2 + 5^2$
  b. $2 + 3(2^2 + 3^2)$ **b.** $3 + 4(4^2 + 5^2)$
  c. $(2 + 3)(2^2 + 3^2)$ **c.** $(3 + 4)(4^2 + 5^2)$

Write each number in scientific notation.

87. 378,000   **88.** 3,780,000   **89.** 4,900   **90.** 490
91. 0.00037   **92.** 0.000037   **93.** 0.00495   **94.** 0.0495

Write each number in expanded form.

95. 5.34×10^3 **96.** 5.34×10^2 **97.** 7.8×10^6 **98.** 7.8×10^4
99. 3.44×10^{-3} **100.** 3.44×10^{-5} **101.** 4.9×10^{-1} **102.** 4.9×10^{-2}

Use the properties of exponents to simplify each of the following expressions. Write all answers in scientific notation.

103. $(4 \times 10^{10})(2 \times 10^{-6})$ **104.** $(3 \times 10^{-12})(3 \times 10^4)$

105. $\dfrac{8 \times 10^{14}}{4 \times 10^5}$ **106.** $\dfrac{6 \times 10^8}{2 \times 10^3}$

107. $\dfrac{(5 \times 10^6)(4 \times 10^{-8})}{8 \times 10^4}$ **108.** $\dfrac{(6 \times 10^{-7})(3 \times 10^9)}{5 \times 10^6}$

109. $\dfrac{(2.4 \times 10^{-3})(3.6 \times 10^{-7})}{(4.8 \times 10^6)(1 \times 10^{-9})}$ **110.** $\dfrac{(7.5 \times 10^{-6})(1.5 \times 10^9)}{(1.8 \times 10^4)(2.5 \times 10^{-2})}$

Simplify. Write answers without using scientific notation, rounding to the nearest whole number.

111. $\dfrac{2.00 \times 10^8}{3.98 \times 10^6}$

112. $\dfrac{2.00 \times 10^8}{3.16 \times 10^5}$

Use a calculator to find each of the following. Write your answer in scientific notation with the first number in each answer rounded to the nearest tenth.

113. $10^{-4.1}$

114. $10^{-5.6}$

Applying the Concepts

115. Large Numbers If you are 20 years old, you have been alive for more than 630,000,000 seconds. Write the number of seconds in scientific notation.

116. Our Galaxy The galaxy in which the Earth resides is called the Milky Way galaxy. It is a spiral galaxy that contains approximately 200,000,000,000 stars (our Sun is one of them). Write the number of stars in words and in scientific notation.

117. Light Year A light year, the distance light travels in 1 year, is approximately 5.9×10^{12} miles. The Andromeda galaxy is approximately 1.7×10^6 light years from our galaxy. Find the distance in miles between our galaxy and the Andromeda galaxy.

118. Distance to the Sun The distance from the Earth to the Sun is approximately 9.3×10^7 miles and light travels 1.2×10^7 miles in 1 minute. How many minutes does it take the light from the Sun to reach the Griffith Observatory in Los Angeles?

Closest Stars to Earth The table below lists the distance (in light-years) of the five closest stars to our solar system:

Common Name	Distance (light-years)
Alpha Centauri C	4.2
Alpha Centauri A, B	4.3
Barnand's Star	5.9
Wolf 359	7.7
Lalande 21185	8.26

Source: space.about.com

Use the table and information given in problems 117 and 118 to help you answer questions 119–122.

119. How far (in miles) is Alpha Centauri C from our solar system? Answer using scientific notation.

120. How far (in miles) is Wolf 359 from our solar system? Answer using scientific notation.

121. It is estimated that our own galaxy, the Milky Way, is 4.13×10^{17} miles wide. How long would it take light to pass through the width of our galaxy?

122. Looking through a telescope in Griffith Observatory, you can see the Sombrero Galaxy, approximately 30 million light years away. How far (in miles) is this galaxy? Answer using scientific notation.

123. TV Shows The chart shows the number of viewers for ABC's top television shows.

> **TOP 5 ABC PRIMETIME SHOWS**
> *www.tvbythenumbers.com, June 15-21, 2009*
>
> | Show | Number of Viewers |
> | --- | --- |
> | Wipeout | 7,995,000 |
> | The Bachelorette | 6,761,000 |
> | I Survived a Japanese Gameshow | 5,345,000 |
> | 20/20 | 4,204,000 |
> | Here Come the Newlyweds | 4,042,000 |
>
> Number of Viewers

For each of the following shows, write the number of viewers in scientific notation.

a. 20/20

b. Wipeout

c. The Bachelorette

124. Fingerprints The FBI has been collecting fingerprint cards since 1924. Their collection has grown to over 200 million cards. When digitized, each fingerprint card turns into about 10 MB of data. (A megabyte [MB] is $2^{20} \approx$ one million bytes.)

a. How many bytes of storage will they need for 200 million cards?

b. A compression routine called the WSQ method will compress the bytes by ratio of 12.9 to 1. Approximately how many bytes of storage will the FBI need for the compressed data from 200 million cards? (Hint: Divide by 12.9.)

125. Credit Card Debt Outstanding credit-card debt in the United States is nearly $800 billion.

a. Write the number 800 billion in scientific notation.

b. If there are approximately 180 million credit card holders, find the average credit-card debt per holder, to the nearest dollar.

126. Cone Nebula Nebulas are collections of dust and gas in space that can be seen by some high-powered telescopes. One well-known nebula, the Cone Nebula, is 2.5 light years or 14,664,240,000,000 miles wide. Round this number to the nearest trillion and then write the result in scientific notation.

127. Computer Science We all use the language of computers to indicate how much memory our computers hold or how much information we can put on a storage device such as a flash drive. Scientific notation gives us a way to compare the actual numbers associated with the words we use to describe data storage in computers. The smallest amount of data that a computer can hold is measured in bits. A byte is the next largest unit and is equal to 8, or 2^3, bits. Fill in the following table:

Number of Bytes		
Unit	Exponential Form	Scientific Notation
Kilobyte	$2^{10} = 1{,}024$	
Megabyte	$2^{20} \approx 1{,}049{,}000$	
Gigabyte	$2^{30} \approx 1{,}074{,}000{,}000$	
Terabyte	$2^{40} \approx 1{,}099{,}500{,}000{,}000$	

Learning Objectives Assessment

The following problems can be used to help assess if you have successfully met the learning objectives for this section.

128. Simplify: $\dfrac{x^{-3}(x^2)^5}{x^4}$.

   a. 1   **b.** x^9 **c.** $\dfrac{1}{x^{34}}$ **d.** x^3

129. Write 0.0000357 in scientific notation.

   a. 3.57×10^{-5} **b.** 3.57×10^5

   c. 357×10^{-7} **d.** 0.357×10^4

130. Simplify: $\dfrac{(4 \times 10^{-5})(2 \times 10^3)}{5 \times 10^4}$.

   a. 1.2×10^{-19} **b.** 1.6×10^{-6}

   c. 1.6×10^2 **d.** 1.2×10^6

Chapter 1 Summary

The numbers in brackets refer to the section(s) in which the topic can be found.

EXAMPLES

Absolute Value [1.1]

1. $|5| = 5$

$|-5| = 5$

The *absolute value* of a real number is its distance from 0 on the number line. If $|x|$ represents the absolute value of x, then

$$|x| = \begin{cases} x & \text{if} & x \geq 0 \\ -x & \text{if} & x < 0 \end{cases}$$

The absolute value of a real number is never negative.

Opposites [1.1]

2. The numbers 5 and -5 are opposites; their sum is 0.

$5 + (-5) = 0$

Any two real numbers the same distance from 0 on the number line, but in opposite directions from 0, are called *opposites,* or *additive inverses.* Opposites always add to 0.

Reciprocals [1.1]

3. The numbers 3 and $\frac{1}{3}$ are reciprocals; their product is 1.

$3\left(\frac{1}{3}\right) = 1$

Any two real numbers whose product is 1 are called *reciprocals.* Every real number has a reciprocal except 0.

Properties of Real Numbers [1.2]

	For Addition	For Multiplication
Commutative	$a + b = b + a$	$ab = ba$
Associative	$a + (b + c) = (a + b) + c$	$a(bc) = (ab)c$
Identity	$a + 0 = a$	$a \cdot 1 = a$
Inverse	$a + (-a) = 0$	$a\left(\frac{1}{a}\right) = 1$
Distributive	$a(b + c) = ab + ac$	

Addition [1.3]

4. $5 + 3 = 8$

$5 + (-3) = 2$

$-5 + 3 = -2$

$-5 + (-3) = -8$

To add two real numbers with

1. *The same sign:* Simply add absolute values and use the common sign.

2. *Different signs:* Subtract the smaller absolute value from the larger absolute value. The answer has the same sign as the number with the larger absolute value.

Subtraction [1.3]

5. $6 - 2 = 6 + (-2) = 4$
$6 - (-2) = 6 + 2 = 8$

If a and b are real numbers,

$$a - b = a + (-b)$$

To subtract b, add the opposite of b.

Multiplication [1.3]

6. $5(4) = 20$
$5(-4) = -20$
$-5(4) = -20$
$-5(-4) = 20$

To multiply two real numbers, simply multiply their absolute values. Like signs give a positive answer. Unlike signs give a negative answer.

Division [1.3]

7. $\frac{12}{-3} = -4$

$\frac{-12}{-3} = 4$

If a and b are real numbers and $b \neq 0$, then

$$\frac{a}{b} = a \cdot \left(\frac{1}{b}\right)$$

To divide by b, multiply by the reciprocal of b.

Order of Operations [1.3]

8. $5^2 - 2 \cdot 6 + 3(8 - 1)$
$= 5^2 - 2 \cdot 6 + 3(7)$
$= 25 - 2 \cdot 6 + 3(7)$
$= 25 - 12 + 21$
$= 34$

⟨∆≠∑⟩ **RULE**   *Order of Operations*

When evaluating a mathematical expression, we will perform the operations in the following order:
1. Begin with the expression in the innermost parentheses or brackets and work your way out.
2. Simplify all numbers with exponents, working from left to right if more than one of these expressions is present.
3. Work all multiplications and divisions left to right.
4. Perform all additions and subtractions left to right.

Properties of Exponents [1.4]

9. These expressions illustrate the properties of exponents.

If a and b represent real numbers and r and s represent integers, then

a. $x^2 \cdot x^3 = x^{2+3} = x^5$

1. $a^r \cdot a^s = a^{r+s}$ Product property for exponents

b. $(x^2)^3 = x^{2 \cdot 3} = x^6$

2. $(a^r)^s = a^{r \cdot s}$ Power property for exponents

c. $(3x)^2 = 3^2 \cdot x^2 = 9x^2$

3. $(ab)^r = a^r \cdot b^r$ Distributive property for exponents over multiplication

d. $2^{-3} = \dfrac{1}{2^3} = \dfrac{1}{8}$

4. $a^{-r} = \dfrac{1}{a^r}$ $(a \neq 0)$ Negative exponent property

e. $\left(\dfrac{x}{5}\right)^2 = \dfrac{x^2}{5^2} = \dfrac{x^2}{25}$

5. $\left(\dfrac{a}{b}\right)^r = \dfrac{a^r}{b^r}$ $(b \neq 0)$ Distributive property for exponents over division

f. $\dfrac{x^7}{x^5} = x^{7-5} = x^2$

6. $\dfrac{a^r}{a^s} = a^{r-s}$ $(a \neq 0)$ Quotient property for exponents

g. $3^1 = 3$
$3^0 = 1$

7. $a^1 = a$ Identity property for exponents

8. $a^0 = 1$ $(a \neq 0)$ Zero exponent property

Scientific Notation [1.4]

10. $49{,}800{,}000 = 4.98 \times 10^7$
$0.00462 = 4.62 \times 10^{-3}$

A number is written in scientific notation when it is written as the product of a number between 1 and 10 and an integer power of 10; that is, when it has the form

$$n \times 10^r$$

where $1 \leq n < 10$ and $r =$ an integer.

⚠ COMMON MISTAKES

1. Interpreting absolute value as changing the sign of the number inside the absolute value symbols. That is, $|-5| = +5, |+5| = -5$. To avoid this mistake, remember, absolute value is defined as a distance and distance is always measured in non-negative units.
2. Confusing $-(-5)$ with $-|-5|$. The first answer is $+5$, but the second answer is -5.

Chapter 1 Test

Write each fraction as an equivalent fraction with denominator 18. [1.1]

1. $\dfrac{1}{2}$

2. $\dfrac{5}{6}$

Insert a $<$ or $>$ to make the statement true. [1.1]

3. $-6 \quad -4$

4. $7 \quad 9$

5. $3 \quad -6$

6. $-\dfrac{6}{7} \quad -\dfrac{2}{5}$

Simplify each expression. [1.1]

7. $|-16|$

8. $-|-9|$

9. Give the opposite and the reciprocal of $\dfrac{2}{7}$. [1.1]

10. Factor 588 into a product of prime factors. [1.1]

Reduce to lowest terms. [1.1]

11. $\dfrac{192}{312}$

12. $\dfrac{162}{459}$

Simplify. [1.2]

13. $4(3x + 2) - 6$

14. $6x - 3 + 4x + 5$

15. $4x + 3(2x + 4y) - 6y$

16. $3a + 2(5b + 4) - 1$

17. $6y + 3(x + 3) - 4y$

18. $18\left(\dfrac{5}{6}y - \dfrac{2}{3}x\right) - 5y$

Add the following fractions. [1.2]

19. $\dfrac{7}{18} + \dfrac{7}{9}$

20. $\dfrac{7}{12} + \dfrac{13}{34}$

Simplify the expression. [1.3]

21. $4 - (2x - 7) + 4x$

22. $(5 - 4x) - (3 - 2x)$

23. $5x - 3(2x + 4) + 15$

24. $\dfrac{(8)(-9) - (6)(-2)}{-(12 - 8)}$

25. $|7 - 10| - |3 - 5|$

26. $6^2 + |3 \cdot 2 - 4| - 3^3$

Simplify as much as possible. Write all answers with positive exponents. [1.4]

27. $(-3x^4)^{-3}$

28. $(4x^2y)^3$

29. $\left(\dfrac{5}{9}\right)^{-2}$

30. $-5x^2y \cdot -7x^3y$

31. $-6xy^3 \cdot 3x^{-2}y^4$

32. $\dfrac{18a^4b}{3ab^3}$

33. $\dfrac{18a^5b^9}{24a^7b^4}$

34. $\left(\dfrac{3x^{-2}y}{9xy^2}\right)^{-3}$

Write in scientific notation. [1.4]

35. $12{,}530{,}000$

36. 0.0052

Write in expanded form. [1.4]

37. 5.26×10^{-3}

38. 4.9×10^5

Simplify. Write in scientific notation. [1.4]

39. $\dfrac{(1.4 \times 10^7)(6.5 \times 10^{-4})}{7.0 \times 10^2}$

40. $\dfrac{(1.8 \times 10^7)(6.8 \times 10^6)}{(2.4 \times 10^9)(3.0 \times 10^{-4})}$

Equations and Inequalities in One Variable

iStockphoto.com © Andresr

Martina is an international student who is planning on taking some of her college courses here in the U.S. Because her country uses the Celsius scale, she is not familiar with temperatures measured in degrees Fahrenheit. The formula

$$F = \frac{9}{5}C + 32$$

gives the relationship between the Celsius and Fahrenheit temperature scales. Using this formula, we can construct a table that shows the Fahrenheit values for a variety of temperatures measured in degrees Celsius.

Degrees Celsius	Degrees Fahrenheit
20°	68°
25°	77°
30°	86°
35°	95°
40°	104°

In this chapter we will see how Martina could use a linear equation or linear inequality to find the Celsius values for temperatures given in degrees Fahrenheit.

© Alex Nikada/iStockPhoto

Success Skills

If you have successfully completed Chapter 1, then you have made a good start at developing the study skills necessary to succeed in all math classes. Some of the study skills for this chapter are a continuation of the skills from Chapter 1, while others are new to this chapter.

1. Continue to Set and Keep a Schedule

Sometimes I find students do well in Chapter 1 and then become overconfident. They will begin to put in less time with their homework. Don't do it. Keep to the same schedule.

2. Increase Effectiveness

You want to become more and more effective with the time you spend on your homework. Increase those activities that are the most beneficial and decrease those that have not given you the results you want.

3. List Difficult Problems

Begin to make lists of problems that give you the most difficulty. These are the problems in which you are repeatedly making mistakes.

4. Begin to Develop Confidence with Word Problems

It seems that the main difference between people who are good at working word problems and those who are not is confidence. People with confidence know that no matter how long it takes them, they will eventually be able to solve the problem. Those without confidence begin by saying to themselves, "I'll never be able to work this problem." If you are in this second category, then instead of telling yourself that you can't do word problems, decide to do whatever it takes to master them. The more word problems you work, the better you will become at them.

5. Organize Your Work

Many of my students keep a notebook that contains everything that they need for the course: class notes, homework, quizzes, tests, and research projects. A three-ring binder with tabs is ideal. Organize your notebook so that you can easily get to any item you want to look at.

Learning Objectives

In this section we will learn how to:

1. Use the Properties of Equality to solve a linear equation in one variable.
2. Solve a linear equation containing fractions or decimals.
3. Recognize a contradiction or an identity.
4. Solve applied problems involving a linear equation in one variable.

iStockPhoto.com/©zorani

Introduction

A recent newspaper article gave the following guideline for college students taking out loans to finance their education: The maximum monthly payment on the amount borrowed should not exceed 8% of their monthly starting salary. In this situation, the maximum monthly payment can be described mathematically with the formula

$$y = 0.08x.$$

From this formula, we find that the solution to the equation

$$300 = 0.08x$$

is the starting salary needed to pay back a student loan at $300 per month. Solving this type of equation is one of the things we will do in this section.

Linear Equations

A ***linear equation in one variable*** is any equation that can be put in the form

$$ax + b = 0$$

where a and b are constants and $a \neq 0$. For example, each of the equations

$$5x + 3 = 0 \qquad 2x = 7 \qquad 2x + 5 = 3$$

are linear because they can be put in the form $ax + b = 0$. In the first equation, $5x$, 3, and 0 are called ***terms*** of the equation: $5x$ is a variable term; 3 and 0 are constant terms.

> (def **DEFINITION** *solution set*
>
> The ***solution set*** for an equation is the set of all numbers that, when used in place of the variable, make the equation a true statement.

> (def **DEFINITION** *equivalent equations*
>
> Two or more equations with the same solution set are called ***equivalent equations***.

The equations $2x - 5 = 9$, $x - 1 = 6$, and $x = 7$ are all equivalent equations because the solution set for each is $\{7\}$.

Properties of Equality

The first property of equality states that adding the same quantity to both sides of an equation preserves equality. Or, more importantly, adding the same amount to both sides of an equation *never changes* the solution set. This property is called the **addition property of equality** and is stated in symbols as follows:

> **⟨Δ≠Σ PROPERTY**   *Addition Property of Equality*
>
> For any three algebraic expressions A, B, and C,
>
> $$\text{if} \qquad A = B$$
> $$\text{then} \qquad A + C = B + C$$
>
> *In words:*   Adding the same quantity to both sides of an equation will not change the solution set.

> *Note*   Because subtraction is defined in terms of addition and division is defined in terms of multiplication, we do not need to introduce separate properties for subtraction and division. The solution set for an equation will never be changed by subtracting the same amount from both sides or by dividing both sides by the same nonzero quantity.

Our second property is called the **multiplication property of equality** and is stated as follows:

> **⟨Δ≠Σ PROPERTY**   *Multiplication Property of Equality*
>
> For any three algebraic expressions A, B, and C, where $C \neq 0$,
>
> $$\text{if} \qquad A = B$$
> $$\text{then} \qquad AC = BC$$
>
> *In words:*   Multiplying both sides of an equation by the same nonzero quantity will not change the solution set.

VIDEO EXAMPLES

SECTION 2.1

> *Note*   We know that multiplication by a number and division by its reciprocal always produce the same result. Because of this fact, instead of multiplying each side of our equation by $\frac{1}{9}$, we could just as easily divide each side by 9. If we did so, the last two lines in our solution would look like this:
>
> $$\frac{9a}{9} = \frac{6}{9}$$
> $$a = \frac{2}{3}$$

◼ EXAMPLE 1   Find the solution set for $3a - 5 = -6a + 1$.

SOLUTION   To solve for a, we must isolate it on one side of the equation. Let's decide to isolate a on the left side. We start by adding $6a$ to both sides of the equation.

$$3a - 5 = -6a + 1$$
$$3a + 6a - 5 = -6a + 6a + 1 \qquad \text{Add } 6a \text{ to both sides}$$
$$9a - 5 = 1$$
$$9a - 5 + 5 = 1 + 5 \qquad \text{Add 5 to both sides}$$
$$9a = 6$$
$$\frac{1}{9}(9a) = \frac{1}{9}(6) \qquad \text{Multiply both sides by } \frac{1}{9}$$
$$a = \frac{2}{3} \qquad \frac{1}{9}(6) = \frac{6}{9} = \frac{2}{3}$$

The solution set is $\left\{ \dfrac{2}{3} \right\}$. ◼

The next example involves fractions. The least common denominator, which is the smallest expression that is divisible by each of the denominators, can be used with the multiplication property of equality to simplify equations containing fractions.

EXAMPLE 2 Solve $\frac{2}{3}x + \frac{1}{2} = -\frac{3}{8}$.

SOLUTION We can solve this equation by applying our properties and working with fractions, or we can begin by eliminating the fractions. Let's work the problem using both methods.

Method 1: *Working with the fractions*

$$\frac{2}{3}x + \frac{1}{2} + \left(-\frac{1}{2}\right) = -\frac{3}{8} + \left(-\frac{1}{2}\right) \qquad \text{Add } -\frac{1}{2} \text{ to each side}$$

$$\frac{2}{3}x = -\frac{7}{8} \qquad\qquad -\frac{3}{8} + \left(-\frac{1}{2}\right) = -\frac{3}{8} + \left(-\frac{4}{8}\right)$$

$$\frac{3}{2}\left(\frac{2}{3}x\right) = \frac{3}{2}\left(-\frac{7}{8}\right) \qquad \text{Multiply each side by } \frac{3}{2}$$

$$x = -\frac{21}{16}$$

Method 2: *Eliminating the fractions in the beginning*

Our original equation has denominators of 3, 2, and 8. The least common denominator, abbreviated LCD, for these three denominators is 24, and it has the property that all three denominators will divide it evenly. Therefore, if we multiply both sides of our equation by 24, each denominator will divide into 24, and we will be left with an equation that does not contain any denominators other than 1.

$$24\left(\frac{2}{3}x + \frac{1}{2}\right) = 24\left(-\frac{3}{8}\right) \qquad \text{Multiply each side by the LCD 24}$$

$$24\left(\frac{2}{3}x\right) + 24\left(\frac{1}{2}\right) = 24\left(-\frac{3}{8}\right) \qquad \text{Distributive Property on the left side}$$

$$16x + 12 = -9 \qquad \text{Multiply}$$

$$16x = -21 \qquad \text{Add } -12 \text{ to each side}$$

$$x = -\frac{21}{16} \qquad \text{Multiply each side by } \frac{1}{16}$$

As the third line above indicates, multiplying each side of the equation by the LCD eliminates all the fractions from the equation. Both methods yield the same solution. ◼

EXAMPLE 3 Solve the equation $0.06x + 0.05(10{,}000 - x) = 560$.

SOLUTION We can solve the equation in its original form by working with the decimals, or we can eliminate the decimals first by using the multiplication property of equality and solve the resulting equation. Here are both methods.

Method 1: *Working with the decimals*

$$0.06x + 0.05(10{,}000 - x) = 560 \qquad \text{Original equation}$$

$$0.06x + 0.05(10{,}000) - 0.05x = 560 \qquad \text{Distributive property}$$

$$0.01x + 500 = 560 \qquad \text{Simplify the left side}$$

$$0.01x = 60 \qquad \text{Add } -500 \text{ to each side}$$

$$\frac{0.01x}{0.01} = \frac{60}{0.01} \qquad \text{Divide each side by } 0.01$$

$$x = 6{,}000$$

Method 2: *Eliminating the decimals in the beginning.*

To move the decimal point two places to the right in $0.06x$ and 0.05, we multiply each side of the equation by 100.

$$0.06x + 0.05(10{,}000 - x) = 560 \qquad \text{Original equation}$$

$$0.06x + 500 - 0.05x = 560 \qquad \text{Distributive Property}$$

$$100(0.06x) + 100(500) - 100(0.05x) = 100(560) \qquad \text{Multiply each side by 100}$$

$$6x + 50{,}000 - 5x = 56{,}000 \qquad \text{Multiply}$$

$$x + 50{,}000 = 56{,}000 \qquad \text{Simplify the left side}$$

$$x = 6{,}000 \qquad \text{Add } -50{,}000 \text{ to each side}$$

Using either method, the solution to our equation is 6,000. We check our work (to be sure we have not made a mistake in applying the properties or an arithmetic mistake) by substituting 6,000 into our original equation and simplifying each side of the result separately.

Check: Substituting 6,000 for x in the original equation, we have

$$0.06(6{,}000) + 0.05(10{,}000 - 6{,}000) \overset{?}{=} 560$$

$$0.06(6{,}000) + 0.05(4{,}000) \overset{?}{=} 560$$

$$360 + 200 \overset{?}{=} 560$$

$$560 = 560 \qquad \text{A true statement} \qquad ▨$$

Note We are placing question marks over the equal signs because we don't know yet if the expressions on the left will be equal to the expressions on the right.

Here is a list of steps to use as a guideline for solving linear equations in one variable.

🔍 HOW TO *Solve Linear Equations in One Variable*

Step 1a: Use the distributive property with grouping symbols to separate terms, if necessary.

1b: If fractions are present, consider multiplying both sides by the LCD to eliminate the fractions. If decimals are present, consider multiplying both sides by a power of 10 to clear the equation of decimals.

1c: Combine similar terms on each side of the equation.

Step 2: Use the addition property of equality to get all variable terms on one side of the equation and all constant terms on the other side. A variable term is a term that contains the variable. A constant term is a term that does not contain the variable (the number 3, for example).

Step 3: Use the multiplication property of equality to get the variable by itself on one side of the equation.

Step 4: Check your solution in the original equation to be sure that you have not made a mistake in the solution process.

As you work through the problems in the problem set, you will see that it is not always necessary to use all four steps when solving equations. The number of steps used depends on the equation. In Example 4, there are no fractions or decimals in the original equation, so step 1b will not be used.

EXAMPLE 4 Solve the equation $8 - 3(4x - 2) + 5x = 35$.

SOLUTION We must begin by distributing the -3 across the quantity $4x - 2$. (It would be a mistake to subtract 3 from 8 first, because the rule for order of operations indicates we are to do multiplication before subtraction.) After we have simplified the left side of our equation, we apply the addition property and the multiplication property. In this example, we will show only the result:

$$8 - 3(4x - 2) + 5x = 35 \qquad \text{Original Equation}$$

Step 1a: $\qquad 8 - 12x + 6 + 5x = 35 \qquad$ Distributive Property

Step 1c: $\qquad\qquad\quad -7x + 14 = 35 \qquad$ Simplify

Step 2: $\qquad\qquad\qquad\quad -7x = 21 \qquad$ Add -14 to each side

Step 3: $\qquad\qquad\qquad\qquad x = -3 \qquad$ Multiply by $-\frac{1}{7}$

Step 4: When x is replaced by -3 in the original equation, a true statement results. Therefore, -3 is the solution to our equation.

Identities and Equations with No Solution

There are two special cases associated with solving linear equations in one variable, which are illustrated in the following examples.

EXAMPLE 5 Solve for x: $2(3x - 4) = 3 + 6x$.

SOLUTION Applying the distributive property to the left side gives us

$$6x - 8 = 3 + 6x \qquad \text{Distributive property}$$

Now, if we add $-6x$ to each side, we are left with

$$-8 = 3$$

which is a false statement. This means that there is no solution to our equation. Any number we substitute for x in the original equation will lead to a similar false statement. We say the original equation is a **contradiction**.

EXAMPLE 6 Solve for x: $-15 + 3x = 3(x - 5)$.

SOLUTION We start by applying the distributive property to the right side.

$$-15 + 3x = 3x - 15 \qquad \text{Distributive property}$$

If we add $-3x$ to each side, we are left with the true statement

$$-15 = -15$$

In this case, our result tells us that any number we use in place of x in the original equation will lead to a true statement. Therefore, all real numbers are solutions to our equation. We say the original equation is an **identity** because the left side is always identically equal to the right side.

Applications

 EXAMPLE 7   In the chapter opener we mentioned that the relationship between temperature in degrees Fahrenheit, F, and degrees Celsius, C, is given by the formula

$$F = \frac{9}{5}C + 32$$

Solve the following equation to find the temperature in degrees Celsius if it is 77 degrees Fahrenheit.

$$77 = \frac{9}{5}C + 32$$

SOLUTION   First, we multiply both sides of the equation by the LCD, which is 5, to eliminate fractions. Then we isolate C.

$$5(77) = 5\left(\frac{9}{5}C + 32\right)$$ Multiply each side by the LCD 5

$$5(77) = 5\left(\frac{9}{5}C\right) + 5(32)$$ Distributive Property on the right side

$$385 = 9C + 160$$ Multiply

$$225 = 9C$$ Add -160 to each side

$$25 = C$$ Multiply each side by $\frac{1}{9}$

A temperature of 77°F corresponds to 25°C.

Getting Ready for Class

After reading through the preceding section, respond in your own words and in complete sentences.

A. Name the constant terms in the equation $5x + 3 = 2$.

B. In your own words, explain how you could use the multiplication property of equality to solve a linear equation that contains fractions.

C. How can you eliminate decimals in a linear equation in one variable?

D. Explain how to tell the difference between a contradiction and an identity.

Each odd/even pair of problems below is matched to an example in the text. If you have any trouble with any of these problems, go to the example that is matched with that problem.

Solve each of the following equations.

1. $7y - 4 = 2y + 11$

2. $5 - 2x = 3x + 1$

3. $-\dfrac{2}{5}x + \dfrac{2}{15} = \dfrac{2}{3}$

4. $\dfrac{1}{2}x + \dfrac{1}{4} = \dfrac{1}{3}x + \dfrac{5}{4}$

5. $0.14x + 0.08(10{,}000 - x) = 1{,}220$

6. $-0.3y + 0.1 = 0.5$

7. $5(y + 2) - 4(y + 1) = 3$

8. $6(y - 3) - 5(y + 2) = 8$

9. $2(4x + 5) = 11 + 8x$

10. $7x + 9 = 3(2x - 1) + x$

11. $4(x + 1) = 7 + 4x - 3$

12. $6(4 - 3x) = 2(12 - 9x)$

Paying Attention to Instructions The next four problems are intended to give you practice reading, and paying attention to, the instructions that accompany the problems you are working. Working these problems is an excellent way to get ready for a test or a quiz.

13. Work each problem according to the instructions given.
  a. Distribute: $3(x - 4)$
  b. Simplify: $3x - 4x$
  c. Solve: $3(x - 4) = 0$
  d. Solve: $3x - 4 = 0$

14. Work each problem according to the instructions given.
  a. Simplify: $(3x + 5) - (7x - 4)$
  b. Solve: $3x + 5 = 7x - 4$
  c. Solve: $3(x + 5) = 7(x - 4)$
  d. Simplify: $3(x + 5) - 7(x - 4)$

15. Solve each equation, if possible.
  a. $9x - 5 = 0$
  b. $9(x - 5) = 0$
  c. $9x - 5 = 13$
  d. $9(x - 5) = 9x$

16. Solve each equation, if possible.
  a. $6(x + 4) = 6x + 4$
  b. $6(x + 4) = 6x + 24$
  c. $2(5 - 3x) = 6x + 1$
  d. $2(5 - 3x) = 1 - 6x$

Now that you have practiced solving a variety of equations, we can turn our attention to the type of equation you will see as you progress through the book. Many of these next equations appear later in the book exactly as you see them here.

Solve each equation.

17. $-3 - 4x = 15$

18. $-\dfrac{3}{5}a + 2 = 8$

19. $0 = 6400a + 70$

20. $.07x = 1.4$

21. $5(2x + 1) = 12$

22. $50 = \dfrac{K}{48}$

23. $100P = 2{,}400$

24. $2x - 3(3x - 5) = -6$

25. $5\left(-\dfrac{19}{15}\right) + 5y = 9$

26. $2\left(-\dfrac{29}{22}\right) - 3y = 4$

27. $\dfrac{3}{4}x - \dfrac{1}{6} = \dfrac{2}{3}$

28. $\dfrac{5}{8} + \dfrac{x}{4} = -\dfrac{1}{2}$

29. $\dfrac{x}{2} - \dfrac{5}{6} = \dfrac{x}{3} + 1$

30. $\dfrac{7}{12}x + 2 = \dfrac{1}{4} - \dfrac{3}{8}x$

31. $5(x + 4) - 3x = 14$ **32.** $3x + (x - 2) \cdot 2 = 6$

33. $15 - 3(x - 1) = x - 2$ **34.** $2(2x - 3) + 2x = 45$

35. $2(20 + x) = 3(20 - x)$ **36.** $2x + 1.5(75 - x) = 127.5$

37. $0.08x + 0.09(9{,}000 - x) = 750$ **38.** $0.12x + 0.10(15{,}000 - x) = 1{,}600$

39. $\frac{1}{2}(4x + 11) = 5x - \frac{2}{3}$ **40.** $9 + \frac{3}{4}(x - 7) = \frac{x}{2}$

41. $\frac{1}{6}(2x - 3) = \frac{7}{12} - \frac{1}{4}(x + 5)$ **42.** $8 - \frac{1}{3}(6x + 1) = \frac{4}{9}(x - 2)$

Solve each equation, if possible.

43. $3x - 6 = 3(x + 4)$ **44.** $4y + 2 - 3y + 5 = 3 + y + 4$

45. $2(4t - 1) + 3 = 5t + 4 + 3t$ **46.** $7x - 3(x - 2) = -4(5 - x)$

47. $7(x + 2) - 4(2x - 1) = 18 - x$

48. $2x + 3 + 4(x - 1) = 11x - 5(x + 2)$

49. Temperature and Altitude As an airplane gains altitude, the temperature outside the plane decreases. The relationship between temperature T and altitude A can be described with the formula

$$T = -0.0035A + 70$$

when the temperature on the ground is 70°F. Solve the equation below to find the altitude at which the temperature outside the plane is -35°F.

$$-35 = -0.0035A + 70$$

50. Retail Price A clothing store is offering a \$10 discount on any item worth at least \$25. After including sales tax, the total cost of an item is given by the formula

$$c = 1.08(x - 10)$$

where x is the original retail price of the item. If Bianca has \$54 to spend, what is the most expensive item she can afford to buy at the clothing store?

Learning Objectives Assessment

The following problems can be used to help assess if you have successfully met the learning objectives for this section.

51. Solve: $11 + 4x = 2x - 3$

     a. 4 **b.** -4 **c.** 7 **d.** -7

52. Solve: $0.07x + 0.05(6{,}000 - x) = 370$

     a. $2{,}500$ **b.** $3{,}000$ **c.** $3{,}500$ **d.** $4{,}000$

53. Which of the following is an identity?

     a. $2x + 3 = 2(x + 1) + 1$ **b.** $2x + 3 = 2(x + 1) - 1$

     c. $2x + 3 = 2(x + 1) + x$ **d.** $2x + 3 = 2(x + 1) - x$

54. Solve the equation $59 = \frac{9}{5}C + 32$ to find the temperature in degrees Celsius corresponding to 59 degrees Fahrenheit.

     a. -11 **b.** 50 **c.** 15 **d.** 29

Getting Ready for the Next Section

Problems under this heading, "Getting Ready for the Next Section", are problems that you must be able to work in order to understand the material in the next section. In this case, the problems below are variations on the type of problems you have already worked in this problem set. They are exactly the type of problems you will see in the explanations and examples in the next section.

Solve each equation.

55. $-4y = -10$

56. $x \cdot 84 = 21$

57. $375 = 300p$

58. $35 = 0.4x$

59. $12 - 4y = 2$

60. $-6 - 3y = 6$

61. $525 = 900 - 300p$

62. $375 = 900 - 300p$

63. $52 = (16 - c) \cdot 4$

64. $48 = (4 - c)3$

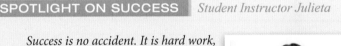 **SPOTLIGHT ON SUCCESS** *Student Instructor Julieta*

Success is no accident. It is hard work, perseverance, learning, studying, sacrifice, and most of all, love of what you are doing or learning to do.
—Pelé

Success really is no accident, nor is it something that happens overnight. Sure you may be sitting there wondering why you don't understand a certain lesson or topic, but you are not alone. There are many others who are sitting in your exact position. Throughout my first year in college (and more specifically in Calculus I) I learned that it is normal for any student to feel stumped every now and then. The students who do well are the ones who keep working, even when they are confused.

Pelé wasn't just born with all that legendary talent. It took dedication and hard work as well. Don't ever feel bad because there's something you don't understand—it's not worth it. Stick with it 100% and just keep working problems; I'm sure you'll be successful with whatever you set your mind to achieve in this course.

Formulas

2.2

Learning Objectives

In this section we will learn how to:

1. Solve a formula for a variable given numerical replacements for the remaining variables.

2. Use the rate equation to find the speed for an object.

3. Solve a formula for a variable without being given numerical replacements for the other variables.

4. Solve applied problems involving formulas.

Introduction

The study of design, architecture, or construction is not complete without an introduction to the ideas presented in the book *A Pattern Language*, published by the Center for Environmental Structure. The book presents a language for design and construction of houses, neighborhoods, and cities, based on what is best for the people who will live there. Here is a passage from that book concerning the size of public places.

> For public squares, courts, pedestrian streets, any place where crowds are drawn together, estimate the mean number of people in the place at any given moment (P), and make the area of the place between $150P$ and $300P$ square feet.

This passage gives us a formula for the area A of a public square that averages P people at any given time. In the language of algebra, the minimum size of such a public square should be

$$A = 150P$$

In this section we review formulas, such as the one shown here. We want to evaluate formulas given specific values for some of its variables, and second, we must learn how to change the form of a formula without changing its meaning.

Formulas

A formula in mathematics is an equation that contains more than one variable. Some formulas are probably already familiar to you. For example, the formula for the area A of a rectangle with length l and width w is $A = lw$.

To begin our work with formulas, we will consider some examples in which we are given numerical replacements for all but one of the variables.

VIDEO EXAMPLES

SECTION 2.2

EXAMPLE 1 Find y when x is 4 in the formula $3x - 4y = 2$.

SOLUTION   We substitute 4 for x in the formula and then solve for y:

When $\qquad\qquad x = 4$

the formula $\qquad 3x - 4y = 2$

becomes $\qquad 3(4) - 4y = 2$

$\qquad\qquad\qquad 12 - 4y = 2 \qquad$ Multiply 3 and 4

$\qquad\qquad\qquad\quad -4y = -10 \qquad$ Add -12 to each side

$\qquad\qquad\qquad\qquad y = \dfrac{5}{2} \qquad$ Divide each side by -4

Note that, in the last line of Example 1, we divided each side of the equation by -4. Remember that this is equivalent to multiplying each side of the equation by $-\frac{1}{4}$.

EXAMPLE 2   A store selling art supplies finds that they can sell x sketch pads each week at a price of p dollars each, according to the formula $x = 900 - 300p$. What price should they charge for each sketch pad if they want to sell 525 pads each week?

SOLUTION   Here we are given a formula, $x = 900 - 300p$, and asked to find the value of p if x is 525. To do so, we simply substitute 525 for x and solve for p:

When $x = 525$

the formula $x = 900 - 300p$

becomes $525 = 900 - 300p$

$$-375 = -300p \qquad \text{Add } -900 \text{ to each side}$$

$$1.25 = p \qquad \text{Divide each side by } -300$$

To sell 525 sketch pads, the store should charge $1.25 for each pad.

Rate Equation and Average Speed

Now we will look at some problems that use what is called the **rate equation**. You use this equation on an intuitive level when you are estimating how long it will take you to drive long distances. For example, if you drive at 50 miles per hour for 2 hours, you will travel 100 miles. Here is the rate equation:

$$\text{Distance} = \text{rate} \cdot \text{time, or } d = r \cdot t$$

EXAMPLE 3   A boat is traveling upstream against a current. If the speed of the boat in still water is r and the speed of the current is c, then the formula for the distance traveled by the boat is $d = (r - c) \cdot t$, where t is the length of time. Find c if $d = 52$ miles, $r = 16$ miles per hour, and $t = 4$ hours.

SOLUTION   Substituting 52 for d, 16 for r, and 4 for t into the formula, we have

$$52 = (16 - c) \cdot 4$$

$$13 = 16 - c \qquad \text{Divide each side by 4}$$

$$-3 = -c \qquad \text{Add } -16 \text{ to each side}$$

$$3 = c \qquad \text{Divide each side by } -1$$

The speed of the current is 3 miles per hour.

The rate equation has two equivalent forms, one of which is obtained by solving for r, while the other is obtained by solving for t. Here they are:

$$r = \frac{d}{t} \quad \text{and} \quad t = \frac{d}{r}$$

The rate in this equation is also referred to as average speed.

The **average speed** of a moving object is defined to be the ratio of distance to time. If you drive your car for 5 hours and travel a distance of 200 miles, then your average rate of speed is

$$\text{Average speed} = \frac{200 \text{ miles}}{5 \text{ hours}} = 40 \text{ miles per hour}$$

Our next example involves both the formula for the circumference of a circle and the rate equation.

EXAMPLE 4 The first Ferris wheel was designed and built by George Ferris in 1893. The diameter of the wheel was 250 feet. It had 36 carriages, equally spaced around the wheel, each of which held a maximum of 40 people. One trip around the wheel took 20 minutes. Find the average speed of a rider on the first Ferris wheel. (Use 3.14 as an approximation for π.)

SOLUTION The distance traveled is the circumference of the wheel, which is

$$C = \pi d = 250\pi = 250(3.14) = 785 \text{ feet}$$

To find the average speed, we divide the distance traveled by the amount of time it took to go once around the wheel. (Round answer to the nearest tenth.)

$$r = \frac{d}{t} = \frac{785 \text{ feet}}{20 \text{ minutes}} \approx 39.3 \text{ feet per minute}$$

For our remaining examples, we will solve a formula for a specified variable without being given numerical replacements for the other variables. When doing so, our solution will not be a numerical value. Rather, our goal is to isolate the specified variable. The result will still be a formula. The given variable should appear by itself on one side of the equation only, and the other side of the equation should be simplified.

Formulas from Geometry

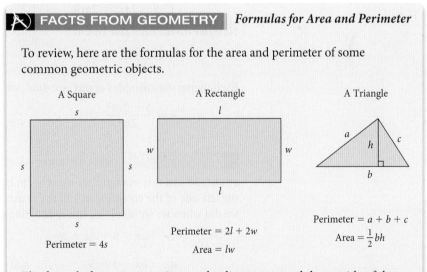

FACTS FROM GEOMETRY *Formulas for Area and Perimeter*

To review, here are the formulas for the area and perimeter of some common geometric objects.

A Square
Perimeter = 4s

A Rectangle
Perimeter = 2l + 2w
Area = lw

A Triangle
Perimeter = a + b + c
Area = $\frac{1}{2}bh$

The formula for perimeter gives us the distance around the outside of the object along its sides, while the formula for area gives us a measure of the amount of surface the object covers.

▨ **EXAMPLE 5**   Given the formula $P = 2w + 2l$, solve for w.

SOLUTION   To solve for w, we must isolate it on one side of the equation. We can accomplish this if we delete the $2l$ term and the coefficient 2 from the right side of the equation.

To begin, we add $-2l$ to both sides:

$$P + (-2l) = 2w + 2l + (-2l)$$

$$P - 2l = 2w$$

To delete the 2 from the right side, we can multiply both sides by $\frac{1}{2}$:

$$\frac{1}{2}(P - 2l) = \frac{1}{2}(2w)$$

$$\frac{P - 2l}{2} = w$$

The two formulas

$$P = 2w + 2l \qquad \text{and} \qquad w = \frac{P - 2l}{2}$$

give the relationship between P, l, and w. They look different, but they both say the same thing about P, l, and w. The first formula gives P in terms of l and w, and the second formula gives w in terms of P and l.   ▨

▨ **EXAMPLE 6**   Solve the formula $S = 2\pi rh + 2\pi r^2$ for h.

SOLUTION   This is the formula for the surface area of a right circular cylinder, with radius r and height h, that is closed at both ends. To isolate h, we first add $-2\pi r^2$ to both sides:

$$S + (-2\pi r^2) = 2\pi rh + 2\pi r^2 + (-2\pi r^2)$$

$$S - 2\pi r^2 = 2\pi rh$$

Next, we divide each side by $2\pi r$.

$$\frac{S - 2\pi r^2}{2\pi r} = h$$

Exchanging the two sides of our equation, we have

$$h = \frac{S - 2\pi r^2}{2\pi r}$$ ▨

▨ **EXAMPLE 7**   Solve for x: $ax - 3 = bx + 5$.

SOLUTION   In this example, we must begin by collecting all the variable terms on the left side of the equation and all the constant terms on the other side (just like we did when we were solving linear equations in Section 2.1):

$$ax - 3 = bx + 5$$

$$ax - bx - 3 = 5 \qquad \text{Add } -bx \text{ to each side}$$

$$ax - bx = 8 \qquad \text{Add 3 to each side}$$

At this point, we need to apply the distributive property to write the left side as $(a - b)x$. After that, we divide each side by $a - b$:

$$(a - b)x = 8 \qquad \text{Distributive property}$$

$$x = \frac{8}{a - b} \qquad \text{Divide each side by } a - b$$ ▨

EXAMPLE 8   Solve for y: $\dfrac{y - b}{x - 0} = m$.

SOLUTION Although we will do more extensive work with formulas of this form later in the book, we need to know how to solve this particular formula for y in order to understand some things in the next chapter. We begin by simplifying the denominator on the left side and then multiplying each side of the formula by x. Doing so makes the rest of the solution process simple.

$$\dfrac{y - b}{x - 0} = m \qquad \text{Original formula}$$

$$\dfrac{y - b}{x} = m \qquad x - 0 = x$$

$$x \cdot \dfrac{y - b}{x} = m \cdot x \qquad \text{Multiply each side by } x$$

$$y - b = mx \qquad \text{Simplify each side}$$

$$y = mx + b \qquad \text{Add } b \text{ to each side}$$

This is our solution. If we look back to line four above, we can justify our result on the left side of the equation this way: Dividing by x is equivalent to multiplying by its reciprocal $\frac{1}{x}$. Here is what it looks like when written out completely:

$$x \cdot \dfrac{y - b}{x} = x \cdot \dfrac{1}{x} \cdot (y - b) = 1(y - b) = y - b$$

EXAMPLE 9   Solve for y: $\dfrac{y - 4}{x - 5} = 3$.

SOLUTION We proceed as we did in the previous example, but this time we clear the formula of fractions by multiplying each side of the formula by $x - 5$.

$$\dfrac{y - 4}{x - 5} = 3 \qquad \text{Original formula}$$

$$(x - 5) \cdot \dfrac{y - 4}{x - 5} = 3 \cdot (x - 5) \qquad \text{Multiply each side by } (x - 5)$$

$$y - 4 = 3x - 15 \qquad \text{Simplify each side}$$

$$y = 3x - 11 \qquad \text{Add 4 to each side}$$

We have solved for y. We can justify our result on the left side of the equation this way: Dividing by $x - 5$ is equivalent to multiplying by its reciprocal $\frac{1}{x-5}$. Here are the details:

$$(x - 5) \cdot \dfrac{y - 4}{x - 5} = (x - 5) \cdot \dfrac{1}{x - 5} \cdot (y - 4) = 1(y - 4) = y - 4$$

Getting Ready for Class

After reading through the preceding section, respond in your own words and in complete sentences.

A. What is a formula in mathematics?

B. Give two equivalent forms of the rate equation $d = rt$.

C. What are the formulas for the area and perimeter of a triangle?

D. Explain in words the formula for the perimeter of a rectangle.

Problem Set 2.2

Use the formula $3x - 4y = 12$ to find y if

1. x is 0   **2.** x is -2 **3.** x is 4   **4.** x is -4

Use the formula $y = 2x - 3$ to find x when

5. y is 0   **6.** y is -3 **7.** y is 5   **8.** y is -5

Problems 9 through 18 are problems that you will see later in the text.

9. If $x - 2y = 4$ and $y = -\dfrac{6}{5}$, find x.

10. If $x - 2y = 4$ and $x = \dfrac{8}{5}$, find y.

11. Let $x = 160$ and $y = 0$ in $y = a(x - 80)^2 + 70$ and solve for a.

12. Let $x = 0$ and $y = 0$ in $y = a(x - 80)^2 + 70$ and solve for a.

13. Use the formula $d = (r - c)t$ to find c if $d = 30$, $r = 12$, and $t = 3$.

14. Use the formula $d = (r - c)t$ to find r if $d = 49$, $c = 4$, and $t = 3.5$.

15. If $y = Kx$, find K if $x = 5$ and $y = 15$.

16. If $d = Kt^2$, find K if $t = 2$ and $d = 64$.

17. If $V = \dfrac{K}{P}$, find K if $P = 48$ and $V = 50$.

18. If $y = Kxz^2$, find K if $x = 5$, $z = 3$, and $y = 180$.

Use the formula $5x - 3y = -15$ to find y if

19. $x = 2$ **20.** $x = -3$ **21.** $x = -\dfrac{1}{5}$ **22.** $x = 3$

Solve each of the following formulas for the indicated variable.

23. $d = rt$ for r **24.** $d = rt$ for t

25. $d = (r + c)t$ for t **26.** $d = (r + c)t$ for r

27. $A = lw$ for l **28.** $A = \dfrac{1}{2}bh$ for b

29. $I = prt$ for t **30.** $I = prt$ for r

31. $PV = nRT$ for T **32.** $PV = nRT$ for R

33. $y = mx + b$ for x **34.** $A = P + Prt$ for t

35. $C = \dfrac{5}{9}(F - 32)$ for F **36.** $F = \dfrac{9}{5}C + 32$ for C

37. $h = vt + 16t^2$ for v **38.** $h = vt - 16t^2$ for v

39. $A = a + (n - 1)d$ for d **40.** $A = a + (n - 1)d$ for n

41. $2x + 3y = 6$ for y **42.** $2x - 3y = 6$ for y

43. $-3x + 5y = 15$ for y **44.** $-2x - 7y = 14$ for y

45. $2x - 6y + 12 = 0$ for y **46.** $7x - 2y - 6 = 0$ for y

47. $ax + 4 = bx + 9$ for x **48.** $ax - 5 = cx - 2$ for x

49. $S = \pi r^2 + 2\pi rh$ for h **50.** $A = P + Prt$ for P

51. $-3x + 4y = 12$ for x **52.** $-3x + 4y = 12$ for y

53. $ax + 3 = cx - 7$ for x **54.** $by - 9 = dy + 3$ for y

Problems 55 through 62 are problems that you will see later in the text. Solve each formula for y.

55. $x = 2y - 3$

56. $x = 4y + 1$

57. $y - 3 = -2(x + 4)$

58. $y - 1 = \dfrac{1}{4}(x - 3)$

59. $y - 3 = -\dfrac{2}{3}(x + 3)$

60. $y + 1 = -\dfrac{2}{3}(x - 3)$

61. $y - 4 = -\dfrac{1}{2}(x + 1)$

62. $y - 2 = \dfrac{1}{3}(x - 1)$

63. Solve for y.

   a. $\dfrac{y + 1}{x - 0} = 4$ **b.** $\dfrac{y + 2}{x - 4} = -\dfrac{1}{2}$ **c.** $\dfrac{y + 3}{x - 7} = 0$

64. Solve for y.

   a. $\dfrac{y - 1}{x - 0} = -3$ **b.** $\dfrac{y - 2}{x - 6} = \dfrac{2}{3}$ **c.** $\dfrac{y - 3}{x - 1} = 0$

Solve for y.

65. $\dfrac{x}{8} + \dfrac{y}{2} = 1$ **66.** $\dfrac{x}{7} + \dfrac{y}{9} = 1$ **67.** $\dfrac{x}{5} + \dfrac{y}{-3} = 1$ **68.** $\dfrac{x}{16} + \dfrac{y}{-2} = 1$

Paying Attention to Instructions The next two problems are intended to give you practice reading, and paying attention to, the instructions that accompany the problems you are working. As we have mentioned previously, working these problems is an excellent way to get ready for a test or a quiz.

69. Work each problem according to the instructions given.

   a. Solve: $-4x + 5 = 20$ **b.** Find the value of $-4x + 5$ when x is 3

   c. Solve for y: $-4x + 5y = 20$ **d.** Solve for x: $-4x + 5y = 20$

70. Work each problem according to the instructions given.

   a. Solve: $2x + 1 = -4$ **b.** Find the value of $2x + 1$ when x is 8

   c. Solve for y: $2x + y = 20$ **d.** Solve for x: $2x + y = 20$

Estimating Vehicle Weight If you can measure the area that the tires on your car contact the ground and know the air pressure in the tires, then you can estimate the weight of your car with the following formula:

$$W = \dfrac{APN}{2{,}000}$$

where W is the vehicle's weight in tons, A is the average tire contact area with a hard surface in square inches, P is the air pressure in the tires in pounds per square inch (psi, or lb/in^2), and N is the number of tires.

XYZ Textbooks/©Matthew Hoy

71. What is the approximate weight of a car if the average tire contact area is a rectangle 6 inches by 5 inches and if the air pressure in the tires is 30 psi?

72. What is the approximate weight of a car if the average tire contact area is a rectangle 5 inches by 4 inches and the tire pressure is 30 psi?

73. Current It takes a boat 2 hours to travel 18 miles upstream against the current. If the speed of the boat in still water is 15 miles per hour, what is the speed of the current?

74. Current It takes a boat 6.5 hours to travel 117 miles upstream against the current. If the speed of the current is 5 miles per hour, what is the speed of the boat in still water?

75. Wind An airplane takes 4 hours to travel 864 miles while flying against the wind. If the speed of the airplane on a windless day is 258 miles per hour, what is the speed of the wind?

76. Wind A cyclist takes 3 hours to travel 39 miles while pedaling against the wind. If the speed of the wind is 4 miles per hour, how fast would the cyclist be able to travel on a windless day?

For problems 77 and 78, use 3.14 as an approximation for π. Round answers to the nearest tenth.

iStockPhoto.com/©Nikada

77. Average Speed A person riding a Ferris wheel with a diameter of 65 feet travels once around the wheel in 30 seconds. What is the average speed of the rider in feet per second?

78. Average Speed A person riding a Ferris wheel with a diameter of 102 feet travels once around the wheel in 3.5 minutes. What is the average speed of the rider in feet per minute?

Temperature The relationship between temperature in degrees Fahrenheit, F, and degrees Celsius, C, is given by the formula

$$F = \frac{9}{5}C + 32$$

Use this formula for problems 79 and 80.

79. Find C if $F = 95$.

80. Find C if $F = 50$.

Digital Video The biggest video download of all time was a *Star Wars* movie trailer.

© Lucasfilm/The Kobal Collection/Hamshere, Keith

The video was compressed so it would be small enough for people to download over the Internet. A formula for estimating the size, in kilobytes, of a compressed video is

$$S = \frac{height \cdot width \cdot fps \cdot time}{35{,}000}$$

where *height* and *width* are in pixels, *fps* is the number of frames per second the video is to play (television plays at 30 fps), and time is given in seconds.

81. Estimate the size in kilobytes of the *Star Wars* trailer that has a height of 480 pixels, a width of 216 pixels, plays at 30 fps, and runs for 150 seconds.

82. Estimate the size in kilobytes of the *Star Wars* trailer that has a height of 320 pixels, a width of 144 pixels, plays at 15 fps, and runs for 150 seconds.

Fermat's Last Theorem This postage stamp shows Fermat's last theorem, which states that if n is an integer greater than 2, then there are no positive integers x, y, and z that will make the formula $x^n + y^n = z^n$ true. Use the formula $x^n + y^n = z^n$ to

83. Find x if $n = 1$, $y = 7$, and $z = 15$.

84. Find y if $n = 1$, $x = 23$, and $z = 37$.

Exercise Physiology In exercise physiology, a person's maximum heart rate, in beats per minute, is found by subtracting his age, in years, from 220. So, if A represents your age in years, then your maximum heart rate is

$$M = 220 - A$$

A person's training heart rate, in beats per minute, is her resting heart rate plus 60% of the difference between her maximum heart rate and her resting heart rate. If resting heart rate is R and maximum heart rate is M, then the formula that gives training heart rate is

$$T = R + 0.6(M - R)$$

85. Training Heart Rate Shar is 46 years old. Her daughter, Sara, is 26 years old. If they both have a resting heart rate of 60 beats per minute, find the training heart rate for each.

86. Training Heart Rate Shane is 30 years old and has a resting heart rate of 68 beats per minute. Her mother, Carol, is 52 years old and has the same resting heart rate. Find the training heart rate for Shane and for Carol.

Learning Objectives Assessment

The following problems can be used to help assess if you have successfully met the learning objectives for this section.

87. If $4x - 3y = 5$, find y if $x = 2$.

    a. $-\dfrac{13}{3}$ **b.** 1 **c.** $\dfrac{11}{4}$ **d.** 9

88. Use the formula $d = (r - c) \cdot t$ to find c if $d = 60$, $r = 20$, and $t = 5$.

    a. 12 **b.** 3 **c.** 4 **d.** 8

89. Solve $4x - 3y = 6$ for y.

    a. $y = 4x - 2$ **b.** $y = 4x + 2$

    c. $y = \dfrac{4}{3}x - 2$ **d.** $y = \dfrac{4}{3}x + 2$

90. A person riding a Ferris wheel with a diameter of 120 feet travels once around the wheel in 40 seconds. What is the average speed of the rider in feet per second? Use 3.14 as an approximation of π.

    a. 6.3 **b.** 18.8 **c.** 9.4 **d.** 4.7

Getting Ready for the Next Section

To understand all of the explanations and examples in the next section you must be able to work the problems below.

Translate into symbols.

91. Three less than twice a number

92. Ten less than four times a number

93. The sum of x and y is 180

94. The sum of a and b is 90

Solve each equation.

95. $x + 2x = 90$

96. $x + 5x = 180$

97. $2(2x - 3) + 2x = 45$

98. $2(4x - 10) + 2x = 12.5$

99. $0.06x + 0.05(10,000 - x) = 560$

100. $x + 0.0725x = 17,481.75$

Applications

Learning Objectives

In this section we will learn how to:

1. Solve applications of linear equations from geometry.

2. Solve applications of linear equations involving percents.

3. Solve applications of linear equations using the rate equation.

4. Use a formula to build a table of paired data.

Introduction

In this section, we begin our work with application problems and use the skills we have developed for solving equations to solve problems written in words. You may find that some of the examples and problems are more realistic than others. Because we are just beginning our work with application problems, even the ones that seem unrealistic are good practice. What is important in this section is the *method* we use to solve application problems, not the applications themselves. The method, or strategy, that we use to solve application problems is called the *Blueprint for Problem Solving.* It is an outline that will overlay the solution process we use on all application problems.

BLUEPRINT FOR PROBLEM SOLVING

Step 1: **Read** the problem, and then mentally *list* the items that are known and the items that are unknown.

Step 2: **Assign a variable** to one of the unknown items. (In most cases, this will amount to letting $x =$ the item that is asked for in the problem.) Then **translate** the other **information** in the problem to expressions involving the variable.

Step 3: **Reread** the problem, and then **write an equation,** using the items and variable listed in steps 1 and 2, that describes the situation.

Step 4: **Solve the equation** found in step 3.

Step 5: **Write your answer** using a complete sentence.

Step 6: **Reread** the problem, and **check** your solution with the original words in the problem.

A number of substeps occur within each of the steps in our blueprint. For instance, with steps 1 and 2 it is always a good idea to draw a diagram or picture if it helps you visualize the relationship among the items in the problem.

Blueprint for Problem Solving

EXAMPLE 1   The length of a rectangle is 3 inches less than twice the width. The perimeter is 45 inches. Find the length and width.

SOLUTION   When working problems that involve geometric figures, a sketch of the figure helps organize and visualize the problem.

Step 1: ***Read and list.***

Known items: The figure is a rectangle. The length is 3 inches less than twice the width. The perimeter is 45 inches.

Unknown items: The length and the width

Step 2: ***Assign a variable and translate information.***

Because the length is given in terms of the width (the length is 3 less than twice the width), we let $x =$ the width of the rectangle. The length is 3 less than twice the width, so it must be $2x - 3$. The diagram in Figure 1 is a visual description of the relationships we have listed so far.

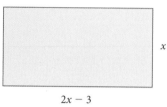

x

$2x - 3$

FIGURE 1

Step 3: ***Reread and write an equation.***

The equation that describes the situation is

Twice the length + twice the width = perimeter

$$2(2x - 3)   +   2x   =   45$$

Step 4: ***Solve the equation.***

$$2(2x - 3) + 2x = 45$$
$$4x - 6 + 2x = 45$$
$$6x - 6 = 45$$
$$6x = 51$$
$$x = 8.5$$

Step 5: ***Write the answer.***

The width is 8.5 inches. The length is $2x - 3 = 2(8.5) - 3 = 14$ inches.

Step 6: ***Reread and check.***

If the length is 14 inches and the width is 8.5 inches, then the perimeter must be $2(14) + 2(8.5) = 28 + 17 = 45$ inches. Also, the length, 14, is 3 less than twice the width.

Remember as you read through the steps in the solutions to the examples in this section that step 1 is done mentally. Read the problem and then *mentally* list the items that you know and the items that you don't know. The purpose of step 1 is to give you direction as you begin to work application problems. Finding the solution to an application problem is a process; it doesn't happen all at once. The first step is to read the problem with a purpose in mind. That purpose is to mentally note the items that are known and the items that are unknown.

███ **EXAMPLE 2**   In April, Pat bought a Ford Mustang with a 5.0-liter engine. The total price, which includes the price of the used car plus sales tax, was $17,481.75. If the sales tax rate was 7.25%, what was the price of the car?

SOLUTION

Step 1: ***Read and list.***

Known items: The total price is $17,481.75. The sales tax rate is 7.25%, which is 0.0725 in decimal form.

Unknown item: The price of the car

Step 2: ***Assign a variable and translate information.***

If we let x = the price of the car, then to calculate the sales tax we multiply the price of the car x by the sales tax rate:

$$\text{Sales tax} = (\text{sales tax rate})(\text{price of the car})$$
$$= 0.0725x$$

Step 3: ***Reread and write an equation.***

$$\text{Car price} + \text{sales tax} = \text{total price}$$
$$x \quad\;\; + 0.0725x = 17{,}481.75$$

Step 4: ***Solve the equation.***

$$x + 0.0725x = 17{,}481.75$$
$$1.0725x = 17{,}481.75$$
$$x = \frac{17{,}481.75}{1.0725}$$
$$= 16{,}300.00$$

Step 5: ***Write the answer.***

The price of the car is $16,300.00.

Step 6: ***Reread and check.***

The price of the car is $16,300.00. The tax is

$$0.0725(16{,}300) = \$1{,}181.75.$$

Adding the retail price and the sales tax we have a total bill of $17,481.75.   ███

⚗ **FACTS FROM GEOMETRY**   *Angles*

An angle is formed by two rays with the same endpoint. The common endpoint is called the ***vertex*** of the angle, and the rays are called the ***sides*** of the angle.

In Figure 2, angle θ (theta) is formed by the two rays OA and OB. The vertex of θ is O. Angle θ is also denoted as angle AOB, where the letter associated with the vertex is always the middle letter in the three letters used to denote the angle.

Degree Measure

The angle formed by rotating a ray through one complete revolution about its endpoint (Figure 3) has a measure of 360 degrees, which we write as 360°.

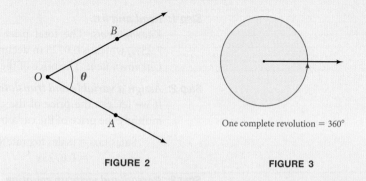

FIGURE 2   FIGURE 3

One degree of angle measure, written 1°, is $\frac{1}{360}$ of a complete rotation of a ray about its endpoint; there are 360° in one full rotation. (The number 360 was decided upon by early civilizations because it was believed that the Earth was at the center of the universe and the sun would rotate once around Earth every 360 days.) Similarly, 180° is half of a complete rotation, and 90° is a quarter of a full rotation. Angles that measure 90° are called **right angles**, and angles that measure 180° are called **straight angles**. If an angle measures between 0° and 90° it is called an **acute angle**, and an angle that measures between 90° and 180° is an **obtuse angle**. Figure 4 illustrates further.

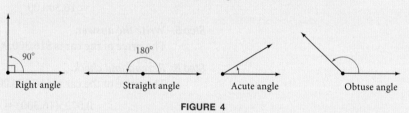

Right angle   Straight angle   Acute angle   Obtuse angle

FIGURE 4

Complementary Angles and Supplementary Angles

If two angles add up to 90°, we call them **complementary angles**, and each is called the **complement** of the other. If two angles have a sum of 180°, we call them **supplementary angles**, and each is called the **supplement** of the other. Figure 5 illustrates the relationship between angles that are complementary and angles that are supplementary.

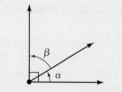

Complementary angles: $\alpha + \beta = 90°$ Supplementary angles: $\alpha + \beta = 180°$

FIGURE 5

EXAMPLE 3 Two complementary angles are such that one is twice as large as the other. Find the two angles.

SOLUTION Applying the Blueprint for Problem Solving, we have:

Step 1: Read and list.
Known items: Two complementary angles. One is twice as large as the other.
Unknown items: The size of the angles

Step 2: Assign a variable and translate information.
Let x = the smaller angle. The larger angle is twice the smaller, so we represent the larger angle with $2x$.

Step 3: Reread and write an equation.
Because the two angles are complementary, their sum is 90. Therefore,

$$x + 2x = 90$$

Step 4: Solve the equation.

$$x + 2x = 90$$
$$3x = 90$$
$$x = 30$$

Step 5: Write the answer.
The smaller angle is 30°, and the larger angle is $2 \cdot 30 = 60°$.

Step 6: Reread and check.
The larger angle is twice the smaller angle, and their sum is 90°.

Suppose we know that the sum of two numbers is 50. If we let x represent one of the two numbers, how can we represent the other? Let's suppose for a moment that x turns out to be 30. Then the other number will be 20, because their sum is 50. That is, if two numbers add up to 50, and one of them is 30, then the other must be $50 - 30 = 20$. Generalizing this to any number x, we see that if two numbers have a sum of 50, and one of the numbers is x, then the other must be $50 - x$. The following table shows some additional examples:

If Two Numbers Have a Sum of	And One of Them Is	Then the Other Must Be
50	x	$50 - x$
10	y	$10 - y$
12	n	$12 - n$

EXAMPLE 4   Suppose a person invests a total of $10,000 in two accounts. One account earns 5% annually, and the other earns 6% annually. If the total interest earned from both accounts in a year is $560, how much is invested in each account?

SOLUTION

Step 1: *Read and list.*

Known items: Two accounts. One pays interest of 5%, and the other pays 6%. The total invested is $10,000. Total interest from both accounts in a year is $560.

Unknown items: The number of dollars invested in each individual account

Step 2: *Assign a variable and translate information.*

If we let $x =$ the amount invested at 6%, then $10,000 - x$ is the amount invested at 5%. The total interest earned from both accounts is $560. The amount of interest earned on x dollars at 6% is $.06x$, whereas the amount of interest earned on $10,000 - x$ dollars at 5% is $0.05(10,000 - x)$.

	Dollars at 6%	Dollars at 5%	Total
Number of	x	$10,000 - x$	10,000
Interest on	$0.06x$	$0.05(10,000 - x)$	560

Step 3: *Reread and write an equation.*

The last line gives us the equation we are after:

$$0.06x + 0.05(10,000 - x) = 560$$

Step 4: *Solve the equation.*

To make the equation a little easier to solve, we begin by multiplying both sides by 100 to move the decimal point two places to the right.

$$6x + 5(10,000 - x) = 56,000$$
$$6x + 50,000 - 5x = 56,000$$
$$x + 50,000 = 56,000$$
$$x = 6,000$$

Step 5: *Write the answer.*

The amount of money invested at 6% is $6,000. The amount of money invested at 5% is $10,000 - $6,000 = $4,000$.

Step 6: *Reread and check.*

To check our results, we find the total interest from the two accounts:

The interest earned on $6,000 at 6% is $0.06(6,000) = 360$
The interest earned on $4,000 at 5% is $0.05(4,000) = 200$

The total interest $= \$560$

(A) FACTS FROM GEOMETRY   *Triangles*

Isosceles Triangles

It is not unusual to have the terms we use in mathematics show up in the descriptions of things we find in the world around us. The flag of Puerto Rico shown here is described on the government website as "Five equal horizontal bands of red (top and bottom) alternating with white; a blue isosceles triangle based on the hoist side bears a large white five-pointed star in the center." An *isosceles triangle* as shown here and in Figure 6, is a triangle with two sides of equal length.

Angles A and B in the isosceles triangle in Figure 6 are called the ***base angles***: they are the angles opposite the two equal sides. In every isosceles triangle, the base angles are equal.

Isosceles Triangle

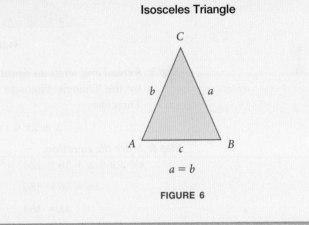

$$a = b$$

FIGURE 6

Note As you can see from Figure 6, one way to label the important parts of a triangle is to label the vertices with capital letters and the sides with small letters: side a is opposite vertex A, side b is opposite vertex B, and side c is opposite vertex C.

Also, because each vertex is the vertex of one of the angles of the triangle, we refer to the three interior angles as A, B, and C.

$\lfloor \Delta \neq \Sigma$ THEOREM   *Triangle Theorem*

In any triangle, the sum of the interior angles is $180°$. For the triangle shown in Figure 7, the relationship is written

$$A + B + C = 180°$$

FIGURE 7

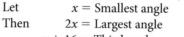 **EXAMPLE 5** The largest angle in a triangle is twice the smallest angle. The third angle is 16° more than the smallest angle. Find the measure of each angle.

SOLUTION

Step 1: *Read and list.*

Known items: A triangle. The largest angle is twice the smallest. The third angle is 16° more than the smallest.

Unknown items: The three angles.

Step 2: *Assign a variable and translate information.*

Let $x =$ Smallest angle

Then $2x =$ Largest angle

$x + 16 =$ Third angle

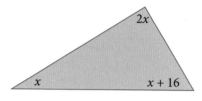

FIGURE 8

Step 3: *Reread and write an equation.*

By the Triangle Theorem, the sum of the interior angles is 180°. Therefore,

$$x + 2x + (x + 16) = 180$$

Step 4: *Solve the equation.*

$$x + 2x + x + 16 = 180$$

$$4x + 16 = 180$$

$$4x = 164$$

$$x = 41$$

Step 5: *Write the answer.*

The smallest angle is 41°, the largest angle is $2 \cdot 41 = 82°$, and the third angle is $41 + 16 = 57°$.

Step 6: *Reread and check.*

The largest angle is twice the smallest angle, the third angle is 16° more than the smallest angle, and the three angles add up to 180°. ▨

███ **EXAMPLE 6**   Allison can run 2 miles per hour faster than Kaitlin, and can cover the same distance in 3 hours that Kaitlin can do in 4 hours. Find the speed at which each girl can run.

SOLUTION

Step 1: *Read and list.*

Known items: Allison can run 2 miles per hour faster than Kaitlin. Allison can run the same distance in 3 hours that Kaitlin can run in 4 hours.

Unknown items: The rate for each girl, and the distance.

Step 2: *Assign a variable and translate information.*

Let r = Kaitlin's rate

Then $r + 2$ = Allison's rate

Also, using the rate equation $d = rt$, we have

$r \cdot 4$ = Kaitlin's distance after 4 hours

$(r + 2) \cdot 3$ = Allison's distance after 3 hours

Step 3: *Reread and write an equation.*

Because Kaitlin can run the same distance in 4 hours that Allison can run in 3 hours,

$$r \cdot 4 = (r + 2) \cdot 3$$

Step 4: *Solve the equation.*

$$4r = 3r + 6$$

$$r = 6$$

Step 5: *Write the answer.*

Kaitlin can run 6 miles per hour, and Allison can run $6 + 2 = 8$ miles per hour.

Step 6: *Reread and check.*

Allison's speed is 2 miles per hour more than Kaitlin's speed. In 3 hours, Allison can run $8 \cdot 3 = 24$ miles, and in 4 hours, Kaitlin can run $6 \cdot 4 = 24$ miles which is the same distance.

███

Table Building

We can use our knowledge of formulas from Section 2.2 to build tables of paired data. As you will see, equations or formulas that contain exactly two variables produce pairs of numbers that can be used to construct tables.

EXAMPLE 7   A piece of string 12 inches long is to be formed into a rectangle. Build a table that gives the length of the rectangle if the width is 1, 2, 3, 4, or 5 inches. Then find the area of each of the rectangles formed.

SOLUTION   Because the formula for the perimeter of a rectangle is $P = 2l + 2w$, and our piece of string is 12 inches long, the formula we will use to find the lengths for the given widths is $12 = 2l + 2w$. To solve this formula for l, we divide each side by 2 and then subtract w. The result is $l = 6 - w$. Table 1 organizes our work so that the formula we use to find l for a given value of w is shown, and we have added a last column to give us the areas of the rectangles formed. The units for the first three columns are inches, and the units for the numbers in the last column are square inches.

12 inches

TABLE 1	Length, Width, and Area		
Width (in.)	Length (in.)		Area (in.²)
w	$l = 6 - w$	l	$A = lw$
1	$l = 6 - 1$	5	5
2	$l = 6 - 2$	4	8
3	$l = 6 - 3$	3	9
4	$l = 6 - 4$	2	8
5	$l = 6 - 5$	1	5

Figures 9 and 10 show two **bar charts** constructed from the information in Table 1.

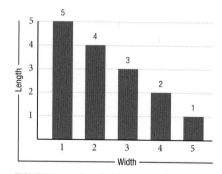

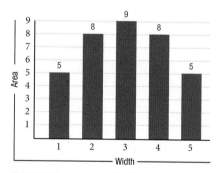

FIGURE 9 *Length and width of rectangles with perimeters fixed at 12 inches*

FIGURE 10 *Area and width of rectangles with perimeters fixed at 12 inches*

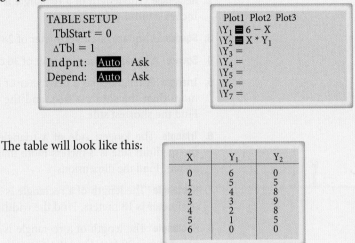

USING TECHNOLOGY   *Graphing Calculators*

A number of graphing calculators have table-building capabilities. We can let the calculator variable X represent the widths of the rectangles in Example 7. To find the lengths, we set variable Y_1 equal to $6 - X$. The area of each rectangle can be found by setting variable Y_2 equal to $X * Y_1$. To have the calculator produce the table automatically, we use a table minimum of 0 and a table increment of 1. Here is a summary of how the graphing calculator is set up:

TABLE SETUP
TblStart = 0
ΔTbl = 1
Indpnt: Auto Ask
Depend: Auto Ask

Plot1 Plot2 Plot3
\Y_1 = 6 − X
\Y_2 = X * Y_1
\Y_3 =
\Y_4 =
\Y_5 =
\Y_6 =
\Y_7 =

The table will look like this:

X	Y_1	Y_2
0	6	0
1	5	5
2	4	8
3	3	9
4	2	8
5	1	5
6	0	0

Getting Ready for Class

After reading through the preceding section, respond in your own words and in complete sentences.

A. What is the first step in solving an application problem?

B. What is the biggest obstacle between you and success in solving application problems?

C. Write an application problem for which the solution depends on solving the equation $2x + 2 \cdot 3 = 18$.

D. What is the last step in solving an application problem? Why is this step important?

Solve each application problem. Be sure to follow the steps in the Blueprint for Problem Solving.

Geometry Problems

1. **Rectangle** A rectangle is twice as long as it is wide. The perimeter is 60 feet. Find the dimensions.

2. **Rectangle** The length of a rectangle is 5 times the width. The perimeter is 48 inches. Find the dimensions.

3. **Square** A square has a perimeter of 28 feet. Find the length of each side.

4. **Square** A square has a perimeter of 36 centimeters. Find the length of each side.

5. **Triangle** A triangle has a perimeter of 23 inches. The medium side is 3 inches more than the shortest side, and the longest side is twice the shortest side. Find the shortest side.

6. **Triangle** The longest side of a triangle is two times the shortest side, while the medium side is 3 meters more than the shortest side. The perimeter is 27 meters. Find the dimensions.

7. **Rectangle** The length of a rectangle is 3 meters less than twice the width. The perimeter is 18 meters. Find the width.

8. **Rectangle** The length of a rectangle is 1 foot more than twice the width. The perimeter is 20 feet. Find the dimensions.

9. **Livestock Pen** A livestock pen is built in the shape of a rectangle that is twice as long as it is wide. The perimeter is 48 feet. If the material used to build the pen is $1.75 per foot for the longer sides and $2.25 per foot for the shorter sides (the shorter sides have gates, which increase the cost per foot), find the cost to build the pen.

10. **Garden** A garden is in the shape of a square with a perimeter of 42 feet. The garden is surrounded by two fences. One fence is around the perimeter of the garden, whereas the second fence is 3 feet from the first fence, as Figure 12 indicates. If the material used to build the two fences is $1.28 per foot, what was the total cost of the fences?

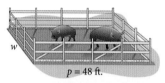

w
$p = 48$ ft.

3 feet

FIGURE 12

Percent Problems

11. **Money** Shane returned from a trip to Las Vegas with $300.00, which was 50% more money than he had at the beginning of the trip. How much money did Shane have at the beginning of his trip?

12. **Items Sold** Every item in the Just a Dollar store is priced at $1.00. When Mary Jo opens the store, there is $125.50 in the cash register. When she counts the money in the cash register at the end of the day, the total is $1,058.60. If the sales tax rate is 8.5%, how many items were sold that day?

13. **Textbook Price** Suppose a college bookstore buys a textbook from a publishing company and then marks up the price they paid for the book 33% and sells it to a student at the marked-up price. If the student pays $115.00 for the textbook, what did the bookstore pay for it? Round your answer to the nearest cent.

14. Hourly Wage A sheet metal worker earns $26.80 per hour after receiving a 4.5% raise. What was the sheet metal worker's hourly pay before the raise? Round your answer to the nearest cent.

15. Movies *Batman Forever* grossed $52.8 million on its opening weekend and had one of the most successful movie launches in history. If *Batman Forever* accounted for approximately 53% of all box office receipts that weekend, what were the total box office receipts? (Round to the nearest tenth of a million.)

16. Fat Content in Milk I was reading the information on the milk carton in Figure 13 at breakfast one morning when I was working on this book. According to the carton, this milk contains 70% less fat than whole milk. The nutrition label on the other side of the carton states that one serving of this milk contains 2.5 grams of fat. How many grams of fat are in an equivalent serving of whole milk? (Round to the nearest hundredth.)

FIGURE 13

More Geometry Problems

17. Angles Two supplementary angles are such that one is eight times larger than the other. Find the two angles.

18. Angles Two complementary angles are such that one is five times larger than the other. Find the two angles.

19. Angles One angle is 12° less than four times another. Find the measure of each angle if
  a. they are complements of each other.
  b. they are supplements of each other.

20. Angles One angle is 4° more than three times another. Find the measure of each angle if
  a. they are complements of each other.
  b. they are supplements of each other.

21. Triangles A triangle is such that the largest angle is three times the smallest angle. The third angle is 9° less than the largest angle. Find the measure of each angle.

22. Triangles The smallest angle in a triangle is half of the largest angle. The third angle is 15° less than the largest angle. Find the measure of all three angles.

23. Triangles The smallest angle in a triangle is one-third of the largest angle. The third angle is 10° more than the smallest angle. Find the measure of all three angles.

24. Isosceles Triangles The third angle in an isosceles triangle is half as large as each of the two base angles. Find the measure of each angle.

25. Isosceles Triangles The third angle in an isosceles triangle is 8° more than twice as large as each of the two base angles. Find the measure of each angle.

26. Isosceles Triangles The third angle in an isosceles triangle is 4° more than one fifth of each of the two base angles. Find the measure of each angle.

Interest Problems

27. Investing A woman has a total of $9,000 to invest. She invests part of the money in an account that pays 8% per year and the rest in an account that pays 9% per year. If the interest earned in the first year is $750, how much did she invest in each account?

	Dollars at 8%	Dollars at 9%	Total
Number of			
Interest on			

28. Investing A man invests $12,000 in two accounts. If one account pays 10% per year and the other pays 7% per year, how much was invested in each account if the total interest earned in the first year was $960?

	Dollars at 10%	Dollars at 7%	Total
Number of			
Interest on			

29. Investing A total of $15,000 is invested in two accounts. One of the accounts earns 12% per year, and the other earns 10% per year. If the total interest earned in the first year is $1,600, how much was invested in each account?

30. Investing A total of $11,000 is invested in two accounts. One of the two accounts pays 9% per year, and the other account pays 11% per year. If the total interest paid in the first year is $1,150, how much was invested in each account?

31. Investing Stacey has a total of $6,000 in two accounts. The total amount of interest she earns from both accounts in the first year is $500. If one of the accounts earns 8% interest per year and the other earns 9% interest per year, how much did she invest in each account?

32. Investing Travis has a total of $6,000 invested in two accounts. The total amount of interest he earns from the accounts in the first year is $410. If one account pays 6% per year and the other pays 8% per year, how much did he invest in each account?

Distance, Rate and Time Problems

33. Distance Eliana and Allegra both belong to a cycling club. Eliana can ride 4 miles per hour faster than Allegra, and can cover the same distance in 2 hours that Allegra can do in 2.5 hours. Find the speed at which each girl can ride.

34. Distance Lucus can roller blade 9 miles per hour faster than he can walk. He can roller blade the same distance in 2 hours that it would take him 8 hours to walk. Find the rates at which Lucus can walk and roller blade.

35. **Distance** A search is being conducted for someone guilty of a hit-and-run felony. In order to set up roadblocks at appropriate points, the police must determine how far the guilty party might have traveled during the past half-hour. Use the formula $d = rt$ with $t = 0.5$ hour to complete the following table.

Speed (miles per hour)	Distance (miles)
20	
30	
40	
50	
60	
70	

36. **Speed** To determine the average speed of a bullet when fired from a rifle, the time is measured from when the gun is fired until the bullet hits a target that is 1,000 feet away. Use the formula $r = \frac{d}{t}$ with $d = 1,000$ feet to complete the following table.

Time (seconds)	Rate (feet per second)
1.00	
0.80	
0.64	
0.50	
0.40	
0.32	

37. **Current** A boat that can travel 10 miles per hour in still water is traveling along a stream with a current of 4 miles per hour. The distance the boat will travel upstream is given by the formula $d = (r - c) \cdot t$, and the distance it will travel downstream is given by the formula $d = (r + c) \cdot t$. Use these formulas with $r = 10$ and $c = 4$ to complete the following table.

Time (hours)	Distance Upstream (miles)	Distance Downstream (miles)
1		
2		
3		
4		
5		
6		

38. **Wind** A plane that can travel 300 miles per hour in still air is traveling in a wind stream with a speed of 20 miles per hour. The distance the plane will travel against the wind is given by the formula $d = (r - w) \cdot t$, and the distance it will travel with the wind is given by the formula $d = (r + w) \cdot t$. Use these formulas with $r = 300$ and $w = 20$ to complete the following table.

Time (hours)	Distance against the wind (miles)	Distance with the wind (miles)
0.50		
1.00		
1.50		
2.00		
2.50		
3.00		

Miscellaneous Problems

39. **Tickets** Tickets for the father-and-son breakfast were $2.00 for fathers and $1.50 for sons. If a total of 75 tickets were sold for $127.50, how many fathers and how many sons attended the breakfast?

40. **Tickets** A Girl Scout troop sells 62 tickets to their mother-and-daughter dinner, for a total of $216. If the tickets cost $4.00 for mothers and $3.00 for daughters, how many of each ticket did they sell?

41. **Sales Tax** A woman owns a small, cash-only business in a state that requires her to charge 6% sales tax on each item she sells. At the beginning of the day, she has $250 in the cash register. At the end of the day, she has $1,204 in the register. How much money should she send to the state government for the sales tax she collected?

42. **Sales Tax** A store is located in a state that requires 6% tax on all items sold. If the store brings in $3,392 in one day, how much of that total was sales tax?

43. **Long-Distance Phone Calls** Patrick goes away to college. The first week he is away from home he calls his girlfriend, using his parents' telephone credit card, and talks for a long time. The telephone company charges 40 cents for the first minute and 30 cents for each additional minute, and then adds on a 50-cent service charge for using the credit card. If his parents receive a bill for $13.80 for Patrick's call, how long did he talk?

44. **Long-Distance Phone Calls** A person makes a long distance person-to-person call to Santa Barbara, California. The telephone company charges 41 cents for the first minute and 32 cents for each additional minute. Because the call is person-to-person, there is also a service charge of $3.00. If the cost of the call is $6.29, how many minutes did the person talk?

Table Building

45. Use $y = 0.08x$ to complete the table.

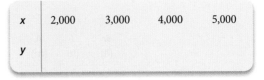

x	2,000	3,000	4,000	5,000
y				

46. Use $h = \dfrac{60}{t}$ to complete the table.

t	4	6	8	10
h				

Coffee Sales Use the information in the chart below to complete the following tables.

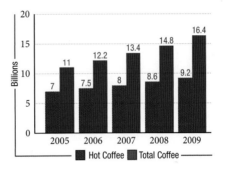

Hot Coffee ■ Total Coffee

47.

Hot Coffee Sales

Year	Sales (billions of dollars)
2005	
2006	
2007	
2008	
2009	

48.

Total Coffee Sales

Year	Sales (billions of dollars)
2005	
2006	
2007	
2008	
2009	

49. Livestock Pen A farmer buys 48 feet of fencing material to build a rectangular livestock pen. Fill in the second column of the table to find the length of the pen if the width is 2, 4, 6, 8, 10, or 12 feet. Then find the area of each of the pens formed.

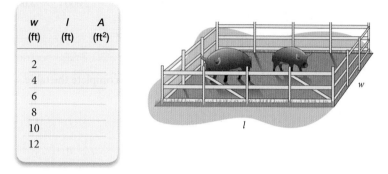

w (ft)	l (ft)	A (ft²)
2		
4		
6		
8		
10		
12		

50. Temperature The relationship between the Fahrenheit scale and Celsius scale for temperature is given by

$$F = \frac{9}{5}C + 32$$

where C is temperature in degrees Celsius and F is temperature in degrees Fahrenheit. Use this formula to complete the following table.

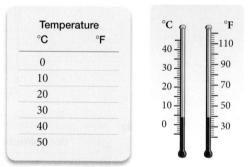

Temperature °C	°F
0	
10	
20	
30	
40	
50	

Maximum Heart Rate In exercise physiology, a person's maximum heart rate, in beats per minute, is found by subtracting his age in years from 220. So, if A represents your age in years, then your maximum heart rate is

$$M = 220 - A$$

Use this formula to complete the following tables.

51.

Age (years)	Maximum Heart Rate (beats per minute)
18	
19	
20	
21	
22	
23	

52.

Age (years)	Maximum Heart Rate (beats per minute)
15	
20	
25	
30	
35	
40	

Training Heart Rate A person's training heart rate, in beats per minute, is his resting heart rate plus 60% of the difference between his maximum heart rate and his resting heart rate. If resting heart rate is R and maximum heart rate is M, then the formula that gives training heart rate is

$$T = R + 0.6(M - R)$$

Use this formula along with the results of Problems 51 and 52 to fill in the following two tables.

53. For a 20-year-old person

Resting Heart Rate (beats per minute)	Training Heart Rate (beats per minute)
60	
62	
64	
68	
70	
72	

54. For a 40-year-old person

Resting Heart Rate (beats per minute)	Training Heart Rate (beats per minute)
60	
62	
64	
68	
70	
72	

Learning Objectives Assessment

The following problems can be used to help assess if you have successfully met the learning objectives for this section.

55. Rectangle The length of a rectangle is 2 inches more than three times the width. The perimeter is 36 inches. Find the width, in inches.

   a. 9   **b.** 4   **c.** 8.5   **d.** 14

56. Investing A total of $1,000 is invested in two accounts, one earning 2% per year, and the other earning 3% per year. If the total interest earned in the first year is $27.50, how much was invested in the account earning 2%?

   a. $150   **b.** $250   **c.** $600   **d.** $750

57. Distance Vance can run 9 miles per hour slower than he can skateboard. He can cover the same distance in 2 hours on his skateboard that he can run in 5 hours. Find the rate at which Vance can skateboard, in miles per hour.

   a. 15   **b.** 12   **c.** 18   **d.** 21

58. Which pair of data could appear in a table for the following formula?

$$\frac{1}{2}bh = 24$$

   a. $b = 8, h = 3$ **b.** $b = 2, h = 6$

   c. $b = 3, h = 16$ **d.** $b = 12, h = 5$

Getting Ready for the Next Section

To understand all of the explanations and examples in the next section you must be able to work the problems below.

Place either $<$ or $>$ between each pair of numbers so that the resulting statement is true.

59. $0 \,\square\, 2$ **60.** $2 \,\square\, 0$ **61.** $-5 \,\square\, -1$ **62.** $\dfrac{1}{2} \,\square\, -\dfrac{1}{2}$

Solve each equation.

63. $-2x - 3 = 7$ **64.** $3x + 3 = 2x - 1$

65. $3(2x - 4) - 7x = -3x$ **66.** $3(2x + 5) = -3x$

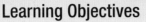

Linear Inequalities in One Variable and Interval Notation

2.4

Learning Objectives

In this section we will learn how to:

1. Use the addition and multiplication properties for inequalities to solve linear inequalities in one variable.

2. Solve linear inequalities in one variable containing decimals or fractions.

3. Use interval notation to express the solution set to a linear inequality in one variable.

4. Solve applications involving linear inequalities in one variable.

Introduction

Wikimedia Commons/©footwarrior

The 'Apapane, a native Hawaiian bird, feeds mainly on the nectar of the 'ohi'a lehua blossom, although the adult diet also includes insects and spiders. The 'Apapane live in high altitude regions where the 'ohi'a blossoms are found and where the birds are protected from mosquitoes which transmit avian malaria and avian pox. Predators of the 'Apapane include the rat, feral cat, mongoose, and owl. According to the *U.S. Geological Survey*:

> Annual survival rates based on 1,584 recaptures of 429 banded individuals:
> 0.72 ± 0.11 for adults and 0.13 ± 0.07 for juveniles.

We can write the survival rate for the adults as an inequality:

$$0.61 \leq r \leq 0.83$$

Inequalities are what we will study in this section.

A linear inequality in one variable is any inequality that can be put in the form

$$ax + b < 0 \qquad (a \text{ and } b \text{ constants}, a \neq 0)$$

where the inequality symbol ($<$) can be replaced with any of the other three inequality symbols ($\leq$, $>$, or $\geq$).

Some examples of *linear inequalities* are

$$x + 9 \leq 0 \qquad 3x - 2 \geq 7 \qquad -5y < 25 \qquad 3(x - 4) > 2x$$

Addition Property for Inequalities

Our first property for inequalities is similar to the addition property we used when solving equations.

Note Because subtraction is defined as addition of the opposite, our new property holds for subtraction as well as addition. That is, we can subtract the same quantity from each side of an inequality and always be sure that we have not changed the solution.

> **[Δ≠Σ] PROPERTY** *Addition Property for Inequalities*
>
> For any algebraic expressions, A, B, and C,
>
> $$\text{if} \qquad A < B$$
>
> $$\text{then} \qquad A + C < B + C$$
>
> *In words*: Adding the same quantity to both sides of an inequality will not change the solution set.

EXAMPLE 1   Solve $3x + 3 < 2x - 1$, and graph the solution set.

SOLUTION   We use the addition property for inequalities to write all the variable terms on one side and all constant terms on the other side:

$$3x + 3 < 2x - 1$$

$$3x + (-2x) + 3 < 2x + (-2x) - 1 \qquad \text{Add } -2x \text{ to each side}$$

$$x + 3 < -1$$

$$x + 3 + (-3) < -1 + (-3) \qquad \text{Add } -3 \text{ to each side}$$

$$x < -4$$

The solution set is all real numbers that are less than -4. To show this we can use **set-builder notation** and write

$$\{x \mid x < -4\}$$

Note We read the expression this way, the set of all numbers x, such that x is less than -4.

We can graph the solution set on the number line using an open circle at -4 to show that -4 is not part of the solution set. This is the format you used when graphing inequalities in beginning algebra.

Here is an equivalent graph that uses a parenthesis opening left, instead of an open circle, to represent the endpoint of the graph.

Interval Notation and Graphing

As we saw in Example 1, the solution set to a linear inequality in one variable is typically a subset of the real numbers, which can be represented graphically by an entire portion of the number line. A continuous segment of the number line is called an **interval**, and we can describe an interval using **interval notation**. In doing so, we always work from left to right on the number line, indicating where the interval begins on the left, followed by where it ends on the right. In general, if an interval begins at a and ends at b, then the notation will look like this:

$$(a, b) \quad [a, b) \quad (a, b] \quad [a, b]$$

An endpoint is included in the interval if a bracket is used.
An endpoint is not included in the interval if a parenthesis is used.

Note The English mathematician John Wallis (1616–1703) was the first person to use the ∞ symbol to represent infinity. When we encounter the interval $(3, \infty)$, we read it as "the interval from 3 to infinity," and we mean the set of real numbers that are greater than three. Likewise, the interval $(-\infty, -4)$ is read "the interval from negative infinity to -4," which is all real numbers less than -4.

If the interval extends indefinitely to the left, the symbol $-\infty$ is used for the left endpoint a. If the interval extends indefinitely to the right, the symbol ∞ is used for the right endpoint b. If the values a and b are included in the solution set, then brackets are used. Otherwise parentheses are used. A parenthesis is always used with $-\infty$ or ∞.

Using interval notation, the solution set for Example 1 would be expressed as $(-\infty, -4)$, indicating that the solution set is all real numbers up to, but not including, -4.

The following table shows the connection between set-builder notation, interval notation, and number line graphs. We have included the graphs with open and closed circles for those of you who have used this type of graph previously. In this book, we will continue to show our graphs using the parentheses/brackets method.

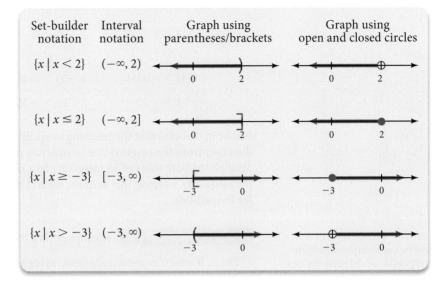

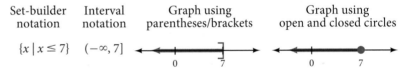 **EXAMPLE 2**   For each inequality, describe the solution set in words. Then write the solution set using set-builder and interval notation, and graph the solution set on the number line.

a. $x \leq 7$ **b.** $x > 0$

SOLUTION

a. The solution set is all real numbers up to and including 7.

> **Note** Because we always describe an interval from left to right on the number line, the notation $[7, -\infty)$ would be incorrect for the solution set to $x \leq 7$.

Set-builder notation	Interval notation	Graph using parentheses/brackets	Graph using open and closed circles
$\{x \mid x \leq 7\}$	$(-\infty, 7]$		

b. The solution set is all real numbers to the right of, but not including, 0.

Set-builder notation	Interval notation	Graph using parentheses/brackets	Graph using open and closed circles
$\{x \mid x > 0\}$	$(0, \infty)$		

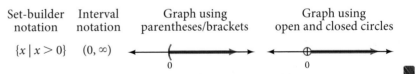

Before we state the multiplication property for inequalities, we will take a look at what happens to an inequality statement when we multiply both sides by a positive number and what happens when we multiply by a negative number.

We begin by writing three true inequality statements:

$$3 < 5 \qquad -3 < 5 \qquad -5 < -3$$

We multiply both sides of each inequality by a positive number, say, 4:

$$4(3) < 4(5) \qquad 4(-3) < 4(5) \qquad 4(-5) < 4(-3)$$

$$12 < 20 \qquad\quad -12 < 20 \qquad\quad -20 < -12$$

Notice in each case that when the resulting inequality symbol points in the same direction as the original inequality symbol, all three statements are still true. Multiplying both sides of an inequality by a positive number preserves the *sense* of the inequality.

Let's take the same three original inequalities and multiply both sides by -4:

$$3 < 5 \qquad\qquad -3 < 5 \qquad\qquad -5 < -3$$

$$-4(3) > -4(5) \qquad -4(-3) > -4(5) \qquad -4(-5) > -4(-3)$$

$$-12 > -20 \qquad\qquad 12 > -20 \qquad\qquad 20 > 12$$

Notice in this case that the resulting inequality symbol must point in the opposite direction from the original one in order to preserve the truth of the statements. Multiplying both sides of an inequality by a negative number *reverses* the sense of the inequality. Keeping this in mind, we will now state the multiplication property for inequalities.

Note Because division is defined as multiplication by the reciprocal, we can apply our new property to division as well as to multiplication. We can divide both sides of an inequality by any nonzero number as long as we reverse the direction of the inequality when the number we are dividing by is negative.

[△≠∑] PROPERTY   *Multiplication Property for Inequalities*

Let A, B, and C represent algebraic expressions.

$$\text{If} \qquad A < B$$
$$\text{then} \quad AC < BC \qquad \text{if} \qquad C \text{ is positive } (C > 0)$$
$$\text{or} \quad AC > BC \qquad \text{if} \qquad C \text{ is negative } (C < 0)$$

In words: Multiplying both sides of an inequality by a positive number always produces an equivalent inequality. Multiplying both sides of an inequality by a negative number reverses the sense of the inequality.

The multiplication property for inequalities does not limit what we can do with inequalities. We are still free to multiply both sides of an inequality by any nonzero number we choose. If the number we multiply by happens to be *negative,* then we *must also reverse* the direction of the inequality.

EXAMPLE 3   Find the solution set for $-2y - 3 < 7$.

SOLUTION   We begin by adding 3 to each side of the inequality:

$$-2y - 3 < 7$$
$$-2y < 10 \qquad\qquad \text{Add 3 to both sides}$$

$$-\frac{1}{2}(-2y) > -\frac{1}{2}(10) \qquad\qquad \text{Multiply by } -\frac{1}{2} \text{ and reverse the direction of the inequality symbol}$$

$$y > -5$$

The solution set is all real numbers that are greater than -5. The following are three equivalent ways to represent this solution set.

Set-Builder Notation	Number Line Graph	Interval Notation
$\{y \mid y > -5\}$		$(-5, \infty)$

Notice how a parenthesis is used with interval notation to show that -5 is not part of the solution set.

When our inequalities become more complicated, we use the same basic steps we used when we were solving equations. That is, we simplify each side of the inequality before we apply the addition property or multiplication property. When we have solved the inequality, we graph the solution on a number line.

Note In Examples 3 and 4, notice that each time we multiplied both sides of the inequality by a negative number we also reversed the direction of the inequality symbol. Failure to do so would cause our graph to lie on the wrong side of the endpoint.

EXAMPLE 4   Solve: $3(2x - 4) - 7x \leq -3x$.

SOLUTION   We begin by using the distributive property to separate terms. Next, simplify both sides.

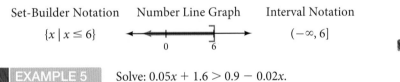

$3(2x - 4) - 7x \leq -3x$	Original inequality
$6x - 12 - 7x \leq -3x$	Distributive property
$-x - 12 \leq -3x$	$6x - 7x = (6 - 7)x = -x$
$-12 \leq -2x$	Add x to both sides
$-\dfrac{1}{2}(-12) \geq -\dfrac{1}{2}(-2x)$	Multiply both sides by $-\dfrac{1}{2}$ and reverse the direction of the inequality symbol
$6 \geq x$	

This last line is equivalent to $x \leq 6$. The solution set can be represented in any of the three following ways.

Set-Builder Notation	Number Line Graph	Interval Notation
$\{x \mid x \leq 6\}$		$(-\infty, 6]$

EXAMPLE 5   Solve: $0.05x + 1.6 > 0.9 - 0.02x$.

SOLUTION   We can solve the inequality in its original form by working with the decimals, or we can eliminate the decimals first using the multiplication property for inequalities. Here are both methods:

Method 1: *Working with the decimals*

$0.05x + 1.6 > 0.9 - 0.02x$	
$0.07x + 1.6 > 0.9$	Add $0.02x$ to both sides
$0.07x > -0.7$	Add -1.6 to both sides
$x > -10$	Divide each side by 0.07

Method 2: *Eliminating the decimals first*

$100(0.05x) + 100(1.6) > 100(0.9) - 100(0.02x)$	
$5x + 160 > 90 - 2x$	Multiply each side by 100
$7x + 160 > 90$	Add $2x$ to both sides
$7x > -70$	Add -160 to both sides
$\dfrac{1}{7}(7x) > \dfrac{1}{7}(-70)$	Multiply both sides by $\dfrac{1}{7}$
$x > -10$	

With either method, the solution set is all real numbers greater than -10. We express the solution set three different ways.

Set-Builder Notation	Number Line Graph	Interval Notation
$\{x \mid x > -10\}$		$(-10, \infty)$

EXAMPLE 6   Solve: $\dfrac{2}{5}(x + 1) \geq \dfrac{8}{15} - \dfrac{x}{3}$.

SOLUTION   Although the inequality can be solved by working with fractions, we will use the multiplication property of inequalities to eliminate fractions by multiplying each side by the LCD, which is 15.

$$15\left(\frac{2}{5}(x + 1)\right) \geq 15\left(\frac{8}{15}\right) - 15\left(\frac{x}{3}\right) \qquad \text{Multiply both sides by the LCD 15}$$

$$6(x + 1) \geq 8 - 5x \qquad \text{Multiply}$$

$$6x + 6 \geq 8 - 5x \qquad \text{Distributive property}$$

$$11x + 6 \geq 8 \qquad \text{Add } 5x \text{ to both sides}$$

$$11x \geq 2 \qquad \text{Add } -6 \text{ to both sides}$$

$$x \geq \frac{2}{11} \qquad \text{Multiply both sides by } \frac{1}{11}$$

The solution set can be expressed in the following equivalent ways.

Set-Builder Notation	Number Line Graph	Interval Notation
$\left\{x \mid x \geq \dfrac{2}{11}\right\}$		$\left[\dfrac{2}{11}, \infty\right)$

Applications

EXAMPLE 7   A company that manufactures ink cartridges for printers finds that they can sell x cartridges each week at a price of p dollars each, according to the formula $x = 1,300 - 100p$. What price should they charge for each cartridge if they want to sell at least 300 cartridges a week?

SOLUTION   Because x is the number of cartridges they sell each week, an inequality that corresponds to selling at least 300 cartridges a week is

$$x \geq 300$$

Substituting $1,300 - 100p$ for x gives us an inequality in the variable p.

$$1,300 - 100p \geq 300$$

$$-100p \geq -1,000 \qquad \text{Add } -1,300 \text{ to each side}$$

$$p \leq 10 \qquad \text{Divide each side by } -100, \text{ and reverse the direction of the inequality symbol}$$

To sell at least 300 cartridges each week, the price per cartridge should be no more than \$10. That is, selling the cartridges for \$10 or less will produce weekly sales of 300 or more cartridges.

Getting Ready for Class

After reading through the preceding section, respond in your own words and in complete sentences.

A. What is the addition property for inequalities?

B. When we use interval notation to denote a section of the real number line, when do we use parentheses ( ) and when do we use brackets [ ]?

C. Explain the difference between the multiplication property of equality and the multiplication property for inequalities.

D. When solving an inequality, when do we change the direction of the inequality symbol?

Problem Set 2.4

For each of the following inequalities, write the solution set in set-builder and interval notation, and graph the solution set on a number line.

1. $x < 6$ **2.** $x \leq -2$ **3.** $x \geq -1$ **4.** $x > 0$

5. $x > \dfrac{3}{2}$ **6.** $x \geq \dfrac{13}{3}$ **7.** $x \leq -\dfrac{5}{4}$ **8.** $x < -\dfrac{7}{6}$

For each of the following graphs, express the interval shown using interval notation.

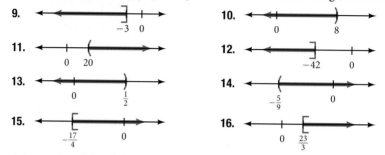

9.

10.

11.

12.

13.

14.

15.

16.

Solve each of the following inequalities. Write the solution set using interval notation, and graph it on a number line.

17. $2x \leq 3$ **18.** $5x \geq -115$ **19.** $\dfrac{1}{2}x > 2$ **20.** $\dfrac{1}{3}x > 4$

21. $-5x \leq 25$ **22.** $-7x \geq 35$ **23.** $-\dfrac{3}{2}x > -6$ **24.** $-\dfrac{2}{3}x < -8$

25. $-12 \leq 2x$ **26.** $-20 \geq 4x$ **27.** $-1 \geq -\dfrac{1}{4}x$ **28.** $-1 \leq -\dfrac{1}{5}x$

Solve each of the following inequalities. Write the solution set using interval notation, and graph it on a number line.

29. $-3x + 1 > 10$ **30.** $-2x - 5 \leq 15$

31. $\dfrac{1}{2} - \dfrac{m}{12} \leq \dfrac{7}{12}$ **32.** $\dfrac{1}{2} - \dfrac{m}{10} > -\dfrac{1}{5}$

33. $\dfrac{1}{2} \geq -\dfrac{1}{6} - \dfrac{2}{9}x$ **34.** $\dfrac{9}{5} > -\dfrac{1}{5} - \dfrac{1}{2}x$

35. $-40 \leq 30 - 20y$ **36.** $-20 > 50 - 30y$

37. $\dfrac{2}{3}x - 3 < 1$ **38.** $\dfrac{3}{4}x - 2 > 7$

39. $10 - \dfrac{1}{2}y \leq 36$ **40.** $8 - \dfrac{1}{3}y \geq 20$

41. $4 - \dfrac{1}{2}x < \dfrac{2}{3}x - 5$ **42.** $5 - \dfrac{1}{3}x > \dfrac{1}{4}x + 2$

43. $0.03x - 0.4 \leq 0.08x + 1.2$ **44.** $2.0 - 0.7x < 1.3 - 0.3x$

45. $3 - \dfrac{x}{5} < 5 - \dfrac{x}{4}$ **46.** $-2 + \dfrac{x}{3} \geq \dfrac{x}{2} - 5$

Simplify each side first, then solve the following inequalities. Write your answers with interval notation.

47. $2(3y + 1) \leq -10$ **48.** $3(2y - 4) > 0$

49. $-(a + 1) - 4a \leq 2a - 8$ **50.** $-(a - 2) - 5a \leq 3a + 7$

51. $\dfrac{1}{3}t - \dfrac{1}{2}(5 - t) < 0$ **52.** $\dfrac{1}{4}t - \dfrac{1}{3}(2t - 5) < 0$

53. $-2 \le 5 - 7(2a + 3)$

54. $1 < 3 - 4(3a - 1)$

55. $-\dfrac{1}{3}(x + 5) \le -\dfrac{2}{9}(x - 1)$

56. $-\dfrac{1}{2}(2x + 1) \le -\dfrac{3}{8}(x + 2)$

57. $5(x - 2) - 7(x + 1) \le -4x + 3$

58. $-3(1 - 2x) - 3(x - 4) < -3 - 4x$

59. $\dfrac{2}{3}x - \dfrac{1}{3}(4x - 5) < 1$

60. $\dfrac{1}{4}x - \dfrac{1}{2}(3x + 1) \ge 2$

61. $20x + 9,300 > 18,000$

62. $20x + 4,800 > 18,000$

Paying Attention to Instructions The next two problems are intended to give you practice reading, and paying attention to, the instructions that accompany the problems you are working.

63. Work each problem according to the instructions given.

    a. Evaluate when $x = 0$: $-\dfrac{1}{2}x + 1$ **b.** Solve: $-\dfrac{1}{2}x + 1 = -7$

    c. Is 0 a solution to $-\dfrac{1}{2}x + 1 < -7$ **d.** Solve: $-\dfrac{1}{2}x + 1 < -7$

64. Work each problem according to the instructions given.

    a. Evaluate when $x = 0$: $-\dfrac{2}{3}x - 5$ **b.** Solve: $-\dfrac{2}{3}x - 5 = 1$

    c. Is 0 a solution to $-\dfrac{2}{3}x - 5 > 1$ **d.** Solve: $-\dfrac{2}{3}x - 5 > 1$

65. Art Supplies A store selling art supplies finds that they can sell x sketch pads each week at a price of p dollars each, according to the formula $x = 900 - 300p$. What price should they charge if they want to sell

    a. At least 300 pads each week?   **b.** More than 600 pads each week?

    c. Less than 525 pads each week?   **d.** At most 375 pads each week?

66. Fuel Efficiency The fuel efficiency (mpg rating) for cars has been increasing steadily since 1980. The formula for a car's fuel efficiency for a given year between 1980 and 1996 is

$$E = 0.36x + 15.9$$

where E is miles per gallon and x is the number of years after 1980.

                 (*Source:* U.S. Federal Highway Administration, *Highway Statistics,* annual)

    a. In what years was the average fuel efficiency for cars less than 17 mpg?

    b. In what years was the average fuel efficiency for cars more than 20 mpg?

67. Student Loan When considering how much debt to incur in student loans, you learn that it is wise to keep your student loan payment to 8% or less of your starting monthly income. Suppose you anticipate a starting annual salary of $24,000. Set up and solve an inequality that represents the amount of monthly debt for student loans that would be considered manageable.

68. Current Nanette can travel 3 miles per hour in her kayak in still water. Use the formula $d = (r - c) \cdot t$ to set up and solve an inequality to determine how fast of a current she can kayak in upstream in order to travel at least 10 miles in 4 hours.

Learning Objectives Assessment

The following problems can be used to help assess if you have successfully met the learning objectives for this section.

69. Solve: $11 - 3x \le 20$. Write the solution set using interval notation.

    a. $(-\infty, -3]$ **b.** $[-3, \infty)$ **c.** $(-\infty, 3]$ **d.** $[3, \infty)$

70. Solve: $\dfrac{x}{5} + 2 > \dfrac{3}{10} - \dfrac{x}{2}$. Write the solution set using interval notation.

    a. $\left(-\infty, -\dfrac{17}{7}\right)$ **b.** $\left(-\dfrac{17}{7}, \infty\right)$ **c.** $\left(\dfrac{1}{7}, \infty\right)$ **d.** $\left(-\infty, \dfrac{1}{7}\right)$

71. Which of the following inequalities has the solution set $(-\infty, 4)$?

    a. $x < 4$ **b.** $x > 4$ **c.** $x \le 4$ **d.** $x \ge 4$

72. Sales Commission Ernesto earns a weekly salary of $600 plus a 15% commission on his sales for the week. What amount of sales must he have to make at least $840 this week?

    a. $9,600   **b.** $240   **c.** $5,600   **d.** $1,600

Getting Ready for the Next Section

To understand all of the explanations and examples in the next section you must be able to work the problems below.

Graph each interval on a number line.

73. $(-\infty, 4)$ **74.** $[-1, \infty)$ **75.** $[-2, \infty)$ **76.** $(-\infty, -1)$

Graph the solution set for each inequality on a number line.

77. $x \ge -2$ **78.** $x > 3$ **79.** $x < 2$ **80.** $x \le -\dfrac{11}{3}$

Solve each inequality.

81. $3(1 - 4x) \le 27$ **82.** $5x - 9 > 2x + 3$

83. $3x + 7 \ge 4$ **84.** $3x + 7 \le -4$

85. $-3 \le 2x - 5$ **86.** $2x - 5 \le 3$

87. $\dfrac{9}{5}x + 32 \le 104$ **88.** $86 \le \dfrac{9}{5}x + 32$

Learning Objectives

In this section we will learn how to:

1. Find the union of two intervals.

2. Find the intersection of two intervals.

3. Solve compound inequalities involving union.

4. Solve compound inequalities involving intersection.

Introduction

A company is about to introduce a new product. Their revenue is given by the formula $R = 15x$ and their cost by the formula $C = 400 + 3x$, where x is the number of units produced and sold each month. To be successful, the company needs a monthly revenue of at least \$12,000, but they must keep their monthly costs to no more than \$5,500. As a result, the company must determine how many units of their product, x, to produce so that

$$15x \geq 12,000 \quad \text{and} \quad 400 + 3x \leq 5,500$$

Solving both inequalities gives us $x \geq 800$ and $x \leq 1,700$. The company can produce between 800 and 1,700 units per month, inclusive, to remain successful. This scenario leads us to the concept of a *compound inequality*. But before we can address compound inequalities, we must first extend our work with sets.

Union and Intersection

> (dĕf **DEFINITION** *union*
>
> The union of two sets A and B is the set of all elements that are in either A or B, and is denoted by $A \cup B$.

The idea behind the union is to merge the two sets into a single set. Everything in A should appear in the union and everything in B should also appear in the union. The word *or* is the key word in the definition.

> (dĕf **DEFINITION** *intersection*
>
> The intersection of two sets A and B is the set of all elements contained in both A and B, and is denoted by $A \cap B$.

With the intersection, we are looking for the things that are common to both sets, or where the two sets overlap. Everything in the intersection must appear in A and also in B. The key word in the definition is *and*.

EXAMPLE 1   If $A = \{2, 4, 6, 8\}$ and $B = \{6, 7, 8, 9\}$, find $A \cup B$ and $A \cap B$.

SOLUTION   For the union, we combine the two sets into a single set.

$$A \cup B = \{2, 4, 6, 7, 8, 9\}$$

Notice that we only list the elements 6 and 8 once.
        The intersection contains the common elements, which are 6 and 8.

$$A \cap B = \{6, 8\}$$

Because intervals are sets of real numbers, we can also find the union and intersection of two intervals. We illustrate this in the next couple of examples.

EXAMPLE 2   If $A = (-\infty, 4)$ and $B = [-1, \infty)$, find $A \cup B$ and $A \cap B$.

SOLUTION   With intervals, it is helpful to graph both intervals on a single number line. We have done this below using different colors for A and B.

For the union, we identify the portion of the number line that is covered by either color. In this case, the whole number line is covered by either blue or green, so

$$A \cup B = (-\infty, \infty)$$

To find the intersection, we look for the overlap of the two colors; that is, the portion of the number line that is covered by both blue and green. We see that

$$A \cap B = [-1, 4)$$

The endpoint -1 is included in the intersection because of the bracket. It is a value contained in both intervals. However, because of the parenthesis, 4 is not an element of set A, and therefore not included in the intersection.

EXAMPLE 3   If $A = [-2, \infty)$ and $B = [3, \infty)$, find $A \cup B$ and $A \cap B$.

SOLUTION   We begin by graphing both intervals on a common number line.

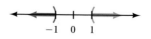

The portion of the number line covered by either color begins at -2, so

$$A \cup B = [-2, \infty)$$

The portion of the number line covered by both colors (the overlap) begins at 3. Therefore,

$$A \cap B = [3, \infty)$$

EXAMPLE 4   If $A = (-\infty, -1)$ and $B = (1, \infty)$, find $A \cup B$ and $A \cap B$.

SOLUTION   The graph of the two intervals is shown below.

We see that

$$A \cup B = (-\infty, -1) \cup (1, \infty)$$

and

$$A \cap B = \varnothing$$

Because the union consists of two separate segments of the number line, it cannot be expressed as a single interval. Also, since the two intervals never overlap, the intersection is the empty set.

Compound Inequalities

We can use the concepts of union and intersection, together with our methods of graphing inequalities, to graph some ***compound inequalities***. Compound inequalities are expressions containing two inequalities together with the word *and* or *or*.

 **EXAMPLE 5**   Solve: $x \geq -2$ or $x > 3$. Graph the solution set.

SOLUTION   We begin by graphing each inequality on a single number line:

The two inequalities are connected by the word *or*, which indicates a union. The compound inequality will be true for any value of x that satisfies either inequality. So the solution set for the compound inequality is the union of the two intervals we have graphed. In interval notation, the union is

$$[-2, \infty)$$

and the graph of the union is

The graph of the solution set is shown below:

EXAMPLE 6   Solve: $x > -1$ and $x < 2$. Graph the solution set.

SOLUTION   We first graph each inequality on a single number line:

Because the two inequalities are connected by the word *and*, an intersection is indicated. The compound inequality will be true for any value of x that satisfies both inequalities. Thus, the solution set is the intersection of the two intervals we have graphed. Using interval notation, we have

$$(-1, 2)$$

The graph of the solution set is shown below:

Here is a summary of the general process for solving a compound inequality.

> **△≠∑ HOW TO   *Solve a Compound Inequality***
>
> **Step 1:** Solve the individual inequalities appearing in the compound inequality.
> **Step 2:** Graph the solution sets of the individual inequalities on a common number line.
> **Step 3:** If the word *or* is used in the compound inequality, find the union of the two solution sets. If the word *and* is used, find the intersection of the two solution sets.
> **Step 4:** Write your answer using interval notation and graph it on a number line.

The next examples illustrate this process.

EXAMPLE 7   Solve: $3t + 7 \leq -4$ or $3t + 7 \geq 4$. Graph the solution set.

SOLUTION   First, we solve each individual inequality separately by isolating the variable.

$$3t + 7 \leq -4 \quad \text{or} \quad 3t + 7 \geq 4$$
$$3t \leq -11 \quad \text{or} \quad 3t \geq -3 \qquad \text{Add } -7$$
$$t \leq -\frac{11}{3} \quad \text{or} \quad t \geq -1 \qquad \text{Multiply by } \frac{1}{3}$$

Now we graph the two solution sets as intervals on a number line.

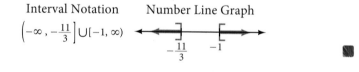

The word *or* indicates a union. We write the union using interval notation and graph it on a number line.

Interval Notation   Number Line Graph

$$\left(-\infty, -\frac{11}{3}\right] \cup [-1, \infty)$$

EXAMPLE 8   Solve: $3(1 - 4x) \leq 27$ and $5x - 9 > 2x + 3$. Graph the solution set.

SOLUTION   We begin by solving the two individual inequalities.

$$3(1 - 4x) \leq 27 \quad \text{and} \quad 5x - 9 > 2x + 3$$
$$3 - 12x \leq 27 \quad \text{and} \quad 3x - 9 > 3$$
$$-12x \leq 24 \quad \text{and} \quad 3x > 12$$
$$x \geq -2 \quad \text{and} \quad x > 4$$

The graphs of the individual solution sets are shown below.

The word *and* indicates an intersection. From the graph, we see that the intersection of these intervals is all real numbers greater than 4. The interval notation and number line graph of the solution set for the compound inequality is

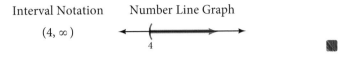

Interval Notation   Number Line Graph

$(4, \infty)$

Continued Inequalities

Sometimes compound inequalities that use the word *and* as the connecting word can be written in a shorter form. For example, the compound inequality $-3 \le x$ and $x \le 4$ can be written $-3 \le x \le 4$. The word *and* does not appear when an inequality is written in this form. It is implied. Inequalities of the form $-3 \le x \le 4$ are called **continued inequalities**. This new notation is useful because writing it takes fewer symbols. The graph of $-3 \le x \le 4$ is

The table below shows the connection between set-builder notation, interval notation, and number line graphs for a variety of continued inequalities. Again, we have included the graphs with open and closed circles for those of you who have used this type of graph previously. Remember, however, that in this book we will be using the parentheses/brackets method of graphing.

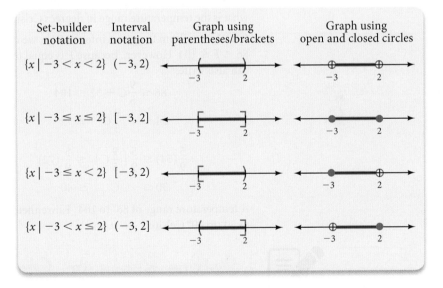

Set-builder notation	Interval notation	Graph using parentheses/brackets	Graph using open and closed circles
$\{x \mid -3 < x < 2\}$	$(-3, 2)$		
$\{x \mid -3 \le x \le 2\}$	$[-3, 2]$		
$\{x \mid -3 \le x < 2\}$	$[-3, 2)$		
$\{x \mid -3 < x \le 2\}$	$(-3, 2]$		

EXAMPLE 9   Solve $-3 \le 2x - 5 \le 3$ and graph the solution set.

SOLUTION   The continued inequality is equivalent to the compound inequality

$$-3 \le 2x - 5 \quad \text{and} \quad 2x - 5 \le 3$$

We could solve this compound inequality using the process we have illustrated in the previous examples. However, with continued inequalities it is much simpler to solve using the original format and not rewriting the problem as a compound inequality.

We can extend our properties for addition and multiplication to cover this situation. If we add a number to the middle expression, we must add the same number to the outside expressions. If we multiply the center expression by a number, we must do the same to the outside expressions, remembering to reverse the direction of the inequality symbols if we multiply by a negative number. We begin by adding 5 to all three parts of the inequality:

$$-3 \leq 2x - 5 \leq 3$$

$$2 \leq \quad 2x \quad \leq 8 \qquad \text{Add 5 to all three members}$$

$$1 \leq \quad x \quad \leq 4 \qquad \text{Multiply through by } \frac{1}{2}$$

The solution set is all real numbers between 1 and 4, inclusive. The interval notation and graph are shown below.

Interval Notation   Number Line Graph

[1, 4]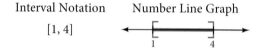

Notice that we did not need to graph individual solution sets or find an intersection. By isolating x in the continued inequality we are led directly to the solution set of the equivalent compound inequality.

EXAMPLE 10   In the chapter opener, we introduced the formula $F = \frac{9}{5}C + 32$, which gives the relationship between the Celsius and Fahrenheit temperature scales. If the temperature range on a certain day is 86° to 104° Fahrenheit, what is the temperature range in degrees Celsius?

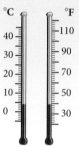

SOLUTION   From the given information, we can write $86 \leq F \leq 104$. However, because F is equal to $\frac{9}{5}C + 32$, we can also write:

$$86 \leq \frac{9}{5}C + 32 \leq 104$$

$$54 \leq \quad \frac{9}{5}C \quad \leq 72 \qquad \text{Add } -32 \text{ to each number}$$

$$\frac{5}{9}(54) \leq \frac{5}{9}\left(\frac{9}{5}C\right) \leq \frac{5}{9}(72) \qquad \text{Multiply each number by } \frac{5}{9}$$

$$30 \leq \quad C \quad \leq 40$$

A temperature range of 86° to 104° Fahrenheit corresponds to a temperature range of 30° to 40° Celsius.

Getting Ready for Class

After reading through the preceding section, respond in your own words and in complete sentences.

A. What is the difference between the union and the intersection of two sets?

B. How do we solve a compound inequality containing the word *or*?

C. If a compound inequality contained the word *and*, how would your answer to B be different?

D. In your own words, explain how to solve a continued inequality.

Find the union, $A \cup B$, for each of the following pairs of sets.

1. $A = \{1, 2, 3\}, B = \{4, 5, 6\}$

2. $A = \{5, 10, 15, 20\}, B = \{10, 20\}$

3. $A = \{5, 6, 7, 8\}, B = \{2, 4, 6, 8\}$

4. $A = \{3, 13, 23\}, B = \varnothing$

5. $A = (1, \infty), B = (6, \infty)$

6. $A = [-5, \infty), B = [2, \infty)$

7. $A = (-\infty, -3], B = (-\infty, -4)$

8. $A = (-\infty, 0), B = \left(-\infty, \dfrac{3}{2}\right]$

9. $A = (-\infty, -7), B = (7, \infty)$

10. $A = \left(-\infty, -\dfrac{1}{4}\right], B = \left[\dfrac{1}{4}, \infty\right)$

11. $A = \left(-\infty, \dfrac{3}{5}\right], B = (-2, \infty)$

12. $A = (-\infty, 9), B = \left(\dfrac{7}{3}, \infty\right)$

Find the intersection, $A \cap B$, for each of the following pairs of sets.

13. $A = \{1, 2, 3\}, B = \{4, 5, 6\}$

14. $A = \{5, 10, 15, 20\}, B = \{10, 20\}$

15. $A = \{5, 6, 7, 8\}, B = \{2, 4, 6, 8\}$

16. $A = \{3, 13, 23\}, B = \varnothing$

17. $A = (1, \infty), B = (6, \infty)$

18. $A = [-5, \infty), B = [2, \infty)$

19. $A = (-\infty, -3], B = (-\infty, -4)$

20. $A = (-\infty, 0), B = \left(-\infty, \dfrac{3}{2}\right]$

21. $A = (-\infty, -7), B = (7, \infty)$

22. $A = \left(-\infty, -\dfrac{1}{4}\right], B = \left[\dfrac{1}{4}, \infty\right)$

23. $A = \left(-\infty, \dfrac{3}{5}\right], B = (-2, \infty)$

24. $A = (-\infty, 9), B = \left(\dfrac{7}{3}, \infty\right)$

Solve the following compound inequalities. Write the solution set using interval notation and graph it on a number line (when possible).

25. $x < 4$ or $x > 1$

26. $x < -1$ or $x > -5$

27. $x \geq -9$ or $x \geq -2$

28. $x \leq -3$ or $x \leq 6$

29. $2x - 9 < 5$ or $5x + 1 \leq 6$

30. $3x + 4 > 13$ or $6x - 2 \geq -8$

31. $10 - x \leq 15$ or $7 + x \geq 4$

32. $3 + x < 12$ or $3 - x > 4$

33. $x + 5 \leq -2$ or $x + 5 \geq 2$

34. $3x + 2 < -3$ or $3x + 2 > 3$

35. $5y + 1 \leq -4$ or $5y + 1 \geq 4$

36. $7y - 5 \leq -2$ or $7y - 5 \geq 2$

37. $5 - 3x > 3x$ or $8x + 1 \geq 2x$

38. $3 - 5x \leq x$ or $4x < x + 2$

39. $2x + 5 < 3x - 1$ or $x - 4 < 2x + 6$

40. $3x - 1 > 2x + 4$ or $5x - 2 > 3x + 4$

41. $2(3y + 1) \geq 3(y - 4)$ or $7(2y + 3) \geq 4(3y - 1)$

42. $5(2y - 1) \leq 4(y + 3)$ or $6(y - 2) \geq 2(4y + 3)$

43. $\dfrac{1}{2} \geq -\dfrac{1}{6} - \dfrac{2}{9}x$ or $4 - \dfrac{1}{2}x < \dfrac{2}{3}x - 5$

44. $\dfrac{9}{5} < -\dfrac{1}{5} - \dfrac{1}{2}x$ or $5 - \dfrac{1}{3}x > \dfrac{1}{4}x + 2$

Solve the following compound inequalities. Write the solution set using interval notation and graph it on a number line (when possible).

45. $x < 3$ and $x > 1$

46. $x \leq -2$ and $x \geq -6$

47. $x \geq -8$ and $x \geq -3$

48. $x < -4$ and $x < 5$

49. $3x - 4 > 2$ and $6x + 2 > -10$

50. $2x + 9 < 1$ and $5x - 1 < 14$

51. $3x + 1 < -8$ and $-2x + 1 \leq -3$ **52.** $2x - 5 \leq -1$ and $-3x - 6 < -15$

53. $x + 3 \geq -1$ and $x + 3 \leq 1$ **54.** $2x - 5 > -3$ and $2x - 5 < 3$

55. $4y - 1 > -2$ and $4y - 1 < 2$ **56.** $5y + 7 \geq -4$ and $5y + 7 \leq 4$

57. $4 - 2x \leq x$ and $7x > x - 5$ **58.** $3x + 5 > 4x$ and $1 - 8x \geq 5x$

59. $3(y - 5) < 4(2y + 3)$ and $2(5y + 1) < 9(y + 2)$

60. $6(y + 1) > 5(3y - 2)$ and $3(4y - 5) > 4(y - 8)$

61. $\dfrac{1}{2} - \dfrac{x}{12} \leq \dfrac{7}{12}$ and $3 - \dfrac{x}{5} < 5 - \dfrac{x}{4}$

62. $\dfrac{1}{2} - \dfrac{x}{10} < -\dfrac{1}{5}$ and $-2 + \dfrac{x}{3} \geq \dfrac{x}{2} - 5$

Solve the following continued inequalities. Use interval notation to write each solution set.

63. $-2 \leq m - 5 \leq 2$ **64.** $-3 \leq m + 1 \leq 3$

65. $-60 < 20a + 20 < 60$ **66.** $-60 < 50a - 40 < 60$

67. $0.5 \leq 0.3a - 0.7 \leq 1.1$ **68.** $0.1 \leq 0.4a + 0.1 \leq 0.3$

69. $3 < \dfrac{1}{2}x + 5 < 6$ **70.** $5 < \dfrac{1}{4}x + 1 < 9$

71. $4 < 6 + \dfrac{2}{3}x < 8$ **72.** $3 < 7 + \dfrac{4}{5}x < 15$

Translate each of the following phrases into an equivalent inequality statement.

73. x is greater than -2 and at most 4   **74.** x is less than 9 and at least -3

75. x is less than -4 or at least 1   **76.** x is at most 1 or more than 6

77. Temperature Range Each of the following temperature ranges is in degrees Fahrenheit. Use the formula $F = \dfrac{9}{5}C + 32$ to find the corresponding temperature range in degrees Celsius.

    a. $95°$ to $113°$ **b.** $68°$ to $86°$ **c.** $-13°$ to $14°$ **d.** $-4°$ to $23°$

78. Survival Rates for 'Apapane Here is what the United States Geological Survey has to say about the survival rates of the 'Apapane, one of the endemic birds of Hawaii.

> Annual survival rates based on 1,584 recaptures of 429 banded individuals 0.72 ± 0.11 for adults and 0.13 ± 0.07 for juveniles.

Write the survival rates using inequalities. Then give the survival rates in terms of percent.

79. Survival Rates for Sea Gulls Here is part of a report concerning the survival rates of Western Gulls that appeared on the web site of Cornell University.

> Survival of eggs to hatching is 70%–80%; of hatched chicks to fledglings 50%–70%; of fledglings to age of first breeding <50%.

Write the survival rates using inequalities without percent.

Learning Objectives Assessment

The following problems can be used to help assess if you have successfully met the learning objectives for this section.

80. Find the union of $[-2, \infty)$ and $(3, \infty)$.

    a. $[-2, 3)$ **b.** $(-\infty, \infty)$ **c.** $(3, \infty)$ **d.** $[-2, \infty)$

81. Find the intersection of $[-2, \infty)$ and $(3, \infty)$.

    a. $[-2, 3)$ **b.** $\varnothing$ **c.** $(3, \infty)$ **d.** $[-2, \infty)$

82. Solve the compound inequality $2x + 5 \leq 3$ or $4 - x \leq 9$.

    a. $[-5, -1]$ **b.** $(-\infty, -5] \cup [-1, \infty)$

    c. $(-\infty, \infty)$ **d.** $\varnothing$

83. Solve the compound inequality $5x - 3 > 2$ and $x + 9 < 4$.

    a. $(-5, 1)$ **b.** $(-\infty, -5) \cup (1, \infty)$

    c. $(-\infty, \infty)$ **d.** $\varnothing$

Getting Ready for the Next Section

To understand all of the explanations and examples in the next section you must be able to work the problems below.

Solve each equation.

84. $2a - 1 = -7$ **85.** $3x - 6 = 9$ **86.** $\dfrac{2}{3}x - 3 = 7$

87. $\dfrac{2}{3}x - 3 = -7$ **88.** $x - 5 = x - 7$ **89.** $x + 3 = x + 8$

90. $x - 5 = -x - 7$ **91.** $x + 3 = -x + 8$

Learning Objectives

In this section we will learn how to:

1. Use the property of absolute value equations to solve an equation containing a single absolute value.

2. Solve an absolute value equation having a single solution or no solution.

3. Solve an equation containing two absolute value expressions.

iStockPhoto.com/©Rob Vomund

Introduction

At one time, Amtrak's annual passenger revenue could be modeled approximately by the formula

$$R = -60|x - 11| + 962$$

where R is the annual revenue in millions of dollars and x is the number of years after 1980 (Association of American Railroads, Washington, DC, *Railroad Facts, Statistics of Railroads of Class 1*, annual). Notice the absolute symbols in the equation.

Equations Containing Absolute Value

You may recall that the ***absolute value*** of x, $|x|$, is the distance between x and 0 on the number line. The absolute value of a number measures its distance from 0.

VIDEO EXAMPLES

SECTION 2.6

EXAMPLE 1   Solve for x: $|x| = 5$.

SOLUTION   Using the definition of absolute value, we can read the equation as, "The distance between x and 0 on the number line is 5." If x is 5 units from 0, then x can be 5 or -5:

$$\text{If } |x| = 5 \quad \text{then} \quad x = 5 \quad \text{or} \quad x = -5$$

In general, then, we can see that any equation of the form $|x| = b$ is equivalent to the equations $x = b$ or $x = -b$, as long as $b > 0$. We generalize this result with the following property.

> **[Δ≠Σ] PROPERTY**   *Absolute Value Equations*
>
> For any algebraic expression A and positive constant b,
>
> $$\text{If} \quad |A| = b$$
> $$\text{then} \quad A = b \quad \text{or} \quad A = -b$$

EXAMPLE 2   Solve $|2a - 1| = 7$.

SOLUTION   We can read this question as "$2a - 1$ is 7 units from 0 on the number line." The quantity $2a - 1$ must be equal to 7 or -7:

$$|2a - 1| = 7$$
$$2a - 1 = 7 \quad \text{or} \quad 2a - 1 = -7$$

We have transformed our absolute value equation into two equations that do not involve absolute value. We can solve each equation using the method in Section 2.1:

$$2a - 1 = 7 \quad \text{or} \quad 2a - 1 = -7$$

$$2a = 8 \quad \text{or} \quad 2a = -6 \qquad \text{Add 1 to both sides}$$

$$a = 4 \quad \text{or} \quad a = -3 \qquad \text{Multiply by } \tfrac{1}{2}$$

Our solution set is $\{-3, 4\}$.

To check our solutions, we put them into the original absolute value equation:

When $a = 4$ When $a = -3$

the equation $|2a - 1| \overset{?}{=} 7$ the equation $|2a - 1| \overset{?}{=} 7$

becomes $|2(4) - 1| \overset{?}{=} 7$ becomes $|2(-3) - 1| \overset{?}{=} 7$

$$|7| \overset{?}{=} 7 \qquad\qquad |-7| \overset{?}{=} 7$$

$$7 = 7 \qquad\qquad\qquad 7 = 7$$

⟮Δ≠Σ⟯ **HOW TO**   *Solve an Absolute Value Equation*

Step 1: Isolate the absolute value on one side of the equation.

Step 2: If the constant term on the other side of the equation is positive, proceed to Step 3. If it is zero or negative, treat the problem as a special case.

Step 3: Use the property of absolute value equations to write two equations that do not involve an absolute value.

Step 4: Solve for the variable in the resulting two equations.

Step 5: Check the solutions in the original equation.

EXAMPLE 3   Solve: $\left| \dfrac{2}{3}x - 3 \right| + 5 = 12.$

SOLUTION   To use the property of absolute value equations to solve this problem, we must isolate the absolute value on the left side of the equal sign. To do so, we add -5 to both sides of the equation to obtain

$$\left| \frac{2}{3}x - 3 \right| = 7$$

Now that the equation is in the correct form, we can write

$$\frac{2}{3}x - 3 = 7 \quad \text{or} \quad \frac{2}{3}x - 3 = -7$$

$$\frac{2}{3}x = 10 \quad \text{or} \quad \frac{2}{3}x = -4 \qquad \text{Add 3 to both sides}$$

$$x = 15 \quad \text{or} \quad x = -6 \qquad \text{Multiply by } \tfrac{3}{2}$$

The solution set is $\{-6, 15\}$.

The next two examples illustrate the special cases where the constant term is zero or negative after the absolute value is isolated.

EXAMPLE 4   Solve: $|3a - 6| + 4 = 0$.

Note Recall that $\varnothing$ is the symbol we use to denote the empty set. When we use it to indicate the solutions to an equation, then we are saying the equation has no solution.

SOLUTION   First, we isolate the absolute value by adding -4 to both sides of the equation to obtain

$$|3a - 6| = -4$$

The solution set is $\varnothing$ because the right side is negative but the left side cannot be negative. No matter what we try to substitute for the variable a, the quantity $|3a - 6|$ will always be positive or zero. It can never be -4.

EXAMPLE 5   Solve: $|2 - 5x| = 0$.

SOLUTION   The absolute value is already isolated, but we have 0 on the right side, which makes this a special case. Because there is no difference between 0 and -0, the property of absolute value equations results in the following single equation:

$$2 - 5x = 0$$
$$-5x = -2$$
$$x = \frac{2}{5}$$

The solution set is $\left\{ \dfrac{2}{5} \right\}$. Notice that we get a single solution in this case.

Consider the statement $|a| = |b|$. What can we say about a and b? We know they are equal in absolute value. By the definition of absolute value, they are the same distance from 0 on the number line. They must be equal to each other or opposites of each other. In symbols, we write:

Note $\Leftrightarrow$ means "if and only if" and "is equivalent to"

$$|a| = |b| \quad \Leftrightarrow \quad a = b \quad \text{or} \quad a = -b$$

| Equal in absolute value | Equals | or | Opposites |

EXAMPLE 6   Solve $|3a + 2| = |2a + 3|$.

SOLUTION   The quantities $3a + 2$ and $2a + 3$ have equal absolute values. They are, therefore, the same distance from 0 on the number line. They must be equals or opposites:

$$|3a + 2| = |2a + 3|$$

Equals		Opposites
$3a + 2 = 2a + 3$	or	$3a + 2 = -(2a + 3)$
$a + 2 = 3$		$3a + 2 = -2a - 3$
$a = 1$		$5a + 2 = -3$
		$5a = -5$
		$a = -1$

The solution set is $\{1, -1\}$.

It makes no difference in the outcome of the problem if we take the opposite of the first or second expression. It is very important, once we have decided which one to take the opposite of, that we take the opposite of both its terms and not just the first term. That is, the opposite of $2a + 3$ is $-(2a + 3)$, which we can think of as $-1(2a + 3)$. Distributing the -1 across *both* terms, we have

$$-1(2a + 3) = -2a - 3$$

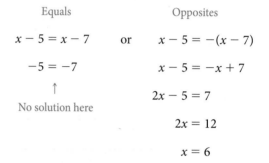 **EXAMPLE 7**   Solve $|x - 5| = |x - 7|$.

SOLUTION   As was the case in Example 6, the quantities $x - 5$ and $x - 7$ must be equal or they must be opposites, because their absolute values are equal:

Equals	Opposites

$$x - 5 = x - 7 \quad \text{or} \quad x - 5 = -(x - 7)$$

$$-5 = -7 \qquad\qquad x - 5 = -x + 7$$

$$\uparrow \qquad\qquad\qquad 2x - 5 = 7$$

No solution here $\qquad\qquad 2x = 12$

$$x = 6$$

Because the first equation leads to a false statement, it will not give us a solution. (If either of the two equations were to reduce to a true statement, it would mean all real numbers would satisfy the original equation.) In this case, our only solution is $x = 6$.

Getting Ready for Class

After reading through the preceding section, respond in your own words and in complete sentences.

A. Why do some of the equations in this section have two solutions instead of one?

B. Translate $|x| = 6$ into words using the definition of absolute value.

C. Explain in words what the equation $|x - 3| = 4$ means with respect to distance on the number line.

D. In your own words, describe the process for solving an equation that contains two absolute value expressions.

Use the property of absolute value equations to solve each of the following problems.

1. $|x| = 4$ **2.** $|x| = 7$ **3.** $2 = |a|$

4. $5 = |a|$ **5.** $|x| = -3$ **6.** $|x| = -4$

7. $|a| + 2 = 3$ **8.** $|a| - 5 = 2$ **9.** $|y| + 4 = 3$

10. $|y| + 3 = 1$ **11.** $|a - 4| = \dfrac{5}{3}$ **12.** $|a + 2| = \dfrac{7}{5}$

13. $\left|\dfrac{3}{5}a + \dfrac{1}{2}\right| = 1$ **14.** $\left|\dfrac{2}{7}a + \dfrac{3}{4}\right| = 1$ **15.** $60 = |20x - 40|$

16. $800 = |400x - 200|$ **17.** $|2x + 1| = -3$ **18.** $|2x - 5| = -7$

19. $\left|\dfrac{3}{4}x - 6\right| = 9$ **20.** $\left|\dfrac{4}{5}x - 5\right| = 15$ **21.** $\left|1 - \dfrac{1}{2}a\right| = 3$

22. $\left|2 - \dfrac{1}{3}a\right| = 10$ **23.** $|2x - 5| = 3$ **24.** $|3x + 1| = 4$

25. $|4 - 7x| = 5$ **26.** $|9 - 4x| = 1$ **27.** $\left|3 - \dfrac{2}{3}y\right| = 5$

28. $\left|-2 - \dfrac{3}{4}y\right| = 6$ **29.** $|3x + 12| = 0$ **30.** $|8 - 6x| = 0$

Solve each equation.

31. $|3x + 4| + 1 = 7$ **32.** $|5x - 3| - 4 = 3$

33. $|3 - 2y| + 4 = 3$ **34.** $|8 - 7y| + 9 = 1$

35. $3 + |4t - 1| = 8$ **36.** $2 + |2t - 6| = 10$

37. $5 + |3a + 2| = 5$ **38.** $|6a - 5| - 11 = -11$

39. $\left|9 - \dfrac{3}{5}x\right| + 6 = 12$ **40.** $\left|4 - \dfrac{2}{7}x\right| + 2 = 14$

41. $5 = \left|\dfrac{2}{7}x + \dfrac{4}{7}\right| - 3$ **42.** $7 = \left|\dfrac{3}{5}x + \dfrac{1}{5}\right| + 2$

43. $2 = -8 + \left|4 - \dfrac{1}{2}y\right|$ **44.** $1 = -3 + \left|2 - \dfrac{1}{4}y\right|$

45. $|3(x + 1)| - 4 = -1$ **46.** $|2(2x + 3)| - 5 = -1$

47. $|1 + 3(2x - 1)| = 5$ **48.** $|3 + 4(3x + 1)| = 7$

49. $3 = -2 + \left|5 - \dfrac{2}{3}a\right|$ **50.** $4 = -1 + \left|6 - \dfrac{4}{5}a\right|$

51. $6 = |7(k + 3) - 4|$ **52.** $5 = |6(k - 2) + 1|$

Solve the following equations.

53. $|3a + 1| = |2a - 4|$ **54.** $|5a + 2| = |4a + 7|$

55. $\left|x - \dfrac{1}{3}\right| = \left|\dfrac{1}{2}x + \dfrac{1}{6}\right|$ **56.** $\left|\dfrac{1}{10}x - \dfrac{1}{2}\right| = \left|\dfrac{1}{5}x + \dfrac{1}{10}\right|$

57. $|y - 2| = |y + 3|$ **58.** $|y - 5| = |y - 4|$

59. $|3x - 1| = |3x + 1|$ **60.** $|5x - 8| = |5x + 8|$

61. $|0.03 - 0.01x| = |0.04 + 0.05x|$ **62.** $|0.07 - 0.01x| = |0.08 - 0.02x|$

63. $|x - 2| = |2 - x|$ **64.** $|x - 4| = |4 - x|$

65. $\left|\dfrac{x}{5} - 1\right| = \left|1 - \dfrac{x}{5}\right|$

66. $\left|\dfrac{x}{3} - 1\right| = \left|1 - \dfrac{x}{3}\right|$

67. $\left|\dfrac{2}{3}b - \dfrac{1}{4}\right| = \left|\dfrac{1}{6}b + \dfrac{1}{2}\right|$

68. $\left|-\dfrac{1}{4}x + 1\right| = \left|\dfrac{1}{2}x - \dfrac{1}{3}\right|$

69. $|0.1a - 0.04| = |0.3a + 0.08|$

70. $|-0.4a + 0.6| = |1.3 - 0.2a|$

71. Paying Attention to Instructions Work each problem according to the instructions given.

　　a. Solve: $4x - 5 = 0$

　　b. Solve: $|4x - 5| = 0$

　　c. Solve: $4x - 5 = 3$

　　d. Solve: $|4x - 5| = 3$

　　e. Solve: $|4x - 5| = |2x + 3|$

72. Paying Attention to Instructions Work each problem according to the instructions given.

　　a. Solve: $3x + 6 = 0$

　　b. Solve: $|3x + 6| = 0$

　　c. Solve: $3x + 6 = 4$

　　d. Solve: $|3x + 6| = 4$

　　e. Solve: $|3x + 6| = |7x + 4|$

Applying the Concepts

iStockPhoto.com/©Rob Vomund

73. Amtrak Amtrak's annual passenger revenue for the years 1985–1995 is modeled approximately by the formula

$$R = -60|x - 11| + 962$$

where R is the annual revenue in millions of dollars and x is the number of years after 1980 (Association of American Railroads, Washington, DC, *Railroad Facts, Statistics of Railroads of Class 1*, annual). In what years was the passenger revenue $722 million?

74. Corporate Profits The corporate profits for various U.S. industries vary from year to year. An approximate model for profits of U.S. "communications companies" during a given year between 1990 and 1997 is given by

$$P = -3400|x - 5.5| + 36000$$

where P is the annual profits (in millions of dollars) and x is the number of years after 1990 (U.S. Bureau of Economic Analysis, Income and Product Accounts of the U.S. (1929–1994), *Survey of Current Business*, September 1998). Use the model to determine the years in which profits of "communications companies" were $31.5 billion ($31,500 million).

Learning Objectives Assessment

The following problems can be used to help assess if you have successfully met the learning objectives for this section.

75. Solve: $|x + 9| - 3 = 5$.

    a. $\{-1, -11\}$ **b.** $\{-1, -17\}$ **c.** $\{-1\}$ **d.** $\varnothing$

76. Solve: $8 + |3x + 7| = 5$.

    a. $\left\{-\dfrac{4}{3}, -\dfrac{10}{3}\right\}$ **b.** $\left\{-\dfrac{20}{3}, -\dfrac{10}{3}\right\}$ **c.** $\varnothing$ **d.** $\left\{-\dfrac{10}{3}\right\}$

77. Solve: $|2a + 3| = |a - 4|$.

    a. $\{-7\}$ **b.** $\left\{-7, -\dfrac{7}{3}\right\}$ **c.** $\left\{-7, \dfrac{1}{3}\right\}$ **d.** $\varnothing$

Getting Ready for the Next Section

To understand all of the explanations and examples in the next section you must be able to work the problems below.

Solve each inequality. Do not graph the solution set.

78. $2x - 5 < 3$ **79.** $-3 < 2x - 5$ **80.** $-4 \leq 3a + 7$

81. $3a + 2 \leq 4$ **82.** $4t - 3 \leq -9$ **83.** $4t - 3 \geq 9$

Learning Objectives

In this section we will learn how to:

1. Use the property of absolute value inequalities to solve an inequality involving an absolute value.

2. Solve special cases of absolute value inequalities.

Introduction

iStockPhoto.com/©bluestocking

In a student survey conducted by the University of Minnesota, it was found that 30% of students were solely responsible for their finances. The survey was reported to have a margin of error plus or minus 3.74%. This means that the difference between the sample estimate of 30% and the actual percent of students who are responsible for their own finances is most likely less than 3.74%. We can write this as an inequality:

$$|x - 0.30| \leq 0.0374$$

where x represents the true percent of students who are responsible for their own finances.

In this section, we will apply the definition of absolute value to solve inequalities involving absolute value. Again, the absolute value of x, which is denoted $|x|$, represents the distance that x is from 0 on the number line. We will begin by considering three absolute value expressions and their verbal translations:

Expression	In Words
$\|x\| = 5$	x is exactly 5 units from 0 on the number line.
$\|a\| < 5$	a is less than 5 units from 0 on the number line.
$\|y\| > 5$	y is greater than 5 units from 0 on the number line.

Once we have translated the expression into words, we can use the translation to graph the original equation or inequality. The graph is then used to write a final equation or inequality that does not involve absolute value.

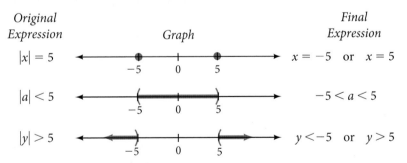

Original Expression	Graph	Final Expression
$\|x\| = 5$		$x = -5$ or $x = 5$
$\|a\| < 5$		$-5 < a < 5$
$\|y\| > 5$		$y < -5$ or $y > 5$

Although we will not always write out the verbal translation of an absolute value inequality, it is important that we understand the translation. Our second expression, $|a| < 5$, means a is within 5 units of 0 on the number line. That is, a must lie *between* the values -5 and 5, as expressed by the continued inequality $-5 < a < 5$. With the third expression, $|y| > 5$, notice that y must lie *outside* the values -5 and 5. That is, y must lie further to the left of -5 or further to the right of 5. The compound inequality $y < -5$ or $y > 5$ expresses this fact.

Absolute Value Inequalities

We can follow this same kind of reasoning to solve more complicated absolute value inequalities by generalizing these results as follows:

⟮Δ≠Σ⟯ PROPERTY   *Absolute Value Inequalities*

For any algebraic expression A and positive constant b,

$$\text{if } |A| < b \text{ then } -b < A < b \quad (A \text{ is between } -b \text{ and } b.)$$

and

$$\text{if } |A| > b \text{ then } A < -b \text{ or } A > b \quad (A \text{ is outside } -b \text{ and } b.)$$

These properties also hold if the strict inequalities are replaced with appropriate inclusive inequality symbols $\leq$ and $\geq$.

VIDEO EXAMPLES

SECTION 2.7

◼ EXAMPLE 1   Graph the solution set: $|2x - 5| < 3$

SOLUTION   The absolute value of $2x - 5$ is the distance that $2x - 5$ is from 0 on the number line. We can translate the inequality as, "$2x - 5$ is less than 3 units from 0 on the number line." That is, $2x - 5$ must appear between -3 and 3 on the number line.

A picture of this relationship is

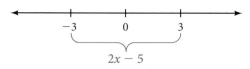

Using the picture, we can write an inequality without absolute value that describes the situation. The value of $2x - 5$ must lie between -3 and 3:

$$-3 < 2x - 5 < 3$$

Next, we solve the continued inequality by first adding 5 to all three members and then multiplying all three by $\frac{1}{2}$.

$$-3 < 2x - 5 < 3$$
$$2 < \quad 2x \quad < 8 \qquad \text{Add 5 to all three expressions}$$
$$1 < \quad x \quad < 4 \qquad \text{Multiply each expression by } \frac{1}{2}$$

We can express the solution set in three ways:

Set-Builder Notation	Number Line Graph	Interval Notation
$\{x \mid 1 < x < 4\}$	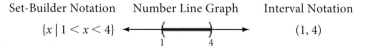	$(1, 4)$

We can see from the solution that for the quantity $2x - 5$ to be within 3 units of 0 on the number line, x must be between 1 and 4.   ◼

EXAMPLE 2   Solve $|x - 3| > 5$, and graph the solution set.

SOLUTION   We interpret the absolute value inequality to mean that $x - 3$ is more than 5 units from 0 on the number line. The values of $x - 3$ must lie outside -5 and 5; that is, to the left of -5 or to the right of 5. Here is a picture of the relationship:

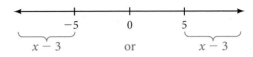

An inequality without absolute value that also describes this situation is

$$x - 3 < -5 \quad \text{or} \quad x - 3 > 5$$

Adding 3 to both sides of each inequality we have

$$x < -2 \quad \text{or} \quad x > 8$$

Here are three ways to write our result:

Set-Builder Notation	Number Line Graph	Interval Notation
$\{x \mid x < -2 \text{ or } x > 8\}$		$(-\infty, -2) \cup (8, \infty)$

We can see from Examples 1 and 2 that to solve an inequality involving absolute value, we must be able to write an equivalent expression that does not involve absolute value.

⟨Δ≠Σ⟩ HOW TO   *Solve an Absolute Value Inequality*

Step 1: Isolate the absolute value on one side of the inequality.
Step 2: If the constant term on the other side of the inequality is positive, proceed to Step 3. If it is zero or negative, treat the problem as a special case.
Step 3: Use the property of absolute value inequalities to write an equivalent expression that does not involve an absolute value.
Step 4: Solve for the variable in the resulting expression.
Step 5: Write the solution set using set-builder or interval notation, and graph it on a number line.

EXAMPLE 3   Solve and graph the solution set for $|3a + 7| + 6 \leq 10$.

SOLUTION   First, we isolate the absolute value by adding -6 to both sides.

$$|3a + 7| \leq 4$$

We can read the resulting inequality as, "The distance between $3a + 7$ and 0 is less than or equal to 4." Or, "$3a + 7$ is within 4 units of 0 on the number line." The value of $3a + 7$ must lie between -4 and 4, inclusive. This relationship can be written without absolute value as:

$$-4 \leq 3a + 7 \leq 4$$

Solving as usual, we have:

$$-4 \leq 3a + 7 \leq 4$$
$$-11 \leq \quad 3a \quad \leq -3 \qquad \text{Add } -7 \text{ to all three members}$$
$$-\frac{11}{3} \leq \quad a \quad \leq -1 \qquad \text{Multiply each expression by } \frac{1}{3}$$

We can now express the solution set in three ways:

Set-Builder Notation   Number Line Graph   Interval Notation

$\left\{a \mid -\frac{11}{3} \le a \le -1\right\}$ 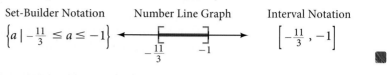 $\left[-\frac{11}{3}, -1\right]$

EXAMPLE 4   Graph the solution set: $2|4t - 3| \ge 18$.

SOLUTION   We must first isolate the absolute value by multiplying both sides of the inequality by $\frac{1}{2}$.

$$|4t - 3| \ge 9$$

The quantity $4t - 3$ is greater than or equal to 9 units from 0. The value of $4t - 3$ must equal -9 or 9, or lie outside of these numbers; that is, to the left of -9 or to the right of 9.

$$4t - 3 \le -9 \qquad \text{or} \qquad 4t - 3 \ge 9$$

$$4t \le -6 \qquad \text{or} \qquad 4t \ge 12 \qquad \text{Add 3}$$

$$t \le -\frac{6}{4} \qquad \text{or} \qquad t \ge \frac{12}{4} \qquad \text{Multiply by } \frac{1}{4}$$

$$t \le -\frac{3}{2} \qquad \text{or} \qquad t \ge 3$$

Here are three ways to write the solution:

Set-Builder Notation   Number Line Graph   Interval Notation

$\left\{t \mid t \le -\frac{3}{2} \text{ or } t \ge 3\right\}$ 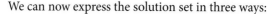 $\left(-\infty, -\frac{3}{2}\right] \cup [3, \infty)$

EXAMPLE 5   Solve and graph the solution set for $|4 - 2t| > 2$.

SOLUTION   Because the absolute value is already isolated, we can use the property of absolute value inequalities to write an equivalent expression without absolute value symbols as

$$4 - 2t < -2 \qquad \text{or} \qquad 4 - 2t > 2$$

To solve these inequalities we begin by adding -4 to each side.

$$4 + (-4) - 2t < -2 + (-4) \quad \text{or} \quad 4 + (-4) - 2t > 2 + (-4)$$

$$-2t < -6 \qquad \text{or} \qquad -2t > -2$$

Next we must multiply both sides of each inequality by $-\frac{1}{2}$. When we do so, we must also reverse the direction of each inequality symbol.

$$-2t < -6 \qquad \text{or} \qquad -2t > -2$$

$$-\frac{1}{2}(-2t) > -\frac{1}{2}(-6) \qquad \text{or} \qquad -\frac{1}{2}(-2t) < -\frac{1}{2}(-2)$$

$$t > 3 \qquad \text{or} \qquad t < 1$$

Although in situations like this we are used to seeing the "less than" symbol written first, the meaning of the solution is clear. We want all real numbers that are either greater than 3 or less than 1. Here are three ways to express this:

Set-Builder Notation	Number Line Graph	Interval Notation
$\{t \mid t < 1 \text{ or } t > 3\}$		$(-\infty, 1) \cup (3, \infty)$

We can use the results of our first few examples and the material in the previous section to summarize the information we have related to absolute value equations and inequalities.

⌈Δ≠Σ⌉ *Rewriting Absolute Value Equations and Inequalities*

If c is a positive real number, then each of the following statements on the left is equivalent to the corresponding statement on the right.

With Absolute Value	Without Absolute Value
$\lvert x \rvert = c$	$x = -c$ or $x = c$
$\lvert x \rvert < c$	$-c < x < c$
$\lvert x \rvert > c$	$x < -c$ or $x > c$
$\lvert ax + b \rvert = c$	$ax + b = -c$ or $ax + b = c$
$\lvert ax + b \rvert < c$	$-c < ax + b < c$
$\lvert ax + b \rvert > c$	$ax + b < -c$ or $ax + b > c$

Special Cases

Because absolute value always results in a nonnegative quantity, we sometimes come across special cases when a negative number or zero appears on the right side of an absolute value inequality after the absolute value has been isolated.

EXAMPLE 6 Solve: $\lvert 7y - 1 \rvert < -2$.

SOLUTION The *left* side is never negative because it is an absolute value. The *right* side is negative. We have a positive quantity (or zero) less than a negative quantity, which is impossible. The solution set is the empty set, $\varnothing$. There is no real number to substitute for y to make this inequality a true statement.

EXAMPLE 7 Solve: $\lvert 6x + 2 \rvert \geq -5$.

SOLUTION This is the opposite case from that in Example 6. No matter what real number we use for x on the *left* side, the result will always be positive, or zero. The *right* side is negative. We have a positive quantity (or zero) greater than or equal to a negative quantity. Every real number we choose for x gives us a true statement. The absolute value will never *equal* -5, but it will always be *greater than* -5. Therefore, the solution set is the set of all real numbers.

EXAMPLE 8   Solve: $|3 - 4a| \leq 0$.

SOLUTION   For any value of a, it is not possible for the absolute value on the left side to be *less than* zero, but it is possible for the absolute value to *equal* zero. Therefore, we must solve the equation

$$|3 - 4a| = 0$$

$$3 - 4a = 0 \qquad \text{Special case}$$

$$-4a = -3 \qquad \text{Add } -3 \text{ to both sides}$$

$$a = \frac{3}{4} \qquad \text{Divide both sides by } -4$$

The solution set is a single value, $\left\{ \frac{3}{4} \right\}$.

Getting Ready for Class

After reading through the preceding section, respond in your own words and in complete sentences.

A. Write an inequality containing absolute value, the solution to which is all the numbers between -5 and 5 on the number line.

B. Translate $|x| \geq 3$ into words using the definition of absolute value.

C. Explain in words what the inequality $|x - 5| < 2$ means with respect to distance on the number line.

D. Why is there no solution to the inequality $|2x - 3| < 0$?

Solve each of the following inequalities using the definition of absolute value. Write your answer using interval notation (when possible), and graph the solution set in each case.

1. $|x| < 3$ **2.** $|x| \leq 7$ **3.** $|x| \geq 2$ **4.** $|x| > 4$

5. $|x| + 2 < 5$ **6.** $|x| - 3 < -1$ **7.** $|t| - 3 > 4$ **8.** $|t| + 5 > 8$

9. $|y| < -5$ **10.** $|y| > -3$ **11.** $|x| \geq -2$ **12.** $|x| \leq -4$

13. $|x - 3| < 7$ **14.** $|x + 4| < 2$ **15.** $|a + 5| \geq 4$ **16.** $|a - 6| \geq 3$

17. $|b + 1| < 0$ **18.** $|b - 9| \leq 0$ **19.** $|b - 2| \geq 0$ **20.** $|b + 8| > 0$

Solve each inequality and graph the solution set.

21. $|a - 1| < -3$ **22.** $|a + 2| \geq -5$ **23.** $|2x - 4| < 6$

24. $|2x + 6| < 2$ **25.** $|3y + 9| \geq 6$ **26.** $|5y - 1| \geq 4$

27. $|2k + 3| \geq 7$ **28.** $|2k - 5| \geq 3$ **29.** $|x - 3| + 2 < 6$

30. $|x + 4| - 3 < -1$ **31.** $|2a + 1| + 4 \geq 7$ **32.** $|2a - 6| - 1 \geq 2$

33. $|3x + 5| - 8 < 5$ **34.** $|6x - 1| - 4 \leq 2$

35. $9 + |5y - 6| < 4$ **36.** $15 + |3y + 7| \geq 1$

Solve each inequality and write your answer using interval notation. Keep in mind that if you multiply or divide both sides of an inequality by a negative number you must reverse the sense of the inequality.

37. $|x - 3| \leq 5$ **38.** $|a + 4| < 6$ **39.** $|3y + 1| < 5$

40. $|2x - 5| \leq 3$ **41.** $|a + 4| \geq 1$ **42.** $|y - 3| > 6$

43. $|2x + 5| > 2$ **44.** $|-3x + 1| \geq 7$ **45.** $|-5x + 3| \leq 8$

46. $|-3x + 4| \leq 7$ **47.** $|-3x + 7| < 2$ **48.** $|-4x + 2| < 6$

Solve each inequality and graph the solution set.

49. $|5 - x| > 3$ **50.** $|7 - x| > 2$ **51.** $\left|3 - \dfrac{2}{3}x\right| \geq 5$

52. $\left|3 - \dfrac{3}{4}x\right| \geq 9$ **53.** $\left|2 - \dfrac{1}{2}x\right| > 1$ **54.** $\left|3 - \dfrac{1}{3}x\right| > 1$

Solve each inequality.

55. $|x - 1| < 0.01$ **56.** $|x + 1| < 0.01$ **57.** $|2x + 1| \geq \dfrac{1}{5}$

58. $|2x - 1| \geq \dfrac{1}{8}$ **59.** $|3x - 2| \leq \dfrac{1}{3}$ **60.** $|2x + 5| < \dfrac{1}{2}$

61. $\left|\dfrac{3x + 1}{2}\right| > \dfrac{1}{2}$ **62.** $\left|\dfrac{2x - 5}{3}\right| \geq \dfrac{1}{6}$ **63.** $\left|\dfrac{4 - 3x}{2}\right| \geq 1$

64. $\left|\dfrac{2x - 3}{4}\right| < 0.35$ **65.** $\left|\dfrac{3x - 2}{5}\right| \leq \dfrac{1}{2}$ **66.** $\left|\dfrac{4x - 3}{2}\right| \leq \dfrac{1}{3}$

67. $\left|2x - \dfrac{1}{5}\right| < 0.3$ **68.** $\left|3x - \dfrac{3}{5}\right| < 0.2$

69. Write the continued inequality $-4 \leq x \leq 4$ as a single inequality involving absolute value.

70. Write the continued inequality $-8 \leq x \leq 8$ as a single inequality involving absolute value.

71. Write $-1 \leq x - 5 \leq 1$ as a single inequality involving absolute value.

72. Write $-3 \leq x + 2 \leq 3$ as a single inequality involving absolute value.

73. Paying Attention to Instructions Work each problem according to the instructions given.
    a. Evaluate when $x = 0$: $|5x + 3|$ **b.** Solve: $|5x + 3| = 7$
    c. Is 0 a solution to $|5x + 3| > 7$ **d.** Solve: $|5x + 3| > 7$

74. Paying Attention to Instructions Work each problem according to the instructions given.
    a. Evaluate when $x = 0$: $|-2x - 5|$ **b.** Solve: $|-2x - 5| = 1$
    c. Is 0 a solution to $|-2x - 5| > 1$ **d.** Solve: $|-2x - 5| > 1$

75. Speed Limits The interstate speed limit for cars is 75 miles per hour in Nebraska, Nevada, New Mexico, Oklahoma, South Dakota, Utah, and Wyoming and is the highest in the United States. To discourage passing, minimum speeds are also posted, so that the difference between the fastest and slowest moving traffic is no more than 20 miles per hour. Write an absolute value inequality that describes the relationship between the minimum allowable speed and a maximum speed of 75 miles per hour.

76. Wavelengths of Light When white light from the sun passes through a prism, it is broken down into bands of light that form colors. The wavelength, v, (in nanometers) of some common colors are:

Blue: $424 < v < 491$
Green: $491 < v < 575$
Yellow: $575 < v < 585$
Orange: $585 < v < 647$
Red: $647 < v < 700$

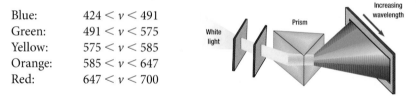

When a fireworks display made of copper is burned, it lets out light with wavelengths, v, that satisfy the relationship $|v - 455| < 23$. Write this inequality without absolute values, find the range of possible values for v, and then using the preceding list of wavelengths, determine the color of that copper fireworks display.

Learning Objectives Assessment

The following problems can be used to help assess if you have successfully met the learning objectives for this section.

77. Solve: $|x - 9| + 5 \le 10$.

    a. $(-\infty, 14]$ **b.** $[-6, 14]$ **c.** $(-\infty, 4] \cup [14, \infty)$ **d.** $[4, 14]$

78. Solve: $|x + 4| > 2$.

    a. $(-2, \infty)$ **b.** $(-6, \infty)$ **c.** $(-\infty, -6) \cup (-2, \infty)$ **d.** $(-6, 2)$

79. Solve: $|5 - y| < -7$.

    a. $\varnothing$ **b.** $(12, \infty)$

    c. $(-\infty, -2) \cup (12, \infty)$ **d.** All real numbers

80. Solve: $|2y + 8| + 11 \ge 9$.

    a. $[-5, \infty)$ **b.** $(-\infty, -5] \cup [-3, \infty)$

    c. $\varnothing$ **d.** All real numbers

Maintaining Your Skills

Simplify each expression. Assume all variables represent nonzero real numbers, and write your answer with positive exponents only.

81. 3^{-2} **82.** $\dfrac{x^6}{x^{-4}}$ **83.** $\dfrac{15x^3y^8}{5xy^{10}}$

84. $(2a^{-3}b^4)^2$ **85.** $\dfrac{(3x^{-3}y^5)^{-2}}{(9xy^{-2})^{-1}}$ **86.** $(3x^4y)^2(5x^3y^4)^3$

Write each number in scientific notation.

87. 54,000   **88.** 0.0359

Write each number in expanded form.

89. 6.44×10^3 **90.** 2.5×10^{-2}

Simplify each expression as much as possible. Write all answers in scientific notation.

91. $(3 \times 10^8)(4 \times 10^{-5})$ **92.** $\dfrac{8 \times 10^5}{2 \times 10^{-8}}$

Chapter 2 Summary

EXAMPLES

Addition Property of Equality [2.1]

1. We can solve

$$x + 3 = 5$$

by adding -3 to both sides:

$$x + 3 + (-3) = 5 + (-3)$$
$$x = 2$$

For algebraic expressions A, B, and C,

$$\text{if} \qquad A = B$$
$$\text{then} \qquad A + C = B + C$$

This property states that we can add the same quantity to both sides of an equation without changing the solution set.

Multiplication Property of Equality [2.1]

2. We can solve $3x = 12$ by multiplying both sides by $\frac{1}{3}$.

$$3x = 12$$
$$\frac{1}{3}(3x) = \frac{1}{3}(12)$$
$$x = 4$$

For algebraic expressions A, B, and C,

$$\text{if} \qquad A = B$$
$$\text{then} \qquad AC = BC \qquad (C \neq 0)$$

Multiplying both sides of an equation by the same nonzero quantity never changes the solution set.

Strategy for Solving Linear Equations in One Variable [2.1]

3. Solve: $3(2x - 1) = 9$.

$$3(2x - 1) = 9$$
$$6x - 3 = 9$$
$$6x - 3 + 3 = 9 + 3$$
$$6x = 12$$
$$\frac{1}{6}(6x) = \frac{1}{6}(12)$$
$$x = 2$$

Step 1a: Use the distributive property to separate terms, if necessary.

1b: If fractions are present, consider multiplying both sides by the LCD to eliminate the fractions. If decimals are present, consider multiplying both sides by a power of 10 to clear the equation of decimals.

1c: Combine similar terms on each side of the equation.

Step 2: Use the addition property of equality to get all variable terms on one side of the equation and all constant terms on the other side. A variable term is a term that contains the variable (for example, $5x$). A constant term is a term that does not contain the variable (the number 3, for example).

Step 3: Use the multiplication property of equality to get the variable by itself on one side of the equation.

Step 4: Check your solution in the original equation to be sure that you have not made a mistake in the solution process.

Formulas [2.2]

4. Solve for w:

$$P = 2l + 2w$$
$$P - 2l = 2w$$
$$\frac{P - 2l}{2} = w$$

A *formula* in algebra is an equation involving more than one variable. To solve a formula for one of its variables, simply isolate that variable on one side of the equation.

Blueprint for Problem Solving [2.3]

5. The perimeter of a rectangle is 32 inches. If the length is 3 times the width, find the dimensions.

Step 1: This step is done mentally.

Step 2: Let x = the width. Then the length is $3x$.

Step 3: The perimeter is 32; therefore

$$2x + 2(3x) = 32$$

Step 4: $8x = 32$
                  $x = 4$

Step 5: The width is 4 inches. The length is $3(4) = 12$ inches.

Step 6: The perimeter is $2(4) + 2(12)$, which is 32. The length is 3 times the width.

Step 1: *Read* the problem, and then mentally *list* the items that are known and the items that are unknown.

Step 2: *Assign a variable* to one of the unknown items. (In most cases this will amount to letting x = the item that is asked for in the problem.) Then *translate* the other *information* in the problem to expressions involving the variable.

Step 3: *Reread* the problem, and then *write an equation,* using the items and variables listed in steps 1 and 2, that describes the situation.

Step 4: *Solve the equation* found in step 3.

Step 5: *Write your answer* using a complete sentence.

Step 6: *Reread* the problem, and *check* your solution with the original words in the problem.

Addition Property for Inequalities [2.4]

6. Adding 5 to both sides of the inequality $x - 5 < -2$ gives

$$x - 5 + 5 < -2 + 5$$
$$x < 3$$

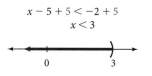

For expressions A, B, and C,

$$\text{if} \qquad\qquad A < B$$
$$\text{then} \qquad A + C < B + C$$

Adding the same quantity to both sides of an inequality never changes the solution set.

Multiplication Property for Inequalities [2.4]

7. Multiplying both sides of $-2x \geq 6$ by $-\frac{1}{2}$ gives

$$-2x \geq 6$$
$$-\frac{1}{2}(-2x) \leq -\frac{1}{2}(6)$$
$$x \leq -3$$

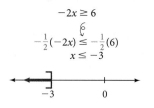

For expressions A, B, and C,

$$\text{if} \qquad A < B$$
$$\text{then} \qquad AC < BC \qquad \text{if} \qquad C > 0 \ (C \text{ is positive})$$
$$\text{or} \qquad AC > BC \qquad \text{if} \qquad C < 0 \ (C \text{ is negative})$$

We can multiply both sides of an inequality by the same nonzero number without changing the solution set as long as each time we multiply by a negative number we also reverse the direction of the inequality symbol.

Compound Inequalities [2.5]

8. Solve: $3x - 1 \leq 5$ and $x + 4 > 1$.

$$3x - 1 \leq 5 \quad \text{and} \quad x + 4 > 1$$
$$3x \leq 6 \quad \text{and} \quad x > -3$$
$$x \leq 2$$

The intersection of $x \leq 2$ and $x > -3$ is $-3 < x \leq 2$, or $(-3, 2]$ using interval notation.

To solve a compound inequality that contains the word "and," solve the two inequalities separately and then find the intersection of the two solution sets.

To solve a compound inequality that contains the word "or," solve the two inequalities separately and then find the union of the two solution sets.

Absolute Value Equations [2.6]

9. To solve

$$|2x - 1| + 2 = 7$$

we first isolate the absolute value on the left side by adding -2 to each side to obtain

$$|2x - 1| = 5$$

$$2x - 1 = 5 \quad \text{or} \quad 2x - 1 = -5$$
$$2x = 6 \quad \text{or} \quad 2x = -4$$
$$x = 3 \quad \text{or} \quad x = -2$$

To solve an equation that involves absolute value, we isolate the absolute value on one side of the equation and then rewrite the absolute value equation as two separate equations that do not involve absolute value. In general, if b is a positive real number, then

$$|A| = b \quad \text{is equivalent to} \quad A = b \quad \text{or} \quad A = -b$$

Absolute Value Inequalities [2.7]

10. To solve

$$|x - 3| + 2 < 6$$

we first add -2 to both sides to obtain

$$|x - 3| < 4$$

which is equivalent to

$$-4 < x - 3 < 4$$
$$-1 < \quad x \quad < 7$$

To solve an inequality that involves absolute value, we first isolate the absolute value on the left side of the inequality symbol. Then we rewrite the absolute value inequality as an equivalent continued or compound inequality that does not contain absolute value symbols. In general, if b is a positive real number, then

$$|A| < b \quad \text{is equivalent to} \quad -b < A < b$$

and

$$|A| > b \quad \text{is equivalent to} \quad A < -b \quad \text{or} \quad A > b$$

> ⚠ **COMMON MISTAKE**
>
> A very common mistake in solving inequalities is to forget to reverse the direction of the inequality symbol when multiplying both sides by a negative number. When this mistake occurs, the graph of the solution set is always drawn on the wrong side of the endpoint.

Chapter 2 Test

Solve the following equations. [2.1]

1. $5 - \dfrac{4}{7}a = -11$

2. $\dfrac{1}{5}x - \dfrac{1}{2} - \dfrac{1}{10}x + \dfrac{2}{5} = \dfrac{3}{10}x + \dfrac{1}{2}$

3. $5(x - 1) - 2(2x + 3) = 5x - 4$

4. $0.07 - 0.02(3x + 1) = -0.04x + 0.01$

Solve for the indicated variable. [2.2]

5. $P = 2l + 2w$; for w

6. $A = \dfrac{1}{2}h(b + B)$; for B

Solve for y. [2.2]

7. $5x - 2y = 10$

8. $\dfrac{y - 5}{x - 4} = 3$

Solve each of the following. [2.3]

9. Geometry A rectangle is twice as long as it is wide. The perimeter is 36 inches. Find the dimensions.

10. Geometry Two angles are supplementary. If the larger angle is $15°$ more than twice the smaller angle, find the measure of each angle.

Solve the following inequalities. Write the solution set using interval notation, then graph the solution set. [2.4]

11. $-5t \le 30$

12. $5 - \dfrac{3}{2}x > -1$

13. $1.6x - 2 < 0.8x + 2.8$

14. $3(2y + 4) \ge 5(y - 8)$

Find the union. [2.5]

15. $\{1, 2, 3\} \cup \{2, 4, 6\}$

16. $(-\infty, -1) \cup (-\infty, 4]$

Find the intersection. [2.5]

17. $\{1, 2, 3\} \cap \{2, 4, 6\}$

18. $(-\infty, -1) \cap (-\infty, 4]$

Solve the compound inequality. [2.5]

19. $x + 5 \le 9$ or $x - 2 > -8$

20. $5x - 3 < 7$ and $4 - x > 7$

Solve the following equations. [2.6]

21. $|x + 6| - 3 = 1$

22. $\left| \dfrac{1}{4}x - 1 \right| = \dfrac{1}{2}$

23. $|3 - 2x| + 5 = 2$

24. $|x - 1| = |5x + 2|$

Solve the following inequalities and graph the solution sets. [2.7]

25. $|6x - 1| > 7$

26. $|3x - 5| - 4 \le 3$

27. $|5 - 4x| \ge -7$

28. $|4t - 1| < -3$

Linear Equations in Two Variables, and Functions

3

Chapter Outline

iStockphoto.com © ishmeriev

A student is working with a spring in a physics lab. As she adds various weights to the spring, the spring stretches by various amounts. She records the weights applied and the corresponding length by which the spring stretches. The table below shows some of the data she collects. The scatter diagram gives a visual representation of the data in the table.

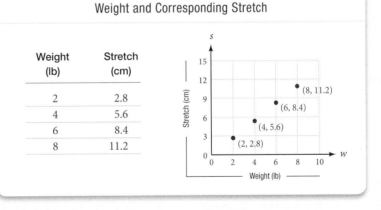

Weight and Corresponding Stretch

Weight (lb)	Stretch (cm)
2	2.8
4	5.6
6	8.4
8	11.2

The exact relationship between the weight and amount of stretch in the spring is given by the following formula, which is an example of a direct variation.

$$s = 1.4w$$

We have three ways to describe the relationship between the weight and amount of stretch in the spring: a table, a graph, and an equation. But, most important to us, we don't need to accept this formula on faith. Later, you will derive the formula from the data in the table above. We will also see that this relationship can be expressed as a function, $f(w) = 1.4w$, because the amount of stretch depends upon the weight applied to the spring.

© Steve Debenport / iStockPhoto

Success Skills

The study skills for this chapter are about attitude. They are points of view that point toward success.

1. **Be Focused, Not Distracted** I have students who begin their assignments by asking themselves, "Why am I taking this class?" If you are asking yourself similar questions, you are distracting yourself from doing the things that will produce the results you want in this course. Don't dwell on questions and evaluations of the class that can be used as excuses for not doing well. If you want to succeed in this course, focus your energy and efforts toward success, rather than distracting yourself from your goals.

2. **Be Resilient** Don't let setbacks keep you from your goals. You want to put yourself on the road to becoming a person who can succeed in this class, or any class in college. Failing a test or quiz, or having a difficult time on some topics, is normal. No one goes through college without some setbacks. Don't let a temporary disappointment keep you from succeeding in this course. A low grade on a test or quiz is simply a signal that you need to reevaluate your study habits.

3. **Intend to Succeed** I have a few students who simply go through the motions of studying without intending to master the material. It is more important to them to look like they are studying than to actually study. You need to study with the intention of being successful in the course. Intend to master the material, no matter what it takes.

The Rectangular Coordinate System and Graphing Lines

Learning Objectives

In this section we will learn how to:

1. Graph an ordered pair on the rectangular coordinate system.

2. Graph a linear equation in two variables.

3. Find the x- and y-intercepts for an equation.

4. Graph horizontal and vertical lines.

Introduction

In this section we place our work with charts and graphs in a more formal setting. Our foundation will be the *rectangular coordinate system*, because it gives us a link between algebra and geometry. With it we notice relationships between certain equations and different lines and curves.

Table 1 gives the net price of a popular intermediate algebra text at the beginning of each year in which a new edition was published. (The net price is the price the bookstore pays for the book, not the price you pay for it.)

TABLE 1	Price of a Textbook	
Edition	Year Published	Net Price ($)
First	1991	30.50
Second	1995	39.25
Third	1999	47.50
Fourth	2003	55.00
Fifth	2007	65.75

The information in Table 1 is represented visually in Figures 1 and 2. The diagram in Figure 1 is called a *bar chart*. The diagram in Figure 2 is called a *line graph*. The data in Table 1 is called *paired data* because each number in the year column is paired with a specific number in the price column.

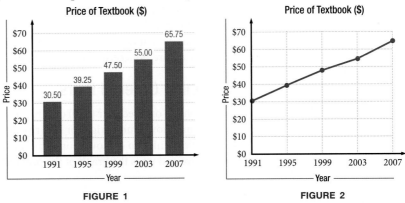

FIGURE 1   FIGURE 2

Ordered Pairs

Paired data play an important role in equations that contain two variables. Working with these equations is easier if we standardize the terminology and notation associated with paired data. So here is a definition that will do just that.

> **ᵈⁿᶠ DEFINITION**
>
> A pair of numbers enclosed in parentheses and separated by a comma, such as $(-2, 1)$, is called an ***ordered pair*** of numbers. The first number in the pair is called the ***x-coordinate*** of the ordered pair; the second number is called the ***y-coordinate***. For the ordered pair $(-2, 1)$, the x-coordinate is -2 and the y-coordinate is 1.

Rectangular Coordinate System

Note A rectangular coordinate system allows us to connect algebra and geometry by associating geometric shapes (the curves shown in the diagrams) with algebraic equations. The French philosopher and mathematician René Descartes (1596 – 1650) is usually credited with the invention of the rectangular coordinate system, which is often referred to as the Cartesian coordinate system in his honor. As a philosopher, Descartes is responsible for the statement, "I think, therefore, I am." Until Descartes invented his coordinate system in 1637, algebra and geometry were treated as separate subjects.

A ***rectangular coordinate system*** is made by drawing two real number lines at right angles to each other. The two number lines, called ***axes***, cross each other at 0. This point is called the ***origin***. Positive directions are to the right and up. Negative directions are to the left and down. The rectangular coordinate system is shown in Figure 3.

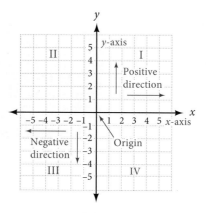

FIGURE 3

The horizontal number line is called the ***x-axis***, and the vertical number line is called the ***y-axis***. The two number lines divide the coordinate system into four quadrants, which we number I through IV in a counterclockwise direction. Points on the axes are not considered as being in any quadrant.

Graphing Ordered Pairs

To graph the ordered pair (a, b) on a rectangular coordinate system, we start at the origin and move a units right or left (right if a is positive, left if a is negative). Then we move b units up or down (up if b is positive, down if b is negative). The point where we end up is the graph of the ordered pair (a, b).

EXAMPLE 1   Plot (graph) the ordered pairs $(2, 5)$, $(-2, 5)$, $(-2, -5)$, and $(2, -5)$.

SOLUTION   To graph the ordered pair $(2, 5)$, we start at the origin and move 2 units to the right, then 5 units up. We are now at the point whose coordinates are $(2, 5)$. We graph the other three ordered pairs in a similar manner (see Figure 4).

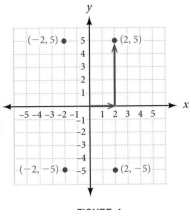

FIGURE 4

Note   From Example 1, we see that any point in quadrant I has both its x- and y-coordinates positive $(+, +)$. Points in quadrant II have negative x-coordinates and positive y-coordinates $(-, +)$. In quadrant III, both coordinates are negative $(-, -)$. In quadrant IV, the form is $(+, -)$.

EXAMPLE 2   Graph the ordered pairs $(1, -3)$, $\left(\frac{1}{2}, 2\right)$, $(3, 0)$, $(0, -2)$, $(-1, 0)$, and $(0, 5)$.

SOLUTION

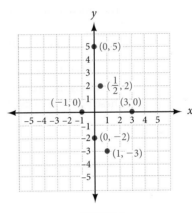

FIGURE 5

From Figure 5, we see that any point on the x-axis has a y-coordinate of 0 (it has no vertical displacement), and any point on the y-axis has an x-coordinate of 0 (no horizontal displacement).

Linear Equations in Two Variables

A **linear equation** in two variables is any equation that can be expressed in the form $ax + by = c$, where a, b, and c are real numbers with a and b not both zero. If we isolate the variable y, the equation will have the form $y = mx + b$. We will define the various forms of linear equation in two variables in more detail later in Section 3.3.

Graphing Linear Equations in Two Variables

To graph a linear equation in two variables, we simply graph its solution set. That is, we plot a point for each ordered pair whose coordinates satisfy the equation. If we do this for every ordered pair in the solution set, the resulting graph will be a line. However, because a line is determined by just two points, we only need to find two ordered pairs that satisfy the equation in order to graph the line. We will often choose to find a third point for "insurance." If all three points do not line up, we have made a mistake.

EXAMPLE 3   Graph the equation $y = 4x$.

SOLUTION   The graph of this equation will be a line. We find three ordered pairs that satisfy the equation. To do so, we can let x equal any number we choose and find the corresponding value of y.

$$\text{Let } x = -1; \quad y = 4(-1) = -4$$

The ordered pair $(-1, -4)$ is one solution.

$$\text{Let } x = 0; \quad y = 4(0) = 0$$

The ordered pair $(0, 0)$ is a second solution.

$$\text{Let } x = 1; \quad y = 4(1) = 4$$

The ordered pair $(1, 4)$ is a third solution.

Plotting the ordered pairs $(-1, -4), (0, 0)$, and $(1, 4)$ and drawing a line through these points, we have the graph of the equation $y = 4x$ as shown in Figure 6.

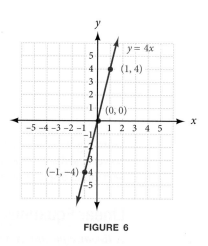

FIGURE 6

Example 3 illustrates again the connection between algebra and geometry that we mentioned earlier in this section. Descartes's rectangular coordinate system allows us to associate the equation $y = 4x$ (an algebraic concept) with a specific line (a geometric concept). The study of the relationship between equations in algebra and their associated geometric figures is called ***analytic geometry***.

EXAMPLE 4   Graph the equation $y = 3x + 2$.

SOLUTION   Again, we need ordered pairs that are solutions to the equation.

Input x	Calculate Using the Equation	Output y	Form Ordered Pairs
-1	$y = 3(-1) + 2 = -3 + 2 =$	-1	$(-1, -1)$
0	$y = 3(0) + 2 = 0 + 2 =$	2	$(0, 2)$
1	$y = 3(1) + 2 = 3 + 2 =$	5	$(1, 5)$

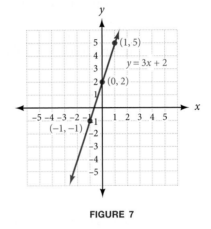

FIGURE 7

Lines Through the Origin

As you can see from Figure 6, the graph of the equation $y = 4x$ is a line that passes through the origin. The same will be true of the graph of any equation that has the same form as $y = 4x$. Here are two more equations that have that form, along with their graphs.

Graph of $y = -2x$ **Graph of $y = \frac{1}{2}x$**

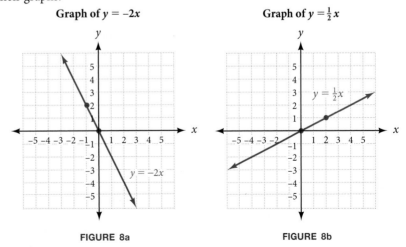

FIGURE 8a   **FIGURE 8b**

Here is a summary of this discussion.

> **⎰Δ≠Σ⎱   *Lines Through the Origin***
>
> The graph of any equation of the form
>
> $$y = mx$$
>
> will be a line passing through the origin and through the point $(1, m)$.

EXAMPLE 5   Graph the equation $y = -\dfrac{1}{3}x$.

SOLUTION   First, we find three ordered pairs that satisfy the equation. Because every value of x we substitute into the equation is going to be multiplied by $-\frac{1}{3}$, let's use numbers that are divisible by 3, like -3, 0, and 3. That way, when we multiply them by $-\frac{1}{3}$, the result will be an integer.

$$\text{Let } x = -3; \quad y = -\frac{1}{3}(-3) = 1$$

$$\text{Let } x = 0; \quad y = -\frac{1}{3}(0) = 0$$

$$\text{Let } x = 3; \quad y = -\frac{1}{3}(3) = -1$$

In table form

x	y
-3	1
0	0
3	-1

Plotting the ordered pairs $(-3, 1)$, $(0, 0)$, and $(3, -1)$ and drawing a line through these points, we have the graph of the equation $y = -\frac{1}{3}x$ as shown in Figure 9.

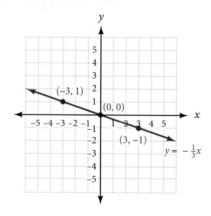

FIGURE 9

Intercepts

Two important points on the graph of a line, if they exist, are the points where the graph crosses the axes.

> (dĕf) **DEFINITION**   *intercepts*
>
> An *x-intercept* of the graph of an equation is the *x*-coordinate of a point where the graph intersects the *x*-axis. The *y-intercept* is defined similarly.

Because any point on the *x*-axis has a *y*-coordinate of 0, we can find the *x*-intercept by letting $y = 0$ and solving the equation for *x*. We find the *y*-intercept by letting $x = 0$ and solving for *y*.

EXAMPLE 6   Find the *x*- and *y*-intercepts for $2x + 3y = 6$; then graph the equation.

SOLUTION   To find the *y*-intercept, we let $x = 0$.

When $$x = 0$$

we have $$2(0) + 3y = 6$$
$$3y = 6$$
$$y = 2$$

The *y*-intercept is 2 so the graph crosses the *y*-axis at the point (0, 2). To find the *x*-intercept, we let $y = 0$.

When $$y = 0$$

we have $$2x + 3(0) = 6$$
$$2x = 6$$
$$x = 3$$

The *x*-intercept is 3, so the graph crosses the *x*-axis at the point (3, 0). We use these results to graph the solution set for $2x + 3y = 6$. The graph is shown in Figure 10.

Note   Graphing lines by finding the intercepts works especially well when the coefficients of *x* and *y* are factors of the constant term.

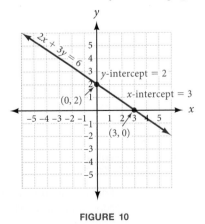

FIGURE 10

EXAMPLE 7   Find the x- and y-intercepts for $y = \frac{3}{2}x - 3$; then graph the equation.

SOLUTION   To find the y-intercept, we let $x = 0$.

When $x = 0$

we have $y = \frac{3}{2}(0) - 3$

$y = -3$

To find the x-intercept, we let $y = 0$.

When $y = 0$

we have $0 = \frac{3}{2}x - 3$

$3 = \frac{3}{2}x$

$\frac{2}{3}(3) = x$

$2 = x$

The graph will cross the y-axis at $(0, -3)$ and the x-axis at $(2, 0)$. Figure 11 shows the graph.

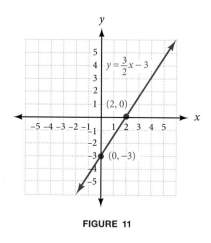

FIGURE 11

In Example 7, notice that the y-intercept, -3, was given by the constant term in the equation. This will be true for any equation that has the same form as $y = \frac{3}{2}x - 3$. We generalize this result below.

[Δ≠Σ] PROPERTY   *y-intercept*

For any equation of the form $y = mx + b$, the y-intercept is given by the constant term, b. That is, the graph will be a line crossing the y-axis at the point $(0, b)$.

Horizontal and Vertical Lines

 **EXAMPLE 8**   Graph each of the following lines.

a. $x = 3$ **b.** $y = -2$

SOLUTION

a. The line $x = 3$ is the set of all points whose x-coordinate is 3. The variable y does not appear in the equation, so the y-coordinate can be any number. Note that we can write our equation as a linear equation in two variables by writing it as $x + 0y = 3$. Because the product of 0 and y will always be 0, y can be any number. The graph of $x = 3$ is the vertical line shown in Figure 12a.

b. The line $y = -2$ is the set of all points whose y-coordinate is -2. The variable x does not appear in the equation, so the x-coordinate can be any number. Again, we can write our equation as a linear equation in two variables by writing it as $0x + y = -2$. Because the product of 0 and x will always be 0, x can be any number. The graph of $y = -2$ is the horizontal line shown in Figure 12b.

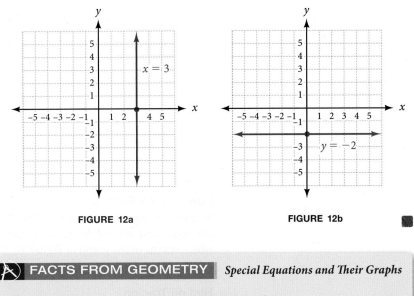

FIGURE 12a   FIGURE 12b

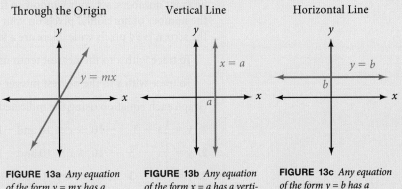

FACTS FROM GEOMETRY   *Special Equations and Their Graphs*

For the equations below, m, a, and b are real numbers.

FIGURE 13a *Any equation of the form $y = mx$ has a graph that passes through the origin.*

FIGURE 13b *Any equation of the form $x = a$ has a vertical line for its graph that passes through $(a, 0)$.*

FIGURE 13c *Any equation of the form $y = b$ has a horizontal line for its graph that passes through $(0, b)$.*

Graphing With Trace and Zoom

All graphing calculators have the ability to graph an equation and then trace over the points on the graph, giving their coordinates. Furthermore, all graphing calculators can zoom in and out on a graph that has been drawn. To graph an equation on a graphing calculator, we first set the graph window. Most calculators call the smallest value of x, Xmin, and the largest value of x, Xmax. The counterpart values of y are Ymin and Ymax. We will use the notation

$$\text{Window:} \quad -5 \leq x \leq 4 \text{ and } -3 \leq y \leq 2$$

to stand for a window in which Xmin $= -5$, Xmax $= 4$, Ymin $= -3$, and Ymax $= 2$

Set your calculator to the following window:

$$\text{Window:} \quad -10 \leq x \leq 10 \text{ and } -10 \leq y \leq 10$$

Graph the equation $Y_1 = -X + 8$ and compare your results with this graph:

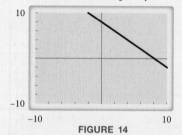

FIGURE 14

Use the Trace feature of your calculator to name three points on the graph. Next, use the Zoom feature of your calculator to zoom out so your window is twice as large.

Solving for y First

To graph the equation from Example 6, $2x + 3y = 6$, on a graphing calculator, you must first solve it for y. When you do so, you will get $y = -\frac{2}{3}x + 2$, which results in the graph in Figure 10. Use this window:

$$\text{Window:} \quad -6 \leq x \leq 6 \text{ and } -6 \leq y \leq 6$$

Hint on Tracing

If you are going to use the Trace feature and you want the x-coordinates to be exact numbers, set your window so the range of X inputs is a multiple of the number of horizontal pixels on your calculator screen. On the TI-83/84, the screen is 94 pixels wide. Here are a few convenient trace windows:

To trace with x to the nearest tenth use $-4.7 \leq x \leq 4.7$ or $0 \leq x \leq 9.4$

To trace with x to the nearest integer use $-47 \leq x \leq 47$ or $0 \leq x \leq 94$

Graph each equation using the indicated window.

1. $y = \frac{1}{2}x - 3$ $\quad -10 \leq x \leq 10$ and $-10 \leq y \leq 10$

2. $y = \frac{1}{2}x^2 - 3$ $\quad -10 \leq x \leq 10$ and $-10 \leq y \leq 10$

3. $y = \frac{1}{2}x^2 - 3$ $\quad -4.7 \leq x \leq 4.7$ and $-10 \leq y \leq 10$

4. $y = x^3$ $\quad -10 \leq x \leq 10$ and $-10 \leq y \leq 10$

5. $y = x^3 - 5$ $\quad -4.7 \leq x \leq 4.7$ and $-10 \leq y \leq 10$

Getting Ready for Class

After reading through the preceding section, respond in your own words and in complete sentences.

A. Explain how you would construct a rectangular coordinate system from two real number lines.

B. Explain in words how you would graph the ordered pair $(2, -3)$.

C. How can you tell if an ordered pair is a solution to the equation $y = 2x - 5$?

D. If you were looking for solutions to the equation $y = \frac{1}{3}x + 5$, why would it be easier to substitute 6 for x than to substitute 5 for x?

SPOTLIGHT ON SUCCESS   *Student Instructor Gordon*

Math takes time. This fact holds true in the smallest of math problems as much as it does in the most math intensive careers. I see proof in each video I make. My videos get progressively better with each take, though I still make mistakes and find aspects I can improve on with each new video. In order to keep trying to improve in spite of any failures or lack of improvement, something else is needed. For me it is the sense of a specific goal in sight, to help me maintain the desire to put in continued time and effort.

When I decided on the number one university I wanted to attend, I wrote the name of that school in bold block letters on my door, written to remind myself daily of my ultimate goal. Stuck in the back of my head, this end result pushed me little by little to succeed and meet all of the requirements for the university I had in mind. And now I can say I'm at my dream school bringing with me that skill.

I recognize that others may have much more difficult circumstances than my own to endure, with the goal of improving or escaping those circumstances, and I deeply respect that. But that fact demonstrates to me how easy but effective it is, in comparison, to "stay with the problems longer" with a goal in mind of something much more easily realized, like a good grade on a test. I've learned to set goals, small or big, and to stick with them until they are realized.

Problem Set 3.1

Graph each of the following ordered pairs on a rectangular coordinate system.

1. a. $(-1, 2)$ **b.** $(-1, -2)$ **c.** $(5, 0)$ **d.** $(0, 2)$ **e.** $(-5, -5)$ **f.** $\left(\dfrac{1}{2}, 2\right)$

2. a. $(1, 2)$ **b.** $(1, -2)$ **c.** $(0, -3)$ **d.** $(4, 0)$ **e.** $(-4, -1)$ **f.** $\left(3, \dfrac{1}{4}\right)$

Give the coordinates of each point.

3.

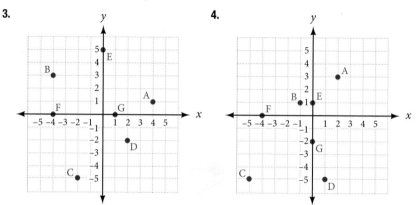

4.

5. Which of the following tables could be produced from the equation $y = 2x - 6$?

a.

x	y
0	6
1	4
2	2
3	0

b.

x	y
0	-6
1	-4
2	-2
3	0

c.

x	y
0	-6
1	-5
2	-4
3	-3

6. Which of the following tables could be produced from the equation $3x - 5y = 15$?

a.

x	y
0	5
-3	0
10	3

b.

x	y
0	-3
5	0
10	3

c.

x	y
0	-3
-5	0
10	-3

7. The graph shown here is the graph of which of the following equations?

a. $y = \dfrac{3}{2}x - 3$

b. $y = \dfrac{2}{3}x - 2$

c. $y = -\dfrac{2}{3}x + 2$

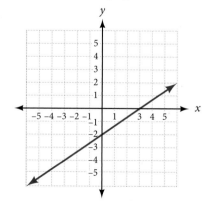

8. The graph shown here is the graph of which of the following equations?

   a. $3x - 2y = 8$

   b. $2x - 3y = 8$

   c. $2x + 3y = 8$

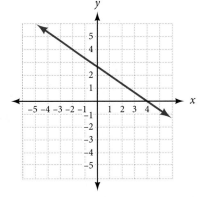

For each of the following equations, find three ordered pairs that satisfy the equation, then use them to graph the line.

9. $y = 2x$ **10.** $y = \frac{1}{3}x$ **11.** $y = -\frac{3}{4}x$ **12.** $y = -4x$

13. $y = x$ **14.** $y = -x$ **15.** $y = \frac{1}{3}x + 1$ **16.** $y = 2x - 5$

17. $y = -3x - 2$ **18.** $y = -\frac{3}{2}x + 2$

Find the intercepts for each graph and write them as ordered pairs. Then use them to help sketch the graph.

19. $3x - 2y = 6$ **20.** $2x + 4y = 8$ **21.** $y + 2x = 4$ **22.** $y - x = 5$

23. $y = 3x$ **24.** $y = -\frac{1}{3}x$ **25.** $y = 2x - 4$ **26.** $y = 4x - 2$

27. $y = \frac{1}{2}x + 1$ **28.** $y = -\frac{1}{2}x + 1$ **29.** $y = x - 3$ **30.** $y = x + 2$

31. $2x + 4y = 5$ **32.** $5x - 3y = 9$

33. Graph the line $0.02x + 0.03y = 0.06$.

34. Graph the line $0.05x - 0.03y = 0.15$.

Graph each of the following lines.

35. $x = -1$ **36.** $x = 5$ **37.** $y = 3$ **38.** $y = -2$

39. $x - \frac{1}{2} = 0$ **40.** $x + \frac{9}{2} = 0$ **41.** $y + \frac{10}{3} = 0$ **42.** $y - \frac{2}{3} = 0$

43. Graph each of the following lines.

   a. $y = 2x$ **b.** $x = -3$ **c.** $y = 2$

44. Graph each of the following lines.

   a. $y = 3x$ **b.** $x = -2$ **c.** $y = 4$

45. Graph each of the following lines.

   a. $y = -\frac{1}{2}x$ **b.** $x = 4$ **c.** $y = -3$

46. Graph each of the following lines.

   a. $y = -\frac{1}{3}x$ **b.** $x = 1$ **c.** $y = -5$

47. Paying Attention to Instructions Work each problem according to the instructions given:

   a. Solve: $4x + 12 = -16$

   b. Find x when y is 0: $4x + 12y = -16$

   c. Find y when x is 0: $4x + 12y = -16$

   d. Graph: $4x + 12y = -16$

   e. Solve for y: $4x + 12y = -16$

48. Paying Attention to Instructions Work each problem according to the instructions given:

   a. Solve: $3x - 8 = -12$

   b. Find x when y is 0: $3x - 8y = -12$

   c. Find y when x is 0: $3x - 8y = -12$

   d. Graph: $3x - 8y = -12$

   e. Solve for y: $3x - 8y = -12$

Applying the Concepts

49. Solar Energy The graph shows the rise in shipments of solar thermal collectors from 1997 to 2006. Use the chart to answer the following questions.

   a. Does the graph contain the point (2000, 7,500)?

   b. Does the graph contain the point (2004, 15,000)?

   c. Does the graph contain the point (2005, 15,000)?

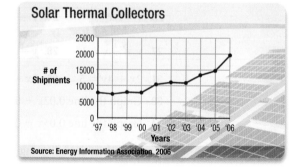

50. Health Care Costs The graph shows the projected rise in the cost of health care from 2002 to 2014. Using years as x and billions of dollars as y, write five ordered pairs that describe the information in the graph.

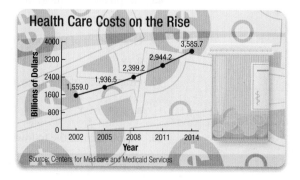

51. **Hourly Wages** Suppose you have a job that pays $7.50 per hour, and you work anywhere from 0 to 40 hours per week. Table 2 gives the amount of money you will earn in 1 week for working various hours. Construct a line graph from the information in Table 2.

TABLE 2	Weekly Wages
Hours Worked	**Pay ($)**
0	0
10	75
20	150
30	225
40	300

52. **Value of a Painting** A piece of abstract art was purchased in 1990 for $125. Table 3 shows the value of the painting at various times, assuming that it doubles in value every 5 years. Construct a bar chart from the information in the table.

TABLE 3	Value of a Painting
Year	**Value ($)**
1990	125
1995	250
2000	500
2005	1,000
2010	2,000

53. **Reading Graphs** The graph shows the number of people in line at a theater box office to buy tickets for a movie that starts at 7:30. The box office opens at 6:45.

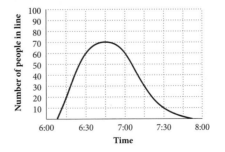

a. How many people are in line at 6:30?

b. How many people are in line when the box office opens?

c. How many people are in line when the show starts?

d. At what times are there 60 people in line?

e. How long after the show starts is there no one left in line?

54. Kentucky Derby The graph gives the monetary bets placed at the Kentucky Derby for specific years. If x represents the year in question and y represents the total wagering for that year, write five ordered pairs that describe the information in the graph.

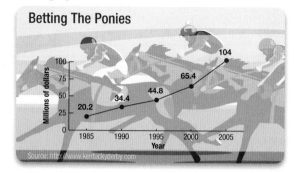

55. Cost of Course Materials The chart below shows the average amount of money college students paid for course materials over a six-year period.

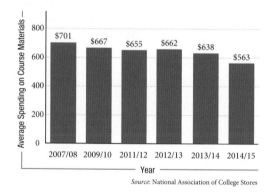

Source: National Association of College Stores

a. Represent the 2007/08 school year with the number 0, the 2008/09 with the number 1, and so on. Then find the equation of the line that connects the information in year 0 with the information in year 7. (You can round the slope to the nearest tenth.)

b. Use the equation you found in part *a* to predict the average amount of money students spent in the 2015/16 school year.

c. What value of x will correspond to the 2017/18 school year.

d. Use the equation you found in Part *a* to predict the average amount of money students spent in the 2017/18 school year.

56. Half-Marathon Popularity Use the chart below to answer questions about the number of people running half-marathons in the United States. If we were to redraw the chart as a scatter diagram, the point (7, 0.75) would indicate that in 2007, 0.75 million people ran a half-marathon.

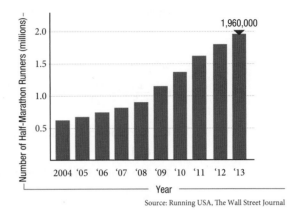

Source: Running USA, The Wall Street Journal

a. If we represent the first bar in the chart with the ordered pair (4, 0.6), how will we represent the last bar in the chart?

b. Find the equation of the line that passes through the points in Part a. Write your answer in slope-intercept form. (Round the slope to the nearest hundredth.)

c. Use the equation from Part b to estimate the number of people running a half-marathon in the year 2016.

Learning Objectives Assessment

The following problems can be used to help assess if you have successfully met the learning objectives for this section.

57. Which ordered pair lies in quadrant IV?

  a. $(3, -5)$ **b.** $(3, 5)$ **c.** $(-3, 5)$ **d.** $(-3, -5)$

58. Graph the equation $y = 2x + 3$.

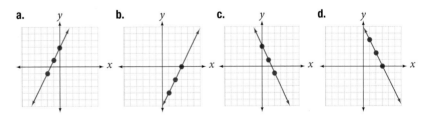

59. Find the x-intercept of $-2x + 5y = 10$.

  a. 2   **b.** -5 **c.** 10   **d.** There isn't one.

60. Which of the following is the equation for a horizontal line passing through $(1, 3)$?

  a. $y = 1$ **b.** $y = 3$ **c.** $x = 1$ **d.** $x = 3$

Getting Ready for the Next Section

61. Write -0.06 as a fraction with denominator 100.

62. Write -0.07 as a fraction with denominator 100.

63. If $y = 2x - 3$, find y when $x = 2$

64. If $y = 2x - 3$, find x when $y = 5$

Simplify.

65. $\dfrac{1 - (-3)}{-5 - (-2)}$ **66.** $\dfrac{-3 - 1}{-2 - (-5)}$ **67.** $\dfrac{-1 - 4}{3 - 3}$ **68.** $\dfrac{-3 - (-3)}{2 - (-1)}$

69. The product of $\dfrac{2}{3}$ and what number will result in

   a. 1?   **b.** -1?

70. The product of 3 and what number will result in

   a. 1?   **b.** -1?

The Slope of a Line

Learning Objectives

In this section we will learn how to:

1. Find the slope of a line geometrically using the graph.

2. Find the slope of a line algebraically using two points on the line.

3. Use slope to identify parallel and perpendicular lines.

4. Interpret slope as an average rate of change.

Introduction

A highway sign tells us we are approaching a 6% downgrade. As we drive down this hill, each 100 feet we travel horizontally is accompanied by a 6-foot drop in elevation.

In mathematics we say the slope of the highway is $-0.06 = -\frac{6}{100} = -\frac{3}{50}$. The slope is the ratio of the vertical change to the accompanying horizontal change.

In defining the slope of a line, we want to associate a number with the line. This does two things. First, we want the slope of a line to measure the "steepness" of the line. That is, in comparing two lines, the slope of the steeper line should have the larger numerical value. Second, we want a line that rises going from left to right to have a *positive* slope. We want a line that falls going from left to right to have a *negative* slope. (A line that neither rises nor falls going from left to right must, therefore, have 0 slope.)

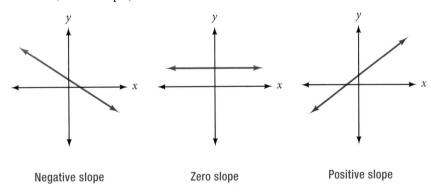

Negative slope   Zero slope   Positive slope

Geometrically, we can define the *slope* of a line as the ratio of the vertical change to the horizontal change encountered when moving from one point to another on the line. The vertical change is sometimes called the *rise*. The horizontal change is called the *run*.

EXAMPLE 1   Find the slope of the line $y = 2x - 3$.

SOLUTION   To use our geometric definition, we first graph $y = 2x - 3$ (Figure 1). We then pick any two convenient points and find the ratio of rise to run. By convenient points we mean points with integer coordinates. If we let $x = 2$ in the equation, then $y = 1$. Likewise, if we let $x = 4$, then y is 5.

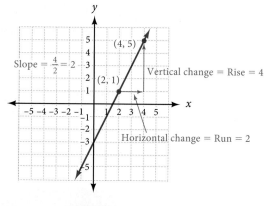

FIGURE 1

The ratio of vertical change to horizontal change is 4 to 2, giving us a slope of $\frac{4}{2} = 2$. Our line has a slope of 2.

Notice that we can measure the vertical change (rise) by subtracting the y-coordinates of the two points shown in Figure 1: $5 - 1 = 4$. The horizontal change (run) is the difference of the x-coordinates: $4 - 2 = 2$. This gives us a second way of defining the slope of a line.

(def) DEFINITION   *slope*

The *slope* of the line passing through the points (x_1, y_1) and (x_2, y_2) is given by

$$\text{Slope} = m = \frac{\text{Rise}}{\text{Run}} = \frac{y_2 - y_1}{x_2 - x_1}$$

↑   ↖

Geometric Form   Algebraic Form

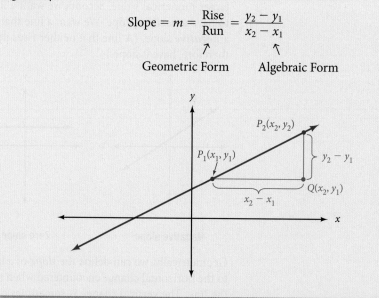

EXAMPLE 2 Find the slope of the line through $(-2, -3)$ and $(-5, 1)$.

SOLUTION If we let $(x_1, y_1) = (-2, -3)$ and $(x_2, y_2) = (-5, 1)$, then

$$m = \frac{y_2 - y_1}{x_2 - x_1} = \frac{1 - (-3)}{-5 - (-2)} = \frac{4}{-3} = -\frac{4}{3}$$

Of course, we can also find the slope geometrically using the graph of the line as shown in Figure 2.

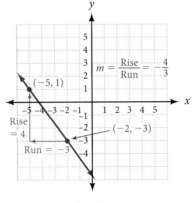

FIGURE 2

We can see the algebraic approach and the geometric approach give the same result.

We should note here that it does not matter which ordered pair we call (x_1, y_1) and which we call (x_2, y_2). If we were to reverse the order of subtraction of both the x- and y-coordinates in the preceding example, we would have

$$m = \frac{-3 - 1}{-2 - (-5)} = \frac{-4}{3} = -\frac{4}{3}$$

which is the same as our previous result.

EXAMPLE 3 Find the slope of the line containing $(3, -1)$ and $(3, 4)$.

SOLUTION Using the definition for slope, we have

$$m = \frac{-1 - 4}{3 - 3} = \frac{-5}{0}$$

The expression $\frac{-5}{0}$ is undefined. That is, there is no real number to associate with it. In this case, we say the line *has no slope*.

Note The two most common mistakes students make when first working with the formula for the slope of a line are

1. Putting the difference of the x-coordinates over the difference of the y-coordinates.

2. Subtracting in one order in the numerator and then subtracting in the opposite order in the denominator. You would make this mistake in Example 2 if you wrote $1 - (-3)$ in the numerator and then $-2 - (-5)$ in the denominator.

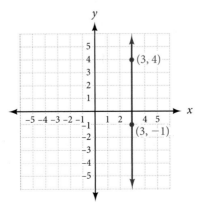

FIGURE 3

The graph of our line is shown in Figure 3. As you can see, the graph is a vertical line. All vertical lines have no slope. (All horizontal lines, as we mentioned earlier, have zero slope.)

Slopes of Parallel and Perpendicular Lines

In geometry, we call lines in the same plane that never intersect parallel. For two lines to be nonintersecting, they must rise or fall at the same rate. In other words, two lines are **parallel** if and only if they have the *same slope*.

Although it is not as obvious, it is also true that two nonvertical lines are **perpendicular** if and only if the *product of their slopes is* −1. This is the same as saying their slopes are negative reciprocals.

We can state these facts with symbols as follows: If line l_1 has slope m_1 and line l_2 has slope m_2, then

$$l_1 \text{ is parallel to } l_2 \Leftrightarrow m_1 = m_2$$
and
$$l_1 \text{ is perpendicular to } l_2 \Leftrightarrow m_1 \cdot m_2 = -1 \text{ or } \left(m_1 = -\frac{1}{m_2}\right)$$

For example, if a line has a slope of $\frac{2}{3}$, then any line parallel to it has a slope of $\frac{2}{3}$. Any line perpendicular to it has a slope of $-\frac{3}{2}$ (the negative reciprocal of $\frac{2}{3}$).

Although we cannot give a formal proof of the relationship between the slopes of perpendicular lines at this level of mathematics, we can offer some justification for the relationship. Figure 4 shows the graphs of two lines. One of the lines has a slope of $\frac{2}{3}$; the other has a slope of $-\frac{3}{2}$. As you can see, the lines are perpendicular.

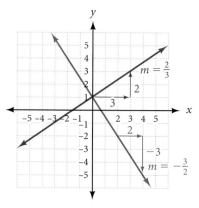

FIGURE 4

Slope and Rate of Change

So far, the slopes we have worked with represent the ratio of the change in y to the corresponding change in x, or, on the graph of the line, the slope is the ratio of vertical change to horizontal change in moving from one point on the line to another. However, when our variables represent quantities from the world around us, slope can have additional interpretations.

EXAMPLE 4 On the chart below, find the slope of the line connecting the first point (1955, 0.29) with the last point (2005, 2.93). Explain the significance of the result.

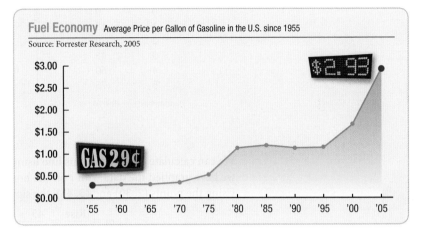

SOLUTION The slope of the line connecting the first point (1955, 0.29) with the last point (2005, 2.93), is

$$m = \frac{2.93 - 0.29}{2005 - 1955} = \frac{2.64}{50} = 0.0528$$

The units are dollars/year. If we write this in terms of cents we have

$$m = 5.28 \text{ cents/year}$$

which is the average change in the price of a gallon of gasoline over a 50-year period of time.

Likewise, if we connect the points (1995, 1.10) and (2005, 2.93), the line that results has a slope of

$$m = \frac{2.93 - 1.10}{2005 - 1995} = \frac{1.83}{10} = 0.183 \text{ dollars/year} = 18.3 \text{ cents/year}$$

which is the average change in the price of a gallon of gasoline over a 10-year period. As you can imagine by looking at the chart, the line connecting the first and last point is not as steep as the line connecting the points from 1995 and 2005, and this is what we are seeing numerically with our slope calculations. If we were summarizing this information for an article in the newspaper, we could say, "Although the price of a gallon of gasoline has increased only 5.28 cents per year over the last 50 years, in the last 10 years the average annual rate of increase has more than tripled, to 18.3 cents per year."

Slope and Average Speed

Previously we introduced the rate equation $d = rt$. Suppose that a boat is traveling at a constant speed of 15 miles per hour in still water. The following table shows the distance the boat will have traveled in the specified number of hours. The graph of this data is shown in Figure 5. Notice that the points all lie along a line.

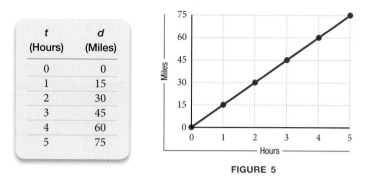

t (Hours)	d (Miles)
0	0
1	15
2	30
3	45
4	60
5	75

FIGURE 5

We can calculate the slope of this line using any two points from the table. Notice we have graphed the data with t on the horizontal axis and d on the vertical axis. Using the points (2, 30) and (3, 45), the slope will be

$$m = \frac{\text{Rise}}{\text{Run}} = \frac{45 - 30}{3 - 2} = \frac{15}{1} = 15$$

The units of the rise are miles and the units of the run are hours, so the slope will be in units of miles per hour. We see that the slope is simply the change in distance divided by the change in time, which is how we compute the average speed. Since the speed is constant, the slope of the line represents the speed of 15 miles per hour.

EXAMPLE 5   A car is traveling at a constant speed. A graph (Figure 6) of the distance the car has traveled over time is shown below. Use the graph to find the speed of the car.

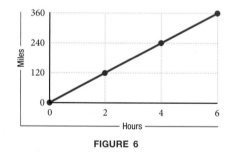

FIGURE 6

SOLUTION   Using the second and third points, we see the rise is $240 - 120 = 120$ miles, and the run is $4 - 2 = 2$ hours. The speed is given by the slope, which is

$$m = \frac{\text{Rise}}{\text{Run}}$$

$$= \frac{120 \text{ miles}}{2 \text{ hours}}$$

$$= 60 \text{ miles per hour}$$

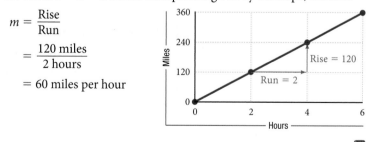

USING TECHNOLOGY *Families of Curves*

We can use a graphing calculator to investigate the effects of the numbers a and b on the graph of $y = ax + b$. To see how the number a affects the graph, we can hold b constant and let a vary. Doing so will give us a *family of curves*. Suppose we set $b = 1$ and then let a take on integer values from -3 to 3.

We will give three methods of graphing this set of equations on a graphing calculator.

Method 1: Y-Variables List

To use the Y-variables list, enter each equation at one of the Y variables, set the graph window, then graph. The calculator will graph the equations in order, starting with Y_1 and ending with Y_7. Following is the Y-variables list, an appropriate window, and a sample of the type of graph obtained (Figure 7).

$$Y_1 = -3x + 1$$
$$Y_2 = -2x + 1$$
$$Y_3 = -x + 1$$
$$Y_4 = 1$$
$$Y_5 = x + 1$$
$$Y_6 = 2x + 1$$
$$Y_7 = 3x + 1$$

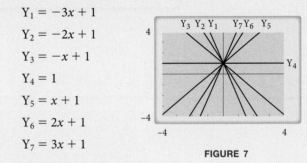

FIGURE 7

Window: X from -4 to 4, Y from -4 to 4

Method 2: Programming

The same result can be obtained by programming your calculator to graph $y = ax + 1$ for $a = -3, -2, -1, 0, 1, 2,$ and 3. Here is an outline of a program that will do this. Check the manual that came with your calculator to find the commands for your calculator.

Step 1: Clear screen
Step 2: Set window for X from -4 to 4 and Y from -4 to 4
Step 3: $-3 \rightarrow A$
Step 4: Label 1
Step 5: Graph Y = AX + 1
Step 6: $A + 1 \rightarrow A$
Step 7: If $A < 4$, go to 1
Step 8: End

Method 3: Using Lists

On the TI-83/84 you can set Y_1 as follows

$$Y_1 = \{-3, -2, -1, 0, 1, 2, 3\} X + 1$$

When you press $\boxed{\text{GRAPH}}$, the calculator will graph each line from $y = -3x + 1$ to $y = 3x + 1$.

Each of the three methods will produce graphs similar to those in Figure 7.

Getting Ready for Class

After reading through the preceding section, respond in your own words and in complete sentences.

A. If you were looking at a graph that described the performance of a stock you had purchased, why would it be better if the slope of the line were positive, rather than negative?

B. Would you rather climb a hill with a slope of $\frac{1}{2}$ or a slope of 3? Explain why.

C. Explain why a vertical line has no slope.

D. Describe how to obtain the slope of a line if you know the coordinates of two points on the line.

Find the slope of each of the following lines from the given graph.

1.

2.

3.

4.

5.

6.

Find the slope of the line through each of the following pairs of points. Then, plot each pair of points, draw a line through them, and indicate the rise and run in the graph in the manner shown in Example 2.

7. $(2, 1), (4, 4)$

8. $(3, 1), (5, 6)$

9. $(1, 4), (5, 2)$

10. $(1, 3), (5, 2)$

11. $(1, -3), (4, 2)$

12. $(2, -4), (5, -9)$

13. $(-3, 5), (-1, -1)$

14. $(-3, 2), (-1, 6)$

15. $(-3, 5), (1, -1)$

16. $(-2, -1), (3, -5)$

17. $(-4, 6), (2, 6)$

18. $(2, -3), (2, 7)$

19. $(-1, -3), (-1, 5)$

20. $(-5, -2), (4, -2)$

Solve for the indicated variable if the line through the two given points has the given slope.

21. $(a, 3)$ and $(2, 6)$, $m = -1$

22. $(a, -2)$ and $(4, -6)$, $m = -3$

23. $(2, b)$ and $(-1, 4b)$, $m = -2$

24. $(-4, y)$ and $(-1, 6y)$, $m = 2$

For each of the equations in Problems 25–28, complete the table, and then use the results to find the slope of the graph of the equation.

25. $2x + 3y = 6$ **26.** $3x - 2y = 6$ **27.** $y = \dfrac{2}{3}x - 5$ **28.** $y = -\dfrac{3}{4}x + 2$

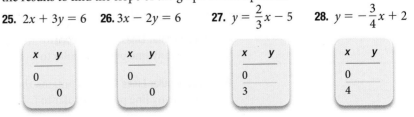

x	y
0	
	0

x	y
0	
	0

x	y
0	
3	

x	y
0	
4	

29. **Finding Slope from Intercepts** Graph the line that has an x-intercept of 3 and a y-intercept of -2. What is the slope of this line?

30. **Finding Slope from Intercepts** Graph the line with x-intercept -4 and y-intercept -2. What is the slope of this line?

31. **Finding Slope from Intercepts** A line has an x-intercept of -1 and no y-intercept. What is the slope of this line?

32. **Finding Slope from Intercepts** A line has a y-intercept of 6 and no x-intercept. What is the slope of this line?

Determine if the lines through each pair of points are parallel, perpendicular, or neither.

33. $(-1, 2)$ and $(1, 3)$, $(2, -2)$ and $(6, 0)$

34. $(-2, -2)$ and $(4, 4)$, $(1, 5)$ and $(3, 3)$

35. $(1, 1)$ and $(2, 4)$, $(-3, 3)$ and $(3, 5)$

36. $(3, -6)$ and $(-1, -3)$, $(-5, 6)$ and $(3, 0)$

37. $(-4, 1)$ and $(5, -5)$, $(-2, 4)$ and $(0, 7)$

38. $(2, 0)$ and $(7, 3)$, $(1, -1)$ and $(-4, 2)$

39. **Parallel Lines** Find the slope of any line parallel to the line through $(2, 3)$ and $(-8, 1)$.

40. **Parallel Lines** Find the slope of any line parallel to the line through $(2, 5)$ and $(5, -3)$.

41. **Perpendicular Lines** Line l contains the points $(5, -6)$ and $(5, 2)$. Give the slope of any line perpendicular to l.

42. **Perpendicular Lines** Line l contains the points $(3, 4)$ and $(-3, 1)$. Give the slope of any line perpendicular to l.

43. **Parallel Lines** Line l contains the points $(-2, 1)$ and $(4, -5)$. Find the slope of any line parallel to l.

44. **Parallel Lines** Line l contains the points $(3, -4)$ and $(-2, -6)$. Find the slope of any line parallel to l.

45. **Perpendicular Lines** Line l contains the points $(-2, -5)$ and $(1, -3)$. Find the slope of any line perpendicular to l.

46. **Perpendicular Lines** Line l contains the points $(6, -3)$ and $(-2, 7)$. Find the slope of any line perpendicular to l.

47. Determine if each of the following tables could represent ordered pairs from an equation of a line.

a.

x	y
0	5
1	7
2	9
3	11

b.

x	y
-2	-5
0	-2
2	0
4	1

48. The following lines have slope 2, $\frac{1}{2}$, 0, and -1. Match each line to its slope value.

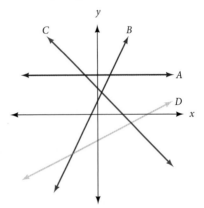

Applying the Concepts

An object is traveling at a constant speed. The distance and time data are shown on the given graph. Use the graph to find the speed of the object.

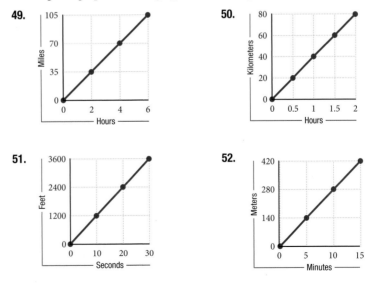

49.

50.

51.

52.

53. **Heating a Block of Ice** A block of ice with an initial temperature of $-20°C$ is heated at a steady rate. The graph shows how the temperature changes as the ice melts to become water and the water boils to become steam and water.

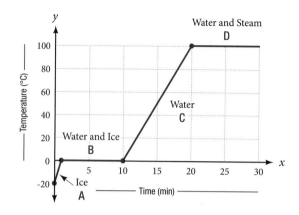

a. How long does it take all the ice to melt?

b. From the time the heat is applied to the block of ice, how long is it before the water boils?

c. Find the slope of the line segment labeled A. What units would you attach to this number?

d. Find the slope of the line segment labeled C. Be sure to attach units to your answer.

e. Is the temperature changing faster during the 1st minute or the 16th minute?

54. **Slope of a Highway** A sign at the top of the Cuesta Grade, outside of San Luis Obispo, reads "7% downgrade next 3 miles." The following diagram is a model of the Cuesta Grade that takes into account the information on that sign.

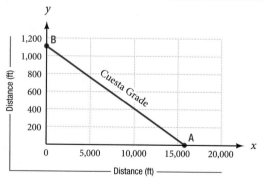

a. At point B, the graph crosses the y-axis at 1,106 feet. How far is it from the origin to point A?

b. What is the slope of the Cuesta Grade?

55. Solar Energy The graph below shows the annual shipments of solar thermal collectors in the United States. Using the graph below, find the slope of the line connecting the first (1997, 8,000) and last (2006, 20,000) endpoints and then explain in words what the slope represents.

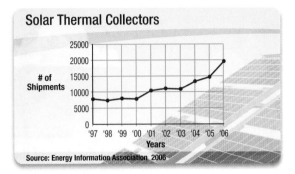

56. Age of New Mothers The graph shows the increase in average age of first time mothers in the U.S. since 1970 Find the slope of the line that connects the points (1975, 21.75) to (1990, 24.25). Round to the nearest hundredth. Explain in words what the slope represents.

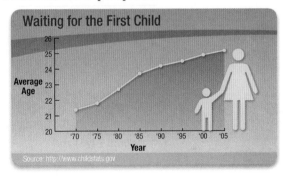

57. Light Bulbs The chart shows a comparison of power usage between incandescent and energy efficient light bulbs. Use the chart to work the following problems involving slope.

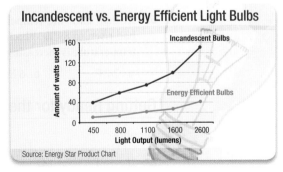

    a. Find the slope of the line for the incandescent bulb from the two endpoints and then explain in words what the slope represents.

    b. Find the slope of the line for the energy efficient bulb from the two endpoints and then explain in words what the slope represents.

    c. Which light bulb is better? Why?

58. Horse Racing The graph shows the amount of money bet on horse racing from 1985 to 2005. Use the chart to work the following problems involving slope.

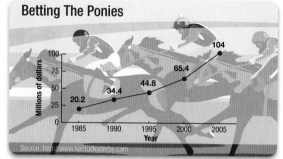

Betting The Ponies

Source: http://www.kentuckyderby.com

a. Find the slope of the line from 1985 to 1990, and then explain in words what the slope represents.

b. Find the slope of the line from 2000 to 2005, and then explain in words what the slope represents.

Learning Objectives Assessment

The following problems can be used to help assess if you have successfully met the learning objectives for this section.

59. Find the slope of the line whose graph is shown in Figure 8.

a. $\dfrac{5}{2}$ **b.** $\dfrac{2}{5}$ **c.** $-\dfrac{5}{2}$ **d.** $-\dfrac{2}{5}$

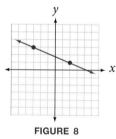

FIGURE 8

60. Find the slope of the line passing through the points $(4, 5)$ and $(-2, -5)$.

a. $\dfrac{3}{5}$ **b.** $\dfrac{5}{3}$ **c.** $-\dfrac{3}{5}$ **d.** $-\dfrac{5}{3}$

61. What would be the slope of a line that is perpendicular to the line passing through $(-4, 0)$ and $(0, 2)$?

a. -2 **b.** 2 **c.** $\dfrac{1}{2}$ **d.** $-\dfrac{1}{2}$

62. At 9:00 a.m. the temperature was $65°F$, and by 3:00 p.m. the temperature had risen to $89°F$. Which of the following gives the average rate at which the temperature is increasing?

a. $24°F$ **b.** 6 hours   **c.** $4°F/hr$ **d.** $0.25°F/hr$

Getting Ready for the Next Section

Simplify.

63. $2\left(-\dfrac{1}{2}\right)$ **64.** $\dfrac{3 - (-1)}{-3 - 3}$ **65.** $-\dfrac{5 - (-3)}{2 - 6}$ **66.** $3\left(-\dfrac{2}{3}x + 1\right)$

Solve for y.

67. $\dfrac{y - b}{x - 0} = m$ **68.** $2x + 3y = 6$

69. $y - 3 = -2(x + 4)$ **70.** $y + 1 = -\dfrac{2}{3}(x - 3)$

71. If $y = -\dfrac{4}{3}x + 5$, find y when x is 0.   **72.** If $y = -\dfrac{4}{3}x + 5$, find y when x is 3.

Learning Objectives

In this section we will learn how to:

1. Use slope-intercept form to graph a line.

2. Use slope-intercept form to find the equation of a line.

3. Use point-slope form to find the equation of a line.

4. Write the equation of a line in standard form.

Introduction

The table and illustrations below show some corresponding temperatures on the Fahrenheit and Celsius temperature scales. For example, water freezes at 32°F and 0°C, and boils at 212°F and 100°C.

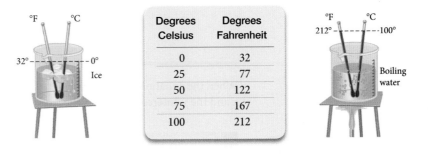

Degrees Celsius	Degrees Fahrenheit
0	32
25	77
50	122
75	167
100	212

If we plot all the points in the table using the x-axis for temperatures on the Celsius scale and the y-axis for temperatures on the Fahrenheit scale, we see that they line up in a straight line (Figure 1).

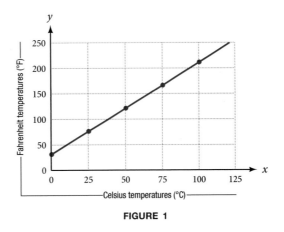

FIGURE 1

This means that a linear equation in two variables will give a perfect description of the relationship between the two scales. That equation is

$$F = \frac{9}{5}C + 32$$

The techniques we use to find the equation of a line from a set of points is what this section is all about.

Suppose line l has slope m and y-intercept b. What is the equation of l? Because the y-intercept is b, we know the point $(0, b)$ is on the line. If (x, y) is any other point on l, then using the definition for slope, we have

$$\frac{y - b}{x - 0} = m \qquad \text{Definition of Slope}$$

$$y - b = mx \qquad \text{Multiply both sides by } x$$

$$y = mx + b \qquad \text{Add } b \text{ to both sides}$$

Slope-Intercept Form

This last equation is known as the ***slope-intercept form*** of the equation of a line.

> ⟨Δ≠Σ⟩ **PROPERTY**   *Slope-Intercept Form of the Equation of a Line*
>
> The equation of any line with slope m and y-intercept b is given by
>
> $$y = mx + b$$
>
>   ↗   ↑
>
>   Slope y-intercept

When the equation is in this form, the ***slope*** of the line is always the ***coefficient*** of x and the y-intercept is always the ***constant term***.

VIDEO EXAMPLES

SECTION 3.3

EXAMPLE 1   Graph the line $y = -\frac{2}{3}x + 2$ using the slope and y-intercept.

SOLUTION   The slope is $m = -\frac{2}{3}$ and the y-intercept is $b = 2$. Therefore, the point $(0, 2)$ is on the graph, and the ratio of rise to run going from $(0, 2)$ to any other point on the line is $-\frac{2}{3}$. If we interpret $-\frac{2}{3}$ as $\frac{-2}{3}$, we can start at $(0, 2)$ and move 2 units down (a rise of -2) and 3 units to the right (a run of 3) to locate another point on the graph. Or, we can interpret $-\frac{2}{3}$ as $\frac{2}{-3}$ and move 2 units up (a rise of 2) and 3 units to the left (a run of -3) from $(0, 2)$ to locate another point. The resulting graph is shown in Figure 2.

Note   As we mentioned earlier in this chapter, the rectangular coordinate system is the tool we use to connect algebra and geometry. Example 1 illustrates this connection, as do the many other examples in this chapter. In Example 1, Descartes's rectangular coordinate system allows us to associate the equation $y = -\frac{2}{3}x + 2$ (an algebraic concept) with the line (a geometric concept) shown in Figure 2.

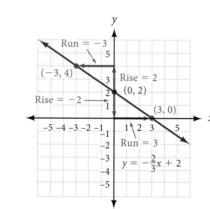

FIGURE 2

EXAMPLE 2   Give the slope and y-intercept for the line $2x - 3y = 5$.

SOLUTION   To use the slope-intercept form, we must solve the equation for y in terms of x:

$$2x - 3y = 5$$
$$-3y = -2x + 5 \qquad \text{Add } -2x \text{ to both sides}$$
$$y = \frac{2}{3}x - \frac{5}{3} \qquad \text{Divide by } -3$$

The last equation has the form $y = mx + b$. The slope must be $m = \frac{2}{3}$ and the y-intercept is $b = -\frac{5}{3}$.   ∎

EXAMPLE 3   Find the equation of the line with slope $-\frac{4}{3}$ and y-intercept 5.

SOLUTION   Substituting $m = -\frac{4}{3}$ and $b = 5$ into the equation $y = mx + b$, we have

$$y = -\frac{4}{3}x + 5$$

Finding the equation from the slope and y-intercept is just that easy. If the slope is m and the y-intercept is b, then the equation is always $y = mx + b$.   ∎

EXAMPLE 4   Find the equation of the line with slope $m = 2$ and passing through the point $(1, -3)$.

SOLUTION   We can find the equation using the slope-intercept form. However, because $(1, -3)$ is not on the y-axis, it is not a y-intercept. That is, we cannot assume $b = -3$. But we can use the slope and given point to solve for b.

Using	$m = 2, x = 1, \text{ and } y = -3$	
in	$y = mx + b$	Slope-intercept form
gives us	$-3 = 2(1) + b$	
	$-3 = 2 + b$	Simplify
	$-5 = b$	Add -2 to both sides

Therefore, the equation of the line is $y = 2x - 5$.   ∎

Point-Slope Form

A second useful form of the equation of a line is point-slope form.

Let line l contain the point (x_1, y_1) and have slope m. If (x, y) is any other point on l, then by the definition of slope we have

$$\frac{y - y_1}{x - x_1} = m$$

Multiplying both sides by $(x - x_1)$ gives us

$$(x - x_1) \cdot \frac{y - y_1}{x - x_1} = m(x - x_1)$$
$$y - y_1 = m(x - x_1)$$

This last equation is known as the ***point-slope form*** of the equation of a line.

> **⟨Δ≠Σ⟩ PROPERTY**   *Point-Slope Form of the Equation of a Line*
>
> The equation of the line through (x_1, y_1) with slope m is given by
>
> $$y - y_1 = m(x - x_1)$$

This form of the equation of a line is used to find the equation of a line, either given one point on the line and the slope, or given two points on the line.

EXAMPLE 5   Find the equation of the line with slope -2 that contains the point $(-4, 3)$. Write the answer in slope-intercept form.

SOLUTION

Using	$(x_1, y_1) = (-4, 3)$ and $m = -2$	
in	$y - y_1 = m(x - x_1)$	Point-slope form
gives us	$y - 3 = -2(x + 4)$	Note: $x - (-4) = x + 4$
	$y - 3 = -2x - 8$	Multiply out right side
	$y = -2x - 5$	Add 3 to each side

Figure 3 is the graph of the line that contains $(-4, 3)$ and has a slope of -2. Notice that the y-intercept on the graph matches that of the equation we found.

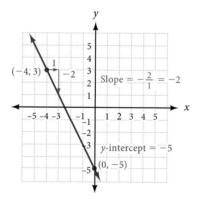

FIGURE 3

Note   We could have used the point $(-3, 3)$ instead of $(3, -1)$ and obtained the same equation. That is, using $(x_1, y_1) = (-3, 3)$ and $m = -\frac{2}{3}$ in

$$y - y_1 = m(x - x_1)$$

gives us

$$y - 3 = -\frac{2}{3}(x + 3)$$

$$y - 3 = -\frac{2}{3}x - 2$$

$$y = -\frac{2}{3}x + 1$$

which is the same result we obtained using $(3, -1)$.

EXAMPLE 6   Find the equation of the line that passes through the points $(-3, 3)$ and $(3, -1)$.

SOLUTION   We begin by finding the slope of the line:

$$m = \frac{3 - (-1)}{-3 - 3} = \frac{4}{-6} = -\frac{2}{3}$$

Using $(x_1, y_1) = (3, -1)$ and $m = -\frac{2}{3}$ in $y - y_1 = m(x - x_1)$ yields

$$y + 1 = -\frac{2}{3}(x - 3)$$

$$y + 1 = -\frac{2}{3}x + 2 \qquad \text{Multiply out right side}$$

$$y = -\frac{2}{3}x + 1 \qquad \text{Add } -1 \text{ to each side}$$

Figure 4 shows the graph of the line that passes through the points $(-3, 3)$ and $(3, -1)$. As you can see, the slope and y-intercept are $-\frac{2}{3}$ and 1, respectively.

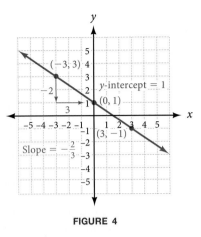

FIGURE 4

Standard Form

The last form of the equation of a line that we will consider in this section is called the **standard form**. It is used mainly to write equations in a form that is free of fractions and is easy to compare with other equations.

> **[Δ≠Σ]** **PROPERTY** *Standard Form for the Equation of a Line*
>
> If a, b, and c are integers with $a \geq 0$, then the equation of a line is in standard form when it has the form
>
> $$ax + by = c$$

If we were to write the equation

$$y = -\frac{2}{3}x + 1$$

in standard form, we would first multiply both sides by 3 to obtain

$$3y = -2x + 3$$

Then we would add $2x$ to each side, yielding

$$2x + 3y = 3$$

which is a linear equation in standard form.

EXAMPLE 7   Give the equation of the line through $(-1, 4)$ whose graph is perpendicular to the graph of $2x - y = -3$. Write the answer in standard form.

SOLUTION   To find the slope of $2x - y = -3$, we solve for y:

$$2x - y = -3$$
$$y = 2x + 3$$

The slope of this line is 2. The line we are interested in is perpendicular to the line with slope 2 and must, therefore, have a slope of $-\frac{1}{2}$.

Using $(x_1, y_1) = (-1, 4)$ and $m = -\frac{1}{2}$, we have

$$y - y_1 = m(x - x_1)$$

$$y - 4 = -\frac{1}{2}(x + 1)$$

Because we want our answer in standard form, we multiply each side by 2.

$$2y - 8 = -1(x + 1)$$

$$2y - 8 = -x - 1$$

$$x + 2y - 8 = -1$$

$$x + 2y = 7$$

The last equation is in standard form.

As a final note, the following summary reminds us that all horizontal lines have equations of the form $y = b$, and slopes of 0. Since they cross the y-axis at b, the y-intercept is b; there is no x-intercept. Vertical lines have no slope, and equations of the form $x = a$. Each will have an x-intercept at a, and no y-intercept. Finally, equations of the form $y = mx$ have graphs that pass through the origin. The slope is always m and both the x-intercept and the y-intercept are 0.

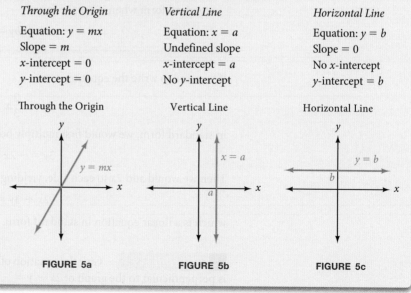

FACTS FROM GEOMETRY

Special Equations: Their Graphs, Slopes, and Intercepts

For the equations below, m, a, and b are real numbers.

Through the Origin	*Vertical Line*	*Horizontal Line*
Equation: $y = mx$	Equation: $x = a$	Equation: $y = b$
Slope $= m$	Undefined slope	Slope $= 0$
x-intercept $= 0$	x-intercept $= a$	No x-intercept
y-intercept $= 0$	No y-intercept	y-intercept $= b$

Through the Origin   Vertical Line   Horizontal Line

FIGURE 5a   FIGURE 5b   FIGURE 5c

EXAMPLE 8   Find the equation of the line passing through the points $(-2, -3)$ and $(-2, 5)$.

SOLUTION   First, we find the slope of the line.

$$m = \frac{5 - (-3)}{-2 - (-2)} = \frac{8}{0}, \text{ which is undefined}$$

Because there is no slope, we cannot use slope-intercept form or point-slope form. A line having no slope must be a vertical line. The graph of the line is shown in Figure 6.

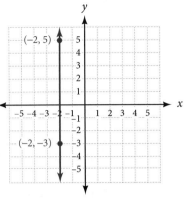

FIGURE 6

The equation of a vertical line is $x = a$. Since $a = -2$, we have $x = -2$.

USING TECHNOLOGY   *Graphing Calculators*

One advantage of using a graphing calculator to graph lines is that a calculator does not care whether the equation has been simplified or not. To illustrate, in Example 5 we found that the equation of the line with slope $-\frac{2}{3}$ that passes through the point $(3, -1)$ is

$$y + 1 = -\frac{2}{3}(x - 3)$$

Normally, to graph this equation we would simplify it first. With a graphing calculator, we add -1 to each side and enter the equation this way:

$$Y_1 = -(2/3)(X - 3) - 1$$

No simplification is necessary. We can graph the equation in this form, and the graph will be the same as the simplified form of the equation, which is $y = -\frac{2}{3}x + 1$. To convince yourself that this is true, graph both the simplified form for the equation and the unsimplified form in the same window. As you will see, the two graphs coincide.

Getting Ready for Class

After reading through the preceding section, respond in your own words and in complete sentences.

A. How would you graph the line $y = \frac{1}{2}x + 3$?

B. What is the slope-intercept form of the equation of a line?

C. Describe how you would find the equation of a line if you knew the slope and the y-intercept of the line.

D. If you had the graph of a line, how would you use it to find the equation of the line?

Problem Set 3.3

Give the equation of the line with the following slope and y-intercept.

1. $m = -4, b = -3$ **2.** $m = -6, b = \dfrac{4}{3}$ **3.** $m = -\dfrac{2}{3}, b = 0$

4. $m = 0, b = \dfrac{3}{4}$ **5.** $m = -\dfrac{2}{3}, b = \dfrac{1}{4}$ **6.** $m = \dfrac{5}{12}, b = -\dfrac{3}{2}$

Find the slope of a line **a.** parallel and **b.** perpendicular to the given line.

7. $y = 3x - 4$ **8.** $y = -4x + 1$ **9.** $3x + y = -2$

10. $2x - y = -4$ **11.** $2x + 5y = -11$ **12.** $3x - 5y = -4$

Give the slope and y-intercept for each of the following equations. Sketch the graph using the slope and y-intercept. Give the slope of any line perpendicular to the given line.

13. $y = 3x - 2$ **14.** $y = 2x + 3$ **15.** $2x - 3y = 12$

16. $3x - 2y = 12$ **17.** $4x + 5y = 20$ **18.** $5x - 4y = 20$

For each of the following lines, name the slope and y-intercept. Then write the equation of the line in slope-intercept form.

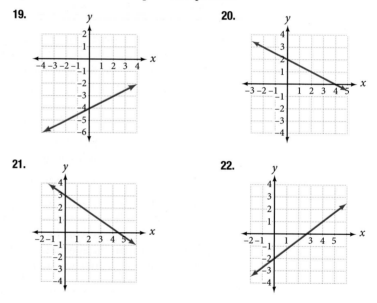

19.

20.

21.

22.

For each of the following problems, the slope and one point on the line are given. In each case, find the equation of that line and write your answer in slope-intercept form.

23. $(-2, -5); m = 2$ **24.** $(-1, -5); m = 2$ **25.** $(-4, 1); m = -\dfrac{1}{2}$

26. $(-2, 1); m = -\dfrac{1}{2}$ **27.** $\left(-\dfrac{1}{3}, 2\right); m = -3$ **28.** $\left(-\dfrac{2}{3}, 5\right); m = -3$

29. $(-4, 2), m = \dfrac{2}{3}$ **30.** $(3, -4), m = -\dfrac{1}{3}$ **31.** $(-5, -2), m = -\dfrac{1}{4}$

32. $(-4, -3), m = \dfrac{1}{6}$ **33.** $(5, 6), m = 0$ **34.** $(-8, -1), m = 0$

Find the equation of the line that passes through each pair of points. Write your answers in standard form.

35. $(3, -2), (-2, 1)$ **36.** $(-4, 1), (-2, -5)$ **37.** $\left(-2, \frac{1}{2}\right), \left(-4, \frac{1}{3}\right)$

38. $(-6, -2), (-3, -6)$ **39.** $\left(\frac{1}{3}, -\frac{1}{5}\right), \left(-\frac{1}{3}, -1\right)$ **40.** $\left(-\frac{1}{2}, -\frac{1}{2}\right), \left(\frac{1}{2}, \frac{1}{10}\right)$

41. $\left(\frac{3}{4}, 2\right), \left(-\frac{1}{6}, 2\right)$ **42.** $\left(1, -\frac{2}{3}\right), \left(-4, -\frac{2}{3}\right)$ **43.** $\left(-\frac{4}{3}, 1\right), \left(-\frac{4}{3}, 3\right)$

44. $(5, 3), (5, -1)$

For each of the following lines, name the coordinates of any two points on the line. Then use those two points to find the equation of the line. Write your answer in slope-intercept form.

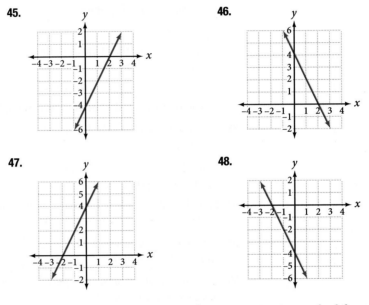

45.

46.

47.

48.

49. The equation $3x - 2y = 10$ is a linear equation in standard form. From this equation, answer the following:
  a. Find the x- and y-intercepts
  b. Find a solution to this equation other than the intercepts in part a.
  c. Write this equation in slope-intercept form
  d. Is the point $(2, 2)$ a solution to the equation?

50. The equation $4x + 3y = 8$ is a linear equation in standard form. From this equation, answer the following:
  a. Find the x- and y-intercepts
  b. Find a solution to this equation other than the intercepts in part a.
  c. Write this equation in slope-intercept form
  d. Is the point $(-3, 2)$ a solution to the equation?

51. Paying Attention to Instructions Work each problem according to the instructions given:

a. Solve: $-2x + 1 = -3$

b. Find x when y is 0: $-2x + y = -3$

c. Find y when x is 0: $-2x + y = -3$

d. Graph: $-2x + y = -3$

e. Solve for y: $-2x + y = -3$

52. Paying Attention to Instructions Work each problem according to the instructions given:

a. Solve: $\frac{x}{3} + \frac{1}{4} = 1$

b. Find x when y is 0: $\frac{x}{3} + \frac{y}{4} = 1$

c. Find y when x is 0: $\frac{x}{3} + \frac{y}{4} = 1$

d. Graph: $\frac{x}{3} + \frac{y}{4} = 1$

e. Solve for y: $\frac{x}{3} + \frac{y}{4} = 1$

53. Graph each of the following lines. In each case, name the slope, the x-intercept, and the y-intercept.

a. $y = \frac{1}{2}x$ **b.** $x = 3$ **c.** $y = -2$

54. Graph each of the following lines. In each case, name the slope, the x-intercept, and the y-intercept.

a. $y = -2x$ **b.** $x = 2$ **c.** $y = -4$

For problems 55-62, write the equation of the line in standard form.

55. Find the equation of the line parallel to the graph of $3x - y = 5$ that contains the point $(-1, 4)$.

56. Find the equation of the line parallel to the graph of $2x - 4y = 5$ that contains the point $(0, 3)$.

57. Line l is perpendicular to the graph of the equation $2x - 5y = 10$ and contains the point $(-4, -3)$. Find the equation for l.

58. Line l is perpendicular to the graph of the equation $-3x - 5y = 2$ and contains the point $(2, -6)$. Find the equation for l.

59. Give the equation of the line perpendicular to the graph of $y = -4x + 2$ that has an x-intercept of -1.

60. Write the equation of the line parallel to the graph of $7x - 2y = 14$ that has an x-intercept of 5.

61. Find the equation of the line with x-intercept 3 and y-intercept 2.

62. Find the equation of the line with x-intercept 2 and y-intercept 3.

Applying the Concepts

63. Deriving the Temperature Equation The table below resembles the table from the introduction to this section. The rows of the table give us ordered pairs (C, F).

Degrees Celsius	Degrees Fahrenheit
C	F
0	32
25	77
50	122
75	167
100	212

a. Use any two of the ordered pairs from the table to derive the equation $F = \frac{9}{5}C + 32$.

b. Use the equation from part *a* to find the Fahrenheit temperature that corresponds to a Celsius temperature of 30°.

64. Maximum Heart Rate The table below gives the maximum heart rate for adults 30, 40, 50, and 60 years old. Each row of the table gives us an ordered pair (A, M).

Age (years)	Maximum Heart Rate (beats per minute)
A	M
30	190
40	180
50	170
60	160

a. Use any two of the ordered pairs from the table to derive the equation $M = 220 - A$, which gives the maximum heart rate M for an adult whose age is A.

b. Use the equation from part *a* to find the maximum heart rate for a 25-year-old adult.

65. Spring Length The table below shows the data from the table in the introduction to this chapter, where w is the weight applied to a spring and s is the amount the spring stretches as a result of that weight. Each row of the table gives us an ordered pair (w, s).

Weight (pounds)	Stretch (cm)
w	s
2	2.8
4	5.6
6	8.4
8	11.2

a. Use any two ordered pairs from the table to derive the equation $s = 1.4w$.

b. Use your equation to find the amount by which the spring will stretch if 11 pounds of weight is added to the spring.

66. Spring Length The table below gives the length of a spring after various weights are added onto the spring. Each row of the table gives us an ordered pair (W, L).

Weight (pounds)	Length of Spring (cm)
W	L
0.5	5.95
1	7.3
1.5	8.65
2	10

a. Use any two ordered pairs from the table to find the equation of the line passing through all four points. Write your answer so that L is expressed in terms of W.

b. What is the normal length of the spring (with no weight attached)?

67. Textbook Cost To produce this textbook, suppose the publisher spent $125,000 for typesetting and $6.50 per book for printing and binding. The total cost to produce and print n books can be written as

$$C = 125{,}000 + 6.5n$$

a. Suppose the number of books printed in the first printing is 10,000. What is the total cost?

b. If the average cost is the total cost divided by the number of books printed, find the average cost of producing 10,000 textbooks.

c. Find the cost to produce one more textbook when you have already produced 10,000 textbooks.

68. Exercise Heart Rate In an aerobics class, the instructor indicates that her students' exercise heart rate is 60% of their maximum heart rate, where maximum heart rate is 220 minus their age.

a. Determine the equation that gives exercise heart rate E in terms of age A.

b. Use the equation to find the exercise heart rate of a 22-year-old student.

c. Sketch the graph of the equation for students from 18 to 80 years of age.

69. Horse Racing The graph shows the total amount of money wagered on the Kentucky Derby. Find an equation for the line segment between 2000 and 2005. Write the equation in slope-intercept form. Round the slope and intercept to the nearest tenth.

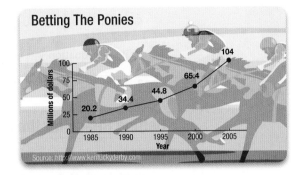

Betting The Ponies

Millions of dollars

Source: http://www.kentuckyderby.com

70. Solar Energy The graph shows the annual number of solar thermal collector shipments in the United States. Find an equation for the line segment that connects the points (2004, 13,750) and (2005, 15,000). Write your answer in slope-intercept form.

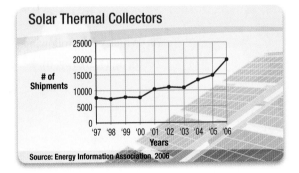

Solar Thermal Collectors

# of Shipments

Source: Energy Information Association, 2006

Learning Objectives Assessment

The following problems can be used to help assess if you have successfully met the learning objectives for this section.

71. In graphing the equation $y = \frac{2}{5}x - 1$, which of the following correctly describes how to locate a second point on the line?

   a. From $(0, -1)$, go up 5 units and right 2 units.

   b. From $(-1, 0)$, go up 5 units and right 2 units.

   c. From $(0, -1)$, go up 2 units and right 5 units.

   d. From $(-1, 0)$, go up 2 units and right 5 units.

72. Use the slope-intercept form to find the equation of the line with slope $m = -3$ and passing through the point $(0, 4)$.

   a. $x = 4y - 3$ **b.** $x = -3y + 4$ **c.** $y = 4x - 3$ **d.** $y = -3x + 4$

73. Use the point-slope form to find the equation of the line passing through $(-2, 3)$ and $(1, -1)$.

a. $y + 1 = -\dfrac{3}{4}(x - 1)$ **b.** $y + 2 = -\dfrac{3}{4}(x - 3)$

c. $y + 2 = -\dfrac{4}{3}(x - 3)$ **d.** $y + 1 = -\dfrac{4}{3}(x - 1)$

74. Write the equation $y = -\dfrac{3}{2}x + \dfrac{5}{4}$ in standard form.

a. $3x + 4y = 5$ **b.** $6x + 4y = 5$ **c.** $4y - 6x = 5$ **d.** $-3x - 4y = 5$

Getting Ready for the Next Section

75. Which of the following are solutions to $x + y \le 4$?

$(0, 0), (4, 0), (2, 3)$

76. Which of the following are solutions to $y < 2x - 3$?

$(0, 0), (3, -2), (-3, 2)$

77. Which of the following are solutions to $y \le \dfrac{1}{2}x$?

$(0, 0), (2, 0), (-2, 0)$

78. Which of the following are solutions to $y > -2x$?

$(0, 0), (2, 0), (-2, 0)$

Linear Inequalities in Two Variables

Learning Objective

In this section we will learn how to:

1. Graph the solution set for a linear inequality in two variables.

Introduction

A small movie theater holds 100 people. The owner charges more for adults than for children, so it is important to know the different combinations of adults and children that can be seated at one time. The shaded region in Figure 1 contains all the seating combinations. The line $x + y = 100$ shows the combinations for a full theater: The y-intercept corresponds to a theater full of adults, and the x-intercept corresponds to a theater full of children. In the shaded region below the line $x + y = 100$ are the combinations that occur if the theater is not full.

Shaded regions like the one shown in Figure 1 are produced by linear inequalities in two variables, which is the topic of this section.

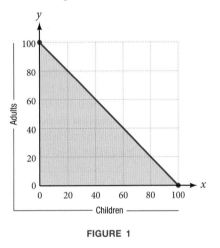

FIGURE 1

Linear Inequalities in Two Variables

A **_linear inequality in two variables_** is any expression that can be put in the form

$$ax + by < c$$

where a, b, and c are real numbers (a and b not both 0). The inequality symbol can be any one of the following four: $<, \leq, >, \geq$.

Some examples of linear inequalities are

$$2x + 3y < 6 \qquad y \geq 2x + 1 \qquad x - y \leq 0$$

Although not all of these examples have the form $ax + by < c$, each one can be put in that form.

The solution set for a linear inequality is a **_section of the coordinate plane_**. The **_boundary_** for the section is found by replacing the inequality symbol with an equal sign and graphing the resulting equation. The boundary is included in the solution set (and is represented with a _solid line_) if the inequality symbol used originally is $\leq$ or $\geq$. The boundary is not included (and is represented with a _broken line_) if the original symbol is $<$ or $>$.

EXAMPLE 1   Graph the solution set for $x + y \leq 4$.

SOLUTION   The boundary for the graph is the graph of $x + y = 4$. The boundary is included in the solution set because the inequality symbol is $\leq$.
Figure 2 is the graph of the boundary:

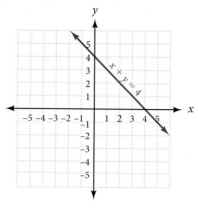

FIGURE 2

The boundary separates the coordinate plane into two regions: the region above the boundary and the region below it. The solution set for $x + y \leq 4$ is one of these two regions along with the boundary. To find the correct region, we simply choose any convenient point that is *not* on the boundary. We then substitute the coordinates of the point into the original inequality $x + y \leq 4$. If the point we choose satisfies the inequality, then it is a member of the solution set, and we can assume that all points on the same side of the boundary as the chosen point are also in the solution set. If the coordinates of our point do not satisfy the original inequality, then the solution set lies on the other side of the boundary.

In this example, a convenient point that is not on the boundary is the origin.

Substituting	$(0, 0)$
into	$x + y \leq 4$
gives us	$0 + 0 \leq 4$
	$0 \leq 4$ A true statement

Because the origin is a solution to the inequality $x + y \leq 4$ and the origin is below the boundary, all other points below the boundary are also solutions. We indicate this by shading the region below the boundary.

Figure 3 is the graph of $x + y \leq 4$.

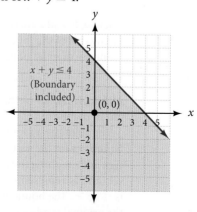

FIGURE 3

Here is a list of steps to follow when graphing the solution sets for linear inequalities in two variables.

> **HOW TO**   *Graph a Linear Inequality in Two Variables*
>
> **Step 1:** Replace the inequality symbol with an equal sign. The resulting equation represents the boundary for the solution set.
> **Step 2:** Graph the boundary found in step 1. Use a *solid line* if the boundary is included in the solution set (i.e., if the original inequality symbol was either $\leq$ or $\geq$). Use a *broken line* to graph the boundary if it is *not* included in the solution set. (It is not included if the original inequality was either $<$ or $>$.)
> **Step 3:** Choose any convenient point not on the boundary and substitute the coordinates into the *original* inequality. If the resulting statement is *true*, the solution set lies on the *same* side of the boundary as the chosen point. If the resulting statement is *false*, the solution set lies on the *opposite* side of the boundary.
> **Step 4:** Shade the region on the appropriate side of the boundary to indicate the solution set.

EXAMPLE 2   Graph the solution set for $y < 2x - 3$.

SOLUTION The boundary is the graph of $y = 2x - 3$, a line with slope 2 and y-intercept -3. The boundary is not included because the original inequality symbol is $<$. We therefore use a broken line to represent the boundary in Figure 4.

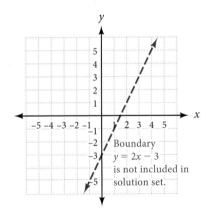

FIGURE 4

A convenient test point is again the origin:

Using	$(0, 0)$
in	$y < 2x - 3$
we have	$0 < 2(0) - 3$
	$0 < -3$ A false statement

Because our test point gives us a false statement and it lies above the boundary, the solution set must lie on the other side of the boundary. Therefore, we shade the region below the boundary. The graph of the solution set is shown in Figure 5.

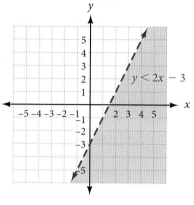

$$y < 2x - 3$$

FIGURE 5

USING TECHNOLOGY   *Graphing Calculators*

Most graphing calculators have a Shade command that allows a portion of a graphing screen to be shaded. With this command we can visualize the solution sets to linear inequalities in two variables. Because most graphing calculators cannot draw a dotted line, however, we are not actually "graphing" the solution set, only visualizing it.

Strategy for Visualizing a Linear Inequality in Two Variables on a Graphing Calculator

Step 1: Solve the inequality for y.

Step 2: Replace the inequality symbol with an equal sign. The resulting equation represents the boundary for the solution set.

Step 3: Graph the equation in an appropriate viewing window.

Step 4: Use the Shade command to indicate the solution set:
For inequalities having the $<$ or $\leq$ sign, use Shade (Xmin, Y_1).
For inequalities having the $>$ or $\geq$ sign, use Shade (Y_1, Xmax).
Figures 6 and 7 show the graphing calculator screens that help us visualize the solution set to the inequality $y < 2x - 3$ that we graphed in Example 2.

Windows: X from -5 to 5, Y from -5 to 5

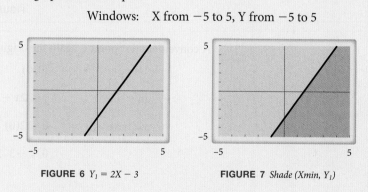

FIGURE 6 $Y_1 = 2X - 3$ **FIGURE 7** *Shade (Xmin, Y_1)*

EXAMPLE 3   Graph the solution set for $x \leq 5$.

SOLUTION   The boundary is $x = 5$, which is a vertical line. All points in Figure 8 to the left have x-coordinates less than 5 and all points to the right have x-coordinates greater than 5.

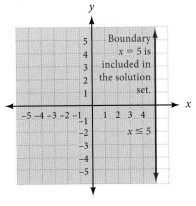

Boundary $x = 5$ is included in the solution set.

$x \leq 5$

FIGURE 8

Getting Ready for Class

After reading through the preceding section, respond in your own words and in complete sentences.

A. When graphing a linear inequality in two variables, how do you find the equation of the boundary line?

B. What is the significance of a broken line in the graph of an inequality?

C. When graphing a linear inequality in two variables, how do you know which side of the boundary line to shade?

D. Describe the set of ordered pairs that are solutions to $x + y < 6$.

Problem Set 3.4

Graph the solution set for each of the following.

1. $x + y < 5$

2. $x + y \leq 5$

3. $x - y \geq -3$

4. $x - y > -3$

5. $2x + 3y < 6$

6. $2x - 3y > -6$

7. $-x + 2y > -4$

8. $-x - 2y < 4$

9. $2x + y < 5$

10. $2x + y < -5$

11. $y < 2x - 1$

12. $y \leq 2x - 1$

13. $3x - 4y < 12$

14. $-2x + 3y < 6$

15. $-5x + 2y \leq 10$

16. $4x - 2y \leq 8$

For each graph shown here, name the linear inequality in two variables that is represented by the shaded region.

17.

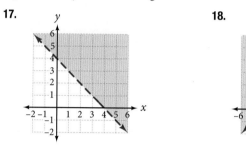

18.

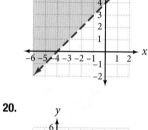

19.

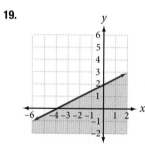

20.

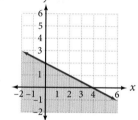

Graph each inequality.

21. $x \geq 3$

22. $x > -2$

23. $y \leq 4$

24. $y > -5$

25. $y < 2x$

26. $y > -3x$

27. $y \geq \dfrac{1}{2}x$

28. $y \leq \dfrac{1}{3}x$

29. $y \geq \dfrac{3}{4}x - 2$

30. $y > -\dfrac{2}{3}x + 3$

31. $\dfrac{x}{3} + \dfrac{y}{2} > 1$

32. $\dfrac{x}{5} + \dfrac{y}{4} < 1$

33. $\dfrac{x}{3} - \dfrac{y}{2} > 1$

34. $-\dfrac{x}{4} - \dfrac{y}{3} > 1$

35. $y \leq -\dfrac{2}{3}x$

36. $y \geq \dfrac{1}{4}x - 1$

37. $5x - 3y < 0$

38. $2x + 3y > 0$

39. $\dfrac{x}{4} + \dfrac{y}{5} \leq 1$

40. $\dfrac{x}{2} + \dfrac{y}{3} < 1$

Applying the Concepts

41. Number of People in a Dance Club A dance club holds a maximum of 200 people. The club charges one price for students and a higher price for nonstudents. If the number of students in the club at any time is x and the number of nonstudents is y, sketch the graph and shade the region in the first quadrant that contains all combinations of students and nonstudents that are in the club at any time.

42. Many Perimeters Suppose you have 500 feet of fencing that you will use to build a rectangular livestock pen. Let x represent the length of the pen and y represent the width. Sketch the graph and shade the region in the first quadrant that contains all possible values of x and y that will give you a rectangle from 500 feet of fencing. (You don't have to use all of the fencing, so the perimeter of the pen could be less than 500 feet.)

43. Gas Mileage You have two cars. The first car travels an average of 12 miles on a gallon of gasoline, and the second averages 22 miles per gallon. Suppose you can afford to buy up to 30 gallons of gasoline this month. If the first car is driven x miles this month, and the second car is driven y miles this month, sketch the graph and shade the region in the first quadrant that gives all the possible values of x and y that will keep you from buying more than 30 gallons of gasoline this month.

44. Student Loan Payments When considering how much debt to incur in student loans, it is advisable to keep your student loan payment after graduation to 8% or less of your starting monthly income. Let x represent your starting monthly salary and let y represent your monthly student loan payment. Write an inequality that describes this situation. Sketch the graph and shade the region in the first quadrant that is a solution to your inequality.

Learning Objectives Assessment

The following problems can be used to help assess if you have successfully met the learning objectives for this section.

45. In graphing the solution set for $2x - 5y > 6$, the boundary should be drawn as a

   a. solid line

   b. broken line

46. The solution set for $2x - 5y > 6$ is the region of the plane

   a. above the boundary line

   b. below the boundary line

Getting Ready for the Next Section

Complete each table using the given equation.

47. $y = 7.5x$

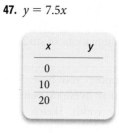

x	y
0	
10	
20	

48. $h = 32t - 16t^2$

t	h
0	
1	
−1	

49. $y = 7.5x$

x	y
0	
$\frac{1}{2}$	
1	

50. $h = 32t - 16t^2$

t	h
−3	
0	
3	

Introduction to Functions

Learning Objectives

In this section we will learn how to:

1. Differentiate between a relation and a function.

2. Find the domain and range for a relation or function.

3. Use the vertical line test to identify if a graph represents a function.

4. Sketch the graph of a relation or function.

Introduction

Kaitlin was given an assignment in her math class. For a particular number, which we will call x, she was to ask her friends two questions:

1. What is the square of x?

2. Can you tell me a number whose square is x?

When Kaitlin performed this experiment on her friends, she found that their answer to the first question was always predictable. For example, if $x = 4$, her friends all gave an answer of 16. However, when she asked for a number whose square is 4, even though most of her friends gave an answer of 2, sometimes they would say -2. She found she could not reliably predict which answer her friends would give to the second question.

If we let y represent the answer given to each question, then we can rephrase the two questions as follows:

1. What is y if $y = x^2$?

2. What is y if $y^2 = x$?

The tables below show the results of Kaitlin's experiment.

TABLE 1 $y = x^2$

x	y
1	1
4	16
9	81

TABLE 2 $y^2 = x$

x	y
1	1
1	-1
4	2
4	-2
9	3
9	-3

Both of these tables express a relationship between values of x and y, and therefore both are an example of a *relation*. The first table is also an example of a *function*. It is the predictable nature of functions that makes them so useful as models in describing relationships in the real world.

Relations

Any relationship between two sets that can be described in terms of ordered pairs is called a relation.

> **(def) DEFINITION** *relation*
>
> A *relation* is a rule of correspondence that pairs each element in one set, called the domain, with **one or more elements** from a second set, called the **range.**

> **(def) ALTERNATE DEFINITION** *relation*
>
> A *relation* is a set of ordered pairs. The set of all first coordinates is the **domain** of the relation. The set of all second coordinates is the **range** of the relation.

VIDEO EXAMPLES

SECTION 3.5

EXAMPLE 1   State the domain and range for the relation

$$\{(1, -1), (1, 1), (4, -2), (4, 2), (9, -3), (9, 3)\}$$

SOLUTION   This relation consists of six ordered pairs. The domain is the set of all first coordinates and the range is the set of all second coordinates. Thus, we have

Domain $= \{1, 4, 9\}$

Range $= \{-3, -2, -1, 1, 2, 3\}$

EXAMPLE 2   Table 3 shows the prices of used Ford Mustangs that were listed in the local newspaper. Graph this relation, and give the domain and range.

TABLE 3   Used Mustang Prices

Year x	Price ($) y
1997	13,925
1997	11,850
1997	9,995
1996	10,200
1996	9,600
1995	9,525
1994	8,675
1994	7,900
1993	6,975

SOLUTION   The graph is shown in Figure 1. This type of graph is called a *scatter diagram*.

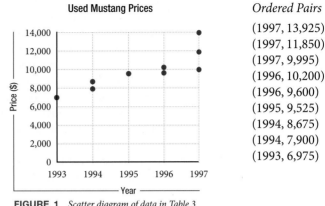

Used Mustang Prices

Ordered Pairs

(1997, 13,925)
(1997, 11,850)
(1997, 9,995)
(1996, 10,200)
(1996, 9,600)
(1995, 9,525)
(1994, 8,675)
(1994, 7,900)
(1993, 6,975)

FIGURE 1   *Scatter diagram of data in Table 3*

The domain is the set of all years listed in Table 3 and the range is the set of the listed prices.

Domain = {1993, 1994, 1995, 1996, 1997}
Range = {6,975, 7,900, 8,675, 9,525, 9,995, 10,200, 11,850, 13,925}

Sometimes a relation can be defined by an equation. The equation acts as the "rule of correspondence," and the relation consists of all ordered pairs that satisfy the equation.

EXAMPLE 3   Sketch the graph of the relation $x = y^2$, and state the domain and range.

SOLUTION   We graph the relation $x = y^2$ by finding a number of ordered pairs that satisfy the equation, plotting these points, then drawing a smooth curve that connects them. Table 4 shows some values for x and y that satisfy the equation. The graph of $x = y^2$ is shown in Figure 2.

TABLE 4

x	y
0	0
1	1
1	−1
4	2
4	−2
9	3
9	−3

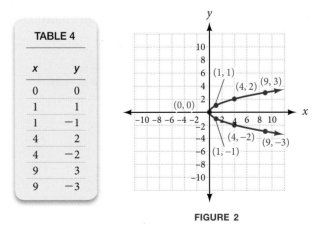

FIGURE 2

To find the domain and range, we must identify *all* values of x and y that satisfy the equation $x = y^2$. We cannot get the domain and range simply by looking at Table 4. For example, the ordered pairs $\left(\frac{1}{4}, \frac{1}{2}\right)$ and $\left(5, \sqrt{5}\right)$ satisfy the equation but are not listed in the table.

From the graph we can see that every point on the graph has a first coordinate that is either positive or zero. In addition, for every real number x greater than or equal to zero, there is a corresponding point on the graph. Therefore, the domain for this relation is $\{x \mid x \geq 0\}$, or $[0, \infty)$ in interval notation.

In contrast, the second coordinates can be positive or negative. Because the graph will continue to widen as we travel further to the right, every real number y will ultimately appear as a second coordinate. (For example, $y = -100$ will appear as a second coordinate in the point $(10{,}000, -100)$, which we could observe if we extended the graph far enough to the right.) Therefore, the range is $\{y \mid y$ is any real number$\}$, which we can write in interval notation as $(-\infty, \infty)$.

Functions

Suppose you have a job that pays $7.50 per hour and that you work anywhere from 0 to 40 hours per week. The amount of money you make in one week *depends* on the number of hours you work that week. In mathematics, we say that your weekly earnings are a ***function*** of the number of hours you work.

If we let the variable x represent hours and the variable y represent the money you make, then the relationship between x and y can be written as

$$y = 7.5x \qquad \text{for} \qquad 0 \leq x \leq 40$$

This equation defines a relation in which each value of x in the interval $[0, 40]$ will correspond to a single value of y. This leads us to the following definition.

> (dĕf **DEFINITION**   *function*
>
> A ***function*** is a relation that pairs each element in the ***domain*** with exactly one element from the ***range.***

If we think of each domain value as an input and the corresponding range value as an output, then a function is a rule for which each input is paired with exactly one output.

Domain:
The set of all inputs

The Function Rule →

Range:
The set of all outputs

EXAMPLE 4 Construct a table and graph for the function $y = 7.5x$ for $0 \leq x \leq 40$, and state the domain and range.

SOLUTION Table 5 gives some of the paired data that satisfy the equation $y = 7.5x$. Figure 3 is the graph of the equation with the restriction $0 \leq x \leq 40$.

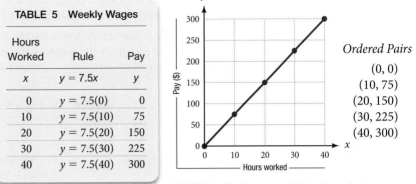

TABLE 5 Weekly Wages

Hours Worked	Rule	Pay
x	$y = 7.5x$	y
0	$y = 7.5(0)$	0
10	$y = 7.5(10)$	75
20	$y = 7.5(20)$	150
30	$y = 7.5(30)$	225
40	$y = 7.5(40)$	300

Ordered Pairs
$(0, 0)$
$(10, 75)$
$(20, 150)$
$(30, 225)$
$(40, 300)$

FIGURE 3 *Weekly wages at $7.50 per hour*

The equation $y = 7.5x$ with the restriction $0 \leq x \leq 40$, Table 5, and Figure 3 are three ways to describe the same relationship between the number of hours you work in one week and your gross pay for that week. In all three, we **input** values of x, and then use the function rule to **output** values of y.

The domain is the set of all inputs, or the values that x can assume. Because of the restriction on the number of hours worked, the domain is $\{x \mid 0 \leq x \leq 40\}$, or $[0, 40]$ in interval notation. From the line graph in Figure 3, we see that the corresponding values of y range from 0 to 300. Therefore, the range is $\{y \mid 0 \leq y \leq 300\}$, or $[0, 300]$ in interval notation.

Function Maps

Another way to visualize the relationship between x and y in Example 4 is with the diagram in Figure 4, which we call a **function map**.

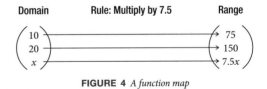

FIGURE 4 *A function map*

Although the diagram in Figure 4 does not show all the values that x and y can assume, it does give us a visual description of how x and y are related. It shows that values of y in the range come from values of x in the domain according to a specific rule (multiply by 7.5 each time).

EXAMPLE 5　Kendra tosses a softball into the air with an underhand motion. The distance of the ball above her hand is given by the function

$$h = 32t - 16t^2 \qquad \text{for} \qquad 0 \le t \le 2$$

where h is the height of the ball in feet and t is the time in seconds. Construct a table that gives the height of the ball at quarter-second intervals, starting with $t = 0$ and ending with $t = 2$, then graph the function.

SOLUTION　We construct Table 6 using the following values of t: 0, $\frac{1}{4}$, $\frac{1}{2}$, $\frac{3}{4}$, 1, $\frac{5}{4}$, $\frac{3}{2}$, $\frac{7}{4}$, 2. Then we construct the graph in Figure 5 from the table. The graph appears only in the first quadrant because neither t nor h can be negative.

TABLE 6　Tossing a Softball into the Air		
Input		Output
Time (sec) t	Function Rule $h = 32t - 16t^2$	Distance (ft) h
0	$h = 32(0) - 16(0)^2 = 0 - 0 = 0$	0
$\frac{1}{4}$	$h = 32\left(\frac{1}{4}\right) - 16\left(\frac{1}{4}\right)^2 = 8 - 1 = 7$	7
$\frac{1}{2}$	$h = 32\left(\frac{1}{2}\right) - 16\left(\frac{1}{2}\right)^2 = 16 - 4 = 12$	12
$\frac{3}{4}$	$h = 32\left(\frac{3}{4}\right) - 16\left(\frac{3}{4}\right)^2 = 24 - 9 = 15$	15
1	$h = 32(1) - 16(1)^2 = 32 - 16 = 16$	16
$\frac{5}{4}$	$h = 32\left(\frac{5}{4}\right) - 16\left(\frac{5}{4}\right)^2 = 40 - 25 = 15$	15
$\frac{3}{2}$	$h = 32\left(\frac{3}{2}\right) - 16\left(\frac{3}{2}\right)^2 = 48 - 36 = 12$	12
$\frac{7}{4}$	$h = 32\left(\frac{7}{4}\right) - 16\left(\frac{7}{4}\right)^2 = 56 - 49 = 7$	7
2	$h = 32(2) - 16(2)^2 = 64 - 64 = 0$	0

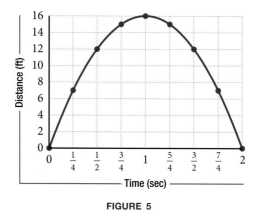

FIGURE 5

Here is a summary of what we know about functions as it applies to this example: We input values of t and output values of h according to the function rule

$$h = 32t - 16t^2 \qquad \text{for} \qquad 0 \le t \le 2$$

The domain is given by the inequality that follows the equation; it is

$$\text{Domain} = \{\, t \mid 0 \le t \le 2 \,\} = [0, 2]$$

The range is the set of all outputs that are possible by substituting the values of t from the domain into the equation. From our table and graph, it seems that the range is

$$\text{Range} = \{\, h \mid 0 \le h \le 16 \,\} = [0, 16]$$

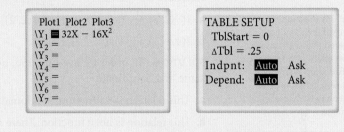

USING TECHNOLOGY   *More About Example 5*

Most graphing calculators can easily produce the information in Table 6. Simply set Y_1 equal to $32X - 16X^2$. Then set up the table so it starts at 0 and increases by an increment of 0.25 each time. (On a TI-82/83, use the TBLSET key to set up the table.)

Plot1 Plot2 Plot3
\Y$_1$ ■ 32X $-$ 16X^2
\Y$_2$ =
\Y$_3$ =
\Y$_4$ =
\Y$_5$ =
\Y$_6$ =
\Y$_7$ =

TABLE SETUP
TblStart = 0
ΔTbl = .25
Indpnt: Auto Ask
Depend: Auto Ask

The table will look like this:

X	Y$_1$	
0	0	
.25	7	
.5	12	
.75	15	
1	16	
1.25	15	
1.5	12	

Graph each equation and build a table as indicated.

1. $y = 64t - 16t^2$ TblStart = 0 ΔTbl = 1
2. $y = \dfrac{1}{2}x - 4$ TblStart = -5 ΔTbl = 1
3. $y = \dfrac{12}{x}$ TblStart = 0.5 ΔTbl = 0.5

Identifying Functions

Because every function is a relation, the function rule also produces ordered pairs of numbers. We use this result to write an alternative definition for a function.

(déf) ALTERNATE DEFINITION   *function*

A ***function*** is a set of ordered pairs in which no two different ordered pairs have the same first coordinate. The set of all first coordinates is the ***domain*** of the function. The set of all second coordinates is the ***range*** of the function.

The restriction on first coordinates in the alternative definition keeps us from assigning a number in the domain to more than one number in the range.

Here are some facts that will help clarify the distinction between relations and functions:

1. Any rule that assigns numbers from one set to numbers in another set is a relation. If that rule makes the assignment so no input has more than one output, then it is also a function.
2. Any set of ordered pairs is a relation. If none of the first coordinates of those ordered pairs is repeated, the set of ordered pairs is also a function.
3. Every function is a relation.
4. Not every relation is a function.

EXAMPLE 6 Determine if each of the following relations is also a function.

a. $\{(-2, 2), (-1, 1), (0, 0), (1, 1), (2, 2)\}$

b. $\{(1, -1), (1, 1), (4, -2), (4, 2)\}$

SOLUTION In order to be a function, no two different ordered pairs can have the same first coordinate.

a. This relation is a function. No domain value is repeated as a first coordinate.

b. This relation is not a function. There are two ordered pairs having the same first coordinate of $x = 1$ (and two ordered pairs having the first coordinate $x = 4$). To be a function, each domain value can appear in only one ordered pair.

Note A function *may* use a *range* value in any number of ordered pairs. This is why the first relation in Example 6 is a function, even though there are different ordered pairs having the same second coordinate.

EXAMPLE 7 In the introduction to this section we described an experiment where students were asked two questions, which can be represented by the equations $y = x^2$ and $y^2 = x$. Determine if these equations represent functions.

SOLUTION Looking at Table 1, we can see that each value of x gets paired with a single value of y. Once we choose an input, the square of that input is uniquely determined. The relation $y = x^2$ is a function because no two ordered pairs will have the same first coordinate.

On the other hand, a quick glance at Table 2 shows us that the equation $y^2 = x$ does not represent a function. For any positive real number x, we can always find two different values of y whose square is x (one positive and one negative). Different ordered pairs can have the same first coordinate, such as $(1, -1)$ and $(1, 1)$.

Vertical Line Test

Look back at the scatter diagram for used Mustang prices shown in Figure 1. Notice that some of the points on the diagram lie above and below each other along vertical lines. This is an indication that the data do not constitute a function. Two data points that lie on the same vertical line must have come from two ordered pairs with the same first coordinates.

Now, look at the graph shown in Figure 2. The reason this graph is the graph of a relation, but not of a function, is that some points on the graph have the same first coordinates, for example, the points $(4, 2)$ and $(4, -2)$. Furthermore, any time two points on a graph have the same first coordinates, those points must lie on a vertical line. [To convince yourself, connect the points $(4, 2)$ and $(4, -2)$ with a straight line. You will see that it must be a vertical line.] This allows us to write the following test that uses the graph to determine whether a relation is also a function.

> **⟨Δ≠Σ⟩ RULE** *Vertical Line Test*
>
> If a vertical line crosses the graph of a relation in more than one place, the relation cannot be a function. If no vertical line can be found that crosses a graph in more than one place, then the graph represents a function.

If we look back to the graph of $h = 32t - 16t^2$ as shown in Figure 5, we see that no vertical line can be found that crosses this graph in more than one place. The graph shown in Figure 5 is therefore the graph of a function.

EXAMPLE 8 Match each relation with its graph, then indicate which relations are functions

a. $y = |x| - 4$ **b.** $y = x^2 - 4$ **c.** $y = 2x + 2$

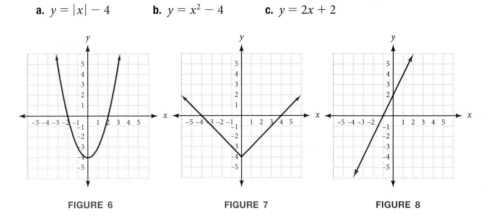

FIGURE 6   FIGURE 7   FIGURE 8

SOLUTION Using the basic graphs for a guide along with our knowledge of translations, we have the following:

a. Figure 7   **b.** Figure 6   **c.** Figure 8

And, since all graphs pass the vertical line test, all are functions.

Getting Ready for Class

After reading through the preceding section, respond in your own words and in complete sentences.

A. What is a function?

B. How is a relation different from a function?

C. What is the vertical line test?

D. Is every line the graph of a function? Explain.

Problem Set 3.5

For each of the following relations, give the domain and range, and indicate which are also functions.

1. $\{(1, 2), (3, 4), (5, 6), (7, 8)\}$ **2.** $\{(2, 1), (4, 3), (6, 5), (8, 7)\}$

3. $\{(2, 5), (3, 4), (1, 4), (0, 6)\}$ **4.** $\{(0, 4), (1, 6), (2, 4), (1, 5)\}$

5. $\{(a, 3), (b, 4), (c, 3), (d, 5)\}$ **6.** $\{(a, 5), (b, 5), (c, 4), (d, 5)\}$

7. $\{(a, 1), (a, 2), (a, 3), (a, 4)\}$ **8.** $\{(a, 1), (b, 1), (c, 1), (d, 1)\}$

State whether each of the following graphs represents a function.

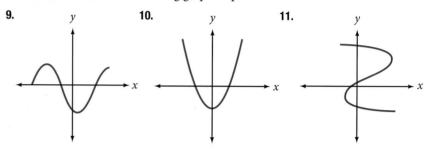

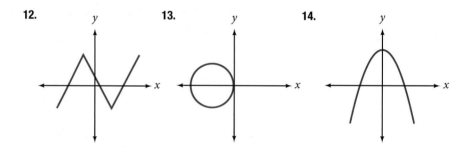

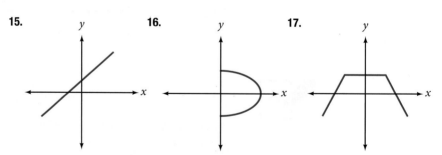

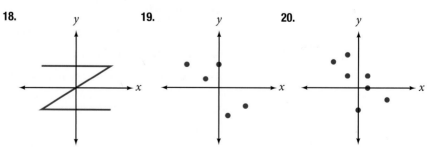

Determine the domain and range of the following functions. Assume the *entire* function is shown.

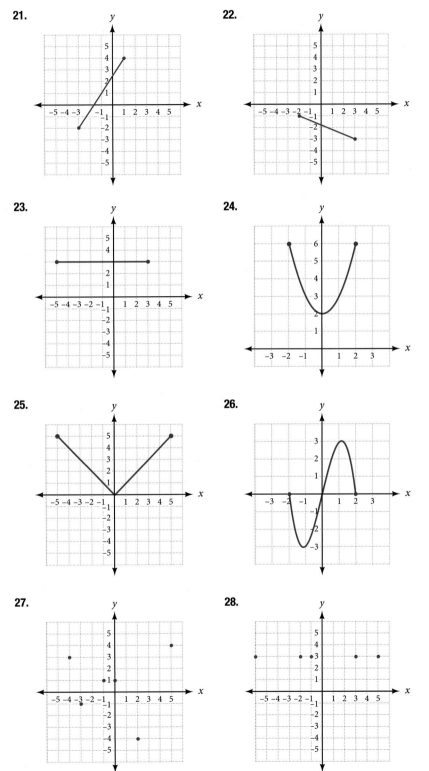

21.

22.

23.

24.

25.

26.

27.

28.

Graph each of the following relations. In each case, use the graph to find the domain and range, and indicate whether the graph is the graph of a function.

29. $y = 2x - 3$ **30.** $y = -\dfrac{1}{3}x + 2$ **31.** $x = 4$ **32.** $y = -5$

33. $y = -|x|$ **34.** $x = |y|$ **35.** $y = x^2 - 1$ **36.** $y = x^2 + 4$

37. $x = 3 - y^2$ **38.** $x = y^2 - 1$

39. Wages Suppose you have a job that pays $8.50 per hour and you work anywhere from 10 to 40 hours per week.

   a. Write an equation, with a restriction on the variable x, that gives the amount of money, y, you will earn for working x hours in one week.

   b. Use the function rule you have written in part a to complete Table 7.

TABLE 7 Weekly Wages		
Hours Worked	Function Rule	Gross Pay ($)
x		y
10		
20		
30		
40		

   c. Construct a line graph from the information in Table 7.

   d. State the domain and range of this function.

   e. What is the minimum amount you can earn in a week with this job? What is the maximum amount?

40. Wages The ad shown here was in the local newspaper. Suppose you are hired for the job described in the ad.

```
   312 HELP WANTED
TTER       ESPRESSO BAR        NEW
LDREN        OPERATOR          DELIV
16 years old   Must be dependable,   Must be 16
h children   honest, serivce-oriented.   Hardw
al, friendly   Coffee exp desired. 15-30   Morning
 hr/week   hrs per wk. $5.25/hour.   10-12 hou
) 444-1237  Start 5/31. Apply in person:   $6.00/hc
5/25      Espresso Yourself      Tribu
          Central Coast Mall     San L
            Deadline 5/23         Dea
```

   a. If x is the number of hours you work per week and y is your weekly gross pay, write the equation for y. (Be sure to include any restrictions on the variable x that are given in the ad.)

b. Use the function rule you have written in part *a* to complete Table 8.

TABLE 8 Weekly Wages		
Hours Worked	Function Rule	Gross Pay ($)
x		*y*
15		
20		
25		
30		

c. Construct a line graph from the information in Table 8.

d. State the domain and range of this function.

e. What is the minimum amount you can earn in a week with this job? What is the maximum amount?

41. Camera Phones The chart shows the estimated number of camera phones and non-camera phones sold from 2004 to 2010. Using the chart, list all the values in the domain and range for the total phones sales.

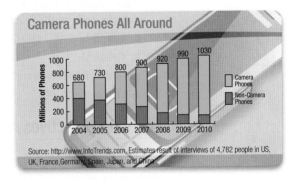

42. Light Bulbs The chart shows a comparison of power usage between incandescent and energy efficient light bulbs. Use the chart to state the domain and range of the function for an energy efficient bulb.

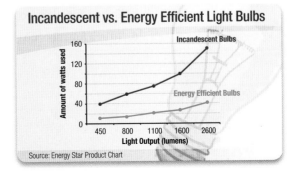

43. **Profits** Match each of the following statements to the appropriate graph indicated by labels I–IV in Figure 9.

    a. Sarah works 25 hours to earn $250.

    b. Justin works 35 hours to earn $560.

    c. Rosemary works 30 hours to earn $360.

    d. Marcus works 40 hours to earn $320.

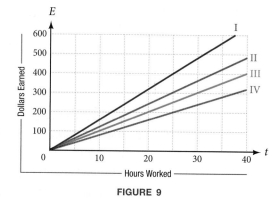

FIGURE 9

44. Find an equation for each of the functions shown in the graph in Figure 9. Show dollars earned, E, as a function of hours worked, t. Then, indicate the domain and range of each function.

    a. Graph I: $E =$ Domain $= \{ t |$ $\}$ Range $= \{ E |$ $\}$

    b. Graph II: $E =$ Domain $= \{ t |$ $\}$ Range $= \{ E |$ $\}$

    c. Graph III: $E =$ Domain $= \{ t |$ $\}$ Range $= \{ E |$ $\}$

    d. Graph IV: $E =$ Domain $= \{ t |$ $\}$ Range $= \{ E |$ $\}$

Learning Objectives Assessment

The following problems can be used to help assess if you have successfully met the learning objectives for this section.

45. Which of the following statements are true? (You may choose more than one.)

    a. Every relation is a function.

    b. Every function is a relation.

    c. To be a function, no two ordered pairs can have the same first coordinate.

    d. To be a function, no two ordered pairs can have the same second coordinate.

46. Find the domain of the relation whose graph is shown in Figure 10.

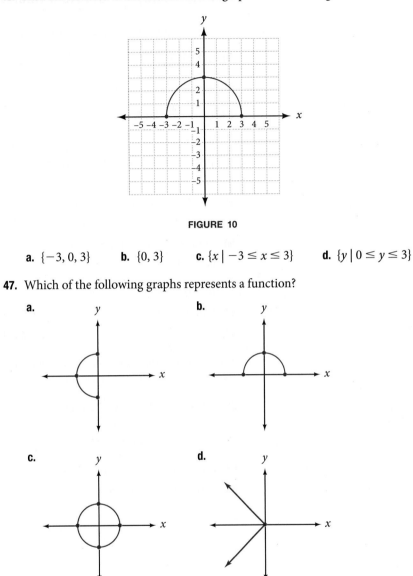

FIGURE 10

 a. $\{-3, 0, 3\}$ **b.** $\{0, 3\}$ **c.** $\{x \mid -3 \le x \le 3\}$ **d.** $\{y \mid 0 \le y \le 3\}$

47. Which of the following graphs represents a function?

48. Which of the following ordered pairs corresponds to a point on the graph of the function $y = x^2 - 9$?

 a. $(-8, 1)$ **b.** $(2, 5)$ **c.** $(9, 0)$ **d.** $(-2, -5)$

Getting Ready for the Next Section

Simplify. Round to the nearest whole number if necessary.

49. $4(3.14)(9)$

50. $\frac{4}{3}(3.14) \cdot 3^3$

51. $4(-2) - 1$

52. $3(3)^2 + 2(3) - 1$

53. If $s = \dfrac{60}{t}$, find s when

    a. $t = 10$ **b.** $t = 8$

54. If $y = 3x^2 + 2x - 1$, find y when

    a. $x = 0$ **b.** $x = -2$

55. Find the value of $x^2 + 2$ for

    a. $x = 5$ **b.** $x = -2$

56. Find the value of $125 \cdot 2^t$ for

    a. $t = 0$ **b.** $t = 1$

For the equation $y = x^2 - 3$:

57. Find y if x is 2.

58. Find y if x is -2.

59. Find y if x is 0.

60. Find y if x is -4.

The problems that follow review some of the more important skills you have learned in previous sections and chapters.

61. If $x - 2y = 4$, and $x = \frac{8}{5}$ find y.

62. If $\frac{x^2}{25} + \frac{y^2}{9} = 1$, find y when x is -4.

63. Let $x = 0$ and $y = 0$ in $y = a(x - 8)^2 + 70$ and solve for a.

64. Find R if $p = 2.5$ and $R = (900 - 300p)p$.

Evaluating Functions

Learning Objectives

In this section we will learn how to:

1. Use function notation to represent an ordered pair for a function.
2. Evaluate a function using an equation.
3. Evaluate a function using a table.
4. Evaluate a function using a graph.

Introduction

Let's return to an example we considered in the previous section. If a job pays $7.50 per hour for working from 0 to 40 hours a week, then the amount of money y earned in one week is a function of the number of hours worked x. The exact relationship between x and y is written

$$y = 7.5x \quad \text{for} \quad 0 \leq x \leq 40$$

Because the amount of money earned y depends on the number of hours worked x, we call y the **dependent variable** and x the **independent variable**. Furthermore, if we let f represent all the ordered pairs produced by the equation, then we can write

$$f = \{(x, y) \,|\, y = 7.5x \quad \text{and} \quad 0 \leq x \leq 40\}$$

Once we have named a function with a letter, we can use an alternative notation to represent the dependent variable y. The alternative notation for y is $f(x)$. It is read "f of x" and can be used instead of the variable y when working with functions. The notation y and the notation $f(x)$ are equivalent. That is,

$$y = 7.5x \Leftrightarrow f(x) = 7.5x$$

When we use the notation $f(x)$ we are using **function notation**.

Function Notation

The benefit of using function notation is that we can write more information with fewer symbols than we can by using just the variable y. For example, in the preceding discussion, asking how much money a person will make for working 20 hours is simply a matter of asking for $f(20)$. Without function notation, we would have to say, "Find the value of y that corresponds to a value of $x = 20$." To illustrate further, using the variable y, we can say "y is 150 when x is 20." Using the notation $f(x)$, we simply say "$f(20) = 150$." Each expression indicates that you will earn $150 for working 20 hours.

> **(dèf) DEFINITION** *function notation*
>
> If we let f represent the rule of correspondence defining y as a function of x, then the notation $y = f(x)$ is used to indicate that y is the range value assigned by the function to the domain value x.

EXAMPLE 1   If the function f is given by

$$f = \{(-2, 0), (3, -1), (2, 4), (7, 5)\}$$

express each ordered pair using function notation.

SOLUTION   We have the following:

$(-2, 0)$	becomes	$f(-2) = 0$
$(3, -1)$	becomes	$f(3) = -1$
$(2, 4)$	becomes	$f(2) = 4$
$(7, 5)$	becomes	$f(7) = 5$

Evaluating Functions

When we use a function to find the value of y for a given value of x, we call this *evaluating the function*. Because a function is often defined by an equation which performs some operations on x, we sometimes refer to a domain value as an ***input*** and the corresponding range value as an ***output***.

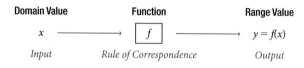

Our next example illustrates this process of finding outputs for given inputs.

EXAMPLE 2   If $f(x) = 7.5x$, find $f(0), f(10),$ and $f(20)$.

SOLUTION   To find $f(0)$, we substitute 0 for x in the expression $7.5x$ and simplify. We find $f(10)$ and $f(20)$ in a similar manner — by substitution.

If $f(x) = 7.5x$

then $f(0) = 7.5(0) = 0$

$f(10) = 7.5(10) = 75$

$f(20) = 7.5(20) = 150$

We can also express the relationships between inputs and outputs from Example 2 using a function map as shown in Figure 1.

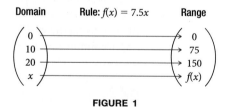

FIGURE 1

EXAMPLE 3   If $f(x) = 3x^2 + 2x - 1$, find $f(0), f(3),$ and $f(-2)$.

SOLUTION   Since $f(x) = 3x^2 + 2x - 1$, we have

$$f(0) = 3(0)^2 + 2(0) - 1 = 0 - 1 = -1$$

$$f(3) = 3(3)^2 + 2(3) - 1 = 27 + 6 - 1 = 32$$

$$f(-2) = 3(-2)^2 + 2(-2) - 1 = 12 - 4 - 1 = 7$$

Keep in mind that a function is a relation, which is a set of ordered pairs. In Example 3, because $f(0) = -1$, $(0, -1)$ is an ordered pair in f. Likewise, $(3, 32)$ and $(-2, 7)$ are also ordered pairs in f.

EXAMPLE 4   If $f(x) = 4x - 1$ and $g(x) = x^2 + 2$, then

$$f(5) = 4(5) - 1 = 19 \qquad \text{and} \qquad g(5) = 5^2 + 2 = 27$$

$$f(-2) = 4(-2) - 1 = -9 \qquad \text{and} \qquad g(-2) = (-2)^2 + 2 = 6$$

$$f(a) = 4a - 1 \qquad \text{and} \qquad g(a) = a^2 + 2$$

$$\begin{aligned} f(x + 3) &= 4(x + 3) - 1 & \text{and} \quad g(x + 3) &= (x + 3)^2 + 2 \\ &= 4x + 12 - 1 & &= (x^2 + 6x + 9) + 2 \\ &= 4x + 11 & &= x^2 + 6x + 11 \end{aligned}$$

$$\begin{aligned} f(x) + 3 &= (4x - 1) + 3 & \text{and} \quad g(x) + 3 &= (x^2 + 2) + 3 \\ &= 4x + 2 & &= x^2 + 5 \end{aligned}$$

$$\begin{aligned} 3\,f(x) &= 3(4x - 1) & \text{and} \quad 3\,g(x) &= 3(x^2 + 2) \\ &= 12x - 3 & &= 3x^2 + 6 \end{aligned}$$

USING TECHNOLOGY   *More About Example 4*

Most graphing calculators can use tables to evaluate functions. To work Example 4 using a graphing calculator table, set Y_1 equal to $4X - 1$ and Y_2 equal to $X^2 + 2$. Then set the independent variable in the table to Ask instead of Auto. Go to your table and input 5, -2, and 0. Under Y_1 in the table, you will find $f(5)$, $f(-2)$, and $f(0)$. Under Y_2, you will find $g(5)$, $g(-2)$, and $g(0)$.

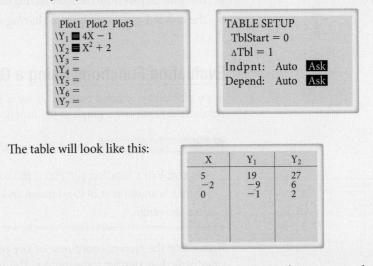

The table will look like this:

X	Y_1	Y_2
5	19	27
-2	-9	6
0	-1	2

Although the calculator asks us for a table increment, the increment doesn't matter because we are inputting the X values ourselves.

EXAMPLE 5   If $f(x) = 2x^2$ and $g(x) = 3x - 1$, find

a. $f[g(2)]$ **b.** $g[f(2)]$

SOLUTION   The expression $f[g(2)]$ is read "f of g of 2."

a. Because $g(2) = 3(2) - 1 = 5$,

$$f[g(2)] = f(5) = 2(5)^2 = 50$$

b. Because $f(2) = 2(2)^2 = 8$,

$$g[f(2)] = g(8) = 3(8) - 1 = 23$$

Evaluating Functions Using a Table

In the previous examples we have seen how to evaluate a function that is defined by an equation. Another way to define a function is using a table. Because a function is simply a set of ordered pairs, we can specify the ordered pairs in table format.

If (x, y) is an ordered pair in f, then the first coordinate is an input (domain value) and the second coordinate is the corresponding output (range value). To evaluate a function from a table, we identify the ordered pair having the given value of x. The y-coordinate of that pair is the function value $f(x)$.

EXAMPLE 6   Given $y = f(x)$, use the table below to find

a. $f(2)$ **b.** $f(-1)$

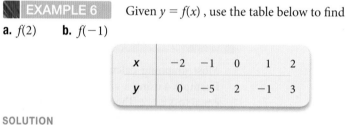

x	−2	−1	0	1	2
y	0	−5	2	−1	3

SOLUTION

a. The ordered pair from the table having an x-coordinate of 2 is $(2, 3)$, so $f(2) = 3$.

b. The ordered pair from the table having an x-coordinate of -1 is $(-1, -5)$, so $f(-1) = -5$.

Evaluating Functions Using a Graph

If we plot every ordered pair (x, y) for a function on a rectangular coordinate system, the result is the graph of the function.

> **(dēf DEFINITION**   *graph of a function*
>
> The **graph** of a function $y = f(x)$ is the set of all points in the plane $(x, f(x))$, where x is an element of the domain and $f(x)$ is the corresponding element from the range.

Notice that the second coordinate of any point on the graph of a function is the *value* of the function for some input x. This means we can evaluate a function from a graph. We locate the point having the given x-value. The y-value of this point will be $f(x)$.

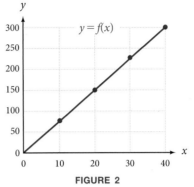

EXAMPLE 7   Figure 2 shows the graph of the function $y = f(x)$ introduced in the beginning of this section. Use the graph to evaluate $f(30)$ and $f(40)$.

FIGURE 2

SOLUTION   From Figure 2, we can see that the point on the graph having an x-coordinate of 30 is (30, 225), so $f(30) = 225$. Likewise, the point on the graph having an x-coordinate of 40 is (40, 300), so $f(40) = 300$. We illustrate this in Figure 3.

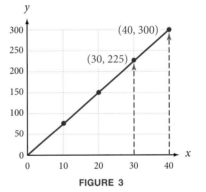

FIGURE 3

Using Function Notation in Applications

The next examples show a variety of ways to use and interpret function notation.

EXAMPLE 8   If it takes Lorena t minutes to run a mile, then her average speed s, in miles per hour, is given by the formula

$$s(t) = \frac{60}{t} \qquad \text{for} \qquad t > 0$$

Find $s(10)$ and $s(8)$, and then explain what they mean.

SOLUTION   To find $s(10)$, we substitute 10 for t in the equation and simplify:

$$s(10) = \frac{60}{10} = 6$$

In words: When Lorena runs a mile in 10 minutes, her average speed is 6 miles per hour.

We calculate $s(8)$ by substituting 8 for t in the equation. Doing so gives us

$$s(8) = \frac{60}{8} = 7.5$$

In words: Running a mile in 8 minutes is running at a rate of 7.5 miles per hour.

◼ EXAMPLE 9   A painting is purchased as an investment for $125. If its value increases continuously so that it doubles every 5 years, then its value is given by the function

$$V(t) = 125 \cdot 2^{t/5} \qquad \text{for} \qquad t \geq 0$$

where t is the number of years since the painting was purchased, and V is its value (in dollars) at time t. Find $V(5)$ and $V(10)$, and explain what they mean.

SOLUTION   The expression $V(5)$ is the value of the painting when $t = 5$ (5 years after it is purchased). We calculate $V(5)$ by substituting 5 for t in the equation $V(t) = 125 \cdot 2^{t/5}$. Here is our work:

$$V(5) = 125 \cdot 2^{5/5} = 125 \cdot 2^1 = 125 \cdot 2 = 250$$

In words: After 5 years, the painting is worth $250.
The expression $V(10)$ is the value of the painting after 10 years. To find this number, we substitute 10 for t in the equation:

$$V(10) = 125 \cdot 2^{10/5} = 125 \cdot 2^2 = 125 \cdot 4 = 500$$

In words: The value of the painting 10 years after it is purchased is $500.   ◼

◼ EXAMPLE 10   A balloon has the shape of a sphere with a radius of 3 inches. Use the following formulas to find the volume and surface area of the balloon.

$$V(r) = \frac{4}{3}\,\pi r^3 \qquad S(r) = 4\pi r^2$$

SOLUTION   As you can see, we have used function notation to write the formulas for volume and surface area, because each quantity is a function of the radius. To find these quantities when the radius is 3 inches, we evaluate $V(3)$ and $S(3)$:

$$V(3) = \frac{4}{3}\pi \cdot 3^3 = \frac{4}{3}\pi \cdot 27$$

$$= 36\pi \text{ cubic inches, or } 113 \text{ cubic inches}$$
$$\text{(to the nearest whole number)}$$

$$S(3) = 4\pi \cdot 3^2$$

$$= 36\pi \text{ square inches, or } 113 \text{ square inches}$$
$$\text{(to the nearest whole number)}$$

The fact that $V(3) = 36\pi$ means that the ordered pair $(3, 36\pi)$ belongs to the function V. Likewise, the fact that $S(3) = 36\pi$ tells us that the ordered pair $(3, 36\pi)$ is a member of function S.   ◼

USING TECHNOLOGY *More About Example 10*

If we look at Example 10, we see that when the radius of a sphere is 3, the numerical values of the volume and surface area are equal. How unusual is this? Are there other values of r for which $V(r)$ and $S(r)$ are equal? We can answer this question by looking at the graphs of both V and S.

To graph the function $V(r) = \frac{4}{3}\pi r^3$, set $Y_1 = 4\pi X^3/3$. To graph $S(r) = 4\pi r^2$, set $Y_2 = 4\pi X^2$. Graph the two functions in each of the following windows:

Window 1: X from -4 to 4, Y from -2 to 10

Window 2: X from 0 to 4, Y from 0 to 50

Window 3: X from 0 to 4, Y from 0 to 150

Then use the Trace and Zoom features of your calculator to locate the point in the first quadrant where the two graphs intersect. How do the coordinates of this point compare with the results in Example 10?

We conclude this section by taking another look at the equation of a line from a function perspective.

Linear Functions

In Section 3.3, we introduced the slope-intercept form of a line, $y = mx + b$. As long as the slope is defined (the line is not vertical), the graph of the line will pass the vertical line test. Therefore, every non-vertical line represents a function. This leads us to the following definition.

DEFINITION *linear function*

A *linear function* is any function that can be expressed in the form

$$f(x) = mx + b$$

The graph of f is a line with slope m passing through the y-axis at $(0, b)$.

EXAMPLE 11   The average price per gallon for gasoline between the years 2002 and 2008 can be modeled by the linear function

$$f(x) = 0.32x + 0.76$$

where x is the number of years from 2000 and $f(x)$ is the average price in dollars per gallon.   (*Source*: U.S. Department of Energy)

a. Graph the function.

b. Evaluate $f(8)$ and explain what it means in practical terms.

c. Based on this model, what would we expect the price of gasoline to be in the year 2020?

SOLUTION

a. The graph, shown in Figure 4, is a line with slope $m = 0.32$ and y-intercept $(0, 0.76)$.

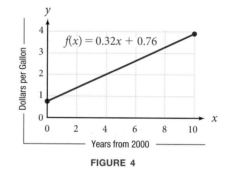

FIGURE 4

b. $f(8) = 0.32(8) + 0.76 = 3.32$

In 2008 (8 years after 2000) the average price of gasoline was $3.32 per gallon.

c. For the year 2020 we use $x = 20$, so

$$f(20) = 0.32(20) + 0.76 = 7.16$$

The expected price of gasoline is $7.16 per gallon.

Getting Ready for Class

After reading through the preceding section, respond in your own words and in complete sentences.

A. Explain what you are calculating when you find $f(2)$ for a given function f.

B. If $f(x) = 7.5x$, how do you find $f(10)$?

C. If $f(2) = 3$ for a function f, what is the relationship between the numbers 2 and 3 and the graph of f?

D. Explain how you would evaluate $f(6)$ using a table.

1. If $f(3) = 8$, then what ordered pair is in f?

2. If $f(8) = 3$, then what ordered pair is in f?

3. If $g(-1) = 5$, then what ordered pair is in g?

4. If $g(5) = -1$, then what ordered pair is in g?

If each of the following ordered pairs belongs to the function $y = f(x)$, express each pair using function notation.

5. $(4, 0)$ **6.** $(0, 3)$ **7.** $(-1, 9)$ **8.** $(2, -2)$

9. $(10, 0.1)$ **10.** $(0.25, 0.5)$ **11.** $\left(-\dfrac{1}{5}, -\dfrac{1}{10}\right)$ **12.** $\left(-\dfrac{1}{3}, \dfrac{1}{9}\right)$

Let $f(x) = 2x - 5$ and $g(x) = x^2 + 3x + 4$. Evaluate the following.

13. $f(2)$ **14.** $f(3)$ **15.** $f(-3)$ **16.** $g(-2)$

17. $g(-1)$ **18.** $f(-4)$ **19.** $g(-3)$ **20.** $g(2)$

21. $g(a)$ **22.** $f(a)$ **23.** $f(a + 6)$ **24.** $g(a + 6)$

Let $f(x) = 3x^2 - 4x + 1$ and $g(x) = 2x - 1$. Evaluate the following.

25. $f(0)$ **26.** $g(0)$ **27.** $g(-4)$ **28.** $f(1)$

29. $f(-1)$ **30.** $g(-1)$ **31.** $g\left(\dfrac{1}{2}\right)$ **32.** $g\left(\dfrac{1}{4}\right)$

33. $f(a)$ **34.** $g(a)$ **35.** $f(a + 2)$ **36.** $g(a + 2)$

Let $f(x) = \dfrac{1}{x + 3}$ and $g(x) = \dfrac{1}{x} + 1$. Evaluate the following.

37. $f\left(\dfrac{1}{3}\right)$ **38.** $g\left(\dfrac{1}{3}\right)$ **39.** $f\left(-\dfrac{1}{2}\right)$ **40.** $g\left(-\dfrac{1}{2}\right)$

41. $f(-3)$ **42.** $g(0)$

Let $f(x) = x^2 - 2x$ and $g(x) = 5x - 4$. Evaluate the following.

43. $f(-2) + g(-1)$ **44.** $f(-1) + g(-2)$ **45.** $f(2) - g(3)$

46. $f(3) - g(2)$ **47.** $f[g(3)]$ **48.** $g[f(3)]$

If $f = \{(1, 4), (-2, 0), \left(3, \dfrac{1}{2}\right), (\pi, 0)\}$ and $g = \{(1, 1), (-2, 2), \left(\dfrac{1}{2}, 0\right)\}$, find each of the following values of f and g.

49. $f(1)$ **50.** $g(1)$ **51.** $g\left(\dfrac{1}{2}\right)$ **52.** $f(3)$

53. $g(-2)$ **54.** $f(\pi)$

If $y = f(x)$, use the table given below to find each value.

x	-3	-2	-1	0	1	2	3
y	4	5	1	-2	-1	0	2

55. $f(0)$ **56.** $f(-3)$ **57.** $f(3)$ **58.** $f(1)$

59. $f(-1)$ **60.** $f(2)$ **61.** $3f(-2)$ **62.** $-2f(3)$

Use the graphs of $y = f(x)$ and $y = g(x)$ shown to find the following values of f and g.

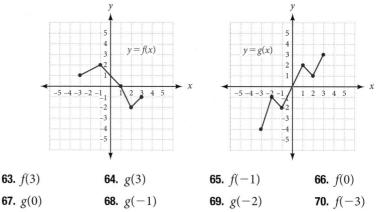

63. $f(3)$ **64.** $g(3)$ **65.** $f(-1)$ **66.** $f(0)$

67. $g(0)$ **68.** $g(-1)$ **69.** $g(-2)$ **70.** $f(-3)$

71. For the function $f(x) = x^2 - 4$, evaluate each of the following expressions.

     a. $f(a) - 3$ **b.** $f(a - 3)$ **c.** $f(x) + 2$ **d.** $f(x + 2)$

     e. $f(a + b)$ **f.** $f(x + h)$ **g.** $f(3x)$ **h.** $3f(x)$

72. For the function $f(x) = 3x^2$, evaluate each of the following expressions.

     a. $f(a) - 2$ **b.** $f(a - 2)$ **c.** $f(x) + 5$ **d.** $f(x + 5)$

     e. $f(a + b)$ **f.** $f(x + h)$ **g.** $f(2x)$ **h.** $2f(x)$

73. Graph the linear function $f(x) = \frac{1}{2}x + 2$. Then use the graph to evaluate $f(4)$.

74. Graph the linear function $f(x) = -\frac{1}{2}x + 6$. Then use the graph to evaluate $f(4)$.

75. Graph the function $f(x) = x^2$. Then use the graph to evaluate $f(1)$, $f(2)$, and $f(3)$.

76. Graph the function $f(x) = x^2 - 2$. Then use the graph to evaluate $f(2)$ and $f(3)$.

77. If f is a linear function, and $f(0) = 2$ and $f(6) = 0$, find the equation for f. Write your answer using function notation.

78. If g is a linear function, and $g(-2) = -4$ and $g(4) = 5$, find the equation for g. Write your answer using function notation.

Applying the Concepts

79. Investing in Art A painting is purchased as an investment for $150. If its value increases continuously so that it doubles every 3 years, then its value is given by the function

$$V(t) = 150 \cdot 2^{t/3} \qquad \text{for} \qquad t \geq 0$$

where t is the number of years since the painting was purchased, and $V(t)$ is its value (in dollars) at time t. Find $V(3)$ and $V(6)$, and then explain what they mean.

80. Average Speed If it takes Minke t minutes to run a mile, then her average speed $s(t)$, in miles per hour, is given by the formula

$$s(t) = \frac{60}{t} \qquad \text{for} \qquad t > 0$$

Find $s(4)$ and $s(5)$, and then explain what they mean.

81. **Antidepressant Sales** Suppose x represents one of the years in the chart. Suppose further that we have three functions f, g, and h that do the following:

    - f pairs each year with the total sales of Zoloft in billions of dollars for that year.
    - g pairs each year with the total sales of Effexor in billions of dollars for that year.
    - h pairs each year with the total sales of Wellbutrin in billions of dollars for that year.

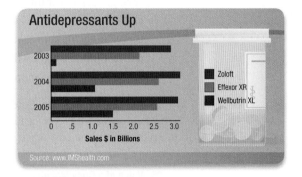

    For each statement below, indicate whether the statement is true or false.

    a. The domain of g is $\{\,2003, 2004, 2005\,\}$

    b. The domain of g is $\{\,x \mid 2003 \leq x \leq 2005\,\}$

    c. $f(2004) > g(2004)$

    d. $h(2005) > 1.5$

    e. $h(2005) > h(2004) > h(2003)$

82. **Mobile Phone Sales** Suppose x represents one of the years in the chart. Suppose further that we have three functions f, g, and h that do the following:

    - f pairs each year with the number of camera phones sold that year.
    - g pairs each year with the number of non-camera phones sold that year.
    - h is such that $h(x) = f(x) + g(x)$.

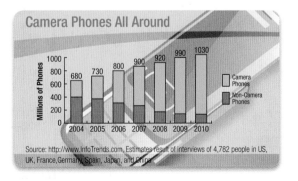

    For each statement below, indicate whether the statement is true or false.

    a. The domain of f is {2004, 2005, 2006, 2007, 2008, 2009, 2010}

    b. $h(2005) = 741,000,000$

    c. $f(2009) > g(2009)$

    d. $f(2004) < f(2005)$

    e. $h(2010) > h(2007) > h(2004)$

83. **Length of a Spring** The linear function $S(x) = 1.4x$ gives the amount of stretch in a spring, $S(x)$, in centimeters, when a weight of x pounds is added to the spring.

    a. How much will the spring stretch if a weight of 2.5 pounds is added?

    b. What amount of weight will cause the spring to stretch 5.6 cm?

    c. Graph the function for $0 \le x \le 10$.

84. **Length of a Spring** The linear function $L(x) = 0.03x + 12$ gives the length of a spring, $L(x)$, in centimeters, when a weight of x grams is added to the spring.

    a. What is the normal length of the spring with no weight attached?

    b. What will be the length of the spring if 200 grams of weight is attached?

    c. How much weight will cause the spring to double in length?

    d. Graph the function for $0 \le x \le 500$.

Straight-Line Depreciation Straight-line depreciation is an accounting method used to help spread the cost of new equipment over a number of years. It takes into account both the cost when new and the salvage value, which is the value of the equipment at the time it gets replaced.

85. **Value of a Copy Machine** The linear function $V(t) = -3{,}300t + 18{,}000$, where V is value and t is time in years, can be used to find the value of a large copy machine during the first 5 years of use.

    a. What is the value of the copier after 3 years and 9 months?

    b. What is the salvage value of this copier if it is replaced after 5 years?

    c. State the domain of this function.

    d. Sketch the graph of this function.

    e. What is the range of this function?

    f. After how many years will the copier be worth only $10,000?

86. **Step Function** Figure 5 shows the graph of the step function C that was used to calculate the first-class postage on a letter weighing x ounces in 2006. Use this graph to answer questions *a* through *d*.

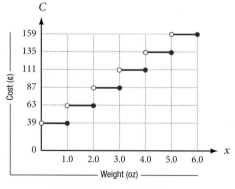

FIGURE 5 *The graph of C(x)*

a. Fill in the following table:

Weight (ounces)	0.6	1.0	1.1	2.5	3.0	4.8	5.0	5.3
Cost (cents)								

b. If a letter cost 87 cents to mail, how much does it weigh? State your answer in words. State your answer as an inequality.

c. If the entire function is shown in Figure 5, state the domain.

d. State the range of the function shown in Figure 5.

Learning Objectives Assessment

The following problems can be used to help assess if you have successfully met the learning objectives for this section.

87. If $(-2, 6)$ is an ordered pair in a function f, express this using function notation.

    a. $-2f = 6$ **b.** $6f = -2$ **c.** $f(-2) = 6$ **d.** $f(6) = -2$

88. If $f(x) = 3x^2 - 4x$, find $f(-2)$.

    a. 4   **b.** 20   **c.** -4 **d.** $-6x^2 + 8x$

89. If $y = f(x)$, use the table below to evaluate $f(1)$.

x	-1	1	2	4
y	-2	-3	1	0

    a. -3 **b.** 2   **c.** -2 **d.** 0

90. Use the graph of $y = g(x)$ shown in Figure 6 to evaluate $g(-1)$.

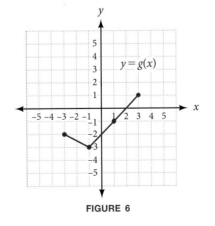

FIGURE 6

    a. 1   **b.** -1 **c.** 3   **d.** -3

Getting Ready for the Next Section

Multiply.

91. $x(35 - 0.1x)$

92. $0.6(M - 70)$

93. $(4x - 3)(x - 1)$

94. $(4x - 3)(4x^2 - 7x + 3)$

Simplify.

95. $(35x - 0.1x^2) - (8x + 500)$

96. $(4x - 3) + (4x^2 - 7x + 3)$

97. $(4x^2 + 3x + 2) - (2x^2 - 5x - 6)$

98. $(4x^2 + 3x + 2) + (2x^2 - 5x - 6)$

99. $4(2)^2 - 3(2)$

100. $4(-1)^2 - 7(-1)$

SPOTLIGHT ON SUCCESS   *Student Instructor CJ*

We are what we repeatedly do.
Excellence, then, is not an act, but a habit.
—Aristotle

Something that has worked for me in college, in addition to completing the assigned homework, is working on some extra problems from each section. Working on these extra problems is a great habit to get into because it helps further your understanding of the material, and you see the many different types of problems that can arise. If you have completed every problem that your book offers, and you still don't feel confident that you have a full grasp of the material, look for more problems. Many problems can be found online or in other books. Your professors may even have some problems that they would suggest doing for extra practice. The biggest benefit to working all the problems in the course's assigned textbook is that often teachers will choose problems either straight from the book or ones similar to problems that were not assigned for tests. Doing this will ensure that you do your best in all your classes.

Algebra and Composition with Functions

3.7

Learning Objectives

In this section we will learn how to:

1. Evaluate the sum, difference, product, or quotient of two functions.

2. Find the formulas for the sum, difference, product, and quotient of two functions.

3. Evaluate a composition of functions.

4. Find the formula for a composition of two functions.

iStockPhoto.com/©Slobo Mitic

Introduction

A company produces and sells copies of an accounting program for home computers. The price they charge for the program is related to the number of copies sold by the demand function

$$p(x) = 35 - 0.1x$$

We find the revenue for this business by multiplying the number of items sold by the price per item. When we do so, we are forming a new function by combining two existing functions. That is, if $n(x) = x$ is the number of items sold and $p(x) = 35 - 0.1x$ is the price per item, then revenue is

$$R(x) = n(x) \cdot p(x) = x(35 - 0.1x) = 35x - 0.1x^2$$

In this case, the revenue function is the product of two functions. When we combine functions in this manner, we are applying our rules for algebra to functions.

To carry this situation further, we know the profit function is the difference between two functions. If the cost function for producing x copies of the accounting program is $C(x) = 8x + 500$, then the profit function is

$$P(x) = R(x) - C(x) = (35x - 0.1x^2) - (8x + 500) = -500 + 27x - 0.1x^2$$

The relationship between these last three functions is represented visually in Figure 1.

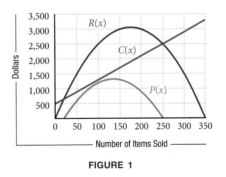

FIGURE 1

Algebra with Functions

Again, when we combine functions in the manner shown, we are applying our rules for algebra to functions. To begin this section, we take a formal look at addition, subtraction, multiplication, and division with functions.

If we are given two functions f and g with a common domain, we can define four other functions as follows.

Operations with Functions

If f and g are functions and x is in the domain of both f and g, then

$(f + g)(x) = f(x) + g(x)$ The function $f + g$ is the sum of the functions f and g.

$(f - g)(x) = f(x) - g(x)$ The function $f - g$ is the difference of the functions f and g.

$(fg)(x) = f(x)g(x)$ The function fg is the product of the functions f and g.

$\left(\dfrac{f}{g}\right)(x) = \dfrac{f(x)}{g(x)}$ The function $\frac{f}{g}$ is the quotient of the functions f and g, where $g(x) \neq 0$.

Notice that, for a given input x, the value of the sum function $(f + g)(x)$ is found by evaluating $f(x)$ and $g(x)$ individually, and then adding the resulting outputs. In other words, when adding two functions, all we are really doing is adding the corresponding range values for a given domain value. The difference function, product function, and quotient function all behave in a similar way.

VIDEO EXAMPLES

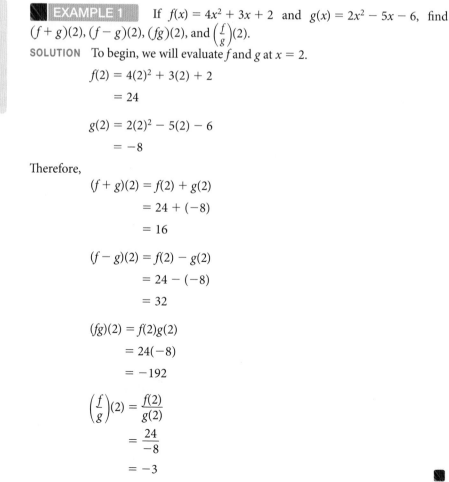

SECTION 3.7

EXAMPLE 1   If $f(x) = 4x^2 + 3x + 2$ and $g(x) = 2x^2 - 5x - 6$, find $(f + g)(2)$, $(f - g)(2)$, $(fg)(2)$, and $\left(\dfrac{f}{g}\right)(2)$.

SOLUTION   To begin, we will evaluate f and g at $x = 2$.

$$f(2) = 4(2)^2 + 3(2) + 2$$
$$= 24$$

$$g(2) = 2(2)^2 - 5(2) - 6$$
$$= -8$$

Therefore,

$$(f + g)(2) = f(2) + g(2)$$
$$= 24 + (-8)$$
$$= 16$$

$$(f - g)(2) = f(2) - g(2)$$
$$= 24 - (-8)$$
$$= 32$$

$$(fg)(2) = f(2)g(2)$$
$$= 24(-8)$$
$$= -192$$

$$\left(\frac{f}{g}\right)(2) = \frac{f(2)}{g(2)}$$
$$= \frac{24}{-8}$$
$$= -3$$

Given formulas for f and g, we can find formulas for the sum, difference, product, and quotient functions by performing the appropriate operation while leaving x as an unknown. We illustrate how this is done in the following example:

EXAMPLE 2 If $f(x) = 4x^2 + 3x + 2$ and $g(x) = 2x^2 - 5x - 6$, write the formulas for the functions $f + g$, $f - g$, fg, and $\frac{f}{g}$.

SOLUTION The function $f + g$ is defined by

$$(f + g)(x) = f(x) + g(x)$$
$$= (4x^2 + 3x + 2) + (2x^2 - 5x - 6)$$
$$= 6x^2 - 2x - 4$$

The function $f - g$ is defined by

$$(f - g)(x) = f(x) - g(x)$$
$$= (4x^2 + 3x + 2) - (2x^2 - 5x - 6)$$
$$= 4x^2 + 3x + 2 - 2x^2 + 5x + 6$$
$$= 2x^2 + 8x + 8$$

The function fg is defined by

$$(fg)(x) = f(x)g(x)$$
$$= (4x^2 + 3x + 2)(2x^2 - 5x - 6)$$
$$= 8x^4 - 20x^3 - 24x^2 + 6x^3 - 15x^2 - 18x + 4x^2 - 10x - 12$$
$$= 8x^4 - 14x^3 - 35x^2 - 28x - 12$$

The function $\frac{f}{g}$ is defined by

$$\left(\frac{f}{g}\right)(x) = \frac{f(x)}{g(x)}$$
$$= \frac{4x^2 + 3x + 2}{2x^2 - 5x - 6}$$

If we have found the formulas for $f + g$, $f - g$, fg, and $\frac{f}{g}$, then we can, of course, use them to evaluate these functions instead of evaluating f and g individually and then performing the corresponding operation on the outputs. Example 3 illustrates this.

EXAMPLE 3 If f and g are the same functions defined in Example 2, evaluate $(f + g)(2)$ and $(fg)(2)$.

SOLUTION We use the formulas for $f + g$ and fg found in Example 2:

$$(f + g)(2) = 6(2)^2 - 2(2) - 4$$
$$= 24 - 4 - 4$$
$$= 16$$

$$(fg)(2) = 8(2)^4 - 14(2)^3 - 35(2)^2 - 28(2) - 12$$
$$= 128 - 112 - 140 - 56 - 12$$
$$= -192$$

Notice that these values are in agreement with those we found in Example 1.

Composition of Functions

In addition to the four operations used to combine functions shown so far in this section, there is a fifth way to combine two functions to obtain a new function. It is called **composition of functions.** To illustrate the concept, recall from Chapter 2 the definition of training heart rate: training heart rate, in beats per minute, is resting heart rate plus 60% of the difference between maximum heart rate and resting heart rate. If your resting heart rate is 70 beats per minute, then your training heart rate is a function of your maximum heart rate M.

$$T(M) = 70 + 0.6(M - 70) = 70 + 0.6M - 42 = 28 + 0.6M$$

But your maximum heart rate is found by subtracting your age in years from 220. So, if x represents your age in years, then your maximum heart rate is

$$M(x) = 220 - x$$

Therefore, if your resting heart rate is 70 beats per minute and your age in years is x, then your training heart rate can be written as a function of x.

$$T(x) = 28 + 0.6(220 - x)$$

This last line is the composition of functions T and M. We input x into function M, which outputs $M(x)$. Then, we input $M(x)$ into function T, which outputs $T(M(x))$, which is the training heart rate as a function of age x. Here is a diagram, called a function map, of the situation:

$$\begin{array}{ccc} \text{Age} & \text{Maximum} & \text{Training} \\ & \text{heart rate} & \text{heart rate} \\ x \xrightarrow{\ M\ } & M(x) \xrightarrow{\ T\ } & T(M(x)) \end{array}$$

FIGURE 2

Now let's generalize the preceding ideas into a formal development of composition of functions. To find the composition of two functions f and g, we first require that the range of g have numbers in common with the domain of f. Then the composition of f with g is defined as follows:

(dĕf **DEFINITION** *function composition*

If f and g are functions, and if x is in the domain of g and $g(x)$ is in the domain of f, then

$$(f \circ g)(x) = f(g(x))$$

To understand this new function, we begin with a number x, and we operate on it with g, giving us $g(x)$. Then we take $g(x)$ and operate on it with f, giving us $f(g(x))$. The diagrams in Figure 3 illustrate the composition of f with g.

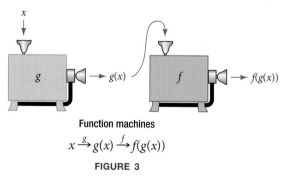

Function machines

$$x \xrightarrow{\ g\ } g(x) \xrightarrow{\ f\ } f(g(x))$$

FIGURE 3

Composition of functions is not commutative. The composition of f with g, $f \circ g$, may therefore be different from the composition of g with f, $g \circ f$.

$$(g \circ f)(x) = g(f(x))$$

The only numbers we can use for the domain of the composition of g with f are numbers in the domain of f, for which $f(x)$ is in the domain of g. The diagrams in Figure 4 illustrate the composition of g with f.

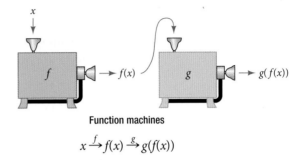

Function machines

$$x \xrightarrow{f} f(x) \xrightarrow{g} g(f(x))$$

FIGURE 4

EXAMPLE 4   If $f(x) = x + 5$ and $g(x) = x^2 - 2x$, find $(f \circ g)(2)$ and $(g \circ f)(2)$.

SOLUTION   To evaluate $(f \circ g)(2)$, we first find $g(2)$.

$$g(2) = (2)^2 - 2(2)$$
$$= 4 - 4$$
$$= 0$$

The output value of 0 from g is now used as an input for f.

$$(f \circ g)(2) = f(g(2))$$
$$= f(0)$$
$$= 0 + 5$$
$$= 5$$

We evaluate $(g \circ f)(2)$ in a similar manner.

Because $f(2) = 2 + 5$
$$= 7$$

we have $(g \circ f)(2) = g(f(2))$
$$= g(7)$$
$$= (7)^2 - 2(7)$$
$$= 49 - 14$$
$$= 35$$

If we repeat the process illustrated in Example 4 but allow x to remain an unknown, then we will find the formulas for the composition functions.

◼ **EXAMPLE 5**   If $f(x) = x + 5$ and $g(x) = x^2 - 2x$, find $(f \circ g)(x)$ and $(g \circ f)(x)$.

SOLUTION   The composition of f with g is

$$(f \circ g)(x) = f(g(x))$$
$$= f(x^2 - 2x)$$
$$= (x^2 - 2x) + 5$$
$$= x^2 - 2x + 5$$

The composition of g with f is

$$(g \circ f)(x) = g(f(x))$$
$$= g(x + 5)$$
$$= (x + 5)^2 - 2(x + 5)$$
$$= (x^2 + 10x + 25) - 2x - 10$$
$$= x^2 + 8x + 15$$ ◼

> *Note* Be careful not to confuse the composition symbol with a multiplication dot. The notation $f \circ g$ does not mean the product of f and g.

Now that we have the formulas from Example 5, we could use them to evaluate $f \circ g$ and $g \circ f$. For instance,

$$(f \circ g)(2) = (2)^2 - 2(2) + 5 = 5$$

and

$$(g \circ f)(2) = (2)^2 + 8(2) + 15 = 35$$

which agree with the values we obtained in Example 4.

Using Tables and Graphs

If we are given tables or graphs for two functions, we can still find values for the sum, difference, product, quotient, or composition of the functions even though we may not know their formulas.

◼ **EXAMPLE 6**   Use the tables shown for functions f and g to find $(f + g)(2)$, $(fg)(1)$, and $(f \circ g)(-2)$.

x	$f(x)$
-2	-3
-1	2
0	6
1	3
2	-1

x	$g(x)$
-2	0
-1	3
0	-4
1	-2
2	5

SOLUTION   From the tables we see that $f(2) = -1$ and $g(2) = 5$. Therefore,

$$(f + g)(2) = f(2) + g(2)$$
$$= -1 + 5$$
$$= 4$$

Also, because $f(1) = 3$ and $g(1) = -2$, we have

$$(fg)(1) = f(1)g(1)$$
$$= 3(-2)$$
$$= -6$$

Since $(f \circ g)(-2) = f(g(-2))$, we first find $g(-2) = 0$. We then evaluate f at $x = 0$, because the output from g becomes the input to f. Thus,

$$(f \circ g)(-2) = f(g(-2))$$
$$= f(0)$$
$$= 6$$

EXAMPLE 7 The graphs of functions f and g are shown in Figure 5. Use the graphs to evaluate $(f - g)(-4)$ and $(g \circ f)(1)$.

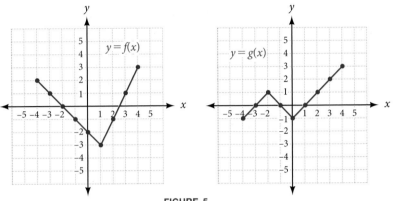

FIGURE 5

SOLUTION From the graphs, we observe that $f(-4) = 2$ and $g(-4) = -1$. Therefore,

$$(f - g)(-4) = f(-4) - g(-4)$$
$$= 2 - (-1)$$
$$= 3$$

Because $f(1) = -3$ and $g(-3) = 0$, we have

$$(g \circ f)(1) = g(f(1))$$
$$= g(-3)$$
$$= 0$$

Getting Ready for Class

After reading through the preceding section, respond in your own words and in complete sentences.

A. How are profit, revenue, and cost related?

B. Explain how you would evaluate the sum function $f + g$ for a given input.

C. For functions f and g, how do you find the composition of f with g?

D. For functions f and g, how do you find the composition of g with f?

Problem Set 3.7

Let $f(x) = 2x + 1$, $g(x) = 4x + 2$, and $h(x) = 4x^2 + 4x + 1$, and find the following.

1. $(f + g)(2)$ **2.** $(f - g)(-1)$ **3.** $(fg)(3)$ **4.** $\left(\dfrac{f}{g}\right)(-3)$

5. $\left(\dfrac{h}{g}\right)(1)$ **6.** $(hg)(1)$ **7.** $(fh)(0)$ **8.** $(h - g)(-4)$

9. $(f + g + h)(2)$ **10.** $(h - f + g)(0)$ **11.** $(h + fg)(3)$ **12.** $(h - fg)(5)$

Let $f(x) = 4x - 3$ and $g(x) = 2x + 5$. Write a formula for each of the following functions.

13. $f + g$ **14.** $f - g$ **15.** $g - f$ **16.** $g + f$

17. fg **18.** $\dfrac{f}{g}$ **19.** $\dfrac{g}{f}$ **20.** ff

If the functions f, g, and h are defined by $f(x) = 3x - 5$, $g(x) = x - 2$ and $h(x) = 3x^2 - 11x + 10$, write a formula for each of the following functions.

21. $g + f$ **22.** $f + h$ **23.** $g + h$ **24.** $f - g$

25. $g - f$ **26.** $h - g$ **27.** fg **28.** gf

29. fh **30.** gh **31.** $\dfrac{h}{f}$ **32.** $\dfrac{h}{g}$

33. $\dfrac{f}{h}$ **34.** $\dfrac{g}{h}$ **35.** $f + g + h$ **36.** $h - g + f$

37. $h + fg$ **38.** $h - fg$

39. If $f(x) = x^2 + 3x$ and $g(x) = 4x - 1$, find
   a. $(f + g)(2)$ by evaluating $f(2)$ and $g(2)$
   b. $(f + g)(x)$
   c. Evaluate your formula from part b at $x = 2$ to verify your answer to part a

40. If $f(x) = 2x - 5$ and $g(x) = 4x^2 + 3$, find
   a. $(f - g)(-1)$ by evaluating $f(-1)$ and $g(-1)$
   b. $(f - g)(x)$
   c. Evaluate your formula from part b at $x = -1$ to verify your answer to part a.

41. If $f(x) = 3x - 2$ and $g(x) = 5x + 4$, find
   a. $(fg)(-2)$
   b. $(fg)(x)$
   c. Evaluate your formula from part b at $x = -2$ to verify your answer to part a.

42. If $f(x) = x^2 - 4$ and $g(x) = x + 2$, find

   a. $\left(\dfrac{f}{g}\right)(3)$ by evaluating $f(3)$ and $g(3)$

   b. $\left(\dfrac{f}{g}\right)(x)$

   c. Evaluate your formula from part b at $x = 3$ to verify your answer to part a.

If $f(x) = 2x + 1$, $g(x) = 4x + 2$, and $h(x) = 4x^2 + 4x + 1$, find the following:

43. $(f \circ g)(-1)$ **44.** $(f \circ h)(-2)$ **45.** $(g \circ h)(1)$ **46.** $(g \circ f)(3)$

47. $(h \circ f)(-3)$ **48.** $(h \circ g)(2)$ **49.** $(g \circ f)(0)$ **50.** $(h \circ f)(0)$

51. Let $f(x) = x^2$ and $g(x) = x + 4$, and find
   a. $(f \circ g)(5)$ **b.** $(g \circ f)(5)$ **c.** $(f \circ g)(x)$ **d.** $(g \circ f)(x)$

52. Let $f(x) = 3 - x$ and $g(x) = x^3 - 1$, and find
   a. $(f \circ g)(0)$ **b.** $(g \circ f)(0)$ **c.** $(f \circ g)(x)$ **d.** $(g \circ f)(x)$

53. Let $f(x) = x^2 + 3x$ and $g(x) = 4x - 1$, and find
   a. $(f \circ g)(0)$ **b.** $(g \circ f)(0)$ **c.** $(f \circ g)(x)$ **d.** $(g \circ f)(x)$

54. Let $f(x) = (x - 2)^2$ and $g(x) = x + 1$, and find the following
   a. $(f \circ g)(-1)$ **b.** $(g \circ f)(-1)$ **c.** $(f \circ g)(x)$ **d.** $(g \circ f)(x)$

For each of the following pairs of functions f and g, show that
$(f \circ g)(x) = (g \circ f)(x) = x$.

55. $f(x) = 5x - 4$ and $g(x) = \dfrac{x + 4}{5}$ **56.** $f(x) = \dfrac{x}{6} - 2$ and $g(x) = 6x + 12$

Use the tables given for functions f and g to find each value in Problems 57–70.

x	-3	-2	-1	0	1	2	3
$f(x)$	4	5	1	-2	-1	0	2

x	-3	-2	-1	0	1	2	3
$g(x)$	3	0	2	-4	-2	1	-1

57. $(f + g)(0)$ **58.** $(g + f)(3)$ **59.** $(g - f)(-2)$ **60.** $(f - g)(1)$

61. $(fg)(-3)$ **62.** $\left(\dfrac{f}{g}\right)(0)$ **63.** $\left(\dfrac{g}{f}\right)(3)$ **64.** $(gf)(2)$

65. $(f \circ g)(-1)$ **66.** $(f \circ g)(-3)$ **67.** $(g \circ f)(2)$ **68.** $(g \circ f)(1)$

69. $(f \circ f)(0)$ **70.** $(g \circ g)(3)$

Use the graphs of $y = f(x)$ and $y = g(x)$ given below to find each value in Problems 71–84.

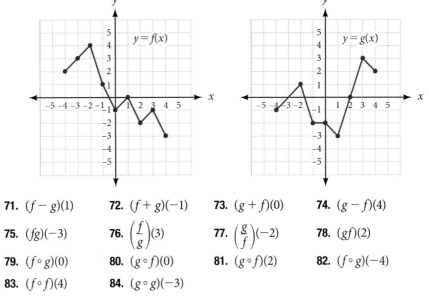

71. $(f - g)(1)$ **72.** $(f + g)(-1)$ **73.** $(g + f)(0)$ **74.** $(g - f)(4)$

75. $(fg)(-3)$ **76.** $\left(\dfrac{f}{g}\right)(3)$ **77.** $\left(\dfrac{g}{f}\right)(-2)$ **78.** $(gf)(2)$

79. $(f \circ g)(0)$ **80.** $(g \circ f)(0)$ **81.** $(g \circ f)(2)$ **82.** $(f \circ g)(-4)$

83. $(f \circ f)(4)$ **84.** $(g \circ g)(-3)$

Applying the Concepts

85. Profit, Revenue, and Cost A company manufactures and sells DVDs. Here are the equations they use in connection with their business.

Number of DVDs sold each day: $n(x) = x$

Selling price for each DVD: $p(x) = 11.5 - 0.05x$

Daily fixed costs: $f(x) = 200$

Daily variable costs: $v(x) = 2x$

Find the following functions.

  a. Revenue $= R(x) =$ the product of the number of DVDs sold each day and the selling price of each DVD.

  b. Cost $= C(x) =$ the sum of the fixed costs and the variable costs.

  c. Profit $= P(x) =$ the difference between revenue and cost.

  d. Average cost $= \overline{C}(x) =$ the quotient of cost and the number of DVDs sold each day.

86. Profit, Revenue, and Cost A company manufactures and sells CDs for home computers. Here are the equations they use in connection with their business.

Number of CDs sold each day: $n(x) = x$

Selling price for each CD: $p(x) = 3 - \dfrac{1}{300}x$

Daily fixed costs: $f(x) = 200$

Daily variable costs: $v(x) = 2x$

Find the following functions.

  a. Revenue $= R(x) =$ the product of the number of CDs sold each day and the selling price of each CD.

  b. Cost $= C(x) =$ the sum of the fixed costs and the variable costs.

  c. Profit $= P(x) =$ the difference between revenue and cost.

  d. Average cost $= \overline{C}(x) =$ the quotient of cost and the number of CDs sold each day.

87. Training Heart Rate Recall the heart rate functions discussed earlier in this section. Find the training heart rate function, $T(M)$, for a person with a resting heart rate of 62 beats per minute, then find the following to the nearest whole number.

  a. Find the maximum heart rate function, $M(x)$, for a person x years of age.

  b. What is the maximum heart rate for a 24-year-old person?

  c. What is the training heart rate for a 24-year-old person with a resting heart rate of 62 beats per minute?

  d. What is the training heart rate for a 36-year-old person with a resting heart rate of 62 beats per minute?

  e. What is the training heart rate for a 48-year-old person with a resting heart rate of 62 beats per minute?

88. **Training Heart Rate** Find the training heart rate function, $T(M)$ for a person with a resting heart rate of 72 beats per minute, then find the following to the nearest whole number.

   a. Find the maximum heart rate function, $M(x)$, for a person x years of age.

   b. What is the maximum heart rate for a 20-year-old person?

   c. What is the training heart rate for a 20-year-old person with a resting heart rate of 72 beats per minute?

   d. What is the training heart rate for a 30-year-old person with a resting heart rate of 72 beats per minute?

   e. What is the training heart rate for a 40-year-old person with a resting heart rate of 72 beats per minute?

Learning Objectives Assessment

The following problems can be used to help assess if you have successfully met the learning objectives for this section.

89. If $f(x) = 3x - 2$ and $g(x) = 4x + 5$, find $(f + g)(-1)$.

   a. -6 **b.** 3 **c.** -4 **d.** 10

90. If $f(x) = 3x - 2$ and $g(x) = 4x + 5$, find $(f - g)(x)$.

   a. $-x - 7$ **b.** -8 **c.** -7 **d.** $x^2 + 7x$

91. If $f(x) = 3x - 2$ and $g(x) = 4x + 5$, find $(f \circ g)(-2)$.

   a. -3 **b.** 24 **c.** -27 **d.** -11

92. If $f(x) = 3x - 2$ and $g(x) = 4x + 5$, find $(g \circ f)(x)$.

   a. $12x - 3$ **b.** $12x + 13$ **c.** $12x^2 + 7x - 10$ **d.** $12x^2 - 3x$

Getting Ready for the Next Section

Simplify.

93. $16(3.5)^2$

94. $\dfrac{2,400}{100}$

95. $\dfrac{180}{45}$

96. $4(2)(4)^2$

97. $\dfrac{0.0005(200)}{(0.25)^2}$

98. $\dfrac{0.2(0.5)^2}{100}$

99. If $y = Kx$, find K if $x = 5$ and $y = 15$.

100. If $d = Kt^2$, find K if $t = 2$ and $d = 64$.

101. If $P = \dfrac{K}{V}$, find K if $P = 48$ and $V = 50$.

102. If $y = Kxz^2$, find K if $x = 5$, $z = 3$, and $y = 180$.

Variation

Learning Objectives

In this section we will learn how to:

1. Solve a direct variation problem.

2. Solve an inverse variation problem.

3. Solve problems involving joint or combined variations.

4. Solve applied problems involving variation.

Introduction

If you are a runner and you average t minutes for every mile you run during one of your workouts, then your speed s in miles per hour is given by the equation and graph shown here. The graph (Figure 1) is shown in the first quadrant only because both t and s are positive.

$$s = \frac{60}{t}$$

Input	Output
t	s
4	15
6	10
8	7.5
10	6
12	5
14	4.3

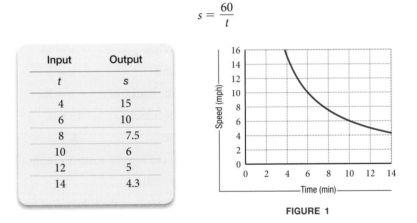

FIGURE 1

You know intuitively that as your average time per mile t increases, your speed s decreases. Likewise, lowering your time per mile will increase your speed. The equation and Figure 1 also show this to be true: Increasing t decreases s, and decreasing t increases s. Quantities that are connected in this way are said to *vary inversely* with each other. Inverse variation is one of the topics we will study in this section.

There are two main types of variation: ***direct variation*** and ***inverse variation***. Variation problems are most common in the sciences, particularly in chemistry and physics.

Direct Variation

When we say the variable y *varies directly* with the variable x, we mean that the relationship can be written in symbols as $y = Kx$, where K is a nonzero constant called the ***constant of variation*** (or *proportionality constant*).

Another way of saying y varies directly with x is to say y is ***directly proportional*** to x.

Study the following list. It gives the mathematical equivalent of some direct variation statements.

Verbal Phrase	Algebraic Equation
y varies directly with x.	$y = Kx$
s varies directly with the square of t.	$s = Kt^2$
y is directly proportional to the cube of z.	$y = Kz^3$
u is directly proportional to the square root of v.	$u = K\sqrt{v}$

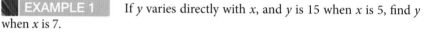 **EXAMPLE 1**   If y varies directly with x, and y is 15 when x is 5, find y when x is 7.

SOLUTION   The first part of the sentence gives us the general relationship between x and y. The equation equivalent to the statement "y varies directly with x" is

$$y = Kx$$

The second part of the sentence in our example gives us the information necessary to evaluate the constant K:

When	$y = 15$
and	$x = 5$
the equation	$y = Kx$
becomes	$15 = K \cdot 5$
or	$K = 3$

The equation can now be written specifically as

$$y = 3x$$

We can now determine the value asked for in the third part of the sentence. Letting $x = 7$, we have

$$y = 3 \cdot 7$$
$$y = 21$$

Notice that the equation $y = 3x$ from Example 1 is a linear equation. The graph of $y = 3x$ is a line with slope $m = 3$ passing through the origin.

In general, the direct variation $y = Kx$ represents a linear function whose graph is a line that passes through the origin. The variation constant K is the slope of this line, or the rate of change in y with respect to x.

The reverse is also true. Any linear function having a y-intercept of 0 will be a direct variation with K given by the slope.

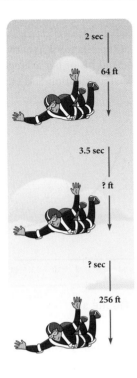

EXAMPLE 2   A skydiver jumps from a plane. Like any object that falls toward earth, the distance the skydiver falls is directly proportional to the square of the time he has been falling, until he reaches his terminal velocity. If the skydiver falls 64 feet in the first 2 seconds of the jump, then

a. How far will he have fallen after 3.5 seconds?
b. Graph the relationship between distance and time.
c. How long will it take him to fall 256 feet?

SOLUTION   We let t represent the time the skydiver has been falling, then we can let $d(t)$ represent the distance he has fallen.

a. Since $d(t)$ is directly proportional to the square of t, we have the general function that describes this situation:

$$d(t) = Kt^2$$

Next, we use the fact that $d(2) = 64$ to find K.

$$64 = K(2)^2$$
$$K = 16$$

The specific equation that describes this situation is

$$d(t) = 16t^2$$

To find how far a skydiver will fall after 3.5 seconds, we find $d(3.5)$,

$$d(3.5) = 16(3.5)^2$$
$$d(3.5) = 196$$

A skydiver will fall 196 feet after 3.5 seconds.

b. To graph this equation, we use a table:

Input	Output
t	$d(t)$
0	0
1	16
2	64
3	144
4	256
5	400

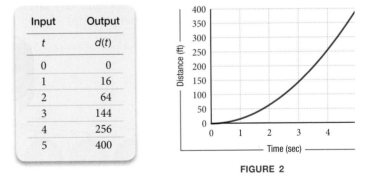

FIGURE 2

c. From the table or the graph (Figure 2), we see that it will take 4 seconds for the skydiver to fall 256 feet.

Inverse Variation

From the introduction to this section, we know that the relationship between the number of minutes t it takes a person to run a mile and his or her average speed in miles per hour s can be described with the following equation and table, and with Figure 3.

$$s = \frac{60}{t}$$

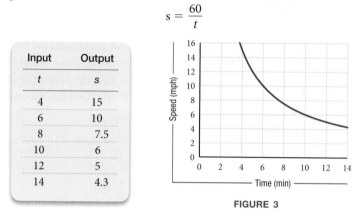

Input	Output
t	s
4	15
6	10
8	7.5
10	6
12	5
14	4.3

FIGURE 3

If t decreases, then s will increase, and if t increases, then s will decrease. The variable s is **inversely proportional** to the variable t. In this case, the *constant of proportionality* is 60.

Another example of inverse variation can be found in photography. If you are familiar with the terminology and mechanics associated with photography, you know that the f-stop for a particular lens will increase as the aperture (the maximum diameter of the opening of the lens) decreases. In mathematics, we say that f-stop and aperture vary inversely with each other. The following diagram illustrates this relationship.

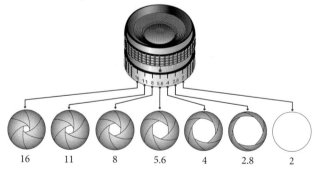

If f is the f-stop and d is the aperture, then their relationship can be written

$$f = \frac{K}{d}$$

In this case, K is the constant of proportionality. (Those of you familiar with photography know that K is also the focal length of the camera lens.)

We generalize this discussion of inverse variation as follows: If y varies inversely with x, then

$$y = K\frac{1}{x} \qquad \text{or} \qquad y = \frac{K}{x}$$

We can also say y is inversely proportional to x. The constant K is again called the constant of variation or proportionality constant.

Verbal Phrase	Algebraic Equation
y is inversely proportional to x.	$y = \dfrac{K}{x}$
s varies inversely with the square of t.	$s = \dfrac{K}{t^2}$
y is inversely proportional to x^4.	$y = \dfrac{K}{x^4}$
z varies inversely with the cube root of t.	$z = \dfrac{K}{\sqrt[3]{t}}$

EXAMPLE 3 The volume of a gas is inversely proportional to the pressure of the gas on its container. If a pressure of 48 pounds per square inch corresponds to a volume of 50 cubic feet, what pressure is needed to produce a volume of 100 cubic feet?

SOLUTION We can represent volume with V and pressure with P:

$$V = \frac{K}{P}$$

Using $P = 48$ and $V = 50$, we have

$$50 = \frac{K}{48}$$

$$K = 50(48)$$

$$K = 2{,}400$$

The equation that describes the relationship between P and V is

$$V = \frac{2{,}400}{P}$$

Here is a graph of this relationship.

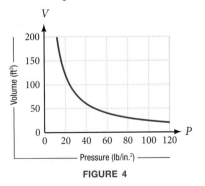

FIGURE 4

Note   The relationship between pressure and volume as given in this example is known as Boyle's law and applies to situations such as those encountered in a piston-cylinder arrangement. It was Robert Boyle (1627–1691) who, in 1662, published the results of some of his experiments that showed, among other things, that the volume of a gas decreases as the pressure increases. This is an example of inverse variation.

Substituting $V = 100$ into our last equation, we get

$$100 = \frac{2{,}400}{P}$$

$$100P = 2{,}400$$

$$P = \frac{2{,}400}{100}$$

$$P = 24$$

A volume of 100 cubic feet is produced by a pressure of 24 pounds per square inch.

Joint Variation and Other Variation Combinations

Many times relationships among different quantities are described in terms of more than two variables. If the variable y varies directly with *two* other variables, say x and z, then we say y varies *jointly* with x and z. In addition to **joint variation**, there are many other combinations of direct and inverse variation involving more than two variables.

The following table is a list of some variation statements and their equivalent mathematical forms:

Verbal Phrase	Algebraic Equation
y varies jointly with x and z.	$y = Kxz$
z varies jointly with r and the square of s.	$z = Krs^2$
V is directly proportional to T and inversely proportional to P.	$V = \dfrac{KT}{P}$
F varies jointly with m_1 and m_2 and inversely with the square of r.	$F = \dfrac{Km_1m_2}{r^2}$

EXAMPLE 4 y varies jointly with x and the square of z. When x is 5 and z is 3, y is 180. Find y when x is 2 and z is 4.

SOLUTION   The general equation is given by

$$y = Kxz^2$$

Substituting $x = 5$, $z = 3$, and $y = 180$, we have

$$180 = K(5)(3)^2$$

$$180 = 45K$$

$$K = 4$$

The specific equation is

$$y = 4xz^2$$

When $x = 2$ and $z = 4$, the last equation becomes

$$y = 4(2)(4)^2$$

$$y = 128$$

EXAMPLE 5 In electricity, the resistance of a cable is directly proportional to its length and inversely proportional to the square of the diameter. If a 100-foot cable 0.5 inch in diameter has a resistance of 0.2 ohm, what will be the resistance of a cable made from the same material if it is 200 feet long with a diameter of 0.25 inch?

SOLUTION Let R = resistance, l = length, and d = diameter. The equation is

$$R = \frac{Kl}{d^2}$$

When $R = 0.2$, $l = 100$, and $d = 0.5$, the equation becomes

$$0.2 = \frac{K(100)}{(0.5)^2}$$

or

$$K = 0.0005$$

Using this value of K in our original equation, the result is

$$R = \frac{0.0005l}{d^2}$$

When $l = 200$ and $d = 0.25$, the equation becomes

$$R = \frac{0.0005(200)}{(0.25)^2}$$

$$R = 1.6 \text{ ohms}$$

Getting Ready for Class

After reading through the preceding section, respond in your own words and in complete sentences.

A. Give an example of a direct variation statement, and then translate it into symbols.

B. Translate the equation $y = \frac{K}{x}$ into words.

C. For the inverse variation equation $y = \frac{3}{x}$ what happens to the values of y as x gets larger?

D. How are direct variation statements and linear equations in two variables related?

Problem Set 3.8

For the following problems, determine whether the equation represents a direct, inverse, or joint variation, and identify the constant of variation, K.

1. $y = \dfrac{3}{x}$

2. $y = 12x$

3. $C = 2\pi r$

4. $L = \dfrac{4}{d^2}$

5. $A = \dfrac{1}{2}bh$

6. $A = \pi r^2$

7. $S = 0.5\sqrt{d}$

8. $F = \dfrac{m_1 m_2}{9}$

Express each of the following sentences symbolically as an equation.

9. z is directly proportional to the square root of x.

10. y varies inversely with the cube of r.

11. F varies directly with the square of m and inversely with d.

12. p is jointly proportional to C and the square of t.

13. A varies jointly with h and the sum of a and b.

14. m varies directly with d and inversely with the square root of l.

For the following problems, y varies directly with x.

15. If y is 10 when x is 2, find y when x is 6.

16. If y is -32 when x is 4, find x when y is -40.

For the following problems, r is inversely proportional to s.

17. If r is -3 when s is 4, find r when s is 2.

18. If r is 8 when s is 3, find s when r is 48.

For the following problems, d varies directly with the square of r.

19. If $d = 10$ when $r = 5$, find d when $r = 10$.

20. If $d = 12$ when $r = 6$, find d when $r = 9$.

For the following problems, y varies inversely with the square of x.

21. If $y = 45$ when $x = 3$, find y when x is 5.

22. If $y = 12$ when $x = 2$, find y when x is 6.

For the following problems, z varies jointly with x and the square of y.

23. If z is 54 when x and y are 3, find z when $x = 2$ and $y = 4$.

24. If z is 27 when $x = 6$ and $y = 3$, find x when $z = 50$ and $y = 4$.

For the following problems, I varies inversely with the cube of w.

25. If $I = 32$ when $w = \dfrac{1}{2}$, find I when $w = \dfrac{1}{3}$.

26. If $I = \dfrac{1}{25}$ when $w = 5$, find I when $w = 10$.

For the following problems, z varies jointly with y and the square of x.

27. If $z = 72$ when $x = 3$ and $y = 2$, find z when $x = 5$ and $y = 3$.

28. If $z = 240$ when $x = 4$ and $y = 5$, find z when $x = 6$ and $y = 3$.

29. If $x = 1$ when $z = 25$ and $y = 5$, find x when $z = 160$ and $y = 8$.

30. If $x = 4$ when $z = 96$ and $y = 2$, find x when $z = 108$ and $y = 1$.

For the following problems, F varies directly with m and inversely with the square of d.

31. If $F = 150$ when $m = 240$ and $d = 8$, find F when $m = 360$ and $d = 3$.

32. If $F = 72$ when $m = 50$ and $d = 5$, find F when $m = 80$ and $d = 6$.

33. If $d = 5$ when $F = 24$ and $m = 20$, find d when $F = 18.75$ and $m = 40$.

34. If $d = 4$ when $F = 75$ and $m = 20$, find d when $F = 200$ and $m = 120$.

Applying the Concepts

35. Length of a Spring The length a spring stretches is directly proportional to the force applied. If a force of 5 pounds stretches a spring 7 inches, how much force is necessary to stretch the same spring 10 inches?

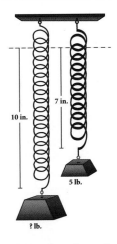

36. Weight and Surface Area The weight of a certain material varies directly with the surface area of that material. If 8 square feet weighs half a pound, how much will 10 square feet weigh?

37. Pressure and Temperature The temperature of a gas varies directly with its pressure. A temperature of 200 K produces a pressure of 50 pounds per square inch.

   a. Find the equation that relates pressure and temperature.

   b. Graph the equation from part a in the first quadrant only.

   c. What pressure will the gas have at 280 K?

38. Circumference and Diameter The circumference of a wheel is directly proportional to its diameter. A wheel has a circumference of 8.5 feet and a diameter of 2.7 feet.

   a. Find the equation that relates circumference and diameter.

   b. Graph the equation from part a in the first quadrant only.

   c. What is the circumference of a wheel that has a diameter of 11.3 feet?

39. Volume and Pressure The volume of a gas is inversely proportional to the pressure. If a pressure of 36 pounds per square inch corresponds to a volume of 25 cubic feet, what pressure is needed to produce a volume of 75 cubic feet?

40. Wave Frequency The frequency of an electromagnetic wave varies inversely with the wavelength. If a wavelength of 200 meters has a frequency of 800 kilocycles per second, what frequency will be associated with a wavelength of 500 meters?

41. f-Stop and Aperture Diameter The relative aperture, or f-stop, for a camera lens is inversely proportional to the diameter of the aperture. An f-stop of 2 corresponds to an aperture diameter of 40 millimeters for the lens on an automatic camera.

  a. Find the equation that relates f-stop and diameter.

  b. Graph the equation from part a in the first quadrant only.

  c. What is the f-stop of this camera when the aperture diameter is 10 millimeters?

42. f-Stop and Aperture Diameter The relative aperture, or f-stop, for a camera lens is inversely proportional to the diameter of the aperture. An f-stop of 2.8 corresponds to an aperture diameter of 75 millimeters for a certain telephoto lens.

  a. Find the equation that relates f-stop and diameter.

  b. Graph the equation from part a. in the first quadrant only.

  c. What aperture diameter corresponds to an f-stop of 5.6?

43. Surface Area of a Cylinder The surface area of a hollow cylinder varies jointly with the height and radius of the cylinder. If a cylinder with radius 3 inches and height 5 inches has a surface area of 94 square inches, what is the surface area of a cylinder with radius 2 inches and height 8 inches?

44. Capacity of a Cylinder The capacity of a cylinder varies jointly with its height and the square of its radius. If a cylinder with a radius of 3 centimeters and a height of 6 centimeters has a capacity of 169.56 cubic centimeters, what will be the capacity of a cylinder with radius 4 centimeters and height 9 centimeters?

45. Electrical Resistance The resistance of a wire varies directly with its length and inversely with the square of its diameter. If 100 feet of wire with diameter 0.01 inch has a resistance of 10 ohms, what is the resistance of 60 feet of the same type of wire if its diameter is 0.02 inch?

46. Volume and Temperature The volume of a gas varies directly with its temperature and inversely with the pressure. If the volume of a certain gas is 30 cubic feet at a temperature of 300 K and a pressure of 20 pounds per square inch, what is the volume of the same gas at 340 K when the pressure is 30 pounds per square inch?

47. Period of a Pendulum The time it takes for a pendulum to complete one period varies directly with the square root of the length of the pendulum. A 100-centimeter pendulum takes 2.1 seconds to complete one period.

  a. Find the equation that relates period and pendulum length.

  b. Graph the equation from part a in quadrant I only.

  c. How long does it take to complete one period if the pendulum hangs 225 centimeters?

48. Intensity of Light Table 4 gives the intensity of light that falls on a surface at various distances from a 100-watt light bulb. Construct a bar chart from the information in Table 4.

TABLE 4	Light intensity from a 100-watt light bulb
Distance Above Surface (ft)	Intensity (lumens/sq ft)
1	120.0
2	30.0
3	13.3
4	7.5
5	4.8
6	3.3

Learning Objectives Assessment

The following problems can be used to help assess if you have successfully met the learning objectives for this section.

49. If y varies directly with the square of x, and y is 450 when x is 5, find y when x is 3.

    a. 270   **b.** 750   **c.** 1,250   **d.** 162

50. If z is inversely proportional to p, and z is 36 when p is 4, find z when p is 6.

    a. 144   **b.** 9   **c.** 54   **d.** 24

51. If d varies jointly with r and the square root of x, and d is 36 when r is 6 and x is 4, find d when r is 5 and x is 9.

    a. 45   **b.** 180   **c.** 22.5   **d.** 6

52. The water temperature in a lake varies inversely with the depth. If the temperature is 10°C at a depth of 200 meters, what is the temperature at a depth of 250 meters?

    a. 16°C   **b.** 6.5°C   **c.** 8°C   **d.** 12.5°C

Maintaining Your Skills

The problems that follow review some of the more important skills you have learned in previous sections and chapters.

Solve the following equations.

53. $x - 5 = 7$ **54.** $3y = -4$ **55.** $5 - \dfrac{4}{7}a = -11$

56. $\dfrac{1}{5}x - \dfrac{1}{2} - \dfrac{1}{10}x + \dfrac{2}{5} = \dfrac{3}{10}x + \dfrac{1}{2}$

57. $5(x - 1) - 2(2x + 3) = 5x - 4$

58. $0.07 - 0.02(3x + 1) = -0.04x + 0.01$

Solve for the indicated variable.

59. $P = 2l + 2w$ for w

60. $A = \frac{1}{2}h(b + B)$ for B

Solve the following inequalities. Write the solution set using interval notation, then graph the solution set.

61. $-5t \leq 30$

62. $5 - \frac{3}{2}x > -1$

63. $1.6x - 2 < 0.8x + 2.8$

64. $3(2y + 4) \geq 5(y - 8)$

Solve the following equations.

65. $\left|\frac{1}{4}x - 1\right| = \frac{1}{2}$

66. $\left|\frac{2}{3}a + 4\right| = 6$

67. $|3 - 2x| + 5 = 2$

68. $5 = |3y + 6| - 4$

Chapter 3 Summary

EXAMPLES

Linear Equations in Two Variables [3.1, 3.3]

1. The equation $3x + 2y = 6$ is an example of a linear equation in two variables.

A *linear equation in two variables* is any equation that can be put in *standard form* $ax + by = c$. The graph of every linear equation is a straight line.

Intercepts [3.1]

2. To find the x-intercept for $3x + 2y = 6$, we let $y = 0$ and get

$$3x = 6$$
$$x = 2$$

In this case the x-intercept is 2, and the graph crosses the x-axis at $(2, 0)$.

The *x-intercept* of an equation is the *x-coordinate* of the point where the graph crosses the *x-axis*. The *y-intercept* is the *y-coordinate* of the point where the graph crosses the *y-axis*. We find the y-intercept by substituting $x = 0$ into the equation and solving for y. The x-intercept is found by letting $y = 0$ and solving for x.

The Slope of a Line [3.2]

3. The slope of the line through $(6, 9)$ and $(1, -1)$ is

$$m = \frac{9 - (-1)}{6 - 1} = \frac{10}{5} = 2$$

The *slope* of the line containing points (x_1, y_1) and (x_2, y_2) is given by

$$\text{Slope} = m = \frac{\text{Rise}}{\text{Run}} = \frac{y_2 - y_1}{x_2 - x_1}$$

Horizontal lines have 0 slope, and vertical lines have no slope.
Parallel lines have equal slopes, and perpendicular lines have slopes that are negative reciprocals.

The Slope-Intercept Form of a Line [3.3]

4. The equation of the line with slope 5 and y-intercept 3 is

$$y = 5x + 3$$

The equation of a line with slope m and y-intercept b is given by

$$y = mx + b$$

The Point-Slope Form of a Line [3.3]

5. The equation of the line through $(3, 2)$ with slope -4 is

$$y - 2 = -4(x - 3)$$

which can be simplified to

$$y = -4x + 14$$

The equation of the line through (x_1, y_1) that has slope m can be written as

$$y - y_1 = m(x - x_1)$$

Linear Inequalities in Two Variables [3.4]

6. The graph of

$$x - y \leq 3$$

is

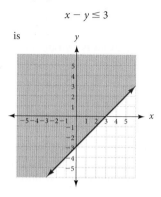

An inequality of the form $ax + by < c$ is a **linear inequality in two variables**. The equation for the boundary of the solution set is given by $ax + by = c$. (This equation is found by simply replacing the inequality symbol with an equal sign.)

To graph a linear inequality, first graph the boundary, using a solid line if the boundary is included in the solution set and a broken line if the boundary is not included in the solution set. Next, choose any point not on the boundary and substitute its coordinates into the original inequality. If the resulting statement is true, the graph lies on the same side of the boundary as the test point. A false statement indicates that the solution set lies on the other side of the boundary.

Relations and Functions [3.5]

7. The relation

$$\{(8, 1), (6, 1), (-3, 0)\}$$

is also a function because no ordered pairs have the same first coordinates. The domain is $\{-3, 6, 8\}$ and the range is $\{0, 1\}$.

A **relation** is any set of ordered pairs. The set of all first coordinates is called the **domain** of the relation, and the set of all second coordinates is the **range** of the relation.

A **function** is a rule that pairs each element in one set, called the **domain**, with exactly one element from a second set, called the **range**. A function is a relation in which no two different ordered pairs have the same first coordinates.

Vertical Line Test [3.5]

8. The graph of $x = y^2$ shown in Figure 2 in Section 3.5 fails the vertical line test. It is not the graph of a function.

If a vertical line crosses the graph of a relation in more than one place, the relation cannot be a function. If no vertical line can be found that crosses the graph in more than one place, the relation must be a function.

Function Notation [3.6]

9. If $f(x) = 5x - 3$ then
$$\begin{aligned} f(0) &= 5(0) - 3 \\ &= -3 \\ f(1) &= 5(1) - 3 \\ &= 2 \\ f(-2) &= 5(-2) - 3 \\ &= -13 \\ f(a) &= 5a - 3 \end{aligned}$$

The alternative notation for y is $f(x)$. It is read "f of x" and can be used instead of the variable y when working with functions. The notation y and the notation $f(x)$ are equivalent; that is, $y = f(x)$.

Linear Function [3.6]

10. The function $f(x) = 2x - 3$ is a linear function. The graph of f is a line with slope $m = 2$ and y-intercept $(0, -3)$.

A **linear function** is any function that can be expressed in the form $f(x) = mx + b$. The graph of f is a line with slope m passing through the y-axis at $(0, b)$.

Algebra with Functions [3.7]

11. If $f(x) = 3x - 1$ and $g(x) = 4x + 2$, then

$$(f + g)(0) = f(0) + g(0)$$
$$= -1 + 2$$
$$= 1$$

and

$$(f + g)(x) = f(x) + g(x)$$
$$= (3x - 1) + (4x + 2)$$
$$= 7x + 1$$

If f and g are any two functions with a common domain, then:

$(f + g)(x) = f(x) + g(x)$ The function $f + g$ is the sum of the functions f and g.

$(f - g)(x) = f(x) - g(x)$ The function $f - g$ is the difference of the functions f and g.

$(fg)(x) = f(x)g(x)$ The function fg is the product of the functions f and g.

$\dfrac{f}{g}(x) = \dfrac{f(x)}{g(x)}$ The function $\dfrac{f}{g}$ is the quotient of the functions f and g, where $g(x) \neq 0$

Composition of Functions [3.7]

12. If $f(x) = 3x - 1$ and $g(x) = 4x + 2$, then

$$(f \circ g)(0) = f(g(0))$$
$$= f(2)$$
$$= 5$$

and

$$(f \circ g)(x) = f(g(x))$$
$$= f(4x + 2)$$
$$= 3(4x + 2) - 1$$
$$= 12x + 5$$

If f and g are two functions for which the range of each has numbers in common with the domain of the other, then we have the following definitions:

The composition of f with g: $(f \circ g)(x) = f[g(x)]$

The composition of g with f: $(g \circ f)(x) = g[f(x)]$

Variation [3.8]

13. If y varies directly with x, then

$$y = Kx$$

Then if y is 18 when x is 6,

$$18 = K \cdot 6$$

or

$$K = 3$$

So the equation can be written more specifically as

$$y = 3x$$

If we want to know what y is when x is 4, we simply substitute:

$$y = 3 \cdot 4$$
$$y = 12$$

If y **varies directly** with x (y is directly proportional to x), then

$$y = Kx$$

If y **varies inversely** with x (y is inversely proportional to x), then

$$y = \frac{K}{x}$$

If z **varies jointly** with x and y (z is directly proportional to both x and y), then

$$z = Kxy$$

In each case, K is called the **constant of variation**.

⚠ COMMON MISTAKE

1. When graphing ordered pairs, the most common mistake is to associate the first coordinate with the y-axis and the second with the x-axis. If you make this mistake you would graph (3, 1) by going up 3 and to the right 1, which is just the reverse of what you should do. Remember, the first coordinate is always associated with the horizontal axis, and the second coordinate is always associated with the vertical axis.

2. The two most common mistakes students make when first working with the formula for the slope of a line are the following:

   a. Putting the difference of the x-coordinates over the difference of the y-coordinates.

   b. Subtracting in one order in the numerator and then subtracting in the opposite order in the denominator.

3. When graphing linear inequalities in two variables, remember to graph the boundary with a broken line when the inequality symbol is $<$ or $>$. The only time you use a solid line for the boundary is when the inequality symbol is $\leq$ or $\geq$.

For each of the following lines, identify the x-intercept, y-intercept, and slope, and sketch the graph. [3.1–3.3]

1. $2x + y = 6$ **2.** $y = -2x - 3$ **3.** $y = \dfrac{3}{2}x + 4$ **4.** $x = -2$

Find the equation for each line. [3.3]

5. Give the equation of the line through $(-1, 3)$ that has slope $m = 2$.

6. Give the equation of the line through $(-3, 2)$ and $(4, -1)$.

7. Line l contains the point $(5, -3)$ and has a graph parallel to the graph of $2x - 5y = 10$. Find the equation for l.

8. Line l contains the point $(-1, -2)$ and has a graph perpendicular to the graph of $y = 3x - 1$. Find the equation for l.

9. Give the equation of the vertical line through $(4, -7)$.

Graph the following linear inequalities. [3.4]

10. $3x - 4y < 12$ **11.** $y \le -x + 2$

State the domain and range for the following relations, and indicate which relations are also functions. [3.5]

12. $\{(-2, 0), (-3, 0), (-2, 1)\}$ **13.** $y = x^2 - 9$

Let $f(x) = x - 2$, $g(x) = 3x + 4$ and $h(x) = 3x^2 - 2x - 8$, and find the following. [3.6, 3.7]

14. $f(3) + g(2)$ **15.** $h(x) - g(x)$ **16.** $(f \circ g)(2)$ **17.** $(g \circ h)(x)$

Solve the following variation problems. [3.8]

18. Direct Variation Quantity y varies directly with the square of x. If y is 50 when x is 5, find y when x is 3.

19. Joint Variation Quantity z varies jointly with x and the cube of y. If z is 15 when x is 5 and y is 2, find z when x is 2 and y is 3.

20. Maximum Load The maximum load (L) a horizontal beam can safely hold varies jointly with the width (w) and the square of the depth (d) and inversely with the length (l). If a 10-foot beam with width 3 feet and depth 4 feet will safely hold up to 800 pounds, how many pounds will a 12-foot beam with width 3 feet and depth 4 feet hold?

Systems of Equations

4

iStockphoto.com © Neustock

Suppose you decide to buy a cellular phone and are trying to decide between two rate plans. Plan A is $18.95 per month plus $0.48 for each minute, or fraction of a minute, that you use the phone. Plan B is $34.95 per month plus $0.36 for each minute, or fraction of a minute. The monthly cost $C(x)$ for each plan can be represented with a linear equation in two variables:

$$\text{Plan A:} \quad C(x) = 0.48x + 18.95$$

$$\text{Plan B:} \quad C(x) = 0.36x + 34.95$$

To compare the two plans, we use the table and graph shown below.

Monthly Cellular Phone Charges

Number of Minutes x	Monthly Cost ($) Plan A	Plan B
0	18.95	34.95
40	38.15	49.35
80	57.35	63.75
120	76.55	78.15
160	95.75	92.55
200	114.95	106.95
240	134.15	121.35

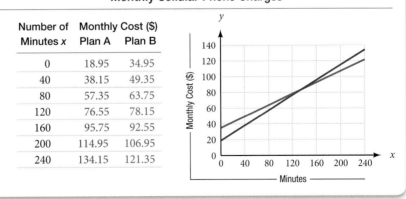

The point of intersection of the two lines in the graph is the point at which the monthly costs of the two plans are equal. In this chapter, we will develop methods of finding that point of intersection.

© pawel.gaul/iStockPhoto

Success Skills

The study skills for this chapter are concerned with getting ready to take an exam.

1. **Getting Ready to Take an Exam** Try to arrange your daily study habits so you have little studying to do the night before your next exam. The next two goals will help you achieve goal number 1.

2. **Review With the Exam in Mind** You should review material that will be covered on the next exam every day. Your review should consist of working problems. Preferably, the problems you work should be problems from your list of difficult problems.

3. **Continue to List Difficult Problems** You should continue to list and rework the problems that give you the most difficulty. It is this list that you will use to study for the next exam. Your goal is to go into the next exam knowing you can successfully work any problem from your list of hard problems.

4. **Pay Attention to Instructions** Taking a test is different from doing homework. When you take a test, the problems will be mixed up. When you do your homework, you usually work a number of similar problems. Sometimes students who do well on their homework become confused when they see the same problems on a test, because they have not paid attention to the instructions on their homework. For example, suppose you see the equation $y = 3x - 2$ on your next test. By itself, the equation is simply a statement. There isn't anything to do unless the equation is accompanied by instructions. Each of the following is a valid instruction with respect to the equation $y = 3x - 2$ and the result of applying the instructions will be different in each case:

> Find x when y is 10.
> Solve for x.
> Graph the equation.
> Find the intercepts.
> Find the slope.

There are many things to do with the equation. If you train yourself to pay attention to the instructions that accompany a problem as you work through the assigned problems, you will not find yourself confused about what to do with a problem when you see it on a test.

Systems of Linear Equations in Two Variables

Learning Objectives

In this section we will learn how to:

1. Solve a system of linear equations in two variables graphically.
2. Solve a system of linear equations in two variables using the addition method.
3. Solve a system of linear equations in two variables using substitution.
4. Recognize an inconsistent system or a system with dependent equations.

Introduction

Previously, we found the graph of an equation of the form $ax + by = c$ to be a line. Because the graph is a line, the equation is said to be a linear equation. Two linear equations considered together form a **linear system** of equations. For example,

$$3x - 2y = 6$$
$$2x + 4y = 20$$

is a linear system. The **solution set** to the system is the set of all ordered pairs that satisfy both equations. If we graph each equation on the same set of axes, we can see the solution set (see Figure 1).

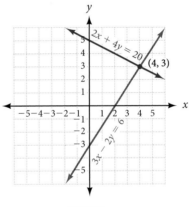

FIGURE 1

The point (4, 3) lies on both lines and therefore must satisfy both equations. It is obvious from the graph that it is the only point that does so. The solution set for the system is {(4, 3)}.

More generally, if $a_1x + b_1y = c_1$ and $a_2x + b_2y = c_2$ are linear equations, then the solution set for the system

$$a_1x + b_1y = c_1$$
$$a_2x + b_2y = c_2$$

can be illustrated through one of the graphs in Figure 2.

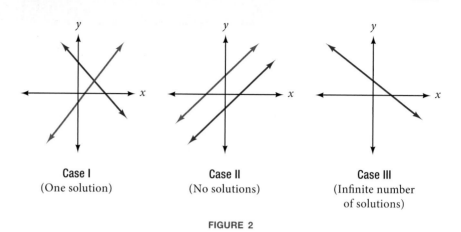

Case I
(One solution)

Case II
(No solutions)

Case III
(Infinite number
of solutions)

FIGURE 2

Case I   The two lines intersect at one and only one point. The coordinates of the point give the solution to the system. This is what usually happens.

Case II   The lines are parallel and therefore have no points in common. The solution set to the system is the empty set, $\varnothing$. In this case, we say the equations are *inconsistent*.

Case III   The lines coincide. That is, their graphs represent the same line. The solution set consists of all ordered pairs that satisfy either equation. In this case, the equations are said to be *dependent*.

Note A system of equations is *consistent* if it has at least one solution. It is *inconsistent* if it has no solution. Two equations are *dependent* if one is a multiple of the other. Otherwise, they are *independent*.

The Addition Method

In the beginning of this section, we found the solution set for the system

$$3x - 2y = 6$$
$$2x + 4y = 20$$

by graphing each equation and then reading the solution set from the graph. Solving a system of linear equations by graphing is the least accurate method. If the coordinates of the point of intersection are not integers, it can be difficult to read the solution set from the graph. There is another method of solving a linear system that does not depend on the graph. It is called the *addition method*.

VIDEO EXAMPLES

SECTION 4.1

EXAMPLE 1   Solve the system.

$$4x + 3y = 10$$
$$2x + y = 4$$

SOLUTION   If we multiply the bottom equation by -3, the coefficients of y in the resulting equation and the top equation will be opposites:

$$4x + 3y = 10 \xrightarrow{\text{No Change}} 4x + 3y = 10$$
$$2x + y = 4 \xrightarrow[\text{Multiply by } -3]{} -6x - 3y = -12$$

Adding the left and right sides of the resulting equations, we have

$$
\begin{array}{rcr}
4x + 3y = & & 10 \\
-6x - 3y = & & -12 \\
\hline
-2x \quad\quad\ = & & -2
\end{array}
$$

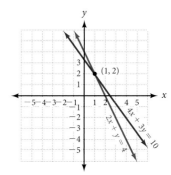

The result is a linear equation in one variable. We have eliminated the variable y from the equations by addition. (It is for this reason we call this method of solving a linear system the **addition method**.) Solving $-2x = -2$ for x, we have

$$x = 1$$

FIGURE 3 *A visual representation of the solution to the system in Example 1*

Note   If we had put $x = 1$ into the first equation in our system, we would have obtained $y = 2$ also:

$$4(1) + 3y = 10$$

$$3y = 6$$

$$y = 2$$

This is the x-coordinate of the solution to our system. To find the y-coordinate, we substitute $x = 1$ into any of the equations containing both the variables x and y. Let's try the second equation in our original system:

$$2(1) + y = 4$$

$$2 + y = 4$$

$$y = 2$$

This is the y-coordinate of the solution to our system. The ordered pair $(1, 2)$ is the solution to the system.

Checking Solutions

We can check our solution by substituting it into both of our equations.

Substituting $x = 1$ and $y = 2$ into $4x + 3y = 10$, we have

$$4(1) + 3(2) \overset{?}{=} 10$$

$$4 + 6 \overset{?}{=} 10$$

$$10 = 10 \quad \text{A true statement}$$

Substituting $x = 1$ and $y = 2$ into $2x + y = 4$, we have

$$2(1) + 2 \overset{?}{=} 4$$

$$2 + 2 \overset{?}{=} 4$$

$$4 = 4 \quad \text{A true statement}$$

Our solution satisfies both equations; therefore, it is a solution to our system of equations.

EXAMPLE 2   Solve the system.

$$3x - 5y = -2$$

$$2x - 3y = 1$$

SOLUTION   We can eliminate either variable. Let's decide to eliminate the variable x. We can do so by multiplying the top equation by 2 and the bottom equation by -3, and then adding the left and right sides of the resulting equations:

$$
\begin{array}{lcl}
3x - 5y = -2 & \xrightarrow{\text{Multiply by } 2} & 6x - 10y = -4 \\
2x - 3y = \ \ 1 & \xrightarrow[\text{Multiply by } -3]{} & -6x + 9y \ = -3 \\
& & \hline \\
& & \qquad\ -y = -7 \\
& & \qquad\ \ \ y = \ \ 7
\end{array}
$$

The y-coordinate of the solution to the system is 7. Substituting this value of y into any of the equations with both x- and y-variables gives $x = 11$. The solution to the system is $(11, 7)$. It is the only ordered pair that satisfies both equations.

Checking Solutions

Checking $(11, 7)$ in each equation looks like this:

Substituting $x = 11$ and $y = 7$ into $3x - 5y = -2$, we have

$$3(11) - 5(7) \overset{?}{=} -2$$
$$33 - 35 \overset{?}{=} -2$$
$$-2 = -2 \qquad \text{A true statement}$$

Substituting $x = 11$ and $y = 7$ into $2x - 3y = 1$, we have

$$2(11) - 3(7) \overset{?}{=} 1$$
$$22 - 21 \overset{?}{=} 1$$
$$1 = 1 \qquad \text{A true statement}$$

Our solution satisfies both equations; therefore, $(11, 7)$ is a solution to our system.

EXAMPLE 3   Solve the system.

$$2x - 3y = 4$$
$$4x + 5y = 3$$

SOLUTION   We can eliminate x by multiplying the top equation by -2 and adding it to the bottom equation:

$$
\begin{array}{ll}
2x - 3y = 4 & \xrightarrow{\text{Multiply by } -2} \quad -4x + 6y = -8 \\
4x + 5y = 3 & \xrightarrow[\text{No Change}]{} \qquad\quad\; 4x + 5y = 3 \\
\hline
& \hspace{4.5cm} 11y = -5 \\
& \hspace{4.5cm} y = -\dfrac{5}{11}
\end{array}
$$

The y-coordinate of our solution is $-\frac{5}{11}$. If we were to substitute this value of y back into either of our original equations, we would find the arithmetic necessary to solve for x cumbersome. For this reason, it is probably best to go back to the original system and solve it a second time—for x instead of y. Here is how we do that:

$$
\begin{array}{ll}
2x - 3y = 4 & \xrightarrow{\text{Multiply by } 5} \quad 10x - 15y = 20 \\
4x + 5y = 3 & \xrightarrow[\text{Multiply by } 3]{} \quad 12x + 15y = 9 \\
\hline
& \hspace{3.8cm} 22x = 29 \\
& \hspace{4.5cm} x = \dfrac{29}{22}
\end{array}
$$

The solution to our system is $\left(\frac{29}{22}, -\frac{5}{11} \right)$.

The main idea in solving a system of linear equations by the addition method is to use the multiplication property of equality on one or both of the original equations, if necessary, to make the coefficients of either variable opposites. The following box shows some steps to follow when solving a system of linear equations by the addition method.

> **HOW TO** *Solve a System of Linear Equations by the Addition Method*
>
> **Step 1:** Decide which variable to eliminate. (In some cases, one variable will be easier to eliminate than the other. With some practice, you will notice which one it is.)
>
> **Step 2:** Use the multiplication property of equality on each equation separately to make the coefficients of the variable that is to be eliminated opposites.
>
> **Step 3:** Add the respective left and right sides of the system together.
>
> **Step 4:** Solve for the remaining variable.
>
> **Step 5:** Substitute the value of the variable from step 4 into an equation containing both variables and solve for the other variable. (Or repeat steps 2–4 to eliminate the other variable.)
>
> **Step 6:** Check your solution in both equations, if necessary.

EXAMPLE 4 Solve the system.

$$5x - 2y = 5$$
$$-10x + 4y = 15$$

SOLUTION We can eliminate y by multiplying the first equation by 2 and adding the result to the second equation:

$$5x - 2y = 5 \xrightarrow{\text{Multiply by 2}} 10x - 4y = 10$$
$$-10x + 4y = 15 \xrightarrow[\text{No Change}]{} \underline{-10x + 4y = 15}$$
$$0 = 25$$

The result is the false statement $0 = 25$, which indicates there is no solution to the system. If we were to graph the two lines, we would find that they are parallel. In a case like this, we say the system is *inconsistent*. Whenever both variables have been eliminated and the resulting statement is false, the solution set for the system will be the empty set, $\varnothing$.

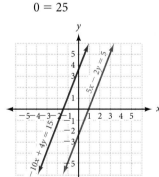

FIGURE 4 *A visual representation of the situation in Example 4 — the two lines are parallel*

EXAMPLE 5   Solve the system.

$$4x + 3y = 2$$

$$8x + 6y = 4$$

SOLUTION   Multiplying the top equation by -2 and adding, we can eliminate the variable x:

$$4x + 3y = 2 \xrightarrow{\text{Multiply by } -2} -8x - 6y = -4$$

$$8x + 6y = 4 \xrightarrow[\text{No Change}]{} \underline{8x + 6y = 4}$$

$$0 = 0$$

Both variables have been eliminated and the resulting statement $0 = 0$ is true. In this case, the lines coincide and the equations are said to be *dependent*. The solution set consists of all ordered pairs that satisfy either equation. We can write the solution set as $\{(x, y) | 4x + 3y = 2\}$ or $\{(x, y) | 8x + 6y = 4\}$.

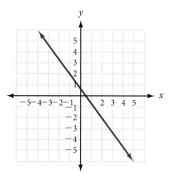

FIGURE 5 *A visual representation of the situation in Example 5 — both equations produce the same graph*

Special Cases

The previous two examples illustrate the two special cases in which the graphs of the equations in the system either coincide or are parallel. In both cases the left-hand sides of the equations were multiples of each other. In the case of the dependent equations the right-hand sides were also multiples. We can generalize these observations for the system

$$a_1 x + b_1 y = c_1$$

$$a_2 x + b_2 y = c_2$$

Inconsistent System

What happens	*Geometric Interpretation*	*Algebraic Interpretation*
Both variables are eliminated, and the resulting statement is false.	The lines are parallel, and there is no solution to the system.	$\dfrac{a_1}{a_2} = \dfrac{b_1}{b_2} \neq \dfrac{c_1}{c_2}$

Dependent Equations

What happens	*Geometric Interpretation*	*Algebraic Interpretation*
Both variables are eliminated, and the resulting statement is true.	The lines coincide, and there are an infinite number of solutions to the system.	$\dfrac{a_1}{a_2} = \dfrac{b_1}{b_2} = \dfrac{c_1}{c_2}$

EXAMPLE 6   Solve the system.

$$\frac{1}{2}x - \frac{1}{3}y = 2$$

$$\frac{1}{4}x + \frac{2}{3}y = 6$$

SOLUTION   Although we could solve this system without clearing the equations of fractions, there is probably less chance for error if we have only integer coefficients to work with. So let's begin by multiplying both sides of the top equation by 6, and both sides of the bottom equation by 12, to clear each equation of fractions:

$$\frac{1}{2}x - \frac{1}{3}y = 2 \xrightarrow{\text{Multiply by 6}} 3x - 2y = 12$$

$$\frac{1}{4}x + \frac{2}{3}y = 6 \xrightarrow[\text{Multiply by 12}]{} 3x + 8y = 72$$

Now we can eliminate x by multiplying the top equation by -1 and leaving the bottom equation unchanged:

$$3x - 2y = 12 \xrightarrow{\text{Multiply by} -1} -3x + 2y = -12$$

$$3x + 8y = 72 \xrightarrow[\text{No Change}]{} \underline{\quad 3x + 8y = \quad 72 \quad}$$

$$10y = 60$$

$$y = 6$$

We can substitute $y = 6$ into any equation that contains both x and y. Let's use $3x - 2y = 12$.

$$3x - 2(6) = 12$$

$$3x - 12 = 12$$

$$3x = 24$$

$$x = \ \ 8$$

The solution to the system is $(8, 6)$.

The Substitution Method

We end this section by considering another method of solving a linear system. The method is called the *substitution method* and is shown in the following examples.

EXAMPLE 7   Solve the system.

$$2x - 3y = -6$$

$$y = 3x - 5$$

SOLUTION   The second equation tells us y is $3x - 5$. Substituting the expression $3x - 5$ for y in the first equation, we have

$$2x - 3(3x - 5) = -6$$

The result of the substitution is the elimination of the variable y. Solving the resulting linear equation in x as usual, we have

$$2x - 9x + 15 = -6$$
$$-7x + 15 = -6$$
$$-7x = -21$$
$$x = 3$$

Putting $x = 3$ into the second equation in the original system, we have

$$y = 3(3) - 5$$
$$= 9 - 5$$
$$= 4$$

The solution to the system is $(3, 4)$.

Checking Solutions

Checking $(3, 4)$ in each equation looks like this:

Substituting $x = 3$ and $y = 4$ into $2x - 3y = -6$, we have

$$2(3) - 3(4) \stackrel{?}{=} -6$$
$$6 - 12 \stackrel{?}{=} -6$$
$$-6 = -6 \qquad \text{A true statement}$$

Substituting $x = 3$ and $y = 4$ into $y = 3x - 5$, we have

$$4 \stackrel{?}{=} 3(3) - 5$$
$$4 \stackrel{?}{=} 9 - 5$$
$$4 = 4 \qquad \text{A true statement}$$

Our solution satisfies both equations; therefore, $(3, 4)$ is a solution to our system.

Here are the steps to use in solving a system of equations by the substitution method.

> **HOW TO** **Solve a System of Equations by the Substitution Method**
>
> **Step 1:** Solve either one of the equations for x or y. (This step is not necessary if one of the equations is already in the correct form, as in Example 7.)
>
> **Step 2:** Substitute the expression for the variable obtained in step 1 into the other equation and solve it for the remaining variable.
>
> **Step 3:** Substitute the solution from step 2 into your result from step 1 and simplify.
>
> **Step 4:** Check your results, if necessary.

Note Both the substitution method and the addition method can be used to solve any system of linear equations in two variables. Systems like the one in Example 7, however, are easier to solve using the substitution method, because one of the variables is already written in terms of the other. A system like the one in Example 6 is easier to solve using the addition method, because solving for one of the variables would lead to an expression involving fractions. The system in Example 8 could be solved easily by either method, because solving the second equation for x is a one-step process.

EXAMPLE 8   Solve by substitution

$$2x + 3y = 5$$
$$x - 2y = 6$$

SOLUTION To use the substitution method, we must solve one of the two equations for x or y. We can solve for x in the second equation by adding $2y$ to both sides:

$$x - 2y = 6$$
$$x = 2y + 6 \qquad \text{Add } 2y \text{ to both sides}$$

Substituting the expression $2y + 6$ for x in the first equation of our system, we have

$$2(2y + 6) + 3y = 5$$
$$4y + 12 + 3y = 5$$
$$7y + 12 = 5$$
$$7y = -7$$
$$y = -1$$

Replacing y with -1 in our first step, we find

$$x = 2(-1) + 6$$
$$= 4$$

The solution is $(4, -1)$.

USING TECHNOLOGY   *Graphing Calculators*

Solving Systems That Intersect in Exactly One Point

A graphing calculator can be used to solve a system of equations in two variables if the equations intersect in exactly one point. To solve the system shown in Example 3, we first solve each equation for y. Here is the result:

$$2x - 3y = 4 \quad \text{becomes} \quad y = \frac{4 - 2x}{-3}$$

$$4x + 5y = 3 \quad \text{becomes} \quad y = \frac{3 - 4x}{5}$$

Graphing these two functions on the calculator gives a diagram similar to the one in Figure 6.

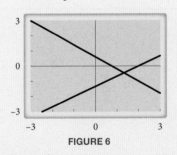

FIGURE 6

Using the Intersect command, we find that the two lines intersect at $x = 1.32$ and $y = -0.45$, which are the decimal equivalents (accurate to the nearest hundredth) of the fractions found in Example 3.

Special Cases

We cannot assume that two lines that look parallel in a calculator widow are in fact parallel. If you graph the functions $y = x - 5$ and $y = 0.99x + 2$ in a window where x and y range from -10 to 10, the lines look parallel. We know this is not the case, however, since their slopes are different. As we zoom out repeatedly, the lines begin to look as if they coincide. We know this is not the case, because the two lines have different y-intercepts. To summarize: If we graph two lines on a calculator and the graphs look as if they are parallel or coincide, we should use algebraic methods, not the calculator, to determine the solution to the system.

Getting Ready for Class

After reading through the preceding section, respond in your own words and in complete sentences.

A. Two linear equations, each with the same two variables, form a system of equations. How do we define a solution to this system? That is, what form will a solution have, and what properties does a solution possess?

B. When would substitution be more efficient than the addition method in solving two linear equations?

C. Explain what an inconsistent system of linear equations looks like graphically and what would result algebraically when attempting to solve the system.

D. When might the graphing method of solving a system of equations be more desirable than the other techniques, and when might it be less desirable?

Solve each system by graphing both equations on the same set of axes and then reading the solution from the graph.

1. $3x - 2y = 6$

$x - y = 1$

2. $5x - 2y = 10$

$x - y = -1$

3. $y = \dfrac{3}{5}x - 3$

$2x - y = -4$

4. $y = \dfrac{1}{2}x - 2$

$2x - y = -1$

5. $y = \dfrac{1}{2}x$

$y = -\dfrac{3}{4}x + 5$

6. $y = \dfrac{2}{3}x$

$y = -\dfrac{1}{3}x + 6$

7. $3x + 3y = -2$

$y = -x + 4$

8. $2x - y = 5$

$y = 2x - 5$

Solve each of the following systems by the addition method.

9. $3x + y = 5$

$3x - y = 3$

10. $-x - y = 4$

$-x + 2y = -3$

11. $-3x - 2y = -1$

$-6x + 4y = -2$

12. $x + 3y = 3$

$2x - 9y = 1$

13. $2x - 5y = 16$

$4x - 3y = 11$

14. $5x - 3y = -11$

$7x + 6y = -12$

15. $6x + 3y = -1$

$9x + 5y = 1$

16. $5x + 4y = -1$

$7x + 6y = -2$

17. $4x + 3y = 14$

$9x - 2y = 14$

18. $7x - 6y = 13$

$6x - 5y = 11$

19. $2x - 5y = 3$

$-4x + 10y = 3$

20. $x + 2y = 0$

$-3x - 6y = 5$

21. $\dfrac{1}{2}x + \dfrac{1}{3}y = 13$

$\dfrac{2}{5}x + \dfrac{1}{4}y = 10$

22. $\dfrac{1}{2}x + \dfrac{1}{3}y = \dfrac{2}{3}$

$\dfrac{2}{3}x + \dfrac{2}{5}y = \dfrac{14}{15}$

23. $\dfrac{1}{2}x - \dfrac{3}{4}y = -\dfrac{1}{2}$

$\dfrac{1}{3}x - \dfrac{1}{2}y = -\dfrac{1}{3}$

24. $\dfrac{1}{2}x - \dfrac{1}{3}y = \dfrac{5}{6}$

$-\dfrac{3}{10}x + \dfrac{1}{5}y = -\dfrac{1}{2}$

Solve each of the following systems by the substitution method.

25. $7x - y = 24$

$x = 2y + 9$

26. $3x - y = -8$

$y = 6x + 3$

27. $6x - y = 10$

$\dfrac{3}{4}x + y = -1$

28. $2x - y = 6$

$\dfrac{4}{3}x + y = 1$

29. $y = 3x - 2$

$y = 4x - 4$

30. $y = 5x - 2$

$y = -2x + 5$

31. $2x - y = 10$

$4x - 2y = 10$

32. $-10x + 8y = -6$

$5x - 4y = 0$

33. $\dfrac{1}{3}x - \dfrac{1}{2}y = 0$

$x = \dfrac{3}{2}y$

34. $\dfrac{2}{5}x - \dfrac{2}{3}y = 0$

$y = \dfrac{3}{5}x$

You may want to read Example 3 again before solving the systems that follow.

35. $4x - 7y = 3$
    $5x + 2y = -3$

36. $3x - 4y = 7$
    $6x - 3y = 5$

37. $9x - 8y = 4$
    $2x + 3y = 6$

38. $4x - 7y = 10$
    $-3x + 2y = -9$

39. $3x - 5y = 2$
    $7x + 2y = 1$

40. $4x - 3y = -1$
    $5x + 8y = 2$

Solve each of the following systems by using either the addition or substitution method. Choose the method that is most appropriate for the problem.

41. $x - 3y = 7$

    $2x + y = -6$

42. $2x - y = 9$

    $x + 2y = -11$

43. $y = \dfrac{1}{2}x + \dfrac{1}{3}$

    $y = -\dfrac{1}{3}x + 2$

44. $y = \dfrac{3}{4}x - \dfrac{4}{5}$

    $y = \dfrac{1}{2}x - \dfrac{1}{2}$

45. $3x - 4y = 12$

    $x = \dfrac{2}{3}y - 4$

46. $-5x + 3y = -15$

    $x = \dfrac{4}{5}y - 2$

47. $4x - 3y = -7$

    $-8x + 6y = -11$

48. $3x - 4y = 8$

    $y = \dfrac{3}{4}x - 2$

49. $3y + z = 17$

    $5y + 20z = 65$

50. $x + y = 850$

    $1.5x + y = 1,100$

51. $\dfrac{3}{4}x - \dfrac{1}{3}y = 1$

    $y = \dfrac{1}{4}x$

52. $-\dfrac{2}{3}x + \dfrac{1}{2}y = -1$

    $y = -\dfrac{1}{3}x$

53. $\dfrac{1}{4}x - \dfrac{1}{2}y = \dfrac{1}{3}$

    $\dfrac{1}{3}x - \dfrac{1}{4}y = -\dfrac{2}{3}$

54. $\dfrac{1}{5}x - \dfrac{1}{10}y = -\dfrac{1}{5}$

    $\dfrac{2}{3}x - \dfrac{1}{2}y = -\dfrac{1}{6}$

55. Work each problem according to the instructions given.
  a. Simplify: $(3x - 4y) - 3(x - y)$
  b. Find y when x is 0 in $3x - 4y = 8$.
  c. Find the y-intercept: $3x - 4y = 8$
  d. Graph: $3x - 4y = 8$
  e. Find the point where the graphs of $3x - 4y = 8$ and $x - y = 2$ cross.

56. Work each problem according to the instructions given.
  a. Solve: $4x - 5 = 20$
  b. Solve for y: $4x - 5y = 20$
  c. Solve for x: $x - y = 5$
  d. Solve the system:

$$4x - 5 = 20$$
$$x - y = 5$$

57. Multiply both sides of the second equation in the following system by 100, and then solve as usual.

$$x + y = 10,000$$
$$0.06x + 0.05y = 560$$

58. What value of c will make the following system a dependent system (one in which the lines coincide)?

$$6x - 9y = 3$$
$$4x - 6y = c$$

59. Where do the graphs of the lines $x + y = 4$ and $x - 2y = 4$ intersect?

60. Where do the graphs of the lines $x = -1$ and $x - 2y = 4$ intersect?

Learning Objectives Assessment

The following problems can be used to help assess if you have successfully met the learning objectives for this section.

61. Figure 7 shows the graphs of two linear equations in two variables. Which of the following could be a solution to the system consisting of the two equations?

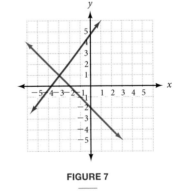

FIGURE 7

  a. $(0, 4)$ **b.** $(-4, 0)$ **c.** $(-3, 1)$ **d.** $(3, -1)$

62. Solve the following system using the addition method. Which value is part of the solution?

$$3x - 2y = 9$$
$$x + 5y = 1$$

  a. $y = \dfrac{1}{5}$ **b.** $y = -\dfrac{6}{17}$ **c.** $x = \dfrac{5}{2}$ **d.** $x = \dfrac{7}{4}$

63. Solve the following system using substitution. Which value is part of the solution?

$$5x + y = 2$$
$$2x - 3y = 4$$

  a. $y = 2$ **b.** $y = -\dfrac{14}{17}$ **c.** $x = \dfrac{2}{5}$ **d.** $x = \dfrac{10}{17}$

64. For which value of c will the following system have an infinite number of solutions?

$$x - 2y = 1$$
$$-2x + 4y = c$$

a. 1 **b.** -1 **c.** 2 **d.** -2

Getting Ready for the Next Section

Simplify.

65. $2 - 2(6)$ **66.** $2(1) - 2 + 3$

67. $(x + 3y) - 1(x - 2z)$ **68.** $(x + y + z) + (2x - y + z)$

Solve.

69. $-9y = -9$ **70.** $30x = 38$

71. $3(1) + 2z = 9$ **72.** $4\left(\dfrac{19}{15}\right) - 2y = 4$

Apply the distributive property, then simplify if possible.

73. $2(5x - z)$ **74.** $-1(x - 2z)$

75. $3(3x + y - 2z)$ **76.** $2(2x - y + z)$

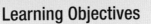

4.2

Learning Objectives

In this section, we will learn how to:

1. Determine if an ordered triple is a solution to a system of linear equations in three variables.

2. Solve a system of three linear equations in three variables.

3. Recognize a system of linear equations in three variables having dependent equations.

4. Recognize an inconsistent system of linear equations in three variables.

Introduction

A solution to an equation in three variables such as

$$2x + y - 3z = 6$$

is an ordered triple of numbers (x, y, z). For example, the ordered triples $(0, 0, -2)$, $(2, 2, 0)$, and $(0, 9, 1)$ are solutions to the equation $2x + y - 3z = 6$, because they produce a true statement when their coordinates are substituted for x, y, and z in the equation.

> **(dĕf) DEFINITION** *solution set*
>
> The *solution set* for a system of three linear equations in three variables is the set of ordered triples that satisfies all three equations.

VIDEO EXAMPLES

SECTION 4.2

EXAMPLE 1 Determine if the given ordered triple is a solution to the system.

$$x + y + 2z = 4$$
$$2x - y + z = -1$$
$$-x - 2y + 3z = 5$$

a. $(2, 4, -1)$ **b.** $(-1, 1, 2)$

SOLUTION All we need to do is see if the ordered triple satisfies all three equations.

a. Substituting $x = 2$, $y = 4$, and $z = -1$ we get

$$2 + 4 + 2(-1) \stackrel{?}{=} 4$$
$$4 = 4 \qquad \text{True}$$

$$2(2) - 4 + (-1) \stackrel{?}{=} -1$$
$$-1 = -1 \qquad \text{True}$$

$$-2 - 2(4) + 3(-1) \stackrel{?}{=} 5$$
$$-13 \neq 5 \qquad \text{False}$$

The ordered triple $(2, 4, -1)$ is not a solution because it does not satisfy the third equation.

b. Substituting $x = -1$, $y = 1$, and $z = 2$ gives us

$$-1 + 1 + 2(2) \stackrel{?}{=} 4$$
$$4 = 4 \qquad \text{True}$$

$$2(-1) - 1 + 2 \stackrel{?}{=} -1$$
$$-1 = -1 \qquad \text{True}$$

$$-(-1) - 2(1) + 3(2) \stackrel{?}{=} 5$$
$$5 = 5 \qquad \text{True}$$

The ordered triple $(-1, 1, 2)$ satisfies all three equations, so it is a solution to the system.

EXAMPLE 2 Solve the system.

$$\begin{aligned} x + y + z &= 6 & (1) \\ 2x - y + z &= 3 & (2) \\ x + 2y - 3z &= -4 & (3) \end{aligned}$$

SOLUTION We want to find the ordered triple (x, y, z) that satisfies all three equations. We have numbered the equations so it will be easier to keep track of where they are and what we are doing.

There are many ways to proceed. The main idea is to take two different pairs of equations and eliminate the same variable from each pair. We begin by adding equations (1) and (2) to eliminate the y-variable. The resulting equation is numbered (4):

$$\begin{aligned} x + y + z &= 6 & (1) \\ \underline{2x - y + z} &= \underline{3} & (2) \\ 3x + 2z &= 9 & (4) \end{aligned}$$

Adding twice equation (2) to equation (3) will also eliminate the variable y. The resulting equation is numbered (5):

$$\begin{aligned} 4x - 2y + 2z &= 6 & \text{Twice (2)} \\ \underline{x + 2y - 3z} &= \underline{-4} & (3) \\ 5x - z &= 2 & (5) \end{aligned}$$

Equations (4) and (5) form a linear system in two variables. By multiplying equation (5) by 2 and adding the result to equation (4), we succeed in eliminating the variable z from the new pair of equations:

$$\begin{aligned} 3x + 2z &= 9 & (4) \\ \underline{10x - 2z} &= \underline{4} & \text{Twice (5)} \\ 13x &= 13 \\ x &= 1 \end{aligned}$$

Substituting $x = 1$ into equation (4), we have

$$3(1) + 2z = 9$$

$$2z = 6$$

$$z = 3$$

Using $x = 1$ and $z = 3$ in equation (1) gives us

$$1 + y + 3 = 6$$

$$y + 4 = 6$$

$$y = 2$$

The solution is the ordered triple $(1, 2, 3)$.

EXAMPLE 3   Solve the system.

$$2x + y - z = 3 \quad (1)$$

$$3x + 4y + z = 6 \quad (2)$$

$$2x - 3y + z = 1 \quad (3)$$

SOLUTION   It is easiest to eliminate z from the equations. The equation produced by adding (1) and (2) is

$$5x + 5y = 9 \quad (4)$$

The equation that results from adding (1) and (3) is

$$4x - 2y = 4 \quad (5)$$

Equations (4) and (5) form a linear system in two variables. We can eliminate the variable y from this system as follows:

$$5x + 5y = 9 \xrightarrow{\text{Multiply by 2}} 10x + 10y = 18$$

$$4x - 2y = 4 \xrightarrow[\text{Multiply by 5}]{} 20x - 10y = 20$$

$$\overline{30x = 38}$$

$$x = \frac{38}{30}$$

$$= \frac{19}{15}$$

Now we substitute $x = \frac{19}{15}$ into equation (5) or equation (4) and solve for y. Using equation (4), we obtain

$$5\left(\frac{19}{15}\right) + 5y = 9$$

$$\frac{19}{3} + 5y = 9 \qquad \text{Reduce}$$

$$19 + 15y = 27 \qquad \text{Multiply both sides by 3}$$

$$15y = 8$$

$$y = \frac{8}{15}$$

Finally, we substitute $x = \frac{19}{15}$ and $y = \frac{8}{15}$ into equation (1), (2), or (3) and solve for z. Using equation (3) gives us

$$2\left(\frac{19}{15}\right) - 3\left(\frac{8}{15}\right) + z = 1$$

$$\frac{38}{15} - \frac{24}{15} + z = 1$$

$$\frac{14}{15} + z = 1$$

$$z = 1 - \frac{14}{15}$$

$$z = \frac{1}{15}$$

The ordered triple that satisfies all three equations is $\left(\frac{19}{15}, \frac{8}{15}, \frac{1}{15}\right)$.

EXAMPLE 4   Solve the system.

$$2x + 3y - z = 5 \quad (1)$$
$$4x + 6y - 2z = 10 \quad (2)$$
$$x - 4y + 3z = 5 \quad (3)$$

SOLUTION   Multiplying equation (1) by -2 and adding the result to equation (2) looks like this:

$$-4x - 6y + 2z = -10 \qquad -2 \text{ times } (1)$$
$$\underline{4x + 6y - 2z = 10} \qquad (2)$$
$$0 = 0$$

All three variables have been eliminated, and we are left with a true statement. This implies that the two equations are dependent. With a system of three equations in three variables, however, a dependent system can have no solution or an infinite number of solutions. After we have concluded the examples in this section, we will discuss the geometry behind these systems. Doing so will give you some additional insight into dependent systems.

EXAMPLE 5   Solve the system.

$$x - 5y + 4z = 8 \quad (1)$$
$$3x + y - 2z = 7 \quad (2)$$
$$-9x - 3y + 6z = 5 \quad (3)$$

SOLUTION   Multiplying equation (2) by 3 and adding the result to equation (3) produces

$$9x + 3y - 6z = 21 \qquad 3 \text{ times } (2)$$
$$\underline{-9x - 3y + 6z = 5} \qquad (3)$$
$$0 = 26$$

In this case, all three variables have been eliminated, and we are left with a false statement. The two equations are inconsistent; there are no ordered triples that satisfy both equations. The solution set for the system is the empty set, $\varnothing$. If equations (2) and (3) have no ordered triples in common, then certainly (1), (2), and (3) do not either. ∎

EXAMPLE 6   Solve the system.

$$
\begin{aligned}
x + 3y &= 5 &(1) \\
6y + z &= 12 &(2) \\
x - 2z &= -10 &(3)
\end{aligned}
$$

SOLUTION   It may be helpful to rewrite the system as

$$
\begin{aligned}
x + 3y &= 5 &(1) \\
6y + z &= 12 &(2) \\
x - 2z &= -10 &(3)
\end{aligned}
$$

Equation (2) does not contain the variable x. If we multiply equation (3) by -1 and add the result to equation (1), we will be left with another equation that does not contain the variable x:

$$
\begin{array}{ll}
x + 3y = 5 & (1) \\
-x + 2z = 10 & -1 \text{ times } (3) \\
\hline
3y + 2z = 15 & (4)
\end{array}
$$

Equations (2) and (4) form a linear system in two variables. Multiplying equation (2) by -2 and adding the result to equation (4) eliminates the variable z:

$$
\begin{array}{lcl}
6y + z = 12 & \xrightarrow{\text{Multiply by } -2} & -12y - 2z = -24 \\
3y + 2z = 15 & \xrightarrow[\text{No Change}]{} & 3y + 2z = 15 \\
& & \hline \\
& & -9y = -9 \\
& & y = 1
\end{array}
$$

Using $y = 1$ in equation (4) and solving for z, we have

$$z = 6$$

Substituting $y = 1$ into equation (1) gives

$$x = 2$$

The ordered triple that satisfies all three equations is $(2, 1, 6)$. ∎

The Geometry Behind Equations in Three Variables

We can graph an ordered triple on a coordinate system with three axes. The graph will be a point in space. The coordinate system is drawn in perspective; you have to imagine that the x-axis comes out of the paper and is perpendicular to both the y-axis and the z-axis. To graph the point $(3, 4, 5)$, we move 3 units in the x-direction, 4 units in the y-direction, and then 5 units in the z-direction, as shown in Figure 1.

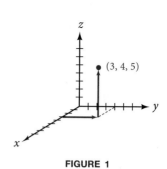

FIGURE 1

Although in actual practice it is sometimes difficult to graph equations in three variables, if we were to graph a linear equation in three variables, we would find that the graph was a plane in space. A system of three equations in three variables is represented by three planes in space.

There are a number of possible ways in which these three planes can intersect, some of which are shown below. And there are still other possibilities that are not among those shown.

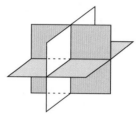

CASE 1 *The three planes have exactly one point in common. In this case we get one solution to our system, as in Examples 2, 3, and 6.*

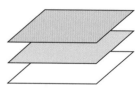

CASE 2 *The three planes have no points in common because they are all parallel to one another. The system they represent is an inconsistent system.*

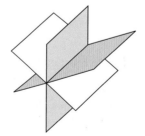

CASE 3 *The three planes intersect in a line. Any point on the line is a solution to the system of equations represented by the planes, so there is an infinite number of solutions to the system. This is an example of a dependent system.*

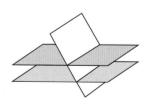

CASE 4 *Two of the planes are parallel; the third plane intersects each of the parallel planes. In this case, the three planes have no points in common. There is no solution to the system; it is an inconsistent system.*

In Example 4, we found that equations (1) and (2) were dependent equations. They represent the same plane. That is, they have all their points in common. But the system of equations that they came from has either no solution or an infinite number of solutions. It all depends on the third plane. If the third plane coincides with the first two, then the solution to the system is a plane. If the third plane is parallel to the first two, then there is no solution to the system. Finally, if the third plane intersects the first two but does not coincide with them, then the solution to the system is that line of intersection.

In Example 5 we found that trying to eliminate a variable from the second and third equations resulted in a false statement. This means that the two planes represented by these equations are parallel. It makes no difference where the third plane is; there is no solution to the system in Example 5. (If we were to graph the three planes from Example 5, we would obtain a diagram similar to Case 2 or Case 4.)

If, in the process of solving a system of linear equations in three variables, we eliminate all the variables from a pair of equations and are left with a false statement, we will say the system is inconsistent. If we eliminate all the variables and are left with a true statement, then we will say the system is a dependent one.

Getting Ready for Class

After reading through the preceding section, respond in your own words and in complete sentences.

A. What is an ordered triple of numbers?

B. Explain what it means for (1, 2, 3) to be a solution to a system of linear equations in three variables.

C. Explain in a general way the procedure you would use to solve a system of three linear equations in three variables.

D. How do you know when a system of linear equations in three variables has no solution?

Problem Set 4.2

Determine if each ordered triple is a solution to the given system.

1. $2x + y - z = 2$
$x + y + z = -1$
$3x - 2y + 2z = 3$

   a. $(2, 1, 3)$ **b.** $(1, -1, -1)$

2. $3x - 2y + z = -4$
$x - y - 2z = 0$
$-2x + y - z = 6$

   a. $(-9, -11, 1)$ **b.** $(-4, -4, 0)$

Solve the following systems.

3. $x + y + z = 4$
$x - y + 2z = 1$
$x - y - 3z = -4$

4. $x - y - 2z = -1$
$x + y + z = 6$
$x + y - z = 4$

5. $x + y + z = 6$
$x - y + 2z = 7$
$2x - y - 4z = -9$

6. $x + y + z = 0$
$x + y - z = 6$
$x - y + 2z = -7$

7. $x + 2y + z = 3$
$2x - y + 2z = 6$
$3x + y - z = 5$

8. $2x + y - 3z = -14$
$x - 3y + 4z = 22$
$3x + 2y + z = 0$

9. $2x + 3y - 2z = 4$
$x + 3y - 3z = 4$
$3x - 6y + z = -3$

10. $4x + y - 2z = 0$
$2x - 3y + 3z = 9$
$-6x - 2y + z = 0$

11. $-x + 4y - 3z = 2$
$2x - 8y + 6z = 1$
$3x - y + z = 0$

12. $4x + 6y - 8z = 1$
$-6x - 9y + 12z = 0$
$x - 2y - 2z = 3$

13. $\dfrac{1}{2}x - y + z = 0$

   $2x + \dfrac{1}{3}y + z = 2$

   $x + y + z = -4$

14. $\dfrac{1}{3}x + \dfrac{1}{2}y + z = -1$

   $x - y + \dfrac{1}{5}z = -1$

   $x + y + z = -5$

15. $2x - y - 3z = 1$
$x + 2y + 4z = 3$
$4x - 2y - 6z = 2$

16. $3x + 2y + z = 3$
$x - 3y + z = 4$
$-6x - 4y - 2z = 1$

17. $2x - y + 3z = 4$
$x + 2y - z = -3$
$4x + 3y + 2z = -5$

18. $6x - 2y + z = 5$
$3x + y + 3z = 7$
$x + 4y - z = 4$

19. $x + y = 9$
$y + z = 7$
$x - z = 2$

20. $x - y = -3$
$x + z = 2$
$y - z = 7$

21. $2x + y = 2$
$y + z = 3$
$4x - z = 0$

22. $2x + y = 6$
$3y - 2z = -8$
$x + z = 5$

23. $2x - 3y = 0$
$6y - 4z = 1$
$x + 2z = 1$

24. $3x + 2y = 3$
$y + 2z = 2$
$6x - 4z = 1$

25.
$$x + y - z = 2$$
$$2x + y + 3z = 4$$
$$x - 2y + 2z = 6$$

26.
$$x + 2y - 2z = 4$$
$$3x + 4y - z = -2$$
$$2x + 3y - 3z = -5$$

27.
$$2x + 3y = -\frac{1}{2}$$
$$4x + 8z = 2$$
$$3y + 2z = -\frac{3}{4}$$

28.
$$3x - 5y = 2$$
$$4x + 6z = \frac{1}{3}$$
$$5y - 7z = \frac{1}{6}$$

29.
$$\frac{1}{3}x + \frac{1}{2}y - \frac{1}{6}z = 4$$
$$\frac{1}{4}x - \frac{3}{4}y + \frac{1}{2}z = \frac{3}{2}$$
$$\frac{1}{2}x - \frac{2}{3}y - \frac{1}{4}z = -\frac{16}{3}$$

30.
$$-\frac{1}{4}x + \frac{3}{8}y + \frac{1}{2}z = -1$$
$$\frac{2}{3}x - \frac{1}{6}y - \frac{1}{2}z = 2$$
$$\frac{3}{4}x - \frac{1}{2}y - \frac{1}{8}z = 1$$

31.
$$x - \frac{1}{2}y - \frac{1}{3}z = -\frac{4}{3}$$
$$\frac{1}{3}x - \frac{1}{2}z = 5$$
$$-\frac{1}{4}x + \frac{2}{3}y - z = -\frac{3}{4}$$

32.
$$x + \frac{1}{3}y - \frac{1}{2}z = -\frac{3}{2}$$
$$\frac{1}{2}x - y + \frac{1}{3}z = 8$$
$$\frac{1}{3}x - \frac{1}{4}y - z = -\frac{5}{6}$$

33. Electric Current In the following diagram of an electrical circuit, x, y, and z represent the amount of current (in amperes) flowing across the 5-ohm, 20-ohm, and 10-ohm resistors, respectively. (In circuit diagrams, resistors are represented by —W— and potential differences by —|⊢.)

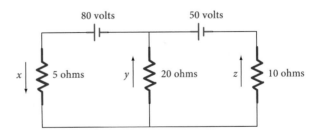

The system of equations used to find the three currents x, y, and z is

$$x - y - z = 0$$
$$5x + 20y = 80$$
$$20y - 10z = 50$$

Solve the system for all variables.

34. Cost of a Rental Car If a car rental company charges \$10 a day and 8¢ a mile to rent one of its cars, then the cost z, in dollars, to rent a car for x, days and drive y miles can be found from the equation

$$z = 10x + 0.08y$$

a. How much does it cost to rent a car for 2 days and drive it 200 miles under these conditions?

b. A second company charges \$12 a day and 6¢ a mile for the same car. Write an equation that gives the cost z, in dollars, to rent a car from this company for x days and drive it y miles.

c. A car is rented from each of the companies mentioned in parts a and b for 2 days. To find the mileage at which the cost of renting the cars from each of the two companies will be equal, solve the following system for y:

$$z = 10x + 0.08y$$
$$z = 12x + 0.06y$$
$$x = 2$$

Learning Objectives Assessment

The following problems can be used to help assess if you have successfully met the learning objectives for this section.

35. Which ordered triple is a solution to the following system?

$$2x - 3y = 1$$
$$x + 4z = 3$$
$$y - 2z = 4$$

a. $(2, 1, -1)$　　　b. $(-1, -1, 1)$　　　c. $\left(5, 3, -\dfrac{1}{2}\right)$　　　d. $\left(8, 5, \dfrac{1}{2}\right)$

36. Solve the system. Which value of z appears in the solution?

$$x - y + z = 2$$
$$-x - y + 2z = -4$$
$$2x - 2y - z = -2$$

a. $z = 0$　　　b. $z = -1$　　　c. $z = 2$　　　d. $z = 4$

37. Which value of c will result in the following system having dependent equations?

$$x + 2y - z = 3$$
$$2x + 4y - 2z = c$$
$$x - y + z = 1$$

a. 3　　　b. 6　　　c. 0　　　d. -3

38. Which value of c will result in the following system being inconsistent?

$$x + y - z = 2$$
$$-x - y + z = c$$
$$2x + y - 3z = 0$$

a. -2　　　b. 2

Getting Ready for the Next Section

Translate into symbols.

39. Two more than 3 times a number

40. One less than twice a number

Simplify.

41. $25 - \dfrac{385}{9}$

42. $0.30(12)$

43. $0.08(4,000)$

44. $500(1.5)$

45. $10(0.2x + 0.5y)$

46. $100(0.09x + 0.08y)$

Solve.

47. $x + (3x + 2) = 26$

48. $5x = 2,500$

Solve each system.

49. $\begin{aligned} -2y - 4z &= -18 \\ -7y + 4z &= 27 \end{aligned}$

50. $\begin{aligned} -x + 2y &= 200 \\ 4x - 2y &= 1,300 \end{aligned}$

There are a lot of word problems in algebra and many of them involve topics that I don't know much about. I am better off solving these problems if I know something about the subject. So, I try to find something I can relate to. For instance, an example may involve the amount of fuel used by a pilot in a jet airplane engine. In my mind, I'd change the subject to something more familiar, like the mileage I'd be getting in my car and the amount spent on fuel, driving from my hometown to my college. Changing these problems to more familiar topics makes math much more interesting and gives me a better chance of getting the problem right. It also helps me to understand how greatly math affects and influences me in my everyday life. We really do use math more than we would like to admit—budgeting our income, purchasing gasoline, planning a day of shopping with friends—almost everything we do is related to math. So the best advice I can give with word problems is to learn how to associate the problem with something familiar to you.

You should know that I have always enjoyed math. I like working out problems and love the challenges of solving equations like individual puzzles. Although there are more interesting subjects to me, and I don't plan on pursuing a career in math or teaching, I do think it's an important subject that will help you in any profession.

Applications

Learning Objectives

In this section, we will learn how to:

1. Solve applications with two unknowns using a system of two linear equations in two variables.

2. Solve applications with three unknowns using a system of three linear equations in three variables.

Introduction

Many times word problems involve more than one unknown quantity. If a problem is stated in terms of two unknowns and we represent each unknown quantity with a different variable, then we must write the relationships between the variables with two equations. The two equations written in terms of the two variables form a system of linear equations that we solve using the methods developed in this chapter. If we find a problem that relates three unknown quantities, then we need three equations to form a linear system we can solve.

Here is our Blueprint for Problem Solving, modified to fit the application problems that you will find in this section.

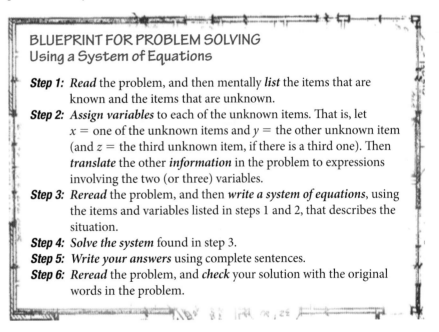

> **BLUEPRINT FOR PROBLEM SOLVING**
> Using a System of Equations
>
> **Step 1:** *Read* the problem, and then mentally *list* the items that are known and the items that are unknown.
> **Step 2:** *Assign variables* to each of the unknown items. That is, let $x =$ one of the unknown items and $y =$ the other unknown item (and $z =$ the third unknown item, if there is a third one). Then *translate* the other *information* in the problem to expressions involving the two (or three) variables.
> **Step 3:** *Reread* the problem, and then *write a system of equations*, using the items and variables listed in steps 1 and 2, that describes the situation.
> **Step 4:** *Solve the system* found in step 3.
> **Step 5:** *Write your answers* using complete sentences.
> **Step 6:** *Reread* the problem, and *check* your solution with the original words in the problem.

EXAMPLE 1 One number is 2 more than 3 times another. Their sum is 26. Find the two numbers.

SOLUTION Applying the steps from our Blueprint, we have:

Step 1: *Read and list.*
We know that we have two numbers, whose sum is 26. One of them is 2 more than 3 times the other. The unknown quantities are the two numbers.

Step 2: *Assign variables and translate information.*
Let $x =$ one of the numbers and $y =$ the other number.

Step 3: *Write a system of equations.*

The first sentence in the problem translates into $y = 3x + 2$. The second sentence gives us a second equation: $x + y = 26$. Together, these two equations give us the following system of equations:

$$x + y = 26$$
$$y = 3x + 2$$

Step 4: *Solve the system.*

Substituting the expression for y from the second equation into the first and solving for x yields

$$x + (3x + 2) = 26$$
$$4x + 2 = 26$$
$$4x = 24$$
$$x = 6$$

Using $x = 6$ in $y = 3x + 2$ gives the second number:

$$y = 3(6) + 2$$
$$y = 20$$

Step 5: *Write answers.*

The two numbers are 6 and 20.

Step 6: *Reread and check.*

The sum of 6 and 20 is 26, and 20 is 2 more than 3 times 6. ◼

EXAMPLE 2   Suppose 850 tickets were sold for a game for a total of $1,100. If adult tickets cost $1.50 and children's tickets cost $1.00, how many of each kind of ticket were sold?

SOLUTION

Step 1: *Read and list.*

The total number of tickets sold is 850. The total income from tickets is $1,100. Adult tickets are $1.50 each. Children's tickets are $1.00 each. We don't know how many of each type of ticket have been sold.

Step 2: *Assign variables and translate information.*

We let $x =$ the number of adult tickets and $y =$ the number of children's tickets.

Step 3: *Write a system of equations.*

The total number of tickets sold is 850, giving us our first equation.

$$x + y = 850$$

Because each adult ticket costs $1.50, and each children's ticket costs $1.00, and the total amount of money paid for tickets was $1,100, a second equation is

$$1.50x + 1.00y = 1,100$$

The same information can also be obtained by summarizing the problem with a table. One such table follows. Notice that the two equations we obtained previously are given by the two rows of the table.

Note Example 2 involves the concept of *value*. In writing the second equation we use the formula

$$P \cdot Q = V$$

where P is the price of each item, Q is the quantity (number) of items, and V is the total value of these items.

	Adult Tickets	Children's Tickets	Total
Number	x	y	850
Value	$1.50x$	$1.00y$	1,100

Whether we use a table to summarize the information in the problem or just talk our way through the problem, the system of equations that describes the situation is

$$x + y = 850$$
$$1.50x + 1.00y = 1,100$$

Step 4: *Solve the system.*

If we multiply the second equation by 10 to clear it of decimals, we have the system

$$x + y = 850$$
$$15x + 10y = 11,000$$

Multiplying the first equation by -10 and adding the result to the second equation eliminates the variable y from the system:

$$-10x - 10y = -8,500$$
$$\underline{15x + 10y = 11,000}$$
$$5x = 2,500$$
$$x = 500$$

The number of adult tickets sold was 500. To find the number of children's tickets, we substitute $x = 500$ into $x + y = 850$ to get

$$500 + y = 850$$
$$y = 350$$

Step 5: *Write answers.*

The number of children's tickets is 350, and the number of adult tickets is 500.

Step 6: *Reread and check.*

The total number of tickets is $350 + 500 = 850$. The amount of money from selling the two types of tickets is

350 children's tickets at \$1.00 each is $350(1.00) = \$350$
500 adult tickets at \$1.50 each is $500(1.50) = \$750$

The total income from ticket sales is \$1,100

EXAMPLE 3   Suppose a person invests a total of \$10,000 in two accounts. One account earns 8% annually, and the other earns 9% annually. If the total interest earned from both accounts in a year is \$860, how much was invested in each account?

SOLUTION

Step 1: *Read and list.*

The total investment is \$10,000 split between two accounts. One account earns 8% annually, and the other earns 9% annually. The interest from both accounts is \$860 in 1 year. We don't know how much is in each account.

Step 2: *Assign variables and translate information.*

We let x equal the amount invested at 9% and y be the amount invested at 8%.

Step 3: *Write a system of equations.*

Because the total investment is $10,000, one relationship between x and y can be written as

$$x + y = 10,000$$

The total interest earned from both accounts is $860. The amount of interest earned on x dollars at 9% is $0.09x$, while the amount of interest earned on y dollars at 8% is $0.08y$. This relationship is represented by the equation

$$0.09x + 0.08y = 860$$

> **Note** The interest earned in each account is found using the simple interest formula
>
> $$R \cdot P = I$$
>
> where R is the interest rate (the percent in decimal form), P is the principal (the amount of money invested), and I is the interest earned after one year.

The two equations we have just written can also be found by first summarizing the information from the problem in a table. Again, the two rows of the table yield the two equations we found previously. Here is the table.

	Dollars at 9%	Dollars at 8%	Total
Number	x	y	10,000
Interest	$0.09x$	$0.08y$	860

The system of equations that describes this situation is given by

$$x + \quad y = 10,000$$
$$0.09x + 0.08y = \quad 860$$

Step 4: *Solve the system.*

Multiplying the second equation by 100 will clear it of decimals. The system that results after doing so is

$$x + \quad y = 10,000$$
$$9x + 8y = 86,000$$

We can eliminate y from this system by multiplying the first equation by -8 and adding the result to the second equation.

$$-8x - 8y = -80,000$$
$$\underline{9x + 8y = \quad 86,000}$$
$$x \quad = \quad 6,000$$

The amount of money invested at 9% is $6,000. Because the total investment was $10,000, the amount invested at 8% must be $4,000.

Step 5: *Write answers.*

The amount invested at 8% is $4,000, and the amount invested at 9% is $6,000.

Step 6: *Reread and check.*

The total investment is $4,000 + $6,000 = $10,000. The amount of interest earned from the two accounts is

In 1 year, $4,000 invested at 8% earns $0.08(4,000) = $320

In 1 year, $6,000 invested at 9% earns $\underline{0.09(6,000) = \$540}$

The total interest from the two accounts is $860

EXAMPLE 4   How much 20% alcohol solution and 50% alcohol solution must be mixed to get 12 gallons of 30% alcohol solution?

SOLUTION   To solve this problem, we must first understand that a 20% alcohol solution is 20% alcohol and 80% water.

Step 1: *Read and list.*

We will mix two solutions to obtain 12 gallons of solution that is 30% alcohol. One of the solutions is 20% alcohol and the other 50% alcohol. We don't know how much of each solution we need.

Step 2: *Assign variables and translate information.*

Let x = the number of gallons of 20% alcohol solution needed, and y = the number of gallons of 50% alcohol solution needed.

Step 3: *Write a system of equations.*

Because we must end up with a total of 12 gallons of solution, one equation for the system is

$$x + y = 12$$

The amount of alcohol in the x gallons of 20% solution is $0.20x$, while the amount of alcohol in the y gallons of 50% solution is $0.50y$. Because the total amount of alcohol in the 20% and 50% solutions must add up to the amount of alcohol in the 12 gallons of 30% solution, the second equation in our system can be written as

$$0.20x + 0.50y = 0.30(12)$$

Again, let's make a table that summarizes the information we have to this point in the problem.

> *Note*   For mixture problems we use the formula
>
> $$P \cdot Q = A$$
>
> where P is the percent of some item (expressed as a decimal), Q is the quantity of the solution, and A is the amount of this item contained in the solution.

	20% Solution	50% Solution	Final Solution
Total number of gallons	x	y	12
Gallons of alcohol	$0.20x$	$0.50y$	$0.30(12)$

Our system of equations is

$$x + y = 12$$
$$0.20x + 0.50y = 0.30(12) = 3.6$$

Step 4: *Solve the system.*

Multiplying the second equation by 10 gives us an equivalent system:

$$x + y = 12$$
$$2x + 5y = 36$$

Multiplying the top equation by -2 to eliminate the x-variable, we have

$$-2x - 2y = -24$$
$$\underline{2x + 5y = 36}$$
$$3y = 12$$
$$y = 4$$

Substituting $y = 4$ into $x + y = 12$, we solve for x:

$$x + 4 = 12$$
$$x = 8$$

> *Note*   In mixture problems, it is a common mistake to forget that the right side of our second equation must also be the *product* of a percent times a quantity, and is not just the percent itself.

Step 5: *Write answers.*

It takes 8 gallons of 20% alcohol solution and 4 gallons of 50% alcohol solution to produce 12 gallons of 30% alcohol solution.

Step 6: *Reread and check.*

If we mix 8 gallons of 20% solution and 4 gallons of 50% solution, we end up with a total of 12 gallons of solution. To check the percentages we look for the total amount of alcohol in the two initial solutions and in the final solution.

In the initial solutions

The amount of alcohol in 8 gallons of 20% solution is $0.20(8) = 1.6$ gallons
The amount of alcohol in 4 gallons of 50% solution is $0.50(4) = 2.0$ gallons
The total amount of alcohol in the initial solutions is 3.6 gallons

In the final solution

The amount of alcohol in 12 gallons of 30% solution is $0.30(12) = 3.6$ gallons.

EXAMPLE 5   It takes 2 hours for a boat to travel 28 miles downstream (with the current). The same boat can travel 18 miles upstream (against the current) in 3 hours. What is the speed of the boat in still water, and what is the speed of the current of the river?

SOLUTION

Step 1: *Read and list.*

A boat travels 18 miles upstream and 28 miles downstream. The trip upstream takes 3 hours. The trip downstream takes 2 hours. We don't know the speed of the boat or the speed of the current.

Step 2: *Assign variables and translate information.*

Let $x =$ the speed of the boat in still water and let $y =$ the speed of the current. The average speed (rate) of the boat upstream is $x - y$, because it is traveling against the current. The rate of the boat downstream is $x + y$, because the boat is traveling with the current.

Step 3: *Write a system of equations.*

Putting the information into a table, we have

	d (distance, miles)	*r* (rate, mph)	*t* (time, h)
Upstream	18	$x - y$	3
Downstream	28	$x + y$	2

The formula for the relationship between distance d, rate r, and time t is $d = rt$ (the rate equation).

Because $d = r \cdot t$, the system we need to solve the problem is

$$18 = (x - y) \cdot 3$$
$$28 = (x + y) \cdot 2$$

which is equivalent to

$$6 = x - y$$
$$14 = x + y$$

> **Note** When setting up motion problems involving current or wind, the rate of the object is always first, followed by the rate of the current or wind. That is, the two expressions for r are given by
>
> $$\begin{matrix} \text{rate of} \\ \text{object} \end{matrix} \pm \begin{matrix} \text{rate of} \\ \text{current/wind} \end{matrix}$$

Step 4: *Solve the system.*

Adding the two equations, we have

$$20 = 2x$$

$$x = 10$$

Substituting $x = 10$ into $14 = x + y$, we see that

$$y = 4$$

Step 5: *Write answers.*

The speed of the boat in still water is 10 miles per hour; the speed of the current is 4 miles per hour.

Step 6: *Reread and check.*

The boat travels at $10 + 4 = 14$ miles per hour downstream, so in 2 hours it will travel $14 \cdot 2 = 28$ miles. The boat travels at $10 - 4 = 6$ miles per hour upstream, so in 3 hours it will travel $6 \cdot 3 = 18$ miles.

EXAMPLE 6 A coin collection consists of 14 coins with a total value of $1.35. If the coins are nickels, dimes, and quarters, and the number of nickels is 3 less than twice the number of dimes, how many of each coin is there in the collection?

SOLUTION This problem will require three variables and three equations.

Step 1: *Read and list.*

We have 14 coins with a total value of $1.35. The coins are nickels, dimes, and quarters. The number of nickels is 3 less than twice the number of dimes. We do not know how many of each coin we have.

Step 2: *Assign variables and translate information.*

Because we have three types of coins, we will have to use three variables. Let's let $x =$ the number of nickels, $y =$ the number of dimes, and $z =$ the number of quarters.

Step 3: *Write a system of equations.*

Because the total number of coins is 14, our first equation is

$$x + y + z = 14$$

Because the number of nickels is 3 less than twice the number of dimes, a second equation is

$$x = 2y - 3 \qquad \text{which is equivalent to} \qquad x - 2y = -3$$

Our last equation is obtained by considering the value of each coin and the total value of the collection. Let's write the equation in terms of cents, so we won't have to clear it of decimals later.

$$5x + 10y + 25z = 135$$

Here is our system, with the equations numbered for reference:

$$
\begin{aligned}
x + y + z &= 14 & (1) \\
x - 2y \phantom{{}+2z} &= -3 & (2) \\
5x + 10y + 25z &= 135 & (3)
\end{aligned}
$$

Step 4: *Solve the system.*

Let's begin by eliminating x from the first and second equations, and the first and third equations. Adding -1 times the second equation to the first equation gives us an equation in only y and z. We call this equation (4).

$$3y + z = 17 \qquad (4)$$

Adding -5 times equation (1) to equation (3) gives us

$$5y + 20z = 65 \qquad (5)$$

We can eliminate z from equations (4) and (5) by adding -20 times (4) to (5). Here is the result:

$$-55y = -275$$

$$y = 5$$

Substituting $y = 5$ into equation (4) gives us $z = 2$. Substituting $y = 5$ and $z = 2$ into equation (1) gives us $x = 7$.

Step 5: *Write answers.*

The collection consists of 7 nickels, 5 dimes, and 2 quarters.

Step 6: *Reread and check.*

The total number of coins is $7 + 5 + 2 = 14$. The number of nickels, 7, is 3 less than twice the number of dimes, 5. To find the total value of the collection, we have

The value of the 7 nickels is	$7(0.05) = \$0.35$
The value of the 5 dimes is	$5(0.10) = \$0.50$
The value of the 2 quarters is	$2(0.25) = \$0.50$

The total value of the collection is $1.35

If you go on to take a chemistry class, you may see the next example (or one much like it).

77°F

EXAMPLE 7   In a chemistry lab, students record the temperature of water at room temperature and find that it is 77° on the Fahrenheit temperature scale and 25° on the Celsius temperature scale. The water is then heated until it boils. The temperature of the boiling water is 212°F and 100°C. Assume that the relationship between the two temperature scales is a linear one, then use the preceding data to find the formula that gives the Celsius temperature, C, in terms of the Fahrenheit temperature, F.

SOLUTION   The data is summarized in the following table.

Corresponding Temperatures	
In Degrees Fahrenheit	In Degrees Celsius
77	25
212	100

If we assume the relationship is linear, then the formula that relates the two temperature scales can be written in slope-intercept form as

$$C = mF + b$$

Substituting $C = 25$ and $F = 77$ into this formula gives us

$$25 = 77m + b$$

Substituting $C = 100$ and $F = 212$ into the formula yields

$$100 = 212m + b$$

Together, the two equations form a system of equations, which we can solve using the addition method.

$$25 = \ 77m + b \ \xrightarrow{\text{Multiply by } -1} \ -25 = -77m - b$$

$$100 = 212m + b \ \xrightarrow[\text{No Change}]{} \ \underline{100 = 212m + b}$$

$$75 = 135m$$

$$m = \frac{75}{135} = \frac{5}{9}$$

To find the value of b, we substitute $m = \frac{5}{9}$ into $25 = 77m + b$ and solve for b.

$$25 = 77\left(\frac{5}{9}\right) + b$$

$$25 = \frac{385}{9} + b$$

$$b = 25 - \frac{385}{9} = \frac{225}{9} - \frac{385}{9} = -\frac{160}{9}$$

The equation that gives C in terms of F is

$$C = \frac{5}{9}F - \frac{160}{9}$$

Getting Ready for Class

After reading through the preceding section, respond in your own words and in complete sentences.

A. If you were to apply the Blueprint for Problem Solving from Section 2.3 to the examples in this section, what would be the first step?

B. If you were to apply the Blueprint for Problem Solving from Section 2.3 to the examples in this section, what would be the last step?

C. When working application problems involving boats moving in rivers, how does the current of the river affect the speed of the boat?

D. Write an application problem for which the solution depends on solving the system of equations:

$$0.05x + 0.06y = 1{,}000$$

$$0.05x + 0.06y = 55$$

Problem Set 4.3

Number Problems

1. One number is 3 more than twice another. The sum of the numbers is 18. Find the two numbers.

2. The sum of two numbers is 32. One of the numbers is 4 less than 5 times the other. Find the two numbers.

3. The difference of two numbers is 6. Twice the smaller is 4 more than the larger. Find the two numbers.

4. The larger of two numbers is 5 more than twice the smaller. If the smaller is subtracted from the larger, the result is 12. Find the two numbers.

5. The sum of three numbers is 8. Twice the smallest is 2 less than the largest, while the sum of the largest and smallest is 5. Use a linear system in three variables to find the three numbers.

6. The sum of three numbers is 14. The largest is 4 times the smallest, while the sum of the smallest and twice the largest is 18. Use a linear system in three variables to find the three numbers.

Ticket and Interest Problems

7. A total of 925 tickets were sold for a game for a total of $1,150. If adult tickets sold for $2.00 and children's tickets sold for $1.00, how many of each kind of ticket were sold?

8. If tickets for a show cost $2.00 for adults and $1.50 for children, how many of each kind of ticket were sold if a total of 300 tickets were sold for $525?

9. Mr. Jones has $20,000 to invest. He invests part at 6% and the rest at 7%. If he earns $1,280 in interest after 1 year, how much did he invest at each rate?

10. A man invests $17,000 in two accounts. One account earns 5% interest per year and the other 6.5%. If his total interest after 1 year is $970, how much did he invest at each rate?

11. Susan invests twice as much money at 7.5% as she does at 6%. If her total interest after 1 year is $840, how much does she have invested at each rate?

12. A woman earns $1,350 in interest from two accounts in 1 year. If she has three times as much invested at 7% as she does at 6%, how much does she have in each account?

13. A man invests $2,200 in three accounts that pay 6%, 8%, and 9% in annual interest, respectively. He has three times as much invested at 9% as he does at 6%. If his total interest for the year is $178, how much is invested at each rate?

14. A student has money in three accounts that pay 5%, 7%, and 8% in annual interest. She has three times as much invested at 8% as she does at 5%. If the total amount she has invested is $1,600 and her interest for the year comes to $115, how much money does she have in each account?

Mixture Problems

15. How many gallons of 20% alcohol solution and 50% alcohol solution must be mixed to get 9 gallons of 30% alcohol solution?

16. How many ounces of 30% hydrochloric acid solution and 80% hydrochloric acid solution must be mixed to get 10 ounces of 50% hydrochloric acid solution?

17. A mixture of 16% disinfectant solution is to be made from 20% and 14% disinfectant solutions. How much of each solution should be used if 15 gallons of the 16% solution are needed?

18. How much 25% antifreeze and 50% antifreeze should be combined to give 40 gallons of 30% antifreeze?

19. Paul mixes nuts worth $1.55 per pound with oats worth $1.35 per pound to get 25 pounds of trail mix worth $1.45 per pound. How many pounds of nuts and how many pounds of oats did he use?

20. A chemist has three different acid solutions. The first acid solution contains 20% acid, the second contains 40%, and the third contains 60%. He wants to use all three solutions to obtain a mixture of 60 liters containing 50% acid, using twice as much of the 60% solution as the 40% solution. How many liters of each solution should be used?

Rate Problems

21. It takes a boat 2 hours to travel 24 miles downstream and 3 hours to travel 18 miles upstream. What is the speed of the boat in still water? What is the speed of the current of the river?

22. A boat on a river travels 20 miles downstream in only 2 hours. It takes the same boat 6 hours to travel 12 miles upstream. What are the speed of the boat and the speed of the current?

23. An airplane flying with the wind can cover a certain distance in 2 hours. The return trip against the wind takes $2\frac{1}{2}$ hours. How fast is the plane and what is the speed of the air, if the distance is 600 miles?

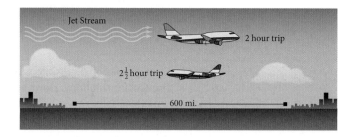

24. An airplane covers a distance of 1,500 miles in 3 hours when it flies with the wind and $3\frac{1}{3}$ hours when it flies against the wind. What is the speed of the plane in still air?

Coin Problems

25. Bob has 20 coins totaling $1.40. If he has only dimes and nickels, how many of each coin does he have?

26. If Amy has 15 coins totaling $2.70, and the coins are quarters and dimes, how many of each coin does she have?

27. A collection of nickels, dimes, and quarters consists of 9 coins with a total value of $1.20. If the number of dimes is equal to the number of nickels, find the number of each type of coin.

28. A coin collection consists of 12 coins with a total value of $1.20. If the collection consists only of nickels, dimes, and quarters, and the number of dimes is two more than twice the number of nickels, how many of each type of coin are in the collection?

29. A collection of nickels, dimes, and quarters amount to $10.00. If there are 140 coins in all and there are twice as many dimes as there are quarters, find the number of nickels.

30. A cash register contains a total of 95 coins consisting of pennies, nickels, dimes, and quarters. There are only 5 pennies and the total value of the coins is $12.05. Also, there are 5 more quarters than dimes. How many of each coin is in the cash register?

Additional Problems

31. **Price and Demand** A manufacturing company finds that they can sell 300 items if the price per item is $2.00, and 400 items if the price is $1.50 per item. If the relationship between the number of items sold x and the price per item p is a linear one, find a formula that gives x in terms of p. Then use the formula to find the number of items they will sell if the price per item is $3.00.

32. **Price and Demand** A company manufactures and sells bracelets. They have found from past experience that they can sell 300 bracelets each week if the price per bracelet is $2.00, but only 150 bracelets are sold if the price is $2.50 per bracelet. If the relationship between the number of bracelets sold x and the price per bracelet p is a linear one, find a formula that gives x in terms of p. Then use the formula to find the number of bracelets they will sell at $3.00 each.

33. **Height of a Ball** A ball is tossed into the air so that the height after 1, 3, and 5 seconds is as given in the following table.

t (sec)	h (ft)
1	128
3	128
5	0

If the relationship between the height of the ball h and the time t is quadratic, then the relationship can be written as

$$h = at^2 + bt + c$$

Use the information in the table to write a system of three equations in three variables a, b, and c. Solve the system to find the exact relationship between h and t.

34. **Height of a Ball** A ball is tossed into the air and its height above the ground after 1, 3, and 4 seconds is recorded as shown in the following table.

t (sec)	h (ft)
1	96
3	64
4	0

The relationship between the height of the ball h and the time t is quadratic and can be written as

$$h = at^2 + bt + c$$

Use the information in the table to write a system of three equations in three variables a, b, and c. Solve the system to find the exact relationship between the variables h and t

Learning Objectives Assessment

The following problems can be used to help assess if you have successfully met the learning objectives for this section.

35. A 25% salt solution and a 60% salt solution are to be mixed to obtain 10 gallons of a 50% salt solution. Which system of equations correctly models this problem?

a. $x + y = 50$
$0.25x + 0.60y = 10$

b. $x + y = 10$
$0.25x + 0.60y = 0.50$

c. $x + y = 10$
$0.25x + 0.60y = 5$

d. $x + y = 0.50$
$25x + 60y = 10$

36. A collection of nickels, dimes, and quarters amount to $2.00. If there are 16 coins in all and there are three more nickels than dimes, how many nickels are in the collection?

a. 7   **b.** 4   **c.** 5   **d.** 9

Getting Ready for the Next Section

37. Does the graph of $x + y < 4$ include the boundary line?

38. Does the graph of $-x + y \leq 3$ include the boundary line?

39. Where do the graphs of the lines $x + y = 4$ and $x - 2y = 4$ intersect?

40. Where do the graphs of the line $x = -1$ and $x - 2y = 4$ intersect?

Solve.

41. $20x + 9{,}300 > 18{,}000$

42. $20x + 4{,}800 > 18{,}000$

Systems of Linear Inequalities

Learning Objectives

In this section, we will learn how to:

1. Determine if an ordered pair is a solution to a system of linear inequalities in two variables.

2. Graph the solution set to a system of linear inequalities in two variables.

3. Use a system of linear inequalities in two variables to solve applied problems.

Introduction

In Section 3.4, we graphed linear inequalities in two variables. To review, we graph the boundary line, using a solid line if the boundary is part of the solution set and a broken line if the boundary is not part of the solution set. Then we test any point that is not on the boundary line in the original inequality. A true statement tells us that the point lies in the solution set, a false statement tells us the solution set is the other region.

Figure 1 shows the graph of the inequality $x + y < 4$. Note that the boundary is not included in the solution set, and is therefore drawn with a broken line. Figure 2 shows the graph of $-x + y \leq 3$. Note that the boundary is drawn with a solid line, because it is part of the solution set.

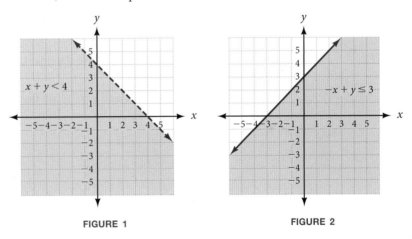

FIGURE 1   **FIGURE 2**

If we form a system of inequalities with the two inequalities, the solution set will be all the points common to both solution sets shown in the two figures above. It is the intersection of the two solution sets. Therefore, the solution set for the system of inequalities

$$x + y < 4$$
$$-x + y \leq 3$$

is all the ordered pairs that satisfy both inequalities. It is the set of points that are below the line $x + y = 4$, and also below (and including) the line $-x + y = 3$. The graph of the solution set to this system is shown in Figure 3. We have written the system in Figure 3 with the word *and* just to remind you that the solution set to a system of equations or inequalities is all the points that satisfy both equations or inequalities.

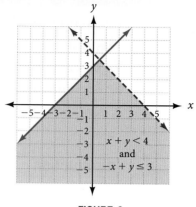

FIGURE 3

EXAMPLE 1   Determine if each ordered pair is a solution to the system of linear inequalities.

$$x + y < 4$$
$$x - 2y \le -2$$

a. $(3, 2)$ **b.** $(1, -1)$ **c.** $(-1, 3)$

SOLUTION   We check each ordered pair to see if it satisfies both inequalities in the system.

a. Because $3 + 2 < 4$ is a false statement, $(3, 2)$ does not satisfy the first inequality. It cannot be a solution.

b. Substituting $x = 1$ and $y = -1$ we find

$$1 + (-1) < 4$$
$$0 < 4 \qquad \text{A true statement}$$
$$\text{and} \quad 1 - 2(-1) \le -2$$
$$3 \le -2 \qquad \text{A false statement}$$

Because $(1, -1)$ does not satisfy the second inequality, it is not a solution to the system.

c. Using $x = -1$ and $y = 3$ we obtain

$$-1 + 3 < 4$$
$$2 < 4 \qquad \text{A true statement}$$
$$\text{and} \quad -1 - 2(3) \le -2$$
$$-7 \le -2 \qquad \text{A true statement}$$

The pair $(-1, 3)$ is a solution because it satisfies both inequalities.

EXAMPLE 2 Graph the solution set for the system of inequalities.

$$y < \frac{1}{2}x + 3$$

$$y \geq \frac{1}{2}x - 2$$

SOLUTION Figures 4 and 5 show the solution set for each of the inequalities separately.

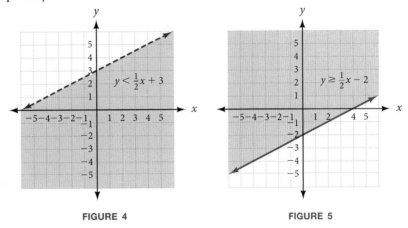

FIGURE 4   FIGURE 5

Figure 6 is the solution set to the system of inequalities. It is the region consisting of points whose coordinates satisfy both inequalities.

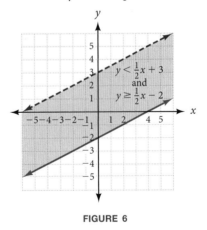

FIGURE 6

EXAMPLE 3   Graph the solution set for the system of inequalities.

$$x + y < 4$$
$$x \geq 0$$
$$y \geq 0$$

SOLUTION   We graphed the first inequality, $x + y < 4$, in Figure 1 at the beginning of this section. The solution set to the inequality $x \geq 0$, shown in Figure 7, is all the points to the right of the y-axis; that is, all the points with x-coordinates that are greater than or equal to 0. Figure 8 shows the graph of $y \geq 0$. It consists of all points with y-coordinates greater than or equal to 0; that is, all points from the x-axis up.

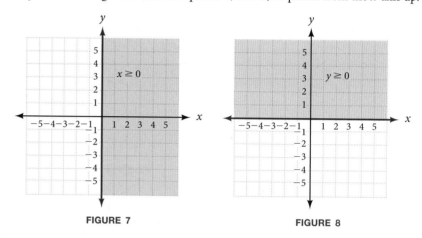

FIGURE 7   **FIGURE 8**

The regions shown in Figures 7 and 8 overlap in the first quadrant. Therefore, putting all three regions together we have the points in the first quadrant that are below the line $x + y = 4$. This region is shown in Figure 9, and it is the solution to our system of inequalities.

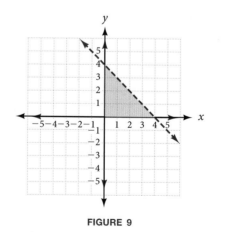

FIGURE 9

EXAMPLE 4 Graph the solution set for the system of inequalities.

$$x \leq 4$$

$$y \geq -3$$

SOLUTION The solution to this system will consist of all points to the left of and including the vertical line $x = 4$ that intersect with all points above and including the horizontal line $y = -3$. The solution set is shown in Figure 10.

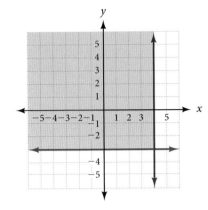

FIGURE 10

EXAMPLE 5 Graph the solution set for the system of inequalities.

$$y \geq 2x + 3$$

$$2x - y > 2$$

SOLUTION The solution set for the first inequality includes all points above and including the line $y = 2x + 3$ as shown in Figure 11. Figure 12 shows the solution set for the second inequality, which consists of all points below the line $2x - y = 2$.

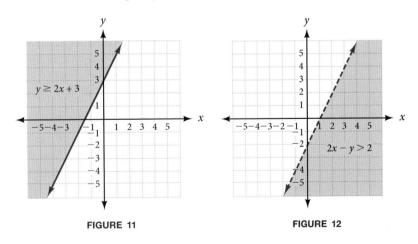

FIGURE 11   **FIGURE 12**

If we isolate y in the second inequality, we obtain $y < 2x - 2$. The lines in Figure 11 and Figure 12 have the same slope, and are therefore parallel. Looking at both graphs, it is evident the two shaded regions will never overlap. There are no ordered pairs that will satisfy both inequalities, so the system has no solution.

EXAMPLE 6   Graph the solution set for the following system.

$$x - 2y \leq 4$$
$$x + y \leq 4$$
$$x \geq -1$$

SOLUTION   We have three linear inequalities, representing three sections of the coordinate plane. The graph of the solution set for this system will be the intersection of these three sections. The graph of $x - 2y \leq 4$ is the section above and including the boundary $x - 2y = 4$. The graph of $x + y \leq 4$ is the section below and including the boundary line $x + y = 4$. The graph of $x \geq -1$ is all the points to the right of, and including, the vertical line $x = -1$. The intersection of these three graphs is shown in Figure 13.

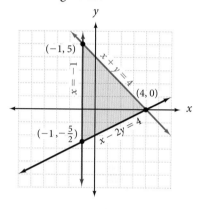

FIGURE 13

EXAMPLE 7   A college basketball arena plans on charging $20 for certain seats and $15 for others. They want to bring in more than $18,000 from all ticket sales and have reserved at least 500 tickets at the $15 rate. Find a system of inequalities describing all possibilities and sketch the graph. If 620 tickets are sold for $15, at least how many tickets are sold for $20?

SOLUTION   Let $x =$ the number of $20 tickets and $y =$ the number of $15 tickets. We need to write a list of inequalities that describe this situation. That list will form our system of inequalities. First of all, we note that we cannot use negative numbers for either x or y. So, we have our first inequalities:

$$x \geq 0$$
$$y \geq 0$$

Next, we note that they are selling at least 500 tickets for $15, so we can replace our second inequality with $y \geq 500$. Now our system is

$$x \geq 0$$
$$y \geq 500$$

Now the amount of money brought in by selling $20 tickets is $20x$, and the amount of money brought in by selling $15 tickets is $15y$. If the total income from ticket sales is to be more than $18,000, then $20x + 15y$ must be greater than 18,000. This gives us our last inequality and completes our system.

$$20x + 15y > 18,000$$
$$x \geq 0$$
$$y \geq 500$$

We have used all the information in the problem to arrive at this system of inequalities. The solution set contains all the values of x and y that satisfy all the conditions given in the problem. Here is the graph of the solution set.

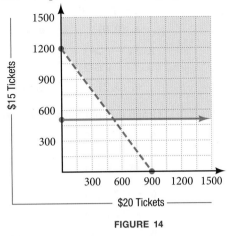

FIGURE 14

If 620 tickets are sold for \$15, then we substitute 620 for y in our first inequality to obtain

$$20x + 15(620) > 18000 \quad \text{Substitute 620 for } y$$
$$20x + 9300 > 18000 \quad \text{Multiply}$$
$$20x > 8700 \quad \text{Add } -9300 \text{ to each side}$$
$$x > 435 \quad \text{Divide each side by 20}$$

If they sell 620 tickets for \$15 each, then they need to sell more than 435 tickets at \$20 each to bring in more than \$18,000.

Getting Ready for Class

After reading through the preceding section, respond in your own words and in complete sentences.

A. What does it mean for an ordered pair to be a solution to a system of linear inequalities in two variables?

B. Explain the process of solving a system of linear inequalities in two variables.

C. What are the circumstances for which a system of linear inequalities in two variables will have no solution?

D. Is it possible for a system of linear inequalities in two variables to have a solution set consisting of the entire xy-plane? Explain why or why not.

Problem Set 4.4

Determine if each ordered pair is a solution to the system of linear inequalities.

1. $x + y \leq -2$
$x - y < \ \ 0$

a. $(0, 0)$ **b.** $(-3, -1)$ **c.** $(1, -4)$ **d.** $(-2, 0)$

2. $3x - y > 3$
$y \leq 2$

a. $(3, 2)$ **b.** $(0, -1)$ **c.** $(4, 3)$ **d.** $(1, -2)$

Graph the solution set for each system of linear inequalities.

3. $x + y < 5$

     $2x - y > 4$

4. $x + y < 5$

     $2x - y < 4$

5. $y < \dfrac{1}{3}x + 4$

     $y \geq \dfrac{1}{3}x - 3$

6. $y < 2x + 4$
$y \geq 2x - 3$

7. $x \geq -3$
$y < -2$

8. $x \leq 4$
$y \geq -2$

9. $1 \leq x \leq 3$
$2 \leq y \leq 4$

10. $-4 \leq x \leq -2$
$1 \leq y \leq \ \ 3$

11. $x + 2y < \ \ 4$
$x + 2y \leq -4$

12. $x \geq 3$
$x > -2$

13. $y > 1$
$y < -3$

14. $x - y \leq -3$
$x - y \geq \ \ 1$

15. $x + y \leq 4$
$x \geq 0$
$y \geq 0$

16. $x - y \leq 2$
$x \geq 0$
$y \leq 0$

17. $x + \ \ y \leq \ \ 3$
$x - 3y \leq \ \ 3$
$x \geq -2$

18. $x - \ \ y \leq \ \ 4$
$x + 2y \leq \ \ 4$
$x \geq -1$

19. $\ \ x + y \leq \ \ 2$
$-x + y \leq \ \ 2$
$y \geq -2$

20. $\ \ x - y \leq \ \ 3$
$-x - y \leq \ \ 3$
$y \leq -1$

21. $x + y < 5$
$y > x$
$y \geq 0$

22. $x + y < 5$
$y > x$
$x \geq 0$

23. $2x + 3y \leq 6$
$x \geq 0$
$y \geq 0$

24. $\ \ x + 2y \leq 10$
$3x + 2y \leq 12$
$x \geq 0$
$y \geq 0$

For each figure below, find a system of inequalities that describes the shaded region.

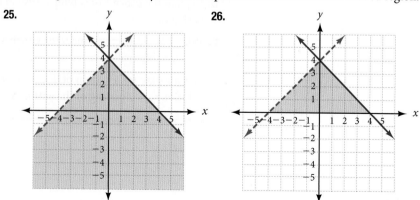

25.

26.

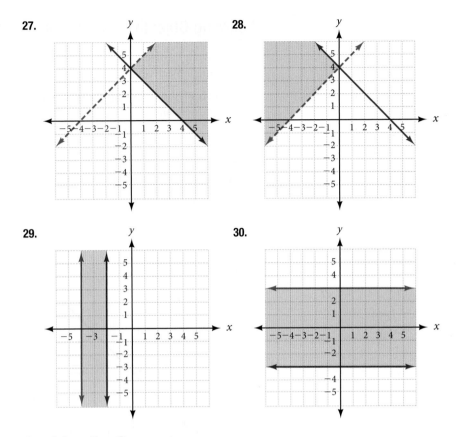

27.

28.

29.

30.

Applying the Concepts

31. **Office Supplies** An office worker wants to purchase some $0.55 postage stamps and also some $0.65 postage stamps totaling no more than $40. It is also desired to have at least twice as many $0.55 stamps and more than 15 $0.55 stamps.

   a. Find a system of inequalities describing all the possibilities and sketch the graph.

   b. If he purchases 20 $0.55 stamps, what is the maximum number of $0.65 stamps he can purchase?

32. **Inventory** A store sells two brands of DVD players. Customer demand indicates that it is necessary to stock at least twice as many DVD players of brand A as of brand B. At least 30 of brand A and 15 of brand B must be on hand. In the store, there is room for not more than 100 DVD players in the store.

   a. Find a system of inequalities describing all possibilities, then sketch the graph.

   b. If there are 35 DVD players of brand A, what is the most number of brand B DVD players on hand?

Learning Objectives Assessment

The following problems can be used to help assess if you have successfully met the learning objectives for this section.

33. Which of the following ordered pairs is a solution to the given system of linear inequalities?

$$x + 3y < 6$$
$$2x - y \geq 4$$

   a. $(0, 0)$ **b.** $(4, 1)$ **c.** $(1, 3)$ **d.** $(3, -1)$

34. Graph the solution set for the system of linear inequalities.

$$x > y$$
$$x \leq 2$$

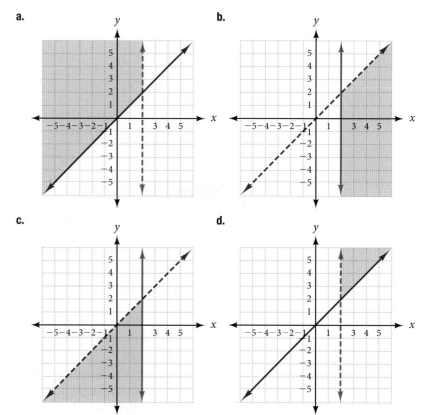

35. You need to purchase some cookies and cupcakes for an office party. Cookies cost $1.25 each and cupcakes cost $1.75 each. You can spend at most $100 and need at least twice as many cookies as cupcakes. Which system of linear inequalities describes all the possibilities?

a. $x + y \leq 100$

$\quad 1.25x \geq 1.75y + 2$

$\quad\quad x \geq 0$

$\quad\quad y \geq 0$

b. $2x + y \leq 100$

$\quad 1.25x \geq 1.75y$

$\quad\quad x \geq 0$

$\quad\quad y \geq 0$

c. $1.25x + 1.75y \leq 100$

$\quad\quad x \geq 2y$

$\quad\quad x \geq 0$

$\quad\quad y \geq 0$

d. $1.25x + 1.75y \leq 100$

$\quad\quad y \geq 2x$

$\quad\quad x \geq 0$

$\quad\quad y \geq 0$

Maintaining Your Skills

For each of the following straight lines, identify the x-intercept, y-intercept, and slope, and sketch the graph.

36. $2x + y = 6$

37. $y = \dfrac{3}{2}x + 4$

38. $x = -2$

Find the equation for each line.

39. Give the equation of the line through $(-1, 3)$ that has slope $m = 2$.

40. Give the equation of the line through $(-3, 2)$ and $(4, -1)$.

41. Line l contains the point $(5, -3)$ and has a graph parallel to the graph of $2x - 5y = 10$. Find the equation for l.

42. Give the equation of the vertical line through $(4, -7)$.

State the domain and range for the following relations, and indicate which relations are also functions.

43. $\{(-2, 0), (-3, 0), (-2, 1)\}$

44. $y = x^2 - 9$

Let $f(x) = x - 2$, $g(x) = 3x + 4$ and $h(x) = 3x^2 - 2x - 8$, and find the following.

45. $f(3) + g(2)$

46. $h(0) + g(0)$

47. $f[g(2)]$

48. $g[f(2)]$

Solve the following variation problems.

49. Direct Variation Quantity y varies directly with the square of x. If y is 50 when x is 5, find y when x is 3.

50. Joint Variation Quantity z varies jointly with x and the cube of y. If z is 15 when x is 5 and y is 2, find z when x is 2 and y is 3.

Chapter 4 Summary

Systems of Linear Equations [4.1, 4.2]

1. The solution to the system

$$x + 2y = 4$$
$$x - y = 1$$

is the ordered pair (2, 1).

A system of linear equations is two or more linear equations considered simultaneously. The solution set to a linear system in two variables is the set of ordered pairs that satisfy both equations. The solution set to a linear system in three variables consists of the ordered triples that satisfy each equation in the system.

To Solve a System by the Addition Method [4.1]

2. We can eliminate the *y*-variable from the system in Example 1 by multiplying both sides of the second equation by 2 and adding the result to the first equation:

$$x + 2y = 4 \xrightarrow{\text{No Change}} x + 2y = 4$$
$$x - y = 1 \xrightarrow[\text{Multiply by 2}]{} 2x - 2y = 2$$
$$\begin{array}{rl} 3x & = 6 \\ x & = 2 \end{array}$$

Substituting $x = 2$ into either of the original two equations gives $y = 1$. The solution is (2, 1).

Step 1: Look the system over to decide which variable will be easier to eliminate.

Step 2: Use the multiplication property of equality on each equation separately, if necessary, to ensure that the coefficients of the variable to be eliminated are opposites.

Step 3: Add the left and right sides of the system produced in step 2, and solve the resulting equation.

Step 4: Substitute the solution from step 3 back into any equation with both *x*- and *y*-variables, and solve.

Step 5: Check your solution in both equations if necessary.

To Solve a System by the Substitution Method [4.1]

3. In Example 1, we solve the second equation for *x* to get

$$x = y + 1$$

Substituting for *x* gives us

$$(y + 1) + 2y = 4$$
$$3y + 1 = 4$$
$$3y = 3$$
$$y = 1$$

Using $y = 1$ in Step 1 gives $x = 2$.

Step 1: Solve either of the equations for one of the variables (this step is not necessary if one of the equations has the correct form already).

Step 2: Substitute the results of step 1 into the other equation, and solve.

Step 3: Substitute the results of step 2 into the equation produced in step 1 and simplify.

Step 4: Check your solution if necessary.

Inconsistent and Dependent Equations [4.1, 4.2]

4. If the lines are parallel and do not coincide, then the system will be inconsistent and the solution is ∅. If the lines coincide, the equations are dependent.

A system of two linear equations that have no solutions in common is said to be an *inconsistent* system, whereas two linear equations that have all their solutions in common are said to be *dependent* equations.

Systems of Linear Inequalities [4.4]

A system of linear inequalities is two or more linear inequalities considered at the same time. To find the solution set to the system, we graph each of the inequalities on the same coordinate system. The solution set is the region that is common to all the regions graphed.

Chapter 4 Test

Solve the following systems by the addition method. [4.1]

1. $2x + 4y = 3$

$-4x - 8y = -6$

2. $4x - 7y = -2$

$-5x + 6y = -3$

3. $\dfrac{1}{3}x - \dfrac{1}{6}y = 3$

$-\dfrac{1}{5}x + \dfrac{1}{4}y = 0$

Solve the following systems by the substitution method. [4.1]

4. $2x - 5y = 14$

$\quad y = 3x + 8$

5. $2x - 5y = -8$

$3x + \ y = \ \ 5$

6. $2x - 4y = 0$

$-x + 2y = 5$

Solve each system. [4.2]

7. $2x - y + z = \ \ 9$
$x + y - 3z = -2$
$3x + y - z = \ \ 6$

8. $2x - y + 3z = 2$
$x - 4y - z = 6$
$3x - 2y + z = 4$

9. $x + 2y - \ z = \ \ 7$
$3x - 4y - 3z = -4$
$2x + \ y - 2z = \ \ 1$

Solve each word problem. [4.3]

10. Number Problem A number is 1 less than twice another. Their sum is 14. Find the two numbers.

11. Investing John invests twice as much money at 6% as he does at 5%. If his investments earn a total of $680 in 1 year, how much does he have invested at each rate?

12. Ticket Cost There were 750 tickets sold for a basketball game for a total of $1,090. If adult tickets cost $2.00 and children's tickets cost $1.00, how many of each kind were sold?

13. Mixture Problem How much 30% alcohol solution and 70% alcohol solution must be mixed to get 16 gallons of 60% solution?

14. Speed of a Boat A boat can travel 20 miles down-stream in 2 hours. The same boat can travel 18 miles upstream in 3 hours. What is the speed of the boat in still water, and what is the speed of the current?

15. Coin Problem A collection of nickels, dimes, and quarters consists of 15 coins with a total value of $1.10. If the number of nickels is 1 less than 4 times the number of dimes, how many of each coin are contained in the collection?

Graph the solution set for each system of linear inequalities. [4.4]

16. $\quad x + 4y \le \ \ 4$

$-3x + 2y > -12$

17. $x + y \le -3$

$y \ge \ \ 2 - x$

18. $y < -\dfrac{1}{2}x + 4$

$x \ge \ \ 0$

$y \ge \ \ 0$

Polynomials and Factoring

5

Chapter Outline

iStockphoto.com © shironosov

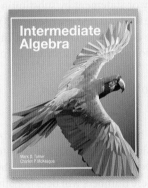

Intermediate Algebra

Mark D. Turner
Charles P. McKeague

XYZ Textbooks publishes the book you are using. If college bookstores buy the book for $75 a copy, then the amount of money XYZ Textbooks receives (revenue) is $R = 75x$, where x is the number of books purchased by bookstores in a year. But XYZ Textbooks does not keep all the revenue. Each copy of the book costs $10 to print, and 15% of each sale is paid to the authors in royalties, plus $3 of each sale goes to sales commissions. In addition to this, the company has expenses for its employees, rent, utilities, and other costs associated with the office. The share of office expenses for this book is $30,000. Therefore, the cost of x books sold to bookstores is

$$C(x) = 10x + 0.15(75)x + 3x + 30{,}000 = 24.25x + 30{,}000$$

The company's profit for selling x books is revenue minus cost.

$$P(x) = 75x - (24.25x + 30{,}000)$$

We can generalize this discussion to produce the most common relationship in business:

$$\text{Profit} = \text{Revenue} - \text{Cost}$$
$$P(x) = R(x) - C(x)$$

As in the discussion above, revenue itself can be broken down into the product of the number of items sold and the price per item. That is,

$$\text{Revenue} = (\text{number of items sold})(\text{price of each item})$$
$$R(x) = xp$$

which in our example is $R(x) = 75x$. Cost can be broken down further into fixed costs and variable costs.

$$C(x) = v(x) + f$$

which in our example is $C(x) = 24.25(x) + 30{,}000$.

In this chapter, these formulas serve as links from the topics in the chapter to the business world.

© Lajos Repasi / iStockPhoto

Success Skills

The study skills for this chapter cover the way you approach new situations in mathematics. The first study skill is a point of view you hold about your natural instincts for what does and doesn't work in mathematics. The second study skill gives you a way of testing your instincts.

1. **Don't Let Your Intuition Fool You** As you become more experienced and more successful in mathematics you will be able to trust your mathematical intuition. For now, though, it can get in the way of your success. For example, if you ask some students to "subtract 3 from -5" they will answer -2 or 2. Both answers are incorrect, even though they may seem intuitively true. Likewise, some students will expand $(a + b)^2$ and arrive at $a^2 + b^2$, which is incorrect. In both cases, intuition leads directly to the wrong answer.

2. **Test Properties of Which You are Unsure** From time to time, you will be in a situation where you would like to apply a property or rule, but you are not sure it is true. You can always test a property or statement by substituting numbers for variables. For instance, I always have students that rewrite $(x + 3)^2$ as $x^2 + 9$, thinking that the two expressions are equivalent. The fact that the two expressions are not equivalent becomes obvious when we substitute 10 for x in each one.

   When $x = 10$, the expression $(x + 3)^2$ is $(10 + 3)^2 = 13^2 = 169$

   When $x = 10$, the expression $x^2 + 9 = 10^2 + 9 = 100 + 9 = 109$

   When you test the equivalence of expressions by substituting numbers for the variable, make it easy on yourself by choosing numbers that are easy to work with, such as 10. Don't try to verify the equivalence of expressions by substituting 0, 1, or 2 for the variable, as using these numbers will occasionally give you false results.

It is not good practice to trust your intuition or instincts in every new situation in algebra. If you have any doubt about the generalizations you are making, test them by replacing variables with numbers and simplifying.

Sums and Differences of Polynomials

5.1

Learning Objectives

In this section, we will learn how to:

1. Give the degree of a polynomial.

2. Add and subtract polynomials.

3. Evaluate a polynomial for a given value of its variable.

Introduction

We begin this section with the definition around which polynomials are defined. Once we have listed all the terminology associated with polynomials, we will show how the distributive property is used to find sums and differences of polynomials.

Polynomials in General

> **(déf) DEFINITION** *term, or monomial*
>
> A *term*, or *monomial*, is a constant or the product of a constant and one or more variables raised to whole-number exponents.

The following are monomials, or terms:

$$-16 \qquad 3x^2 \qquad -\frac{2}{5}a^3b^2c \qquad xy^2z$$

The numerical part of each monomial is called the ***numerical coefficient***, or just *coefficient*. For the preceding terms, the coefficients are -16, 3, $-\frac{2}{5}$, and 1. Notice that the coefficient for xy^2z is understood to be 1.

A term consisting only of a constant is called a ***constant term***. For example, in the preceding terms, -16 is a constant term.

> **(déf) DEFINITION** *degree of a term*
>
> A constant term has degree 0. For a term containing a single variable, the *degree* is the exponent on the variable.
>
> If the term contains multiple variables, the *degree* is the sum of the exponents appearing on the variables.

For example, in the terms listed previously, -16 has degree 0, $3x^2$ has degree 2, and xy^2z has degree 4.

> **(déf) DEFINITION** *polynomial*
>
> A *polynomial* is any finite sum of terms. Because subtraction can be written in terms of addition, finite differences are also included in this definition.

The following are polynomials:

$$2x^2 - 6x + 3 \qquad -5x^2y + 2xy^2 \qquad 4a - 5b + 6c + 7d$$

Polynomials can be classified further according to the number of terms present. If a polynomial consists of two terms, it is said to be a **binomial**. If it has three terms, it is called a **trinomial**. And, as stated, a polynomial with only one term is said to be a **monomial**.

Degree of a Polynomial

> (dĕf **DEFINITION** *degree of a polynomial*
>
> The *degree* of a polynomial is the highest degree of its terms. For a polynomial with one variable, the degree is the highest power to which the variable is raised in any one term.

VIDEO EXAMPLES

SECTION 5.1

> ▊ **EXAMPLE 1**
>
> **a.** $6x^2 + 2x - 1$ — A trinomial of degree 2
> **b.** $5x^3y - 3xy^2$ — A binomial of degree 4
> **c.** $7x^6 - 5x^3 + 2x - 4$ — A polynomial of degree 6
> **d.** $-7x^4$ — A monomial of degree 4
> **e.** 15 — A monomial of degree 0

Polynomials in one variable are usually written in decreasing powers of the variable. When this is the case, the first term is called the **leading term**, and the coefficient of the first term is called the **leading coefficient**. In Example 1a, the leading coefficient is 6. In Example 1c, it is 7. However, if a polynomial is not written in decreasing powers of the variable, the leading coefficient is attached to the variable with the highest degree.

> (dĕf **DEFINITION** *similar, or like, terms*
>
> Two or more terms that differ only in their numerical coefficients are called *similar*, or *like*, terms. Since similar terms can differ only in their coefficients, they have identical variable parts, which includes identical powers on the variables.

For example, here are some pairs of like terms and some pairs of unlike terms.

Like Terms	Unlike Terms
$2x$ and $5x$	$2x$ and $5y$
x^2 and $4x^2$	x and x^2
$3x^2y$ and x^3y	$3x^3y$ and xy^3

Addition and Subtraction of Polynomials

To add two polynomials, we simply apply the commutative and associative properties to group similar terms together, and then use the distributive property as we have in the following example:

> *Note* In practice it is not necessary to show all the steps shown in Example 2. It is important to understand that addition of polynomials is equivalent to combining similar terms.

EXAMPLE 2 Find the sum of $5x^2 - 4x + 2$ and $3x^2 + 9x - 6$.

SOLUTION $(5x^2 - 4x + 2) + (3x^2 + 9x - 6)$

$$= (5x^2 + 3x^2) + (-4x + 9x) + (2 - 6) \quad \text{Commutative and associative properties}$$

$$= (5 + 3)x^2 + (-4 + 9)x + (2 - 6) \quad \text{Distributive property}$$

$$= 8x^2 + 5x + (-4)$$

$$= 8x^2 + 5x - 4$$

EXAMPLE 3 Add: $(-8x^3 + 7x^2 - 6x + 5) + (10x^3 + 3x^2 - 2x - 6)$.

SOLUTION We can add the two polynomials using the method of Example 2, or we can arrange similar terms in columns and add vertically. Using the column method, we have

$$\begin{array}{r} -8x^3 + 7x^2 - 6x + 5 \\ \underline{10x^3 + 3x^2 - 2x - 6} \\ 2x^3 + 10x^2 - 8x - 1 \end{array}$$

To find the difference of two polynomials, we need to use the fact that the opposite of a sum is the sum of the opposites; that is,

$$-(a + b) = -a + (-b)$$

One way to remember this is to observe that $-(a + b)$ is equivalent to

$$-1(a + b) = (-1)a + (-1)b = -a + (-b)$$

If a negative sign directly precedes the parentheses surrounding a polynomial, we may remove the parentheses and the preceding negative sign by changing the sign of each term within the parentheses. For example,

$$-(3x + 4) = -3x + (-4) = -3x - 4$$

$$-(5x^2 - 6x + 9) = -5x^2 + 6x - 9$$

$$-(-x^2 + 7x - 3) = x^2 - 7x + 3$$

◼ **EXAMPLE 4**   Find the difference of $9x^2 - 3x + 5$ and $4x^2 + 2x - 3$.

SOLUTION   We subtract by adding the opposite of each term in the polynomial that follows the subtraction sign.

$$(9x^2 - 3x + 5) - (4x^2 + 2x - 3)$$

$$= 9x^2 - 3x + 5 + (-4x^2) + (-2x) + 3 \qquad \text{The opposite of a sum is the sum of the opposites}$$

$$= (9x^2 - 4x^2) + (-3x - 2x) + (5 + 3) \qquad \text{Commutative and associative property}$$

$$= 5x^2 - 5x + 8 \qquad \text{Combine similar terms}$$

◼

◼ **EXAMPLE 5**   Subtract: $(-3x^2 + 5x - 2) - (4x^2 - 9x + 1)$.

SOLUTION   Again, to subtract, we add the opposite.

$$(-3x^2 + 5x - 2) - (4x^2 - 9x + 1) = -3x^2 + 5x - 2 - 4x^2 + 9x - 1$$

$$= (-3x^2 - 4x^2) + (5x + 9x) + (-2 - 1)$$

$$= -7x^2 + 14x - 3 \qquad ◼$$

◼ **EXAMPLE 6**   Simplify: $4x - 3[2 - (3x + 4)]$.

SOLUTION   Removing the innermost parentheses first, we have

$$4x - 3[2 - (3x + 4)] = 4x - 3(2 - 3x - 4)$$

$$= 4x - 3(-3x - 2)$$

$$= 4x + 9x + 6$$

$$= 13x + 6 \qquad ◼$$

◼ **EXAMPLE 7**   Simplify: $(2x + 3) - [(3x + 1) - (x - 7)]$.

SOLUTION $(2x + 3) - [(3x + 1) - (x - 7)] = (2x + 3) - (3x + 1 - x + 7)$

$$= (2x + 3) - (2x + 8)$$

$$= 2x + 3 - 2x - 8$$

$$= -5 \qquad ◼$$

Evaluating Polynomials

In the example that follows we will find the value of a polynomial for a given value of the variable.

EXAMPLE 8   Find the value of $5x^3 - 3x^2 + 4x - 5$ when x is 2.

SOLUTION   We begin by substituting 2 for x in the original polynomial.

$$\begin{aligned}
&\text{When} && x = 2 \\
&\text{the polynomial} && 5x^3 - 3x^2 + 4x - 5 \\
&\text{becomes} && 5 \cdot 2^3 - 3 \cdot 2^2 + 4 \cdot 2 - 5 \\
& && = 5 \cdot 8 - 3 \cdot 4 + 4 \cdot 2 - 5 \\
& && = 40 - 12 + 8 - 5 \\
& && = 31
\end{aligned}$$

Example 8 can be restated using function notation by calling the polynomial $P(x)$ and asking for $P(2)$. The solution would look like this:

$$\begin{aligned}
\text{If} \quad & P(x) = 5x^3 - 3x^2 + 4x - 5 \\
\text{then} \quad & P(2) = 5 \cdot 2^3 - 3 \cdot 2^2 + 4 \cdot 2 - 5 \\
& \quad\quad\;\; = 31
\end{aligned}$$

Revenue, Cost, and Profit

Three functions that occur very frequently in business and economics classes are profit $P(x)$, revenue $R(x)$, and cost functions $C(x)$. If a company manufactures and sells x items, then the revenue $R(x)$ is the total amount of money obtained by selling all x items. The cost $C(x)$ is the total amount of money it costs the company to manufacture the x items. The profit $P(x)$ obtained by selling all x items is the difference between the revenue and the cost and is given by the equation

$$P(x) = R(x) - C(x)$$

EXAMPLE 9   The total cost (in dollars) for a company to manufacture and sell x items per month is $C(x) = 20x + 1,300$. If the revenue brought in by selling all x items is $R(x) = 75x - 0.03x^2$, find the profit equation $P(x)$. How much profit will be made by producing and selling 300 items each month?

SOLUTION   Because $P(x) = R(x) - C(x)$, we have

$$\begin{aligned}
P(x) &= (75x - 0.03x^2) - (20x + 1,300) \\
&= 75x - 0.03x^2 - 20x - 1,300 \\
&= -0.03x^2 + 55x - 1,300
\end{aligned}$$

Evaluating $P(x)$ at $x = 300$ give us

$$\begin{aligned}
P(300) &= -0.03(300)^2 + 55(300) - 1,300 \\
&= 12,500
\end{aligned}$$

A profit of $12,500 will be made by producing and selling 300 items each month.

Getting Ready for Class

After reading through the preceding section, respond in your own words and in complete sentences.

A. Is $3x^2 + 2x - \frac{1}{x}$ a polynomial? Explain.

B. What is the degree of a polynomial?

C. What are similar terms?

D. Explain in words how you subtract one polynomial from another.

Identify those of the following that are monomials, binomials, or trinomials. Give the degree of each, and name the leading coefficient.

1. $5x^2 - 3x + 2$ **2.** $2x^2 + 4x - 1$ **3.** $3x - 5$

4. $5y + 3$ **5.** $8a^2b^3 + 3ab^2 - 5b$ **6.** $9a^2b^2 - 8ab - 4$

7. $4x^3 - 6x^2 + 5x - 3$ **8.** $9x^4 + 4x^3 - 2x^2 + x$ **9.** $-\dfrac{3}{4}$

10. -16 **11.** $4x - 5 + 6x^3$ **12.** $9x + 2 + 3x^3$

Simplify each of the following by combining similar terms.

13. $2x^2 - 3x + 10x - 15$ **14.** $6x^2 - 4x - 15x + 10$

15. $12a^2 + 8ab - 15ab - 10b^2$ **16.** $28a^2 - 8ab + 7ab - 2b^2$

Add or subtract as indicated.

17. $(4x + 2) + (3x - 1)$ **18.** $(8x - 5) + (-5x + 4)$

19. $(5x^2 - 6x + 1) - (4x^2 + 7x - 2)$ **20.** $(11x^2 - 8x) - (4x^2 - 2x - 7)$

21. $\left(\dfrac{1}{2}x^2 - \dfrac{1}{3}x - \dfrac{1}{6}\right) - \left(\dfrac{1}{4}x^2 + \dfrac{7}{12}x\right) + \left(\dfrac{1}{3}x - \dfrac{1}{12}\right)$

22. $\left(\dfrac{2}{3}x^2 - \dfrac{1}{2}x\right) - \left(\dfrac{1}{4}x^2 + \dfrac{1}{6}x\right) + \dfrac{1}{12} - \left(\dfrac{1}{2}x^2 + \dfrac{1}{4}\right)$

23. $(y^3 - 2y^2 - 3y + 4) - (2y^3 - y^2 + y - 3)$

24. $(8y^3 - 3y^2 + 7y + 2) - (-4y^3 + 6y^2 - 5y - 8)$

25. $(5x^3 - 4x^2) - (3x + 4) + (5x^2 - 7) - (3x^3 + 6)$

26. $(x^3 - x) - (x^2 + x) + (x^3 - 1) - (-3x + 2)$

27. $\left(\dfrac{4}{7}x^2 - \dfrac{1}{7}xy + \dfrac{1}{14}y^2\right) - \left(\dfrac{1}{2}x^2 - \dfrac{2}{7}xy - \dfrac{9}{14}y^2\right)$

28. $\left(\dfrac{1}{5}x^2 - \dfrac{1}{2}xy + \dfrac{1}{10}y^2\right) - \left(-\dfrac{3}{10}x^2 + \dfrac{2}{5}xy - \dfrac{1}{2}y^2\right)$

29. $(3a^3 + 2a^2b + ab^2 - b^3) - (6a^3 - 4a^2b + 6ab^2 - b^3)$

30. $(a^3 - 3a^2b + 3ab^2 - b^3) - (a^3 + 3a^2b + 3ab^2 + b^3)$

31. Find the sum of $x^2 - 6xy + y^2$ and $2x^2 - 6xy - y^2$.

32. Find the sum of $9x^3 - 6x^2 + 2$ and $3x^2 - 5x + 4$.

33. Find the sum of $11a^2 + 3ab + 2b^2$, $9a^2 - 2ab + b^2$, and $-6a^2 - 3ab + 5b^2$.

34. Find the sum of $a^2 - ab - b^2$, $a^2 + ab - b^2$, and $a^2 + 2ab + b^2$.

35. Find the difference of $2x^2 - 7x$ and $2x^2 - 4x$.

36. Find the difference of $-3x + 9$ and $-3x + 6$.

37. Let $p(x) = -8x^5 - 4x^3 + 6$ and $q(x) = 9x^5 - 4x^3 - 6$. Find $q(x) - p(x)$.

38. Let $p(x) = 4x^4 - 3x^3 - 2x^2$ and $q(x) = 2x^4 + 3x^3 + 4x^2$. Find $q(x) - p(x)$.

Simplify each of the following. Begin by working within the innermost parentheses.

39. $-[2 - (4 - x)]$

40. $-[-3 - (x - 6)]$

41. $-5[-(x - 3) - (x + 2)]$

42. $-6[(2x - 5) - 3(8x - 2)]$

43. $4x - 5[3 - (x - 4)]$

44. $x - 7[3x - (2 - x)]$

45. $-(3x - 4y) - [(4x + 2y) - (3x + 7y)]$

46. $(8x - y) - [-(2x + y) - (-3x - 6y)]$

47. $4a - (3a + 2[a - 5(a + 1) + 4])$

48. $6a - (-2a - 6[2a + 3(a - 1) - 6])$

Evaluate the following polynomials.

49. Find the value of $2x^2 - 3x - 4$ when x is 2.

50. Find the value of $4x^2 + 3x - 2$ when x is -1.

51. If $P(x) = \dfrac{3}{2}x^2 - \dfrac{3}{4}x + 1$, find

    a. $P(12)$ **b.** $P(-8)$

52. If $P(x) = \dfrac{2}{5}x^2 - \dfrac{1}{10}x + 2$, find

    a. $P(10)$ **b.** $P(-10)$

53. If $Q(x) = x^3 - x^2 + x - 1$, find

    a. $Q(4)$ **b.** $Q(-2)$

54. If $Q(x) = x^3 + x^2 + x - 1$, find

    a. $Q(5)$ **b.** $Q(-2)$

55. If $R(x) = 11.5x - 0.05x^2$, find

    a. $R(10)$ **b.** $R(-10)$

56. If $R(x) = 11.5x - 0.01x^2$, find

    a. $R(10)$ **b.** $R(-10)$

57. If $P(x) = 600 + 1{,}000x - 100x^2$, find

    a. $P(-4)$ **b.** $P(4)$

58. If $P(x) = 500 + 800x - 100x^2$, find

    a. $P(-6)$ **b.** $P(8)$

Applying the Concepts

59. Caffeine The chart shows the number of milligrams of caffeine in different sodas. If Mary drinks three Dr. Peppers, a Mountain Dew, and two Barq's Root Beers, write an expression that describes the amount of caffeine she drinks, and simplify it.

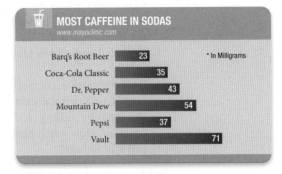

60. Sports Fields The chart shows the minimum acreage needed for different sports. A school has twice as many basketball courts as it does baseball fields and half as many soccer fields as it does baseball fields. Write an expression that describes the minimum acreage of sports fields, then simplify the expression if the school has two baseball fields.

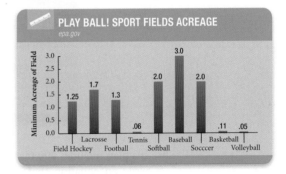

Problems 61–66 may be solved using a graphing calculator.

61. Height of an Object If an object is thrown straight up into the air with a velocity of 128 feet/second, then its height $h(t)$ above the ground t seconds later is given by the formula

$$h(t) = -16t^2 + 128t$$

Find the height after 3 seconds and after 5 seconds. [Find $h(3)$ and $h(5)$.]

62. Height of an Object The formula for the height of an object that has been thrown straight up with a velocity of 64 feet/second is

$$h(t) = -16t^2 + 64t$$

Find the height after 1 second and after 3 seconds. [Find $h(1)$ and $h(3)$.]

63. Profit The total cost (in dollars) for a company to manufacture and sell x items per week is $C(x) = 60x + 300$. If the revenue brought in by selling all x items is $R(x) = 100x - 0.5x^2$, find the weekly profit. How much profit will be made by producing and selling 60 items each week?

64. Profit The total cost (in dollars) for a company to produce and sell x items per week is $C(x) = 200x + 1{,}600$. If the revenue brought in by selling all x items is $R(x) = 300x - 0.6x^2$, find the weekly profit. How much profit will be made by producing and selling 50 items each week?

65. Profit Suppose it costs a company selling sewing patterns $C(x) = 800 + 6.5x$ dollars to produce and sell x patterns a month. If the revenue obtained by selling x patterns is $R(x) = 10x - 0.002x^2$, what is the profit equation? How much profit will be made if 1,000 patterns are produced and sold in May?

66. Profit Suppose a company manufactures and sells x picture frames each month with a total cost of $C(x) = 1{,}200 + 3.5x$ dollars. If the revenue obtained by selling x frames is $R(x) = 9x - 0.003x^2$, find the profit equation. How much profit will be made if 1,000 frames are manufactured and sold in June?

Text Messages and Voice Minutes The chart shows the average number of text messages sent and voice minutes used by different age groups. Use the chart to answer Problems 67 and 68.

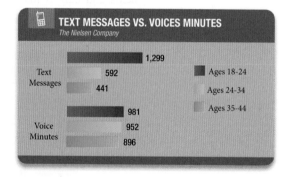

67. If a plan includes 400 text messages for $6, and 5¢ for each additional text, write an equation that describes the cost for text messages for any age group. Then, find how much the average 24–34 year old would owe.

68. A couple in the 35–44 age range have a family plan that includes 700 free text messages and 5¢ for each additional text. Write an expression that describes how much they would owe for text messages and solve.

Learning Objectives Assessment

The following problems can be used to help assess if you have successfully met the learning objectives for this section.

69. Find the degree of the polynomial $8x^4y + 2xy^2 - y^3$.

    a. 5   **b.** 4   **c.** 3   **d.** 8

70. Subtract: $(5x^3 - 3x^2 + 2) - (x^3 + 4x^2 - 9)$.

    a. $6x^3 + x^2 + 11$ **b.** $-4x^5 + 2x - 7$

    c. $4x^3 - 7x^2 + 11$ **d.** $4x^3 + x^2 - 7$

71. Evaluate $5x^2 - 3x - 1$ for $x = -2$.

    a. 25   **b.** 13   **c.** -15 **d.** -27

Getting Ready for the Next Section

Simplify.

72. $2x^2 - 3x + 10x - 15$

73. $12a^2 + 8ab - 15ab - 10b^2$

74. $(6x^3 - 2x^2y + 8xy^2) + (-9x^2y + 3xy^2 - 12y^3)$

75. $(3x^3 - 15x^2 + 18x) + (2x^2 - 10x + 12)$

76. $4x^3(-3x)$ **77.** $5x^2(-4x)$

78. $4x^3(5x^2)$ **79.** $5x^2(3x^2)$

80. $(a^3)^2$ **81.** $(a^4)^2$

82. $11.5(130) - 0.05(130)^2$ **83.** $-0.05(130)^2 + 9.5(130) - 200$

SPOTLIGHT ON SUCCESS   *Student Instructor Stefanie*

Never confuse a single defeat with a final defeat.
—F. Scott Fitzgerald

The idea that has worked best for my success in college, and more specifically in my math courses, is to stay positive and be resilient. I have learned that a 'bad' grade doesn't make me a failure; if anything it makes me strive to do better. That is why I never let a bad grade on a test or even in a class get in the way of my overall success.

By sticking with this positive attitude, I have been able to achieve my goals. My grades have never represented how well I know the material. This is because I have struggled with test anxiety and it has consistently lowered my test scores in a number of courses. However, I have not let it defeat me. When I applied to graduate school, I did not meet the grade requirements for my top two schools, but that did not stop me from applying.

One school asked that I convince them that my knowledge of mathematics was more than my grades indicated. If I had let my grades stand in the way of my goals, I wouldn't have been accepted to both of my top two schools, and will be attending one of them in the Fall, on my way to becoming a mathematics teacher.

Multiplication of Polynomials 5.2

Learning Objectives

In this section, we will learn how to:

1. Multiply polynomials.

2. Multiply binomials using the FOIL method.

3. Find the square of a binomial.

4. Multiply binomials to find the difference of two squares.

Introduction

In the previous section, we found the relationship between profit, revenue, and cost to be

$$P(x) = R(x) - C(x)$$

Revenue itself can be broken down further by another formula common in the business world. The revenue obtained from selling all x items is the product of the number of items sold and the price per item; that is,

$$\text{Revenue} = (\text{Number of items sold})(\text{Price of each item})$$

$$R = xp$$

Many times, x and p are polynomials, which means that the expression xp is the product of two polynomials. In this section, we learn how to multiply polynomials, and in so doing, increase our understanding of the equations and formulas that describe business applications.

Multiplying Polynomials

EXAMPLE 1 Find the product of $4x^3$ and $5x^2 - 3x + 1$.

SOLUTION To multiply, we apply the distributive property.

$$4x^3(5x^2 - 3x + 1)$$

$$= 4x^3(5x^2) + 4x^3(-3x) + 4x^3(1) \qquad \text{Distributive property}$$

$$= 20x^5 - 12x^4 + 4x^3$$

Notice that we multiply coefficients and add exponents.

EXAMPLE 2 Multiply $2x - 3$ and $x + 5$.

SOLUTION Distributing the $2x - 3$ across the sum $x + 5$ gives us

$$(2x - 3)(x + 5)$$

$$= (2x - 3)x + (2x - 3)5 \qquad \text{Distributive property}$$

$$= 2x(x) + (-3)x + 2x(5) + (-3)5 \quad \text{Distributive property}$$

$$= 2x^2 - 3x + 10x - 15$$

$$= 2x^2 + 7x - 15 \qquad \text{Combine like terms}$$

Notice the third line in the previous example. It consists of all possible products of terms in the first binomial and those of the second binomial. We can generalize this into a rule for multiplying two polynomials.

> **[△≠Σ] RULE** *Multiplication of Polynomials*
>
> To multiply two polynomials, multiply each term in the first polynomial by every term in the second polynomial.

Multiplying polynomials can be accomplished by a method that looks very similar to long multiplication with whole numbers.

EXAMPLE 3 Multiply $(2x - 3y)$ and $(3x^2 - xy + 4y^2)$ vertically.

SOLUTION

$$
\begin{array}{r}
3x^2 - xy + 4y^2 \\
2x - 3y \\
\hline
6x^3 - 2x^2y + 8xy^2 \\
- 9x^2y + 3xy^2 - 12y^3 \\
\hline
6x^3 - 11x^2y + 11xy^2 - 12y^3
\end{array}
$$

Multiply $(3x^2 - xy + 4y^2)$ by $2x$

Multiply $(3x^2 - xy + 4y^2)$ by $-3y$

Add similar terms

Note The vertical method of multiplying polynomials does not directly show the use of the distributive property. It is, however, very useful since it always gives the correct result and is easy to remember.

Multiplying Binomials—The FOIL Method

Consider the product of $(2x - 5)$ and $(3x - 2)$. Distributing $(3x - 2)$ over $2x$ and -5, we have

$$(2x - 5)(3x - 2) = (2x)(3x - 2) + (-5)(3x - 2)$$

$$= (2x)(3x) + (2x)(-2) + (-5)(3x) + (-5)(-2)$$

$$= 6x^2 - 4x - 15x + 10$$

$$= 6x^2 - 19x + 10$$

Looking closely at the second and third lines, we notice the following:

1. $6x^2$ comes from multiplying the *first* terms in each binomial.

$(2x - 5)(3x - 2)$ $2x(3x) = 6x^2$ *First* terms

2. $-4x$ comes from multiplying the *outside* terms in the product.

$(2x - 5)(3x - 2)$ $2x(-2) = -4x$ *Outside* terms

3. $-15x$ comes from multiplying the *inside* terms in the product.

$(2x - 5)(3x - 2)$ $-5(3x) = -15x$ *Inside* terms

4. 10 comes from multiplying the *last* two terms in the product.

$(2x - 5)(3x - 2)$ $-5(-2) = 10$ *Last* terms

Note The FOIL method does not show the properties used in multiplying two binomials. It is simply a way of finding products of binomials quickly. Remember, the FOIL method applies only to products of two binomials. The vertical method applies to all products of polynomials with two or more terms.

Once we know where the terms in the answer come from, we can reduce the number of steps used in finding the product.

$$(2x - 5)(3x - 2) = 6x^2 - 4x - 15x + 10 = 6x^2 - 19x + 10$$

First Outside Inside Last

EXAMPLE 4 Multiply $(4a - 5b)(3a + 2b)$ using the FOIL method.

SOLUTION $(4a - 5b)(3a + 2b) = 12a^2 + 8ab - 15ab - 10b^2$

F   O   I   L

$$= 12a^2 - 7ab - 10b^2$$

EXAMPLE 5 Multiply $(3 - 2t)(4 + 7t)$ using the FOIL method.

SOLUTION $(3 - 2t)(4 + 7t) = 12 + 21t - 8t - 14t^2$

F   O   I   L

$$= 12 + 13t - 14t^2$$

EXAMPLE 6 If $p(x) = 2x + \frac{1}{2}$ and $q(x) = 4x - \frac{1}{2}$, find the product $p(x) \cdot q(x)$.

SOLUTION $p(x) \cdot q(x) = \left(2x + \frac{1}{2}\right)\left(4x - \frac{1}{2}\right)$

$$= 8x^2 - x + 2x - \frac{1}{4}$$

F   O   I   L

$$= 8x^2 + x - \frac{1}{4}$$

EXAMPLE 7 Multiply $(a^5 + 3)(a^5 - 7)$ using the FOIL method.

SOLUTION $(a^5 + 3)(a^5 - 7) = a^{10} - 7a^5 + 3a^5 - 21$

F   O   I   L

$$= a^{10} - 4a^5 - 21$$

EXAMPLE 8 Multiply $(2x + 3)(5y - 4)$ using the FOIL method.

SOLUTION $(2x + 3)(5y - 4) = 10xy - 8x + 15y - 12$

F   O   I   L

The Square of a Binomial

We can use the FOIL method to find the square of a binomial by writing the expression as a product.

EXAMPLE 9 Multiply: $(4x - 6)^2$.

SOLUTION Applying the definition of exponents and then the FOIL method, we have

$$(4x - 6)^2 = (4x - 6)(4x - 6)$$

$$= 16x^2 - 24x - 24x + 36$$

F   O   I   L

$$= 16x^2 - 48x + 36$$

Another way to find the square of a binomial involves using a formula. Consider Figure 1 as a representation of $(a + b)^2$.

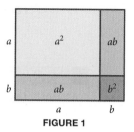

FIGURE 1

This type of product occurs frequently enough in algebra that we have special formulas for it. Here are the formulas for binomial squares:

$$(a + b)^2 = (a + b)(a + b) \qquad\qquad (a - b)^2 = (a - b)(a - b)$$
$$= a^2 + ab + ab + b^2 \qquad\qquad = a^2 - ab - ab + b^2$$
$$= a^2 + 2ab + b^2 \qquad\qquad = a^2 - 2ab + b^2$$

Observing the results in both cases, we have the following rule:

Note Be careful not to mistake $(a + b)^2$ as the same as $a^2 + b^2$. We must multiply both terms of the first binomial by both terms of the second to get $a^2 - 2ab + b^2$. Again, $(a + b)^2 \neq a^2 + b^2$.

$\lceil\Delta{\neq}\Sigma$ RULE   *Squaring of a Binomial*

The square of a binomial is the sum of the square of the first term, twice the product of the two inside terms, and the square of the last term. Or:

$$(a + b)^2 = \qquad a^2 \qquad + \qquad 2ab \qquad + \qquad b^2$$

	Square of first term	Twice the product of the two terms	Square of last term

$$(a - b)^2 = \qquad a^2 \qquad - \qquad 2ab \qquad + \qquad b^2$$

EXAMPLE 10   Use the formulas in the rule for expanding binomial squares to expand the following.

a. $(x + 7)^2 = x^2 + 2(x)(7) + 7^2$

$$= x^2 + 14x + 49$$

b. $(3t - 5)^2 = (3t)^2 - 2(3t)(5) + 5^2$

$$= 9t^2 - 30t + 25$$

c. $(4x + 2y)^2 = (4x)^2 + 2(4x)(2y) + (2y)^2$

$$= 16x^2 + 16xy + 4y^2$$

d. $(5 - a^3)^2 = 5^2 - 2(5)(a^3) + (a^3)^2$

$$= 25 - 10a^3 + a^6$$

Products Resulting in the Difference of Two Squares

Another frequently occurring kind of product is found when multiplying two binomials that differ only in the sign between their terms.

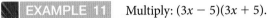 **EXAMPLE 11** Multiply: $(3x - 5)(3x + 5)$.

SOLUTION Applying the FOIL method, we have

$$(3x - 5)(3x + 5) = 9x^2 + 15x - 15x - 25 \quad \text{Two middle terms add to 0}$$
$$\qquad\qquad\qquad\quad \text{F} \quad \text{O} \quad \text{I} \quad \text{L}$$

$$= 9x^2 - 25$$

The outside and inside products in Example 11 are opposites and therefore add to 0. Here it is in general:

$$(a - b)(a + b) = a^2 + ab - ab - b^2 \quad \text{Two middle terms add to 0}$$

$$= a^2 - b^2$$

The expression $a^2 - b^2$ is called the *difference of two squares.*

> ⟨Δ≠Σ⟩ **RULE** *Difference of Two Squares*
>
> To multiply two binomials that differ only in the sign between their two terms, simply subtract the square of the second term from the square of the first term:
> $$(a + b)(a - b) = a^2 - b^2$$

EXAMPLE 12 Find the following products:

a. $(x - 5)(x + 5) = x^2 - 5^2$
$$= x^2 - 25$$

b. $(2a - 3)(2a + 3) = (2a)^2 - 3^2$
$$= 4a^2 - 9$$

c. $(x^2 + 4)(x^2 - 4) = (x^2)^2 - 4^2$
$$= x^4 - 16$$

d. $(x^3 - 2a)(x^3 + 2a) = (x^3)^2 - (2a)^2$
$$= x^6 - 4a^2$$

More About Revenue

From the introduction to this chapter, we know that the revenue obtained from selling x items at p dollars per item is

$$R = \text{Revenue} = xp \qquad \text{(The number of items} \times \text{price per item)}$$

For example, if a store sells 100 items at \$4.50 per item, the revenue is $100(4.50) = \$450$. If we have an equation that gives the relationship between x and p, then we can write the revenue in terms of x or in terms of p. With function notation, we would write the revenue as either $R(x)$ or $R(p)$, where

> $R(x)$ is the revenue function that gives the revenue R in terms of the number of items x that are sold.
>
> $R(p)$ is the revenue function that gives the revenue R in terms of the price per item p.

With function notation we can see exactly which variables we want our formulas written in terms of.

EXAMPLE 13   A company manufactures and sells DVDs. They find that they can sell x DVDs each day at p dollars per disc, according to the equation $x = 230 - 20p$. Find $R(p)$ and $R(x)$.

SOLUTION   The notation $R(p)$ tells us we are to write the revenue equation in terms of the variable p. To do so, we use the formula $R(p) = xp$ and substitute $230 - 20p$ for x to obtain

$$R(p) = xp = (230 - 20p)p = 230p - 20p^2$$

The notation $R(x)$ indicates that we are to write the revenue in terms of the variable x. We need to solve the equation $x = 230 - 20p$ for p. Let's begin by interchanging the two sides of the equation.

$$230 - 20p = x$$

$$-20p = -230 + x \qquad \text{Add } -230 \text{ to each side}$$

$$p = \frac{-230 + x}{-20} \qquad \text{Divide each side by } -20$$

$$p = 11.5 - 0.05x \qquad \tfrac{230}{20} = 11.5 \text{ and } \tfrac{1}{20} = 0.05$$

Now we can find $R(x)$ by substituting $11.5 - 0.05x$ for p in the formula $R(x) = xp$.

$$R(x) = xp = x(11.5 - 0.05x) = 11.5x - 0.05x^2$$

Our two revenue functions are actually equivalent. To offer some justification for this, suppose that the company decides to sell each disc for \$5. The equation $x = 230 - 20p$ indicates that, at \$5 per disc, they will sell $x = 230 - 20(5) = 230 - 100 = 130$ discs per day. To find the revenue from selling the discs for \$5 each, we use $R(p)$ with $p = 5$.

$$\begin{aligned}
\text{If} \qquad & p = 5 \\
\text{then} \qquad & R(p) = R(5) \\
& = 230(5) - 20(5)^2 \\
& = 1{,}150 - 500 \\
& = \$650
\end{aligned}$$

However, to find the revenue from selling 130 discs, we use $R(x)$ with $x = 130$.

$$\text{If} \qquad x = 130$$

$$\text{then} \qquad R(x) = R(130)$$

$$= 11.5(130) - 0.05(130)^2$$

$$= 1{,}495 - 845$$

$$= \$650$$

 EXAMPLE 14 Suppose the daily cost function for the DVDs in Example 13 is $C(x) = 200 + 2x$. Find the profit function $P(x)$ and then find $P(130)$.

SOLUTION Since profit is equal to the difference of the revenue and the cost, we have

$$P(x) = R(x) - C(x)$$

$$= 11.5x - 0.05x^2 - (200 + 2x)$$

$$= -0.05x^2 + 9.5x - 200$$

Notice that we used the formula for $R(x)$ from Example 13 instead of the formula for $R(p)$. We did so because we were asked to find $P(x)$, meaning we want the profit P only in terms of the variable x.

Next, we use the formula we just obtained to find $P(130)$.

$$P(130) = -0.05(130)^2 + 9.5(130) - 200$$

$$= -0.05(16{,}900) + 9.5(130) - 200$$

$$= -845 + 1{,}235 - 200$$

$$= \$190$$

Because $P(130) = \$190$, the company will make a profit of $190 per day by selling 130 discs per day.

Getting Ready for Class

After reading through the preceding section, respond in your own words and in complete sentences.

A. In words, state the rule for multiplying polynomials.

B. How would you use the FOIL method to multiply two binomials?

C. Explain why $(a + b)^2 \neq a^2 + b^2$.

D. Discuss the rule for a difference of two squares.

Problem Set 5.2

Multiply the following by applying the distributive property.

1. $2x(6x^2 - 5x + 4)$ **2.** $-3x(5x^2 - 6x - 4)$ **3.** $-3a^2(a^3 - 6a^2 + 7)$

4. $4a^3(3a^2 - a + 1)$ **5.** $2a^2b(a^3 - ab + b^3)$ **6.** $5a^2b^2(8a^2 - 2ab + b^2)$

Multiply.

7. $(x - 5)(x + 3)$ **8.** $(x + 4)(x + 6)$ **9.** $(2x^2 - 3)(3x^2 - 5)$

10. $(3x^2 + 4)(2x^2 - 5)$ **11.** $(x + 3)(x^2 + 6x + 5)$ **12.** $(x - 2)(x^2 - 5x + 7)$

13. $(a - b)(a^2 + ab + b^2)$ **14.** $(a + b)(a^2 - ab + b^2)$

15. $(2x + y)(4x^2 - 2xy + y^2)$ **16.** $(x - 3y)(x^2 + 3xy + 9y^2)$

17. $(2a - 3b)(a^2 + ab + b^2)$ **18.** $(5a - 2b)(a^2 - ab - b^2)$

Multiply the following using the FOIL method.

19. $(x - 2)(x + 3)$ **20.** $(x + 2)(x - 3)$ **21.** $(2a + 3)(3a + 2)$

22. $(5a - 4)(2a + 1)$ **23.** $(5 - 3t)(4 + 2t)$ **24.** $(7 - t)(6 - 3t)$

25. $(x^3 + 3)(x^3 - 5)$ **26.** $(x^3 + 4)(x^3 - 7)$ **27.** $(5x - 6y)(4x + 3y)$

28. $(6x - 5y)(2x - 3y)$ **29.** $\left(3t + \dfrac{1}{3}\right)\left(6t - \dfrac{2}{3}\right)$ **30.** $\left(5t - \dfrac{1}{5}\right)\left(10t + \dfrac{3}{5}\right)$

31. Let $p(x) = 4x - 3$ and $q(x) = 2x + 1$. Find

    a. $p(x) - q(x)$ **b.** $p(x) + q(x)$ **c.** $p(x) \cdot q(x)$

32. Let $p(x) = 5x + 2$ and $q(x) = 3x - 7$. Find

    a. $p(x) - q(x)$ **b.** $p(x) + q(x)$ **c.** $p(x) \cdot q(x)$

Find the following special products.

33. $(5x + 2y)^2$ **34.** $(3x - 4y)^2$

35. $(5 - 3t^3)^2$ **36.** $(7 - 2t^4)^2$

37. $(2a + 3b)(2a - 3b)$ **38.** $(6a - 1)(6a + 1)$

39. $(3r^2 + 7s)(3r^2 - 7s)$ **40.** $(5r^2 - 2s)(5r^2 + 2s)$

41. $\left(y + \dfrac{3}{2}\right)^2$ **42.** $\left(y - \dfrac{7}{2}\right)^2$

43. $\left(a - \dfrac{1}{2}\right)^2$ **44.** $\left(a - \dfrac{5}{2}\right)^2$

45. $\left(x + \dfrac{1}{4}\right)^2$ **46.** $\left(x - \dfrac{3}{8}\right)^2$

47. $\left(t + \dfrac{1}{3}\right)^2$ **48.** $\left(t - \dfrac{2}{5}\right)^2$

49. $\left(\dfrac{1}{3}x - \dfrac{2}{5}\right)\left(\dfrac{1}{3}x + \dfrac{2}{5}\right)$ **50.** $\left(\dfrac{3}{4}x - \dfrac{1}{7}\right)\left(\dfrac{3}{4}x + \dfrac{1}{7}\right)$

Find the following products.

51. $(x - 2)^3$

52. $(4x + 1)^3$

53. $\left(x - \dfrac{1}{2}\right)^3$

54. $\left(x + \dfrac{1}{4}\right)^3$

55. $3(x - 1)(x - 2)(x - 3)$

56. $2(x + 1)(x + 2)(x + 3)$

57. $(b^2 + 8)(a^2 + 1)$

58. $(b^2 + 1)(a^4 - 5)$

59. $(x + 1)^2 + (x + 2)^2 + (x + 3)^2$

60. $(x - 1)^2 + (x - 2)^2 + (x - 3)^2$

61. $(2x + 3)^2 - (2x - 3)^2$

62. $(x - 3)^3 - (x + 3)^3$

Simplify.

63. $(x + 3)^2 - 2(x + 3) - 8$

64. $(x - 2)^2 - 3(x - 2) - 10$

65. $(2a - 3)^2 - 9(2a - 3) + 20$

66. $(3a - 2)^2 + 2(3a - 2) - 3$

67. $2(4a + 2)^2 - 3(4a + 2) - 20$

68. $6(2a + 4)^2 - (2a + 4) - 2$

69. Let $a = 2$ and $b = 3$, and evaluate each of the following expressions.
  a. $a^4 - b^4$
  b. $(a - b)^4$
  c. $(a^2 + b^2)(a + b)(a - b)$

70. Let $a = 2$ and $b = 3$, and evaluate each of the following expressions.
  a. $a^3 + b^3$
  b. $(a + b)^3$
  c. $a^3 + 3a^2b + 3ab^2 + b^3$

Applying the Concepts

71. Revenue A store selling art supplies finds that it can sell x sketch pads per week at p dollars each, according to the formula $x = 900 - 300p$. Write formulas for $R(p)$ and $R(x)$. Then find the revenue obtained by selling the pads for $1.60 each.

72. Revenue A company selling CDs finds that it can sell x CDs per day at p dollars per CD, according to the formula $x = 800 - 100p$. Write formulas for $R(p)$ and $R(x)$. Then find the revenue obtained by selling the CDs for $3.80 each.

73. Revenue A company sells an inexpensive accounting program for home computers. If it can sell x programs per week at p dollars per program, according to the formula $x = 350 - 10p$, find formulas for $R(p)$ and $R(x)$. How much will the weekly revenue be if it sells 65 programs?

74. Revenue A company sells boxes of greeting cards through the mail. It finds that it can sell x boxes of cards each week at p dollars per box, according to the formula $x = 1,475 - 250p$. Write formulas for $R(p)$ and $R(x)$. What revenue will it bring in each week if it sells 200 boxes of cards?

75. Profit If the cost to produce the x programs in Problem 73 is $C(x) = 5x + 500$, find $P(x)$ and $P(60)$.

76. Profit If the cost to produce the x CDs in Problem 72 is $C(x) = 2x + 200$, find $P(x)$ and $P(40)$.

77. Surrounding Area A pool 10 feet wide and 40 feet long is to be surrounded by a pathway x feet wide. Find an expression for the area of the pool deck as a function of x. What area will the pool deck have if the pathway is 5 feet wide? What about 8 feet wide?

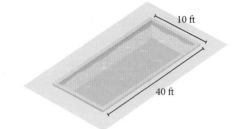

10 ft

40 ft

78. Surrounding Area A lap pool 10 times as long as it is wide. A 2-meter cement pathway is built around the entire pool. Find an expression for the area of the pool deck as a function w, the width of the pool. What will the area of the pool deck be if the path is 3 meters wide? What if the path is 4 meters wide?

79. Volume A rectangular box is built by cutting out square corners from a 9" by 12" piece of cardboard, then folding the resulting flaps up to form the height. Let x represent the sides of the square corners being cut out. Express the volume of the box as a function of x. What will the volume be if 2" squares are cut out?

x

12 in

9 in

80. Volume A rectangular box is built by cutting out square corners from a 9 cm by 12 cm piece of cardboard, then folding the resulting flaps up to form the height. Let x represent the sides of the square corners being cut out. Express the volume of the box as a function of x. What will the volume be if 3 cm squares are cut out? What is the domain of the volume function you found? (Hint: What is the largest square that can be cut out from the sides?)

81. Interest If you deposit $100 in an account with an interest rate r that is compounded annually, then the amount of money in that account at the end of 4 years is given by the formula $A = 100(1 + r)^4$. Expand the right side of this formula.

82. **Interest** If you deposit P dollars in an account with an annual interest rate r that is compounded twice a year, then at the end of a year the amount of money in that account is given by the formula

$$A = P\left(1 + \frac{r}{2}\right)^2$$

Expand the right side of this formula.

83. **Inflation** The future price of an item priced at P dollars is given by

$$F = P(1 + r)^t$$

where r is the inflation rate (expressed as a decimal) and t is the time (in years). If a gallon of gas costs $4.00 today, what will its cost be in 10 years assuming an inflation rate of 8% per year?

84. **Gas** A politician is quoted as saying, "At this rate, we will be paying $20.00 for a gallon of gas in twenty years." Is the quote from the politician reasonable? Use the formula in the previous problem to answer this question.

85. **Tuition** Tuition fees for California State University were $5,131 during the 2010 – 2011 academic year. Historically, tuition has increased at an 8% per year rate. Use the formula from Problem 83 to find what the tuition was 20 years ago and 30 years ago. Hint: First solve for P, then substitute the given value for F. Round to the nearest dollar.

86. **Tuition** Repeat the previous problem for University of California, where the 2010 – 2011 yearly tuition was $11,300. Round to the nearest dollar.

Learning Objectives Assessment

The following problems can be used to help assess if you have successfully met the learning objectives for this section.

87. Multiply: $(x + 2)(3x^2 - 4x + 5)$.

    a. $9x^2 - 7x + 10$ **b.** $3x^3 + 2x^2 - 3x + 10$

    c. $3x^3 + 10x^2 + 13x + 10$ **d.** $3x^3 + 6x^2 + 5x + 10$

88. Multiply $(2x - 3)(3x + 5)$ using the FOIL method.

    a. $6x^2 - 15$ **b.** $5x^2 + x + 2$

    c. $6x^2 + x - 15$ **d.** $7x - 15$

89. Find the product: $(2a - 3b)^2$.

    a. $4a^2 - 9b^2$ **b.** $4a^2 + 9b^2$

    c. $4a^2 - 12ab + 9b^2$ **d.** $4a^2 - 6ab + 9b^2$

90. Multiply: $(3t + 2)(3t - 2)$.

    a. $9t^2 - 4$ **b.** $9t^2 + 4$

    c. $9t^2 + 12t - 4$ **d.** $9t^2 - 12t - 4$

Getting Ready for the Next Section

Divide.

91. $\dfrac{10x^2}{5x^2}$

92. $\dfrac{-15x^4}{5x^2}$

93. $\dfrac{4x^4y^3}{-2x^2y}$

94. $\dfrac{10a^4b^2}{4a^2b^2}$

95. $4{,}628 \div 25$

96. $7{,}546 \div 35$

Multiply.

97. $2x^2(2x - 4)$

98. $3x^2(x - 2)$

99. $(2x - 4)(2x^2 + 4x + 5)$

100. $(x - 2)(3x^2 + 6x + 15)$

Subtract.

101. $(2x^2 - 7x + 9) - (2x^2 - 4x)$

102. $(x^2 - 6xy - 7y^2) - (x^2 + xy)$

Learning Objectives

In this section, we will learn how to:

1. Divide a polynomial by a monomial.

2. Divide polynomials using long division.

3. Divide polynomials using synthetic division.

4. Find the average cost function.

Introduction

First Bank of San Luis Obispo charges $2.00 per month and $0.15 per check for a regular checking account. So, if you write x checks in one month, the total monthly cost of the checking account will be $C(x) = 2.00 + 0.15x$. From this formula, we see that the more checks we write in a month, the more we pay for the account. But it is also true that the more checks we write in a month, the lower the cost per check. To find the cost per check, we use the *average cost* function. To find the average cost function, we divide the total cost by the number of checks written.

$$\text{Average Cost} = \overline{C}(x) = \frac{C(x)}{x} = \frac{2.00 + 0.15x}{x}$$

This last expression gives us the average cost per check for each of the x checks written. To work with this last expression, we need to know something about division with polynomials, and that is what we will cover in this section.

We begin this section by considering division of a polynomial by a monomial. This is the simplest kind of polynomial division. The rest of the section is devoted to division of a polynomial by a polynomial.

Dividing a Polynomial by a Monomial

To divide a polynomial by a monomial, we use the definition of division and apply the distributive property. This method is sometimes referred to as *monomial division*. The following example illustrates the procedure.

VIDEO EXAMPLES

SECTION 5.3

EXAMPLE 1   Divide: $\dfrac{10x^5 - 15x^4 + 20x^3}{5x^2}$.

SOLUTION

$$\frac{10x^5 - 15x^4 + 20x^3}{5x^2} = (10x^5 - 15x^4 + 20x^3) \cdot \frac{1}{5x^2} \qquad \text{Dividing by } 5x^2 \text{ is the same as multiplying by } \tfrac{1}{5x^2}$$

$$= 10x^5 \cdot \frac{1}{5x^2} - 15x^4 \cdot \frac{1}{5x^2} + 20x^3 \cdot \frac{1}{5x^2} \qquad \text{Distributive property}$$

$$= \frac{10x^5}{5x^2} - \frac{15x^4}{5x^2} + \frac{20x^3}{5x^2} \qquad \text{Multiplying by } \tfrac{1}{5x^2} \text{ is the same as dividing by } 5x^2$$

$$= 2x^3 - 3x^2 + 4x$$

Notice that division of a polynomial by a monomial is accomplished by dividing each term of the polynomial by the monomial. The first two steps are usually not shown in a problem like this. They are part of Example 1 to justify distributing $5x^2$ under all three terms of the polynomial $10x^5 - 15x^4 + 20x^3$.

Here are some more examples of monomial division.

EXAMPLE 2   Divide: $\dfrac{8x^3y^5 - 16x^2y^2 + 4x^4y^3}{-2x^2y}$.

SOLUTION

$$\frac{8x^3y^5 - 16x^2y^2 + 4x^4y^3}{-2x^2y} = \frac{8x^3y^5}{-2x^2y} + \frac{-16x^2y^2}{-2x^2y} + \frac{4x^4y^3}{-2x^2y}$$

$$= -4xy^4 + 8y - 2x^2y^2$$

EXAMPLE 3   Divide: $\dfrac{10a^4b^2 + 8ab^3 - 12a^3b + 6ab}{4a^2b^2}$. Write the result with positive exponents.

SOLUTION

$$\frac{10a^4b^2 + 8ab^3 - 12a^3b + 6ab}{4a^2b^2} = \frac{10a^4b^2}{4a^2b^2} + \frac{8ab^3}{4a^2b^2} - \frac{12a^3b}{4a^2b^2} + \frac{6ab}{4a^2b^2}$$

$$= \frac{5a^2}{2} + \frac{2b}{a} - \frac{3a}{b} + \frac{3}{2ab}$$

Notice in Example 3 that the result is not a polynomial because of the last three terms. If we were to write each as a product, some of the variables would have negative exponents. For example, the second term would be

$$\frac{2b}{a} = 2a^{-1}b$$

The divisor in each of the preceding examples was a monomial. We now want to turn our attention to division of polynomials in which the divisor has two or more terms.

Long Division

When the divisor is a polynomial with two or more terms, the previous method of monomial division will not work. Instead, we use a process called *long division*. Because this new method is very similar to long division with whole numbers, we will review the method of long division here.

EXAMPLE 4   Divide: $25\overline{)4{,}628}$.

SOLUTION

$$\begin{array}{r} 1 \\ 25\overline{)4{,}628} \\ \underline{25} \\ 21 \end{array}$$ Estimate 25 into 46

     Multiply $1 \times 25 = 25$
     Subtract $46 - 25 = 21$

$$\begin{array}{r} 1 \\ 25\overline{)4{,}628} \\ \underline{25\downarrow} \\ 212 \end{array}$$

     Bring down the 2

These are the four basic steps in long division: estimate, multiply, subtract, and bring down the next term. To complete the problem, we simply perform the same four steps:

$$\begin{array}{r} 18 \\ 25\overline{)4{,}628} \\ 25 \\ \hline 2\,12 \\ 2\,00 \\ \hline 128 \end{array}$$

Estimate 25 into 212 (we try 8)

Multiply 8×25 to get 200
Subtract to get 12, then bring down the 8

One more time:

$$\begin{array}{r} 185 \\ 25\overline{)4{,}628} \\ 25 \\ \hline 2\,12 \\ 2\,00 \\ \hline 128 \\ 125 \\ \hline 3 \end{array}$$

5 is the estimate of 25 into 128

Multiply 5×25 to get 125
Subtract to get 3

Because 3 is less than 25 and we have no more terms to bring down, we have our answer. The quotient is 185 and the reminder is 3. We write the result in the form

$$\text{Quotient} + \frac{\text{Reminder}}{\text{Divisor}}$$

Here is how it looks:

$$\frac{4{,}628}{25} = 185 + \frac{3}{25}$$

To check our answer, we multiply 185 by 25 and then add 3 to the result:

$$25(185) + 3 = 4{,}625 + 3 = 4{,}628$$

Long division with polynomials is similar to long division with whole numbers. Both use the same four basic steps: estimate, multiply, subtract, and bring down the next term. Here is an example.

EXAMPLE 5 Divide: $\dfrac{2x^2 - 7x + 9}{x - 2}$.

SOLUTION

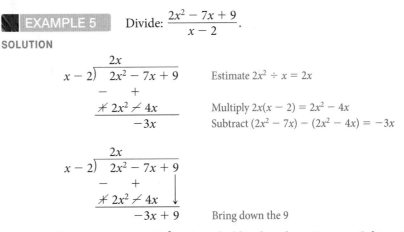

Estimate $2x^2 \div x = 2x$

Multiply $2x(x - 2) = 2x^2 - 4x$
Subtract $(2x^2 - 7x) - (2x^2 - 4x) = -3x$

Bring down the 9

Notice we change the signs on $2x^2 - 4x$ and add in the subtraction step. Subtracting a polynomial is equivalent to adding its opposite, and this method usually results in fewer mistakes while doing the subtraction.

We repeat the four steps.

$$
\begin{array}{r}
2x - 3 \\
x - 2{\overline{\smash{\big)}\,2x^2 - 7x + 9}}
\end{array}
$$

-3 is the estimate: $-3x \div x = -3$

$$
\begin{array}{r}
2x^2 - 4x \\
\hline
-3x + 9 \\
+- \\
-3x + 6 \\
\hline
3
\end{array}
$$

Multiply $-3(x - 2) = -3x + 6$
Subtract $(-3x + 9) - (-3x + 6) = 3$

Because we have no other term to bring down, we have our answer. The quotient is $2x - 3$ and the remainder is 3. We express our answer as follows:

$$\frac{2x^2 - 7x + 9}{x - 2} = 2x - 3 + \frac{3}{x - 2}$$

To check, we multiply $(2x - 3)(x - 2)$ to get $2x^2 - 7x + 6$; then, adding the remainder 3 to this result, we have $2x^2 - 7x + 9$.

In setting up a division problem involving two polynomials, we must remember two things: (1) Both polynomials should be in decreasing powers of the variable, and (2) neither should skip any powers from the highest power down to the constant term. If there are any missing terms, they can be filled in using a coefficient of 0. This will insure that we are always subtracting similar terms. The next example illustrates how this can be done.

EXAMPLE 6 Divide: $2x - 4{\overline{\smash{\big)}\,4x^3 - 6x - 11}}$.

SOLUTION Because the dividend is missing an x^2 term, we can fill it in with $0x^2$:

$$4x^3 - 6x - 11 = 4x^3 + 0x^2 - 6x - 11$$

Adding $0x^2$ does not change our original problem.

Note Adding the $0x^2$ term gives us a column in which to write $-8x^2$. We cannot subtract $-8x^2$ from $-6x$.

$$
\begin{array}{r}
2x^2 + 4x + 5 \\
2x - 4{\overline{\smash{\big)}\,4x^3 + 0x^2 - 6x - 11}} \\
4x^3 - 8x^2 \\
\hline
+ 8x^2 - 6x \\
8x^2 - 16x \\
\hline
+ 10x - 11 \\
10x - 20 \\
\hline
+ 9
\end{array}
$$

The quotient is $2x^2 + 4x + 5$ and the remainder is 9.

$$\frac{4x^3 - 6x - 11}{2x - 4} = 2x^2 + 4x + 5 + \frac{9}{2x - 4}$$

To check this result, we multiply $2x - 4$ and $2x^2 + 4x + 5$:

$$
\begin{array}{r}
2x^2 + 4x \ + \ 5 \\
\times \qquad 2x \ - \ 4 \\
\hline
4x^3 + 8x^2 + 10x \\
- \ 8x^2 - 16x - 20 \\
\hline
4x^3 \qquad - \ 6x - 20
\end{array}
$$

Adding 9 (the remainder) to this result gives us the polynomial $4x^3 - 6x - 11$. Our answer checks. ∎

For our next example, we look at a case where the remainder is zero.

EXAMPLE 7   Divide: $\dfrac{x^2 - 6xy - 7y^2}{x + y}$.

SOLUTION

$$
\begin{array}{r}
x \ - 7y \\
x + y \overline{)\ x^2 - \ 6xy - 7y^2} \\
\end{array}
\qquad x^2 \div x = x
$$

$$
\begin{array}{r}
\cancel{+}\, x^2 \cancel{+}\ xy \\
\hline
- \ 7xy - 7y^2 \\
+ \qquad + \\
\cancel{-}\, 7xy \cancel{-}\, 7y^2 \\
\hline
0
\end{array}
\qquad -7xy \div x = -7y
$$

The quotient is $x - 7y$ and the remainder is 0. When the remainder is zero we do not need to write the fraction containing the remainder in our answer. Thus, we have

$$
\frac{x^2 - 6xy - 7y^2}{x + y} = x - 7y
$$

which is easy to check because

$$
(x + y)(x - 7y) = x^2 - 6xy - 7y^2
$$
∎

EXAMPLE 8   Divide: $\dfrac{4x^3 + 2x^2 - 5x - 3}{2x^2 - 3x + 1}$.

SOLUTION

$$
\begin{array}{r}
2x \ + 4 \\
2x^2 - 3x + 1 \overline{)\ 4x^3 + 2x^2 - 5x \ - 3} \\
- \quad + \quad - \\
\cancel{+}\, 4x^3 \cancel{-}\, 6x^2 \cancel{+}\, 2x \\
\hline
+ \ 8x^2 - 7x \ - 3 \\
- \quad + \quad - \\
\cancel{+}\, 8x^2 \cancel{-}\, 12x \cancel{+}\, 4 \\
\hline
5x - 7
\end{array}
$$

In this case, the quotient is $2x + 4$ and the remainder is the binomial $5x - 7$. We write our answer in the form

$$
\frac{4x^3 + 2x^2 - 5x - 3}{2x^2 - 3x + 1} = 2x + 4 + \frac{5x - 7}{2x^2 - 3x + 1}
$$
∎

Synthetic Division

Synthetic division is a short form of long division with polynomials. We will consider synthetic division only for those cases in which the divisor is of the form $x + k$, where k is a constant.

Let's begin by looking over an example of long division with polynomials.

$$
\begin{array}{r}
3x^2 - 2x + 4 \\
x + 3\overline{)3x^3 + 7x^2 - 2x - 4} \\
\underline{3x^3 + 9x^2} \\
-2x^2 - 2x \\
\underline{-2x^2 - 6x} \\
4x - 4 \\
\underline{4x + 12} \\
-16
\end{array}
$$

We can rewrite the problem without showing the variable since the variable is written in descending powers and similar terms are in alignment. It looks like this:

$$
\begin{array}{r}
3 \quad -2 \quad 4 \\
1 + 3\overline{)3 \quad 7 \quad -2 \quad -4} \\
\underline{(3) + 9} \\
-2 \;\; (-2) \\
\underline{(-2) \;\; -6} \\
4 \;\; (-4) \\
\underline{(4) \quad 12} \\
-16
\end{array}
$$

We have used parentheses to enclose the numbers that are repetitions of the numbers above them. We can compress the problem by eliminating all repetitions except the first one:

$$
\begin{array}{r}
3 \quad -2 \quad 4 \\
1 + 3\overline{)3 \quad 7 \quad -2 \quad -4} \\
\underline{9 \quad -6 \quad 12} \\
3 \quad -2 \quad 4 \;\; -16
\end{array}
$$

The top line is the same as the first three terms of the bottom line, so we eliminate the top line. Also, the 1 that was the coefficient of x in the original problem can be eliminated since we will consider only division problems where the divisor is of the form $x + k$. The following is the most compact form of the original division problem:

$$
\begin{array}{r}
+3\overline{)3 \quad 7 \quad -2 \quad -4} \\
\underline{9 \quad -6 \quad 12} \\
3 \quad -2 \quad 4 \;\; -16
\end{array}
$$

If we check over the problem, we find that the first term in the bottom row is exactly the same as the first term in the top row—and it always will be in problems of this type. Also, the last three terms in the bottom row come from multiplication by $+3$ and then subtraction. We can get an equivalent result by multiplying by -3 and adding. The problem would then look like this:

$$
\begin{array}{r|rrrr}
-3 & 3 & 7 & -2 & -4 \\
 & \downarrow & -9 & 6 & -12 \\
\hline
 & 3 & -2 & 4 & \boxed{-16}
\end{array}
$$

We have used the brackets ⌋ and ⌊ to separate the divisor and the remainder. This last expression is synthetic division. It is an easy process to remember. Simply change the sign of the constant term in the divisor, then bring down the first term of the dividend. The process is then just a series of multiplications and additions, as indicated in the following diagram by the arrows:

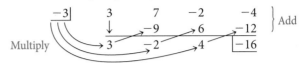

The last term of the bottom row is always the remainder. The numbers to the left of this term are the coefficients of the terms for the quotient.

Here are some additional examples of synthetic division with polynomials.

EXAMPLE 9   Divide $x^4 - 2x^3 + 4x^2 - 6x + 2$ by $x - 2$.

SOLUTION   To begin, we change the sign of the constant term in the divisor to get $+2$ and write this on the left, followed by the coefficients of the dividend. Next, we bring the left-most coefficient straight down. Here is how it looks:

Now we multiply the 1 in the bottom line by 2, and write the result below the next coefficient. Then we add to obtain 0.

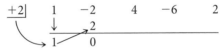

We repeat this procedure with each new number in the bottom row. Here is the result in individual steps:

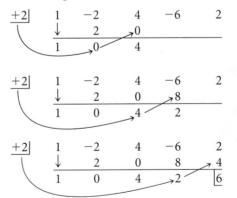

From the last line we have the answer. The first term in the quotient will always have a degree that is one less than the degree of the leading term in the dividend. In this case, the quotient will begin with an x^3 term.

$$1x^3 + 0x^2 + 4x + 2 + \frac{6}{x - 2}$$

Further simplifying the result, we get:

$$x^3 + 4x + 2 + \frac{6}{x - 2}$$

EXAMPLE 10   Divide: $\dfrac{3x^3 - 4x + 5}{x + 4}$.

SOLUTION   Since we cannot skip any powers of the variable in the dividend $3x^3 - 4x + 5$, we rewrite it as $3x^3 + 0x^2 - 4x + 5$ and proceed as we did in Example 9:

$$
\begin{array}{r|rrrr}
-4 & 3 & 0 & -4 & 5 \\
 & \downarrow & -12 & 48 & -176 \\
\hline
 & 3 & -12 & 44 & \boxed{-171}
\end{array}
$$

From the synthetic division, we have

$$\frac{3x^3 - 4x + 5}{x + 4} = 3x^2 - 12x + 44 + \frac{-171}{x + 4}$$

EXAMPLE 11   Divide: $\dfrac{x^3 - 1}{x - 1}$.

SOLUTION   Writing the numerator as $x^3 + 0x^2 + 0x - 1$ and using synthetic division, we have

$$
\begin{array}{r|rrrr}
+1 & 1 & 0 & 0 & -1 \\
 & \downarrow & 1 & 1 & 1 \\
\hline
 & 1 & 1 & 1 & \boxed{0}
\end{array}
$$

which indicates

$$\frac{x^3 - 1}{x - 1} = x^2 + x + 1$$

Average Cost

EXAMPLE 12   First Bank of San Luis Obispo charges $2.00 per month and $0.15 per check for a regular checking account. As we mentioned in the introduction to this section, the total monthly cost of this account is $C(x) = 2.00 + 0.15x$. To find the average cost of each of the x checks, we divide the total cost by the number of checks written. That is,

$$\overline{C}(x) = \frac{C(x)}{x}$$

© bluestocking/iStockPhoto

a. Find the formula for the average cost function, $\overline{C}(x)$.

b. Use the average cost function to fill in the following table.

x	1	5	10	15	20
$\overline{C}(x)$					

c. What happens to the average cost as more checks are written?

d. Assume that you write at least 1 check a month, but never more than 20 checks per month, and graph both $y = C(x)$ and $y = \overline{C}(x)$ on the same set of axes.

SOLUTION

a. The average cost function is

$$\overline{C}(x) = \frac{C(x)}{x}$$

$$= \frac{2.00 + 0.15x}{x}$$

$$= 0.15 + \frac{2.00}{x}$$

b. Using the formula from part *a* we have

x	1	5	10	15	20
$\overline{C}(x)$	2.15	0.55	0.35	0.28	0.25

c. The average cost decreases as more checks are written.

d.

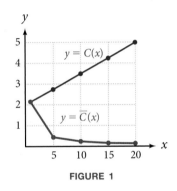

FIGURE 1

Getting Ready for Class

After reading through the preceding section, respond in your own words and in complete sentences.

A. Explain how monomial division is performed.

B. What are the four steps used in long division with polynomials?

C. What does it mean to have a remainder of 0?

D. When must long division be performed, and when can synthetic division be used instead?

Problem Set 5.3

Use monomial division to find the following quotients.

1. $\dfrac{4x^3 - 8x^2 + 6x}{2x}$

2. $\dfrac{6x^3 + 12x^2 - 9x}{3x}$

3. $\dfrac{10x^4 + 15x^3 - 20x^2}{-5x^2}$

4. $\dfrac{12x^5 - 18x^4 - 6x^3}{6x^3}$

5. $\dfrac{8y^5 + 10y^3 - 6y}{4y^3}$

6. $\dfrac{6y^4 - 3y^3 + 18y^2}{9y^2}$

7. $\dfrac{5x^3 - 8x^2 - 6x}{-2x^2}$

8. $\dfrac{-9x^5 + 10x^3 - 12x}{-6x^4}$

9. $\dfrac{28a^3b^5 + 42a^4b^3}{7a^2b^2}$

10. $\dfrac{a^2b + ab^2}{ab}$

11. $\dfrac{10x^3y^2 - 20x^2y^3 - 30x^3y^3}{-10x^2y}$

12. $\dfrac{9x^4y^4 + 18x^3y^4 - 27x^2y^4}{-9xy^3}$

Divide using the long division method.

13. $\dfrac{x^2 - 5x - 7}{x + 2}$

14. $\dfrac{x^2 + 4x - 8}{x - 3}$

15. $\dfrac{6x^2 + 7x - 18}{3x - 4}$

16. $\dfrac{8x^2 - 26x - 9}{2x - 7}$

17. $\dfrac{2x^3 - 3x^2 - 4x + 5}{x + 1}$

18. $\dfrac{3x^3 - 5x^2 + 2x - 1}{x - 2}$

19. $\dfrac{2y^3 - 9y^2 - 17y + 39}{2y - 3}$

20. $\dfrac{3y^3 - 19y^2 + 17y + 4}{3y - 4}$

21. $\dfrac{2x^3 - 9x^2 + 11x - 6}{2x^2 - 3x + 2}$

22. $\dfrac{6x^3 + 7x^2 - x + 3}{3x^2 - x + 1}$

23. $\dfrac{6y^3 - 8y + 5}{2y - 4}$

24. $\dfrac{9y^3 - 6y^2 + 8}{3y - 3}$

25. $\dfrac{a^4 - 2a + 5}{a - 2}$

26. $\dfrac{a^4 + a^3 - 1}{a + 2}$

27. $\dfrac{y^4 - 16}{y - 2}$

28. $\dfrac{y^4 - 81}{y - 3}$

29. $\dfrac{x^4 + x^3 - 3x^2 - x + 2}{x^2 + 3x + 2}$

30. $\dfrac{2x^4 + x^3 + 4x - 3}{2x^2 - x + 3}$

Divide using synthetic division.

31. $\dfrac{x^2 - 5x + 6}{x + 2}$

32. $\dfrac{x^2 + 8x - 12}{x - 3}$

33. $\dfrac{3x^2 - 4x + 1}{x - 1}$

34. $\dfrac{4x^2 - 2x - 6}{x + 1}$

35. $\dfrac{x^3 + 2x^2 + 3x + 4}{x - 2}$

36. $\dfrac{x^3 - 2x^2 - 3x - 4}{x - 2}$

37. $\dfrac{3x^3 - x^2 + 2x + 5}{x - 3}$

38. $\dfrac{2x^3 - 5x^2 + x + 2}{x - 2}$

39. $\dfrac{2x^3 + x - 3}{x - 1}$

40. $\dfrac{3x^3 - 2x + 1}{x - 5}$

41. $\dfrac{x^4 + 2x^2 + 1}{x + 4}$

42. $\dfrac{x^4 - 3x^2 + 1}{x - 4}$

43. $\dfrac{x^5 - 2x^4 + x^3 - 3x^2 - x + 1}{x - 2}$

44. $\dfrac{2x^5 - 3x^4 + x^3 - x^2 + 2x + 1}{x + 2}$

45. $\dfrac{x^2 + x + 1}{x - 1}$

46. $\dfrac{x^2 + x + 1}{x + 1}$

47. $\dfrac{x^4 - 1}{x + 1}$

48. $\dfrac{x^4 + 1}{x - 1}$

49. $\dfrac{x^3 - 1}{x - 1}$

50. $\dfrac{x^3 - 1}{x + 1}$

Applying the Concepts

51. The Remainder Theorem Find $P(-2)$ if $P(x) = x^2 - 5x - 7$. Compare it with the remainder in Problem 13.

52. The Remainder Theorem The remainder theorem of algebra states that if a polynomial, $P(x)$, is divided by $x - a$, then the remainder is $P(a)$. Verify the remainder theorem by showing that when $P(x) = x^2 - x + 3$ is divided by $x - 2$ the remainder is 5, and that $P(2) = 5$.

53. Checking Account Trust Bank of Arroyo Grande charges $3.00 per month and $0.10 per check for a regular checking account. As we mentioned in the introduction to this section, the total monthly cost of this account is $C(x) = 3.00 + 0.10x$. To find the average cost of each of the x checks, we divide the total cost by the number of checks written. That is,

$$\overline{C}(x) = \frac{C(x)}{x}$$

a. Use the total cost function to fill in the following table.

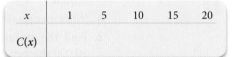

x	1	5	10	15	20
$C(x)$					

b. Find the formula for the average cost function, $\overline{C}(x)$.

c. Use the average cost function to fill in the following table.

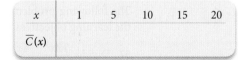

x	1	5	10	15	20
$\overline{C}(x)$					

d. What happens to the average cost as more checks are written?

54. Average Cost A company that manufactures computer diskettes uses the function $C(x) = 200 + 2x$ to represent the daily cost of producing x diskettes.

a. Find the average cost function, $\overline{C}(x)$.

b. Use the average cost function to fill in the following table:

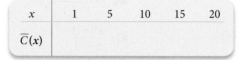

x	1	5	10	15	20
$\overline{C}(x)$					

c. What happens to the average cost as more items are produced?

d. Graph the function $y = \overline{C}(x)$ for $x > 0$.

55. Average Cost For long distance service, a particular phone company charges a monthly fee of $4.95 plus $0.07 per minute of calling time used. The relationship between the number of minutes of calling time used, m, and the amount of the monthly phone bill $T(m)$ is given by the function $T(m) = 4.95 + 0.07m$.

a. Find the total cost when 100, 400, and 500 minutes of calling time is used in 1 month.

b. Find a formula for the average cost per minute function $\overline{T}(m)$.

c. Find the average cost per minute of calling time used when 100, 400, and 500 minutes are used in 1 month.

56. Average Cost A company manufactures electric pencil sharpeners. Each month they have fixed costs of $40,000 and variable costs of $8.50 per sharpener. Therefore, the total monthly cost to manufacture x sharpeners is given by the function $C(x) = 40,000 + 8.5x$.

a. Find the total cost to manufacture 1,000, 5,000, and 10,000 sharpeners a month.

b. Write an expression for the average cost per sharpener function $\overline{C}(x)$.

c. Find the average cost per sharpener to manufacture 1,000, 5,000, and 10,000 sharpeners per month.

Learning Objectives Assessment

The following problems can be used to help assess if you have successfully met the learning objectives for this section.

57. Divide: $\dfrac{12x^5 - 4x^3 + 10x^2}{2x^2}$.

a. $12x^5 - 4x^3 + 5$ **b.** $2x^3 + 10x^2$

c. $6x^3 - 2x + 5$ **d.** $6x^5 - 2x^3 + 10$

58. Use long division to divide $\dfrac{x^3 + 8}{x + 2}$.

a. $x^2 + 4$ **b.** $x^2 - 2x + 4$

c. $x^2 + 2x - 4 + \dfrac{16}{x + 2}$ **d.** $x^2 + 2x + 4$

59. Divide $\dfrac{x^2 - x + 6}{x + 2}$ using synthetic division.

   a. $x + 1 + \dfrac{8}{x + 2}$ **b.** $x - 3 + \dfrac{12}{x + 2}$

   c. $-2x + 3$ **d.** $x^2 + 2$

60. The function $C(x) = 500 + 3x$ gives the total daily cost for a company to manufacture x widgets. Find the average daily cost function $\overline{C}(x)$.

   a. $\overline{C}(x) = \dfrac{500}{x} + 3$ **b.** $\overline{C}(x) = 500x + 3x^2$

   c. $\overline{C}(x) = 500 + 2x$ **d.** $\overline{C}(x) = 503$

Getting Ready for the Next Section

Simplify. Write answers with positive exponents only.

61. $\dfrac{8a^3}{a}$ **62.** $\dfrac{-8a^2}{a}$ **63.** $\dfrac{-48a}{a}$ **64.** $\dfrac{-32a}{a}$

65. $\dfrac{16a^5b^4}{8a^2b^3}$ **66.** $\dfrac{12x^4y^5}{3x^3y^3}$ **67.** $\dfrac{-24a^5b^5}{8a^5b^3}$ **68.** $\dfrac{-15x^5y^3}{3x^3y^3}$

Multiply.

69. $8a(a^2 - a - 6)$ **70.** $5x^3(5x^2 + 4x - 6)$

71. $(x + y)(5 + x)$ **72.** $(a^2 + 1)(b^2 + 8)$

Greatest Common Factor and Factoring by Grouping

5.4

Learning Objectives

In this section, we will learn how to:

1. Factor by factoring out the greatest common factor.

2. Factor by grouping.

Introduction

In professional football, "hang time" refers to the amount of time the ball is in the air when punted. The term came into use during the tenure of legendary NFL punter Ray Guy, who could keep the ball in the air as long as 6 seconds.

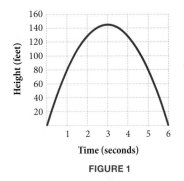

FIGURE 1

Hang time depends on only one variable: the initial vertical velocity imparted to the ball by the kicker's foot. For example, if the punter punts the ball with an initial vertical velocity of 96 feet per second, the height of the ball is given by $h = 96t - 16t^2$. We find hang time by solving the equation

$$96t - 16t^2 = 0$$

Factoring, which we present in this section, is the key to solving this equation. Writing the left side in factored form looks like this

$$16t(6 - t) = 0$$

The second factor, $6 - t$, tells us that the hang time is 6 seconds.

Factoring

In this section we develop techniques that allow us to factor a variety of polynomials. In general, factoring is the reverse of multiplication. The diagram below illustrates the relationship between factoring and multiplication. Reading from left to right, we say the product of 3 and 7 is 21. Reading in the other direction, from right to left, we say 21 factors into 3 times 7. Or 3 and 7 are factors of 21.

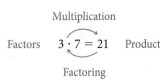

Greatest Common Factor

> (def **DEFINITION**   *greatest common factor*
>
> The greatest common factor (GCF) for a polynomial is the largest monomial that divides (is a factor of) each term of the polynomial.

The greatest common factor for the polynomial $25x^5 + 20x^4 - 30x^3$ is $5x^3$ since it is the largest monomial that is a factor of each term. We can apply the distributive property and write

$$25x^5 + 20x^4 - 30x^3 = 5x^3(5x^2) + 5x^3(4x) - 5x^3(6)$$

$$= 5x^3(5x^2 + 4x - 6)$$

The last line is written in factored form. Notice that the three terms in the grouping symbol are the result of dividing each term in the original polynomial by the GCF. That is,

$$\frac{25x^5}{5x^3} = 5x^2, \qquad \frac{20x^4}{5x^3} = 4x, \qquad \text{and} \qquad \frac{-30x^3}{5x^3} = -6$$

When we factor a greatest common factor from a polynomial, we are essentially dividing. The GCF is written first, followed by a grouping symbol containing the quotients that result when each term is divided by the GCF.

VIDEO EXAMPLES

SECTION 5.4

EXAMPLE 1   Factor the greatest common factor from $8a^3 - 8a^2 - 48a$.

SOLUTION   The greatest common factor is $8a$. It is the largest monomial that divides each term of our polynomial. We can write each term in our polynomial as the product of $8a$ and another monomial. Then, we apply the distributive property to factor $8a$ from each term.

$$8a^3 - 8a^2 - 48a = 8a(a^2) - 8a(a) - 8a(6)$$

$$= 8a(a^2 - a - 6)$$

EXAMPLE 2   Factor the greatest common factor from the expression

$$-16a^5b^4 - 24a^2b^5 - 8a^3b^3$$

SOLUTION   The largest monomial that divides each term is $8a^2b^3$. However, because each term is negative, we include the negative (a factor of -1) in our GCF. We write each term of the original polynomial as a product containing $-8a^2b^3$ and apply the distributive property to write the polynomial in factored form.

$$-16a^5b^4 - 24a^2b^5 - 8a^3b^3 = -8a^2b^3(2a^3b) - 8a^2b^3(3b^2) - 8a^2b^3(a)$$

$$= -8a^2b^3(2a^3b + 3b^2 + a)$$

EXAMPLE 3   Factor the greatest common factor from the expression

$$5x^2(a + b) - 6x(a + b) - 7(a + b)$$

SOLUTION   The greatest common factor is the binomial $a + b$. Factoring it from each term, we have

$$5x^2(a + b) - 6x(a + b) - 7(a + b) = (a + b)(5x^2 - 6x - 7)$$

Factoring by Grouping

The polynomial $5x + 5y + x^2 + xy$ can be factored by noticing that the first two terms have a 5 in common, whereas the last two have an x in common. Applying the distributive property twice, we have

$$5x + 5y + x^2 + xy = 5(x + y) + x(x + y)$$

This last expression can be thought of as having two terms, $5(x + y)$ and $x(x + y)$, each of which has a common factor $(x + y)$. We apply the distributive property again to factor $(x + y)$ from each term.

$$5(x + y) + x(x + y) = (x + y)(5 + x)$$

EXAMPLE 4   Factor: $a^2b^2 + b^2 + 8a^2 + 8$.

SOLUTION   The first two terms have b^2 in common; the last two have 8 in common.

$$a^2b^2 + b^2 + 8a^2 + 8 = b^2(a^2 + 1) + 8(a^2 + 1)$$
$$= (a^2 + 1)(b^2 + 8)$$

EXAMPLE 5   Factor by grouping: $x^3 + 2x^2 + 9x + 18$.

SOLUTION   We begin by factoring x^2 from the first two terms and 9 from the second two terms.

$$x^3 + 2x^2 + 9x + 18 = x^2(x + 2) + 9(x + 2)$$
$$= (x + 2)(x^2 + 9)$$

EXAMPLE 6   Factor: $15 + x^3y^4 - 5y^4 - 3x^3$.

SOLUTION   Notice that the first two terms don't have anything in common (except for a factor of 1). The last two terms only have a -1 in common. If we leave the terms in this order, grouping will not work. Sometimes we can overcome this situation by rearranging terms. For example, if we move the x^3y^4 term to the end, we have

$$15 - 5y^4 - 3x^3 + x^3y^4$$

Now we can factor 5 from the first two terms and $-x^3$ from the last two terms.

$$15 - 5y^4 - 3x^3 + x^3y^4 = 5(3 - y^4) - x^3(3 - y^4)$$
$$= (3 - y^4)(5 - x^3)$$

Getting Ready for Class

After reading through the preceding section, respond in your own words and in complete sentences.

A. What is the relationship between multiplication and factoring?

B. What is the greatest common factor for a polynomial?

C. After factoring a polynomial, how can you check your result?

D. When would you try to factor by grouping?

Problem Set 5.4

Factor the greatest common factor from each of the following.

1. $10x^3 - 15x^2$ **2.** $12x^5 + 18x^7$ **3.** $9y^6 + 18y^3$

4. $24y^4 - 8y^2$ **5.** $9a^2b - 6ab^2$ **6.** $30a^3b^4 + 20a^4b^3$

7. $21xy^4 + 7x^2y^2$ **8.** $14x^6y^3 - 6x^2y^4$ **9.** $3a^2 - 21a + 33$

10. $3a^2 - 3a + 6$ **11.** $4x^3 - 16x^2 + 20x$ **12.** $2x^3 - 14x^2 + 28x$

13. $10x^4y^2 + 20x^3y^3 + 30x^2y^4$ **14.** $6x^4y^2 + 18x^3y^3 + 24x^2y^4$

15. $-x^2y + xy^2 - x^2y^2$ **16.** $-x^3y^2 - x^2y^3 - x^2y^2$

17. $4x^3y^2z - 8x^2y^2z^2 + 6xy^2z^3$ **18.** $7x^4y^3z^2 - 21x^2y^2z^2 - 14x^2y^3z^4$

19. $20a^2b^2c^2 - 30ab^2c + 25a^2bc^2$ **20.** $8a^3bc^5 - 48a^2b^4c + 16ab^3c^5$

21. $5x(a - 2b) - 3y(a - 2b)$ **22.** $3a(x - y) - 7b(x - y)$

23. $3x^2(x + y)^2 - 6y^2(x + y)^2$ **24.** $10x^3(2x - 3y) - 15x^2(2x - 3y)$

25. $2x^2(x + 5) + 7x(x + 5) + 8(x + 5)$ **26.** $2x^2(x + 2) + 13x(x + 2) + 17(x + 2)$

Factor each of the following by grouping.

27. $3xy + 3y + 2ax + 2a$ **28.** $5xy^2 + 5y^2 + 3ax + 3a$

29. $x^2y + x + 3xy + 3$ **30.** $x^3y^3 + 2x^3 + 5x^2y^3 + 10x^2$

31. $3xy^2 - 6y^2 + 4x - 8$ **32.** $8x^2y - 4x^2 + 6y - 3$

33. $x^2 - ax - bx + ab$ **34.** $ax - x^2 - bx + ab$

35. $ab + 5a - b - 5$ **36.** $x^2 - xy - ax + ay$

37. $a^4b^2 + a^4 - 5b^2 - 5$ **38.** $2a^2 - a^2b - bc^2 + 2c^2$

39. $x^3 + 3x^2 + 4x + 12$ **40.** $x^3 + 5x^2 + 4x + 20$

41. $x^3 + 2x^2 + 25x + 50$ **42.** $x^3 + 4x^2 + 9x + 36$

43. $2x^3 + 3x^2 + 8x + 12$ **44.** $3x^3 + 2x^2 + 27x + 18$

45. $4x^3 + 12x^2 + 9x + 27$ **46.** $9x^3 + 18x^2 + 4x + 8$

47. The greatest common factor of the binomial $3x - 9$ is 3. The greatest common factor of the binomial $6x - 2$ is 2. What is the greatest common factor of their product, $(3x - 9)(6x - 2)$, when it has been multiplied out?

48. The greatest common factors of the binomials $5x - 10$ and $2x + 4$ are 5 and 2, respectively. What is the greatest common factor of their product, $(5x - 10)(2x + 4)$, when it has been multiplied out?

Applying the Concepts

49. Investing If P dollars are placed in a savings account in which the rate of interest r is compounded yearly, then at the end of one year the amount of money in the account can be written as $P + Pr$. At the end of two years the amount of money in the account is

$$P + Pr + (P + Pr)r$$

Use factoring by grouping to show that this last expression can be written as $P(1 + r)^2$.

50. Investing At the end of 3 years, the amount of money in the savings account in Problem 49 will be

$$P(1 + r)^2 + P(1 + r)^2 r$$

Use factoring to show that this last expression can be written as $P(1 + r)^3$.

Price and Revenue In Section 5.2, we introduced the following formula for revenue:

$$\text{Revenue} = (\text{Number of items sold})(\text{Price of each item}) = xp$$

Use this formula for Problems 51–52

51. The weekly revenue equation for a company selling an inexpensive accounting program for home computers is given by the equation

$$R(x) = 35x - 0.1x^2$$

where x is the number of programs they sell per week. What price p should they charge if they want to sell 65 programs per week?

52. The weekly revenue equation for a small mail-order company selling boxes of greeting cards is

$$R(x) = 5.9x - 0.004x^2$$

where x is the number of boxes they sell per week. What price p should they charge if they want to sell 200 boxes each week?

Learning Objectives Assessment

The following problems can be used to help assess if you have successfully met the learning objectives for this section.

53. Factor out the greatest common factor from $12x^6y^3 - 18x^4y + 4x^2y^2$.

    a. $4x^2y(3x^3y^2 - 6x^2 + y)$ **b.** $2x^2y(6x^4y^2 - 9x^2 + 2y)$

    c. $18x^6y^3(6 - x^2y^2 + 14x^4y)$ **d.** $2x^2y(10x^3y^3 - 16x^2y + 2xy^2)$

54. Factor $3x^3 - 6x^2y - 4x + 8y$ by grouping.

    a. $(x - 2y)(3x^2 - 4)$ **b.** $(x - 2y)(3x^2 + 4)$

    c. $(3x^2 - 4)(x - 2y)(x + 2y)$ **d.** $3x^2(x - 2y) - 4(x + 2y)$

Getting Ready for the Next Section

Factor out the greatest common factor.

55. $3x^4 - 9x^3y - 18x^2y^2$ **56.** $5x^2 + 10x + 30$

57. $2x^2(x - 3) - 4x(x - 3) - 3(x - 3)$ **58.** $3x^2(x - 2) - 8x(x - 2) + 2(x - 2)$

Multiply.

59. $(x + 2)(3x - 1)$ **60.** $(x - 2)(3x + 1)$ **61.** $(x - 1)(3x - 2)$

62. $(x + 1)(3x + 2)$ **63.** $(x + 2)(x + 3)$ **64.** $(x - 2)(x - 3)$

65. $(2y + 5)(3y - 7)$ **66.** $(2y - 5)(3y + 7)$ **67.** $(4 - 3a)(5 - a)$

68. $(4 - 3a)(5 + a)$ **69.** $(5 + 2x)(5 - 2x)$ **70.** $(3 + 2x)^2$

71. Complete the following table.

Two Numbers a and b	Their Product ab	Their Sum a + b
1, −24		
−1, 24		
2, −12		
−2, 12		
3, −8		
−3, 8		
4, −6		
−4, 6		

72. Complete the following table.

Two Numbers a and b	Their Product ab	Their Sum a + b
1, −54		
−1, 54		
2, −27		
−2, 27		
3, −18		
−3, 18		
6, −9		
−6, 9		

73. Find two numbers whose product is 36 and sum is 13.

74. Find two numbers whose product is 40 and sum is −13.

75. Find two numbers whose product is −40 and sum is 3.

76. Find two numbers whose product is −36 and sum is −9.

77. Find two numbers whose product is −48 and sum is −8.

78. Find two numbers whose product is −48 and sum is 13.

Learning Objectives

In this section, we will learn how to:

1. Factor trinomials in which the leading coefficient is 1.
2. Factor trinomials in which the leading coefficient is a number other than 1.

Factoring Trinomials with a Leading Coefficient of 1

Earlier in this chapter, we multiplied the following binomials:

$$(x - 2)(x + 3) = x^2 + x - 6$$
$$(x + 5)(x + 2) = x^2 + 7x + 10$$

In each case the product of two binomials is a trinomial. The first term in the resulting trinomial is obtained by multiplying the first term in each binomial. The middle term comes from adding the product of the two inside terms with the product of the two outside terms. The last term is the product of the last terms in each binomial.

In general,

$$(x + a)(x + b) = x^2 + ax + bx + ab$$
$$= x^2 + (a + b)x + ab$$

Writing this as a factoring problem, we have

$$x^2 + (a + b)x + ab = (x + a)(x + b)$$

To factor a trinomial with a leading coefficient of 1, we simply find the two numbers a and b whose sum is the coefficient of the middle term and whose product is the constant term.

VIDEO EXAMPLES

SECTION 5.5

EXAMPLE 1 $\quad$ Factor: $x^2 + 8x + 15$.

SOLUTION Since the leading coefficient is 1, we need two integers whose product is 15 and whose sum is 8. The integers are 5 and 3.

$$x^2 + 8x + 15 = (x + 5)(x + 3)$$

In the preceding example, we found factors of $x + 5$ and $x + 3$. These are the only two such factors for $x^2 + 8x + 15$. There is no other pair of binomials $x + a$ and $x + b$ whose product is $x^2 + 8x + 15$.

EXAMPLE 2 $\quad$ Factor: $x^2 - xy - 12y^2$.

SOLUTION We need two numbers whose product is -12 and whose sum is -1. The numbers are -4 and 3. Because the last term in the trinomial also contains y^2, the last term in each of the binomials will need to contain a factor of y.

$$x^2 - xy - 12y^2 = (x - 4y)(x + 3y)$$

Checking this result gives

$$(x - 4y)(x + 3y) = x^2 + 3xy - 4xy - 12y^2$$
$$= x^2 - xy - 12y^2$$

Based on our first two examples, we have the following rules for factoring a trinomial.

> **RULE**   *Factoring Trinomials*
>
> If the third term of the trinomial is positive, then both binomials will have the ***same sign*** as the middle term.
>
> If the third term of the trinomial is negative, then the binomials will have ***opposite signs***.

EXAMPLE 3   Factor: $x^2 - 8x + 6$.

SOLUTION   Since there is no pair of integers whose product is 6 and whose sum is -8, the trinomial $x^2 - 8x + 6$ is not factorable. We say it is a ***prime polynomial***.

Factoring When the Leading Coefficient is not 1

EXAMPLE 4   Factor: $3x^4 - 15x^3y - 18x^2y^2$.

SOLUTION   The leading coefficient is not 1. There is a common factor of $3x^2$, however. Factoring this out to begin with we have

$$3x^4 - 15x^3y - 18x^2y^2 = 3x^2(x^2 - 5xy - 6y^2)$$

Factoring the resulting trinomial as in the previous examples gives

$$3x^2(x^2 - 5xy - 6y^2) = 3x^2(x - 6y)(x + y)$$

Note   As a general rule, it is best to factor out the greatest common factor first.

Factoring Other Trinomials by Trial and Error

We want to turn our attention now to trinomials with leading coefficients other than 1 and with no greatest common factor other than 1.

Suppose we want to factor $3x^2 - x - 2$. The factors will be a pair of binomials having opposite signs. The product of the first terms will be $3x^2$, and the product of the last terms will be -2. We can list all the possible factors along with their products as follows:

Possible Factors	First Term	Middle Term	Last Term
$(x + 2)(3x - 1)$	$3x^2$	$-x + 6x = +5x$	-2
$(x - 2)(3x + 1)$	$3x^2$	$x - 6x = -5x$	-2
$(x + 1)(3x - 2)$	$3x^2$	$-2x + 3x = +x$	-2
$(x - 1)(3x + 2)$	$3x^2$	$2x - 3x = -x$	-2

From the last line we see that the factors of $3x^2 - x - 2$ are $(x - 1)(3x + 2)$. That is,

$$3x^2 - x - 2 = (x - 1)(3x + 2)$$

To factor trinomials with leading coefficients other than 1, when the greatest common factor is 1, we must use trial and error or list all the possible factors. In either case the idea is this: look only at pairs of binomials whose products give the correct first and last terms, then use the FOIL method to find the combination that will give the correct middle term.

EXAMPLE 5   Factor: $2x^2 + 13xy + 15y^2$.

SOLUTION   Because the second and third terms are positive, both binomials must contain + symbols. Listing all such possible factors, the product of whose first terms is $2x^2$ and the product of whose last terms is $15y^2$, yields

Possible Factors	Middle Term of Product
$(2x + 5y)(x + 3y)$	$6xy + 5xy = 11xy$
$(2x + 3y)(x + 5y)$	$10xy + 3xy = 13xy$
$(2x + 15y)(x + y)$	$2xy + 15xy = 17xy$
$(2x + y)(x + 15y)$	$30xy + xy = 31xy$

The second line has the correct middle term:

$$2x^2 + 13xy + 15y^2 = (2x + 3y)(x + 5y)$$

There are other ways to reduce the number of possible factors to consider. For example, if we were to factor the trinomial $2x^2 - 11x + 12$, we would not have to consider the pair of possible factors $(2x - 4)(x - 3)$. If the original trinomial has no greatest common factor other than 1, then neither of its binomial factors will either. The trinomial $2x^2 - 11x + 12$ has a greatest common factor of 1, but the possible factor $2x - 4$ has a greatest common factor of 2: $2x - 4 = 2(x - 2)$. Therefore, we do not need to consider $2x - 4$ as a possible factor.

EXAMPLE 6   Factor: $12x^4 + 17x^2 + 6$.

SOLUTION   This is a trinomial in x^2. Both binomials must contain a positive sign. The first terms in the binomials could be $12x^2$ and x^2, $6x^2$ and $2x^2$, or $4x^2$ and $3x^2$. The last terms could be 1 and 6, or 2 and 3. By eliminating all possible pairs of factors that contain a common factor, we are left with

Possible Factors	Middle Term of Product
$(12x^2 + 1)(x^2 + 6)$	$72x^2 + x^2 = 73x^2$
$(4x^2 + 3)(3x^2 + 2)$	$8x^2 + 9x^2 = 17x^2$

The second line has the correct middle term:

$$12x^4 + 17x^2 + 6 = (4x^2 + 3)(3x^2 + 2)$$

EXAMPLE 7   Factor: $2x^2(x - 3) - 5x(x - 3) - 3(x - 3)$.

SOLUTION We begin by factoring out the greatest common factor $(x - 3)$. Then we factor the trinomial that remains.

$$2x^2(x - 3) - 5x(x - 3) - 3(x - 3) = (x - 3)(2x^2 - 5x - 3)$$
$$= (x - 3)(2x + 1)(x - 3)$$
$$= (x - 3)^2(2x + 1)$$

Factoring Trinomials By Grouping

As an alternative to the trial-and-error method of factoring trinomials, we present the following method. The new method does not require as much trial and error. To use this new method, we must rewrite our original trinomial in such a way that the factoring by grouping method can be applied.

HOW TO   *Factor $ax^2 + bx + c$*

Step 1: Form the product ac.

Step 2: Find a pair of numbers whose product is ac and whose sum is b.

Step 3: Rewrite the polynomial to be factored so that the middle term bx is written as the sum of two terms whose coefficients are the two numbers found in step 2.

Step 4: Factor by grouping.

EXAMPLE 8   Factor $3x^2 - 10x - 8$ by grouping.

SOLUTION The trinomial $3x^2 - 10x - 8$ has the form $ax^2 + bx + c$, where $a = 3$, $b = -10$, and $c = -8$.

Step 1: The product ac is $3(-8) = -24$.

Step 2: We need to find two numbers whose product is -24 and whose sum is -10. Let's list all the pairs of numbers whose product is -24 to find the pair whose sum is -10. Because ac is negative, the numbers must be of opposite signs, and because b is negative, the negative number must have a larger absolute value.

Product	*Sum*
$1(-24) = -24$	$1 + (-24) = -23$
$2(-12) = -24$	$2 + (-12) = -10$
$3(-8) = -24$	$3 + (-8) = -5$
$4(-6) = -24$	$4 + (-6) = -2$

As you can see, only 2 and -12 have a sum of -10.

Step 3: We now rewrite our original trinomial so the middle term $-10x$ is written as the sum of $-12x$ and $2x$:

$$3x^2 - 10x - 8 = 3x^2 - 12x + 2x - 8$$

Step 4: Factoring by grouping, we have

$$3x^2 - 12x + 2x - 8 = 3x(x - 4) + 2(x - 4)$$

$$= (x - 4)(3x + 2)$$

You can see that this method works by multiplying $x - 4$ and $3x + 2$ to get

$$3x^2 - 10x - 8$$

Note It does not matter in which order we write the terms $-12x$ and $2x$. If we had written

$$3x^2 - 10x - 8$$
$$= 3x^2 + 2x - 12x - 8$$

the grouping method would work just as well.

▨ **EXAMPLE 9**   Factor: $9x^2 + 15x + 4$.

SOLUTION   In this case, $a = 9$, $b = 15$, and $c = 4$. The product ac is $9 \cdot 4 = 36$. Listing all the pairs of numbers whose product is 36 with their corresponding sums, we have

Product	Sum
$1(36) = 36$	$1 + 36 = 37$
$2(18) = 36$	$2 + 18 = 20$
$3(12) = 36$	$3 + 12 = 15$
$4(9) = 36$	$4 + 9 = 13$
$6(6) = 36$	$6 + 6 = 12$

Notice we list only positive numbers since both the product and sum we are looking for are positive. The numbers 3 and 12 are the numbers we are looking for. Their product is 36, and their sum is 15. We now rewrite the original polynomial $9x^2 + 15x + 4$ with the middle term written as $3x + 12x$. We then factor by grouping.

$$9x^2 + 15x + 4 = 9x^2 + 3x + 12x + 4$$

$$= 3x(3x + 1) + 4(3x + 1)$$

$$= (3x + 1)(3x + 4)$$

The polynomial $9x^2 + 15x + 4$ factors into the product

$$(3x + 1)(3x + 4)$$

▨ **EXAMPLE 10**   Factor: $8x^2 - 2x - 15$.

SOLUTION   The product ac is $8(-15) = -120$. There are many pairs of numbers whose product is -120. We are looking for the pair whose sum is also -2. The numbers are -12 and 10. Writing $-2x$ as $-12x + 10x$ and then factoring by grouping, we have

$$8x^2 - 2x - 15 = 8x^2 - 12x + 10x - 15$$

$$= 4x(2x - 3) + 5(2x - 3)$$

$$= (2x - 3)(4x + 5)$$

Getting Ready for Class

After reading through the preceding section, respond in your own words and in complete sentences.

A. What is a prime polynomial?

B. When factoring trinomials, what should you look for first?

C. How can you check to see that you have factored a trinomial correctly?

D. Describe how to determine the binomial factors of $6x^2 + 5x - 25$.

Factor each of the following trinomials.

1. $x^2 + 7x + 12$
2. $x^2 - 7x + 12$
3. $x^2 - x - 12$
4. $x^2 + x - 12$
5. $y^2 + y - 6$
6. $y^2 - y - 6$
7. $16 - 6x - x^2$
8. $3 + 2x - x^2$
9. $12 + 8x + x^2$
10. $15 - 2x - x^2$
11. $16 - x^2$
12. $30 - x - x^2$
13. $x^2 + 3xy + 2y^2$
14. $x^2 - 5xy - 24y^2$
15. $a^2 + 3ab - 18b^2$
16. $a^2 - 8ab - 9b^2$
17. $x^2 - 2xa - 48a^2$
18. $x^2 + 14xa + 48a^2$
19. $x^2 - 12xb + 36b^2$
20. $x^2 + 10xb + 25b^2$

Factor completely by first factoring out the greatest common factor and then factoring the trinomial that remains.

21. $3a^2 - 21a + 30$
22. $3a^2 - 3a - 6$
23. $4x^3 - 16x^2 - 20x$
24. $2x^3 - 14x^2 + 20x$

Factor completely. Be sure to factor out the greatest common factor first if it is other than 1.

25. $3x^2 - 6xy - 9y^2$
26. $5x^2 + 25xy + 20y^2$
27. $2a^5 + 4a^4b + 4a^3b^2$
28. $3a^4 - 18a^3b + 27a^2b^2$
29. $10x^4y^2 + 20x^3y^3 - 30x^2y^4$
30. $6x^4y^2 + 18x^3y^3 - 24x^2y^4$
31. $2x^2 + 7x - 15$
32. $2x^2 - 7x - 15$
33. $2x^2 + x - 15$
34. $2x^2 - x - 15$
35. $2x^2 - 13x + 15$
36. $2x^2 + 13x + 15$
37. $2x^2 - 11x + 15$
38. $2x^2 + 11x + 15$
39. $2x^2 + 7x + 15$
40. $2x^2 + x + 15$
41. $2 + 7a + 6a^2$
42. $2 - 7a + 6a^2$
43. $60y^2 - 15y - 45$
44. $72y^2 + 60y - 72$
45. $6x^4 - x^3 - 2x^2$
46. $3x^4 + 2x^3 - 5x^2$
47. $40r^3 - 120r^2 + 90r$
48. $40r^3 + 200r^2 + 250r$
49. $4x^2 - 11xy - 3y^2$
50. $3x^2 + 19xy - 14y^2$
51. $10x^2 - 3xa - 18a^2$
52. $9x^2 + 9xa - 10a^2$
53. $18a^2 + 3ab - 28b^2$
54. $6a^2 - 7ab - 5b^2$
55. $600 + 800t - 800t^2$
56. $200 - 600t - 350t^2$
57. $9y^4 + 9y^3 - 10y^2$
58. $4y^5 + 7y^4 - 2y^3$
59. $24a^2 - 2a^3 - 12a^4$
60. $8x^4y^2 - 2x^3y^3 - 6x^2y^4$
61. $8x^4y^2 - 47x^3y^3 - 6x^2y^4$
62. $600x^4 - 100x^2 - 200$
63. $20a^4 + 37a^2 + 15$
64. $9 + 21r^2 + 12r^4$
65. $2 - 4r^2 - 30r^4$

Factor each of the following by first factoring out the greatest common factor and then factoring the trinomial that remains.

66. $2x^2(x + 5) + 7x(x + 5) + 6(x + 5)$

67. $2x^2(x + 2) + 13x(x + 2) + 15(x + 2)$

68. $x^2(2x + 3) + 7x(2x + 3) + 10(2x + 3)$

69. $2x^2(x + 1) + 7x(x + 1) + 6(x + 1)$

70. $3x^2(x - 3) + 7x(x - 3) - 20(x - 3)$

71. $4x^2(x + 6) + 23x(x + 6) + 15(x + 6)$

72. $6x^2(x - 2) - 17x(x - 2) + 12(x - 2)$

73. $10x^2(x + 4) - 33x(x + 4) - 7(x + 4)$

74. $12x^2(x + 3) + 7x(x + 3) - 45(x + 3)$

75. $24x^2(x - 6) + 38x(x - 6) + 15(x - 6)$

76. $6x^2(5x - 2) - 11x(5x - 2) - 10(5x - 2)$

77. $14x^2(3x + 4) - 39x(3x + 4) + 10(3x + 4)$

78. $20x^2(2x + 3) + 47x(2x + 3) + 21(2x + 3)$

79. $15x^2(4x - 5) - 2x(4x - 5) - 24(4x - 5)$

80. What polynomial, when factored, gives $(3x + 5y)(3x - 5y)$?

81. What polynomial, when factored, gives $(7x + 2y)(7x - 2y)$?

82. One factor of the trinomial $a^2 + 260a + 2{,}500$ is $a + 10$. What is the other factor?

83. One factor of the trinomial $a^2 - 75a - 2{,}500$ is $a + 25$. What is the other factor?

84. One factor of the trinomial $12x^2 - 107x + 210$ is $x - 6$. What is the other factor?

85. One factor of the trinomial $36x^2 + 134x - 40$ is $2x + 8$. What is the other factor?

86. One factor of the trinomial $54x^2 + 111x + 56$ is $6x + 7$. What is the other factor?

87. One factor of the trinomial $63x^2 + 110x + 48$ is $7x + 6$. What is the other factor?

88. One factor of the trinomial $35x^2 + 19x - 24$ is $5x - 3$. What is the other factor?

89. One factor of the trinomial $36x^2 + 43x - 35$ is $4x + 7$. What is the other factor?

90. Factor the right side of the equation $y = 4x^2 + 18x - 10$, and then use the result to find y when x is $\frac{1}{2}$, when x is -5, and when x is 2.

91. Factor the right side of the equation $y = 9x^2 + 33x - 12$, and use the result to find y when x is $\frac{1}{3}$, when x is -4, and when x is 3.

Learning Objectives Assessment

The following problems can be used to help assess if you have successfully met the learning objectives for this section.

92. Factor: $x^2 + 2x - 24$. Which of the following factors appears in the answer?

    a. $(x + 2)$ **b.** $(x + 8)$ **c.** $(x - 4)$ **d.** $(x - 6)$

93. Factor: $12x^2 - x - 6$. Which of the following factors appears in the answer?

    a. $(4x + 3)$ **b.** $(3x + 2)$ **c.** $(x - 3)$ **d.** $(2x - 3)$

Getting Ready for the Next Section

For each problem below, place a number or expression inside the parentheses so that the resulting statement is true.

94. $\dfrac{25}{64} = (\quad)^2$ **95.** $\dfrac{4}{9} = (\quad)^2$ **96.** $x^6 = (\quad)^2$

97. $x^8 = (\quad)^2$ **98.** $16x^4 = (\quad)^2$ **99.** $81y^4 = (\quad)^2$

Write as a perfect cube.

100. $\dfrac{1}{8} = (\quad)^3$ **101.** $\dfrac{1}{27} = (\quad)^3$ **102.** $x^6 = (\quad)^3$

103. $x^{12} = (\quad)^3$ **104.** $27x^3 = (\quad)^3$ **105.** $125y^3 = (\quad)^3$

106. $8y^3 = (\quad)^3$ **107.** $1000x^3 = (\quad)^3$

Learning Objectives

In this section, we will learn how to:

1. Factor perfect square trinomials.

2. Factor the difference of two squares.

3. Factor the sum or difference of two cubes.

Perfect Square Trinomials

In Section 5.2, we listed some special products found in multiplying polynomials. Two of the formulas looked like this:

$$(a + b)^2 = a^2 + 2ab + b^2$$
$$(a - b)^2 = a^2 - 2ab + b^2$$

If we exchange the left and right sides of each formula, we have two special formulas for factoring:

$$a^2 + 2ab + b^2 = (a + b)^2$$
$$a^2 - 2ab + b^2 = (a - b)^2$$

The left side of each formula is called a **perfect square trinomial**. The right sides are binomial squares. Perfect square trinomials can always be factored using the usual methods for factoring trinomials. However, if we notice that the first and last terms of a trinomial are positive and perfect squares, it is wise to see whether the trinomial factors as a binomial square before attempting to factor by the usual method.

VIDEO EXAMPLES

SECTION 5.6

▧ **EXAMPLE 1**   Factor: $x^2 - 6x + 9$.

SOLUTION   Since the first and last terms are positive and perfect squares, we attempt to factor according to the preceding formulas.

$$x^2 - 6x + 9 = (x)^2 - 6x + (3)^2$$

If we let $a = x$ and $b = 3$, then $-2ab = -2(x)(3) = -6x$, which agrees with the middle term in the trinomial. Thus, we have

$$x^2 - 6x + 9 = (x - 3)^2$$

▧

▧ **EXAMPLE 2**   Factor each of the following perfect square trinomials.

a. $16a^2 + 40ab + 25b^2 = (4a)^2 + 40ab + (5b)^2$

$\qquad\qquad\qquad\quad = (4a + 5b)^2$ $\qquad\qquad 2(4a)(5b) = 40ab$

b. $49 - 14t + t^2 = (7)^2 - 14t + (t)^2$

$\qquad\qquad\qquad = (7 - t)^2$ $\qquad\qquad -2(7)(t) = -14t$

c. $9x^4 - 12x^2 + 4 = (3x^2)^2 - 12x^2 + (2)^2$

$\qquad\qquad\qquad\quad = (3x^2 - 2)^2$ $\qquad\qquad -2(3x^2)(2) = -12x^2$

d. $(y + 3)^2 + 10(y + 3) + 25 = [(y + 3) + 5]^2$

$\qquad\qquad\qquad\qquad\qquad\quad = (y + 8)^2$

▧

███ **EXAMPLE 3**   Factor: $8x^2 - 24xy + 18y^2$.

SOLUTION   We begin by factoring the greatest common factor 2 from each term.

$$8x^2 - 24xy + 18y^2 = 2(4x^2 - 12xy + 9y^2)$$
$$= 2((2x)^2 - 12xy + (3y)^2) \qquad -2(2x)(3y) = -12xy$$
$$= 2(2x - 3y)^2$$

███

The Difference of Two Squares

Recall from Section 5.2 the formula that results in the difference of two squares:

$$(a + b)(a - b) = a^2 - b^2$$

Writing this as a factoring formula, we have

$$a^2 - b^2 = (a + b)(a - b)$$

███ **EXAMPLE 4**   Each of the following is the difference of two squares. Use the formula $a^2 - b^2 = (a + b)(a - b)$ to factor each one.

SOLUTION

a. $x^2 - 25 = x^2 - 5^2$
$$= (x + 5)(x - 5)$$

b. $49 - t^2 = 7^2 - t^2$
$$= (7 + t)(7 - t)$$

c. $81a^2 - 25b^2 = (9a)^2 - (5b)^2$
$$= (9a + 5b)(9a - 5b)$$

d. $x^2 - \dfrac{4}{9} = x^2 - \left(\dfrac{2}{3}\right)^2$
$$= \left(x + \dfrac{2}{3}\right)\left(x - \dfrac{2}{3}\right)$$

e. $x^2 + 36$ represents the sum of two squares and cannot be factored. We say that it is prime.

███

> *Note*   The sum of two squares never factors into the product of two binomials; that is, if we were to attempt to factor $(x^2 + 36)$ in Example 4, we would be unable to find two binomials (or any other polynomials) whose product is $x^2 + 36$.

As our next example shows, the difference of two fourth powers can be factored as the difference of two squares.

███ **EXAMPLE 5**   Factor: $16x^4 - 81y^4$.

SOLUTION   The first and last terms are perfect squares. We factor according to the preceding formula.

$$16x^4 - 81y^4 = (4x^2)^2 - (9y^2)^2$$
$$= (4x^2 + 9y^2)(4x^2 - 9y^2)$$

Notice that the second factor is also the difference of two squares. Factoring completely, we have

$$16x^4 - 81y^4 = (4x^2 + 9y^2)((2x)^2 - (3y)^2)$$
$$= (4x^2 + 9y^2)(2x + 3y)(2x - 3y)$$

███

Here is another example of the difference of two squares.

EXAMPLE 6   Factor: $(x - 3)^2 - 25$.

SOLUTION This example has the form $a^2 - b^2$, where a is $x - 3$ and b is 5. We factor it according to the formula for the difference of two squares:

$$(x - 3)^2 - 25 = (x - 3)^2 - 5^2 \qquad \text{Write 25 as } 5^2$$
$$= [(x - 3) + 5][(x - 3) - 5] \qquad \text{Factor}$$
$$= (x + 2)(x - 8) \qquad \text{Simplify}$$

Notice in this example we could have expanded $(x - 3)^2$, subtracted 25, and then factored to obtain the same result.

$$(x - 3)^2 - 25 = x^2 - 6x + 9 - 25 \qquad \text{Expand } (x - 3)^2$$
$$= x^2 - 6x - 16 \qquad \text{Simplify}$$
$$= (x - 8)(x + 2) \qquad \text{Factor}$$

EXAMPLE 7   Factor: $x^2 - 10x + 25 - y^2$.

SOLUTION Notice the first three items form a perfect square trinomial; that is, $x^2 - 10x + 25 = (x - 5)^2$. If we replace the first three terms by $(x - 5)^2$, the expression that results has the form $a^2 - b^2$. We can factor as we did in Example 6.

$$x^2 - 10x + 25 - y^2 = (x^2 - 10x + 25) - y^2 \qquad \text{Group first three terms together}$$
$$= (x - 5)^2 - y^2 \qquad \text{This has the form } a^2 - b^2$$
$$= [(x - 5) + y][(x - 5) - y] \qquad \begin{array}{l}\text{Factor according to the}\\ \text{formula } a^2 - b^2 = (a + b)\\ (a - b)\end{array}$$
$$= (x - 5 + y)(x - 5 - y) \qquad \text{Simplify}$$

We could check this result by multiplying the two factors together. (You may want to do that to convince yourself that we have the correct result.)

EXAMPLE 8   Factor completely: $x^3 + 2x^2 - 9x - 18$.

SOLUTION We use factoring by grouping to begin and then factor the difference of two squares.

$$x^3 + 2x^2 - 9x - 18 = x^2(x + 2) - 9(x + 2)$$
$$= (x + 2)(x^2 - 9)$$
$$= (x + 2)(x + 3)(x - 3)$$

The Sum and Difference of Two Cubes

Here are the formulas for factoring the sum and difference of two cubes:

$$a^3 + b^3 = (a + b)(a^2 - ab + b^2)$$
$$a^3 - b^3 = (a - b)(a^2 + ab + b^2)$$

Since these formulas are unfamiliar, it is important that we verify them.

■ EXAMPLE 9   Verify the two formulas.

SOLUTION   We verify the formulas by multiplying the right sides and comparing the results with the left sides:

$$
\begin{array}{r}
a^2 - ab + b^2 \\
a + b \\
\hline
a^3 - a^2b + ab^2 \\
a^2b - ab^2 + b^3 \\
\hline
a^3 \qquad\quad + b^3
\end{array}
\qquad\qquad
\begin{array}{r}
a^2 + ab + b^2 \\
a - b \\
\hline
a^3 + a^2b + ab^2 \\
- a^2b - ab^2 - b^3 \\
\hline
a^3 \qquad\quad - b^3
\end{array}
$$

The first formula is correct.   The second formula is correct.   ■

Here are some examples using the formulas for factoring the sum and difference of two cubes:

■ EXAMPLE 10   Factor: $64 + t^3$.

SOLUTION   The first term is the cube of 4 and the second term is the cube of t. Therefore,

$$64 + t^3 = 4^3 + t^3$$
$$= (4 + t)(4^2 - 4t + t^2)$$
$$= (4 + t)(16 - 4t + t^2)$$ ■

■ EXAMPLE 11   Factor: $27x^3 + 125y^3$.

SOLUTION   Writing both terms as perfect cubes, we have

$$27x^3 + 125y^3 = (3x)^3 + (5y)^3$$
$$= (3x + 5y)((3x)^2 - (3x)(5y) + (5y)^2)$$
$$= (3x + 5y)(9x^2 - 15xy + 25y^2)$$ ■

EXAMPLE 12   Factor: $a^3 - \dfrac{1}{8}$.

SOLUTION The first term is the cube of a, whereas the second term is the cube of $\dfrac{1}{2}$.

$$a^3 - \frac{1}{8} = a^3 - \left(\frac{1}{2}\right)^3$$

$$= \left(a - \frac{1}{2}\right)\left(a^2 + \frac{1}{2}a + \left(\frac{1}{2}\right)^2\right)$$

$$= \left(a - \frac{1}{2}\right)\left(a^2 + \frac{1}{2}a + \frac{1}{4}\right)$$

Getting Ready for Class

After reading through the preceding section, respond in your own words and in complete sentences.

A. What is a perfect square trinomial?

B. Is it possible to factor the sum of two squares?

C. Write the formula you use to factor the sum of two cubes.

D. Write a problem that uses the formula for the difference of two cubes.

Problem Set 5.6

Factor each perfect square trinomial.

1. $x^2 - 6x + 9$ **2.** $x^2 + 10x + 25$ **3.** $a^2 - 12a + 36$

4. $36 - 12a + a^2$ **5.** $25 - 10t + t^2$ **6.** $64 + 16t + t^2$

7. $\dfrac{1}{9}x^2 + 2x + 9$ **8.** $\dfrac{1}{4}x^2 - 2x + 4$ **9.** $4y^4 - 12y^2 + 9$

10. $9y^4 + 12y^2 + 4$ **11.** $16a^2 + 40ab + 25b^2$ **12.** $25a^2 - 40ab + 16b^2$

13. $\dfrac{1}{25} + \dfrac{1}{10}t^2 + \dfrac{1}{16}t^4$ **14.** $\dfrac{1}{9} - \dfrac{1}{3}t^3 + \dfrac{1}{4}t^6$ **15.** $y^2 + 3y + \dfrac{9}{4}$

16. $y^2 - 7y + \dfrac{49}{4}$ **17.** $a^2 - a + \dfrac{1}{4}$ **18.** $a^2 - 5a + \dfrac{25}{4}$

19. $x^2 - \dfrac{1}{2}x + \dfrac{1}{16}$ **20.** $x^2 - \dfrac{3}{4}x + \dfrac{9}{64}$ **21.** $t^2 + \dfrac{2}{3}t + \dfrac{1}{9}$

22. $t^2 - \dfrac{4}{5}t + \dfrac{4}{25}$ **23.** $16x^2 - 48x + 36$ **24.** $36x^2 + 48x + 16$

25. $75a^3 + 30a^2 + 3a$ **26.** $45a^4 - 30a^3 + 5a^2$

27. $(x + 2)^2 + 6(x + 2) + 9$ **28.** $(x + 5)^2 + 4(x + 5) + 4$

Factor each as the difference of two squares. Be sure to factor completely.

29. $x^2 - 9$ **30.** $x^2 - 16$ **31.** $49x^2 - 64y^2$ **32.** $81x^2 - 49y^2$

33. $4a^2 - \dfrac{1}{4}$ **34.** $25a^2 - \dfrac{1}{25}$ **35.** $x^2 - \dfrac{9}{25}$ **36.** $x^2 - \dfrac{25}{36}$

37. $9x^2 - 16y^2$ **38.** $25x^2 - 49y^2$ **39.** $250 - 10t^2$ **40.** $640 - 10t^2$

41. $x^4 - 81$ **42.** $x^4 - 16$ **43.** $3x^4 - 243$

44. $5x^4 - 80$ **45.** $16a^4 - 81$ **46.** $81a^4 - 16b^4$

47. $\dfrac{1}{81} - \dfrac{y^4}{16}$ **48.** $\dfrac{1}{25} - \dfrac{y^4}{64}$ **49.** $\dfrac{x^4}{16} - \dfrac{16}{81}$

50. $81a^4 - 256$ **51.** $a^4 - \dfrac{81}{256}$ **52.** $16a^4 - 625b^4$

Factor completely.

53. $x^6 - y^6$ **54.** $x^6 - 1$

55. $2a^7 - 128a$ **56.** $128a^8 - 2a^2$

57. $(x - 2)^2 - 9$ **58.** $(x + 2)^2 - 9$

59. $(y + 4)^2 - 16$ **60.** $(y - 4)^2 - 16$

61. $x^2 - 10x + 25 - y^2$ **62.** $x^2 - 6x + 9 - y^2$

63. $a^2 + 8a + 16 - b^2$ **64.** $a^2 + 12a + 36 - b^2$

65. $x^2 + 2xy + y^2 - a^2$ **66.** $a^2 + 2ab + b^2 - y^2$

67. $x^3 + 3x^2 - 4x - 12$ **68.** $x^3 + 5x^2 - 4x - 20$

69. $x^3 + 2x^2 - 25x - 50$

70. $x^3 + 4x^2 - 9x - 36$

71. $2x^3 + 3x^2 - 8x - 12$

72. $3x^3 + 2x^2 - 27x - 18$

73. $4x^3 + 12x^2 - 9x - 27$

74. $9x^3 + 18x^2 - 4x - 8$

75. $(2x - 5)^2 - 100$

76. $(7a + 5)^2 - 64$

77. $(a - 3)^2 - (4b)^2$

78. $(2x - 5)^2 - (6y)^2$

79. $a^2 - 6a + 9 - 16b^2$

80. $x^2 - 10x + 25 - 9y^2$

81. $x^2(x + 4) - 6x(x + 4) + 9(x + 4)$

82. $x^2(x - 6) + 8x(x - 6) + 16(x - 6)$

Factor each of the following as the sum or difference of two cubes.

83. $x^3 - y^3$ **84.** $x^3 + y^3$ **85.** $a^3 + 8$ **86.** $a^3 - 8$

87. $27 + x^3$ **88.** $27 - x^3$ **89.** $y^3 - 1$ **90.** $y^3 + 1$

91. $10r^3 - 1{,}250$ **92.** $10r^3 + 1{,}250$ **93.** $64 + 27a^3$ **94.** $27 - 64a^3$

95. $8x^3 - 27y^3$ **96.** $27x^3 - 8y^3$ **97.** $t^3 + \dfrac{1}{27}$ **98.** $t^3 - \dfrac{1}{27}$

99. $27x^3 - \dfrac{1}{27}$ **100.** $8x^3 + \dfrac{1}{8}$ **101.** $64a^3 + 125b^3$ **102.** $125a^3 - 27b^3$

103. Find two values of b that will make $9x^2 + bx + 25$ a perfect square trinomial.

104. Find a value of c that will make $49x^2 - 42x + c$ a perfect square trinomial.

105. Find a value of c that will make $25x^2 - 90x + c$ a perfect square trinomial.

106. Find two values of b that will make $16x^2 + bx + 25$ a perfect square trinomial.

Learning Objectives Assessment

The following problems can be used to help assess if you have successfully met the learning objectives for this section.

107. Factor: $16x^2 - 40x + 25$.

  a. $(4x - 5)^2$ **b.** $(4x + 5)^2$

  c. $(4x + 5)(4x - 5)$ **d.** $(8x + 5)(2x - 5)$

108. Factor: $16x^2 - 25$.

  a. Prime   **b.** $(4x - 5)^2$

  c. $(4x + 5)^2$ **d.** $(4x + 5)(4x - 5)$

109. Factor: $8x^3 - 27$.

  a. $(2x - 3)^3$ **b.** $(2x - 3)(4x^2 + 12x + 9)$

  c. $(2x - 3)(4x^2 - 6x + 9)$ **d.** $(2x - 3)(4x^2 + 6x + 9)$

Getting Ready for the Next Section

Factor out the greatest common factor.

110. $y^4 + 36y^2$

111. $2ab^5 + 8ab^4 + 2ab^3$

112. $3a^2b^3 + 6a^2b^2 - 3a^2b$

Factor by grouping.

113. $4x^2 - 6x + 2ax - 3a$

114. $6x^2 - 4x + 3ax - 2a$

115. $15ax - 10a + 12x - 8$

116. $15bx + 3b - 25x - 5$

Factor the difference of squares.

117. $x^2 - 4$ **118.** $x^2 - 9$ **119.** $A^2 - 25$ **120.** $25x^2 - 36y^2$

Factor the perfect square trinomial.

121. $x^2 - 6x + 9$

122. $x^2 - 10x + 25$

123. $x^2 + 8xy + 16y^2$

124. $a^2 - 12ab + 36b^2$

Factor.

125. $6a^2 - 11a + 4$

126. $6x^2 - x - 15$

127. $12x^2 - 32x - 35$

128. $12a^2 - 7a - 10$

Factor the sum or difference of cubes.

129. $x^3 + 8$ **130.** $x^3 - 27$ **131.** $8x^3 - 27$ **132.** $27a^3 - 8b^3$

Learning Objective

In this section, we will learn how to:

1. Factor a variety of polynomials.

In this section, we will review the different methods of factoring that we have presented in the previous sections of this chapter. This section is important because it will give you an opportunity to factor a variety of polynomials.

We begin this section by listing the steps that can be used to factor polynomials of any type.

> **HOW TO** *Factor a Polynomial*
>
> **Step 1:** If the polynomial has a greatest common factor other than 1, then factor out the greatest common factor.
>
> **Step 2:** If the polynomial has two terms (a binomial), then check to see if it is the difference of two squares, or the sum or difference of two cubes, then factor accordingly. (*Note:* If it is the **sum** of two squares, it will not factor.)
>
> **Step 3:** If the polynomial has three terms (a trinomial), then either it is a perfect square trinomial, which will factor into the square of a binomial, or it is not a perfect square trinomial, in which case you use the other methods we presented in Section 5.5.
>
> **Step 4:** If the polynomial has more than three terms, try to factor it by grouping.
>
> **Step 5:** As a final check, see if any of the factors you have written can be factored further. If you have overlooked a common factor, you can catch it here.

Here are some examples illustrating how we use the steps in our list. There are no new factoring techniques in this section. The problems here are all similar to the problems you have seen before. What is different is that they are not all of the same type.

VIDEO EXAMPLES

SECTION 5.7

EXAMPLE 1   Factor: $2x^5 - 8x^3$.

SOLUTION   First we check to see if the greatest common factor is other than 1. Since the greatest common factor is $2x^3$, we begin by factoring it out. Once we have done so, we notice that the binomial that remains is the difference of two squares, which we factor according to the formula $a^2 - b^2 = (a + b)(a - b)$.

$$2x^5 - 8x^3 = 2x^3(x^2 - 4) \qquad \text{Factor out the greatest common factor, } 2x^3$$

$$= 2x^3(x + 2)(x - 2) \quad \text{Factor the difference of two squares}$$

EXAMPLE 2   Factor: $3x^4 - 18x^3 + 27x^2$.

SOLUTION   Step 1 is to factor out the greatest common factor $3x^2$. After we have done so, we notice that the trinomial that remains is a perfect square trinomial, which will factor as the square of a binomial.

$$3x^4 - 18x^3 + 27x^2 = 3x^2(x^2 - 6x + 9) \quad \text{Factor out } 3x^2$$
$$= 3x^2(x - 3)^2 \qquad x^2 - 6x + 9 \text{ is the square of } x - 3$$

EXAMPLE 3   Factor: $y^3 + 25y$.

SOLUTION   We begin by factoring out the y that is common to both terms. The binomial that remains after we have done so is the sum of two squares, which does not factor, so after the first step, we are finished.

$$y^3 + 25y = y(y^2 + 25)$$

EXAMPLE 4   Factor: $6a^2 - 11a + 4$.

SOLUTION   Here we have a trinomial that does not have a greatest common factor other than 1. Since it is not a perfect square trinomial, we factor it by trial and error. Without showing all the different possibilities, here is the answer.

$$6a^2 - 11a + 4 = (3a - 4)(2a - 1)$$

EXAMPLE 5   Factor: $2x^4 + 16x$.

SOLUTION   This binomial has a greatest common factor of $2x$. The binomial that remains after the $2x$ has been factored from each term is the sum of two cubes, which we factor according to the formula $a^3 + b^3 = (a + b)(a^2 - ab + b^2)$.

$$2x^4 + 16x = 2x(x^3 + 8) \qquad \text{Factor } 2x \text{ from each term}$$
$$= 2x(x + 2)(x^2 - 2x + 4) \quad \text{The sum of two cubes}$$

EXAMPLE 6   Factor: $2ab^5 + 8ab^4 + 2ab^3$.

SOLUTION   The greatest common factor is $2ab^3$. We begin by factoring it from each term. After that we find that the trinomial that remains cannot be factored further.

$$2ab^5 + 8ab^4 + 2ab^3 = 2ab^3(b^2 + 4b + 1)$$

EXAMPLE 7   Factor: $4x^2 - 6x + 2ax - 3a$.

SOLUTION   Our polynomial has four terms, so we factor by grouping.

$$4x^2 - 6x + 2ax - 3a = 2x(2x - 3) + a(2x - 3)$$
$$= (2x - 3)(2x + a)$$

EXAMPLE 8 Factor: $x^6 - y^6$.

SOLUTION We have a choice of how we want to write the two terms to begin. We can write the expression as the difference of two squares, $(x^3)^2 - (y^3)^2$, or as the difference of two cubes, $(x^2)^3 - (y^2)^3$. It is better to use the difference of two squares if we have a choice. The resulting two binomials can then both be factored as a sum or difference of cubes.

$$x^6 - y^6 = (x^3)^2 - (y^3)^2$$

$$= (x^3 - y^3)(x^3 + y^3)$$

$$= (x - y)(x^2 + xy + y^2)(x + y)(x^2 - xy + y^2)$$

Try this example again writing the first line as the difference of two cubes instead of the difference of two squares. It will become apparent why it is better to use the difference of two squares first.

Getting Ready for Class

After reading through the preceding section, respond in your own words and in complete sentences.

A. How do you know when you've factored completely?

B. What is the first step in factoring a polynomial?

C. If a polynomial has four terms, what method of factoring should you try?

D. Other than a greatest common factor, how do we choose which factoring technique to try?

Factor each of the following polynomials completely. None of the factors in your final answer should still be factorable. Also, note that the even-numbered problems are not necessarily similar to the odd-numbered problems that precede them in this problem set.

1. $x^2 - 81$

2. $x^2 - 18x + 81$

3. $x^2 + 2x - 15$

4. $15x^2 + 13x - 6$

5. $x^2(x + 2) + 6x(x + 2) + 9(x + 2)$

6. $12x^2 - 11x + 2$

7. $x^2y^2 + 2y^2 + x^2 + 2$

8. $21y^2 - 25y - 4$

9. $2a^3b + 6a^2b + 2ab$

10. $6a^2 - ab - 15b^2$

11. $x^2 + x + 1$

12. $x^2y + 3y + 2x^2 + 6$

13. $12a^2 - 75$

14. $18a^2 - 50$

15. $9x^2 - 12xy + 4y^2$

16. $x^3 - x^2$

17. $25 - 10t + t^2$

18. $t^2 + 4t + 4 - y^2$

19. $4x^3 + 16xy^2$

20. $16x^2 + 49y^2$

21. $2y^3 + 20y^2 + 50y$

22. $x^2 + 5bx - 2ax - 10ab$

23. $a^7 + 8a^4b^3$

24. $5a^2 - 45b^2$

25. $t^2 + 6t + 9 - x^2$

26. $36 + 12t + t^2$

27. $x^3 + 5x^2 - 9x - 45$

28. $x^3 + 5x^2 - 16x - 80$

29. $5a^2 + 10ab + 5b^2$

30. $3a^3b^2 + 15a^2b^2 + 3ab^2$

31. $x^2 + 49$

32. $16 - x^4$

33. $3x^2 + 15xy + 18y^2$

34. $3x^2 + 27xy + 54y^2$

35. $9a^2 + 2a + \dfrac{1}{9}$

36. $18 - 2a^2$

37. $x^2(x - 3) - 14x(x - 3) + 49(x - 3)$

38. $x^2 + 3ax - 2bx - 6ab$

39. $x^2 - 64$

40. $9x^2 - 4$

41. $8 - 14x - 15x^2$

42. $5x^4 + 14x^2 - 3$

43. $49a^7 - 9a^5$

44. $a^6 - b^6$

45. $r^2 - \dfrac{1}{25}$

46. $27 - r^3$

47. $49x^2 + 9y^2$

48. $12x^4 - 62x^3 + 70x^2$

49. $100x^2 - 100x - 600$

50. $100x^2 - 100x - 1{,}200$

51. $25a^3 + 20a^2 + 3a$

52. $16a^5 - 54a^2$

53. $3x^4 - 14x^2 - 5$

54. $8 - 2x - 15x^2$

55. $24a^5b - 3a^2b$

56. $18a^4b^2 - 24a^3b^3 + 8a^2b^4$

57. $64 - r^3$

58. $r^2 - \dfrac{1}{9}$

59. $20x^4 - 45x^2$

60. $16x^3 + 16x^2 + 3x$

61. $400t^2 - 900$

62. $900 - 400t^2$

63. $16x^5 - 44x^4 + 30x^3$

64. $16x^2 + 16x - 1$

65. $y^6 - 1$

66. $25y^7 - 16y^5$

67. $50 - 2a^2$

68. $4a^2 + 2a + \dfrac{1}{4}$

69. $12x^4y^2 + 36x^3y^3 + 27x^2y^4$ **70.** $16x^3y^2 - 4xy^2$

71. $x^2 - 4x + 4 - y^2$ **72.** $x^2 - 12x + 36 - b^2$

73. $a^2 - \dfrac{4}{3}ab + \dfrac{4}{9}b^2$ **74.** $a^2 + \dfrac{3}{2}ab + \dfrac{9}{16}b^2$

75. $x^2 - \dfrac{4}{5}xy + \dfrac{4}{25}y^2$ **76.** $x^2 - \dfrac{8}{7}xy + \dfrac{16}{49}y^2$

77. $a^2 - \dfrac{5}{3}ab + \dfrac{25}{36}b^2$ **78.** $a^2 + \dfrac{5}{4}ab + \dfrac{25}{64}b^2$

79. $x^2 - \dfrac{8}{5}xy + \dfrac{16}{25}y^2$ **80.** $a^2 + \dfrac{3}{5}ab + \dfrac{9}{100}b^2$

81. $2x^2(x + 2) - 13x(x + 2) + 15(x + 2)$ **82.** $5x^2(x - 4) - 14x(x - 4) - 3(x - 4)$

83. $(x - 4)^3 + (x - 4)^4$ **84.** $(2x - 7)^5 + (2x - 7)^6$

85. $2y^3 - 54$ **86.** $81 + 3y^3$

87. $2a^3 - 128b^3$ **88.** $128a^3 + 2b^3$

89. $2x^3 + 432y^3$ **90.** $432x^3 - 2y^3$

Learning Objectives Assessment

The following problems can be used to help assess if you have successfully met the learning objectives for this section.

91. Which factoring technique should always be tried first?
    a. Trial and error   **b.** Greatest common factor
    c. Grouping   **d.** Difference of squares

92. Factor completely: $16x^4 - 81y^4$.
    a. $(2x - 3y)^4$ **b.** $(4x^2 + 9y^2)(4x^2 - 9y^2)$
    c. $(2x - 3y)(8x^3 + 6xy + 27y^3)$ **d.** $(4x^2 + 9y^2)(2x + 3y)(2x - 3y)$

Getting Ready for the Next Section

Simplify.

93. $x^2 + (x + 1)^2$ **94.** $x^2 + (x + 3)^2$

95. $\dfrac{16t^2 - 64t + 48}{16}$ **96.** $\dfrac{100p^2 - 1{,}300p + 4{,}000}{100}$

Factor each of the following.

97. $x^2 - 2x - 24$ **98.** $x^2 - x - 6$

99. $2x^3 - 5x^2 - 3x$ **100.** $3x^3 - 5x^2 - 2x$

101. $x^3 + 2x^2 - 9x - 18$ **102.** $x^3 + 5x^2 - 4x - 20$

103. $x^3 + 2x^2 - 5x - 10$ **104.** $2x^3 + 3x^2 - 18x - 27$

Solve.

105. $x - 6 = 0$ **106.** $x + 4 = 0$ **107.** $2x + 1 = 0$ **108.** $3x + 1 = 0$

Solving Equations by Factoring

Learning Objectives

In this section, we will learn how to:

1. Solve quadratic equations by factoring.

2. Solve polynomial equations by factoring.

3. Solve applications of polynomial equations.

Introduction

> (def) **DEFINITION** *quadratic equation*
>
> Any equation that can be written in the form
>
> $$ax^2 + bx + c = 0$$
>
> where a, b, and c are constants and a is not 0 ($a \neq 0$) is called a *quadratic equation*. The form $ax^2 + bx + c = 0$ is called *standard form* for quadratic equations.

Note The third equation is clearly a quadratic equation since it is in standard form. (Notice that a is 4, b is -3, and c is 2.) The first two equations are also quadratic because they could be put in the form $ax^2 + bx + c = 0$ by using the addition property of equality.

Each of the following is a quadratic equation:

$$2x^2 = 5x \qquad 5x^2 = 75 \qquad 4x^2 - 3x + 2 = 0$$

Notation For a quadratic equation written in standard form, the first term ax^2 is called the *quadratic term*; the second term bx is the *linear term*; and the last term c is called the *constant term*.

In the past we have noticed that the number 0 is a special number. There is another property of 0 that is the key to solving quadratic equations. It is called the *zero-factor property*.

Note What the zero-factor property says in words is that we can't multiply and get 0 without multiplying by 0; that is, if we multiply two numbers and get 0, then one or both of the original two numbers we multiplied must have been 0.

> [Δ≠Σ] **PROPERTY** *Zero-Factor Property*
>
> For all real numbers r and s,
>
> $$r \cdot s = 0 \qquad \text{if and only if} \qquad r = 0 \qquad \text{or} \qquad s = 0 \quad \text{(or both)}$$

Solving Quadratic Equations by Factoring

VIDEO EXAMPLES

SECTION 5.8

EXAMPLE 1 Solve: $x^2 - 2x - 24 = 0$.

SOLUTION We begin by factoring the left side as $(x - 6)(x + 4)$ and get

$$(x - 6)(x + 4) = 0$$

Now both $(x - 6)$ and $(x + 4)$ represent real numbers. We notice that their product is 0. By the zero-factor property, one or both of them must be 0.

$$x - 6 = 0 \qquad \text{or} \qquad x + 4 = 0$$

We have used factoring and the zero-factor property to rewrite our original second-degree equation as two first-degree equations connected by the word *or*. Completing the solution, we solve the two linear equations.

$$x - 6 = 0 \qquad \text{or} \qquad x + 4 = 0$$
$$x = 6 \qquad\qquad\qquad x = -4$$

We check our solutions in the original equation as follows:

Check $x = 6$	Check $x = -4$
$6^2 - 2(6) - 24 \stackrel{?}{=} 0$	$(-4)^2 - 2(-4) - 24 \stackrel{?}{=} 0$
$36 - 12 - 24 \stackrel{?}{=} 0$	$16 + 8 - 24 \stackrel{?}{=} 0$
$0 = 0$	$0 = 0$

In both cases the result is a true statement, which means that both 6 and -4 are solutions to the original equation. ▰

To generalize the preceding example, here are the steps used in solving an equation by factoring.

> ### HOW TO   *Solve an Equation by Factoring*
>
> **Step 1:** Write the equation in standard form.
>
> **Step 2:** Factor the left side.
>
> **Step 3:** Use the zero-factor property to set each factor equal to 0.
>
> **Step 4:** Solve the resulting linear equations.

▰ **EXAMPLE 2**   Solve: $100x^2 = 300x$.

SOLUTION   We begin by writing the equation in standard form and factoring.

$$100x^2 = 300x$$
$$100x^2 - 300x = 0 \qquad \text{Standard form}$$
$$100x(x - 3) = 0 \qquad \text{Factor}$$

Using the zero-factor property to set each factor to 0, we have

$$100x = 0 \quad \text{or} \quad x - 3 = 0$$
$$x = 0 \qquad\qquad x = 3$$

The two solutions are 0 and 3. ▰

▰ **EXAMPLE 3**   Solve: $(x - 2)(x + 1) = 4$.

SOLUTION   We begin by multiplying the two factors on the left side. (Notice that it would be incorrect to set each of the factors on the left side equal to 4. The fact that the product is 4 does not imply that either of the factors must be 4.)

$$(x - 2)(x + 1) = 4$$
$$x^2 - x - 2 = 4 \qquad \text{Multiply the left side}$$
$$x^2 - x - 6 = 0 \qquad \text{Standard form}$$
$$(x - 3)(x + 2) = 0 \qquad \text{Factor}$$
$$x - 3 = 0 \quad \text{or} \quad x + 2 = 0 \qquad \text{Zero-factor property}$$
$$x = 3 \qquad\qquad x = -2$$

▰

EXAMPLE 4   Solve: $(x + 2)(3x - 1) = (x + 2)(x + 6)$.

SOLUTION   We begin by multiplying the factors on each side.

$$(x + 2)(3x - 1) = (x + 2)(x + 6)$$

$$3x^2 + 5x - 2 = x^2 + 8x + 12 \qquad \text{FOIL both sides}$$

$$2x^2 - 3x - 14 = 0 \qquad \text{Standard form}$$

$$(2x - 7)(x + 2) = 0 \qquad \text{Factor}$$

$$2x - 7 = 0 \quad \text{or} \quad x + 2 = 0 \qquad \text{Zero-factor Property}$$

$$2x = 7 \qquad\qquad x = -2$$

$$x = \frac{7}{2}$$

We have two solutions: $\frac{7}{2}$ and -2.

Solving Polynomial Equations by Factoring

We can use the zero-factor property to solve polynomial equations of higher degree. The process is similar to the method we have used to solve quadratic equations.

EXAMPLE 5   Solve: $x^3 + 2x^2 - 9x - 18 = 0$.

SOLUTION   We start with factoring by grouping.

$$x^3 + 2x^2 - 9x - 18 = 0$$

$$\left. \begin{array}{c} x^2(x + 2) - 9(x + 2) = 0 \\ (x + 2)(x^2 - 9) = 0 \end{array} \right\} \quad \text{Factor by grouping}$$

$$(x + 2)(x - 3)(x + 3) = 0 \qquad \text{The difference of two squares}$$

$$x + 2 = 0 \quad \text{or} \quad x - 3 = 0 \quad \text{or} \quad x + 3 = 0 \qquad \text{Zero-facor property}$$

$$x = -2 \qquad x = 3 \qquad x = -3$$

We have three solutions: -2, 3, and -3.

EXAMPLE 6   Solve: $\frac{1}{3}x^3 = \frac{5}{6}x^2 + \frac{1}{2}x$.

SOLUTION   We can simplify our work if we clear the equation of fractions. Multiplying both sides by the LCD, 6, we have

$$6 \cdot \frac{1}{3}x^3 = 6 \cdot \frac{5}{6}x^2 + 6 \cdot \frac{1}{2}x$$

$$2x^3 = 5x^2 + 3x$$

Next we add $-5x^2$ and $-3x$ to each side so that the right side will become 0.

$$2x^3 - 5x^2 - 3x = 0 \qquad \text{Standard form}$$

We factor the left side and then use the zero-factor property to set each factor to 0.

$$x(2x^2 - 5x - 3) = 0 \qquad \text{Factor out the greatest common factor}$$

$$x(2x + 1)(x - 3) = 0 \qquad \text{Continue factoring}$$

$$x = 0 \quad \text{or} \quad 2x + 1 = 0 \quad \text{or} \quad x - 3 = 0 \qquad \text{Zero-factor property}$$

Solving each of the resulting equations, we have

$$x = 0 \qquad \text{or} \qquad x = -\frac{1}{2} \qquad \text{or} \qquad x = 3$$

Applications

We conclude this section by looking at some applications of polynomial equations.

EXAMPLE 7   The sum of the squares of two consecutive integers is 25. Find the two integers.

SOLUTION   We apply the Blueprint for Problem Solving to solve this application problem. Remember, step 1 in the blueprint is done mentally.

Step 1: Read and list.

> *Known items:* Two consecutive integers. If we add their squares, the result is 25.
> *Unknown items:* The two integers

Step 2: Assign a variable and translate information.

> Let x = the first integer
> Then $x + 1$ = the next consecutive integer.

Step 3: Reread and write an equation.

> Since the sum of the squares of the two consecutive integers is 25, the equation that describes the situation is
> $$x^2 + (x + 1)^2 = 25$$

Step 4: Solve the equation.

$x^2 + (x + 1)^2 = 25$	
$x^2 + (x^2 + 2x + 1) = 25$	Multiply $(x + 1)^2$
$2x^2 + 2x - 24 = 0$	Standard form
$x^2 + x - 12 = 0$	Divide both sides by 2
$(x + 4)(x - 3) = 0$	Factor
$x = -4$ or $x = 3$	Zero-factor property

Step 5: Write the answer.

> If $x = -4$, then $x + 1 = -3$. If $x = 3$, then $x + 1 = 4$. The two integers are -4 and -3, or the two integers are 3 and 4.

Step 6: Reread and check.

> The two integers in each pair are consecutive integers, and the sum of the squares of either pair is 25.   ◼

Another application of quadratic equations involves the Pythagorean Theorem, an important theorem from geometry. The theorem gives the relationship between the sides of any right triangle (a triangle with a 90-degree angle). We state it here without proof.

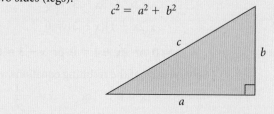

FACTS FROM GEOMETRY   *The Pythagorean Theorem*

In any right triangle, the square of the length of the longest side (hypotenuse) is equal to the sum of the squares of the lengths of the other two sides (legs).

$$c^2 = a^2 + b^2$$

EXAMPLE 8   The lengths of the three sides of a right triangle are given by three consecutive integers. Find the lengths of the three sides.

SOLUTION

Step 1: Read and list.

Known items: A right triangle. The three sides are three consecutive integers.

Unknown items: The three sides

Step 2: Assign a variable and translate information.

Let x = first integer (shortest side)

Then $x + 1$ = next consecutive integer

$x + 2$ = last consecutive integer (longest side)

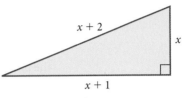

Step 3: Reread and write an equation.

By the Pythagorean Theorem, we have

$$(x + 2)^2 = (x + 1)^2 + x^2$$

Step 4: Solve the equation.

$$x^2 + 4x + 4 = x^2 + 2x + 1 + x^2$$

$$x^2 - 2x - 3 = 0$$

$$(x - 3)(x + 1) = 0$$

$$x = 3 \quad \text{or} \quad x = -1$$

Step 5: Write the answer.

Since x is the length of a side in a triangle, it must be a positive number. Therefore, $x = -1$ cannot be used. The shortest side is 3. The other two sides are 4 and 5.

Step 6: Reread and check.

The three sides are given by consecutive integers. The square of the longest side is equal to the sum of the squares of the two shorter sides.

EXAMPLE 9   A rectangular garden measures 40 meters by 30 meters. A lawn of uniform width surrounds the garden and has an area of 456 square meters. Find the width of the lawn that surrounds the garden.

SOLUTION

Step 1: Read and list.

Known items: The length and width of the garden

The area of the lawn

Unknown items: The width of the lawn

Step 2: Assign a variable and translate information.

Let w = the width of the lawn that surrounds the garden.

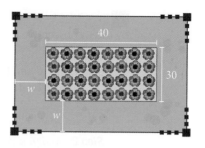

Step 3: *Reread and write an equation.*

The area of the garden is $30(40) = 1{,}200$ m^2.

The area of the lawn is 456 m^2.

The length of the garden plus the lawn is $40 + 2w$.

The width of the garden plus the lawn is $30 + 2w$.

The product of these expressions is equal to the sum of the areas.
$$(40 + 2w)(30 + 2w) = 1{,}200 + 456$$

Step 4: *Solve the equation.*

$$\begin{aligned}
(40 + 2w)(30 + 2w) &= 1{,}200 + 456 \\
1200 + 140w + 4w^2 &= 1{,}656 && \text{FOIL} \\
4w^2 + 140w - 456 &= 0 && \text{Standard form} \\
w^2 + 35w - 114 &= 0 && \text{Divide both sides by 4} \\
(w - 3)(w + 38) &= 0 && \text{Factor} \\
w = 3 \quad \text{or} \quad w &= -38 && \text{Zero-factor property}
\end{aligned}$$

Step 5: *Write the answer.*

Because w is measuring length, it must be a positive number. Therefore, $w = -38$ cannot be used. The width of the lawn is 3 meters.

Step 6: *Reread and check.*

$$\begin{aligned}
(40 + 2(3))(30 + 2(3)) &\overset{?}{=} 1{,}200 + 456 \\
46 \cdot 36 &\overset{?}{=} 1{,}656 \\
1{,}656 &= 1{,}656 && \text{A true statement} \quad \blacksquare
\end{aligned}$$

Our last two examples involve formulas that are quadratic.

EXAMPLE 10   If an object is projected into the air with an initial vertical velocity of v feet/second, its height h, in feet, above the ground after t seconds will be given by
$$h = vt - 16t^2$$
Find t if $v = 64$ feet/second and $h = 48$ feet.

SOLUTION   Substituting $v = 64$ and $h = 48$ into the preceding formula, we have
$$48 = 64t - 16t^2$$
which is a quadratic equation. We write it in standard form and solve by factoring.

$$16t^2 - 64t + 48 = 0$$

$$t^2 - 4t + 3 = 0 \qquad \text{Divide each side by 16}$$

$$(t - 1)(t - 3) = 0$$

$$t - 1 = 0 \quad \text{or} \quad t - 3 = 0$$

$$t = 1 \qquad\qquad t = 3$$

Here is how we interpret our results: If an object is projected upward with an initial vertical velocity of 64 feet/second, it will be 48 feet above the ground after 1 second and after 3 seconds; that is, it passes 48 feet going up and also coming down.

EXAMPLE 11 A manufacturer of headphones knows that the number of headphones she can sell each week is related to the price of the headphones by the equation $x = 1{,}300 - 100p$, where x is the number of headphones and p is the price per set. What price should she charge for each set of headphones if she wants the weekly revenue to be $4,000?

SOLUTION The formula for total revenue is $R = xp$. Since we want R in terms of p, we substitute $1{,}300 - 100p$ for x in the equation $R = xp$.

If $\qquad R = xp$

and $\qquad x = 1{,}300 - 100p$

then $\qquad R = (1{,}300 - 100p)p$

We want to find p when R is 4,000. Substituting 4,000 for R in the formula gives us

$$4{,}000 = (1{,}300 - 100p)p$$

$$4{,}000 = 1{,}300p - 100p^2$$

which is a quadratic equation. To write it in standard form, we add $100p^2$ and $-1{,}300p$ to each side, giving us

$$100p^2 - 1{,}300p + 4{,}000 = 0$$

$$p^2 - 13p + 40 = 0 \qquad \text{Divide each side by 100}$$

$$(p - 5)(p - 8) = 0$$

$$p - 5 = 0 \quad \text{or} \quad p - 8 = 0$$

$$p = 5 \qquad\qquad p = 8$$

If she sells the headphones for $5 each or for $8 each she will have a weekly revenue of $4,000.

Getting Ready for Class

After reading through the preceding section, respond in your own words and in complete sentences.

A. What is standard form for a quadratic equation?

B. Describe the zero-factor property in your own words.

C. What is the first step in solving an equation by factoring?

D. Explain the Pythagorean theorem in words.

Problem Set 5.8

Solve each quadratic equation.

1. $x^2 - 5x - 6 = 0$

2. $x^2 + 5x - 6 = 0$

3. $3y^2 + 11y - 4 = 0$

4. $3y^2 - y - 4 = 0$

5. $60x^2 - 130x + 60 = 0$

6. $90x^2 + 60x - 80 = 0$

7. $\dfrac{1}{10}t^2 - \dfrac{5}{2} = 0$

8. $\dfrac{2}{7}t^2 - \dfrac{7}{2} = 0$

9. $\dfrac{1}{5}y^2 - 2 = -\dfrac{3}{10}y$

10. $\dfrac{1}{2}y^2 + \dfrac{5}{3} = \dfrac{17}{6}y$

11. $9x^2 - 12x = 0$

12. $4x^2 + 4x = 0$

13. $0.02r + 0.01 = 0.15r^2$

14. $0.02r - 0.01 = -0.08r^2$

15. $-100x = 10x^2$

16. $800x = 100x^2$

17. $(x + 6)(x - 2) = -7$

18. $(x - 7)(x + 5) = -20$

19. $(y - 4)(y + 1) = -6$

20. $(y - 6)(y + 1) = -12$

21. $(x + 1)^2 = 3x + 7$

22. $(x + 2)^2 = 9x$

23. $(2r + 3)(2r - 1) = -(3r + 1)$

24. $(3r + 2)(r - 1) = -(7r - 7)$

25. $3x^2 + x = 10$

26. $y^2 + y - 20 = 2y$

27. $12(x + 3) + 12(x - 3) = 3(x^2 - 9)$

28. $8(x + 2) + 8(x - 2) = 3(x^2 - 4)$

29. $(y + 3)^2 + y^2 = 9$

30. $(2y + 4)^2 + y^2 = 4$

31. $(x + 3)^2 + 1 = 2$

32. $(x - 3)^2 + (-1)^2 = 10$

33. $(3x + 1)(x - 4) = (x - 3)(x + 3)$

34. $(x + 3)(x + 6) = (3x + 1)(x - 4)$

35. $(3x - 2)(x + 1) = (x - 4)^2$

36. $(x - 3)(x - 2) = (3x - 2)(x + 1)$

37. $(2x - 3)(x - 5) = (x + 1)(x - 3)$

38. $(3x + 5)(x + 1) = (x - 5)^2$

Use factoring to solve each polynomial equation.

39. $x^3 - 5x^2 + 6x = 0$

40. $x^3 + 5x^2 + 6x = 0$

41. $100x^4 = 400x^3 + 2{,}100x^2$

42. $100x^4 = -400x^3 + 2{,}100x^2$

43. $x^3 + 3x^2 - 4x - 12 = 0$

44. $x^3 + 5x^2 - 4x - 20 = 0$

45. $x^3 + 2x^2 - 25x - 50 = 0$

46. $x^3 + 4x^2 - 9x - 36 = 0$

47. $9a^3 = 16a$

48. $16a^3 = 25a$

49. $2x^3 + 3x^2 - 8x - 12 = 0$

50. $3x^3 + 2x^2 - 27x - 18 = 0$

51. $4x^3 + 12x^2 - 9x - 27 = 0$

52. $9x^3 + 18x^2 - 4x - 8 = 0$

53. Paying Attention to Instructions Work each problem according to the instructions given.

    a. Solve: $8x - 5 = 0$

    b. Add: $(8x - 5) + (2x - 3)$

    c. Multiply: $(8x - 5)(2x - 3)$

    d. Solve: $16x^2 - 34x + 15 = 0$

54. Paying Attention to Instructions Work each problem according to the instructions given.

    a. Subtract: $(3x + 5) - (7x - 4)$

    b. Solve: $3x + 5 = 7x - 4$

    c. Multiply: $(3x + 5)(7x - 4)$

    d. Solve: $21x^2 + 23x - 20 = 0$

55. Solve each equation.

  a. $9x - 25 = 0$ **b.** $9x^2 - 25 = 0$

  c. $9x^2 - 25 = 56$ **d.** $9x^2 - 25 = 30x - 50$

56. Solve each equation.

  a. $5x - 6 = 0$ **b.** $(5x - 6)^2 = 0$

  c. $25x^2 - 36 = 0$ **d.** $25x^2 - 36 = 28$

57. Let $f(x) = \left(x + \frac{3}{2} \right)^2$. Find all values for the variable x, for which $f(x) = 0$.

58. Let $f(x) = \left(x - \frac{5}{2} \right)^2$. Find all values for the variable x, for which $f(x) = 0$.

59. Let $f(x) = (x - 3)^2 - 25$. Find all values for the variable x, for which $f(x) = 0$.

60. Let $f(x) = 9x^3 + 18x^2 - 4x - 8$. Find all values for the variable x, for which $f(x) = 0$.

Let $f(x) = x^2 + 6x + 3$. Find all values for the variable x, for which $f(x) = g(x)$.

61. $g(x) = -6$ **62.** $g(x) = 19$ **63.** $g(x) = 10$ **64.** $g(x) = -2$

Let $h(x) = x^2 - 5x$. Find all values for the variable x, for which $h(x) = f(x)$.

65. $f(x) = 0$ **66.** $f(x) = -6$

67. $f(x) = 2x + 8$ **68.** $f(x) = -2x + 10$

Find all values for the variable x such that $f(x) = x$.

69. $f(x) = x^2$ **70.** $f(x) = 4x - 7$

71. $f(x) = \frac{1}{3}x + 1$ **72.** $f(x) = x^2 - 2$

Applying the Concepts

73. The product of two consecutive odd integers is 99. Find the two integers.

74. The product of two consecutive integers is 132. Find the two integers.

75. The sum of two numbers is 14. Their product is 48. Find the numbers.

76. The sum of two numbers is 12. Their product is 32. Find the numbers.

77. The dimensions of a rectangular photograph are consecutive even integers. The product of the integers is 10 less than 5 times their sum. Find the two integers.

78. The dimensions of a rectangular photograph are consecutive even integers. The sum of the integers squared is 100. Find the integers.

79. In May 2012, fiction authors John Grisham and Toni Morrison had books on the New York Times Hardcover Fiction Best Sellers list. The books, *Calico Joe* and *Home* respectively were ranked as two consecutive odd integers. The product of the integers is 1 less than 4 times their sum. Find the two integers.

80. On Billboard's 2011 year-end Hot 100 Songs list, Katy Perry had songs with rankings that were consecutive integers. The square of the sum of the integers is 49. Find the two integers.

81. Area The area of the rectangle below is 126 cm². Find the length and width using the given measurements.

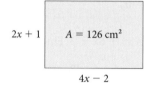

82. Area The area of the rectangle is 120 ft². Find the length and width.

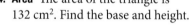

83. Area The area of the triangle is 80 m². Find the base and height.

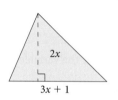

84. Area The area of the triangle is 132 cm². Find the base and height.

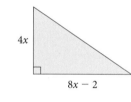

85. Area The area of the triangle is 63 in². Find the base and height.

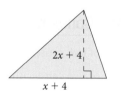

86. Area The area of the triangle is 35 cm². Find the base and height.

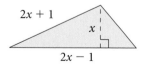

87. Pythagorean Theorem Find the value of x.

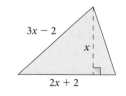

88. Pythagorean Theorem Find the value of x.

89. **Right Triangle** The lengths of the three sides of a right triangle are given by three consecutive even integers. Find the lengths of the three sides.

90. **Right Triangle** The longest side of a right triangle is 3 less than twice the shortest side. The third side measures 12 inches. Find the length of the shortest side.

91. **Geometry** The length of a rectangle is 2 meters more than 3 times the width. If the area is 16 square meters, find the width and the length.

92. **Geometry** The length of a rectangle is 4 yards more than twice the width. If the area is 70 square yards, find the width and the length.

93. **Geometry** The base of a triangle is 2 cm more than 4 times the height. If the area is 36 cm^2, find the base and the height.

94. **Geometry** The height of a triangle is 4 feet less than twice the base. If the area is 48 square feet, find the base and the height.

95. **Surrounding Area** A rectangular garden measures 40 yards by 35 yards. A lawn of uniform width surrounds the garden and has an area of 316 square yards. Find the width of the lawn that surrounds the garden.

96. **Surrounding Area** A rectangular garden measures 100 feet by 50 feet. A lawn of uniform width surrounds the garden and has an area of 1,600 square feet. Find the width of the lawn that surrounds the garden.

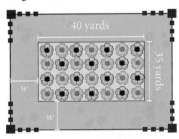

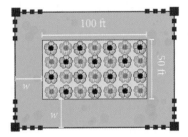

97. **Surrounding Area** A pool that is 75 feet long and 35 feet wide is surrounded by a cement path. The area of the path and the pool combined is 2,975 square feet. What is the area of the path?

98. **Surrounding Area** A pool that is 100 feet long and 50 feet wide is surrounded by a cement path. The area of the path and the pool combined is 8,400 square feet. How wide is the path?

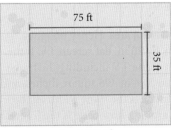

A = 2,975 ft^2

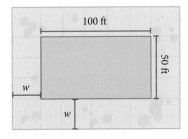

99. Surrounding Area A standard tennis court is 78 feet by 36 feet. There is a path that surrounds the court as shown in the diagram. If the total area of the court and path combined is 7,560 square feet, how wide are the sides of the path? (Solve for x.)

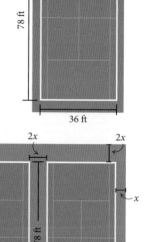

100. Surrounding Area

Two standard tennis courts, 78 feet by 36 feet, are side by side with paths surrounding each court as shown. The total area of the courts and paths combined is 15,120 square feet. How wide are the side paths? (Solve for x.)

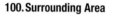

101. Surrounding Area When you open a book, you'll notice the type doesn't touch the edges of the page. The text block is surrounded by margins on all four sides. In the book pictured below, the pages are each 10 inches by 7 inches and the margins are as shown. If there is 23.25 square inches of total margin, how wide is the left margin? (Solve for x.)

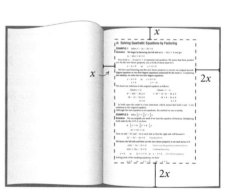

102. Surrounding Area Henry bought a 16 in. by 20 in. flat screen TV. If the front of the TV including the frame has an area of 396 square inches, and the frame is the same width all the way around, how wide is the frame?

The formula $h = vt - 16t^2$ gives the height h, in feet, of an object projected into the air with an initial vertical velocity v, in feet per second, after t seconds.

103. **Projectile Motion** If an object is projected upward with an initial velocity of 48 feet per second, at what times will it reach a height of 32 feet above the ground?

104. **Projectile Motion** If an object is projected upward into the air with an initial velocity of 80 feet per second, at what times will it reach a height of 64 feet above the ground?

105. **Projectile Motion** An object is projected into the air with a vertical velocity of 24 feet per second. At what times will the object be on the ground? (It is on the ground when h is 0.)

106. **Projectile Motion** An object is projected into the air with a vertical velocity of 20 feet per second. At what times will the object be on the ground?

107. **Height of a Bullet** A bullet is fired into the air with an initial upward velocity of 80 feet per second from the top of a building 96 feet high. The equation that gives the height of the bullet at any time t is $h = 96 + 80t - 16t^2$. At what times will the bullet be 192 feet in the air?

108. **Height of an Arrow** An arrow is shot into the air with an upward velocity of 48 feet per second from a hill 32 feet high. The equation that gives the height of the arrow at any time t is $h = 32 + 48t - 16t^2$. Find the times at which the arrow will be 64 feet above the ground.

Learning Objectives Assessment

The following problems can be used to help assess if you have successfully met the learning objectives for this section.

109. Solve: $x^2 - x - 12 = 0$.

    a. $-3, 4$ **b.** $3, -4$ **c.** $2, -6$ **d.** $-2, 6$

110. In solving $(x + 5)(x - 4) = 10$, which of the following is a valid step?

    a. $x + 5 = 10$ or $x - 4 = 10$ **b.** $x + 5 = 0$ or $x - 4 = 0$

    c. $x^2 + x - 20 = 10$ **d.** $x^2 - 20 = 10$

111. Solve: $2x^3 + x^2 - 18x - 9 = 0$.

    a. $-\dfrac{1}{2}$ **b.** $-3, \dfrac{1}{2}$ **c.** $-3, -\dfrac{1}{2}, 3$ **d.** $3, \dfrac{1}{2}$

112. **Projectile Motion** If an object is thrown straight up into the air with an initial velocity of 32 feet per second, then its height above the ground at any time t is given by the formula $h = 32t - 16t^2$. Find the times at which the object is on the ground by letting $h = 0$ in the equation and solving for t.

    a. 2 seconds   **b.** 0 and 2 seconds

    c. 0 and 4 seconds   **d.** 4 seconds

Maintaining Your Skills

Solve each system.

113. $2x - 5y = -8$
$3x + y = 5$

114. $4x - 7y = -2$
$-5x + 6y = -3$

115. $\dfrac{1}{3}x - \dfrac{1}{6}y = 3$

$-\dfrac{1}{5}x + \dfrac{1}{4}y = 0$

116. $x - 5y = 16$
$y = 3x + 8$

Graph the solution set for each system.

117. $3x + 2y < 6$
$-2x + 3y < 6$

118. $y \le x + 3$
$y > x - 4$

119. $x \le 4$
$y < 2$

120. $2x + y < 4$
$x \ge 0$
$y \ge 0$

Chapter 5 Summary

EXAMPLES

Addition of Polynomials [5.1]

1. $(3x^2 + 2x - 5) + (4x^2 - 7x + 2)$

$= 7x^2 - 5x - 3$

To add two polynomials, simply combine the coefficients of similar terms.

Negative Signs Preceding Parentheses [5.1]

2. $-(2x^2 - 8x - 9)$
$= -2x^2 + 8x + 9$

If there is a negative sign directly preceding the parentheses surrounding a polynomial, we may remove the parentheses and preceding negative sign by changing the sign of each term within the parentheses. (This procedure is actually just another application of the distributive property.)

Multiplication of Polynomials [5.2]

3. $(3x - 5)(x + 2)$
$= 3x^2 + 6x - 5x - 10$
$= 3x^2 + x - 10$

To multiply two polynomials, multiply each term in the first by each term in the second.

Special Products [5.2]

4. The following are examples of the three special products.
$(x + 3)^2 = x^2 + 6x + 9$
$(5 - x)^2 = 25 - 10x + x^2$
$(x + 7)(x - 7) = x^2 - 49$

$$\left.\begin{array}{l} (a + b)^2 = a^2 + 2ab + b^2 \\[2mm] (a - b)^2 = a^2 - 2ab + b^2 \end{array}\right\} \text{Binomial squares}$$

$(a + b)(a - b) = a^2 - b^2$ \qquad Difference of two squares

Business Applications [5.1, 5.2]

5. A company makes x items each week and sells them for p dollars each, according to the equation $p = 35 - 0.1x$. Then, the revenue is $R = x(35 - 0.1x) = 35x - 0.1x^2$ If the total cost to make all x items is $C = 8x + 500$, then the profit gained by selling the x items is

$P(x) = 35x - 0.1x^2 - (8x + 500)$

$= -500 + 27x - 0.1x^2$

If a company manufacturers and sells x items at p dollars per item, then the revenue R is given by the formula

$$R(x) = xp$$

If the total cost to manufacture all x items is $C(x)$, then the profit, $P(x)$, obtained from selling all x items is

$$P(x) = R(x) - C(x)$$

Dividing a Polynomial by a Monomial [5.3]

6. $\dfrac{15x^3 - 20x^2 + 10x}{5x}$

$\quad = 3x^2 - 4x + 2$

To divide a polynomial by a monomial, divide each term of the polynomial by the monomial.

Long Division with Polynomials [5.3]

7.
$$\begin{array}{r} x - 2 \\ x - 3\overline{)\,x^2 - 5x + 8} \\ \cancel{+}x^2 \,\cancel{+}\, 3x \\ \hline -2x + 8 \\ \cancel{+}\,2x \,\cancel{+}\, 6 \\ \hline 2 \end{array}$$

If division with polynomials cannot be accomplished by dividing out factors common to the numerator and denominator, then we use a process similar to long division with whole numbers. The steps in the process are estimate, multiply, subtract, and bring down the next term.

Synthetic Division [5.3]

8.

$$\begin{array}{r} 3\,|\ \ 1 \quad -5 \quad \ \ 8 \\ \downarrow \qquad 3 \quad -6 \\ \hline \ \ 1 \quad -2 \quad \boxed{2} \end{array}$$

$$\dfrac{x^2 - 5x + 8}{x - 3} = x - 2 + \dfrac{2}{x - 3}$$

Synthetic division is a short form of long division with polynomials. We can use synthetic division when the divisor is of the form $x + k$, where k is a constant. The process is just a series of multiplications and additions.

Greatest Common Factor [5.4]

9. The greatest common factor of $10x^5 - 15x^4 + 30x^3$ is $5x^3$. Factoring it out of each term, we have

$$5x^3(2x^2 - 3x + 6)$$

The greatest common factor of a polynomial is the largest monomial (the monomial with the largest coefficient and highest exponent) that divides each term of the polynomial. The first step in factoring a polynomial is to factor the greatest common factor (if it is other than 1) out of each term.

Factoring Trinomials [5.5]

10. $x^2 + 5x + 6 = (x + 2)(x + 3)$

$\quad x^2 - 5x + 6 = (x - 2)(x - 3)$

$\quad x^2 + x - 6 = (x - 2)(x + 3)$

$\quad x^2 - x - 6 = (x + 2)(x - 3)$

We factor a trinomial by writing it as the product of two binomials. (This refers to trinomials whose greatest common factor is 1.) Each factorable trinomial has a unique set of factors. Finding the factors is sometimes a matter of trial and error.

Special Factoring [5.6]

11. Here are some binomials that have been factored this way:

$\quad x^2 + 6x + 9 = (x + 3)^2$

$\quad x^2 - 6x + 9 = (x - 3)^2$

$\quad x^2 - 9 = (x + 3)(x - 3)$

$\quad x^3 - 27 = (x - 3)(x^2 + 3x + 9)$

$\quad x^3 + 27 = (x + 3)(x^2 - 3x + 9)$

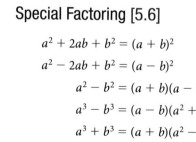

$$a^2 + 2ab + b^2 = (a + b)^2$$
$$a^2 - 2ab + b^2 = (a - b)^2$$

Perfect square trinomials

$$a^2 - b^2 = (a + b)(a - b) \qquad \text{Difference of two squares}$$
$$a^3 - b^3 = (a - b)(a^2 + ab + b^2) \qquad \text{Difference of two cubes}$$
$$a^3 + b^3 = (a + b)(a^2 - ab + b^2) \qquad \text{Sum of two cubes}$$

A sum of two squares does not result in the product of two binomials; that is, there is no binomial factorization for a sum of two squares.

To Factor Polynomials in General [5.7]

12. Factor completely.

   a. $3x^3 - 6x^2 = 3x^2(x - 2)$

   b. $x^2 - 9 = (x + 3)(x - 3)$
     $x^3 - 8 = (x - 2)(x^2 + 2x + 4)$
     $x^3 + 1 = (x + 1)(x^2 - x + 1)$

   c. $x^2 - 6x + 9 = (x - 3)^2$
     $6x^2 - 7x - 5 = (2x + 1)(3x - 5)$

   d. $x^2 + ax + bx + ab$
     $= x(x + a) + b(x + a)$
     $= (x + a)(x + b)$

Step 1: If the polynomial has a greatest common factor other than 1, then factor out the greatest common factor.

Step 2: If the polynomial has two terms (it is a binomial), then see if it is the difference of two squares, or the sum or difference of two cubes, and then factor accordingly. Remember, if it is the sum of two squares it will not factor.

Step 3: If the polynomial has three terms (a trinomial), then it is either a perfect square trinomial, which will factor into the square of a binomial, or it is not a perfect square trinomial, in which case you use one of the methods developed in Section 5.5.

Step 4: If the polynomial has more than three terms, then try to factor it by grouping.

Step 5: As a final check, see if any of the factors you have written can be factored further. If you have overlooked a common factor, you can catch it here.

To Solve an Equation by Factoring [5.8]

13. Solve $x^2 - 5x = -6$.

     $x^2 - 5x + 6 = 0$

    $(x - 3)(x - 2) = 0$

  $x - 3 = 0$ or $x - 2 = 0$

    $x = 3$ $x = 2$

Step 1: Write the equation in standard form.

Step 2: Factor the left side.

Step 3: Use the zero-factor property to set each factor equal to zero.

Step 4: Solve the resulting linear equations.

> ⚠ **COMMON MISTAKES**
>
> When we subtract one polynomial from another, it is common to forget to add the opposite of each term in the second polynomial. For example
>
> $$(6x - 5) - (3x + 4) = 6x - 5 - 3x + 4 \qquad \text{Mistake}$$
>
> This mistake occurs if the negative sign outside the second set of parentheses is not distributed over all terms inside the parentheses. To avoid this mistake, remember: the opposite of a sum is the sum of the opposites, or,
>
> $$-(3x + 4) = -3x + (-4)$$

Chapter 5 Test

Simplify the following expressions. [5.1]

1. $\left(\dfrac{6}{5}x^3 - 2x - \dfrac{3}{5}\right) - \left(\dfrac{6}{5}x^2 - \dfrac{2}{5}x + \dfrac{3}{5}\right)$

2. $5 - 7[9(2x + 1) - 16x]$

Profit, Revenue, and Cost A company making ceramic coffee cups finds that it can sell x cups per week at p dollars each, according to the formula $p(x) = 36 - 0.3x$. If the total cost to produce and sell x coffee cups is $C(x) = 4x + 50$, find the following. [5.1, 5.2]

3. An equation for the revenue that gives the revenue in terms of x

4. The profit equation

5. The revenue brought in by selling 100 coffee cups

6. The cost of producing 100 coffee cups

7. The profit obtained by making and selling 100 coffee cups

Multiply. [5.2]

8. $(x + 7)(-5x + 4)$

9. $(3x - 2)(2x^2 + 6x - 5)$

10. $(3a^4 - 7)^2$

11. $(2x + 3)(2x - 3)$

12. $x(x - 7)(3x + 4)$

13. $\left(2x - \dfrac{1}{7}\right)\left(7x + \dfrac{1}{2}\right)$

Divide. [5.3]

14. $\dfrac{24x^3y + 12x^2y^2 - 16xy^3}{4xy}$

15. $\dfrac{2x^3 - 9x^2 + 10}{2x - 1}$

Divide using synthetic division. [5.3]

16. $\dfrac{x^3 + 2x^2 - 25x - 50}{x + 5}$

17. $\dfrac{y^3 + 16}{y - 1}$

Factor the following expressions. [5.4, 5.5, 5.6, 5.7]

18. $x^2 - 6x + 5$

19. $15x^4 + 33x^2 - 36$

20. $81x^4 - 16y^4$

21. $6ax - ay + 18b^2x - 3b^2y$

22. $y^3 - \dfrac{1}{27}$

23. $3x^4y^4 + 15x^3y^5 - 72x^2y^6$

24. $a^2 - 2ab - 36 + b^2$

25. $16 - x^4$

Solve each equation. [5.8]

26. $\dfrac{1}{4}x^2 = -\dfrac{21}{8}x - \dfrac{5}{4}$

27. $243x^3 = 81x^4$

28. $(x + 5)(x - 2) = 8$

29. $x^3 + 5x^2 - 9x - 45 = 0$

Let $f(x) = x^2 - 2x - 15$. Find all values for the variable x for which $f(x) = g(x)$. [5.8]

30. $g(x) = 0$

31. $g(x) = 5 - 3x$

32. Area Find the value of the variable x in the following figure. [5.8]

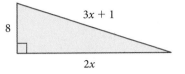

33. Area The area of the figure below is 12 square inches. Find the value of x. [5.8]

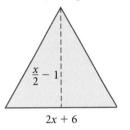

34. Area The area of the figure below is 12 square centimeters. Find the value of x. [5.8]

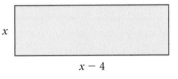

35. Projectile Motion An object is projected into the air with an initial velocity of 64 feet per second. Its height at any time t is given by the formula $h = 64t - 16t^2$. Find the times at which the object is on the ground. [5.8]

6

Rational Expressions and Rational Functions

Chapter Outline

iStockphoto.com © furabolo

I f you have ever put yourself on a weight loss diet, you know that you lose more weight at the beginning of the diet than you do later. If we let $W(x)$ represent a person's weight after x weeks on the diet, then the rational function

$$W(x) = \frac{80(2x + 15)}{x + 6}$$

is a mathematical model of the person's weekly progress on a diet intended to take them from 200 pounds to about 160 pounds. Rational functions are good models for quantities that fall off rapidly to begin with, and then level off over time. The table shows some values for this function, along with the graph of this function.

Weekly Weight Loss

Weeks Since Starting Diet	Weight (Nearest Pound)
0	200
4	184
8	177
12	173
16	171
20	169
24	168

As you progress through this chapter, you will acquire an intuitive feel for these types of functions, and as a result, you will see why they are good models for situations such as dieting.

© clu/iStockPhoto

Success Skills

This is the last chapter in which we will mention study skills. You know by now what works best for you and what you have to do to achieve your goals for this course. From now on, it is simply a matter of sticking with the things that work for you and avoiding the things that do not. It seems simple, but as with anything that takes effort, it is up to you to see that you maintain the skills that get you where you want to be in the course.

If you intend to take more classes in mathematics and want to ensure your success in those classes, then you can work toward this goal: ***Become the type of student who can learn mathematics on his or her own.*** Most people who have degrees in mathematics were students who could learn mathematics on their own. This doesn't mean that you must learn it all on your own, or that you study alone, or that you don't ask questions. It means that you know your resources, both internal and external, and you can count on those resources when you need them. Attaining this goal gives you independence and puts you in control of your success in any math class you take.

Learning Objectives

In this section, we will learn how to:

1. Evaluate a rational expression.
2. Determine when a rational expression is undefined.
3. Reduce a rational expression to lowest terms.
4. Reduce a rational expression containing factors that are opposites.

Introduction

We will begin this section with the definition of a rational expression. Recall from Chapter 1 that a **rational number** is any number that can be expressed as the ratio of two integers:

$$\text{Rational numbers} = \left\{ \frac{a}{b} \mid a \text{ and } b \text{ are integers, } b \neq 0 \right\}$$

We define a rational expression in a similar fashion.

(def) **DEFINITION** *rational expression*

A **rational expression** is any expression that can be written in the form

$$\frac{P}{Q}$$

where P and Q are polynomials and $Q \neq 0$.

Some examples of rational expressions are

$$\frac{2x - 3}{x + 5} \qquad \frac{x^2 - 5x - 6}{x^2 - 1} \qquad \frac{a - b}{b - a}$$

Evaluating Rational Expressions

To *evaluate* a rational expression means to find its value when any variables in the expression are replaced by specific numbers. We simply substitute the given values for each variable into the expression and then simplify following the order of operations.

VIDEO EXAMPLES

SECTION 6.1

EXAMPLE 1 Evaluate $\dfrac{2x - 3}{x + 5}$ if $x = 8$ and if $x = -5$.

SOLUTION If $x = 8$, we have

$$\frac{2(8) - 3}{8 + 5} = \frac{13}{13} = 1$$

If $x = -5$, we have

$$\frac{2(-5) - 3}{-5 + 5} = -\frac{13}{0}$$

which is undefined.

Notice in Example 1 that the value $x = -5$ caused the denominator of the expression to be zero, making the rational expression undefined. As we will see throughout this chapter, being aware of any such values of the variable can be important when working with rational expressions.

Determining when Rational Expressions are Undefined

A rational expression will be undefined for any value of the variable that makes the denominator equal to zero. We can find these values by setting the denominator equal to zero and then solving the resulting equation.

EXAMPLE 2   Determine any values of the variable for which

$$\frac{x^2 - 5x - 6}{x^2 - 1}$$

is undefined.

Note Technically, when $x = -1$, the expression $\frac{x^2 - 5x - 6}{x^2 - 1}$ becomes $\frac{0}{0}$, which is considered an *indeterminant form*. For our purposes, however, we will simply say the expression is undefined.

SOLUTION   We set the denominator equal to zero and solve for x.

$$x^2 - 1 = 0$$

$$(x + 1)(x - 1) = 0$$

$$x + 1 = 0 \quad \text{or} \quad x - 1 = 0$$

$$x = -1 \qquad\qquad x = 1$$

Therefore, the expression is undefined for $x = \pm 1$.

EXAMPLE 3   Determine any values of the variable for which

$$\frac{x - 3}{x^2 + 2}$$

is undefined.

SOLUTION   In order for the expression to be undefined, we need

$$x^2 + 2 = 0$$

which implies

$$x^2 = -2$$

But x^2 will always be positive or zero for any real number x, so this equation has no real solutions. Therefore, the rational expression is defined for all real numbers.

From this point forward, we assume any problems involving rational expressions are limited to values of the variable for which the expression is defined (meaning values that result in a nonzero denominator).

Basic Property

For rational expressions, multiplying the numerator and denominator by the same nonzero expression may change the form of the rational expression, but it will always produce an expression equivalent to the original one.

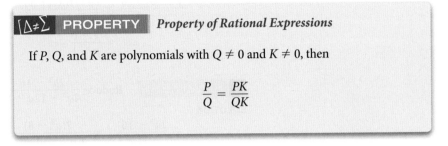

> $[\Delta \neq \Sigma]$ **PROPERTY**   *Property of Rational Expressions*
>
> If P, Q, and K are polynomials with $Q \neq 0$ and $K \neq 0$, then
>
> $$\frac{P}{Q} = \frac{PK}{QK}$$

This property implies that the reverse is also true, meaning that we can divide both the numerator and denominator by a common factor K to obtain an equivalent rational expression. Using the property in this way allows us to reduce a rational expression to lowest terms.

Reducing to Lowest Terms

The fraction $\frac{6}{8}$ can be reduced because both 6 and 8 have a common factor of 2. The process is shown here:

$$\frac{6}{8} = \frac{3 \cdot \cancel{2}}{4 \cdot \cancel{2}} = \frac{3}{4}$$

Reducing $\frac{6}{8}$ to $\frac{3}{4}$ involves dividing the numerator and denominator by 2, which we have indicated by drawing green lines through the common factor. Because 3 and 4 have no common factor other than 1, we say that the fraction $\frac{3}{4}$ is expressed in lowest terms.

In like manner, we reduce rational expressions to lowest terms by first factoring the numerator and denominator, and then dividing both numerator and denominator by any factors they have in common.

EXAMPLE 4   Reduce $\dfrac{x^2 - 9}{x - 3}$ to lowest terms.

SOLUTION   Factoring, we have

$$\frac{x^2 - 9}{x - 3} = \frac{(x + 3)(x - 3)}{x - 3}$$

The numerator and denominator have the factor $x - 3$ in common. Dividing the numerator and denominator by $x - 3$, we have

$$\frac{(x + 3)\cancel{(x - 3)}}{\cancel{x - 3}} = \frac{x + 3}{1} = x + 3 \qquad ∎$$

Note   In Example 4, the original rational expression is undefined if $x = 3$. This is still true even though the expression can be reduced. Thus, when we state the relationship

$$\frac{x^2 - 9}{x - 3} = x + 3$$

we are assuming that it is true for all values of x except $x = 3$.

In the previous example, you may be tempted to divide the x into the x^2, or the 3 into the 9. It is very important to remember that we are only allowed to divide by factors that are common to both the numerator and denominator. Factors are connected to one another by multiplication, not addition or subtraction. The x and 3 in $x - 3$ are terms, not factors, of the denominator and therefore cannot be reduced individually.

Here are some other examples of reducing rational expressions to lowest terms.

EXAMPLE 5   Reduce $\dfrac{y^2 - 5y - 6}{y^2 - 1}$ to lowest terms.

SOLUTION

$$\frac{y^2 - 5y - 6}{y^2 - 1} = \frac{(y - 6)(y + 1)}{(y - 1)(y + 1)}$$

$$= \frac{y - 6}{y - 1}$$

EXAMPLE 6   Reduce $\dfrac{2a^3 - 16}{4a^2 - 12a + 8}$ to lowest terms.

SOLUTION

$$\frac{2a^3 - 16}{4a^2 - 12a + 8} = \frac{2(a^3 - 8)}{4(a^2 - 3a + 2)}$$

$$= \frac{2(a - 2)(a^2 + 2a + 4)}{2 \cdot 2(a - 2)(a - 1)}$$

$$= \frac{a^2 + 2a + 4}{2(a - 1)}$$

EXAMPLE 7   Reduce $\dfrac{x^2 - 3x + ax - 3a}{x^2 - ax - 3x + 3a}$ to lowest terms.

SOLUTION

$$\frac{x^2 - 3x + ax - 3a}{x^2 - ax - 3x + 3a} = \frac{x(x - 3) + a(x - 3)}{x(x - a) - 3(x - a)}$$

$$= \frac{(x - 3)(x + a)}{(x - a)(x - 3)}$$

$$= \frac{x + a}{x - a}$$

The answer to Example 7 cannot be reduced further. It is a fairly common mistake to attempt to divide out an x or an a in this last expression. Remember, we can divide out only the factors common to the numerator and denominator of a rational expression.

In these last examples we consider some situations where the rational expression can be reduced, even though it may not appear so at first.

EXAMPLE 8   Reduce to lowest terms: $\dfrac{x + 2}{2 + x}$.

SOLUTION   Because addition is commutative, $x + 2 = 2 + x$. Therefore,

$$\frac{x + 2}{2 + x} = \frac{x + 2}{x + 2}$$

$$= 1$$

EXAMPLE 9   Reduce to lowest terms: $\dfrac{a - b}{b - a}$.

SOLUTION   In this case, $a - b \neq b - a$. However, there is a relationship between $a - b$ and $b - a$ in that they are opposites. We can show this fact by factoring -1 from each term in the numerator:

$$\frac{a - b}{b - a} = \frac{-b + a}{b - a} \qquad \text{Reverse the order of the terms in the numerator}$$

$$= \frac{-1(b - a)}{b - a} \qquad \text{Factor } -1 \text{ from each term in the numerator}$$

$$= -1 \qquad \text{Divide out common factor } b - a$$

EXAMPLE 10   Reduce to lowest terms: $\dfrac{x^2 - 25}{5 - x}$.

SOLUTION   We begin by factoring the numerator:

$$\frac{x^2 - 25}{5 - x} = \frac{(x - 5)(x + 5)}{5 - x}$$

The factors $x - 5$ and $5 - x$ are once again opposites. We can reverse the order of either by factoring -1 from it.
   That is, $5 - x = -x + 5 = -1(x - 5)$.

$$\frac{(x - 5)(x + 5)}{5 - x} = \frac{(x - 5)(x + 5)}{-1(x - 5)}$$

$$= \frac{x + 5}{-1}$$

$$= -(x + 5)$$

$$= -x - 5$$

Dividing a Polynomial by a Polynomial

EXAMPLE 11   Divide: $\dfrac{x^2 - 6xy - 7y^2}{x + y}$.

SOLUTION   In the previous chapter we used long division to work this problem. As an alternative, we can factor the numerator and perform the division by simply reducing the rational expression to lowest terms.

$$\frac{x^2 - 6xy - 7y^2}{x + y} = \frac{(x + y)(x - 7y)}{x + y}$$

$$= x - 7y$$

Getting Ready for Class

After reading through the preceding section, respond in your own words and in complete sentences.

A. What is a rational expression?

B. When is a rational expression undefined?

C. Explain how to determine if a rational expression is in "lowest terms."

D. Explain why the answer to Example 5 cannot be reduced further by dividing out a y.

Evaluate each rational expression if $x = -2$ and if $x = 3$.

1. $\dfrac{x - 3}{x + 1}$

2. $\dfrac{x + 2}{x + 3}$

3. $\dfrac{x}{x^2 - 5}$

4. $\dfrac{x^2}{x^2 + 4}$

Evaluate each rational expression if $x = 2$ and $y = -1$.

5. $\dfrac{x^2 - xy}{y - 2x}$

6. $\dfrac{x + 5y}{x^2 - 3xy - 4y^2}$

Determine any values of the variable for which the rational expression is undefined.

7. $\dfrac{x - 4}{x + 2}$

8. $\dfrac{x}{x - 1}$

9. $\dfrac{x + 5}{3x}$

10. $\dfrac{x - 6}{x^2}$

11. $\dfrac{2x}{x^2 - 4}$

12. $\dfrac{x + 1}{x^2 - 9}$

13. $\dfrac{x - 3}{x^2 - 3x - 10}$

14. $\dfrac{2x + 2}{x^2 - x - 12}$

15. $\dfrac{x^2 - x}{x^2 - 3x + 2}$

16. $\dfrac{x^2 - 4}{x^2 + 2x}$

17. $\dfrac{5x}{x^2 + 1}$

18. $\dfrac{x - 2}{x^2 + 3}$

Reduce each rational expression to lowest terms.

19. $\dfrac{x^2 - 16}{6x + 24}$

20. $\dfrac{12x - 9y}{3x^2 + 3xy}$

21. $\dfrac{a^4 - 81}{a - 3}$

22. $\dfrac{a^2 - 4a - 12}{a^2 + 8a + 12}$

23. $\dfrac{20y^2 - 45}{10y^2 - 5y - 15}$

24. $\dfrac{20x^2 - 93x + 34}{4x^2 - 9x - 34}$

25. $\dfrac{12y - 2xy - 2x^2y}{6y - 4xy - 2x^2y}$

26. $\dfrac{250a + 100ax + 10ax^2}{50a - 2ax^2}$

27. $\dfrac{(x - 3)^2(x + 2)}{(x + 2)^2(x - 3)}$

28. $\dfrac{(x - 4)^3(x + 3)}{(x + 3)^2(x - 4)}$

29. $\dfrac{x^3 + 1}{x^2 - 1}$

30. $\dfrac{x^3 - 1}{x^2 - 1}$

31. $\dfrac{4am - 4an}{3n - 3m}$

32. $\dfrac{ad - ad^2}{d - 1}$

33. $\dfrac{ab - a + b - 1}{ab + a + b + 1}$

34. $\dfrac{6cd - 4c - 9d + 6}{6d^2 - 13d + 6}$

35. $\dfrac{21x^2 - 23x + 6}{21x^2 + x - 10}$

36. $\dfrac{36x^2 - 11x - 12}{20x^2 - 39x + 18}$

37. $\dfrac{8x^2 - 6x - 9}{8x^2 - 18x + 9}$

38. $\dfrac{42x^2 + 23x - 10}{14x^2 + 45x - 14}$

39. $\dfrac{4x^2 + 29x + 45}{8x^2 - 10x - 63}$

40. $\dfrac{30x^2 - 61x + 30}{60x^2 + 22x - 60}$

41. $\dfrac{a^3 + b^3}{a^2 - b^2}$

42. $\dfrac{a^2 - b^2}{a^3 - b^3}$

43. $\dfrac{8x^4 - 8x}{4x^4 + 4x^3 + 4x^2}$

44. $\dfrac{6x^5 - 48x^3}{12x^3 + 24x^2 + 48x}$

45. $\dfrac{ax + 2x + 3a + 6}{ay + 2y - 4a - 8}$

46. $\dfrac{x^2 - 3ax - 2x + 6a}{x^2 - 3ax + 2x - 6a}$

47. $\dfrac{x^3 + 3x^2 - 4x - 12}{x^2 + x - 6}$

48. $\dfrac{x^3 + 5x^2 - 4x - 20}{x^2 + 7x + 10}$

49. $\dfrac{x^3 - 8}{x^2 - 4}$

50. $\dfrac{y^2 - 9}{y^3 + 27}$

51. $\dfrac{8x^3 - 27}{4x^2 - 9}$

52. $\dfrac{25y^2 - 4}{125y^3 + 8}$

Reduce the rational expression to lowest terms, if possible.

53. $\dfrac{x + 7}{7 + x}$

54. $\dfrac{2x + 5}{5x + 2}$

55. $\dfrac{x + 3y}{x - 3y}$

56. $\dfrac{a + b}{4b + 4a}$

Refer to Examples 9 and 10 in this section, and reduce the following to lowest terms.

57. $\dfrac{x - 4}{4 - x}$

58. $\dfrac{6 - x}{x - 6}$

59. $\dfrac{y^2 - 36}{6 - y}$

60. $\dfrac{1 - y}{y^2 - 1}$

61. $\dfrac{1 - 9a^2}{9a^2 - 6a + 1}$

62. $\dfrac{1 - a^2}{a^2 - 2a + 1}$

63. $\dfrac{x^2 - 7x + 12}{6 + x - x^2}$

64. $\dfrac{x^2 - 16}{4x - x^2}$

Simplify each expression.

65. $\dfrac{(3x - 5) - (3a - 5)}{x - a}$

66. $\dfrac{(2x + 3) - (2a + 3)}{x - a}$

67. $\dfrac{(x^2 - 4) - (a^2 - 4)}{x - a}$

68. $\dfrac{(x^2 - 1) - (a^2 - 1)}{x - a}$

The following problems are similar to those you solved in the previous chapter. In this case, perform each division by reducing the expression to lowest terms.

69. $\dfrac{x^2 - x - 6}{x - 3}$

70. $\dfrac{x^2 - x - 6}{x + 2}$

71. $\dfrac{2a^3 - 3a - 9}{2a + 3}$

72. $\dfrac{5x^2 - 14xy - 24y^2}{x - 4y}$

73. $\dfrac{x^3 + 8}{x + 2}$

74. $\dfrac{x^3 + 2x^2 - 25x - 50}{x - 5}$

75. **Paying Attention to Instructions** Work each problem according to the instructions given.

    a. Evaluate $\dfrac{2x - 2}{x^2 - x}$ if $x = 3$.

    b. Reduce $\dfrac{2x - 2}{x^2 - x}$ to lowest terms.

    c. Determine any values of x for which $\dfrac{2x - 2}{x^2 - x}$ is undefined.

    d. Find the value of $\dfrac{2x - 2}{x^2 - x}$ when $x = -1$.

76. **Paying Attention to Instructions** Work each problem according to the instructions given.

    a. Determine any values of x for which $\dfrac{x^2 - 3x - 4}{4 - x}$ is undefined.

    b. Evaluate $\dfrac{x^2 - 3x - 4}{4 - x}$ if $x = -4$.

    c. Reduce $\dfrac{x^2 - 3x - 4}{4 - x}$ to lowest terms.

    d. Divide $\dfrac{x^2 - 3x - 4}{4 - x}$ using long division.

Learning Objectives Assessment

The following problems can be used to help assess if you have successfully met the learning objectives for this section.

77. Evaluate $\dfrac{x^2 - 4}{x^2 - 2x}$ if $x = 1$.

    a. $-\dfrac{1}{3}$ b. $\dfrac{1}{3}$ c. 3 d. -3

78. For which value of x is $\dfrac{x^2 - 4}{x^2 - 2x}$ undefined?

    a. -2 b. 4 c. 0 d. -1

79. Reduce $\dfrac{x^2 - 4}{x^2 - 2x}$ to lowest terms.

    a. $\dfrac{2}{x}$ b. $\dfrac{x + 2}{x - 2}$ c. $\dfrac{x - 2}{x}$ d. $\dfrac{x + 2}{x}$

80. Which expression reduces to -1?

    a. $\dfrac{3 - x}{x + 3}$ b. $\dfrac{3 - x}{x - 3}$ c. $\dfrac{3 + x}{x + 3}$ d. $\dfrac{x + 3}{x - 3}$

Getting Ready for the Next Section

Multiply or divide, as indicated.

81. $\dfrac{6}{7} \cdot \dfrac{14}{18}$

82. $\dfrac{6}{8} \div \dfrac{3}{5}$

83. $5y^2 \cdot 4x^2$

84. $4y^3 \cdot 3x^2$

85. $9x^4 \cdot 8y^5$

86. $6x^4 \cdot 12y^5$

Factor.

87. $x^2 - 4$

88. $x^2 - 6x + 9$

89. $x^3 - x^2y$

90. $a^2 - 5a + 6$

91. $2y^2 - 2$

92. $xa + xb + ya + yb$

Multiplication and Division of Rational Expressions

6.2

Learning Objectives

In this section, we will learn how to:

1. Multiply rational expressions.

2. Divide rational expressions.

Introduction

In Section 6.1, we found the process of reducing rational expressions to lowest terms to be the same process used in reducing fractions to lowest terms. The similarity also holds for the process of multiplication or division of rational expressions.

Multiplication with fractions is the simplest of the four basic operations. To multiply two fractions, we simply multiply numerators and multiply denominators. That is, if a, b, c, and d are real numbers, with $b \neq 0$ and $d \neq 0$, then

$$\frac{a}{b} \cdot \frac{c}{d} = \frac{ac}{bd}$$

Multiplication with Rational Expressions

VIDEO EXAMPLES

SECTION 6.2

EXAMPLE 1   Multiply: $\dfrac{6}{7} \cdot \dfrac{14}{18}$.

SOLUTION

$$\frac{6}{7} \cdot \frac{14}{18} = \frac{6(14)}{7(18)} \qquad \text{Multiply numerators and denominators}$$

$$= \frac{2 \cdot 3(2 \cdot 7)}{7(2 \cdot 3 \cdot 3)} \qquad \text{Factor}$$

$$= \frac{2}{3} \qquad \text{Divide out common factors}$$

Our next example is similar to Example 1, except that the fractions are now rational expressions. We multiply fractions whose numerators and denominators are monomials by multiplying numerators and multiplying denominators and then reducing to lowest terms. Here is how it looks.

EXAMPLE 2   Multiply: $\dfrac{8x^3}{27y^8} \cdot \dfrac{9y^3}{12x^2}$.

SOLUTION   We multiply numerators and denominators without actually carrying out the multiplication:

$$\frac{8x^3}{27y^8} \cdot \frac{9y^3}{12x^2} = \frac{8 \cdot 9x^3y^3}{27 \cdot 12x^2y^8} \qquad \begin{array}{l}\text{Multiply numerators}\\\text{Multiply denominators}\end{array}$$

$$= \frac{4 \cdot 2 \cdot 9 \cdot x \cdot x^2 \cdot y^3}{9 \cdot 3 \cdot 4 \cdot 3 \cdot x^2 \cdot y^3 \cdot y^5} \qquad \text{Factor coefficients}$$

$$= \frac{2x}{9y^5} \qquad \text{Divide out common factors}$$

The product of two rational expressions is the product of their numerators over the product of their denominators.

Once again, we should mention that the little slashes we have drawn through the factors are simply used to denote the factors we have divided out of the numerator and denominator.

EXAMPLE 3   Multiply: $\dfrac{x-3}{x^2-4} \cdot \dfrac{x+2}{x^2-6x+9}$.

SOLUTION   We begin by multiplying numerators and denominators. We then factor all polynomials and divide out factors common to the numerator and denominator:

$$\dfrac{x-3}{x^2-4} \cdot \dfrac{x+2}{x^2-6x+9} = \dfrac{(x-3)(x+2)}{(x^2-4)(x^2-6x+9)} \qquad \text{Multiply}$$

$$= \dfrac{\cancel{(x-3)}\cancel{(x+2)}}{\cancel{(x+2)}(x-2)\cancel{(x-3)}(x-3)} \qquad \text{Factor}$$

$$= \dfrac{1}{(x-2)(x-3)}$$

The first two steps can be combined to save time. We can perform the multiplication and factoring steps together.

EXAMPLE 4   Multiply: $\dfrac{2y^2-4y}{2y^2-2} \cdot \dfrac{y^2-2y-3}{y^2-5y+6}$.

SOLUTION

$$\dfrac{2y^2-4y}{2y^2-2} \cdot \dfrac{y^2-2y-3}{y^2-5y+6} = \dfrac{2y\cancel{(y-2)}\cancel{(y-3)}\cancel{(y+1)}}{2\cancel{(y+1)}(y-1)\cancel{(y-3)}\cancel{(y-2)}}$$

$$= \dfrac{y}{y-1}$$

Notice in both of the preceding examples that we did not actually multiply the polynomials as we did in Chapter 5. It would be senseless to do that because we would then have to factor each of the resulting products to reduce them to lowest terms.

Division with Rational Expressions

The quotient of two rational expressions is the product of the first and the reciprocal of the second. That is, we find the quotient of two rational expressions the same way we find the quotient of two fractions. Here is an example that reviews division with fractions.

EXAMPLE 5   Divide: $\dfrac{6}{8} \div \dfrac{3}{5}$.

SOLUTION

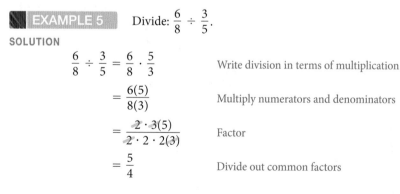

$$\dfrac{6}{8} \div \dfrac{3}{5} = \dfrac{6}{8} \cdot \dfrac{5}{3} \qquad \text{Write division in terms of multiplication}$$

$$= \dfrac{6(5)}{8(3)} \qquad \text{Multiply numerators and denominators}$$

$$= \dfrac{\cancel{2}\cdot 3(5)}{\cancel{2}\cdot 2 \cdot 2(3)} \qquad \text{Factor}$$

$$= \dfrac{5}{4} \qquad \text{Divide out common factors}$$

To divide one rational expression by another, we use the definition of division to multiply by the reciprocal of the expression that follows the division symbol.

EXAMPLE 6 Divide: $\dfrac{8x^3}{5y^2} \div \dfrac{4x^2}{10y^6}$.

SOLUTION First, we rewrite the problem in terms of multiplication. Then we multiply.

$$\frac{8x^3}{5y^2} \div \frac{4x^2}{10y^6} = \frac{8x^3}{5y^2} \cdot \frac{10y^6}{4x^2}$$

$$= \frac{\overset{2}{\cancel{8}} \cdot \overset{2}{\cancel{10}}x^{\overset{x}{\cancel{3}}}y^{\overset{y^4}{\cancel{6}}}}{\cancel{4} \cdot \cancel{5}x^2 y^2}$$

$$= 4xy^4$$

EXAMPLE 7 Divide: $\dfrac{x^2 - y^2}{x^2 - 2xy + y^2} \div \dfrac{x^3 + y^3}{x^3 - x^2 y}$.

SOLUTION We begin by writing the problem as the product of the first and the reciprocal of the second and then proceed as in the previous two examples:

$$\frac{x^2 - y^2}{x^2 - 2xy + y^2} \div \frac{x^3 + y^3}{x^3 - x^2 y}$$ Multiply by the reciprocal of the divisor

$$= \frac{x^2 - y^2}{x^2 - 2xy + y^2} \cdot \frac{x^3 - x^2 y}{x^3 + y^3}$$

$$= \frac{\cancel{(x - y)}(x + y)(x^2)\cancel{(x - y)}}{\cancel{(x - y)}\cancel{(x - y)}\cancel{(x + y)}(x^2 - xy + y^2)}$$ Factor and multiply

$$= \frac{x^2}{x^2 - xy + y^2}$$ Divide out common factors

Here are some more examples of multiplication and division with rational expressions.

EXAMPLE 8 Perform the indicated operations.

$$\frac{a^2 - 8a + 15}{a + 4} \cdot \frac{a + 2}{a^2 - 5a + 6} \div \frac{a^2 - 3a - 10}{a^2 + 2a - 8}$$

SOLUTION First, we rewrite the division as multiplication by the reciprocal. Then we proceed as usual.

$$\frac{a^2 - 8a + 15}{a + 4} \cdot \frac{a + 2}{a^2 - 5a + 6} \div \frac{a^2 - 3a - 10}{a^2 + 2a - 8}$$ Change division to multiplication by the reciprocal

$$= \frac{(a^2 - 8a + 15)(a + 2)(a^2 + 2a - 8)}{(a + 4)(a^2 - 5a + 6)(a^2 - 3a - 10)}$$ Factor

$$= \frac{\cancel{(a - 5)}\cancel{(a - 3)}\cancel{(a + 2)}\cancel{(a + 4)}\cancel{(a - 2)}}{\cancel{(a + 4)}\cancel{(a - 3)}\cancel{(a - 2)}\cancel{(a - 5)}\cancel{(a + 2)}}$$ Divide out common factors

$$= 1$$

Our next example involves factoring by grouping. As you may have noticed, working the problems in this chapter gives you a very detailed review of factoring.

EXAMPLE 9   Multiply: $\dfrac{xa + xb + ya + yb}{xa - xb - ya + yb} \cdot \dfrac{xa + xb - ya - yb}{xa - xb + ya - yb}$.

SOLUTION   We will factor each polynomial by grouping, which takes two steps.

$$\dfrac{xa + xb + ya + yb}{xa - xb - ya + yb} \cdot \dfrac{xa + xb - ya - yb}{xa - xb + ya - yb}$$

$$= \dfrac{x(a + b) + y(a + b)}{x(a - b) - y(a - b)} \cdot \dfrac{x(a + b) - y(a + b)}{x(a - b) + y(a - b)} \qquad \text{Factor by grouping}$$

$$= \dfrac{(a + b)(x + y)(a + b)(x - y)}{(a - b)(x - y)(a - b)(x + y)}$$

$$= \dfrac{(a + b)^2}{(a - b)^2}$$

EXAMPLE 10   Multiply: $(4x^2 - 36) \cdot \dfrac{12}{4x + 12}$.

SOLUTION   We can think of $4x^2 - 36$ as having a denominator of 1. Thinking of it in this way allows us to proceed as we did in the previous examples.

$$(4x^2 - 36) \cdot \dfrac{12}{4x + 12}$$

$$= \dfrac{4x^2 - 36}{1} \cdot \dfrac{12}{4x + 12} \qquad \text{Write } 4x^2 - 36 \text{ with denominator 1}$$

$$= \dfrac{4(x - 3)(x + 3)12}{4(x + 3)} \qquad \text{Factor}$$

$$= 12(x - 3) \qquad \text{Divide out common factors}$$

EXAMPLE 11   Multiply: $3(x - 2)(1 - x) \cdot \dfrac{5}{x^2 - 3x + 2}$.

SOLUTION   This problem is very similar to the problem in Example 10. Writing the first rational expression with a denominator of 1, we have

$$\dfrac{3(x - 2)(1 - x)}{1} \cdot \dfrac{5}{x^2 - 3x + 2} = \dfrac{3(x - 2)(-1)(x - 1)5}{(x - 2)(x - 1)}$$

$$= 3(-1)5$$

$$= -15$$

Notice that we factored -1 out of $1 - x$ because $1 - x$ and $x - 1$ are opposites.

Getting Ready for Class

After reading through the preceding section, respond in your own words and in complete sentences.

A. Summarize the steps used to multiply fractions.

B. What is the first step in multiplying two rational expressions?

C. Why is factoring important when multiplying and dividing rational expressions?

D. How is division with rational expressions different than multiplication of rational expressions?

Perform the indicated operations.

1. $\dfrac{2}{9} \cdot \dfrac{3}{4}$

2. $\dfrac{5}{6} \cdot \dfrac{7}{8}$

3. $\dfrac{3}{4} \div \dfrac{1}{3}$

4. $\dfrac{3}{8} \div \dfrac{5}{4}$

5. $\dfrac{3}{7} \cdot \dfrac{14}{24} \div \dfrac{1}{2}$

6. $\dfrac{6}{5} \cdot \dfrac{10}{36} \div \dfrac{3}{4}$

7. $\dfrac{10x^2}{5y^2} \cdot \dfrac{15y^3}{2x^4}$

8. $\dfrac{8x^3}{7y^4} \cdot \dfrac{14y^6}{16x^2}$

9. $\dfrac{11a^2b}{5ab^2} \div \dfrac{22a^3b^2}{10ab^4}$

10. $\dfrac{8ab^3}{9a^2b} \div \dfrac{16a^2b^2}{18ab^3}$

11. $\dfrac{6x^2}{5y^3} \cdot \dfrac{11z^2}{2x^2} \div \dfrac{33z^5}{10y^8}$

12. $\dfrac{4x^3}{7y^2} \cdot \dfrac{6z^5}{5x^6} \div \dfrac{24z^2}{35x^6}$

Perform the indicated operations. Be sure to write all answers in lowest terms.

13. $\dfrac{x^2 - 9}{x^2 - 4} \cdot \dfrac{x - 2}{x - 3}$

14. $\dfrac{x^2 - 16}{x^2 - 25} \cdot \dfrac{x - 5}{x - 4}$

15. $\dfrac{y^2 - 1}{y + 2} \cdot \dfrac{y^2 + 5y + 6}{y^2 + 2y - 3}$

16. $\dfrac{y - 1}{y^2 - y - 6} \cdot \dfrac{y^2 + 5y + 6}{y^2 - 1}$

17. $\dfrac{3x - 12}{x^2 - 4} \cdot \dfrac{x^2 + 6x + 8}{x - 4}$

18. $\dfrac{x^2 + 5x + 1}{4x - 4} \cdot \dfrac{x - 1}{x^2 + 5x + 1}$

19. $\dfrac{xy}{xy + 1} \div \dfrac{x}{y}$

20. $\dfrac{y}{x} \div \dfrac{xy}{xy - 1}$

21. $\dfrac{1}{x^2 - 9} \div \dfrac{1}{x^2 + 9}$

22. $\dfrac{1}{x^2 - 9} \div \dfrac{1}{(x - 3)^2}$

23. $\dfrac{y - 3}{y^2 - 6y + 9} \cdot \dfrac{y - 3}{4}$

24. $\dfrac{y - 3}{y^2 - 6y + 9} \div \dfrac{y - 3}{4}$

25. $\dfrac{5x + 2y}{25x^2 - 5xy - 6y^2} \cdot \dfrac{20x^2 - 7xy - 3y^2}{4x + y}$

26. $\dfrac{7x + 3y}{42x^2 - 17xy - 15y^2} \cdot \dfrac{12x^2 - 4xy - 5y^2}{2x + y}$

27. $\dfrac{a^2 - 5a + 6}{a^2 - 2a - 3} \div \dfrac{5 - a}{a^2 + 3a + 2}$

28. $\dfrac{a^2 + 7a + 12}{5 - a} \div \dfrac{a^2 + 9a + 18}{a^2 - 7a + 10}$

29. $\dfrac{4t^2 - 1}{6t^2 + t - 2} \div \dfrac{8t^3 + 1}{27t^3 + 8}$

30. $\dfrac{9t^2 - 1}{6t^2 + 7t - 3} \div \dfrac{27t^3 + 1}{8t^3 + 27}$

31. $\dfrac{2x^2 - 5x - 12}{4x^2 + 8x + 3} \div \dfrac{x^2 - 16}{2x^2 + 7x + 3}$

32. $\dfrac{x^2 - 2x + 1}{3x^2 + 7x - 20} \div \dfrac{x^2 + 3x - 4}{3x^2 - 2x - 5}$

33. $\dfrac{2a^2 - 21ab - 36b^2}{a^2 - 11ab - 12b^2} \div \dfrac{10a + 15b}{b^2 - a^2}$

34. $\dfrac{3a^2 + 7ab - 20b^2}{a^2 + 5ab + 4b^2} \div \dfrac{3a^2 - 17ab + 20b^2}{12b - 3a}$

35. $\dfrac{6c^2 - c - 15}{9c^2 - 25} \cdot \dfrac{15c^2 + 22c - 5}{6c^2 + 5c - 6}$ **36.** $\dfrac{m^2 + 4m - 21}{m^2 - 12m + 27} \cdot \dfrac{m^2 - 7m + 12}{m^2 + 3m - 28}$

37. $\dfrac{6a^2b + 2ab^2 - 20b^3}{4a^2b - 16b^3} \cdot \dfrac{10a^2 - 22ab + 4b^2}{27a^3 - 125b^3}$

38. $\dfrac{12a^2b - 3ab^2 - 42b^3}{9a^2 - 36b^2} \cdot \dfrac{6a^2 - 15ab + 6b^2}{8a^3b - b^4}$

39. $\dfrac{360x^3 - 490x}{36x^2 + 84x + 49} \cdot \dfrac{30x^2 + 83x + 56}{150x^3 + 65x^2 - 280x}$

40. $\dfrac{490x^2 - 640}{49x^2 - 112x + 64} \cdot \dfrac{28x^2 - 95x + 72}{56x^3 - 62x^2 - 144x}$

41. $\dfrac{x^5 - x^2}{5x^2 - 5x} \cdot \dfrac{10x^4 - 10x^2}{2x^4 + 2x^3 + 2x^2}$ **42.** $\dfrac{2x^4 - 16x}{3x^6 - 48x^2} \cdot \dfrac{6x^5 + 24x^3}{4x^4 + 8x^3 + 16x^2}$

43. $\dfrac{a^2 - 16b^2}{a^2 - 8ab + 16b^2} \cdot \dfrac{a^2 - 9ab + 20b^2}{a^2 - 7ab + 12b^2} \div \dfrac{a^2 - 25b^2}{a^2 - 6ab + 9b^2}$

44. $\dfrac{a^2 - 6ab + 9b^2}{a^2 - 4b^2} \cdot \dfrac{a^2 - 5ab + 6b^2}{(a - 3b)^2} \div \dfrac{a^2 - 9b^2}{a^2 - ab - 6b^2}$

45. $\dfrac{2y^2 - 7y - 15}{42y^2 - 29y - 5} \cdot \dfrac{12y^2 - 16y + 5}{7y^2 - 36y + 5} \div \dfrac{9 - 4y^2}{49y^2 - 1}$

46. $\dfrac{8y^2 + 18y - 5}{21y^2 - 16y + 3} \cdot \dfrac{35y^2 - 22y + 3}{6y^2 + 17y + 5} \div \dfrac{16y^2 - 1}{1 - 9y^2}$

47. $\dfrac{xy - 2x + 3y - 6}{xy + 2x - 4y - 8} \cdot \dfrac{xy + x - 4y - 4}{xy - x + 3y - 3}$

48. $\dfrac{ax + bx + 2a + 2b}{ax - 3a + bx - 3b} \cdot \dfrac{ax - bx - 3a + 3b}{ax - bx - 2a + 2b}$

49. $\dfrac{xy^2 - y^2 + 4xy - 4y}{xy - 3y + 4x - 12} \div \dfrac{xy^3 + 2xy^2 + y^3 + 2y^2}{xy^2 - 3y^2 + 2xy - 6y}$

50. $\dfrac{4xb - 8b + 12x - 24}{xb^2 + 3b^2 + 3xb + 9b} \div \dfrac{4xb - 8b - 8x + 16}{xb^2 + 3b^2 - 2xb - 6b}$

51. $\dfrac{2x^3 + 10x^2 - 8x - 40}{x^3 + 4x^2 - 9x - 36} \cdot \dfrac{x^2 + x - 12}{2x^2 + 14x + 20}$

52. $\dfrac{x^3 + 2x^2 - 9x - 18}{x^4 + 3x^3 - 4x^2 - 12x} \cdot \dfrac{x^3 + 5x^2 + 6x}{x^2 - x - 6}$

53. $\dfrac{w^3 - w^2 x}{wy - w} \div \left(\dfrac{x - w}{y - 1}\right)^2$

54. $\dfrac{a^3 - a^2 b}{a - ac} \div \left(\dfrac{a - b}{c - 1}\right)^2$

55. $\dfrac{mx + my + 2x + 2y}{6x^2 - 5xy - 4y^2} \div \dfrac{2mx - 4x + my - 2y}{3mx - 6x - 4my + 8y}$

56. $\dfrac{ax - 2a + 2xy - 4y}{ax + 2a - 2xy - 4y} \div \dfrac{ax + 2a + 2xy + 4y}{ax - 2a - 2xy + 4y}$

57. $(3x - 6) \cdot \dfrac{x}{x - 2}$

58. $(4x + 8) \cdot \dfrac{x}{x + 2}$

59. $(x^2 - 25) \cdot \dfrac{2}{x - 5}$

60. $(x^2 - 49) \cdot \dfrac{5}{x + 7}$

61. $(x^2 - 3x + 2) \cdot \dfrac{3}{3x - 3}$

62. $(x^2 - 3x + 2) \cdot \dfrac{-1}{x - 2}$

63. $(y - 3)(y - 4)(y + 3) \cdot \dfrac{-1}{y^2 - 9}$

64. $(y + 1)(y + 4)(y - 1) \cdot \dfrac{3}{y^2 - 1}$

65. $a(a + 5)(a - 5) \cdot \dfrac{a + 1}{a^2 + 5a}$

66. $a(a + 3)(a - 3) \cdot \dfrac{a - 1}{a^2 - 3a}$

Paying Attention to Instructions The next two problems are intended to give you practice reading, and paying attention to, the instructions that accompany the problems you are working. Working these problems is an excellent way to get ready for a test or a quiz.

67. Work each problem according to the instructions given.

   a. Simplify: $\dfrac{16 - 1}{64 - 1}$

   b. Reduce: $\dfrac{25x^2 - 9}{125x^3 - 27}$

   c. Multiply: $\dfrac{25x^2 - 9}{125x^3 - 27} \cdot \dfrac{5x - 3}{5x + 3}$

   d. Divide: $\dfrac{25x^2 - 9}{125x^3 - 27} \div \dfrac{5x - 3}{25x^2 + 15x + 9}$

68. Work each problem according to the instructions given.

   a. Simplify: $\dfrac{64 - 49}{64 + 112 + 49}$

   b. Reduce: $\dfrac{9x^2 - 49}{9x^2 + 42x + 49}$

   c. Multiply: $\dfrac{9x^2 - 49}{9x^2 + 42x + 49} \cdot \dfrac{3x + 7}{3x - 7}$

   d. Divide: $\dfrac{9x^2 - 49}{9x^2 + 42x + 49} \div \dfrac{3x + 7}{3x - 7}$

Learning Objectives Assessment

The following problems can be used to help assess if you have successfully met the learning objectives for this section.

69. Multiply: $\dfrac{x^2 - 1}{x^2 + x - 6} \cdot \dfrac{x + 3}{1 - x}$.

a. $\dfrac{3}{x - 6}$ b. $\dfrac{x + 1}{x - 2}$ c. $\dfrac{x - 1}{x + 2}$ d. $-\dfrac{x + 1}{x - 2}$

70. Divide: $\dfrac{x + 1}{x^2 - x - 6} \div \dfrac{x^2 + x}{x^2 - 5x + 6}$.

a. $\dfrac{1}{x}$ b. $\dfrac{x - 2}{x(x + 2)}$ c. $-\dfrac{1}{x}$ d. $\dfrac{x}{(x - 2)(x + 2)}$

Getting Ready for the Next Section

Combine.

71. $\dfrac{4}{9} + \dfrac{2}{9}$ **72.** $\dfrac{3}{8} + \dfrac{1}{8}$ **73.** $\dfrac{3}{14} + \dfrac{7}{30}$ **74.** $\dfrac{3}{10} + \dfrac{11}{42}$

Multiply.

75. $-1(7 - x)$ **76.** $-1(3 - x)$

Factor.

77. $x^2 - 1$ **78.** $x^2 - 2x - 3$ **79.** $2x + 10$

80. $x^2 + 4x + 3$ **81.** $a^3 - b^3$ **82.** $8y^3 - 27$

Learning Objectives

In this section, we will learn how to:

1. Add or subtract rational expressions having the same denominator.

2. Find the LCD of rational expressions with different denominators.

3. Add or subtract rational expressions having different denominators.

Introduction

This section is concerned with addition and subtraction of rational expressions. In the first part of this section, we will look at addition of expressions that have the same denominator. In the second part of this section, we will look at addition of expressions that have different denominators.

Addition and Subtraction with the Same Denominator

To add two expressions that have the same denominator, we simply add numerators and put the sum over the common denominator. Because the process we use to add and subtract rational expressions is the same process used to add and subtract fractions, we will begin with an example involving fractions.

VIDEO EXAMPLES

SECTION 6.3

 **EXAMPLE 1** Add: $\dfrac{4}{9} + \dfrac{2}{9}$.

SOLUTION We add fractions with the same denominator by using the distributive property. Here is a detailed look at the steps involved.

$$\frac{4}{9} + \frac{2}{9} = 4\left(\frac{1}{9}\right) + 2\left(\frac{1}{9}\right)$$

$$= (4 + 2)\left(\frac{1}{9}\right) \qquad \text{Distributive property}$$

$$= 6\left(\frac{1}{9}\right)$$

$$= \frac{6}{9}$$

$$= \frac{2}{3} \qquad \text{Divide numerator and denominator by common factor 3}$$

Note that the important thing about the fractions in this example is that they each have a denominator of 9. If they did not have the same denominator, we could not have written them as two terms with a factor of $\frac{1}{9}$ in common. Without the $\frac{1}{9}$ common to each term, we couldn't apply the distributive property. Without the distributive property, we would not have been able to add the two fractions in this form. ∎

In the following examples, we will not show all the steps we showed in Example 1. The steps are shown in Example 1 so you will see why both fractions must have the same denominator before we can add them. In practice, we simply add numerators and place the result over the common denominator.

We add and subtract rational expressions with the same denominator by combining numerators and writing the result over the common denominator. Then we reduce the result to lowest terms, if possible. Example 2 shows this process in detail. If you see the similarities between operations on rational numbers and operations on rational expressions, this chapter will look like an extension of rational numbers rather than a completely new set of topics.

EXAMPLE 2   Add: $\dfrac{x}{x^2 - 1} + \dfrac{1}{x^2 - 1}$.

SOLUTION Because the denominators are the same, we simply add numerators:

$$\dfrac{x}{x^2 - 1} + \dfrac{1}{x^2 - 1} = \dfrac{x + 1}{x^2 - 1} \qquad \text{Add numerators}$$

$$= \dfrac{\cancel{x + 1}}{(x - 1)\cancel{(x + 1)}} \qquad \text{Factor denominator}$$

$$= \dfrac{1}{x - 1} \qquad \text{Divide out common factor } x + 1$$

Our next example involves subtraction of rational expressions. Pay careful attention to what happens to the signs of the terms in the numerator of the second expression when we subtract it from the first expression.

EXAMPLE 3   Subtract: $\dfrac{2x - 5}{x - 2} - \dfrac{x - 3}{x - 2}$.

SOLUTION Because each expression has the same denominator, we simply subtract the numerator in the second expression from the numerator in the first expression and write the difference over the common denominator $x - 2$. We must be careful, however, that we subtract both terms in the second numerator. To ensure that we do, we will enclose that numerator in parentheses.

$$\dfrac{2x - 5}{x - 2} - \dfrac{x - 3}{x - 2} = \dfrac{2x - 5 - (x - 3)}{x - 2} \qquad \text{Subtract numerators}$$

$$= \dfrac{2x - 5 - x + 3}{x - 2} \qquad \text{Remove parentheses}$$

$$= \dfrac{\cancel{x - 2}}{\cancel{x - 2}} \qquad \begin{array}{l}\text{Combine similar terms in}\\ \text{the numerator}\end{array}$$

$$= 1 \qquad \text{Reduce (or divide)}$$

Note the $+3$ in the numerator of the second step. It is a common mistake to write this as -3, by forgetting to subtract both terms in the numerator of the second expression. Whenever the expression we are subtracting has two or more terms in its numerator, we have to watch for this mistake.

Next we consider addition and subtraction of fractions and rational expressions that have different denominators.

Addition and Subtraction With Different Denominators

Before we look at an example of addition of fractions with different denominators, we need to review the definition for the least common denominator (LCD).

> **(def DEFINITION** *least common denominator*
>
> The *least common denominator* for a set of denominators is the smallest expression that is divisible by each of the denominators.

The first step in combining two fractions is to find the LCD. Once we have the common denominator, we rewrite each fraction as an equivalent fraction with the common denominator. After that, we simply add or subtract as we did in our first three examples. We provide a more detailed outline of these steps in the following box.

> **HOW TO** *Add or Subtract Fractions with Different Denominators*
>
> **Step 1:** Factor each denominator completely.
>
> **Step 2:** Find the least common denominator (LCD) of the fractions. The LCD will be a product containing each factor the *most* number of times it appears in any single denominator.
>
> **Step 3:** Rewrite each fraction as an equivalent fraction with the common denominator by multiplying both numerator and denominator by the factors from the LCD that are missing in the denominator.
>
> **Step 4:** Add or subtract the numerators and keep the common denominator.
>
> **Step 5:** Reduce to lowest terms, if possible.

Example 4 reviews this step-by-step procedure in adding two rational numbers.

EXAMPLE 4   Add: $\dfrac{3}{14} + \dfrac{7}{30}$.

SOLUTION

Step 1: *Factor each denominator.*
We factor both denominators into prime factors.

$$\text{Factor 14:} \quad 14 = 2 \cdot 7$$
$$\text{Factor 30:} \quad 30 = 2 \cdot 3 \cdot 5$$

Step 2: *Find the LCD.*

Because the LCD must be divisible by 14, it must have factors of 2 and 7. It must also be divisible by 30 and, therefore, have factors of 2, 3, and 5. We do not need to repeat the 2 that appears in both the factors of 14 and those of 30. Therefore,

$$\text{LCD} = 2 \cdot 3 \cdot 5 \cdot 7 = 210$$

Step 3: *Change to equivalent fractions.*

Because we want each fraction to have a denominator of 210 and at the same time keep its original value, we multiply each by 1 in the appropriate form.

Change $\frac{3}{14}$ to a fraction with denominator 210.

$$\frac{3}{14} \cdot \frac{15}{15} = \frac{45}{210}$$ 14 contains a factor of 2 and 7, but is missing the factors 3 and 5.

Change $\frac{7}{30}$ to a fraction with denominator 210.

$$\frac{7}{30} \cdot \frac{7}{7} = \frac{49}{210}$$ 30 contains the factors 2, 3, and 5, but is missing the factor of 7.

Step 4: *Add numerators of equivalent fractions found in step 3.*

$$\frac{45}{210} + \frac{49}{210} = \frac{94}{210}$$

Step 5: *Reduce to lowest terms, if necessary.*

$$\frac{94}{210} = \frac{47}{105}$$

Before applying this entire process to rational expressions, it might be helpful to have some practice in finding the LCD when the denominators are different.

EXAMPLE 5   Find the LCD for each pair of rational expressions.

a. $\dfrac{5}{x}$ and $\dfrac{x}{x+2}$ **b.** $\dfrac{x-1}{x^2+4x+4}$ and $\dfrac{3x}{x^2-4}$

SOLUTION

a. The first denominator contains a single factor of x, and the second denominator contains a single factor of $(x+2)$. Therefore,

$$\text{LCD} = x(x+2)$$

It is important to recognize that the x in $x+2$ is a **term** and not a **factor**. We cannot say LCD $= x+2$ because this would not include x as a factor.

b. Factoring both denominators gives us

$$x^2 + 4x + 4 = (x+2)(x+2) = (x+2)^2$$

$$x^2 - 4 = (x+2)(x-2)$$

The LCD requires two (not three) factors of $(x+2)$ and one factor of $(x-2)$.

$$\text{LCD} = (x+2)^2(x-2)$$

The main idea in adding fractions is to write each fraction again with the LCD for a denominator. In doing so, we must be sure not to change the value of either of the original fractions.

EXAMPLE 6   Add: $\dfrac{-2}{x^2 - 2x - 3} + \dfrac{3}{x^2 - 9}$.

SOLUTION

Step 1: *Factor each denominator.*

$$x^2 - 2x - 3 = (x - 3)(x + 1)$$
$$x^2 - 9 \quad\; = (x - 3)(x + 3)$$

Step 2: *Find the LCD.*

The LCD must contain one factor each of $(x - 3)$, $(x + 3)$, and $(x + 1)$, so we have

$$\text{LCD} = (x - 3)(x + 3)(x + 1)$$

Step 3: *Change each rational expression to an equivalent expression that has the LCD for a denominator.*

$$\frac{-2}{x^2 - 2x - 3} = \frac{-2}{(x - 3)(x + 1)} \cdot \frac{(x + 3)}{(x + 3)} = \frac{-2x - 6}{(x - 3)(x + 3)(x + 1)}$$

$$\frac{3}{x^2 - 9} = \frac{3}{(x - 3)(x + 3)} \cdot \frac{(x + 1)}{(x + 1)} = \frac{3x + 3}{(x - 3)(x + 3)(x + 1)}$$

Step 4: *Add numerators of the rational expressions found in step 3.*

$$\frac{-2x - 6}{(x - 3)(x + 3)(x + 1)} + \frac{3x + 3}{(x - 3)(x + 3)(x + 1)} = \frac{x - 3}{(x - 3)(x + 3)(x + 1)}$$

Step 5: *Reduce to lowest terms by dividing out the common factor $x - 3$.*

$$\frac{\cancel{x - 3}}{\cancel{(x - 3)}(x + 3)(x + 1)} = \frac{1}{(x + 3)(x + 1)}$$

EXAMPLE 7   Subtract: $\dfrac{x + 4}{2x + 10} - \dfrac{5}{x^2 - 25}$.

SOLUTION   We begin by factoring each denominator.

$$\frac{x + 4}{2x + 10} - \frac{5}{x^2 - 25} = \frac{x + 4}{2(x + 5)} - \frac{5}{(x + 5)(x - 5)}$$

The LCD is $2(x + 5)(x - 5)$. Completing the problem, we have

$$= \frac{x + 4}{2(x + 5)} \cdot \frac{(x - 5)}{(x - 5)} - \frac{5}{(x + 5)(x - 5)} \cdot \frac{2}{2}$$

$$= \frac{x^2 - x - 20}{2(x + 5)(x - 5)} - \frac{10}{2(x + 5)(x - 5)}$$

$$= \frac{x^2 - x - 30}{2(x + 5)(x - 5)}$$

To see if this expression will reduce, we factor the numerator into $(x - 6)(x + 5)$.

$$= \frac{(x - 6)\cancel{(x + 5)}}{2\cancel{(x + 5)}(x - 5)}$$

$$= \frac{x - 6}{2(x - 5)}$$

EXAMPLE 8 Subtract: $\dfrac{2x-2}{x^2+4x+3} - \dfrac{x-1}{x^2+5x+6}$.

SOLUTION We factor each denominator and build the LCD from those factors:

$$\frac{2x-2}{x^2+4x+3} - \frac{x-1}{x^2+5x+6}$$

$$= \frac{2x-2}{(x+3)(x+1)} - \frac{x-1}{(x+3)(x+2)}$$

$$= \frac{2x-2}{(x+3)(x+1)} \cdot \frac{(x+2)}{(x+2)} - \frac{x-1}{(x+3)(x+2)} \cdot \frac{(x+1)}{(x+1)} \qquad \text{The LCD is } (x+1)(x+2)(x+3)$$

$$= \frac{2x^2+2x-4}{(x+1)(x+2)(x+3)} - \frac{x^2-1}{(x+1)(x+2)(x+3)} \qquad \text{Multiply out each numerator}$$

$$= \frac{(2x^2+2x-4)-(x^2-1)}{(x+1)(x+2)(x+3)} \qquad \text{Subtract numerators}$$

$$= \frac{x^2+2x-3}{(x+1)(x+2)(x+3)} \qquad \text{Factor numerator to see if we can rdeuce}$$

$$= \frac{(x+3)(x-1)}{(x+1)(x+2)(x+3)} \qquad \text{Reduce}$$

$$= \frac{x-1}{(x+1)(x+2)} \qquad \blacksquare$$

EXAMPLE 9 Add: $\dfrac{x^2}{x-7} + \dfrac{6x+7}{7-x}$.

SOLUTION In Section 6.1, we were able to reverse the terms in a factor such as $7-x$ by factoring -1 from each term. In a problem like this, the same result can be obtained by multiplying the numerator and denominator by -1.

$$\frac{x^2}{x-7} + \frac{6x+7}{7-x} \cdot \frac{-1}{-1} = \frac{x^2}{x-7} + \frac{-6x-7}{x-7}$$

$$= \frac{x^2-6x-7}{x-7} \qquad \text{Add numerators}$$

$$= \frac{(x-7)(x+1)}{(x-7)} \qquad \text{Factor numerator}$$

$$= x+1 \qquad \text{Divide out } x-7 \qquad \blacksquare$$

For our next example, we will look at a problem in which we combine a whole number and a rational expression.

EXAMPLE 10 Subtract: $2 - \dfrac{9}{3x + 1}$.

SOLUTION To subtract these two expressions, we think of 2 as a rational expression with a denominator of 1.

$$2 - \frac{9}{3x + 1} = \frac{2}{1} - \frac{9}{3x + 1}$$

The LCD is $3x + 1$. Multiplying the numerator and denominator of the first expression by $3x + 1$ gives us a rational expression equivalent to 2, but with a denominator of $3x + 1$.

$$\frac{2}{1} \cdot \frac{(3x + 1)}{(3x + 1)} - \frac{9}{3x + 1} = \frac{6x + 2 - 9}{3x + 1}$$

$$= \frac{6x - 7}{3x + 1}$$

The numerator and denominator of this last expression do not have any factors in common other than 1, so the expression is in lowest terms.

EXAMPLE 11 Write an expression for the sum of a number and twice its reciprocal. Then, simplify that expression.

SOLUTION If x is the number, then its reciprocal is $\frac{1}{x}$. Twice its reciprocal is $\frac{2}{x}$. The sum of the number and twice its reciprocal is

$$x + \frac{2}{x}$$

To combine these two expressions, we think of the first term x as a rational expression with a denominator of 1. The LCD is x.

$$x + \frac{2}{x} = \frac{x}{1} + \frac{2}{x}$$

$$= \frac{x}{1} \cdot \frac{x}{x} + \frac{2}{x}$$

$$= \frac{x^2 + 2}{x}$$

Getting Ready for Class

After reading through the preceding section, respond in your own words and in complete sentences.

A. Briefly describe how you would add two rational expressions that have the same denominator.

B. Why is factoring important in finding a least common denominator?

C. What is the last step in adding or subtracting two rational expressions?

D. Explain how you would change the fraction $\dfrac{5}{x - 3}$ to an equivalent fraction with denominator $x^2 - 9$.

Problem Set 6.3

Combine the following fractions.

1. $\dfrac{3}{4} + \dfrac{1}{2}$ **2.** $\dfrac{5}{6} + \dfrac{1}{3}$ **3.** $\dfrac{2}{5} - \dfrac{1}{15}$ **4.** $\dfrac{5}{8} - \dfrac{1}{4}$

5. $\dfrac{5}{6} + \dfrac{7}{8}$ **6.** $\dfrac{3}{4} + \dfrac{2}{3}$ **7.** $\dfrac{9}{48} - \dfrac{3}{54}$ **8.** $\dfrac{6}{28} - \dfrac{5}{42}$

9. $\dfrac{3}{4} - \dfrac{1}{8} + \dfrac{2}{3}$ **10.** $\dfrac{1}{3} - \dfrac{5}{6} + \dfrac{5}{12}$

Combine the following rational expressions. Reduce all answers to lowest terms.

11. $\dfrac{x}{x+3} + \dfrac{3}{x+3}$ **12.** $\dfrac{5x}{5x+2} + \dfrac{2}{5x+2}$ **13.** $\dfrac{4}{y-4} - \dfrac{y}{y-4}$

14. $\dfrac{8}{y+8} + \dfrac{y}{y+8}$ **15.** $\dfrac{x}{x^2-y^2} - \dfrac{y}{x^2-y^2}$ **16.** $\dfrac{x}{x^2-y^2} + \dfrac{y}{x^2-y^2}$

17. $\dfrac{2x-3}{x-2} - \dfrac{x+1}{x-2}$ **18.** $\dfrac{2x-4}{x+2} - \dfrac{x+6}{x+2}$ **19.** $\dfrac{7x-2}{2x+1} - \dfrac{5x-3}{2x+1}$

20. $\dfrac{7x-1}{3x+2} - \dfrac{4x-3}{3x+2}$ **21.** $\dfrac{3x-11}{x^2+x-6} + \dfrac{2x+1}{x^2+x-6}$

22. $\dfrac{x+7}{x^2+5x+6} + \dfrac{3x+5}{x^2+5x+6}$

23. $\dfrac{2x^2-x+6}{x^2-3x-4} - \dfrac{x^2-5x+3}{x^2-3x-4}$

24. $\dfrac{3x^2-2x-1}{4x^2-9} - \dfrac{x^2-3x+2}{4x^2-9}$

Find the LCD for each pair of rational expressions.

25. $\dfrac{2}{x}$ and $\dfrac{5}{x-3}$

26. $\dfrac{x}{x+1}$ and $\dfrac{1}{4x}$

27. $\dfrac{3x}{x+4}$ and $\dfrac{6}{x-4}$

28. $\dfrac{x+2}{x^2-x}$ and $\dfrac{x-2}{x^2-1}$

29. $\dfrac{x^2+5}{x^2+6x+9}$ and $\dfrac{x-9}{x^2-x-12}$

30. $\dfrac{x^2+3x-1}{2x^2-4x+2}$ and $\dfrac{3x^2+7}{4x^2+4x-8}$

Combine the following rational expressions. Reduce all answers to lowest terms.

31. $\dfrac{1}{a} + \dfrac{2}{a^2} - \dfrac{3}{a^3}$ **32.** $\dfrac{3}{a} + \dfrac{2}{a^2} - \dfrac{1}{a^3}$

33. $\dfrac{3x+1}{2x-6} - \dfrac{x+2}{x-3}$ **34.** $\dfrac{x+1}{x-2} - \dfrac{4x+7}{5x-10}$

35. $\dfrac{6x+5}{5x-25} - \dfrac{x+2}{x-5}$ **36.** $\dfrac{4x+2}{3x+12} - \dfrac{x-2}{x+4}$

37. $\dfrac{2}{x} + \dfrac{5}{x - 3}$

38. $\dfrac{x}{x + 1} - \dfrac{1}{4x}$

39. $\dfrac{3}{x + 2} + \dfrac{2}{x + 3}$

40. $\dfrac{x}{x + 6} + \dfrac{4}{x - 6}$

41. $\dfrac{x + 1}{2x - 2} - \dfrac{2}{x^2 - 1}$

42. $\dfrac{x + 7}{2x + 12} + \dfrac{6}{x^2 - 36}$

43. $\dfrac{1}{a - b} - \dfrac{3ab}{a^3 - b^3}$

44. $\dfrac{1}{a + b} + \dfrac{3ab}{a^3 + b^3}$

45. $\dfrac{1}{2y - 3} - \dfrac{18y}{8y^3 - 27}$

46. $\dfrac{1}{3y - 2} - \dfrac{18y}{27y^3 - 8}$

47. $\dfrac{x}{x^2 - 5x + 6} - \dfrac{3}{3 - x}$

48. $\dfrac{x}{x^2 + 4x + 4} - \dfrac{2}{2 + x}$

49. $\dfrac{2}{4t - 5} + \dfrac{9}{8t^2 - 38t + 35}$

50. $\dfrac{3}{2t - 5} + \dfrac{21}{8t^2 - 14t - 15}$

51. $\dfrac{1}{a^2 - 5a + 6} + \dfrac{3}{a^2 - a - 2}$

52. $\dfrac{-3}{a^2 + a - 2} + \dfrac{5}{a^2 - a - 6}$

53. $\dfrac{1}{8x^3 - 1} - \dfrac{1}{4x^2 - 1}$

54. $\dfrac{1}{27x^3 - 1} - \dfrac{1}{9x^2 - 1}$

55. $\dfrac{4}{4x^2 - 9} - \dfrac{6}{8x^2 - 6x - 9}$

56. $\dfrac{9}{9x^2 + 6x - 8} - \dfrac{6}{9x^2 - 4}$

57. $\dfrac{4a}{a^2 + 6a + 5} - \dfrac{3a}{a^2 + 5a + 4}$

58. $\dfrac{3a}{a^2 + 7a + 10} - \dfrac{2a}{a^2 + 6a + 8}$

59. $\dfrac{2x - 1}{x^2 + x - 6} - \dfrac{x + 2}{x^2 + 5x + 6}$

60. $\dfrac{4x + 1}{x^2 + 5x + 4} - \dfrac{x + 3}{x^2 + 4x + 3}$

61. $\dfrac{2x - 8}{3x^2 + 8x + 4} + \dfrac{x + 3}{3x^2 + 5x + 2}$

62. $\dfrac{5x + 3}{2x^2 + 5x + 3} - \dfrac{3x + 9}{2x^2 + 7x + 6}$

63. $\dfrac{2}{x^2 + 5x + 6} - \dfrac{4}{x^2 + 4x + 3} + \dfrac{3}{x^2 + 3x + 2}$

64. $\dfrac{-5}{x^2 + 3x - 4} + \dfrac{5}{x^2 + 2x - 3} + \dfrac{1}{x^2 + 7x + 12}$

65. $\dfrac{2x + 8}{x^2 + 5x + 6} - \dfrac{x + 5}{x^2 + 4x + 3} - \dfrac{x - 1}{x^2 + 3x + 2}$

66. $\dfrac{2x + 11}{x^2 + 9x + 20} - \dfrac{x + 1}{x^2 + 7x + 12} - \dfrac{x + 6}{x^2 + 8x + 15}$

67. $2 + \dfrac{3}{2x + 1}$

68. $3 - \dfrac{2}{2x + 3}$

69. $5 + \dfrac{2}{4 - t}$

70. $7 + \dfrac{3}{5 - t}$

71. $x - \dfrac{4}{2x + 3}$

72. $x - \dfrac{5}{3x + 4} + 1$

73. $\dfrac{x}{x + 2} + \dfrac{1}{2x + 4} - \dfrac{3}{x^2 + 2x}$

74. $\dfrac{x}{x + 3} + \dfrac{7}{3x + 9} - \dfrac{2}{x^2 + 3x}$

75. $\dfrac{1}{x} + \dfrac{x}{2x + 4} - \dfrac{2}{x^2 + 2x}$

76. $\dfrac{1}{x} + \dfrac{x}{3x + 9} - \dfrac{3}{x^2 + 3x}$

77. Let $f(x) = \dfrac{2}{x + 4}$ and $g(x) = \dfrac{x - 1}{x^2 + 3x - 4}$; find $f(x) + g(x)$

78. Let $f(t) = \dfrac{5}{3t - 2}$ and $g(t) = \dfrac{t - 3}{3t^2 + 7t - 6}$; find $f(t) - g(t)$

79. Let $f(x) = \dfrac{7}{x^2 - x - 12}$ and $g(x) = \dfrac{5}{x^2 + x - 6}$; find $f(x) - g(x)$

80. Let $f(x) = \dfrac{2x}{x^2 - x - 2}$ and $g(x) = \dfrac{5}{x^2 + x - 6}$; find $f(x) + g(x)$

81. Paying Attention to Instructions Work each problem according to the instructions given.

    a. Multiply: $\dfrac{3}{8} \cdot \dfrac{1}{6}$.

    b. Divide: $\dfrac{3}{8} \div \dfrac{1}{6}$.

    c. Add: $\dfrac{3}{8} + \dfrac{1}{6}$.

    d. Multiply: $\dfrac{x + 3}{x - 3} \cdot \dfrac{5x + 15}{x^2 - 9}$.

    e. Divide: $\dfrac{x + 3}{x - 3} \div \dfrac{5x + 15}{x^2 - 9}$.

    f. Subtract: $\dfrac{x + 3}{x - 3} - \dfrac{5x + 15}{x^2 - 9}$.

82. Paying Attention to Instructions Work each problem according to the instructions given.

    a. Multiply: $\dfrac{16}{49} \cdot \dfrac{1}{28}$.

    b. Divide: $\dfrac{16}{49} \div \dfrac{1}{28}$.

    c. Subtract: $\dfrac{16}{49} - \dfrac{1}{28}$.

    d. Multiply: $\dfrac{3x - 2}{3x + 2} \cdot \dfrac{15x + 6}{9x^2 - 4}$.

    e. Divide: $\dfrac{3x - 2}{3x + 2} \div \dfrac{15x + 6}{9x^2 - 4}$.

    f. Subtract: $\dfrac{3x + 2}{3x - 2} - \dfrac{15x + 6}{9x^2 - 4}$.

Applying the Concepts

83. Optometry The formula

$$P = \frac{1}{a} + \frac{1}{b}$$

is used by optometrists to help determine how strong to make the lenses for a pair of eyeglasses. If a is 10 and b is 0.2, find the corresponding value of P.

84. Quadratic Formula Later in the book we will work with the quadratic formula. The derivation of the formula requires that you can add the fractions below. Add the fractions.

$$\frac{-c}{a} + \frac{b^2}{(2a)^2}$$

85. Number Problem Write an expression for the sum of a number and 4 times its reciprocal. Then, simplify that expression.

86. Number Problem Write an expression for the sum of a number and 3 times its reciprocal. Then, simplify that expression.

87. Number Problem Write an expression for the sum of the reciprocals of two consecutive integers. Then, simplify that expression.

88. Number Problem Write an expression for the sum of the reciprocals of two consecutive even integers. Then, simplify that expression.

Learning Objectives Assessment

The following problems can be used to help assess if you have successfully met the learning objectives for this section.

89. Add $\dfrac{2x + 1}{x^2 - 4} + \dfrac{3}{x^2 - 4}$ and reduce to lowest terms if possible.

a. $\dfrac{2}{x - 2}$ b. $\dfrac{2}{x + 2}$ c. $\dfrac{1}{x - 2}$ d. $\dfrac{1}{x + 2}$

90. Subtract $\dfrac{6x + 1}{x^2 + 3x - 10} - \dfrac{5x - 4}{x^2 + 3x - 10}$ and reduce to lowest terms if possible.

a. $\dfrac{x - 3}{(x + 5)(x - 2)}$ b. $\dfrac{11x + 5}{(x + 5)(x - 2)}$

c. $\dfrac{1}{x - 2}$ d. $\dfrac{x - 2}{x + 5}$

91. Find the LCD of $\dfrac{3}{x^2 + 2x + 1}$ and $\dfrac{1}{x^2 + 4x + 3}$.

a. $(x + 1)^2(x + 3)$ b. $(x + 1)(x + 3)$

c. $(x + 1)^3(x + 3)$ d. $x + 1$

92. Add: $\dfrac{3}{x + 2} + \dfrac{1}{x - 3}$.

a. $\dfrac{4}{2x - 1}$ b. $\dfrac{4}{(x + 2)(x - 3)}$

c. $\dfrac{4x + 3}{(x + 2)(x - 3)}$ d. $\dfrac{4x - 7}{(x + 2)(x - 3)}$

Getting Ready for the Next Section

Divide.

93. $\dfrac{3}{4} \div \dfrac{5}{8}$ **94.** $\dfrac{2}{3} \div \dfrac{5}{6}$

Multiply.

95. $x\left(1 + \dfrac{2}{x}\right)$ **96.** $3\left(x + \dfrac{1}{3}\right)$ **97.** $3x\left(\dfrac{1}{x} - \dfrac{1}{3}\right)$ **98.** $3x\left(\dfrac{1}{x} + \dfrac{1}{3}\right)$

Factor.

99. $x^2 - 4$ **100.** $x^2 - x - 6$

> *You never fail until you stop trying.*
> —Albert Einstein

Coming to the United States at the age of 10 and not knowing how to speak English was a very difficult hurdle to overcome. However, with hard work and dedication I was able to rise above those obstacles. When I came to the U.S. our school did not have a strong English development program as it was known at that time, English as a Second Language (ESL). The approach back then was "sink or swim." When my self-esteem was low, my mom and my three older sisters were always there for me and they would always encourage me to do well. My mom was a single parent, and her number one priority was that we would receive a good education. My mother's perseverance is what has made me the person I am today. At a young age I was able to see that she had overcome more than what my situation was, and I would always tell myself, "if Mom can do it, I could also do it." Not only did she not have an education, but she also saved us from a civil war that was happening in my home country of El Salvador.

When things in school got hard, I would always reflect on all the hard work, sacrifice and effort of my mother. I would just tell myself that I should not have any excuses and that I needed to keep going. If my mother, who worked as a housekeeper, could send all four of her kids to college doesn't motivate you, I don't know what does. It definitely motivated me. The day everything began to change for me was when I was in eighth grade. I was sitting in my biology class not paying attention to the teacher because I was really focusing on a piece of paper on the wall. It said, "You never fail until you stop trying." I read it over and over, trying to digest what the quote meant. With my limited English I was doing my best to translate what it meant in my native language. It finally clicked! I was able to figure out what those seven words meant. I memorized the quote and began to apply it to my academics and to real-life situations. I began to really focus in my studies. I wanted to do well in school, and most important I wanted to improve my English. To this day I always reflect to that quote when I feel I can't do something.

I was able to finish junior high successfully. Going to high school was a lot easier and I ended up with very good grades and eventually I was accepted to an excellent college. I was never the smartest student on campus, but I always did well because I never quit. I earned my college degree and now I teach at a dual immersion elementary school. I have that same quote in my classroom and I constantly remind my students to never stop trying.

Complex Fractions

Learning Objectives

In this section, we will learn how to:

1. Simplify complex fractions using the division method.

2. Simplify complex fractions using the LCD method.

Introduction

The quotient of two fractions or two rational expressions is called a ***complex fraction***. This section is concerned with the simplification of complex fractions.

Division Method

There are generally two methods that can be used to simplify complex fractions. If the numerator and denominator of the complex fraction are both single fractions, then the most effective way to simplify is to rewrite the expression as a division problem in horizontal format. We can then perform the division using the techniques we learned in Section 6.2.

Our first example illustrates how this is done with a quotient of two rational numbers.

VIDEO EXAMPLES

SECTION 6.4

EXAMPLE 1 Simplify: $\dfrac{\frac{3}{4}}{\frac{5}{8}}$.

SOLUTION Notice that the numerator and denominator of the complex fraction are both single fractions $\left(\frac{3}{4} \text{ and } \frac{5}{8}\right)$. We write the quotient as a division of two fractions and then carry out the division.

$$\frac{\frac{3}{4}}{\frac{5}{8}} = \frac{3}{4} \div \frac{5}{8}$$

$$= \frac{3}{\underset{1}{4}} \cdot \frac{\overset{2}{8}}{5} \qquad \text{Multiply by the reciprocal}$$

$$= \frac{6}{5} \qquad \text{Reduce to lowest terms}$$

The process is similar when the numerator and denominator are rational expressions as seen in the following example.

EXAMPLE 2   Simplify: $\dfrac{\dfrac{x-2}{x^2-9}}{\dfrac{x^2-4}{x+3}}$.

SOLUTION   Using the division method, we have

$$\frac{\dfrac{x-2}{x^2-9}}{\dfrac{x^2-4}{x+3}} = \frac{x-2}{x^2-9} \div \frac{x^2-4}{x+3}$$

$$= \frac{x-2}{x^2-9} \cdot \frac{x+3}{x^2-4}$$

$$= \frac{(x-2)(x+3)}{(x+3)(x-3)(x+2)(x-2)}$$

$$= \frac{1}{(x-3)(x+2)}$$

LCD Method

If the numerator or denominator contains a sum or difference of two or more fractions, then we can use the LCD to simplify the complex fraction more efficiently. With the LCD method, we multiply both numerator and denominator by the entire LCD of all fractions appearing in the larger complex fraction. This will allow us to clear all of the smaller fractions from the expression because each individual denominator will divide out.

Our next example illustrates this process.

EXAMPLE 3   Simplify: $\dfrac{\dfrac{1}{x}+\dfrac{1}{y}}{\dfrac{1}{x}-\dfrac{1}{y}}$.

SOLUTION   Because the numerator and denominator are not both single fractions, we use the LCD method instead of the division method. We begin by multiplying both the numerator and denominator by the quantity xy, which is the LCD for all the fractions. To clarify, we have written the LCD over a 1 in both cases.

$$\frac{\dfrac{1}{x}+\dfrac{1}{y}}{\dfrac{1}{x}-\dfrac{1}{y}} = \frac{\left(\dfrac{1}{x}+\dfrac{1}{y}\right)\cdot\dfrac{xy}{1}}{\left(\dfrac{1}{x}-\dfrac{1}{y}\right)\cdot\dfrac{xy}{1}}$$ Multiply numerator and denominator by the entire LCD

$$= \frac{\dfrac{1}{x}\cdot\dfrac{xy}{1}+\dfrac{1}{y}\cdot\dfrac{xy}{1}}{\dfrac{1}{x}\cdot\dfrac{xy}{1}-\dfrac{1}{y}\cdot\dfrac{xy}{1}}$$ Apply the distributive property to distribute xy over both terms in the numerator and denominator.

$$= \frac{y+x}{y-x}$$ Reduce in each product.

EXAMPLE 4 Simplify: $\dfrac{1 - \dfrac{4}{x^2}}{1 - \dfrac{1}{x} - \dfrac{6}{x^2}}$.

SOLUTION The easiest way to simplify this complex fraction is to multiply the numerator and denominator by the LCD, which is x^2:

$$\dfrac{1 - \dfrac{4}{x^2}}{1 - \dfrac{1}{x} - \dfrac{6}{x^2}} = \dfrac{\dfrac{x^2}{1}\left(1 - \dfrac{4}{x^2}\right)}{\dfrac{x^2}{1}\left(1 - \dfrac{1}{x} - \dfrac{6}{x^2}\right)} \qquad \text{Multiply numerator and denominator by } x^2$$

$$= \dfrac{\dfrac{x^2}{1} \cdot \dfrac{1}{1} - \dfrac{x^2}{1} \cdot \dfrac{4}{x^2}}{\dfrac{x^2}{1} \cdot \dfrac{1}{1} - \dfrac{x^2}{1} \cdot \dfrac{1}{x} - \dfrac{x^2}{1} \cdot \dfrac{6}{x^2}} \qquad \text{Distributive property}$$

$$= \dfrac{x^2 - 4}{x^2 - x - 6} \qquad \text{Simplify each product}$$

$$= \dfrac{(x - 2)(x + 2)}{(x - 3)(x + 2)} \qquad \text{Factor}$$

$$= \dfrac{x - 2}{x - 3} \qquad \text{Reduce}$$

EXAMPLE 5 Simplify: $2 - \dfrac{3}{x + \dfrac{1}{3}}$.

SOLUTION First, we simplify the expression that follows the subtraction sign.

$$2 - \dfrac{3}{x + \dfrac{1}{3}} = 2 - \dfrac{3 \cdot 3}{3\left(x + \dfrac{1}{3}\right)}$$

$$= 2 - \dfrac{9}{3x + 1}$$

Now we subtract by rewriting the first term, 2, with the LCD, $3x + 1$.

$$2 - \dfrac{9}{3x + 1} = \dfrac{2}{1} \cdot \dfrac{3x + 1}{3x + 1} - \dfrac{9}{3x + 1}$$

$$= \dfrac{6x + 2 - 9}{3x + 1}$$

$$= \dfrac{6x - 7}{3x + 1}$$

 **EXAMPLE 6** Simplify: $\dfrac{\dfrac{x+1}{x+4}+\dfrac{x}{x-2}}{\dfrac{2x}{x-2}-\dfrac{x+3}{x+4}}$.

SOLUTION Once again we will use the LCD method. In this case the LCD is $(x+4)(x-2)$.

$$\dfrac{\dfrac{x+1}{x+4}+\dfrac{x}{x-2}}{\dfrac{2x}{x-2}-\dfrac{x+3}{x+4}}=\dfrac{\left(\dfrac{x+1}{x+4}+\dfrac{x}{x-2}\right)\cdot\dfrac{(x+4)(x-2)}{1}}{\left(\dfrac{2x}{x-2}-\dfrac{x+3}{x+4}\right)\cdot\dfrac{(x+4)(x-2)}{1}}$$ Multiply numerator and denominator by LCD

$$=\dfrac{\dfrac{x+1}{x+4}\cdot\dfrac{(x+4)(x-2)}{1}+\dfrac{x}{x-2}\cdot\dfrac{(x+4)(x-2)}{1}}{\dfrac{2x}{x-2}\cdot\dfrac{(x+4)(x-2)}{1}-\dfrac{x+3}{x+4}\cdot\dfrac{(x+4)(x-2)}{1}}$$ Distribute

$$=\dfrac{(x+1)(x-2)+x(x+4)}{2x(x+4)-(x+3)(x-2)}$$ Reduce

$$=\dfrac{x^2-x-2+x^2+4x}{2x^2+8x-(x^2+x-6)}$$ Multiply

$$=\dfrac{2x^2+3x-2}{x^2+7x+6}$$ Simplify

$$=\dfrac{(2x-1)(x+2)}{(x+1)(x+6)}$$ Factor

Getting Ready for Class

After reading through the preceding section, respond in your own words and in complete sentences.

A. What is a complex fraction?

B. Explain how some complex fractions can be converted to division problems. When is it more efficient to convert a complex fraction to a division problem of rational expressions?

C. Explain how a least common denominator can be used to simplify a complex fraction.

D. Explain how you choose which method to use.

Simplify each of the following as much as possible. Use whichever method you feel is most appropriate.

1. $\dfrac{\dfrac{3}{4}}{\dfrac{2}{3}}$

2. $\dfrac{\dfrac{5}{9}}{\dfrac{7}{12}}$

3. $\dfrac{\dfrac{1}{3} - \dfrac{1}{4}}{\dfrac{1}{2} + \dfrac{1}{8}}$

4. $\dfrac{\dfrac{1}{6} - \dfrac{1}{3}}{\dfrac{1}{4} - \dfrac{1}{8}}$

5. $\dfrac{3 + \dfrac{2}{5}}{1 - \dfrac{3}{7}}$

6. $\dfrac{2 + \dfrac{5}{6}}{1 - \dfrac{7}{8}}$

7. $\dfrac{\dfrac{1}{x}}{1 + \dfrac{1}{x}}$

8. $\dfrac{1 - \dfrac{1}{x}}{\dfrac{1}{x}}$

9. $\dfrac{1 + \dfrac{1}{a}}{1 - \dfrac{1}{a}}$

10. $\dfrac{1 - \dfrac{2}{a}}{1 - \dfrac{3}{a}}$

11. $\dfrac{\dfrac{1}{x} - \dfrac{1}{y}}{\dfrac{1}{x} + \dfrac{1}{y}}$

12. $\dfrac{\dfrac{1}{x} + \dfrac{2}{y}}{\dfrac{2}{x} + \dfrac{1}{y}}$

13. $\dfrac{\dfrac{x - 5}{x^2 - 4}}{\dfrac{x^2 - 25}{x + 2}}$

14. $\dfrac{\dfrac{3x + 1}{x^2 - 49}}{\dfrac{9x^2 - 1}{x - 7}}$

15. $\dfrac{\dfrac{4a}{2a^3 + 2}}{\dfrac{8a}{4a + 4}}$

16. $\dfrac{\dfrac{2a}{3a^3 - 3}}{\dfrac{4a}{6a - 6}}$

17. $\dfrac{1 - \dfrac{9}{x^2}}{1 - \dfrac{1}{x} - \dfrac{6}{x^2}}$

18. $\dfrac{4 - \dfrac{1}{x^2}}{4 + \dfrac{4}{x} + \dfrac{1}{x^2}}$

19. $\dfrac{2 + \dfrac{5}{a} - \dfrac{3}{a^2}}{2 - \dfrac{5}{a} + \dfrac{2}{a^2}}$

20. $\dfrac{3 + \dfrac{5}{a} - \dfrac{2}{a^2}}{3 - \dfrac{10}{a} + \dfrac{3}{a^2}}$

21. $\dfrac{2 + \dfrac{3}{x} - \dfrac{18}{x^2} - \dfrac{27}{x^3}}{2 + \dfrac{9}{x} + \dfrac{9}{x^2}}$

22. $\dfrac{3 + \dfrac{5}{x} - \dfrac{12}{x^2} - \dfrac{20}{x^3}}{3 + \dfrac{11}{x} + \dfrac{10}{x^2}}$

23. $\dfrac{1 + \dfrac{1}{x + 3}}{1 - \dfrac{1}{x + 3}}$

24. $\dfrac{1 + \dfrac{1}{x - 2}}{1 - \dfrac{1}{x - 2}}$

25. $\dfrac{1 + \dfrac{1}{x + 3}}{1 + \dfrac{7}{x - 3}}$

26. $\dfrac{1 + \dfrac{1}{x - 2}}{1 - \dfrac{3}{x + 2}}$

27. $\dfrac{1 - \dfrac{1}{a + 1}}{1 + \dfrac{1}{a - 1}}$

28. $\dfrac{\dfrac{1}{a - 1} + 1}{\dfrac{1}{a + 1} - 1}$

29. $\dfrac{\dfrac{1}{x + 3} + \dfrac{1}{x - 3}}{\dfrac{1}{x + 3} - \dfrac{1}{x - 3}}$

30. $\dfrac{\dfrac{1}{x + a} + \dfrac{1}{x - a}}{\dfrac{1}{x + a} - \dfrac{1}{x - a}}$

31. $\dfrac{\dfrac{y+1}{y-1} + \dfrac{y-1}{y+1}}{\dfrac{y+1}{y-1} - \dfrac{y-1}{y+1}}$

32. $\dfrac{\dfrac{y-1}{y+1} - \dfrac{y+1}{y-1}}{\dfrac{y-1}{y+1} + \dfrac{y+1}{y-1}}$

33. $1 - \dfrac{x}{1 - \dfrac{1}{x}}$

34. $x - \dfrac{1}{x - \dfrac{1}{2}}$

35. $1 + \dfrac{1}{1 + \dfrac{1}{1+1}}$

36. $1 - \dfrac{1}{1 - \dfrac{1}{1 - \dfrac{1}{2}}}$

37. $\dfrac{1 - \dfrac{1}{x + \dfrac{1}{2}}}{1 + \dfrac{1}{x + \dfrac{1}{2}}}$

38. $\dfrac{2 + \dfrac{1}{x - \dfrac{1}{3}}}{2 - \dfrac{1}{x - \dfrac{1}{3}}}$

39. $\dfrac{\dfrac{1}{x+h} - \dfrac{1}{x}}{h}$

40. $\dfrac{\dfrac{1}{(x+h)^2} - \dfrac{1}{x^2}}{h}$

41. $\dfrac{\dfrac{3}{ab} + \dfrac{4}{bc} - \dfrac{2}{ac}}{\dfrac{5}{abc}}$

42. $\dfrac{\dfrac{x}{yz} - \dfrac{y}{xz} + \dfrac{z}{xy}}{\dfrac{1}{x^2y^2} - \dfrac{1}{x^2z^2} + \dfrac{1}{y^2z^2}}$

43. $\dfrac{\dfrac{t^2 - 2t - 8}{t^2 + 7t + 6}}{\dfrac{t^2 - t - 6}{t^2 + 2t + 1}}$

44. $\dfrac{\dfrac{y^2 - 5y - 14}{y^2 + 3y - 10}}{\dfrac{y^2 - 8y + 7}{y^2 + 6y + 5}}$

45. $\dfrac{5 + \dfrac{4}{b-1}}{\dfrac{7}{b+5} - \dfrac{3}{b-1}}$

46. $\dfrac{\dfrac{6}{x+5} - 7}{\dfrac{8}{x+5} - \dfrac{9}{x+3}}$

47. $\dfrac{\dfrac{3}{x^2 - x - 6}}{\dfrac{2}{x+2} - \dfrac{4}{x-3}}$

48. $\dfrac{\dfrac{9}{a-7} + \dfrac{8}{2a+3}}{\dfrac{10}{2a^2 - 11a - 21}}$

49. $\dfrac{\dfrac{1}{m-4} + \dfrac{1}{m-5}}{\dfrac{1}{m^2 - 9m + 20}}$

50. $\dfrac{\dfrac{1}{k^2 - 7k + 12}}{\dfrac{1}{k-3} + \dfrac{1}{k-4}}$

Applying the Concepts

51. Doppler Effect The change in the pitch of a sound (such as a train whistle) as an object passes is called the Doppler effect, named after C. J. Doppler (1803–1853). A person will *hear* a sound with a frequency, h, according to the formula

$$h = \dfrac{f}{1 + \dfrac{v}{s}}$$

where f is the actual frequency of the sound being produced, s is the speed of sound (about 740 miles per hour), and v is the velocity of the moving object.

a. Examine this fraction, and then explain why h and f approach the same value as v becomes smaller and smaller.

b. Solve this formula for v.

52. Work Problem A water storage tank has two drains. It can be shown that the time it takes to empty the tank if both drains are open is given by the formula

$$\frac{1}{\dfrac{1}{a} + \dfrac{1}{b}}$$

where a = time it takes for the first drain to empty the tank, and b = time for the second drain to empty the tank.

a. Simplify this complex fraction.

b. Find the amount of time needed to empty the tank using both drains if, used alone, the first drain empties the tank in 4 hours and the second drain can empty the tank in 3 hours.

Learning Objectives Assessment

The following problems can be used to help assess if you have successfully met the learning objectives for this section.

53. Simplify $\dfrac{\dfrac{x-1}{x^2+2x}}{\dfrac{2x^2-2x}{x^2-4}}$ using the division method.

    a. $\dfrac{x-2}{2x^2}$ **b.** $\dfrac{2}{x-2}$ **c.** $\dfrac{1}{2(x-4)}$ **d.** $\dfrac{1}{2x}$

54. Simplify $\dfrac{\dfrac{3}{x} - \dfrac{2}{x+1}}{\dfrac{1}{x} + \dfrac{4}{x+1}}$ using the LCD method.

    a. $\dfrac{1}{5}$ **b.** $-\dfrac{2x+1}{5}$ **c.** $\dfrac{5}{x(x+1)}$ **d.** $\dfrac{x+3}{5x+1}$

Getting Ready for the Next Section

Multiply.

55. $x(y-2)$ **56.** $x(y-1)$ **57.** $6\left(\dfrac{x}{2} - 3\right)$

58. $6\left(\dfrac{x}{3} + 1\right)$ **59.** $xab \cdot \dfrac{1}{x}$ **60.** $xab\left(\dfrac{1}{b} + \dfrac{1}{a}\right)$

Factor.

61. $y^2 - 25$ **62.** $x^2 - 3x + 2$ **63.** $xa + xb$ **64.** $xy - y$

Solve.

65. $5x - 4 = 6$ **66.** $y^2 + y - 20 = 2y$

Equations Involving Rational Expressions

Learning Objectives

In this section, we will learn how to:

1. Use the LCD to solve an equation containing rational expressions.

2. Recognize extraneous solutions for rational equations.

3. Solve formulas involving rational expressions.

Introduction

The first step in solving an equation that contains one or more rational expressions is to find the LCD for all denominators in the equation. We then multiply both sides of the equation by the LCD to clear the equation of all fractions. That is, after we have multiplied through by the LCD, each term in the resulting equation will have a denominator of 1.

VIDEO EXAMPLES

SECTION 6.5

EXAMPLE 1 Solve: $\dfrac{x}{2} - 3 = \dfrac{2}{3}$.

SOLUTION The LCD for 2 and 3 is 6. Multiplying both sides by 6, we have

$$\frac{6}{1}\left(\frac{x}{2} - 3\right) = \frac{6}{1}\left(\frac{2}{3}\right)$$

$$\frac{6}{1}\left(\frac{x}{2}\right) - 6(3) = \frac{6}{1}\left(\frac{2}{3}\right)$$

$$3x - 18 = 4$$

$$3x = 22$$

$$x = \frac{22}{3}$$

Multiplying both sides of an equation by the LCD clears the equation of fractions because the LCD has the property that all the denominators divide it evenly.

EXAMPLE 2 Solve: $\dfrac{6}{a - 4} = \dfrac{3}{8}$.

SOLUTION The LCD for $a - 4$ and 8 is $8(a - 4)$. Multiplying both sides by this quantity yields

$$\frac{8(a - 4)}{1} \cdot \frac{6}{a - 4} = \frac{8(a - 4)}{1} \cdot \frac{3}{8}$$

$$48 = (a - 4) \cdot 3$$

$$48 = 3a - 12$$

$$60 = 3a$$

$$20 = a$$

The solution is 20, which checks in the original equation.

When we multiply both sides of an equation by an expression containing the variable, we must be sure to check our solutions. The multiplication property of equality does not allow multiplication by zero. If the expression we multiply by contains the variable, then it has the possibility of being zero. In the last example, we multiplied both sides by $8(a - 4)$. This step is only valid if $a \neq 4$, and so we must restrict any solutions to values other than 4.

EXAMPLE 3 Solve: $\dfrac{x}{x-2} + \dfrac{2}{3} = \dfrac{2}{x-2}$.

SOLUTION The LCD is $3(x-2)$. We are assuming $x \neq 2$ when we multiply both sides of the equation by $3(x-2)$:

$$\frac{3(x-2)}{1} \cdot \left(\frac{x}{x-2} + \frac{2}{3}\right) = \frac{3(x-2)}{1} \cdot \frac{2}{x-2}$$

$$3x + (x-2) \cdot 2 = 3 \cdot 2$$

$$3x + 2x - 4 = 6$$

$$5x - 4 = 6$$

$$5x = 10$$

$$x = 2$$

Note In the process of solving the equation, we multiplied both sides by $3(x-2)$, solved for x, and got $x = 2$ for our solution. But when x is 2, the quantity $3(x-2) = 3(2-2) = 3(0) = 0$, which means we multiplied both sides of our equation by 0, which is not allowed under the multiplication property of equality.

The only possible solution is $x = 2$. Checking this value back in the original equation gives

$$\frac{2}{2-2} + \frac{2}{3} \stackrel{?}{=} \frac{2}{2-2}$$

$$\frac{2}{0} + \frac{2}{3} \stackrel{?}{=} \frac{2}{0}$$

The first and last terms are undefined. The proposed solution, $x = 2$, does not check in the original equation. The solution set is the empty set. There is no solution to the original equation. ◼

When the proposed solution to an equation is not actually a solution, it is called an *extraneous* solution. In the last example, $x = 2$ is an extraneous solution.

EXAMPLE 4 Solve: $\dfrac{5}{x^2 - 3x + 2} - \dfrac{1}{x-2} = \dfrac{1}{3x-3}$.

SOLUTION Writing the equation again with the denominators in factored form, we have

$$\frac{5}{(x-2)(x-1)} - \frac{1}{x-2} = \frac{1}{3(x-1)}$$

Note We can check the proposed solution in any of the equations obtained before multiplying through by the LCD. We cannot check the proposed solution in an equation obtained after multiplying through by the LCD because, if we have multiplied by 0, the resulting equations will not be equivalent to the original one.

The LCD is $3(x-2)(x-1)$. Assuming $x \neq 1$ and $x \neq 2$, we multiply through by the LCD to obtain

$$\frac{3(x-2)(x-1)}{1} \cdot \frac{5}{(x-2)(x-1)} - \frac{3(x-2)(x-1)}{1} \cdot \frac{1}{(x-2)}$$

$$= \frac{3(x-2)(x-1)}{1} \cdot \frac{1}{3(x-1)}$$

$$3 \cdot 5 - 3(x-1) \cdot 1 = (x-2) \cdot 1$$

$$15 - 3x + 3 = x - 2$$

$$-3x + 18 = x - 2$$

$$-4x + 18 = -2$$

$$-4x = -20$$

$$x = 5$$

Checking the proposed solution $x = 5$ in the original equation yields a true statement. ◼

EXAMPLE 5 Solve: $3 + \dfrac{1}{x} = \dfrac{10}{x^2}$.

SOLUTION To clear the equation of denominators, we multiply both sides by x^2 (thereby assuming $x \neq 0$):

$$\frac{x^2}{1}\left(3 + \frac{1}{x}\right) = \frac{x^2}{1}\left(\frac{10}{x^2}\right)$$

$$3(x^2) + \left(\frac{1}{x}\right)\frac{x^2}{1} = \left(\frac{10}{x^2}\right)\frac{x^2}{1}$$

$$3x^2 + x = 10$$

Rewrite in standard form, and solve:

$$3x^2 + x - 10 = 0$$

$$(3x - 5)(x + 2) = 0$$

$$3x - 5 = 0 \quad \text{or} \quad x + 2 = 0$$

$$x = \frac{5}{3} \quad \text{or} \quad x = -2$$

The solution set is $\left\{-2, \frac{5}{3}\right\}$. Both solutions check in the original equation. Remember: We have to check all solutions any time we multiply both sides of the equation by an expression that contains the variable, just to be sure we haven't multiplied by 0.

EXAMPLE 6 Solve: $\dfrac{y - 4}{y^2 - 5y} = \dfrac{2}{y^2 - 25}$.

SOLUTION Factoring each denominator,

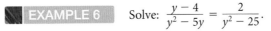

$$\frac{y - 4}{y(y - 5)} = \frac{2}{(y - 5)(y + 5)}$$

Note When multiplying by the LCD in Example 6, we are assuming y is not equal to -5, 0, or 5.

we find the LCD is $y(y - 5)(y + 5)$. Multiplying each side of the equation by the LCD clears the equation of denominators and leads us to our possible solutions:

$$\frac{y(y - 5)(y + 5)}{1} \cdot \frac{y - 4}{y(y - 5)} = \frac{2}{(y - 5)(y + 5)} \cdot \frac{y(y - 5)(y + 5)}{1}$$

$$(y + 5)(y - 4) = 2y$$

$$y^2 + y - 20 = 2y \qquad \text{Multiply out the left side}$$

$$y^2 - y - 20 = 0 \qquad \text{Add } -2y \text{ to each side}$$

$$(y - 5)(y + 4) = 0$$

$$y - 5 = 0 \quad \text{or} \quad y + 4 = 0$$

$$y = 5 \quad \text{or} \quad y = -4$$

The two possible solutions are 5 and -4. If we substitute -4 for y in the original equation, we find that it leads to a true statement. It is therefore a solution. On the other hand, if we substitute 5 for y in the original equation, we find that both sides of the equation are undefined. The only solution to our original equation is $y = -4$. The other possible solution $y = 5$ is extraneous.

Formulas

Remember that a formula is an equation containing more than one variable. In solving a formula we are not trying to find values for a variable. Rather, we want to isolate the specified variable.

If the formula contains rational expressions, we can multiply both sides of the equation by the LCD of all denominators to clear the equation of fractions as we have been doing in the previous examples.

EXAMPLE 7   Solve for y: $x = \dfrac{y - 4}{y - 2}$

SOLUTION   To solve for y, we first multiply each side by $y - 2$ to obtain

$$x(y - 2) = y - 4$$

$$xy - 2x = y - 4 \qquad \text{Distributive property}$$

$$xy - y = 2x - 4 \qquad \text{Collect all terms containing } y \text{ on the left side}$$

$$y(x - 1) = 2x - 4 \qquad \text{Factor } y \text{ from each term on the left side}$$

$$y = \frac{2x - 4}{x - 1} \qquad \text{Divide each side by } x - 1$$

EXAMPLE 8   Solve the formula $\dfrac{1}{x} = \dfrac{1}{b} + \dfrac{1}{a}$ for x.

SOLUTION   We begin by multiplying both sides by the least common denominator xab. As you can see from our previous examples, multiplying both sides of an equation by the LCD is equivalent to multiplying each term of both sides by the LCD:

$$\frac{xab}{1} \cdot \frac{1}{x} = \frac{1}{b} \cdot \frac{xab}{1} + \frac{1}{a} \cdot \frac{xab}{1}$$

$$ab = xa + xb$$

$$ab = (a + b)x \qquad \text{Factor } x \text{ from the right side}$$

$$\frac{ab}{a + b} = x$$

We know we are finished because the variable we were solving for is alone on one side of the equation and does not appear on the other side.

Getting Ready for Class

After reading through the preceding section, respond in your own words and in complete sentences.

A. Explain how a least common denominator can be used to solve an equation containing rational expressions.

B. Why does multiplying both sides of an equation by the LCD sometimes place restrictions on the variable?

C. What is an extraneous solution?

D. How can we determine if a solution is extraneous?

Solve each of the following equations. Be sure to check for any extraneous solutions.

1. $\dfrac{x}{5} + 4 = \dfrac{5}{3}$

2. $\dfrac{x}{5} = \dfrac{x}{2} - 9$

3. $\dfrac{a}{3} + 2 = \dfrac{4}{5}$

4. $\dfrac{a}{4} + \dfrac{1}{2} = \dfrac{2}{3}$

5. $\dfrac{y}{2} + \dfrac{y}{4} + \dfrac{y}{6} = 3$

6. $\dfrac{y}{3} - \dfrac{y}{6} + \dfrac{y}{2} = 1$

7. $\dfrac{5}{2x} = \dfrac{1}{x} + \dfrac{3}{4}$

8. $\dfrac{1}{2a} = \dfrac{2}{a} - \dfrac{3}{8}$

9. $\dfrac{1}{x} = \dfrac{1}{3} - \dfrac{2}{3x}$

10. $\dfrac{5}{2x} = \dfrac{2}{x} - \dfrac{1}{12}$

11. $\dfrac{2x}{x - 3} + 2 = \dfrac{2}{x - 3}$

12. $\dfrac{2}{x + 5} = \dfrac{2}{5} - \dfrac{x}{x + 5}$

13. $1 - \dfrac{1}{x} = \dfrac{12}{x^2}$

14. $2 + \dfrac{5}{x} = \dfrac{3}{x^2}$

15. $y - \dfrac{4}{3y} = -\dfrac{1}{3}$

16. $\dfrac{y}{2} - \dfrac{4}{y} = -\dfrac{7}{2}$

17. $\dfrac{x + 2}{x + 1} = \dfrac{1}{x + 1} + 2$

18. $\dfrac{x + 6}{x + 3} = \dfrac{3}{x + 3} + 2$

19. $\dfrac{3}{a - 2} = \dfrac{2}{a - 3}$

20. $\dfrac{5}{a + 1} = \dfrac{4}{a + 2}$

21. $6 - \dfrac{5}{x^2} = \dfrac{7}{x}$

22. $10 - \dfrac{3}{x^2} = -\dfrac{1}{x}$

23. $\dfrac{1}{x - 1} - \dfrac{1}{x + 1} = \dfrac{3x}{x^2 - 1}$

24. $\dfrac{5}{x - 1} + \dfrac{2}{x - 1} = \dfrac{4}{x + 1}$

25. $\dfrac{2}{x - 3} + \dfrac{x}{x^2 - 9} = \dfrac{4}{x + 3}$

26. $\dfrac{2}{x + 5} + \dfrac{3}{x + 4} = \dfrac{2x}{x^2 + 9x + 20}$

27. $\dfrac{3}{2} - \dfrac{1}{x - 4} = \dfrac{-2}{2x - 8}$

28. $\dfrac{2}{x} - \dfrac{1}{x + 1} = \dfrac{-2}{5x + 5}$

29. $\dfrac{t - 4}{t^2 - 3t} = \dfrac{-2}{t^2 - 9}$

30. $\dfrac{t + 3}{t^2 - 2t} = \dfrac{10}{t^2 - 4}$

31. $\dfrac{3}{y - 4} - \dfrac{2}{y + 1} = \dfrac{5}{y^2 - 3y - 4}$

32. $\dfrac{1}{y + 2} - \dfrac{2}{y - 3} = \dfrac{-2y}{y^2 - y - 6}$

33. $\dfrac{2}{1 + a} = \dfrac{3}{1 - a} + \dfrac{5}{a}$

34. $\dfrac{1}{a + 3} - \dfrac{a}{a^2 - 9} = \dfrac{2}{3 - a}$

35. $\dfrac{3}{2x - 6} - \dfrac{x + 1}{4x - 12} = 4$

36. $\dfrac{2x - 3}{5x + 10} + \dfrac{3x - 2}{4x + 8} = 1$

37. $\dfrac{y + 2}{y^2 - y} - \dfrac{6}{y^2 - 1} = 0$

38. $\dfrac{y + 3}{y^2 - y} - \dfrac{8}{y^2 - 1} = 0$

39. $\dfrac{4}{2x - 6} - \dfrac{12}{4x + 12} = \dfrac{12}{x^2 - 9}$

40. $\dfrac{1}{x + 2} + \dfrac{1}{x - 2} = \dfrac{4}{x^2 - 4}$

41. $\dfrac{2}{y^2 - 7y + 12} - \dfrac{1}{y^2 - 9} = \dfrac{4}{y^2 - y - 12}$

42. $\dfrac{1}{y^2 + 5y + 4} + \dfrac{3}{y^2 - 1} = \dfrac{-1}{y^2 + 3y - 4}$

43. Solve each equation.

    a. $6x - 2 = 0$ **b.** $\dfrac{6}{x} - 2 = 0$ **c.** $\dfrac{x}{6} - 2 = -\dfrac{1}{2}$

    d. $\dfrac{6}{x} - 2 = -\dfrac{1}{2}$ **e.** $\dfrac{6}{x^2} + 6 = \dfrac{20}{x}$

44. Solve each equation.

    a. $5x - 2 = 0$ **b.** $5 - \dfrac{2}{x} = 0$ **c.** $\dfrac{x}{2} - 5 = -\dfrac{3}{4}$

    d. $\dfrac{2}{x} - 5 = -\dfrac{3}{4}$ **e.** $-\dfrac{3}{x} + \dfrac{2}{x^2} = 5$

45. Paying Attention to Instructions Work each problem according to the instructions given.

    a. Divide: $\dfrac{6}{x^2 - 2x - 8} \div \dfrac{x + 3}{x + 2}$

    b. Add: $\dfrac{6}{x^2 - 2x - 8} + \dfrac{x + 3}{x + 2}$

    c. Solve: $\dfrac{6}{x^2 - 2x - 8} + \dfrac{x + 3}{x + 2} = 2$

46. Paying Attention to Instructions Work each problem according to the instructions given.

    a. Divide: $\dfrac{-10}{x^2 - 25} \div \dfrac{x - 4}{x - 5}$

    b. Add: $\dfrac{-10}{x^2 - 25} + \dfrac{x - 4}{x - 5}$

    c. Solve: $\dfrac{-10}{x^2 - 25} + \dfrac{x - 4}{x - 5} = \dfrac{4}{5}$

Solve for y.

47. $x = \dfrac{y - 3}{y - 1}$ **48.** $x = \dfrac{y - 2}{y - 3}$ **49.** $x = \dfrac{2y + 1}{3y + 1}$ **50.** $x = \dfrac{3y + 2}{5y + 1}$

Solve each formula for the indicated variable.

51. $P = \dfrac{h}{\lambda}$ for λ **52.** $x = \dfrac{1}{wc}$ for c **53.** $R = \dfrac{wa^2}{2l}$ for l

54. $R = \dfrac{wa}{2l}(2l - a)$ for l **55.** $\dfrac{a}{b} = \dfrac{x}{y}$ for b **56.** $\dfrac{2}{b} = \dfrac{x - 1}{3}$ for b

57. $\dfrac{a}{a - b} = c$ for b **58.** $\dfrac{a}{a - b} = c$ for a **59.** $\dfrac{x + 3}{a} = \dfrac{x}{b}$ for x

60. $2m - \dfrac{1}{n} = P$ for n **61.** $\dfrac{1}{x} = \dfrac{1}{b} - \dfrac{1}{a}$ for x **62.** $\dfrac{1}{x} = \dfrac{1}{b} - \dfrac{1}{a}$ for a

63. $\dfrac{x}{y} - \dfrac{2}{z} = \dfrac{m}{3}$ for y **64.** $\dfrac{a}{b} + \dfrac{c}{d} = \dfrac{e}{f}$ for d

Applying the Concepts

65. Geometry From plane geometry and the principle of similar triangles, the relationship between y_1, y_2, and h shown in Figure 3 can be expressed as

$$\frac{1}{h} = \frac{1}{y_1} + \frac{1}{y_2}$$

Two poles are 12 feet high and 8 feet high. If a cable is attached to the top of each one and stretched to the bottom of the other, what is the height above the ground at which the two wires will meet?

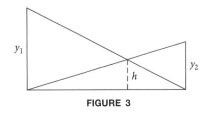

FIGURE 3

66. Kayak Race In a kayak race, the participants must paddle a kayak 450 meters down a river and then return 450 meters up the river to the starting point (Figure 4). Susan has correctly deduced that the total time t (in seconds) depends on the speed c (in meters per second) of the water according to the following expression:

$$t = \frac{450}{v + c} + \frac{450}{v - c}$$

where v is the speed of the kayak relative to the water (the speed of the kayak in still water).

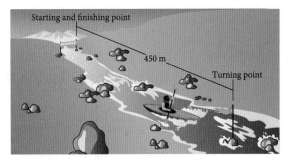

FIGURE 4

a. Fill in the following table.

Time	Speed of Kayak Relative to the Water	Current of the River
t(sec)	v(m/sec)	c(m/sec)
240		1
300		2
	4	3
	3	1
540	3	
	3	3

b. If the kayak race were conducted in the still waters of a lake, do you think that the total time of a given participant would be greater than, equal to, or smaller than the time in the river? Justify your answer.

c. Suppose Peter can drive his kayak at 4.1 meters per second and that the speed of the current is 4.1 meters per second. What will happen when Peter makes the turn and tries to come back up the river? How does this situation show up in the equation for total time?

Learning Objectives Assessment

The following problems can be used to help assess if you have successfully met the learning objectives for this section.

67. Solve: $\dfrac{3}{x+1} - \dfrac{1}{2} = \dfrac{x}{x+1}$.

    a. 2   **b.** $\dfrac{5}{3}$ **c.** $\dfrac{5}{2}$ **d.** $\varnothing$

68. In solving Problem 67, what restriction is placed on the variable?

    a. None   **b.** $x \neq 0$ **c.** $x \neq 1$ **d.** $x \neq -1$

69. Which value is an extraneous solution to the following equation?

$$\frac{x}{x+2} + \frac{3}{x-2} = \frac{12}{x^2-4}$$

    a. -3 **b.** 0   **c.** -2 **d.** 2

70. Solve $\dfrac{x}{y} + z = \dfrac{w}{r}$ for y.

    a. $y = \dfrac{r(x+z)}{w}$ **b.** $y = \dfrac{xr}{w-rz}$ **c.** $y = \dfrac{x(r-z)}{w}$ **d.** $y = \dfrac{rx+z}{w}$

Getting Ready for the Next Section

Multiply.

71. $39.3 \cdot 60$ **72.** $1{,}100 \cdot 60 \cdot 60$

Divide. Round to the nearest tenth, if necessary.

73. $65{,}000 \div 5{,}280$ **74.** $3{,}960{,}000 \div 5{,}280$

Multiply.

75. $2x\left(\dfrac{1}{x} + \dfrac{1}{2x}\right)$ **76.** $3x\left(\dfrac{1}{x} + \dfrac{1}{3x}\right)$

Solve.

77. $12(x+3) + 12(x-3) = 3(x^2-9)$ **78.** $40 + 2x = 60 - 3x$

79. $\dfrac{1}{10} - \dfrac{1}{12} = \dfrac{1}{x}$ **80.** $\dfrac{1}{x} + \dfrac{1}{2x} = 2$

Applications

Learning Objectives

In this section, we will learn how to:

1. Solve number problems.

2. Solve rate problems.

3. Solve work problems.

Introduction

We begin this section with some application problems, the solutions to which involve equations that contain rational expressions. As you will see, the solutions to the examples show only the essential steps from our Blueprint for Problem Solving. Recall that Step 1 was done mentally; we read the problem and mentally list the items that are known and the items that are unknown. This is an essential part of problem solving. Now that you have had experience with application problems, however, you are doing Step 1 automatically.

Number Problems

VIDEO EXAMPLES

SECTION 6.6

EXAMPLE 1   One number is twice another. The sum of their reciprocals is 2. Find the numbers.

SOLUTION   Let x be the smaller number. The larger number is $2x$. Their reciprocals are $\frac{1}{x}$ and $\frac{1}{2x}$. The equation that describes the situation is

$$\frac{1}{x} + \frac{1}{2x} = 2$$

Multiplying both sides by the LCD $2x$, we have

$$2x \cdot \frac{1}{x} + 2x \cdot \frac{1}{2x} = 2x(2)$$

$$2 + 1 = 4x$$

$$3 = 4x$$

$$x = \frac{3}{4}$$

The smaller number is $\frac{3}{4}$. The larger is $2\left(\frac{3}{4}\right) = \frac{6}{4} = \frac{3}{2}$. Adding their reciprocals, we have

$$\frac{4}{3} + \frac{2}{3} = \frac{6}{3} = 2$$

The sum of the reciprocals of $\frac{3}{4}$ and $\frac{3}{2}$ is 2. ∎

Rate Problems

EXAMPLE 2   Two families from the same neighborhood plan a ski trip together. The first family is driving a newer vehicle and makes the 455-mile trip at a speed 5 miles per hour faster than the second family who is traveling in an older vehicle. The second family takes a half-hour longer to make the trip. What are the speeds of the two families?

SOLUTION The following table will be helpful in finding the equation necessary to solve this problem.

	d(distance)	r(rate)	t(time)
First Family			
Second Family			

If we let x be the speed of the second family, then the speed of the first family will be $x + 5$. Both families travel the same distance of 455 miles. Putting this information into the table we have

	d	r	t
First Family	455	$x + 5$	
Second Family	455	x	

To fill in the last two spaces in the table, we use the relationship $d = r \cdot t$. Since the last column of the table is the time, we solve the equation $d = r \cdot t$ for t and get

$$t = \frac{d}{r}$$

Taking the distance and dividing by the rate (speed) for each family, we complete the table.

	d	r	t
First Family	455	$x + 5$	$\frac{455}{x + 5}$
Second Family	455	x	$\frac{455}{x}$

Reading the problem again, we find that the time for the second family is longer than the time for the first family by one-half hour. In other words, the time for the second family can be found by adding one-half hour to the time for the first family, or

Time for first family + One-half hour = Time for second family

$$\frac{455}{x + 5} + \frac{1}{2} = \frac{455}{x}$$

Multiplying both sides by the LCD of $2x(x + 5)$ gives

$$2x \cdot (455) + x(x + 5) \cdot 1 = 455 \cdot 2(x + 5)$$

$$910x + x^2 + 5x = 910x + 4550$$

$$x^2 + 5x - 4550 = 0$$

$$(x + 70)(x - 65) = 0$$

$$x = -70 \quad \text{or} \quad x = 65$$

Since we cannot have a negative speed, the only solution is $x = 65$. Then

$$x + 5 = 65 + 5 = 70$$

The speed of the first family is 70 miles per hour, and the speed of the second family is 65 miles per hour.

EXAMPLE 3   The current of a river is 3 miles per hour. It takes a motorboat a total of 3 hours to travel 12 miles upstream and return 12 miles downstream. What is the speed of the boat in still water?

SOLUTION   This time we let x be the speed of the boat in still water. If there was no current, the rate for the boat would be x in both directions. However, because of the current, the boat will go slower upstream (against the current) and faster downstream (with the current), and this decrease/increase in speed will depend upon how fast the current is. Thus, we have

Upstream rate $= x - 3$ Boat goes 3 miles per hour slower

Downstream rate $= x + 3$ Boat goes 3 miles per hour faster

We can now fill in our table with both distances and both rates.

	d	r	t
Upstream	12	$x - 3$	
Downstream	12	$x + 3$	

The last two boxes can be filled in using the relationship

$$t = \frac{d}{r}$$

	d	r	t
Upstream	12	$x - 3$	$\frac{12}{x - 3}$
Downstream	12	$x + 3$	$\frac{12}{x + 3}$

The total time for the trip up and back is 3 hours:

Time upstream + Time downstream = Total time

$$\frac{12}{x - 3} \quad + \quad \frac{12}{x + 3} \quad = \quad 3$$

Multiplying both sides by $(x - 3)(x + 3)$, we have

$$12(x + 3) + 12(x - 3) = 3(x^2 - 9)$$
$$12x + 36 + 12x - 36 = 3x^2 - 27$$
$$3x^2 - 24x - 27 = 0$$
$$x^2 - 8x - 9 = 0 \qquad \text{Divide both sides by 3}$$
$$(x - 9)(x + 1) = 0$$
$$x = 9 \quad \text{or} \quad x = -1$$

The speed of the motorboat in still water is 9 miles per hour. (We don't use $x = -1$ because the speed of the motorboat cannot be a negative number.)

Work Problems

If you are able to complete some task in x units of time by working at a steady rate, then in one unit of time you will have completed only a fraction of the task, with this fraction given by $\frac{1}{x}$.

For instance, I can sweep the floor at my house in 10 minutes. This means that in one minute, I will have swept just one-tenth of the floor. In other words, my rate of work (sweeping the floor) is $\frac{1}{10}$ floor per minute.

The next two examples show how we can solve work problems by considering the fraction of the task completed in each unit of time.

EXAMPLE 4   Allison can wash the car in 30 minutes. If she and Kaitlin work together, then it only takes them 20 minutes to wash the car. How long would it take Kaitlin to wash the car by herself?

SOLUTION   Let x be the time it takes for Kaitlin to wash the car on her own. By herself, Allison will do $\frac{1}{30}$ the job in one minute. If Kaitlin is helping, then in that same minute she will do an additional $\frac{1}{x}$ the job. Because we know they will complete $\frac{1}{20}$ the job each minute while working together, we have the following equation:

$$\text{In 1 minute}$$

$$\begin{bmatrix} \text{Amount done} \\ \text{by Allison} \end{bmatrix} + \begin{bmatrix} \text{Amount done} \\ \text{by Kaitlin} \end{bmatrix} = \begin{bmatrix} \text{Fraction of car} \\ \text{washed by both} \end{bmatrix}$$

$$\frac{1}{30} \qquad + \qquad \frac{1}{x} \qquad = \qquad \frac{1}{20}$$

Multiplying by the LCD $= 60x$, we have

$$60x \cdot \frac{1}{30} + 60x \cdot \frac{1}{x} = 60x \cdot \frac{1}{20}$$

$$2x + 60 = 3x$$

$$60 = x$$

It would take Kaitlin 60 minutes to wash the car by herself.

EXAMPLE 5   An inlet pipe can fill a pool in 10 hours, while the drain can empty it in 12 hours. If the pool is empty and both the inlet pipe and drain are open, how long will it take to fill the pool?

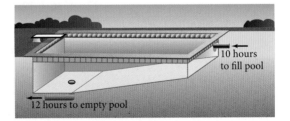

SOLUTION Let x be the time it takes to fill the pool with both pipes open. If the inlet pipe can fill the pool in 10 hours, then in 1 hour it is $\frac{1}{10}$ full. If the outlet pipe empties the pool in 12 hours, then in 1 hour it is $\frac{1}{12}$ empty. If the pool can be filled in x hours with both the inlet pipe and the drain open, then in 1 hour it is $\frac{1}{x}$ full when both pipes are open.

Here is the equation:

In 1 hour

$$\begin{bmatrix} \text{Amount filled} \\ \text{by inlet pipe} \end{bmatrix} - \begin{bmatrix} \text{Amount emptied} \\ \text{by the drain} \end{bmatrix} = \begin{bmatrix} \text{Fraction of pool} \\ \text{filled with both pipes} \end{bmatrix}$$

$$\frac{1}{10} \qquad - \qquad \frac{1}{12} \qquad = \qquad \frac{1}{x}$$

Multiplying through by $60x$, we have

$$60x \cdot \frac{1}{10} - 60x \cdot \frac{1}{12} = 60x \cdot \frac{1}{x}$$

$$6x - 5x = 60$$

$$x = 60$$

It takes 60 hours to fill the pool if both the inlet pipe and the drain are open. ∎

Getting Ready for Class

After reading through the preceding section, respond in your own words and in complete sentences.

A. Briefly list the steps in the Blueprint for Problem Solving that you have used previously to solve application problems.

B. If one number is twice another, write an expression for the sum of their reciprocals.

C. Write a formula for the relationship between distance, rate, and time.

D. Why did we add fractions in Example 4, but subtract them in Example 5?

Problem Set 6.6

Solve the following word problems. Be sure to show the equation in each case.

Number Problems

1. One number is 3 times another. The sum of their reciprocals is $\frac{20}{3}$. Find the numbers.

2. One number is 3 times another. The sum of their reciprocals is $\frac{4}{9}$. Find the numbers.

3. The sum of a number and its reciprocal is $\frac{10}{3}$. Find the number.

4. The sum of a number and twice its reciprocal is $\frac{27}{5}$. Find the number.

5. The sum of the reciprocals of two consecutive integers is $\frac{7}{12}$. Find the two integers.

6. Find two consecutive even integers, the sum of whose reciprocals is $\frac{3}{4}$.

7. If a certain number is added to the numerator and denominator of $\frac{7}{9}$, the result is $\frac{5}{6}$. Find the number.

8. Find the number you would add to both the numerator and denominator of $\frac{8}{11}$ so that the result would be $\frac{6}{7}$.

Rate Problems

9. The speed of a boat in still water is 5 miles per hour. If the boat travels 3 miles downstream in the same amount of time it takes to travel 1.5 miles upstream, what is the speed of the current?

   a. Let x be the speed of the current. Complete the distance and rate columns in the table.

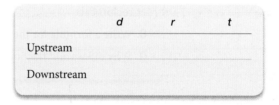

	d	r	t
Upstream			
Downstream			

   b. Now use the distance and rate information to complete the time column.

   c. What does the problem tell us about the two times? Use this fact to write an equation involving the two expressions for time.

   d. Solve the equation. Write your answer as a complete sentence.

10. A boat, which moves at 18 miles per hour in still water, travels 14 miles downstream in the same amount of time it takes to travel 10 miles upstream. Find the speed of the current.

   a. Let x be the speed of the current. Complete the distance and rate columns in the table.

	d	r	t
Upstream			
Downstream			

   b. Now use the distance and rate information to complete the time column.

   c. What does the problem tell us about the two times? Use this fact to write an equation involving the two expressions for time.

   d. Solve the equation. Write your answer as a complete sentence.

11. The current of a river is 2 miles per hour. A boat travels to a point 8 miles upstream and back again in 3 hours. What is the speed of the boat in still water?

12. A motorboat travels at 4 miles per hour in still water. It goes 12 miles upstream and 12 miles back again in a total of 8 hours. Find the speed of the current of the river.

13. Train A has a speed 15 miles per hour greater than that of train B. If train A travels 150 miles in the same time train B travels 120 miles, what are the speeds of the two trains?

   a. Let x be the speed of the train B. Complete the distance and rate columns in the table.

	d	r	t
Train A			
Train B			

   b. Now use the distance and rate information to complete the time column.

   c. What does the problem tell us about the two times? Use this fact to write an equation involving the two expressions for time.

   d. Solve the equation. Write your answer as a complete sentence.

14. A train travels 30 miles per hour faster than a car. If the train covers 120 miles in the same time the car covers 80 miles, what are the speeds of each of them?

    a. Let x be the speed of the car. Complete the distance and rate columns in the table.

	d	r	t
Car			
Train			

    b. Now use the distance and rate information to complete the time column.

    c. What does the problem tell us about the two times? Use this fact to write an equation involving the two expressions for time.

    d. Solve the equation. Write your answer as a complete sentence.

15. A small airplane flies 810 miles from Los Angeles, California, to Portland, Oregon, with an average speed of 270 miles per hour. An hour and a half after the plane leaves, a Boeing 747 leaves Los Angeles for Portland. Both planes arrive in Portland at the same time. What was the average speed of the 747?

16. Lou leaves for a cross-country excursion on a bicycle traveling at 20 miles per hour. His friends are driving the trip and will meet him at several rest stops along the way. The first stop is scheduled 30 miles from the original starting point. If the people driving leave 15 minutes after Lou from the same place, how fast will they have to drive to reach the first rest stop at the same time as Lou?

17. A tour bus leaves Sacramento every Friday evening at 5:00 P.M. for a 270-mile trip to Las Vegas. This week, however, the bus leaves at 5:30 P.M. To arrive in Las Vegas on time, the driver drives 6 miles per hour faster than usual. What is the bus' usual speed?

18. A bakery delivery truck leaves the bakery at 5:00 A.M. each morning on its 140-mile route. One day the driver gets a late start and does not leave the bakery until 5:30 A.M. To finish her route on time the driver drives 5 miles per hour faster than usual. At what speed does she usually drive?

Work Problems

19. Doug can paint the living room in 4 hours. It would take Elaine 6 hours to paint the same room. How long would it take Doug and Elaine, working together, to paint the living room?

20. It takes Andy 45 minutes to clean the kitchen. His sister Anita can do the job in 30 minutes. How long would it take Andy and Anita to clean the kitchen if they worked together?

21. Hector can detail a car in 60 minutes. If Alonzo helps him, together they can finish the job in 36 minutes. How long would it take Alonzo to detail a car by himself?

22. Jesse can build a wood deck in 8 hours. With Grecia's help, they can build the deck in only 5 hours. If Grecia were working alone, how long would she need to build the deck?

23. A water tank can be filled by an inlet pipe in 8 hours. It takes twice that long for the outlet pipe to empty the tank. How long will it take to fill the tank if both pipes are open?

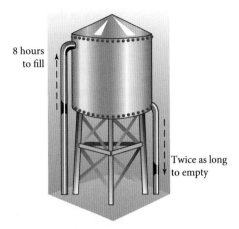

8 hours
to fill

Twice as long
to empty

24. A sink can be filled from the faucet in 5 minutes. It takes only 3 minutes to empty the sink when the drain is open. If the sink is full and both the faucet and the drain are open, how long will it take to empty the sink?

25. It takes 10 hours to fill a pool with the inlet pipe. It can be emptied in 15 hours with the outlet pipe. If the pool is half full to begin with, how long will it take to fill it from there if both pipes are open?

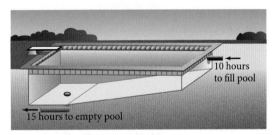

10 hours
to fill pool

15 hours to empty pool

26. A sink is one-quarter full when both the faucet and the drain are opened. The faucet alone can fill the sink in 6 minutes, while it takes 8 minutes to empty it with the drain. How long will it take to fill the remaining three quarters of the sink?

27. A sink has two faucets: one for hot water and one for cold water. The sink can be filled by a cold water faucet in 3.5 minutes. If both faucets are open, the sink is filled in 2.1 minutes. How long does it take to fill the sink with just the hot water faucet open?

28. A water tank is being filled by two inlet pipes. Pipe A can fill the tank in $4\frac{1}{2}$ hours, but both pipes together can fill the tank in 2 hours. How long does it take to fill the tank using only pipe B?

Miscellaneous Problems

29. Rhind Papyrus Nearly 4,000 years ago, Egyptians worked mathematical exercises involving reciprocals. The ***Rhind Papyrus*** contains a wealth of such problems, and one of them is as follows:

"A quantity and its two thirds are added together, one third of this is added, then one third of the sum is taken, and the result is 10."

Write an equation and solve this exercise.

30. Photography For clear photographs, a camera must be properly focused. Professional photographers use a mathematical relationship relating the distance from the camera lens to the object being photographed, a; the distance from the lens to the film, b; and the focal length of the lens, f. These quantities, a, b, and f, are related by the equation

$$\frac{1}{a} + \frac{1}{b} = \frac{1}{f}$$

A camera has a focal length of 3 inches. If the lens is 5 inches from the film, how far should the lens be placed from the object being photographed for the camera to be perfectly focused?

The Periodic Table If you take a chemistry class, you will work with the Periodic Table of Elements. Figure 3 shows three of the elements listed in the periodic table. As you can see, the bottom number in each figure is the molecular weight of the element. In chemistry, a mole is the amount of a substance that will give the weight in grams equal to the molecular weight. For example, 1 mole of lead is 207.2 grams.

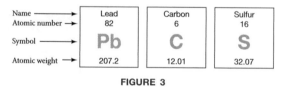

FIGURE 3

31. Chemistry For the element carbon, 1 mole = 12.01 grams.

   a. To the nearest gram, how many grams of carbon are in 2.5 moles of carbon?

   b. How many moles of carbon are in 39 grams of carbon? Round to the nearest hundredth.

32. Chemistry For the element sulfur, 1 mole = 32.07 grams.

   a. How many grams of sulfur are in 3 moles of sulfur?

   b. How many moles of sulfur are found in 80.2 grams of sulfur?

Learning Objectives Assessment

The following problems can be used to help assess if you have successfully met the learning objectives for this section.

33. If the difference between a number and its reciprocal is $\frac{3}{2}$, which of the following could be the number?

a. 3 **b.** $\frac{1}{3}$ **c.** -2 **d.** $-\frac{1}{2}$

34. The current of a stream is 1 mile per hour. If x is the speed of a kayak in still water, then which of the following represents the rate of the kayak while traveling upstream?

a. x **b.** $x + 1$ **c.** $1 - x$ **d.** $x - 1$

35. It takes 10 hours to fill a pool with the inlet pipe, but if a garden hose is used it takes 15 hours to fill it. How long would it take to fill the pool if both the inlet pipe and garden hose are running?

a. 6 hours   **b.** 5 hours   **c.** 30 hours   **d.** 12.5 hours

Getting Ready for the Next Section

36. Evaluate $\dfrac{x - 4}{x - 2}$ if $x = -2$.

Determine any values of x for which the expression is undefined.

37. $\dfrac{x - 4}{x - 2}$ **38.** $\dfrac{2}{x + 4}$

Solve for x.

39. $x^2 - 9 = 0$ **40.** $t^2 - 25 = 0$

Reduce to lowest terms.

41. $\dfrac{x^2 - a^2}{x - a}$ **42.** $\dfrac{2xh + h^2}{h}$

Find the slope of the line passing through the given points.

43. $(1, 4), (3, 9)$ **44.** $(-2, 3), (1, -3)$

45. If $f(x) = x^2 + 5$, find $f(0)$ and $f(-1)$.

46. If $g(x) = 3x - 5$, find $g(2)$ and $g(a)$.

Find the intercepts for each graph.

47. $y = 2x - 4$ **48.** $y = -\dfrac{1}{2}x + 1$

Learning Objectives

In this section, we will learn how to:

1. Evaluate a rational function.

2. Find the domain for a rational function.

3. Find and simplify a difference quotient.

4. Graph a rational function.

Introduction

If you have ever taken a home videotape to be transferred to DVD, you know the amount you pay for the transfer depends on the number of copies you have made: The more copies you have made, the lower the charge per copy. The following demand function gives the price (in dollars) per tape $p(x)$ a company charges for making x DVDs.

$$p(x) = \frac{2(x + 60)}{x + 5}$$

The graph in Figure 1 shows this function from $x = 0$ to $x = 100$. As you can see, the more copies that are made, the lower the price per copy.

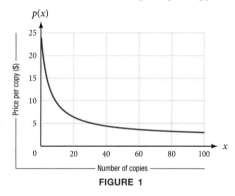

FIGURE 1

This is an example of a rational function. We can extend our knowledge of rational expressions to rational functions with the following definition:

(dĕf) DEFINITION *rational function*

A *rational function* is any function that can be written in the form

$$f(x) = \frac{P(x)}{Q(x)}$$

where $P(x)$ and $Q(x)$ are polynomials and $Q(x) \neq 0$.

VIDEO EXAMPLES

SECTION 6.7

EXAMPLE 1   For the rational function $f(x) = \dfrac{x-4}{x-2}$, find $f(0)$, $f(-4)$, $f(4)$, $f(-2)$, and $f(2)$.

SOLUTION   To find these function values, we substitute the given value of x into the rational expression, and then simplify if possible.

$$f(0) = \frac{0-4}{0-2} = \frac{-4}{-2} = 2$$

$$f(-4) = \frac{-4-4}{-4-2} = \frac{-8}{-6} = \frac{4}{3}$$

$$f(4) = \frac{4-4}{4-2} = \frac{0}{2} = 0$$

$$f(-2) = \frac{-2-4}{-2-2} = \frac{-6}{-4} = \frac{3}{2}$$

$$f(2) = \frac{2-4}{2-2} = \frac{-2}{0} \qquad \text{Undefined}$$

Because the rational function in Example 1 is not defined when x is 2, the domain of that function does not include 2. We have more to say about the domain of a rational function next.

The Domain of a Rational Function

If the domain of a rational function is not specified, it is assumed to be all real numbers for which the function is defined. That is, the **domain** of the rational function

$$f(x) = \frac{P(x)}{Q(x)}$$

is all x for which $Q(x)$ is nonzero.

EXAMPLE 2   Find the domain for each function.

a. $f(x) = \dfrac{x-4}{x-2}$ **b.** $g(x) = \dfrac{x^2+5}{x+1}$ **c.** $h(x) = \dfrac{x}{x^2-9}$

SOLUTION

a. The domain for $f(x) = \dfrac{x-4}{x-2}$ is $\{x \mid x \neq 2\}$.

b. The domain for $g(x) = \dfrac{x^2+5}{x+1}$ is $\{x \mid x \neq -1\}$.

c. The domain for $h(x) = \dfrac{x}{x^2-9}$ is $\{x \mid x \neq -3, x \neq 3\}$.

Notice that, for these functions, $f(2), g(-1), h(-3)$, and $h(3)$ are all undefined, and that is why the domains are written as shown.

Difference Quotients

The diagram in Figure 2 is an important diagram from calculus. Although it may look complicated, the point of it is simple: The slope of the line passing through the points P and Q is given by the formula

$$\text{Slope of line through } PQ = m = \frac{f(x) - f(a)}{x - a}$$

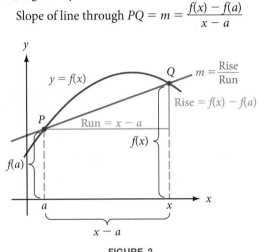

FIGURE 2

The expression $\frac{f(x) - f(a)}{x - a}$ is called a **difference quotient**. When $f(x)$ is a polynomial, it will be a rational expression.

EXAMPLE 3   If $f(x) = 3x - 5$, find $\dfrac{f(x) - f(a)}{x - a}$.

SOLUTION

$$\frac{f(x) - f(a)}{x - a} = \frac{(3x - 5) - (3a - 5)}{x - a}$$
$$= \frac{3x - 3a}{x - a}$$
$$= \frac{3(x - a)}{x - a}$$
$$= 3$$

EXAMPLE 4   If $f(x) = x^2 - 4$, find $\dfrac{f(x) - f(a)}{x - a}$ and simplify.

SOLUTION   Because $f(x) = x^2 - 4$ and $f(a) = a^2 - 4$, we have

$$\frac{f(x) - f(a)}{x - a} = \frac{(x^2 - 4) - (a^2 - 4)}{x - a}$$
$$= \frac{x^2 - 4 - a^2 + 4}{x - a}$$
$$= \frac{x^2 - a^2}{x - a}$$
$$= \frac{(x + a)(x - a)}{x - a} \qquad \text{Factor and divide out common factor}$$
$$= x + a$$

The diagram in Figure 3 is similar to the one in Figure 2. The main difference is in how we label the points. From Figure 3, we can see another difference quotient that gives us the slope of the line through the points P and Q.

$$\text{Slope of line through } PQ = m = \frac{f(x + h) - f(x)}{h}$$

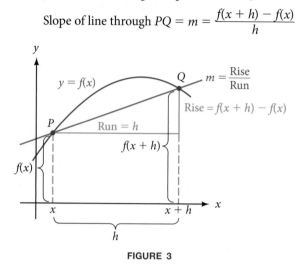

FIGURE 3

Examples 5 and 6 use the same functions used in Examples 1 and 2, but this time the alternative difference quotient is used.

EXAMPLE 5   If $f(x) = 3x - 5$, find $\dfrac{f(x + h) - f(x)}{h}$.

SOLUTION   The expression $f(x + h)$ is given by

$$f(x + h) = 3(x + h) - 5$$
$$= 3x + 3h - 5$$

Using this result gives us

$$\frac{f(x + h) - f(x)}{h} = \frac{(3x + 3h - 5) - (3x - 5)}{h}$$
$$= \frac{3h}{h}$$
$$= 3$$

EXAMPLE 6   If $f(x) = x^2 - 4$, find $\dfrac{f(x + h) - f(x)}{h}$.

SOLUTION   The expression $f(x + h)$ is given by

$$f(x + h) = (x + h)^2 - 4$$
$$= x^2 + 2xh + h^2 - 4$$

Using this result gives us

$$\frac{f(x + h) - f(x)}{h} = \frac{(x^2 + 2xh + h^2 - 4) - (x^2 - 4)}{h}$$
$$= \frac{2xh + h^2}{h}$$
$$= \frac{h(2x + h)}{h}$$
$$= 2x + h$$

Graphing Rational Functions

In our next example, we investigate the graph of a rational function.

EXAMPLE 7 Graph the rational function $f(x) = \dfrac{6}{x-2}$.

SOLUTION To find the y-intercept, we let x equal 0.

$$\text{When } x = 0: \qquad y = \frac{6}{0-2} = \frac{6}{-2} = -3 \qquad y\text{-intercept}$$

The graph will not cross the x-axis. If it did, we would have a solution to the equation

$$0 = \frac{6}{x-2}$$

which has no solution because there is no number to divide 6 by to obtain 0.

The graph of our equation is shown in Figure 4 along with a table giving values of x and y that satisfy the equation. Notice that y is undefined when x is 2. This means that the graph will not cross the vertical line $x = 2$. (If it did, there would be a value of y for $x = 2$.) The line $x = 2$ is called a ***vertical asymptote*** of the graph. The graph will get very close to the vertical asymptote, but will never touch or cross it.

x	y
-4	-1
-1	-2
0	-3
1	-6
2	Undefined
3	6
4	3
5	2

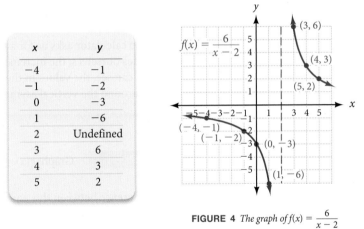

FIGURE 4 *The graph of $f(x) = \dfrac{6}{x-2}$*

If you were to graph $y = \frac{6}{x}$ on the coordinate system in Figure 4, you would see that the graph of $y = \frac{6}{x-2}$ is the graph of $y = \frac{6}{x}$ with all points shifted 2 units to the right.

USING TECHNOLOGY *More About Example 7*

We know the graph of $f(x) = \frac{6}{x-2}$ will not cross the vertical asymptote $x = 2$ because replacing x with 2 in the equation gives us an undefined expression, meaning there is no value of y to associate with $x = 2$. We can use a graphing calculator to explore the behavior of this function when x gets closer and closer to 2 by using the table function on the calculator. We want to put our own values for X into the table, so we set the independent variable to Ask. (On a TI-83/84, use the $\boxed{\text{TBLSET}}$ key to set up the table.)

USING TECHNOLOGY   *More About Example 7 continued*

To see how the function behaves as x gets close to 2, we let X take on values of 1.9, 1.99, and 1.999. Then we move to the other side of 2 and let X become 2.1, 2.01, and 2.001.

TABLE SETUP
TblStart = 0
ΔTbl = 1
Indpnt: Auto **Ask**
Depend: **Auto** Ask

Plot1 Plot2 Plot3
\Y₁ ∎ 6/(X − 2)
\Y₂ =
\Y₃ =
\Y₄ =
\Y₅ =
\Y₆ =
\Y₇ =

The table will look like this:

X	Y₁	
1.9	−60	
1.99	−600	
1.999	−6000	
2.1	60	
2.01	600	
2.001	6000	

Again, the calculator asks us for a table increment. Because we are inputting the x values ourselves, the increment value does not matter.

As you can see, the values in the table support the shape of the curve in Figure 4 around the vertical asymptote $x = 2$.

EXAMPLE 8   Graph: $g(x) = \dfrac{6}{x + 2}$.

SOLUTION

The only difference between this equation and the equation in Example 7 is in the denominator. This graph will have the same shape as the graph in Example 7, but the vertical asymptote will be $x = -2$ instead of $x = 2$. Figure 5 shows the graph.

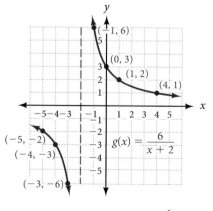

FIGURE 5 *The graph of $g(x) = \dfrac{6}{x + 2}$*

Notice that the graphs shown in Figures 4 and 5 are both graphs of functions because no vertical line will cross either graph in more than one place. Notice the similarities and differences in our two functions,

$$f(x) = \frac{6}{x-2} \quad \text{and} \quad g(x) = \frac{6}{x+2}$$

and their graphs. The vertical asymptotes shown in Figures 4 and 5 correspond to the fact that both $f(2)$ and $g(-2)$ are undefined. The domain for the function f is all real numbers except $x = 2$, while the domain for g is all real numbers except $x = -2$.

We continue our investigation of the graphs of rational functions by considering the graph of a rational function with binomials in the numerator and denominator.

EXAMPLE 9 Graph the rational function $y = \dfrac{x-4}{x-2}$.

SOLUTION In addition to making a table to find some points on the graph, we can analyze the graph as follows:

1. The graph will have a y-intercept of 2, because when $x = 0$, $y = \dfrac{-4}{-2} = 2$.

2. To find the x-intercept, we let $y = 0$ to get

$$0 = \frac{x-4}{x-2}$$

The only way this expression can be 0 is if the numerator is 0, which happens when $x = 4$. (If you want to solve this equation, multiply both sides by $x - 2$. You will get the same solution, $x = 4$.)

3. The graph will have a vertical asymptote at $x = 2$, because $x = 2$ will make the denominator of the function 0, meaning y is undefined when x is 2.

4. The graph will have a **horizontal asymptote** at $y = 1$ because for very large values of x, $\frac{x-4}{x-2}$ is very close to 1. The larger x is, the closer $\frac{x-4}{x-2}$ is to 1. The same is true for very small values of x, such as $-1,000$ and $-10,000$.

Putting this information together with the ordered pairs in the table next to the figure, we have the graph shown in Figure 6.

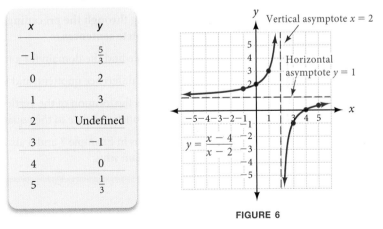

x	y
−1	$\frac{5}{3}$
0	2
1	3
2	Undefined
3	−1
4	0
5	$\frac{1}{3}$

FIGURE 6

USING TECHNOLOGY *More About Example 9*

In the previous example, we used technology to explore the graph of a rational function around a vertical asymptote. This time, we are going to explore the graph near the horizontal asymptote. In Figure 6, the horizontal asymptote is at $y = 1$. To show that the graph approaches this line as x becomes very large, we use the table function on our graphing calculator, with X taking values of 100, 1,000, and 10,000. To show that the graph approaches the line $y = 1$ on the left side of the coordinate system, we let X become -100, $-1,000$ and $-10,000$.

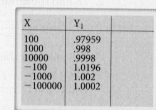

TABLE SETUP
TblStart = 0
ΔTbl = 1
Indpnt: Auto **Ask**
Depend: **Auto** Ask

Plot1 Plot2 Plot3
\Y₁ ▤ (X − 4)(X − 2)
\Y₂ =
\Y₃ =
\Y₄ =
\Y₅ =
\Y₆ =
\Y₇ =

The table will look like this:

X	Y₁	
100	.97959	
1000	.998	
10000	.9998	
−100	1.0196	
−1000	1.002	
−100000	1.0002	

As you can see, as x becomes very large in the positive direction, the graph approaches the line $y = 1$ from below. As x becomes very small in the negative direction, the graph approaches the line $y = 1$ from above.

Getting Ready for Class

After reading through the preceding section, respond in your own words and in complete sentences.

A. How do we find the domain for a rational function?

B. What is a difference quotient, and what does it represent?

C. How does the location of the vertical asymptote in the graph of a rational function relate to the equation of the function?

D. The graphs in Example 7 and 8 also have a horizontal asymptote. Where is it? Explain why.

1. If $g(x) = \frac{x+3}{x-1}$, find $g(0), g(-3), g(3), g(-1)$, and $g(1)$, if possible.

2. If $g(x) = \frac{x-2}{x-1}$, find $g(0), g(-2), g(2), g(-1)$, and $g(1)$, if possible.

3. If $h(t) = \frac{t-3}{t+1}$, find $h(0), h(-3), h(3), h(-1)$, and $h(1)$, if possible.

4. If $h(t) = \frac{t-2}{t+1}$, find $h(0), h(-2), h(2), h(-1)$, and $h(1)$, if possible.

State the domain for each rational function.

5. $f(x) = \frac{x-3}{x-1}$ **6.** $f(x) = \frac{x+4}{x-2}$ **7.** $g(x) = \frac{x^2-4}{x-2}$

8. $g(x) = \frac{x^2-9}{x-3}$ **9.** $h(t) = \frac{t-4}{t^2-16}$ **10.** $h(t) = \frac{t-5}{t^2-25}$

The graphs of two rational functions are given in Figures 7 and 8. Use the graphs to find the following.

11. a. $f(2)$ **b.** $f(-1)$ **c.** $f(0)$ **d.** $g(3)$

12. a. $g(6)$ **b.** $g(-1)$ **c.** $f(g(6))$ **d.** $g(f(-2))$

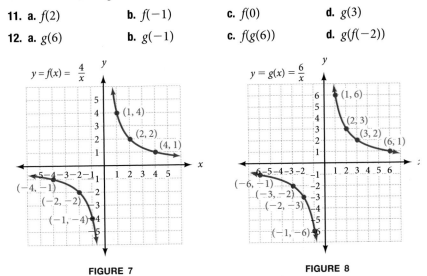

FIGURE 7   **FIGURE 8**

For the functions below, evaluate

     a. $\dfrac{f(x) - f(a)}{x - a}$ **b.** $\dfrac{f(x + h) - f(x)}{h}$

13. $f(x) = 4x$ **14.** $f(x) = -3x$

15. $f(x) = 5x + 3$ **16.** $f(x) = 6x - 5$

17. $f(x) = x^2$ **18.** $f(x) = 3x^2$

19. $f(x) = x^2 + 1$ **20.** $f(x) = x^2 - 3$

21. $f(x) = x^2 - 3x + 4$ **22.** $f(x) = x^2 + 4x - 7$

23. Difference Quotient For each rational function below, find the difference quotient

$$\frac{f(x) - f(a)}{x - a}$$

a. $f(x) = \dfrac{4}{x}$ **b.** $f(x) = \dfrac{1}{x + 1}$ **c.** $f(x) = \dfrac{1}{x^2}$

24. Difference Quotient For each rational function below, find the difference quotient

$$\frac{f(x + h) - f(x)}{h}$$

a. $f(x) = \dfrac{4}{x}$ **b.** $f(x) = \dfrac{1}{x + 1}$ **c.** $f(x) = \dfrac{1}{x^2}$

25. Let $f(x) = \dfrac{1}{x - 3}$ and $g(x) = \dfrac{1}{x + 3}$ and find x if

a. $f(x) + g(x) = \dfrac{5}{8}$ **b.** $\dfrac{f(x)}{g(x)} = 5$ **c.** $f(x) = g(x)$

26. Let $f(x) = \dfrac{4}{x + 2}$ and $g(x) = \dfrac{4}{x - 2}$ and find x if

a. $f(x) - g(x) = -\dfrac{4}{3}$ **b.** $\dfrac{g(x)}{f(x)} = -7$ **c.** $f(x) = -g(x)$

Graph each function. Show the vertical asymptote.

27. $f(x) = \dfrac{1}{x - 3}$ **28.** $f(x) = \dfrac{1}{x + 3}$ **29.** $f(x) = \dfrac{4}{x + 2}$ **30.** $f(x) = \dfrac{4}{x - 2}$

31. $g(x) = \dfrac{2}{x - 4}$ **32.** $g(x) = \dfrac{2}{x + 4}$ **33.** $g(x) = \dfrac{6}{x + 1}$ **34.** $g(x) = \dfrac{6}{x - 1}$

Graph each rational function. In each case, show the vertical asymptote, the horizontal asymptote, and any intercepts that exist.

35. $f(x) = \dfrac{x - 3}{x - 1}$ **36.** $f(x) = \dfrac{x + 4}{x - 2}$ **37.** $f(x) = \dfrac{x + 3}{x - 1}$

38. $f(x) = \dfrac{x - 2}{x - 1}$ **39.** $g(x) = \dfrac{x - 3}{x + 1}$ **40.** $g(x) = \dfrac{x - 2}{x + 1}$

Applying the Concepts

41. Diet The following rational function is the one we mentioned in the introduction to this chapter. The quantity $W(x)$ is the weight (in pounds) of the person after x weeks of dieting. Use the function to fill in the table. Then compare your results with the graph in the chapter introduction.

$$W(x) = \frac{80(2x + 15)}{x + 6}$$

Weeks	Weight (lb)
x	$W(x)$
0	
1	
4	
12	
24	

42. Drag Racing The following rational function gives the speed $V(x)$, in miles per hour, of a dragster at each second x during a quarter-mile race.

Use the function to fill in the table.

$$V(x) = \frac{340x}{x + 3}$$

Time (sec)	Speed (mi/hr)
x	$V(x)$
0	
1	
2	
3	
4	
5	
6	

Learning Objectives Assessment

43. If $f(x) = \dfrac{x + 5}{x - 4}$, find $f(-2)$.

a. $-\dfrac{3}{2}$ **b.** $-\dfrac{1}{2}$ **c.** $\dfrac{7}{6}$ **d.** Undefined

44. Find the domain for $f(x) = \dfrac{x - 2}{x^2 - 4}$.

a. $\{x \mid x \neq 0\}$ **b.** $\{x \mid x \neq -2\}$ **c.** $\{x \mid x \neq 2\}$ **d.** $\{x \mid x \neq 2, x \neq -2\}$

45. If $f(x) = x^2 - 5$, find and simplify $\dfrac{f(x + h) - f(x)}{h}$.

a. 1   **b.** h **c.** $\dfrac{h - 10}{h}$ **d.** $2x + h$

46. The graph of $f(x) = \dfrac{x + 5}{x - 4}$ would have which of the following as a vertical asymptote?

a. $x = -5$ **b.** $x = 4$ **c.** $y = 1$ **d.** $y = 0$

Maintaining Your Skills

Reviewing these problems will help clarify the different methods we have used in this chapter.

Perform the indicated operations.

47. $\dfrac{2a + 10}{a^3} \cdot \dfrac{a^2}{3a + 15}$

48. $\dfrac{4a + 8}{a^2 - a - 6} \div \dfrac{a^2 + 7a + 12}{a^2 - 9}$

49. $(x^2 - 9)\left(\dfrac{x + 2}{x + 3}\right)$

50. $\dfrac{1}{x + 4} + \dfrac{8}{x^2 - 16}$

51. $\dfrac{2x - 7}{x - 2} - \dfrac{x - 5}{x - 2}$

52. $2 + \dfrac{25}{5x - 1}$

Simplify each expression.

53. $\dfrac{\dfrac{1}{x} - \dfrac{1}{3}}{\dfrac{1}{x} + \dfrac{1}{3}}$

54. $\dfrac{1 - \dfrac{9}{x^2}}{1 - \dfrac{1}{x} - \dfrac{6}{x^2}}$

Solve each equation.

55. $\dfrac{x}{x - 3} + \dfrac{3}{2} = \dfrac{3}{x - 3}$

56. $1 - \dfrac{3}{x} = \dfrac{-2}{x^2}$

Chapter 6 Summary

EXAMPLES

Rational Numbers and Expressions [6.1]

1. $\frac{3}{4}$ is a rational number. $\frac{x-3}{x^2-9}$ is a rational expression.

A *rational number* is any number that can be expressed as the ratio of two integers:

$$\text{Rational numbers} = \left\{ \frac{a}{b} \,\middle|\, a \text{ and } b \text{ are integers}, b \neq 0 \right\}$$

A *rational expression* is any quantity that can be expressed as the ratio of two polynomials:

$$\text{Rational expressions} = \left\{ \frac{P}{Q} \,\middle|\, P \text{ and } Q \text{ are polynomials}, Q \neq 0 \right\}$$

Rational Expressions that are Undefined [6.1]

2. $\frac{x-3}{x^2-9}$ is undefined when

$$
\begin{aligned}
x^2 - 9 &= 0 \\
(x+3)(x-3) &= 0 \\
x + 3 = 0 \quad &\text{or} \quad x - 3 = 0 \\
x = -3 \quad & \qquad x = 3
\end{aligned}
$$

A rational expression will be undefined for any value of the variable that makes its denominator equal to zero. To find these values, set the denominator equal to zero and solve the resulting equation.

Property of Rational Expressions [6.1]

If P, Q, and K are polynomials with $Q \neq 0$ and $K \neq 0$, then

$$\frac{P}{Q} = \frac{PK}{QK}$$

which is to say that multiplying or dividing the numerator and denominator of a rational expression by the same nonzero quantity always produces an equivalent rational expression.

Reducing to Lowest Terms [6.1]

3. $\dfrac{x-3}{x^2-9} = \dfrac{x-3}{(x-3)(x+3)}$

$= \dfrac{1}{x+3}$

To reduce a rational expression to lowest terms, we first factor the numerator and denominator and then divide the numerator and denominator by any factors they have in common.

Multiplication [6.2]

4. $\dfrac{x+1}{x^2-4} \cdot \dfrac{x+2}{3x+3}$

$= \dfrac{(x+1)(x+2)}{(x-2)(x+2)(3)(x+1)}$

$= \dfrac{1}{3(x-2)}$

To multiply two rational numbers or rational expressions, multiply numerators and multiply denominators. In symbols,

$$\frac{P}{Q} \cdot \frac{R}{S} = \frac{PR}{QS} \qquad (Q \neq 0 \text{ and } S \neq 0)$$

In practice, we don't really multiply, but rather, we factor and then divide out common factors.

Division [6.2]

5. $\dfrac{x^2 - y^2}{x^3 + y^3} \div \dfrac{x - y}{x^2 - xy + y^2}$

$= \dfrac{x^2 - y^2}{x^3 + y^3} \cdot \dfrac{x^2 - xy + y^2}{x - y}$

$= \dfrac{(x + y)(x - y)(x^2 - xy + y^2)}{(x + y)(x^2 - xy + y^2)(x - y)}$

$= 1$

To divide one rational expression by another, we use the definition of division to rewrite our division problem as an equivalent multiplication problem. To divide by a rational expression we multiply by its reciprocal. In symbols,

$$\frac{P}{Q} \div \frac{R}{S} = \frac{P}{Q} \cdot \frac{S}{R} = \frac{PS}{QR} \qquad (Q \neq 0, S \neq 0, R \neq 0)$$

Least Common Denominator [6.3]

6. The LCD for $\dfrac{2}{x - 3}$ and $\dfrac{3}{5}$ is $5(x - 3)$.

The *least **common denominator**,* LCD, for a set of denominators is the smallest quantity divisible by each of the denominators.

Addition and Subtraction [6.3]

7. $\dfrac{2}{x - 3} + \dfrac{3}{5}$

$= \dfrac{2}{x - 3} \cdot \dfrac{5}{5} + \dfrac{3}{5} \cdot \dfrac{x - 3}{x - 3}$

$= \dfrac{3x + 1}{5(x - 3)}$

If P, Q, and R represent polynomials, $R \neq 0$, then

$$\frac{P}{R} + \frac{Q}{R} = \frac{P + Q}{R} \quad \text{and} \quad \frac{P}{R} - \frac{Q}{R} = \frac{P - Q}{R}$$

When adding or subtracting rational expressions with different denominators, we must find the LCD for all denominators and change each rational expression to an equivalent expression that has the LCD.

Complex Fractions [6.4]

8. $\dfrac{\dfrac{1}{x} + \dfrac{1}{y}}{\dfrac{1}{x} - \dfrac{1}{y}} = \dfrac{xy\left(\dfrac{1}{x} + \dfrac{1}{y}\right)}{xy\left(\dfrac{1}{x} - \dfrac{1}{y}\right)}$

$= \dfrac{y + x}{y - x}$

A rational expression that contains, in its numerator or denominator, other rational expressions is called a ***complex fraction***. One method of simplifying a complex fraction is to multiply the numerator and denominator by the LCD for all denominators.

Equations Involving Rational Expressions [6.5]

9. Solve $\dfrac{x}{2} + 3 = \dfrac{1}{3}$.

$6\left(\dfrac{x}{2}\right) + 6 \cdot 3 = 6 \cdot \dfrac{1}{3}$

$3x + 18 = 2$

$x = -\dfrac{16}{3}$

To solve an equation involving rational expressions, we first find the LCD for all denominators appearing on either side of the equation. We then multiply both sides by the LCD to clear the equation of all fractions and solve as usual.

Rational Function [6.7]

10. $f(x) = \frac{x + 4}{x + 2}$ is a rational function. Its domain is $\{x \mid x \neq -2\}$.

A **rational function** is any function that can be written in the form

$$f(x) = \frac{P(x)}{Q(x)}$$

where $P(x)$ and $Q(x)$ are polynomials and $Q(x) \neq 0$. The domain of a rational function is $\{x \mid Q(x) \neq 0\}$.

Difference Quotient [6.7]

11. If $f(x) = 5x + 3$, then

$$\frac{f(x) - f(a)}{x - a} = \frac{(5x + 3) - (5a + 3)}{x - a}$$

$$= \frac{5x - 5a}{x - a}$$

$$= \frac{5(x - a)}{x - a}$$

$$= 5$$

The expressions

$$\frac{f(x) - f(a)}{x - a} \quad \text{and} \quad \frac{f(x + h) - f(x)}{h}$$

are called **difference quotients**. They represent the slope of a line passing through two points on the graph of $f(x)$. If $f(x)$ is a polynomial, then these expressions will be rational expressions.

⚠ COMMON MISTAKES

1. Attempting to divide the numerator and denominator of a rational expression by a quantity that is not a factor of both. Like this:

$$\frac{x^2 - \overset{3}{9x} + \overset{2}{20}}{\underset{1}{x^2} - \underset{1}{3x} - 10} \quad \text{Mistake}$$

This makes no sense at all. The numerator and denominator must be factored completely before any factors they have in common can be recognized:

$$\frac{x^2 - 9x + 20}{x^2 - 3x - 10} = \frac{(x - 5)(x - 4)}{(x - 5)(x + 2)}$$

$$= \frac{x - 4}{x + 2}$$

2. Forgetting to check solutions to equations involving rational expressions. When we multiply both sides of an equation by a quantity containing the variable, we must be sure to check for extraneous solutions.

Chapter 6 Test

Find any values for which the rational expression is undefined. [6.1]

1. $\dfrac{x+1}{x^2+3x-10}$

2. $\dfrac{4x}{x^2+9}$

Reduce to lowest terms. [6.1]

3. $\dfrac{x^2-y^2}{x-y}$

4. $\dfrac{2x^2-5x+3}{2x^2-x-3}$

5. $\dfrac{7-x}{2x-14}$

6. $\dfrac{4-a^2}{a^2-4a+4}$

Multiply and divide as indicated. [6.2]

7. $\dfrac{a^2-16}{5a-15}\cdot\dfrac{10(a-3)^2}{a^2-7a+12}$

8. $\dfrac{a^4-81}{a^2+9}\div\dfrac{a^2-8a+15}{4a-20}$

9. $\dfrac{x^3-8}{2x^2-9x+10}\div\dfrac{x^2+2x+4}{2x^2+x-15}$

Add and subtract as indicated. [6.3]

10. $\dfrac{4}{21}+\dfrac{6}{35}$

11. $\dfrac{3}{4}-\dfrac{1}{2}+\dfrac{5}{8}$

12. $\dfrac{a}{a^2-9}+\dfrac{3}{a^2-9}$

13. $\dfrac{1}{x}+\dfrac{2}{x-3}$

14. $\dfrac{4x}{x^2+6x+5}-\dfrac{3x}{x^2+5x+4}$

15. $\dfrac{2x+8}{x^2+4x+3}-\dfrac{x+4}{x^2+5x+6}$

Simplify each complex fraction. [6.4]

16. $\dfrac{3-\dfrac{1}{a+3}}{3+\dfrac{1}{a+3}}$

17. $\dfrac{1-\dfrac{9}{x^2}}{1+\dfrac{1}{x}-\dfrac{6}{x^2}}$

Solve each of the following equations. [6.5]

18. $\dfrac{1}{x}+3=\dfrac{4}{3}$

19. $\dfrac{x}{x-3}+3=\dfrac{3}{x-3}$

20. $\dfrac{y+3}{2y}+\dfrac{5}{y-1}=\dfrac{1}{2}$

21. $1-\dfrac{1}{x}=\dfrac{6}{x^2}$

Solve the following applications. Be sure to show the equation in each case. [6.6]

22. Number Problem What number must be subtracted from the denominator of $\frac{10}{23}$ to make the result $\frac{1}{3}$?

23. Speed of a Boat The current of a river is 2 miles per hour. It takes a motorboat a total of 3 hours to travel 8 miles upstream and return 8 miles downstream. What is the speed of the boat in still water?

24. Filling a Pool An inlet pipe can fill a pool in 10 hours, and the drain can empty it in 15 hours. If the pool is half full and both the inlet pipe and the drain are left open, how long will it take to fill the pool the rest of the way?

State the domain for each rational function. [6.7]

25. $f(x)=\dfrac{x-3}{x+2}$

26. $h(t)=\dfrac{t-6}{t^2-36}$

27. If $f(x)=4x-3$, find and simplify the difference quotient $\dfrac{f(x)-f(a)}{x-a}$. [6.7]

28. If $f(x)=x^2+2x-5$, find and simplify the difference quotient $\dfrac{f(x+h)-f(x)}{h}$. [6.7]

29. Graph $f(x)=\dfrac{x+4}{x-1}$. [6.7]

Roots and Rational Exponents

7

Chapter Outline

iStockphoto.com © trait2lumiere

Ecology and conservation are topics that interest most college students. If our rivers and oceans are to be preserved for future generations, we need to work to eliminate pollution from our waters. If a river is flowing at 1 meter per second and a pollutant is entering the river at a constant rate, the shape of the pollution plume can often be modeled by the simple equation

$$y = \sqrt{x}$$

The following table and graph were produced from the equation.

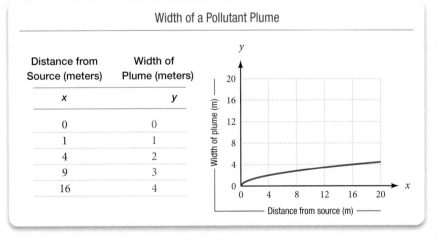

Width of a Pollutant Plume

Distance from Source (meters)	Width of Plume (meters)
x	y
0	0
1	1
4	2
9	3
16	4

To visualize how the graph models the pollutant plume, imagine that the river is flowing from left to right, parallel to the x-axis, with the x-axis as one of its banks. The pollutant is entering the river from the bank at $(0, 0)$.

By modeling pollution with mathematics, we can use our knowledge of mathematics to help control and eliminate pollution.

Success Skills

iStockphoto.com © IPGGutenbergUKLtd

If you have made it this far, then you have the study skills necessary to be successful in this course. Success skills are more general in nature and will help you with all your classes and ensure your success in college as well.

Let's start with a question:

Question: What quality is most important for success in any college course?

Answer: Independence. You want to become an independent learner.

We all know people like this. They are generally happy. They don't worry about getting the right instructor, or whether or not things work out every time. They have a confidence that comes from knowing that they are responsible for their success or failure in the goals they set for themselves.

Here are some of the qualities of an independent learner:

- Intends to succeed.
- Doesn't let setbacks deter them.
- Knows their resources.
  - Instructor's office hours
  - Math lab
  - Student Solutions Manual
  - Group study
  - Internet
- Doesn't mistake activity for achievement.
- Has a positive attitude.

There are other traits as well. The first step in becoming an independent learner is doing a little self-evaluation and then making of list of traits that you would like to acquire. What skills do you have that align with those of an independent learner? What attributes do you have that keep you from being an independent learner? What qualities would you like to obtain that you don't have now?

Roots and Radical Functions

Learning Objectives

In this section, we will learn how to:

1. Find the square root of a number.

2. Find the nth root of a number.

3. Graph a simple root function.

4. Find the domain for a root function.

Introduction

In Chapter 1, we developed notation (exponents) to give us the square, cube, or any other power of a number. For instance, if we wanted the square of 3, we wrote $3^2 = 9$. If we wanted the cube of 3, we wrote $3^3 = 27$. In this section, we will develop notation that will take us in the reverse direction, that is, from the square of a number, say 25, back to the original number, 5.

Figure 1 shows a square in which each of the four sides is 1 inch long. To find the square of the length of the diagonal c, we apply the Pythagorean theorem:

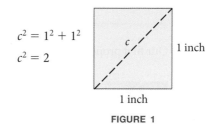

$$c^2 = 1^2 + 1^2$$
$$c^2 = 2$$

FIGURE 1

Note The Pythagoreans, whose motto was "All is number," believed everything in the world was constructed from whole numbers. The discovery that $\sqrt{2}$ is an irrational number (which cannot be expressed as a ratio of whole numbers) was a terrible shock to their society and endangered their beliefs. Therefore, it was kept as a closely guarded secret. According to legend, Hippasus was murdered for divulging this secret.

Because we know that c is positive and that its square is 2, we call c the ***positive square root*** of 2, and we write $c = \sqrt{2}$. This leads us to the following definition.

> (def **DEFINITION** *positive square root*
>
> If x is a nonnegative real number, then the expression $\sqrt{x}$ is called the ***positive square root*** of x and is such that
> $$(\sqrt{x})^2 = x$$
> *In words:* $\sqrt{x}$ is the positive number we square to get x.
> **Note:** $\sqrt{x}$ is sometimes referred to as the ***principal square root*** of x.

The negative square root of x, denoted as $-\sqrt{x}$, is the negative number we square to get x.

VIDEO EXAMPLES

SECTION 7.1

EXAMPLE 1 The positive square root of 64 is 8 because 8 is the positive number with the property $8^2 = 64$. The negative square root of 64 is -8 because -8 is the negative number whose square is 64. We can summarize both these facts by saying

$$\sqrt{64} = 8 \quad \text{and} \quad -\sqrt{64} = -8$$

It is a common mistake to assume that an expression like $\sqrt{25}$ indicates both square roots, 5 and -5. The expression $\sqrt{25}$ indicates only the positive square root of 25, which is 5. If we want the negative square root, we must use a negative sign: $-\sqrt{25} = -5$.

The higher roots, cube roots, fourth roots, and so on, are defined by definitions similar to that of square roots.

Note We have restricted the even roots in this definition to nonnegative numbers. Even roots of negative numbers exist, but are not represented by real numbers. That is, $\sqrt{-4}$ is not a real number because there is no real number whose square is -4.

(def DEFINITION

If x is a real number and n is a positive integer, then

Positive square root of x, $\sqrt{x}$, is such that $(\sqrt{x})^2 = x$ $x \geq 0$

Cube root of x, $\sqrt[3]{x}$, is such that $(\sqrt[3]{x})^3 = x$

Positive fourth root of x, $\sqrt[4]{x}$, is such that $(\sqrt[4]{x})^4 = x$ $x \geq 0$

Fifth root of x, $\sqrt[5]{x}$, is such that $(\sqrt[5]{x})^5 = x$

$$\cdot \qquad \cdot \;\; \cdot$$
$$\cdot \qquad \cdot \;\; \cdot$$
$$\cdot \qquad \cdot \;\; \cdot$$

The nth root of x, $\sqrt[n]{x}$, is such that $(\sqrt[n]{x})^n = x$ $x \geq 0$ if n is even

Our first property for radicals is a direct result of the previous definition.

[Δ≠Σ PROPERTY   *Property 1 for Radicals*

If a is a real number and n is a positive integer, then

$$(\sqrt[n]{a})^n = a, a \geq 0 \text{ if } n \text{ is even}$$

EXAMPLES   Use Property 1 to simplify each expression.

2. $(\sqrt{3})^2 = 3$

3. $(\sqrt[3]{-11})^3 = -11$

4. $(\sqrt[5]{2y})^5 = 2y$

Notation An expression like $\sqrt[3]{8}$ that involves a root is called a ***radical expression***. In the expression $\sqrt[3]{8}$, the ^{3} is called the ***index***, the $\sqrt{}$ is the ***radical sign***, and 8 is called the ***radicand***. The index of a radical must be a positive integer greater than 1. If no index is written, it is assumed to be 2.

The following is a table of the most common roots used in this book. Any of the roots that are unfamiliar should be memorized.

Square Roots		Cube Roots	Fourth Roots
$\sqrt{0} = 0$	$\sqrt{49} = 7$	$\sqrt[3]{0} = 0$	$\sqrt[4]{0} = 0$
$\sqrt{1} = 1$	$\sqrt{64} = 8$	$\sqrt[3]{1} = 1$	$\sqrt[4]{1} = 1$
$\sqrt{4} = 2$	$\sqrt{81} = 9$	$\sqrt[3]{8} = 2$	$\sqrt[4]{16} = 2$
$\sqrt{9} = 3$	$\sqrt{100} = 10$	$\sqrt[3]{27} = 3$	$\sqrt[4]{81} = 3$
$\sqrt{16} = 4$	$\sqrt{121} = 11$	$\sqrt[3]{64} = 4$	
$\sqrt{25} = 5$	$\sqrt{144} = 12$	$\sqrt[3]{125} = 5$	
$\sqrt{36} = 6$	$\sqrt{169} = 13$		

Roots and Negative Numbers

When dealing with negative numbers and radicals, the only restriction concerns negative numbers under even roots. We can have negative signs in front of radicals and negative numbers under odd roots and still obtain real numbers. Here are some examples to help clarify this. In the last section of this chapter, we will see how to deal with square roots of negative numbers.

EXAMPLES   Simplify each expression, if possible.

5. $\sqrt[3]{-8} = -2$ because $(-2)^3 = -8$.

6. $\sqrt{-4}$ is not a real number because there is no real number whose square is -4.

7. $\sqrt[5]{-32} = -2$ because $(-2)^5 = -32$.

8. $\sqrt[4]{-81}$ is not a real number because there is no real number we can raise to the fourth power and obtain -81.

When there is a negative sign in front of a radical, we can simply interpret the negative as indicating the "opposite of" a value, as shown in the next two examples.

EXAMPLES   Simplify each expression, if possible.

9. $-\sqrt{25} = -(\sqrt{25})$

$= -(5)$

$= -5$

10. $-\sqrt[3]{-64} = -(\sqrt[3]{-64})$

$= -(-4)$

$= 4$

Variables Under a Radical

From the preceding examples, it is clear that we must be careful that we do not try to take an even root of a negative number. For this reason, we will assume that all variables appearing under a radical sign represent nonnegative numbers.

EXAMPLES Assume all variables represent nonnegative numbers, and simplify each expression as much as possible.

11. $\sqrt{25a^4b^6} = 5a^2b^3$ because $(5a^2b^3)^2 = 25a^4b^6$.

12. $\sqrt[3]{x^6y^{12}} = x^2y^4$ because $(x^2y^4)^3 = x^6y^{12}$.

13. $\sqrt[4]{81r^8s^{20}} = 3r^2s^5$ because $(3r^2s^5)^4 = 81r^8s^{20}$. ∎

The Spiral of Roots

FACTS FROM GEOMETRY *The Pythagorean Theorem (Again)*

Now that we have had some experience working with square roots, we can rewrite the Pythagorean theorem using a square root. If triangle ABC is a right triangle with $C = 90°$, then the length of the longest side is the *positive square root* of the sum of the squares of the other two sides (see Figure 2).

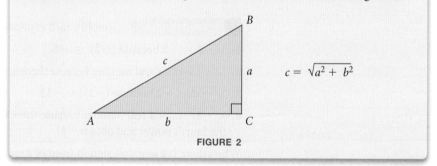

$$c = \sqrt{a^2 + b^2}$$

FIGURE 2

In the introduction to this section, we showed how the Pythagorean theorem can be used to represent $\sqrt{2}$ geometrically as the length of the diagonal of a square. In a similar manner, the Pythagorean theorem can be used to construct the attractive spiral shown in Figure 3.

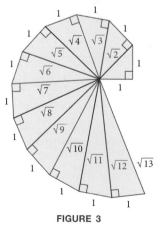

FIGURE 3

This spiral is called the Spiral of Roots because each of the diagonals is the positive square root of one of the positive integers. To construct the spiral, we begin by drawing two line segments, each of length 1, at right angles to each other. Then we use the Pythagorean theorem to find the length of the diagonal. Figure 4 illustrates this procedure.

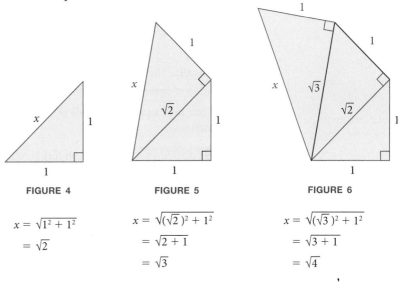

FIGURE 4   **FIGURE 5**   **FIGURE 6**

$x = \sqrt{1^2 + 1^2}$ $x = \sqrt{(\sqrt{2})^2 + 1^2}$ $x = \sqrt{(\sqrt{3})^2 + 1^2}$

$\quad = \sqrt{2}$ $\quad = \sqrt{2 + 1}$ $\quad = \sqrt{3 + 1}$

$\quad\quad\quad\quad = \sqrt{3}$ $\quad\quad\quad = \sqrt{4}$

Next, we construct a second triangle by connecting a line segment of length 1 to the end of the first diagonal so that the angle formed is a right angle. We find the length of the second diagonal using the Pythagorean theorem. Figure 5 illustrates this procedure. Continuing to draw new triangles by connecting line segments of length 1 to the end of each new diagonal, so that the angle formed is a right angle, the spiral of roots begins to appear (Figure 6).

USING TECHNOLOGY

As our preceding discussion indicates, the length of each diagonal in the spiral of roots is used to calculate the length of the next diagonal. The ANS key on a graphing calculator can be used effectively in a situation like this. To begin, we store the number 1 in the variable ANS. Next, we key in the formula used to produce each diagonal using ANS for the variable. After that, it is simply a matter of pressing ENTER, as many times as we like, to produce the lengths of as many diagonals as we like. Here is a summary of what we do:

Enter This	Display Shows
1 ENTER	1.000
√ (ANS² + 1) ENTER	1.414
ENTER	1.732
ENTER	2.000
ENTER	2.236

If you continue to press the ENTER key, you will produce decimal approximations for as many of the diagonals in the spiral of roots as you like.

Root Functions

If we use a root as the formula for a function, the result is called a *root function*. Here is the definition.

> (d͞ef   **DEFINITION**   *root function*
>
> For any positive integer n, the function given by
> $$f(x) = \sqrt[n]{x}$$
> is called the ***nth root function***.

In our next example we will sketch the graph of the square root and cube root functions.

EXAMPLE 14   Graph $f(x) = \sqrt{x}$ and $g(x) = \sqrt[3]{x}$.

SOLUTION   The graphs are shown in Figures 7 and 8. Notice that the graph of $f(x) = \sqrt{x}$ appears in the first quadrant only, because in the equation $y = \sqrt{x}$, x and y cannot be negative.

The graph of $g(x) = \sqrt[3]{x}$ appears in Quadrants 1 and 3 because the cube root of a positive number is also a positive number, and the cube root of a negative number is a negative number. That is, when x is positive, y will be positive, and when x is negative, y will be negative.

The graphs of both equations will contain the origin, because $y = 0$ when $x = 0$ in both equations.

x	y = f(x)
-4	Not real
-1	Not real
0	0
1	1
4	2
9	3
16	4

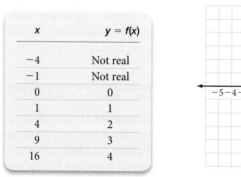

FIGURE 7

x	y = g(x)
-27	-3
-8	-2
-1	-1
0	0
1	1
8	2
27	3

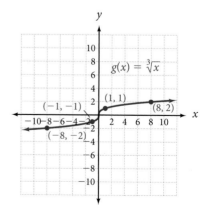

FIGURE 8

Domain of a Root Function

As we observed in Example 14, the square root function $y = \sqrt{x}$ will only give real number outputs if $x \geq 0$, but with the cube root function $y = \sqrt[3]{x}$, x can be any real number. This leads us to the following property:

> **[△≠Σ] PROPERTY** *Domain of a Root Function*
>
> The domain of $f(x) = \sqrt[n]{x}$ is
>
> | **Set-Builder Notation** | **Interval Notation** | |
> | --- | --- | --- |
> | $\{x \mid x \geq 0\}$ | $[0, \infty)$ | if n is even |
> | $\{x \mid x \text{ is any real number}\}$ | $(-\infty, \infty)$ | if n is odd |

EXAMPLES

15. The domain of $f(x) = \sqrt{x}$ is $[0, \infty)$.

16. The domain of $f(x) = \sqrt[3]{x} + 3$ is $(-\infty, \infty)$.

If the radical contains some expression involving x, we must make sure the radicand represents a nonnegative number if the index of the radical is even. We illustrate how this is done in our last example.

EXAMPLE 17   Find the domain for $f(x) = \sqrt{x + 2}$.

SOLUTION   The value of the radicand, $x + 2$, must be nonnegative. We can set up an appropriate inequality and solve for x.

$$\text{Radicand} \geq 0$$

$$x + 2 \geq 0$$

$$x \geq -2$$

The domain is $\{x \mid x \geq -2\}$.

Getting Ready for Class

After reading through the preceding section, respond in your own words and in complete sentences.

A. Every real number has two square roots. Explain the notation we use to tell them apart. Use the square roots of 3 for examples.

B. Explain why the square root of -4 is not a real number.

C. Explain why $(\sqrt[10]{2})^{10} = 2$.

D. When can the radicand of a root function be negative?

Problem Set 7.1

Find each of the following roots, if possible.

1. $\sqrt{144}$ **2.** $-\sqrt{144}$ **3.** $\sqrt{-144}$ **4.** $\sqrt{-49}$

5. $-\sqrt{49}$ **6.** $\sqrt{49}$ **7.** $\sqrt[3]{-27}$ **8.** $-\sqrt[3]{27}$

9. $-\sqrt[3]{-27}$ **10.** $\sqrt[4]{16}$ **11.** $-\sqrt[4]{16}$ **12.** $\sqrt[4]{-16}$

13. $\sqrt{0.04}$ **14.** $\sqrt{0.81}$ **15.** $\sqrt[3]{0.008}$ **16.** $\sqrt[3]{0.125}$

17. $\sqrt{\dfrac{1}{36}}$ **18.** $\sqrt{\dfrac{9}{25}}$ **19.** $\sqrt[3]{\dfrac{1}{8}}$ **20.** $\sqrt[3]{\dfrac{27}{64}}$

Use Property 1 for radicals to simplify each expression. Assume all variables represent nonnegative numbers.

21. $(\sqrt{5})^2$ **22.** $(\sqrt{8})^2$ **23.** $(\sqrt[3]{2})^3$ **24.** $(\sqrt[3]{-6})^3$

25. $(\sqrt[4]{10})^4$ **26.** $(\sqrt[5]{16})^5$ **27.** $(\sqrt{7x})^2$ **28.** $(\sqrt[3]{4a^2})^3$

Simplify each expression. Assume all variables represent nonnegative numbers.

29. $\sqrt{36a^8}$ **30.** $\sqrt{49a^{10}}$ **31.** $\sqrt[3]{27a^{12}}$ **32.** $\sqrt[3]{8a^{15}}$

33. $\sqrt[3]{x^3y^6}$ **34.** $\sqrt[3]{x^6y^3}$ **35.** $\sqrt[5]{32x^{10}y^5}$ **36.** $\sqrt[5]{32x^5y^{10}}$

37. $\sqrt[4]{16a^{12}b^{20}}$ **38.** $\sqrt[4]{81a^{24}b^8}$

Simplify. Assume all variables are nonnegative.

39. a. $\sqrt{25}$ **b.** $\sqrt{0.25}$ **c.** $\sqrt{2500}$ **d.** $\sqrt{0.0025}$

40. a. $\sqrt[3]{8}$ **b.** $\sqrt[3]{0.008}$ **c.** $\sqrt[3]{8,000}$ **d.** $\sqrt[3]{8 \times 10^{-6}}$

41. a. $\sqrt{16a^4b^8}$ **b.** $\sqrt[4]{16a^4b^8}$

42. a. $\sqrt[3]{64x^6y^{18}}$ **b.** $\sqrt[6]{64x^6y^{18}}$

Graph each root function.

43. $f(x) = 2\sqrt{x}$ **44.** $f(x) = -2\sqrt{x}$ **45.** $f(x) = \sqrt{x} - 2$

46. $f(x) = \sqrt{x} + 2$ **47.** $f(x) = \sqrt{x - 2}$ **48.** $f(x) = \sqrt{x + 2}$

49. $f(x) = 3\sqrt[3]{x}$ **50.** $f(x) = -3\sqrt[3]{x}$ **51.** $f(x) = \sqrt[3]{x} + 3$

52. $f(x) = \sqrt[3]{x} - 3$ **53.** $f(x) = \sqrt[3]{x + 3}$ **54.** $f(x) = \sqrt[3]{x - 3}$

Find the domain for each function.

55. $f(x) = \sqrt{x + 3}$ **56.** $f(x) = \sqrt{x - 3}$

57. $f(x) = \sqrt{x} + 3$ **58.** $f(x) = \sqrt{3x}$

59. $f(x) = \sqrt{2x - 10}$ **60.** $f(x) = \sqrt{3x + 6}$

61. $f(x) = \sqrt{5 - x}$ **62.** $f(x) = \sqrt{8 - x}$

63. $f(x) = \sqrt[3]{x + 4}$ **64.** $f(x) = \sqrt[3]{x - 4}$

65. $f(x) = \sqrt[3]{4x}$ **66.** $f(x) = \sqrt[3]{x} - 4$

Applying the Concepts

67. Chemistry Figure 9 shows part of a model of a magnesium oxide (MgO) crystal. Each corner of the square is at the center of one oxygen ion (O^{2-}), and the center of the middle ion is at the center of the square. The radius for each oxygen ion is 150 picometers (pm), and the radius for each magnesium ion (Mg^{2+}) is 60 picometers.

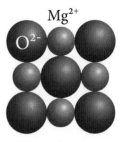

FIGURE 9

   a. Find the length of the side of the square. Write your answer in picometers.

   b. Find the length of the diagonal of the square. Write your answer in picometers.

   c. If 1 meter is 10^{12} picometers, give the length of the diagonal of the square in meters.

68. Geometry The length of each side of the cube shown in Figure 10 is 1 inch.

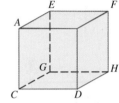

FIGURE 10

   a. Find the length of the diagonal CH.

   b. Find the length of the diagonal CF.

69. Spiral of Roots Construct your own spiral of roots by using a ruler. Draw the first triangle by using two 1-inch lines. The first diagonal will have a length of $\sqrt{2}$ inches. Each new triangle will be formed by drawing a 1-inch line segment at the end of the previous diagonal so the angle formed is 90°.

70. Spiral of Roots Construct a spiral of roots by using line segments of length 2 inches. The length of the first diagonal will be $2\sqrt{2}$ inches. The length of the second diagonal will be $2\sqrt{3}$ inches.

Learning Objectives Assessment

The following problems can be used to help assess if you have successfully met the learning objectives for this section.

71. Find: $-\sqrt{64}$.

   a. 8   **b.** -8 **c.** -4 **d.** Not a real number

72. Simplify: $\sqrt[3]{-64}$.

   a. -4 **b.** -8 **c.** 4   **d.** Not a real number

73. Which point lies on the graph of $f(x) = \sqrt{x + 4}$?

   a. $(4, 2)$ **b.** $(0, 4)$ **c.** $(-3, 1)$ **d.** $(-5, 1)$

74. Find the domain of $f(x) = \sqrt{x + 4}$.

   a. $\{x \mid x \geq 4\}$ **b.** $\{x \mid x \geq -4\}$ **c.** $\{x \mid x \geq 0\}$ **d.** $\{x \mid x \text{ is any real number}\}$

Getting Ready for the Next Section

Simplify. Assume all variables represent positive real numbers.

75. $\sqrt{25}$ **76.** $\sqrt{4}$ **77.** $\sqrt{16x^4y^2}$ **78.** $\sqrt{4x^6y^8}$

79. $\sqrt[3]{27}$ **80.** $\sqrt[3]{-8}$ **81.** $\sqrt[3]{8a^3b^3}$ **82.** $\sqrt[3]{64a^6b^3}$

83. -5^2 **84.** $(-5)^2$ **85.** 5^{-2} **86.** $\left(\dfrac{1}{5}\right)^{-2}$

87. $x^2 \cdot x^5$ **88.** $(x^2)^5$ **89.** $\dfrac{x^5}{x^2}$ **90.** $\left(\dfrac{x^2}{y^5}\right)^3$

Rational Exponents

Learning Objectives

In this section, we will learn how to:

1. Write a root using a rational exponent.
2. Write a rational exponent in radical notation.
3. Use rational exponents to simplify roots.
4. Simplify expressions containing rational exponents.

Introduction

We will now develop a second kind of notation involving exponents that will allow us to designate square roots, cube roots, and so on in another way.

Consider the equation $x = 8^{1/3}$. Although we have not encountered fractional exponents before, let's assume that all the properties of exponents hold in this case. Cubing both sides of the equation, we have

$$x^3 = (8^{1/3})^3$$
$$x^3 = 8^{(1/3)(3)}$$
$$x^3 = 8^1$$
$$x^3 = 8$$

The last line tells us that x is the number whose cube is 8. It must be true, then, that x is the cube root of 8, $x = \sqrt[3]{8}$. Because we started with $x = 8^{1/3}$, it follows that

$$8^{1/3} = \sqrt[3]{8}$$

It seems reasonable, then, to define fractional exponents as indicating roots. Here is the formal definition.

> **(def) DEFINITION**
>
> If x is a real number and n is a positive integer greater than 1, then
> $$x^{1/n} = \sqrt[n]{x} \qquad (x \geq 0 \text{ when } n \text{ is even})$$
> *In words:* The quantity $x^{1/n}$ is the nth root of x.

With this definition, we have a way of representing roots with exponents. Here are some examples.

VIDEO EXAMPLES

SECTION 7.2

EXAMPLES Write each expression as a root and then simplify, if possible.

1. $8^{1/3} = \sqrt[3]{8} = 2$

2. $36^{1/2} = \sqrt{36} = 6$

3. $-25^{1/2} = -\sqrt{25} = -5$

4. $(-25)^{1/2} = \sqrt{-25}$, which is not a real number

5. $\left(\dfrac{4}{9}\right)^{1/2} = \sqrt{\dfrac{4}{9}} = \dfrac{2}{3}$

The properties of exponents developed in Chapter 1 were applied to integer exponents only. We will now extend these properties to include rational exponents also. We do so without proof.

[Δ≠Σ] PROPERTY   *Properties of Exponents*

If a and b are real numbers and r and s are rational numbers, and a and b are nonnegative whenever r and s indicate even roots, then

1. $a^r \cdot a^s = a^{r+s}$ **4.** $a^{-r} = \dfrac{1}{a^r}$ $(a \neq 0)$

2. $(a^r)^s = a^{rs}$ **5.** $\left(\dfrac{a}{b}\right)^r = \dfrac{a^r}{b^r}$ $(b \neq 0)$

3. $(ab)^r = a^r b^r$ **6.** $\dfrac{a^r}{a^s} = a^{r-s}$ $(a \neq 0)$

Sometimes rational exponents can simplify our work with radicals. Here are Examples 12 and 13 from Section 7.1 again, but this time we will work them using rational exponents.

EXAMPLES   Write each radical with a rational exponent, then simplify.

6. $\sqrt[3]{x^6 y^{12}} = (x^6 y^{12})^{1/3}$

$= (x^6)^{1/3}(y^{12})^{1/3}$

$= x^2 y^4$

7. $\sqrt[4]{81 r^8 s^{20}} = (81 r^8 s^{20})^{1/4}$

$= 81^{1/4}(r^8)^{1/4}(s^{20})^{1/4}$

$= 3 r^2 s^5$

So far, the numerators of all the rational exponents we have encountered have been 1. The next theorem extends the work we can do with rational exponents to rational exponents with numerators other than 1.

We can extend our properties of exponents with the following theorem.

[Δ≠Σ] *Theorem 7.1*

If a is a nonnegative real number, m is an integer, and n is a positive integer, then

$$a^{m/n} = (a^{1/n})^m = (\sqrt[n]{a})^m \quad \text{and} \quad a^{m/n} = (a^m)^{1/n} = \sqrt[n]{a^m}$$

With rational exponents, the numerator always represents a power and the denominator represents the index of a root.

Proof   We can prove Theorem 7.1 using the properties of exponents. Because $m/n = m(1/n)$, we have

$$a^{m/n} = a^{m(1/n)} \qquad a^{m/n} = a^{(1/n)(m)}$$

$$= (a^m)^{1/n} \qquad \quad = (a^{1/n})^m$$

$$= \sqrt[n]{a^m} \qquad \quad = (\sqrt[n]{a})^m$$

Here are some examples that illustrate how we use this theorem.

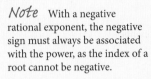 **EXAMPLES**   Simplify as much as possible.

8. $8^{2/3} = (8^{1/3})^2$ Theorem 7.1

$= (\sqrt[3]{8})^2$ Definition of fractional exponents

$= 2^2$ The cube root of 8 is 2

$= 4$ The square of 2 is 4

9. $25^{(3/2)} = (25^{1/2})^3$ Theorem 7.1

$= (\sqrt{25})^3$ Definition of fractional exponents

$= 5^3$ The square root of 25 is 5

$= 125$ The cube of 5 is 125

10. $9^{-3/2} = (9^{1/2})^{-3}$ Theorem 7.1

$= (\sqrt{9})^{-3}$ Definition of fractional exponents

$= 3^{-3}$ The square root of 9 is 3

$= \dfrac{1}{3^3}$ Property 4 for exponents

$= \dfrac{1}{27}$ The cube of 3 is 27

11. $\left(\dfrac{27}{8}\right)^{-4/3} = \left[\left(\dfrac{27}{8}\right)^{1/3}\right]^{-4}$ Theorem 7.1

$= \left[\sqrt[3]{\dfrac{27}{8}}\right]^{-4}$ Definition of fractional exponents

$= \left(\dfrac{3}{2}\right)^{-4}$ Evaluate the cube root

$= \left(\dfrac{2}{3}\right)^{4}$ Property 4 for exponents

$= \dfrac{16}{81}$ The fourth power of $\frac{2}{3}$ is $\frac{16}{81}$

USING TECHNOLOGY   *Graphing Calculators —*
A Word of Caution

Some graphing calculators give surprising results when evaluating expressions such as $(-8)^{2/3}$. As you know from reading this section, the expression $(-8)^{2/3}$ simplifies to 4, either by taking the cube root first and then squaring the result, or by squaring the base first and then taking the cube root of the result. Here are three different ways to evaluate this expression on your calculator:

1. $(-8)\wedge(2/3)$ To evaluate $(-8)^{2/3}$
2. $((-8)\wedge2)\wedge(1/3)$ To evaluate $((-8)^2)^{1/3}$
3. $((-8)\wedge(1/3))\wedge2$ To evaluate $((-8)^{1/3})^2$

Note any differences in the results.
  Next, graph each of the following functions, one at a time.

1. $Y_1 = X^{2/3}$ **2.** $Y_2 = (X^2)^{1/3}$ **3.** $Y_3 = (X^{1/3})^2$

The correct graph is shown in Figure 1. Note which of your graphs match the correct graph.
  Different calculators evaluate exponential expressions in different ways. You should use the method (or methods) that gave you the correct graph.

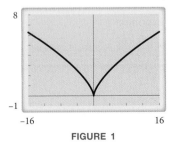

FIGURE 1

The following examples show the application of the properties of exponents to rational exponents.

EXAMPLES   Assume all variables represent positive quantities, and simplify as much as possible.

12. $x^{1/3} \cdot x^{5/6} = x^{1/3\,+\,5/6}$ Property 1 for exponents

$\qquad\qquad = x^{2/6\,+\,5/6}$ LCD is 6

$\qquad\qquad = x^{7/6}$ Add fractions

13. $(y^{2/3})^{3/4} = y^{(2/3)(3/4)}$ Property 2 for exponents

$\qquad\qquad = y^{1/2}$ Multiply fractions: $\frac{2}{3} \cdot \frac{3}{4} = \frac{6}{12} = \frac{1}{2}$

14. $\dfrac{z^{1/3}}{z^{1/4}} = z^{1/3\,-\,1/4}$ Property 6 for exponents

$\qquad\qquad = z^{4/12\,-\,3/12}$ LCD is 12

$\qquad\qquad = z^{1/12}$ Subtract fractions

15. $\left(\dfrac{a^{-1/3}}{b^{1/2}}\right)^6 = \dfrac{(a^{-1/3})^6}{(b^{1/2})^6}$ Property 5 for exponents

$\qquad\qquad = \dfrac{a^{-2}}{b^3}$ Property 2 for exponents

$\qquad\qquad = \dfrac{1}{a^2 b^3}$ Property 4 for exponents

16. $\dfrac{(x^{-3}y^{1/2})^4}{x^{10}y^{3/2}} = \dfrac{(x^{-3})^4(y^{1/2})^4}{x^{10}y^{3/2}}$ Property 3 for exponents

$\qquad\qquad = \dfrac{x^{-12}y^2}{x^{10}y^{3/2}}$ Property 2 for exponents

$\qquad\qquad = x^{-22}y^{1/2}$ Property 6 for exponents

$\qquad\qquad = \dfrac{y^{1/2}}{x^{22}}$ Property 4 for exponents

Getting Ready for Class

After reading through the preceding section, respond in your own words and in complete sentences.

A. What does $5^{1/2}$ represent?

B. Explain why $-9^{1/2}$ and $(-9)^{1/2}$ give different results.

C. For the expression $a^{m/n}$, explain the significance of the numerator m and the significance of the denominator n in the exponent.

D. Why must we interpret $8^{-2/3}$ as $8^{(-2)/3}$ and not $8^{2/(-3)}$?

Use the definition of rational exponents to write each of the following with the appropriate root. Then simplify.

1. $36^{1/2}$ **2.** $49^{1/2}$ **3.** $-9^{1/2}$ **4.** $-16^{1/2}$

5. $8^{1/3}$ **6.** $-8^{1/3}$ **7.** $(-8)^{1/3}$ **8.** $-27^{1/3}$

9. $32^{1/5}$ **10.** $81^{1/4}$ **11.** $\left(\dfrac{81}{25}\right)^{1/2}$ **12.** $\left(\dfrac{9}{16}\right)^{1/2}$

13. $\left(\dfrac{64}{125}\right)^{1/3}$ **14.** $\left(\dfrac{8}{27}\right)^{1/3}$

Use the definition of rational exponents to write each radical expression using a rational exponent.

15. $\sqrt{5}$ **16.** $\sqrt{2}$ **17.** $\sqrt{3x}$ **18.** $\sqrt{11ab}$

19. $\sqrt[3]{9}$ **20.** $\sqrt[3]{15}$ **21.** $\sqrt[3]{4x^2}$ **22.** $\sqrt[3]{10bc}$

Use Theorem 7.1 to simplify each of the following as much as possible.

23. $27^{2/3}$ **24.** $8^{4/3}$ **25.** $25^{3/2}$ **26.** $9^{3/2}$

27. $16^{3/4}$ **28.** $81^{3/4}$

Simplify each expression. Remember, negative exponents give reciprocals.

29. $27^{-1/3}$ **30.** $9^{-1/2}$ **31.** $81^{-3/4}$ **32.** $4^{-3/2}$

33. $\left(\dfrac{25}{36}\right)^{-1/2}$ **34.** $\left(\dfrac{16}{49}\right)^{-1/2}$ **35.** $\left(\dfrac{81}{16}\right)^{-3/4}$ **36.** $\left(\dfrac{27}{8}\right)^{-2/3}$

37. $16^{1/2} + 27^{1/3}$ **38.** $25^{1/2} + 100^{1/2}$ **39.** $8^{-2/3} + 4^{-1/2}$ **40.** $49^{-1/2} + 25^{-1/2}$

Use the properties of exponents to simplify each of the following as much as possible. Assume all bases are positive.

41. $x^{3/5} \cdot x^{1/5}$ **42.** $x^{3/4} \cdot x^{5/4}$ **43.** $y^{1/2} \cdot y^{1/4}$ **44.** $y^{2/3} \cdot y^{3/5}$

45. $(a^{3/4})^{4/3}$ **46.** $(a^{2/3})^{3/4}$ **47.** $\dfrac{x^{1/5}}{x^{3/5}}$ **48.** $\dfrac{x^{2/7}}{x^{5/7}}$

49. $\dfrac{x^{5/6}}{x^{2/3}}$ **50.** $\dfrac{x^{7/8}}{x^{8/7}}$ **51.** $(x^{3/5}y^{5/6}z^{1/3})^{3/5}$ **52.** $(x^{3/4}y^{1/8}z^{5/6})^{4/5}$

53. $\dfrac{a^{3/4}b^2}{a^{7/8}b^{1/4}}$ **54.** $\dfrac{a^{1/3}b^4}{a^{3/5}b^{1/3}}$ **55.** $\dfrac{(y^{2/3})^{3/4}}{(y^{1/3})^{3/5}}$ **56.** $\dfrac{(y^{5/4})^{2/5}}{(y^{1/4})^{4/3}}$

57. $\dfrac{x \cdot x^{2/3}}{(x^{5/6})^3}$ **58.** $\dfrac{x^3 \cdot x^{1/4}}{(x^{2/5})^2}$ **59.** $\left(\dfrac{a^{-1/4}}{b^{1/2}}\right)^8$ **60.** $\left(\dfrac{a^{-1/5}}{b^{1/3}}\right)^{15}$

Use rational exponents to simplify each expression. Assume all variables represent nonnegative numbers.

61. $\sqrt{25a^6}$ **62.** $\sqrt{64b^8}$ **63.** $\sqrt{x^2y^{10}}$ **64.** $\sqrt{x^4y^{12}}$

65. $\sqrt[3]{27b^9}$ **66.** $\sqrt[3]{8a^{15}}$ **67.** $\sqrt[3]{x^6y^{21}}$ **68.** $\sqrt[3]{x^3y^{18}}$

69. $\sqrt[4]{81a^8b^{20}}$ **70.** $\sqrt[5]{32a^{10}b^{20}}$

71. Show that the expression $(a^{1/2} + b^{1/2})^2$ is not equal to $a + b$ by replacing a with 9 and b with 4 in both expressions and then simplifying each.

72. Show that the statement $(a^2 + b^2)^{1/2} = a + b$ is not, in general, true by replacing a with 3 and b with 4 and then simplifying both sides.

73. You may have noticed, if you have been using a calculator to find roots, that you can find the fourth root of a number by pressing the square root button twice. Written in symbols, this fact looks like this:

$$\sqrt{\sqrt{a}} = \sqrt[4]{a} \qquad (a \geq 0)$$

Show that this statement is true by rewriting each side with exponents instead of radical notation and then simplifying the left side.

74. Show that the statement is true by rewriting each side with exponents instead of radical notation and then simplifying the left side.

$$\sqrt[3]{\sqrt{a}} = \sqrt[6]{a} \qquad (a \geq 0)$$

Applying the Concepts

75. Maximum Speed The maximum speed (v) that an automobile can travel around a curve of radius r without skidding is given by the equation

$$v = \left(\frac{5r}{2}\right)^{1/2}$$

where v is in miles per hour and r is measured in feet. What is the maximum speed a car can travel around a curve with a radius of 250 feet without skidding?

76. Relativity The equation

$$L = \left(1 - \frac{v^2}{c^2}\right)^{1/2}$$

gives the relativistic length of a 1-foot ruler traveling with velocity v. Find L if

$$\frac{v}{c} = \frac{3}{5}$$

Learning Objectives Assessment

The following problems can be used to help assess if you have successfully met the learning objectives for this section.

77. Write $\sqrt{8x}$ using a rational exponent.

    a. $(8x)^{1/2}$ **b.** $8x^{1/2}$ **c.** $(8x)^2$ **d.** $8x^2$

78. Write $-8^{2/3}$ as a root.

    a. $-\sqrt[3]{8^2}$ **b.** $\left(\sqrt[3]{-8}\right)^2$ **c.** $\left(\sqrt{-8}\right)^3$ **d.** $-\sqrt{8^3}$

79. Simplify $\sqrt{64x^6y^{12}}$ using rational exponents.

    a. $4x^3y^6$ **b.** $8x^3y^6$ **c.** $4x^2y^4$ **d.** $8x^2y^4$

80. Simplify: $y^{1/2} \cdot y^{2/3}$.

    a. $y^{1/3}$ **b.** $y^{7/6}$ **c.** $y^{3/5}$ **d.** $y^{-1/6}$

Getting Ready for the Next Section

Simplify. Assume all variables represent positive real numbers.

81. $\sqrt{6^2}$ **82.** $\sqrt{3^2}$ **83.** $\sqrt{(5y)^2}$ **84.** $\sqrt{(8x^3)^2}$

85. $\sqrt[3]{2^3}$ **86.** $\sqrt[3]{(-5)^3}$

Fill in the blank.

87. $50 = \underline{\hspace{0.5cm}} \cdot 2$

88. $12 = \underline{\hspace{0.5cm}} \cdot 3$

89. $48x^4y^3 = \underline{\hspace{0.5cm}} \cdot y$

90. $40a^5b^4 = \underline{\hspace{0.5cm}} \cdot 5a^2b$

91. $12x^7y^6 = \underline{\hspace{0.5cm}} \cdot 3x$

92. $54a^6b^2c^4 = \underline{\hspace{0.5cm}} \cdot 2b^2c$

SPOTLIGHT ON SUCCESS   *Instructor Octabio*

*The best thing about the future
is that it comes one day at a time.*
—Abraham Lincoln

For my family, education was always the way to go. Education would move us ahead, but the path through education was not always clear. My parents had immigrated to this country and had not had the opportunity to continue in education. Luckily, with the help of school counselors and the A.V.I.D. (Advancement Via Individual Determination) program in our school district, my older sister and brother were able to get into some of their top colleges. Later, with A.V.I.D. and the guidance of my siblings, I was able to take the right courses and was lucky enough to be accepted at my dream university.

Math has been my favorite subject ever since I can remember. When I got to higher level math classes, however, I struggled more than I had with previous levels of math. This struggle initially stopped me from enjoying the class, but as my understanding grew, I became more and more interested in seeing how things connected. I have found these connections at all levels of mathematics. These connections continue to be a source of satisfaction for me.

Simplified Form for Radicals

Learning Objectives

In this section, we will learn how to:

1. Simplify a radical expression by writing it as a product.
2. Simplify a radical expression by writing it as a quotient.
3. Simplify fractions containing radicals.
4. Simplify nth roots of nth powers.

Introduction

In this section, we will use radical notation instead of rational exponents. We will begin by stating two more properties of radicals. Following this, we will give a definition for simplified form for radical expressions. The examples in this section show how we use the properties of radicals to write radical expressions in simplified form.

Here are the next two properties of radicals. For these two properties, we will assume a and b are nonnegative real numbers whenever n is an even number.

> **PROPERTY** *Property 2 for Radicals*
>
> $$\sqrt[n]{ab} = \sqrt[n]{a}\sqrt[n]{b}$$
>
> *In words:* The nth root of a product is the product of the nth roots.
>
> **Proof of Property 2**
>
> | | |
> |---|---|
> | $\sqrt[n]{ab} = (ab)^{1/n}$ | Definition of fractional exponents |
> | $= a^{1/n}b^{1/n}$ | Exponents distribute over products |
> | $= \sqrt[n]{a}\sqrt[n]{b}$ | Definition of fractional exponents |

Note There is not a property for radicals that says the nth root of a sum is the sum of the nth roots. That is,

$$\sqrt[n]{a + b} \neq \sqrt[n]{a} + \sqrt[n]{b}$$

> **PROPERTY** *Property 3 for Radicals*
>
> $$\sqrt[n]{\frac{a}{b}} = \frac{\sqrt[n]{a}}{\sqrt[n]{b}} \quad (b \neq 0)$$
>
> *In words:* The nth root of a quotient is the quotient of the nth roots.

The proof of Property 3 is similar to the proof of Property 2.

These two properties of radicals allow us to change the form of and simplify radical expressions without changing their value.

> ### 【△≠∑】 RULE *Simplified Form for Radical Expressions*
>
> A radical expression is in **simplified form** if
>
> 1. None of the factors of the radicand (the quantity under the radical sign) can be written as powers greater than or equal to the index — that is, no perfect squares can be factors of the quantity under a square root sign, no perfect cubes can be factors of what is under a cube root sign, and so forth.
> 2. There are no fractions under the radical sign.
> 3. There are no radicals in the denominator.

Satisfying the first condition for simplified form actually amounts to taking as much out from under the radical sign as possible. The following examples illustrate the first condition for simplified form.

VIDEO EXAMPLES

SECTION 7.3

■ EXAMPLE 1   Write $\sqrt{50}$ in simplified form.

SOLUTION   This radical expression does not meet the first condition for simplified form because the radicand contains a perfect square. The largest perfect square that divides 50 is 25. We write 50 as $25 \cdot 2$ and apply Property 2 for radicals:

$$\sqrt{50} = \sqrt{25 \cdot 2} \qquad 50 = 25 \cdot 2$$
$$= \sqrt{25}\sqrt{2} \qquad \text{Property 2}$$
$$= 5\sqrt{2} \qquad \sqrt{25} = 5$$

We have taken as much as possible out from under the radical sign — in this case, factoring 25 from 50 and then writing $\sqrt{25}$ as 5.   ■

■ EXAMPLE 2   Write $\sqrt[3]{24}$ in simplified form.

SOLUTION   Once again, our expression does not meet the first condition for simplified form because 24 contains a perfect cube. We write 24 as $8 \cdot 3$ and simplify as we did in Example 1.

$$\sqrt[3]{24} = \sqrt[3]{8 \cdot 3} \qquad 24 = 8 \cdot 3$$
$$= \sqrt[3]{8}\sqrt[3]{3} \qquad \text{Property 2}$$
$$= 2\sqrt[3]{3} \qquad \sqrt[3]{8} = 2$$ ■

Note   In Example 2, we do not want to write 24 as $4 \cdot 6$, because the problem involves a cube root, not a square root. Neither 4 nor 6 are perfect cubes.

As we progress through this chapter you will see more and more expressions that involve the product of a number and a radical. Here are some examples:

$$3\sqrt{2} \qquad \frac{1}{2}\sqrt{5} \qquad 5\sqrt[3]{7} \qquad 3x\sqrt{2x} \qquad 2a^2b\sqrt[3]{5a}$$

All of these are products. The first expression $3\sqrt{2}$ is the product of 3 and $\sqrt{2}$. That is,

$$3\sqrt{2} = 3 \cdot \sqrt{2}$$

The 3 and the $\sqrt{2}$ are not stuck together in some mysterious way. The expression $3\sqrt{2}$ is simply the product of two numbers, one of which is rational, and the other is irrational.

EXAMPLE 3 Write in simplified form: $\sqrt{48x^4y^3}$, where $x, y \geq 0$

SOLUTION The largest perfect square that is a factor of the radicand is $16x^4y^2$. Applying Property 2, we have

$$\sqrt{48x^4y^3} = \sqrt{16x^4y^2 \cdot 3y}$$
$$= \sqrt{16x^4y^2}\sqrt{3y}$$
$$= 4x^2y\sqrt{3y}$$

EXAMPLE 4 Write $\sqrt[3]{40a^5b^4}$ in simplified form.

SOLUTION We now want to factor the largest perfect cube from the radicand. We write $40a^5b^4$ as $8a^3b^3 \cdot 5a^2b$ and proceed as in previous examples.

$$\sqrt[3]{40a^5b^4} = \sqrt[3]{8a^3b^3 \cdot 5a^2b}$$
$$= \sqrt[3]{8a^3b^3}\sqrt[3]{5a^2b}$$
$$= 2ab\sqrt[3]{5a^2b}$$

Fractions and Radical Expressions

We now consider some examples that involve fractions and simplified form for radicals.

EXAMPLE 5 Simplify each expression.

a. $\dfrac{\sqrt{12}}{6}$ b. $\dfrac{5\sqrt{18}}{15}$ c. $\dfrac{6 + \sqrt{8}}{2}$ d. $\dfrac{-1 + \sqrt{45}}{2}$

SOLUTION These expressions are not yet simplified because they do not meet the first condition for simplified form. In each case, we simplify the radical first, then we factor and reduce to lowest terms.

a. $\dfrac{\sqrt{12}}{6} = \dfrac{2\sqrt{3}}{6}$ Simplify the radical $\sqrt{12} = \sqrt{4 \cdot 3} = \sqrt{4}\sqrt{3} = 2\sqrt{3}$

$= \dfrac{2\sqrt{3}}{2 \cdot 3}$ Factor denominator

$= \dfrac{\sqrt{3}}{3}$ Divide out common factors

b. $\dfrac{5\sqrt{18}}{15} = \dfrac{5 \cdot 3\sqrt{2}}{15}$ $\sqrt{18} = \sqrt{9 \cdot 2} = \sqrt{9}\sqrt{2} = 3\sqrt{2}$

$= \dfrac{5 \cdot 3\sqrt{2}}{3 \cdot 5}$ Factor denominator

$= \sqrt{2}$ Divide out common factors

c. $\dfrac{6 + \sqrt{8}}{2} = \dfrac{6 + 2\sqrt{2}}{2}$ $\sqrt{8} = \sqrt{4 \cdot 2} = \sqrt{4}\sqrt{2} = 2\sqrt{2}$

$= \dfrac{2(3 + \sqrt{2})}{2}$ Factor numerator

$= 3 + \sqrt{2}$ Divide out common factors

d. $\dfrac{-1 + \sqrt{45}}{2} = \dfrac{-1 + 3\sqrt{5}}{2}$ $\sqrt{45} = \sqrt{9 \cdot 5} = \sqrt{9}\sqrt{5} = 3\sqrt{5}$

This expression cannot be simplified further because $-1 + 3\sqrt{5}$ and 2 have no factors in common.

Note In Example 5d, we cannot combine $-1 + 3\sqrt{5}$ in the numerator. That is,

$$-1 + 3\sqrt{5} \neq 2\sqrt{5}$$

The reason is that -1 and $3\sqrt{5}$ are not similar terms. We will define similar terms for radicals in the next section.

▓ EXAMPLE 6   Simplify: $\sqrt{\dfrac{3}{4}}$.

SOLUTION In this case, we want to eliminate the fraction under the radical sign in order to satisfy condition 2 for simplified form. Applying Property 3 for radicals, we have

$$\sqrt{\frac{3}{4}} = \frac{\sqrt{3}}{\sqrt{4}} \qquad \text{Property 3}$$

$$= \frac{\sqrt{3}}{2} \qquad \sqrt{4} = 2$$

The last expression is now simplified because it satisfies all three conditions for simplified form. ▓

▓ EXAMPLE 7   Simplify: $\sqrt[3]{\dfrac{40}{27y^3}}$.

SOLUTION As we did in Example 6, we use Property 3 to avoid having a fraction under the radical sign. Then we simplify each radical.

$$\sqrt[3]{\frac{40}{27y^3}} = \frac{\sqrt[3]{40}}{\sqrt[3]{27y^3}} \qquad \text{Property 3}$$

$$= \frac{\sqrt[3]{8}\sqrt[3]{5}}{3y} \qquad \text{Property 2; } \sqrt[3]{27y^3} = 3y$$

$$= \frac{2\sqrt[3]{5}}{3y} \qquad \sqrt[3]{8} = 2 \qquad ▓$$

▓ EXAMPLE 8   Simplify: $\dfrac{\sqrt{10}}{\sqrt{2}}$.

SOLUTION Neither $\sqrt{10}$ nor $\sqrt{2}$ can be simplified. But because of the radical in the denominator, this expression does not meet the third condition for simplified form.

Notice both radicands contain a common factor. By using Property 3 for radicals we can write the expression as a single radical, which will then allow us to reduce the resulting fraction.

$$\frac{\sqrt{10}}{\sqrt{2}} = \sqrt{\frac{10}{2}} \qquad \text{Property 3}$$

$$= \sqrt{5} \qquad \text{Reduce} \qquad ▓$$

The Golden Ratio

In Section 7.1, we used the Pythagorean theorem to show that the diagonal of a square had a length of $\sqrt{2}$. Associating numbers, such as $\sqrt{2}$, with the diagonal of a square or rectangle allows us to analyze some interesting items from geometry. One particularly interesting geometric object is shown in Figure 1.

The Golden Rectangle

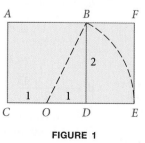

FIGURE 1

Its origins can be traced back over 2,000 years to the Greek civilization that produced Pythagoras, Socrates, Plato, Aristotle, and Euclid. The most important mathematical work to come from that Greek civilization was Euclid's *Elements,* an elegantly written summary of all that was known about geometry at that time in history. Euclid's *Elements,* according to Howard Eves, an authority on the history of mathematics, exercised a greater influence on scientific thinking than any other work. Here is how we construct a golden rectangle from a square of side 2, using the same method that Euclid used in his *Elements.*

Constructing a Golden Rectangle From a Square of Side 2

Step 1: Draw a square with a side of length 2. Connect the midpoint of side *CD* to corner *B.* (Note that we have labeled the midpoint of segment *CD* with the letter *O.*)

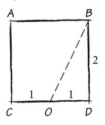

Step 2: Drop the diagonal from step 1 down so it aligns with side *CD.*

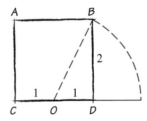

Step 3: Form rectangle *ACEF.* This is a golden rectangle.

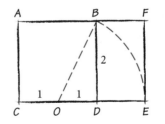

All golden rectangles are constructed from squares. Every golden rectangle, no matter how large or small it is, will have the same shape. To associate a number with the shape of the golden rectangle, we use the ratio of its length to its width. This ratio is called the ***golden ratio.***

To calculate the golden ratio, we must first find the length of the diagonal we used to construct the golden rectangle. Figure 2 shows the golden rectangle we constructed from a square of side 2. The length of the diagonal OB is found by applying the Pythagorean theorem to triangle OBD.

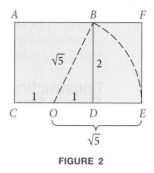

FIGURE 2

The length of segment OE is equal to the length of diagonal OB; both are $\sqrt{5}$. Because the distance from C to O is 1, the length CE of the golden rectangle is $1 + \sqrt{5}$. Now we can find the golden ratio:

$$\text{Golden ratio} = \frac{\text{length}}{\text{width}} = \frac{CE}{EF} = \frac{1 + \sqrt{5}}{2}$$

EXAMPLE 9   Construct a golden rectangle from a square of side 4. Then show that the ratio of the length to the width is the golden ratio $\frac{1 + \sqrt{5}}{2}$.

SOLUTION   Figure 3 shows the golden rectangle constructed from a square of side 4. The length of the diagonal OB is found from the Pythagorean theorem.

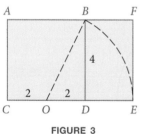

FIGURE 3

$$OB = \sqrt{2^2 + 4^2}$$
$$= \sqrt{4 + 16}$$
$$= \sqrt{20}$$
$$= 2\sqrt{5}$$

The ratio of the length to the width for the rectangle is the golden ratio.

$$\text{Golden ratio} = \frac{CE}{EF}$$
$$= \frac{2 + OE}{4}$$
$$= \frac{2 + OB}{4}$$
$$= \frac{2 + 2\sqrt{5}}{4}$$
$$= \frac{2(1 + \sqrt{5})}{2 \cdot 2}$$
$$= \frac{1 + \sqrt{5}}{2}$$

As you can see, showing that the ratio of length to width in this rectangle is the golden ratio depends on our ability to write $\sqrt{20}$ as $2\sqrt{5}$ and our ability to reduce to lowest terms by factoring and then dividing out the common factor 2 from the numerator and denominator.

Square Root of a Perfect Square

So far in this chapter, we have assumed that all our variables are nonnegative when they appear under a square root symbol. There are times, however, when this is not the case.

Consider $\sqrt{x^2}$, where x is allowed to represent any real number. Let's evaluate this expression for $x = 3$ and $x = -3$.

If $x = 3$ If $x = -3$

then $\sqrt{3^2} = \sqrt{9} = 3$ then $\sqrt{(-3)^2} = \sqrt{9} = 3$

Whether we operate on 3 or -3, the result is the same: Both expressions simplify to 3. When we used $x = 3$, the result is equal to our value of x. But when we used $x = -3$, the result is the opposite of our x value. The other operation we have worked with in the past that produces the same result is absolute value. That is,

$$|3| = 3 \quad \text{and} \quad |-3| = 3$$

This leads us to the fourth property.

> ⎾Δ≠Σ⏌ **PROPERTY**   *Property 4 for Square Roots*
>
> If a is a real number, then $\sqrt{a^2} = |a|$.

The result of this discussion and Property 4 is simply this:

If we know a is positive, then $\sqrt{a^2} = a$.

If we know a is negative, then $\sqrt{a^2} = -a$.

If we don't know if a is positive or negative, then $\sqrt{a^2} = |a|$.

EXAMPLES   Simplify each expression. Do *not* assume the variables represent positive numbers.

10. $\sqrt{9x^2} = \sqrt{9}\sqrt{x^2}$ Property 2

$\qquad = 3|x|$ Property 4

11. $\sqrt{x^3} = \sqrt{x^2}\sqrt{x}$ Property 2

$\qquad = |x|\sqrt{x}$ Property 4

12. $\sqrt{x^2 - 6x + 9} = \sqrt{(x - 3)^2} = |x - 3|$

13. $\sqrt{x^3 - 5x^2} = \sqrt{x^2(x - 5)} = |x|\sqrt{x - 5}$

*n*th Root of an *n*th Power

As you can see, we must use absolute value symbols when we take a square root of a perfect square, unless we know the base of the perfect square is a positive number. The same idea holds for higher even roots, but not for odd roots. With odd roots, no absolute value symbols are necessary.

▧ EXAMPLES Simplify each expression.

14 $\sqrt[3]{2^3} = \sqrt[3]{8} = 2$

15. $\sqrt[3]{(-2)^3} = \sqrt[3]{-8} = -2$

We can extend this discussion to all roots as follows:

▧≠Σ PROPERTY *Property 4 for Radicals*

If a is a real number, then

$$\sqrt[n]{a^n} = |a| \qquad \text{if} \qquad n \text{ is even}$$

$$\sqrt[n]{a^n} = a \qquad \text{if} \qquad n \text{ is odd}$$

▧ EXAMPLES Simplify each expression. Do *not* assume the variables represent positive numbers.

16. $\sqrt[3]{(x + 2)^3} = x + 2$

17. $\sqrt[3]{-64y^3} = -4y$

18. $\sqrt[4]{81b^4} = 3|b|$

19. $\sqrt[5]{32x^5} = 2x$

Getting Ready for Class

After reading through the preceding section, respond in your own words and in complete sentences.

A. Explain why this statement is false: "The square root of a sum is the sum of the square roots."

B. What is simplified form for an expression that contains a square root?

C. Describe two different ways we used Property 3 for radicals in this section.

D. Why is it not necessarily true that $\sqrt{a^2} = a$?

Use Property 2 for radicals to write each of the following expressions in simplified form. Assume all variables represent positive numbers.

1. $\sqrt{8}$ **2.** $\sqrt{32}$ **3.** $\sqrt{98}$ **4.** $\sqrt{75}$

5. $\sqrt{288}$ **6.** $\sqrt{128}$ **7.** $\sqrt{80}$ **8.** $\sqrt{200}$

9. $\sqrt{48}$ **10.** $\sqrt{27}$ **11.** $\sqrt{675}$ **12.** $\sqrt{972}$

13. $\sqrt[3]{54}$ **14.** $\sqrt[3]{24}$ **15.** $\sqrt[3]{128}$ **16.** $\sqrt[3]{162}$

17. $\sqrt[3]{432}$ **18.** $\sqrt[3]{1,536}$ **19.** $\sqrt[5]{64}$ **20.** $\sqrt[4]{48}$

21. $\sqrt{18x^3}$ **22.** $\sqrt{27x^5}$ **23.** $\sqrt[4]{32y^7}$ **24.** $\sqrt[5]{32y^7}$

25. $\sqrt[3]{40x^4y^7}$ **26.** $\sqrt[3]{128x^6y^2}$ **27.** $\sqrt{48a^2b^3c^4}$ **28.** $\sqrt{72a^4b^3c^2}$

29. $\sqrt[3]{48a^2b^3c^4}$ **30.** $\sqrt[3]{72a^4b^3c^2}$ **31.** $\sqrt[5]{64x^8y^{12}}$ **32.** $\sqrt[4]{32x^9y^{10}}$

33. $\sqrt[5]{243x^7y^{10}z^5}$ **34.** $\sqrt[5]{64x^8y^4z^{11}}$

Substitute the given numbers into the expression $\sqrt{b^2 - 4ac}$, and then simplify, if possible.

35. $a = 2, b = -6, c = 3$ **36.** $a = 6, b = 7, c = -5$

37. $a = 1, b = 2, c = 6$ **38.** $a = 2, b = 5, c = 3$

39. $a = \dfrac{1}{2}, b = -\dfrac{1}{2}, c = -\dfrac{5}{4}$ **40.** $a = \dfrac{7}{4}, b = -\dfrac{3}{4}, c = -2$

Simplify each expression using Property 2 for radicals.

41. $\dfrac{\sqrt{20}}{4}$ **42.** $\dfrac{3\sqrt{20}}{15}$ **43.** $\dfrac{\sqrt{12}}{4}$ **44.** $\dfrac{2\sqrt{32}}{8}$

45. $\dfrac{4 + \sqrt{12}}{2}$ **46.** $\dfrac{2 + \sqrt{9}}{5}$ **47.** $\dfrac{9 + \sqrt{27}}{3}$ **48.** $\dfrac{-6 - \sqrt{64}}{2}$

49. $\dfrac{10 + \sqrt{75}}{5}$ **50.** $\dfrac{-6 + \sqrt{45}}{3}$ **51.** $\dfrac{-2 - \sqrt{27}}{6}$ **52.** $\dfrac{12 - \sqrt{12}}{6}$

53. $\dfrac{-4 - \sqrt{8}}{2}$ **54.** $\dfrac{6 - \sqrt{48}}{8}$

Use Property 3 for radicals to simplify each expression. Assume all variables represent positive numbers.

55. $\sqrt{\dfrac{7}{25}}$ **56.** $\sqrt{\dfrac{13}{64}}$ **57.** $\sqrt{\dfrac{5x}{36}}$ **58.** $\sqrt{\dfrac{3ab}{16}}$

59. $\sqrt[3]{\dfrac{3}{64}}$ **60.** $\sqrt[3]{\dfrac{5}{8}}$ **61.** $\sqrt[3]{\dfrac{2a}{b^3}}$ **62.** $\sqrt[3]{\dfrac{11x^2}{27}}$

63. $\sqrt[4]{\dfrac{9}{16}}$ **64.** $\sqrt[5]{\dfrac{3y^3}{32}}$ **65.** $\dfrac{\sqrt{15}}{\sqrt{3}}$ **66.** $\dfrac{\sqrt{30}}{\sqrt{5}}$

67. $\dfrac{\sqrt[3]{12}}{\sqrt[3]{4}}$ **68.** $\dfrac{\sqrt[3]{18}}{\sqrt[3]{2}}$

Simplify each expression. Assume all variables represent positive numbers.

69. $\sqrt{\dfrac{12x^2}{25}}$ **70.** $\sqrt{\dfrac{5ab^4}{36}}$ **71.** $\sqrt{\dfrac{3x^3}{4y^6}}$ **72.** $\sqrt{\dfrac{28a^7}{9b^2}}$

73. $\sqrt[3]{\dfrac{15b^4}{8a^3}}$ **74.** $\sqrt[3]{\dfrac{81x^2y^5}{64z^3}}$ **75.** $\sqrt[4]{\dfrac{9x^6y^{10}}{16z^8}}$ **76.** $\sqrt[4]{\dfrac{48a^3b^9}{625c^{12}}}$

Simplify each expression. Do *not* assume the variables represent positive numbers.

77. $\sqrt{25x^2}$ **78.** $\sqrt{49x^2}$ **79.** $\sqrt{27x^3y^2}$ **80.** $\sqrt{40x^3y^2}$

81. $\sqrt{x^2 - 10x + 25}$ **82.** $\sqrt{x^2 - 16x + 64}$

83. $\sqrt{4x^2 + 12x + 9}$ **84.** $\sqrt{16x^2 + 40x + 25}$

85. $\sqrt{4a^4 + 16a^3 + 16a^2}$ **86.** $\sqrt{9a^4 + 18a^3 + 9a^2}$

87. $\sqrt{4x^3 - 8x^2}$ **88.** $\sqrt{18x^3 - 9x^2}$

89. Show that the statement $\sqrt{a + b} = \sqrt{a} + \sqrt{b}$ is not true by replacing a with 9 and b with 16 and simplifying both sides.

90. Find a pair of values for a and b that will make the statement $\sqrt{a + b} = \sqrt{a} + \sqrt{b}$ true.

Applying the Concepts

91. Diagonal Distance The distance d between opposite corners of a rectangular room with length l and width w is given by

$$d = \sqrt{l^2 + w^2}$$

How far is it between opposite corners of a living room that measures 10 by 15 feet?

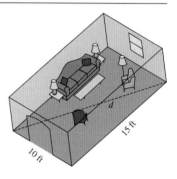

92. Distance to the Horizon If you are at a point k miles above the surface of the Earth, the distance you can see, in miles, is approximated by the equation $d = \sqrt{8000k + k^2}$.

    a. How far can you see from a point that is 1 mile above the surface of the Earth?

    b. How far can you see from a point that is 2 miles above the surface of the Earth?

    c. How far can you see from a point that is 3 miles above the surface of the Earth?

93. Isosceles Right Triangles A triangle is isosceles if it has two equal sides, and a triangle is a right triangle if it has a right angle in it. Sketch an isosceles right triangle, and find the ratio of the hypotenuse to a leg.

94. Equilateral Triangles A triangle is equilateral if it has three equal sides. The triangle in the figure is equilateral with each side of length $2x$. Find the ratio of the height to a side.

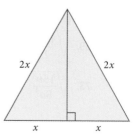

95. **Pyramids** The following solid is called a regular square pyramid because its base is a square and all eight edges are the same length, 5. It is also true that the vertex, *V*, is directly above the center of the base.

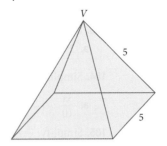

    a. Find the ratio of a diagonal of the base to the length of a side.

    b. Find the ratio of the area of the base to the diagonal of the base.

    c. Find the ratio of the area of the base to the perimeter of the base.

96. **Pyramids** Refer to this diagram of a square pyramid. Find the ratio of the height *h* of the pyramid to the slant height *a*.

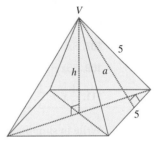

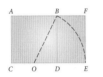

97. **Golden Ratio** The golden ratio is the ratio of the length to the width in any golden rectangle. The exact value of this number is $\frac{1 + \sqrt{5}}{2}$. Use a calculator to find a decimal approximation to this number and round it to the nearest thousandth.

98. **Golden Ratio** The reciprocal of the golden ratio is $\frac{2}{1 + \sqrt{5}}$. Find a decimal approximation to this number that is accurate to the nearest thousandth.

99. **Golden Rectangle** Construct a golden rectangle from a square of side 8. Then show that the ratio of the length to the width is the golden ratio $\frac{1 + \sqrt{5}}{2}$.

100. **Golden Rectangle** Construct a golden rectangle from a square of side 10. Then show that the ratio of the length to the width is the golden ratio $\frac{1 + \sqrt{5}}{2}$.

101. **Golden Rectangle** To show that all golden rectangles have the same ratio of length to width, construct a golden rectangle from a square of side 2*x*. Then show that the ratio of the length to the width is the golden ratio.

102. **Golden Rectangle** To show that all golden rectangles have the same ratio of length to width, construct a golden rectangle from a square of side *x*. Then show that the ratio of the length to the width is the golden ratio.

Learning Objectives Assessment

The following problems can be used to help assess if you have successfully met the learning objectives for this section.

103. Simplify: $\sqrt{48}$.

   a. $2\sqrt{12}$ **b.** $4\sqrt{3}$ **c.** $2\sqrt{6}$ **d.** $24\sqrt{2}$

104. Simplify: $\sqrt{\dfrac{27x^2}{100}}$. Assume x is a positive number.

   a. $\dfrac{3x}{10}$ **b.** $\dfrac{3x\sqrt{3}}{10}$ **c.** $\dfrac{9x\sqrt{3}}{10}$ **d.** $\dfrac{3x\sqrt{x}}{5\sqrt{2}}$

105. Simplify: $\dfrac{2 + \sqrt{48}}{6}$.

   a. $\dfrac{1 + 2\sqrt{3}}{3}$ **b.** $\sqrt{3}$ **c.** $\dfrac{1 + \sqrt{6}}{3}$ **d.** $\dfrac{1 + 4\sqrt{3}}{3}$

106. Simplify $\sqrt{(x - 4)^2}$ for any real number x.

   a. $x - 4$ **b.** $x + 4$ **c.** $|x| - 4$ **d.** $|x - 4|$

Getting Ready for the Next Section

Simplify the following.

107. $5x - 4x + 6x$ **108.** $12x + 8x - 7x$

109. $35xy^2 - 8xy^2$ **110.** $20a^2b + 33a^2b$

111. $\dfrac{1}{2}x + \dfrac{1}{3}x$ **112.** $\dfrac{2}{3}x + \dfrac{5}{8}x$

Write in simplified form for radicals.

113. $\sqrt{18}$ **114.** $\sqrt{8}$

115. $\sqrt{75xy^3}$ **116.** $\sqrt{12xy}$

117. $\sqrt[3]{8a^4b^2}$ **118.** $\sqrt[3]{27ab^2}$

Learning Objectives

In this section, we will learn how to:

1. Identify similar radicals.

2. Combine similar radicals.

Introduction

In Chapter 1, we found we could add similar terms when combining polynomials. The same idea applies to addition and subtraction of radical expressions.

> (dĕf) **DEFINITION** *similar radicals*
>
> Two radicals are said to be *similar radicals* if they have the same index and the same radicand.

The expressions $5\sqrt[3]{7}$ and $-8\sqrt[3]{7}$ are similar since they both have an index of 3 and a radicand of 7. The expressions $3\sqrt[4]{5}$ and $7\sqrt[3]{5}$ are not similar because they have different indices, and the expressions $2\sqrt[5]{8}$ and $3\sqrt[5]{9}$ are not similar because the radicands are not the same.

We add and subtract radical expressions in the same way we add and subtract polynomials — by combining similar terms under the distributive property.

VIDEO EXAMPLES

SECTION 7.4

EXAMPLE 1 Combine: $5\sqrt{3} - 4\sqrt{3} + 6\sqrt{3}$.

SOLUTION All three radicals are similar. We apply the distributive property to get

$$5\sqrt{3} - 4\sqrt{3} + 6\sqrt{3} = (5 - 4 + 6)\sqrt{3}$$
$$= 7\sqrt{3}$$

EXAMPLE 2 Simplify: $7\sqrt{5} + 2\sqrt{7} - \sqrt{5} + 5\sqrt{7}$.

SOLUTION First, we group pairs of similar radicals. Then we combine terms in each group using the distributive property.

$$7\sqrt{5} + 2\sqrt{7} - \sqrt{5} + 5\sqrt{7} = (7\sqrt{5} - \sqrt{5}) + (2\sqrt{7} + 5\sqrt{7})$$

$$= (7 - 1)\sqrt{5} + (2 + 5)\sqrt{7}$$

$$= 6\sqrt{5} + 7\sqrt{7}$$

We cannot simplify any further because $6\sqrt{5}$ and $7\sqrt{7}$ are not similar radicals. They have different radicands and so they cannot be combined.

EXAMPLE 3   Combine: $3\sqrt{8} + 5\sqrt{18}$.

SOLUTION The two radicals do not seem to be similar. We must write each in simplified form before applying the distributive property.

$$
\begin{aligned}
3\sqrt{8} + 5\sqrt{18} &= 3\sqrt{4 \cdot 2} + 5\sqrt{9 \cdot 2} \\
&= 3\sqrt{4}\,\sqrt{2} + 5\sqrt{9}\,\sqrt{2} \\
&= 3 \cdot 2\,\sqrt{2} + 5 \cdot 3\,\sqrt{2} \\
&= 6\,\sqrt{2} + 15\,\sqrt{2} \\
&= (6 + 15)\,\sqrt{2} \\
&= 21\,\sqrt{2}
\end{aligned}
$$

The result of Example 3 can be generalized to the following rule for sums and differences of radical expressions.

[Δ≠Σ] RULE

To add or subtract radical expressions, put each in simplified form and apply the distributive property, if possible. We can only add similar radicals. We must write each expression in simplified form for radicals before we can tell if the radicals are similar.

EXAMPLE 4   Combine $7\sqrt{75xy^3} - 4y\sqrt{12xy}$, where $x, y \geq 0$.

SOLUTION We write each expression in simplified form and combine similar radicals:

$$
\begin{aligned}
7\sqrt{75xy^3} - 4y\sqrt{12xy} &= 7\sqrt{25y^2}\,\sqrt{3xy} - 4y\sqrt{4}\,\sqrt{3xy} \\
&= 7 \cdot 5y\sqrt{3xy} - 4y \cdot 2\sqrt{3xy} \\
&= 35y\sqrt{3xy} - 8y\sqrt{3xy} \\
&= (35y - 8y)\sqrt{3xy} \\
&= 27y\sqrt{3xy}
\end{aligned}
$$

EXAMPLE 5   Combine: $10\sqrt[3]{8a^4b^2} + 11a\sqrt[3]{27ab^2}$.

SOLUTION Writing each radical in simplified form and combining similar terms, we have

$$
\begin{aligned}
10\sqrt[3]{8a^4b^2} + 11a\sqrt[3]{27ab^2} &= 10\sqrt[3]{8a^3}\,\sqrt[3]{ab^2} + 11a\sqrt[3]{27}\,\sqrt[3]{ab^2} \\
&= 10 \cdot 2a\sqrt[3]{ab^2} + 11a \cdot 3\sqrt[3]{ab^2} \\
&= 20a\sqrt[3]{ab^2} + 33a\sqrt[3]{ab^2} \\
&= 53a\sqrt[3]{ab^2}
\end{aligned}
$$

EXAMPLE 6   Combine: $\dfrac{\sqrt{3}}{2} + \dfrac{\sqrt{12}}{6}$.

SOLUTION   We begin by writing the second term in simplified form.

$$\dfrac{\sqrt{3}}{2} + \dfrac{\sqrt{12}}{6} = \dfrac{\sqrt{3}}{2} + \dfrac{\sqrt{4}\sqrt{3}}{6}$$

$$= \dfrac{\sqrt{3}}{2} + \dfrac{2\sqrt{3}}{6}$$

$$= \dfrac{\sqrt{3}}{2} + \dfrac{\sqrt{3}}{3}$$

$$= \dfrac{1}{2}\sqrt{3} + \dfrac{1}{3}\sqrt{3}$$

$$= \left(\dfrac{1}{2} + \dfrac{1}{3}\right)\sqrt{3}$$

$$= \dfrac{5}{6}\sqrt{3}$$

$$= \dfrac{5\sqrt{3}}{6}$$

EXAMPLE 7   Simplify: $3x\sqrt{54} - 4y\sqrt{24}$.

SOLUTION   First, we simplify each radical expression.

$$3x\sqrt{54} - 4y\sqrt{24} = 3x\sqrt{9}\sqrt{6} - 4y\sqrt{4}\sqrt{6}$$

$$= 9x\sqrt{6} - 8y\sqrt{6}$$

Although the two expressions contain similar radicals, they are not similar terms because of the different variable factors. They cannot be combined, though we can still apply the distributive property to factor out the similar radical.

$$9x\sqrt{6} - 8y\sqrt{6} = (9x - 8y)\sqrt{6}$$

Getting Ready for Class

After reading through the preceding section, respond in your own words and in complete sentences.

A. What are similar radicals?

B. When can we add two radical expressions?

C. What is the first step when adding or subtracting expressions containing radicals?

D. Explain why $2 + 4\sqrt{5} \neq 6\sqrt{5}$.

Problem Set 7.4

Simplify the following expressions by combining similar radicals. Assume any variables under an even root are nonnegative.

1. $3\sqrt{5} + 4\sqrt{5}$

2. $6\sqrt{3} - 5\sqrt{3}$

3. $3x\sqrt{7} - 4x\sqrt{7}$

4. $6y\sqrt{a} + 7y\sqrt{a}$

5. $5\sqrt[3]{10} - 4\sqrt[3]{10}$

6. $6\sqrt[4]{2} + 9\sqrt[4]{2}$

7. $8\sqrt[5]{6} - 2\sqrt[5]{6} + 3\sqrt[5]{6}$

8. $7\sqrt[6]{7} - \sqrt[6]{7} + 4\sqrt[6]{7}$

9. $3x\sqrt{2} - 4x\sqrt{2} + x\sqrt{2}$

10. $5x\sqrt{6} - 3x\sqrt{6} - 2x\sqrt{6}$

11. $4\sqrt{2} + \sqrt{3} + 3\sqrt{2} + 2\sqrt{3}$

12. $\sqrt{5} - 2\sqrt{6} + 9\sqrt{5} + 6\sqrt{6}$

13. $6\sqrt{x} - 5\sqrt{y} - 4\sqrt{x} - 7\sqrt{y}$

14. $\sqrt[3]{2a} + 2\sqrt[3]{2a} + 8\sqrt[3]{3b} - 7\sqrt[3]{2a}$

15. $5\sqrt[3]{3} + 4\sqrt[3]{3} + \sqrt[3]{3} - 3\sqrt[3]{3}$

16. $3\sqrt[3]{2} - \sqrt[3]{2} - 6\sqrt[3]{2} + 9\sqrt{2}$

17. $\sqrt{20} - \sqrt{80} + \sqrt{45}$

18. $\sqrt{8} - \sqrt{32} - \sqrt{18}$

19. $4\sqrt{8} - 2\sqrt{50} - 5\sqrt{72}$

20. $\sqrt{48} - 3\sqrt{27} + 2\sqrt{75}$

21. $5x\sqrt{8} + 3\sqrt{32x^2} - 5\sqrt{50x^2}$

22. $2\sqrt{50x^2} - 8x\sqrt{18} - 3\sqrt{72x^2}$

23. $5\sqrt[3]{16} - 4\sqrt[3]{54}$

24. $\sqrt[3]{81} + 3\sqrt[3]{24}$

25. $\sqrt[3]{x^4y^2} + 7x\sqrt[3]{xy^2}$

26. $2\sqrt[3]{x^8y^6} - 3y^2\sqrt[3]{8x^8}$

27. $5a^2\sqrt{27ab^3} - 6b\sqrt{12a^5b}$

28. $9a\sqrt{20a^3b^2} + 7b\sqrt{45a^5}$

29. $b\sqrt[3]{24a^5b} + 3a\sqrt[3]{81a^2b^4}$

30. $7\sqrt[3]{a^4b^3c^2} - 6ab\sqrt[3]{ac^2}$

31. $5x\sqrt[4]{3y^5} + y\sqrt[4]{243x^4y} + \sqrt[4]{48x^4y^5}$

32. $x\sqrt[4]{5xy^8} + y\sqrt[4]{405x^5y^4} + y^2\sqrt[4]{80x^5}$

33. $\dfrac{\sqrt{3}}{2} + \dfrac{\sqrt{27}}{2}$

34. $\dfrac{\sqrt{5}}{3} - \dfrac{\sqrt{20}}{3}$

35. $\sqrt{\dfrac{5}{36}} + \dfrac{\sqrt{45}}{6}$

36. $\sqrt{\dfrac{18}{25}} - \dfrac{\sqrt{8}}{5}$

37. $\dfrac{\sqrt{x}}{3} - \dfrac{\sqrt{x}}{2}$

38. $\dfrac{3\sqrt{a}}{5} + \dfrac{\sqrt{a}}{3}$

39. $\dfrac{\sqrt{18}}{6} + \sqrt{\dfrac{2}{9}}$

40. $\dfrac{\sqrt{12}}{6} + \sqrt{\dfrac{3}{16}}$

41. $2x\sqrt{8} + 3y\sqrt{50}$

42. $4a\sqrt{12} - 3b\sqrt{27}$

43. $2\sqrt[3]{16x^3} - \sqrt[3]{54}$

44. $\sqrt[3]{125x^3y} + 2\sqrt[3]{8y^4}$

45. Use a calculator to find a decimal approximation for $\sqrt{2} + \sqrt{3}$ and for $\sqrt{5}$.

46. Use a calculator to find decimal approximations for $\sqrt{7} - \sqrt{5}$ and $\sqrt{2}$.

47. Use a calculator to find a decimal approximation for $\sqrt{8} + \sqrt{18}$. Is it equal to the decimal approximation for $\sqrt{26}$ or $\sqrt{50}$?

48. Use a calculator to find a decimal approximation for $\sqrt{3} + \sqrt{12}$. Is it equal to the decimal approximation for $\sqrt{15}$ or $\sqrt{27}$?

Each of the following statements is false. Correct the right side of each one to make the statement true.

49. $3\sqrt{2x} + 5\sqrt{2x} = 8\sqrt{4x}$

50. $5\sqrt{3} - 7\sqrt{3} = -2\sqrt{9}$

51. $\sqrt{9 + 16} = 3 + 4$

52. $\sqrt{36 + 64} = 6 + 8$

Learning Objectives Assessment

The following problems can be used to help assess if you have successfully met the learning objectives for this section.

53. Which pair of radicals are similar?

    a. $4x\sqrt[3]{5}, \dfrac{1}{2}x\sqrt[3]{5}$ **b.** $4x\sqrt[3]{5}, 2\sqrt[3]{5y}$ **c.** $4\sqrt{5}, 2\sqrt[3]{5}$ **d.** $4x\sqrt{2}, x\sqrt{3}$

54. Combine: $\sqrt{48} + \sqrt{75}$.

    a. $\sqrt{123}$ **b.** $9\sqrt{6}$ **c.** $41\sqrt{3}$ **d.** $9\sqrt{3}$

Getting Ready for the Next Section

Simplify the following.

55. $3 \cdot 2$ **56.** $5 \cdot 7$

57. $(x + y)(4x - y)$ **58.** $(2x + y)(x - y)$

59. $(x + 3)^2$ **60.** $(3x - 2y)^2$

61. $(x - 2)(x + 2)$ **62.** $(2x + 5)(2x - 5)$

Simplify the following expressions.

63. $2\sqrt{18}$ **64.** $5\sqrt{36}$

65. $(\sqrt{6})^2$ **66.** $(\sqrt{2})^2$

67. $(3\sqrt{x})^2$ **68.** $(2\sqrt{y})^2$

Multiplication and Division of Radical Expressions

7.5

Learning Objectives

In this section, we will learn how to:

1. Multiply radicals.

2. Simplify products involving radicals.

3. Divide radicals.

4. Rationalize the denominator.

Introduction

In Section 7.3, we introduced the golden rectangle. An example of a golden rectangle is shown in Figure 1.

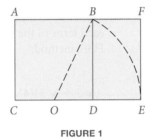

FIGURE 1

By now you know that, in any golden rectangle constructed from a square (of any size), the ratio of the length to the width will be

$$\frac{1 + \sqrt{5}}{2}$$

which we call the golden ratio. What is interesting is that the smaller rectangle on the right, *BFED*, shown in Figure 1, is also a golden rectangle. We will use the mathematics developed in this section to confirm this fact.

In this section, we will look at multiplication and division of expressions that contain radicals. As you will see, multiplication of expressions that contain radicals is very similar to multiplication of polynomials. The division problems in this section are an extension of the work we did previously in Section 7.3.

Multiplication

EXAMPLE 1 Multiply $(3\sqrt{5})(2\sqrt{7})$.

SOLUTION We can rearrange the order and grouping of the numbers in this product by applying the commutative and associative properties. Following this, we apply Property 2 for radicals and multiply:

$$(3\sqrt{5})(2\sqrt{7}) = (3 \cdot 2)(\sqrt{5}\sqrt{7}) \qquad \text{Communicative and associative properties}$$

$$= (3 \cdot 2)(\sqrt{5 \cdot 7}) \qquad \text{Property 2 for radicals}$$

$$= 6\sqrt{35} \qquad \text{Multiplication}$$

In practice, it is not necessary to show the first two steps.

EXAMPLE 2 Multiply $\sqrt{3}(2\sqrt{6} - 5\sqrt{12})$.

SOLUTION Applying the distributive property, we have

$$\sqrt{3}(2\sqrt{6} - 5\sqrt{12}) = \sqrt{3} \cdot 2\sqrt{6} - \sqrt{3} \cdot 5\sqrt{12}$$
$$= 2\sqrt{18} - 5\sqrt{36}$$

Writing each radical in simplified form gives

$$2\sqrt{18} - 5\sqrt{36} = 2\sqrt{9}\sqrt{2} - 5\sqrt{36}$$
$$= 2 \cdot 3\sqrt{2} - 5 \cdot 6$$
$$= 6\sqrt{2} - 30$$

EXAMPLE 3 Multiply $(\sqrt{3} + \sqrt{5})(4\sqrt{3} - \sqrt{5})$.

SOLUTION The same principle that applies when multiplying two binomials applies to this product. We must multiply each term in the first expression by each term in the second one. Any convenient method can be used. Let's use the FOIL method.

$$(\sqrt{3} + \sqrt{5})(4\sqrt{3} - \sqrt{5}) = \overset{F}{\sqrt{3} \cdot 4\sqrt{3}} - \overset{O}{\sqrt{3} \cdot \sqrt{5}} + \overset{I}{\sqrt{5} \cdot 4\sqrt{3}} - \overset{L}{\sqrt{5} \cdot \sqrt{5}}$$
$$= 4 \cdot 3 - \sqrt{15} + 4\sqrt{15} - 5$$
$$= 12 + 3\sqrt{15} - 5$$
$$= 7 + 3\sqrt{15}$$

EXAMPLE 4 Expand and simplify $(\sqrt{x} + 3)^2$.

SOLUTION 1 We can write this problem as a multiplication problem and proceed as we did in Example 3:

$$(\sqrt{x} + 3)^2 = (\sqrt{x} + 3)(\sqrt{x} + 3)$$
$$= \overset{F}{\sqrt{x} \cdot \sqrt{x}} + \overset{O}{3\sqrt{x}} + \overset{I}{3\sqrt{x}} + \overset{L}{3 \cdot 3}$$
$$= x + 3\sqrt{x} + 3\sqrt{x} + 9$$
$$= x + 6\sqrt{x} + 9$$

SOLUTION 2 We can obtain the same result by applying the formula for the square of a sum: $(a + b)^2 = a^2 + 2ab + b^2$.

$$(\sqrt{x} + 3)^2 = (\sqrt{x})^2 + 2(\sqrt{x})(3) + 3^2$$
$$= x + 6\sqrt{x} + 9$$

EXAMPLE 5 Expand $(3\sqrt{x} - 2\sqrt{y})^2$ and simplify the result.

SOLUTION Let's apply the formula for the square of a difference, $(a - b)^2 = a^2 - 2ab + b^2$.

$$(3\sqrt{x} - 2\sqrt{y})^2 = (3\sqrt{x})^2 - 2(3\sqrt{x})(2\sqrt{y}) + (2\sqrt{y})^2$$
$$= 9x - 12\sqrt{xy} + 4y$$

EXAMPLE 6   Expand and simplify $(\sqrt{x+2} - 1)^2$.

SOLUTION   Applying the formula $(a - b)^2 = a^2 - 2ab + b^2$, we have

$$(\sqrt{x+2} - 1)^2 = (\sqrt{x+2})^2 - 2\sqrt{x+2}(1) + 1^2$$

$$= x + 2 - 2\sqrt{x+2} + 1$$

$$= x + 3 - 2\sqrt{x+2}$$

EXAMPLE 7   Multiply $(\sqrt{6} + \sqrt{2})(\sqrt{6} - \sqrt{2})$.

SOLUTION   We notice the product is of the form $(a + b)(a - b)$, which always gives the difference of two squares, $a^2 - b^2$:

$$(\sqrt{6} + \sqrt{2})(\sqrt{6} - \sqrt{2}) = (\sqrt{6})^2 - (\sqrt{2})^2$$

$$= 6 - 2$$

$$= 4$$

In Example 7, the two expressions $(\sqrt{6} + \sqrt{2})$ and $(\sqrt{6} - \sqrt{2})$ are called **conjugates**. In general, the conjugate of $\sqrt{a} + \sqrt{b}$ is $\sqrt{a} - \sqrt{b}$. If a and b are positive rational numbers, multiplying conjugates of this form always produces a rational number. That is, if a and b are positive rational numbers, then

$$(\sqrt{a} + \sqrt{b})(\sqrt{a} - \sqrt{b}) = \sqrt{a}\sqrt{a} - \sqrt{a}\sqrt{b} + \sqrt{a}\sqrt{b} - \sqrt{b}\sqrt{b}$$

$$= a - \sqrt{ab} + \sqrt{ab} - b$$

$$= a - b$$

which is rational if a and b are positive rational numbers. Later in this section, we will see how the conjugate can be used to simplify certain quotients containing radicals.

Division

EXAMPLE 8   Divide: $\dfrac{6\sqrt{14}}{2\sqrt{7}}$.

SOLUTION   We begin by writing the expression as a product of two fractions. Then we can use Property 3 to simplify the fraction containing the radicals.

$$\frac{6\sqrt{14}}{2\sqrt{7}} = \frac{6}{2} \cdot \frac{\sqrt{14}}{\sqrt{7}}$$

$$= \frac{6}{2} \cdot \sqrt{\frac{14}{7}} \qquad \text{Property 3}$$

$$= 3\sqrt{2} \qquad \text{Reduce}$$

EXAMPLE 9   Divide: $\dfrac{3\sqrt{60}}{4\sqrt{3}}$.

SOLUTION   Following our process from the previous example, we have

$$\frac{3\sqrt{60}}{4\sqrt{3}} = \frac{3}{4} \cdot \frac{\sqrt{60}}{\sqrt{3}}$$

$$= \frac{3}{4} \cdot \sqrt{20}$$

Now we simplify the remaining radical expression and reduce if possible.

$$\frac{3\sqrt{20}}{4} = \frac{3 \cdot 2\sqrt{5}}{2 \cdot 2} \qquad \sqrt{20} = \sqrt{4}\sqrt{5} = 2\sqrt{5}$$

$$= \frac{3\sqrt{5}}{2} \qquad \text{Reduce}$$

Rationalizing the Denominator

Next we look at some examples that involve radicals in quotients. When simplifying these expressions, our motivation is to satisfy the second and third conditions for simplified form for radicals that we introduced in Section 7.3.

EXAMPLE 10 Write $\sqrt{\dfrac{5}{6}}$ in simplified form.

SOLUTION Using Property 3 for radicals, we have

$$\sqrt{\frac{5}{6}} = \frac{\sqrt{5}}{\sqrt{6}}$$

Note The phrase "rationalizing the denominator" is used because we have changed the denominator from $\sqrt{6}$, which is an irrational number, into $\sqrt{36} = 6$, which is a rational number.

The resulting expression satisfies the second condition for simplified form because neither radical contains a fraction. It does, however, violate Condition 3 because it has a radical in the denominator. Getting rid of the radical in the denominator is called ***rationalizing the denominator*** and is accomplished by turning the radicand in the denominator into a perfect square. In this case, we can do this by multiplying the numerator and denominator by $\sqrt{6}$:

$$\frac{\sqrt{5}}{\sqrt{6}} = \frac{\sqrt{5}}{\sqrt{6}} \cdot \frac{\sqrt{6}}{\sqrt{6}} \qquad \text{Equivalent fraction}$$

$$= \frac{\sqrt{30}}{\sqrt{36}} \qquad \text{Property 2 for radicals}$$

$$= \frac{\sqrt{30}}{6}$$

EXAMPLES Rationalize the denominator.

11. $\dfrac{4}{\sqrt{3}} = \dfrac{4}{\sqrt{3}} \cdot \dfrac{\sqrt{3}}{\sqrt{3}}$

$$= \frac{4\sqrt{3}}{\sqrt{3^2}}$$

$$= \frac{4\sqrt{3}}{3}$$

12. $\dfrac{2\sqrt{3x}}{\sqrt{5y}} = \dfrac{2\sqrt{3x}}{\sqrt{5y}} \cdot \dfrac{\sqrt{5y}}{\sqrt{5y}}$

$$= \frac{2\sqrt{15xy}}{\sqrt{(5y)^2}}$$

$$= \frac{2\sqrt{15xy}}{5y}$$

When the denominator involves a cube root, we must multiply by a radical that will produce a perfect cube under the cube root sign in the denominator, as Example 13 illustrates.

EXAMPLE 13 Rationalize the denominator in $\dfrac{7}{\sqrt[3]{4}}$.

SOLUTION Because $4 = 2^2$, we can multiply both numerator and denominator by $\sqrt[3]{2}$ and obtain $\sqrt[3]{2^3}$ in the denominator.

$$\frac{7}{\sqrt[3]{4}} = \frac{7}{\sqrt[3]{2^2}}$$

$$= \frac{7}{\sqrt[3]{2^2}} \cdot \frac{\sqrt[3]{2}}{\sqrt[3]{2}}$$

$$= \frac{7\sqrt[3]{2}}{\sqrt[3]{2^3}}$$

$$= \frac{7\sqrt[3]{2}}{2}$$

EXAMPLE 14 Simplify: $\sqrt{\dfrac{12x^5y^3}{5z}}$. Assume all variables represent positive numbers.

SOLUTION We use Property 3 to write the numerator and denominator as two separate radicals:

$$\sqrt{\frac{12x^5y^3}{5z}} = \frac{\sqrt{12x^5y^3}}{\sqrt{5z}}$$

Simplifying the numerator, we have

$$\frac{\sqrt{12x^5y^3}}{\sqrt{5z}} = \frac{\sqrt{4x^4y^2}\sqrt{3xy}}{\sqrt{5z}}$$

$$= \frac{2x^2y\sqrt{3xy}}{\sqrt{5z}}$$

To rationalize the denominator, we multiply the numerator and denominator by $\sqrt{5z}$:

$$\frac{2x^2y\sqrt{3xy}}{\sqrt{5z}} \cdot \frac{\sqrt{5z}}{\sqrt{5z}} = \frac{2x^2y\sqrt{15xyz}}{\sqrt{(5z)^2}}$$

$$= \frac{2x^2y\sqrt{15xyz}}{5z}$$

EXAMPLE 15 Rationalize the denominator: $\dfrac{6}{\sqrt{5} - \sqrt{3}}$.

SOLUTION Because the product of two conjugates is a rational number, we multiply the numerator and denominator by the conjugate of the denominator.

$$\frac{6}{\sqrt{5} - \sqrt{3}} = \frac{6}{\sqrt{5} - \sqrt{3}} \cdot \frac{(\sqrt{5} + \sqrt{3})}{(\sqrt{5} + \sqrt{3})}$$

$$= \frac{6\sqrt{5} + 6\sqrt{3}}{(\sqrt{5})^2 - (\sqrt{3})^2}$$

$$= \frac{6\sqrt{5} + 6\sqrt{3}}{5 - 3}$$

$$= \frac{6\sqrt{5} + 6\sqrt{3}}{2}$$

The numerator and denominator of this last expression have a factor of 2 in common.

We can reduce to lowest terms by factoring 2 from the numerator and then dividing both the numerator and denominator by 2:

$$= \frac{2(3\sqrt{5} + 3\sqrt{3})}{2}$$

$$= 3\sqrt{5} + 3\sqrt{3}$$

EXAMPLE 16   Rationalize the denominator: $\dfrac{\sqrt{5} - 2}{\sqrt{5} + 2}$.

SOLUTION   To rationalize the denominator, we multiply the numerator and denominator by the conjugate of the denominator:

$$\frac{\sqrt{5} - 2}{\sqrt{5} + 2} = \frac{\sqrt{5} - 2}{\sqrt{5} + 2} \cdot \frac{(\sqrt{5} - 2)}{(\sqrt{5} - 2)}$$

$$= \frac{5 - 2\sqrt{5} - 2\sqrt{5} + 4}{(\sqrt{5})^2 - 2^2}$$

$$= \frac{9 - 4\sqrt{5}}{5 - 4}$$

$$= \frac{9 - 4\sqrt{5}}{1}$$

$$= 9 - 4\sqrt{5}$$

Applications

EXAMPLE 17   A golden rectangle constructed from a square of side 2 is shown in Figure 2. Show that the smaller rectangle *BDEF* is also a golden rectangle by finding the ratio of its length to its width.

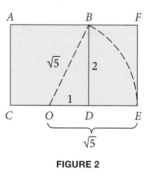

FIGURE 2

SOLUTION   First, find expressions for the length and width of the smaller rectangle.

Length = $EF = 2$

Width = $DE = \sqrt{5} - 1$

Next, we find the ratio of length to width.

Ratio of length to width $= \dfrac{EF}{DE} = \dfrac{2}{\sqrt{5} - 1}$

To show that the small rectangle is a golden rectangle, we must show that the ratio of length to width is the golden ratio. We do so by rationalizing the denominator.

$$\frac{2}{\sqrt{5} - 1} = \frac{2}{\sqrt{5} - 1} \cdot \frac{\sqrt{5} + 1}{\sqrt{5} + 1}$$

$$= \frac{2(\sqrt{5} + 1)}{5 - 1}$$

$$= \frac{2(\sqrt{5} + 1)}{4}$$

$$= \frac{\sqrt{5} + 1}{2} \qquad \text{Divide out common factor 2}$$

Because addition is commutative, this last expression is the golden ratio. Therefore, the small rectangle in Figure 2 is a golden rectangle.

Getting Ready for Class

After reading through the preceding section, respond in your own words and in complete sentences.

A. Explain why $(\sqrt{5} + \sqrt{2})^2 \neq 5 + 2$.

B. What are conjugates?

C. What result is guaranteed when multiplying radical expressions that are conjugates?

D. How is rationalizing the denominator for $\frac{3}{\sqrt{5} - 2}$ different than for $\frac{3}{\sqrt{5}}$?

Problem Set 7.5

Assume all expressions appearing under a square root symbol represent positive numbers throughout this problem set.

Multiply.

1. $\sqrt{6}\sqrt{3}$ **2.** $\sqrt{6}\sqrt{2}$

3. $(2\sqrt{3})(5\sqrt{7})$ **4.** $(3\sqrt{5})(2\sqrt{7})$

5. $(4\sqrt{6})(2\sqrt{15})(3\sqrt{10})$ **6.** $(4\sqrt{35})(2\sqrt{21})(5\sqrt{15})$

7. $(3\sqrt[3]{3})(6\sqrt[3]{9})$ **8.** $(2\sqrt[3]{2})(6\sqrt[3]{4})$

9. $\sqrt{3}(\sqrt{2} - 3\sqrt{3})$ **10.** $\sqrt{2}(5\sqrt{3} + 4\sqrt{2})$

11. $6\sqrt[3]{4}(2\sqrt[3]{2} + 1)$ **12.** $7\sqrt[3]{5}(3\sqrt[3]{25} - 2)$

13. $(\sqrt{3} + \sqrt{2})(3\sqrt{3} - \sqrt{2})$ **14.** $(\sqrt{5} - \sqrt{2})(3\sqrt{5} + 2\sqrt{2})$

15. $(\sqrt{x} + 5)(\sqrt{x} - 3)$ **16.** $(\sqrt{x} + 4)(\sqrt{x} + 2)$

17. $(3\sqrt{6} + 4\sqrt{2})(\sqrt{6} + 2\sqrt{2})$ **18.** $(\sqrt{7} - 3\sqrt{3})(2\sqrt{7} - 4\sqrt{3})$

19. $(\sqrt{3} + 4)^2$ **20.** $(\sqrt{5} - 2)^2$

21. $(\sqrt{x} - 3)^2$ **22.** $(\sqrt{x} + 4)^2$

23. $(2\sqrt{a} - 3\sqrt{b})^2$ **24.** $(5\sqrt{a} - 2\sqrt{b})^2$

25. $(\sqrt{x - 4} + 2)^2$ **26.** $(\sqrt{x - 3} + 2)^2$

27. $(\sqrt{x - 5} - 3)^2$ **28.** $(\sqrt{x - 3} - 4)^2$

29. $(\sqrt{3} - \sqrt{2})(\sqrt{3} + \sqrt{2})$ **30.** $(\sqrt{5} - \sqrt{2})(\sqrt{5} + \sqrt{2})$

31. $(\sqrt{a} + 7)(\sqrt{a} - 7)$ **32.** $(\sqrt{a} + 5)(\sqrt{a} - 5)$

33. $(5 - \sqrt{x})(5 + \sqrt{x})$ **34.** $(3 - \sqrt{x})(3 + \sqrt{x})$

35. $(\sqrt{x - 4} + 2)(\sqrt{x - 4} - 2)$ **36.** $(\sqrt{x + 3} + 5)(\sqrt{x + 3} - 5)$

37. $(\sqrt{3} + 1)^3$ **38.** $(\sqrt{5} - 2)^3$

Divide.

39. $\dfrac{\sqrt{30}}{\sqrt{6}}$ **40.** $\dfrac{\sqrt{21}}{\sqrt{3}}$ **41.** $\dfrac{6\sqrt{10}}{3\sqrt{2}}$ **42.** $\dfrac{8\sqrt{15}}{2\sqrt{5}}$

43. $\dfrac{\sqrt[3]{12}}{\sqrt[3]{3}}$ **44.** $\dfrac{\sqrt[3]{60}}{\sqrt[3]{4}}$ **45.** $\dfrac{2\sqrt[3]{18}}{12\sqrt[3]{9}}$ **46.** $\dfrac{4\sqrt[3]{42}}{10\sqrt[3]{6}}$

47. $\dfrac{3x\sqrt{24}}{9\sqrt{2}}$ **48.** $\dfrac{5\sqrt{54x^2}}{35\sqrt{3}}$ **49.** $\dfrac{11\sqrt{40ab^3}}{2\sqrt{5ab}}$ **50.** $\dfrac{3\sqrt{120x^2y^3}}{4x\sqrt{6y}}$

Rationalize the denominator in each of the following expressions.

51. $\dfrac{2}{\sqrt{3}}$ **52.** $\dfrac{3}{\sqrt{2}}$ **53.** $\dfrac{5}{\sqrt{6}}$ **54.** $\dfrac{7}{\sqrt{5}}$

55. $\sqrt{\dfrac{1}{2}}$ **56.** $\sqrt{\dfrac{1}{3}}$ **57.** $\sqrt{\dfrac{1}{5}}$ **58.** $\sqrt{\dfrac{1}{6}}$

59. $\dfrac{4}{\sqrt[3]{2}}$ **60.** $\dfrac{5}{\sqrt[3]{3}}$ **61.** $\dfrac{2}{\sqrt[3]{9}}$ **62.** $\dfrac{3}{\sqrt[3]{4}}$

63. $\sqrt[4]{\dfrac{3}{2x^2}}$ **64.** $\sqrt[4]{\dfrac{5}{3x^2}}$ **65.** $\sqrt[4]{\dfrac{8}{y}}$ **66.** $\sqrt[4]{\dfrac{27}{y}}$

67. $\sqrt[3]{\dfrac{4x}{3y}}$ **68.** $\sqrt[3]{\dfrac{7x}{6y}}$ **69.** $\sqrt[3]{\dfrac{2x}{9y}}$ **70.** $\sqrt[3]{\dfrac{5x}{4y}}$

Write each of the following in simplified form.

71. $\sqrt{\dfrac{27x^3}{5y}}$ **72.** $\sqrt{\dfrac{12x^5}{7y}}$ **73.** $\sqrt{\dfrac{75x^3y^2}{2z}}$ **74.** $\sqrt{\dfrac{50x^2y^3}{3z}}$

Rationalize the denominator.

75. a. $\dfrac{1}{\sqrt{2}}$ **b.** $\dfrac{1}{\sqrt[3]{2}}$ **c.** $\dfrac{1}{\sqrt[4]{2}}$

76. a. $\dfrac{1}{\sqrt{3}}$ **b.** $\dfrac{1}{\sqrt[3]{9}}$ **c.** $\dfrac{1}{\sqrt[4]{27}}$

Simplify the following expressions by combining similar radicals. Assume any variables under an even root are positive.

77. $\dfrac{\sqrt{2}}{2} + \dfrac{1}{\sqrt{2}}$ **78.** $\dfrac{\sqrt{3}}{3} + \dfrac{1}{\sqrt{3}}$

79. $\dfrac{\sqrt{5}}{3} + \dfrac{1}{\sqrt{5}}$ **80.** $\dfrac{\sqrt{6}}{2} + \dfrac{1}{\sqrt{6}}$

81. $\sqrt{x} - \dfrac{1}{\sqrt{x}}$ **82.** $\sqrt{x} + \dfrac{1}{\sqrt{x}}$

83. $\dfrac{\sqrt{18}}{6} + \sqrt{\dfrac{1}{2}} + \dfrac{\sqrt{2}}{2}$ **84.** $\dfrac{\sqrt{12}}{6} + \sqrt{\dfrac{1}{3}} + \dfrac{\sqrt{3}}{3}$

85. $\sqrt{6} - \sqrt{\dfrac{2}{3}} + \sqrt{\dfrac{1}{6}}$ **86.** $\sqrt{15} - \sqrt{\dfrac{3}{5}} + \sqrt{\dfrac{5}{3}}$

87. $\sqrt[3]{25} + \dfrac{3}{\sqrt[3]{5}}$ **88.** $\sqrt[4]{8} + \dfrac{1}{\sqrt[4]{2}}$

Rationalize the denominator in each of the following.

89. $\dfrac{\sqrt{2}}{\sqrt{6} - \sqrt{2}}$ **90.** $\dfrac{\sqrt{5}}{\sqrt{5} + \sqrt{3}}$ **91.** $\dfrac{\sqrt{5}}{\sqrt{5} + 1}$ **92.** $\dfrac{\sqrt{7}}{\sqrt{7} - 1}$

93. $\dfrac{\sqrt{x}}{\sqrt{x} - 3}$ **94.** $\dfrac{\sqrt{x}}{\sqrt{x} + 2}$ **95.** $\dfrac{\sqrt{5}}{2\sqrt{5} - 3}$ **96.** $\dfrac{\sqrt{7}}{3\sqrt{7} - 2}$

97. $\dfrac{3}{\sqrt{x} - \sqrt{y}}$ **98.** $\dfrac{2}{\sqrt{x} + \sqrt{y}}$ **99.** $\dfrac{\sqrt{6} + \sqrt{2}}{\sqrt{6} - \sqrt{2}}$ **100.** $\dfrac{\sqrt{5} - \sqrt{3}}{\sqrt{5} + \sqrt{3}}$

101. $\dfrac{\sqrt{7} - 2}{\sqrt{7} + 2}$ **102.** $\dfrac{\sqrt{11} + 3}{\sqrt{11} - 3}$

103. Paying Attention to Instructions Work each problem according to the instructions given.

    a. Add: $(\sqrt{x} + 2) + (\sqrt{x} - 2)$ **b.** Multiply: $(\sqrt{x} + 2)(\sqrt{x} - 2)$

    c. Square: $(\sqrt{x} + 2)^2$ **d.** Divide: $\dfrac{\sqrt{x} + 2}{\sqrt{x} - 2}$

104. Paying Attention to Instructions Work each problem according to the instructions given.

    a. Add: $(5 + \sqrt{2}) + (5 - \sqrt{2})$ **b.** Multiply: $(5 + \sqrt{2})(5 - \sqrt{2})$

    c. Square: $(5 + \sqrt{2})^2$ **d.** Divide: $\dfrac{5 + \sqrt{2}}{5 - \sqrt{2}}$

105. Paying Attention to Instructions Work each problem according to the instructions given.

   a. Add: $\sqrt{2} + (\sqrt{6} + \sqrt{2})$ **b.** Multiply: $\sqrt{2}(\sqrt{6} + \sqrt{2})$

   c. Divide: $\dfrac{\sqrt{6} + \sqrt{2}}{\sqrt{2}}$ **d.** Divide: $\dfrac{\sqrt{2}}{\sqrt{6} + \sqrt{2}}$

106. Paying Attention to Instructions Work each problem according to the instructions given.

   a. Add: $\left(\dfrac{1 + \sqrt{5}}{2}\right) + \left(\dfrac{1 - \sqrt{5}}{2}\right)$ **b.** Multiply: $\left(\dfrac{1 + \sqrt{5}}{2}\right)\left(\dfrac{1 - \sqrt{5}}{2}\right)$

107. Show that the product below is 5:

$$(\sqrt[3]{2} + \sqrt[3]{3})(\sqrt[3]{4} - \sqrt[3]{6} + \sqrt[3]{9})$$

108. Show that the product below is $x + 8$:

$$(\sqrt[3]{x} + 2)(\sqrt[3]{x^2} - 2\sqrt[3]{x} + 4)$$

Each of the following statements below is false. Correct the right side of each one to make it true.

109. $5(2\sqrt{3}) = 10\sqrt{15}$ **110.** $3(2\sqrt{x}) = 6\sqrt{3x}$ **111.** $(\sqrt{x} + 3)^2 = x + 9$

112. $(\sqrt{x} - 7)^2 = x - 49$ **113.** $(5\sqrt{3})^2 = 15$ **114.** $(3\sqrt{5})^2 = 15$

Applying the Concepts

115. Gravity If an object is dropped from the top of a 100-foot building, the amount of time t (in seconds) that it takes for the object to be h feet from the ground is given by the formula

$$t = \dfrac{\sqrt{100 - h}}{4}$$

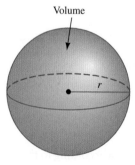

Volume

How long does it take before the object is 50 feet from the ground? How long does it take to reach the ground? (When it is on the ground, h is 0.)

116. Radius of a Sphere The radius r of a sphere with volume V can be found by using the formula

$$r = \sqrt[3]{\dfrac{3V}{4\pi}}$$

Find the radius of a sphere with volume 9 cubic feet. Write your answer in simplified form. $\left(\text{Use } \tfrac{22}{7} \text{ for } \pi.\right)$

117. Golden Rectangle Rectangle *ACEF* in Figure 3 is a golden rectangle. If side *AC* is 6 inches, show that the smaller rectangle *BDEF* is also a golden rectangle.

118. Golden Rectangle Rectangle *ACEF* in Figure 3 is a golden rectangle. If side *AC* is 1 inch, show that the smaller rectangle *BDEF* is also a golden rectangle.

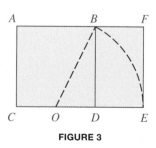

FIGURE 3

119. Golden Rectangle If side AC in Figure 3 is $2x$, show that rectangle $BDEF$ is a golden rectangle.

120. Golden Rectangle If side AC in Figure 3 is x, show that rectangle $BDEF$ is a golden rectangle.

Learning Objectives Assessment

The following problems can be used to help assess if you have successfully met the learning objectives for this section.

121. Multiply: $(3\sqrt{6})(4\sqrt{10})$.

    a. $24\sqrt{15}$ **b.** $7\sqrt{60}$ **c.** $12\sqrt{15}$ **d.** 48

122. Multiply: $(\sqrt{3} + \sqrt{7})^2$.

    a. 10 **b.** $10 + 2\sqrt{21}$ **c.** $12\sqrt{21}$ **d.** $2\sqrt{3} + 2\sqrt{7}$

123. Divide: $\dfrac{8\sqrt{12}}{12\sqrt{2}}$.

    a. $\dfrac{2\sqrt{6}}{3}$ **b.** 4 **c.** $\dfrac{8}{\sqrt{2}}$ **d.** $\dfrac{\sqrt{10}}{4}$

124. Rationalize the denominator: $\dfrac{3}{\sqrt[3]{2}}$.

    a. $\dfrac{3\sqrt[3]{2}}{2}$ **b.** $\dfrac{27}{2}$ **c.** $\dfrac{3\sqrt[3]{4}}{2}$ **d.** $\dfrac{3\sqrt{2}}{2}$

Getting Ready for the Next Section

Simplify.

125. $(t + 5)^2$ **126.** $(x - 4)^2$ **127.** $\sqrt{x} \cdot \sqrt{x}$ **128.** $\sqrt{3x} \cdot \sqrt{3x}$

Solve.

129. $3x + 4 = 5^2$ **130.** $4x - 7 = 3^2$

131. $t^2 + 7t + 12 = 0$ **132.** $x^2 - 3x - 10 = 0$

133. $t^2 + 10t + 25 = t + 7$ **134.** $x^2 - 4x + 4 = x - 2$

135. $(x + 4)^2 = x + 6$ **136.** $(x - 6)^2 = x - 4$

137. Is $x = 7$ a solution to $\sqrt{3x + 4} = 5$?

138. Is $x = 4$ a solution to $\sqrt{4x - 7} = -3$?

139. Is $t = -6$ a solution to $t + 5 = \sqrt{t + 7}$?

140. Is $t = -3$ a solution to $t + 5 = \sqrt{t + 7}$?

Equations Involving Radicals 7.6

Learning Objectives

In this section, we will learn how to:

1. Solve equations containing square roots.

2. Identify extraneous solutions when squaring both sides of an equation.

3. Solve equations involving nth roots.

4. Solve equations containing rational exponents.

Introduction

This section is concerned with solving equations that involve one or more radicals. The first step in solving an equation that contains a radical is to eliminate the radical from the equation. To do so, we need an additional property.

> **PROPERTY** *Squaring Property of Equality*
>
> If both sides of an equation are squared, the solutions to the original equation are solutions to the resulting equation.

We will never lose solutions to our equations by squaring both sides. We may, however, introduce **extraneous solutions**. Extraneous solutions satisfy the equation obtained by squaring both sides of the original equation, but do not satisfy the original equation.

We know that if two real numbers a and b are equal, then so are their squares:

$$\text{If} \qquad a = b$$
$$\text{then} \qquad a^2 = b^2$$

On the other hand, extraneous solutions are introduced when we square opposites. That is, even though opposites are not equal, their squares are. For example,

$$5 = -5 \qquad \text{A false statement}$$
$$(5)^2 = (-5)^2 \qquad \text{Square both sides}$$
$$25 = 25 \qquad \text{A true statement}$$

We are free to square both sides of an equation any time it is convenient. We must be aware, however, that doing so may introduce extraneous solutions. We must, therefore, check all our solutions in the original equation if at any time we square both sides of the original equation.

VIDEO EXAMPLES

SECTION 7.6

EXAMPLE 1 Solve for x: $\sqrt{3x + 4} = 5$.

SOLUTION We square both sides and proceed as usual:

$$\sqrt{3x + 4} = 5$$
$$(\sqrt{3x + 4})^2 = 5^2$$
$$3x + 4 = 25$$
$$3x = 21$$
$$x = 7$$

Checking $x = 7$ in the original equation, we have

$$\sqrt{3(7) + 4} \stackrel{?}{=} 5$$

$$\sqrt{21 + 4} \stackrel{?}{=} 5$$

$$\sqrt{25} \stackrel{?}{=} 5$$

$$5 = 5$$

The solution $x = 7$ satisfies the original equation.

EXAMPLE 2 Solve: $\sqrt{4x - 7} = -3$.

SOLUTION Squaring both sides, we have

$$\sqrt{4x - 7} = -3$$

$$(\sqrt{4x - 7})^2 = (-3)^2$$

$$4x - 7 = 9$$

$$4x = 16$$

$$x = 4$$

Checking $x = 4$ in the original equation gives

$$\sqrt{4(4) - 7} \stackrel{?}{=} -3$$

$$\sqrt{16 - 7} \stackrel{?}{=} -3$$

$$\sqrt{9} \stackrel{?}{=} -3$$

$$3 = -3 \qquad \text{A false statement}$$

Note The fact that there is no solution to the equation in Example 2 was obvious to begin with. Notice that the left side of the equation is the positive square root of $4x - 7$, which must be a positive number or 0. The right side of the equation is -3. Because we cannot have a number that is either positive or zero equal to a negative number, there is no solution to the equation.

The solution $x = 4$ produces a false statement when checked in the original equation. Because $x = 4$ was the only possible solution, there is no solution to the original equation. The possible solution $x = 4$ is an extraneous solution. It satisfies the equation obtained by squaring both sides of the original equation, but does not satisfy the original equation. In this case, the solution set is $\varnothing$.

EXAMPLE 3 Solve: $\sqrt{5x - 1} + 3 = 7$.

SOLUTION We must isolate the radical on the left side of the equation. If we attempt to square both sides without doing so, the resulting equation will also contain a radical. Adding -3 to both sides, we have

$$\sqrt{5x - 1} + 3 = 7$$

$$\sqrt{5x - 1} = 4$$

We can now square both sides and proceed as usual:

$$(\sqrt{5x - 1})^2 = 4^2$$

$$5x - 1 = 16$$

$$5x = 17$$

$$x = \frac{17}{5}$$

Checking $x = \dfrac{17}{5}$, we have

$$\sqrt{5\left(\dfrac{17}{5}\right) - 1} + 3 \overset{?}{=} 7$$

$$\sqrt{17 - 1} + 3 \overset{?}{=} 7$$

$$\sqrt{16} + 3 \overset{?}{=} 7$$

$$4 + 3 \overset{?}{=} 7$$

$$7 = 7$$

EXAMPLE 4　　Solve: $t + 5 = \sqrt{t + 7}$.

SOLUTION　This time, squaring both sides of the equation results in a quadratic equation:

$$(t + 5)^2 = (\sqrt{t + 7})^2 \qquad \text{Square both sides}$$

$$t^2 + 10t + 25 = t + 7$$

$$t^2 + 9t + 18 = 0 \qquad \text{Standard form}$$

$$(t + 3)(t + 6) = 0 \qquad \text{Factor the left side}$$

$$t + 3 = 0 \quad \text{or} \quad t + 6 = 0 \qquad \text{Set factors equal to 0}$$

$$t = -3 \qquad\qquad t = -6$$

We must check each solution in the original equation:

Check $t = -3$	Check $t = -6$
$-3 + 5 \overset{?}{=} \sqrt{-3 + 7}$	$-6 + 5 \overset{?}{=} \sqrt{-6 + 7}$
$2 \overset{?}{=} \sqrt{4}$	$-1 \overset{?}{=} \sqrt{1}$
$2 = 2$　A true statement	$-1 = 1$　A false statement

Because $t = -6$ does not check, our only solution is $t = -3$.

EXAMPLE 5　　Solve: $\sqrt{x - 3} = \sqrt{x} - 3$.

SOLUTION　We begin by squaring both sides. Note carefully what happens when we square the right side of the equation, and compare the square of the right side with the square of the left side. You must convince yourself that these results are correct. (The note in the margin will help if you are having trouble convincing yourself that what is written below is true.)

$$(\sqrt{x - 3})^2 = (\sqrt{x} - 3)^2$$

$$x - 3 = x - 6\sqrt{x} + 9$$

Now we still have a radical in our equation, so we will have to square both sides again. Before we do, though, let's isolate the remaining radical.

$$x - 3 = x - 6\sqrt{x} + 9$$

$$-3 = -6\sqrt{x} + 9 \qquad \text{Add } -x \text{ to each side}$$

$$-12 = -6\sqrt{x} \qquad \text{Add } -9 \text{ to each side}$$

$$2 = \sqrt{x} \qquad \text{Divide each side by } -6$$

$$4 = x \qquad \text{Square each side}$$

Note　It is very important that you realize that the square of $(\sqrt{x} - 3)$ is not $x + 9$. Remember, when we square a difference with two terms, we use the formula

$$(a - b)^2 = a^2 - 2ab + b^2$$

Applying this formula to $(\sqrt{x} - 3)^2$ we have

$$(\sqrt{x} - 3)^2 =$$
$$(\sqrt{x})^2 - 2(\sqrt{x})(3) + 3^2$$
$$= x - 6\sqrt{x} + 9$$

Our only possible solution is $x = 4$, which we check in our original equation as follows:

$$\sqrt{4-3} \stackrel{?}{=} \sqrt{4} - 3$$

$$\sqrt{1} \stackrel{?}{=} 2 - 3$$

$$1 = -1 \qquad \text{A false statement}$$

Substituting 4 for x in the original equation yields a false statement. Because 4 was our only possible solution, there is no solution to our equation.   ◼

Here is another example of an equation for which we must apply our squaring property twice before all radicals are eliminated.

◼ **EXAMPLE 6**   Solve: $\sqrt{x+1} = 1 - \sqrt{2x}$.

SOLUTION   This equation has two separate terms involving radical signs. Squaring both sides gives

$$x + 1 = 1 - 2\sqrt{2x} + 2x$$

$$-x = -2\sqrt{2x} \qquad \text{Add } -2x \text{ and } -1 \text{ to both sides}$$

$$x^2 = 4(2x) \qquad \text{Square both sides}$$

$$x^2 - 8x = 0 \qquad \text{Standard form}$$

Our equation is a quadratic equation in standard form. To solve for x, we factor the left side and set each factor equal to 0:

$$x(x - 8) = 0 \qquad \text{Factor left side}$$

$$x = 0 \quad \text{or} \quad x - 8 = 0$$

$$x = 8 \qquad \text{Set factors equal to 0}$$

Because we squared both sides of our equation, we have the possibility that one or both of the solutions are extraneous. We must check each one in the original equation:

Check $x = 8$	Check $x = 0$
$\sqrt{8+1} \stackrel{?}{=} 1 - \sqrt{2 \cdot 8}$	$\sqrt{0+1} \stackrel{?}{=} 1 - \sqrt{2 \cdot 0}$
$\sqrt{9} \stackrel{?}{=} 1 - \sqrt{16}$	$\sqrt{1} \stackrel{?}{=} 1 - \sqrt{0}$
$3 \stackrel{?}{=} 1 - 4$	$1 \stackrel{?}{=} 1 - 0$
$3 = -3$ A false statement	$1 = 1$ A true statement

Because $x = 8$ does not check, it is an extraneous solution. Our only solution is $x = 0$.   ◼

EXAMPLE 7 Solve: $\sqrt{x+1} = \sqrt{x+2} - 1$.

SOLUTION Squaring both sides we have

$$(\sqrt{x+1})^2 = (\sqrt{x+2} - 1)^2$$
$$x + 1 = x + 2 - 2\sqrt{x+2} + 1$$

Once again, we are left with a radical in our equation. Before we square each side again, we must isolate the radical on the right side of the equation.

$$x + 1 = x + 3 - 2\sqrt{x+2} \qquad \text{Simplify the right side}$$
$$1 = 3 - 2\sqrt{x+2} \qquad \text{Add } -x \text{ to each side}$$
$$-2 = -2\sqrt{x+2} \qquad \text{Add } -3 \text{ to each side}$$
$$1 = \sqrt{x+2} \qquad \text{Divide each side by } -2$$
$$1 = x + 2 \qquad \text{Square both sides}$$
$$-1 = x \qquad \text{Add } -2 \text{ to each side}$$

Checking our only possible solution, $x = -1$, in our original equation, we have

$$\sqrt{-1+1} \stackrel{?}{=} \sqrt{-1+2} - 1$$
$$\sqrt{0} \stackrel{?}{=} \sqrt{1} - 1$$
$$0 \stackrel{?}{=} 1 - 1$$
$$0 = 0 \qquad \text{A true statement}$$

Our solution checks.

It is also possible to raise both sides of an equation to powers greater than 2. We only need to check for extraneous solutions when we raise both sides of an equation to an even power. Raising both sides of an equation to an odd power will not produce extraneous solutions.

EXAMPLE 8 Solve: $\sqrt[3]{4x+5} = 3$.

SOLUTION Cubing both sides, we have

$$(\sqrt[3]{4x+5})^3 = 3^3$$
$$4x + 5 = 27$$
$$4x = 22$$
$$x = \frac{22}{4}$$
$$x = \frac{11}{2}$$

We do not need to check $x = \frac{11}{2}$ because we raised both sides to an odd power.

EXAMPLE 9   Solve: $(x + 3)^{2/3} = 4$.

SOLUTION   We begin by writing the rational exponent as a radical.

$$(x + 3)^{2/3} = \sqrt[3]{(x + 3)^2}$$

Now we cube both sides to eliminate the cube root.

$$\sqrt[3]{(x + 3)^2} = 4$$

$$\left(\sqrt[3]{(x + 3)^2}\right)^3 = 4^3 \qquad \text{Cube both sides}$$

$$(x + 3)^2 = 64$$

$$x^2 + 6x + 9 = 64 \qquad \text{FOIL}$$

$$x^2 + 6x - 55 = 0 \qquad \text{Standard form}$$

$$(x + 11)(x - 5) = 0 \qquad \text{Factor left side}$$

$$x + 11 = 0 \quad \text{or} \quad x - 5 = 0$$

$$x = -11 \qquad\qquad x = 5$$

Because we raised both sides to an odd power, we do not need to check our solutions.

Getting Ready for Class

After reading through the preceding section, respond in your own words and in complete sentences.

A. What is the squaring property of equality?

B. Under what conditions do we obtain extraneous solutions to equations that contain radical expressions?

C. If we have raised both sides of an equation to a power, when is it not necessary to check for extraneous solutions?

D. When will you need to apply the squaring property of equality twice in the process of solving an equation containing radicals?

Solve each of the following equations.

1. $\sqrt{2x + 1} = 3$ **2.** $\sqrt{3x + 1} = 4$ **3.** $\sqrt{4x + 1} = -5$

4. $\sqrt{6x + 1} = -5$ **5.** $\sqrt{2y - 1} = 3$ **6.** $\sqrt{3y - 1} = 2$

7. $\sqrt{5x - 7} = -1$ **8.** $\sqrt{8x + 3} = -6$ **9.** $\sqrt{2x - 3} - 2 = 4$

10. $\sqrt{3x + 1} - 4 = 1$ **11.** $\sqrt{4a + 1} + 3 = 2$ **12.** $\sqrt{5a - 3} + 6 = 2$

13. $\sqrt[4]{3x + 1} = 2$ **14.** $\sqrt[4]{4x + 1} = 3$ **15.** $\sqrt[3]{2x - 5} = 1$

16. $\sqrt[3]{5x + 7} = 2$ **17.** $\sqrt[3]{3a + 5} = -3$ **18.** $\sqrt[3]{2a + 7} = -2$

19. $\sqrt{y - 3} = y - 3$ **20.** $\sqrt{y + 3} = y - 3$ **21.** $\sqrt{a + 2} = a + 2$

22. $\sqrt{a + 10} = a - 2$ **23.** $\sqrt{2x + 4} = \sqrt{1 - x}$

24. $\sqrt{3x + 4} = -\sqrt{2x + 3}$ **25.** $\sqrt{4a + 7} = -\sqrt{a + 2}$

26. $\sqrt{7a - 1} = \sqrt{2a + 4}$ **27.** $\sqrt[4]{5x - 8} = \sqrt[4]{4x - 1}$

28. $\sqrt[4]{6x + 7} = \sqrt[4]{x + 2}$ **29.** $x + 1 = \sqrt{5x + 1}$

30. $x - 1 = \sqrt{6x + 1}$ **31.** $t + 5 = \sqrt{2t + 9}$

32. $t + 7 = \sqrt{2t + 13}$ **33.** $\sqrt{y - 8} = \sqrt{8 - y}$

34. $\sqrt{2y + 5} = \sqrt{5y + 2}$ **35.** $\sqrt[3]{3x + 5} = \sqrt[3]{5 - 2x}$

36. $\sqrt[3]{4x + 9} = \sqrt[3]{3 - 2x}$

The following equations will require that you square both sides twice before all the radicals are eliminated. Solve each equation using the methods shown in Examples 5, 6, and 7.

37. $\sqrt{x - 8} = \sqrt{x} - 2$ **38.** $\sqrt{x + 3} = \sqrt{x} - 3$

39. $\sqrt{x + 1} = \sqrt{x} + 1$ **40.** $\sqrt{x - 1} = \sqrt{x} - 1$

41. $\sqrt{x + 8} = \sqrt{x - 4} + 2$ **42.** $\sqrt{x + 5} = \sqrt{x - 3} + 2$

43. $\sqrt{x - 5} - 3 = \sqrt{x - 8}$ **44.** $\sqrt{x - 3} - 4 = \sqrt{x - 3}$

45. Solve each equation.

    a. $\sqrt{y} - 4 = 6$ **b.** $\sqrt{y - 4} = 6$

    c. $\sqrt{y - 4} = -6$ **d.** $\sqrt{y - 4} = y - 6$

46. Solve each equation.

    a. $\sqrt{2y} + 15 = 7$ **b.** $\sqrt{2y + 15} = 7$

    c. $\sqrt{2y + 15} = y$ **d.** $\sqrt{2y + 15} = y + 6$

47. Solve each equation.

    a. $x - 3 = 0$ **b.** $\sqrt{x} - 3 = 0$

    c. $\sqrt{x - 3} = 0$ **d.** $\sqrt{x} + 3 = 0$

    e. $\sqrt{x} + 3 = 5$ **f.** $\sqrt{x} + 3 = -5$

    g. $x - 3 = \sqrt{5 - x}$

48. Solve each equation.

    a. $x - 2 = 0$ **b.** $\sqrt{x} - 2 = 0$ **c.** $\sqrt{x} + 2 = 0$

    d. $\sqrt{x + 2} = 0$ **e.** $\sqrt{x} + 2 = 7$ **f.** $x - 2 = \sqrt{2x - 1}$

Solve each of the following equations by writing the rational exponent as a radical.

49. $(5x - 1)^{1/2} = 3$

50. $(2x - 5)^{1/2} = 7$

51. $(4x + 1)^{1/3} = 2$

52. $(3x + 4)^{1/3} = -5$

53. $(x + 2)^{2/3} = 1$

54. $(x - 4)^{2/3} = 4$

Applying the Concepts

55. Solving a Formula Solve the following formula for h:

$$t = \frac{\sqrt{100 - h}}{4}$$

56. Solving a Formula Solve the following formula for h:

$$t = \sqrt{\frac{2h - 40t}{g}}$$

57. Pendulum Clock The length of time (T) in seconds it takes the pendulum of a clock to swing through one complete cycle is given by the formula

$$T = 2\pi\sqrt{\frac{L}{32}}$$

where L is the length, in feet, of the pendulum, and π is approximately $\frac{22}{7}$. How long must the pendulum be if one complete cycle takes 2 seconds?

58. Pollution A long straight river, 100 meters wide, is flowing at 1 meter per second. A pollutant is entering the river at a constant rate from one of its banks. As the pollutant disperses in the water, it forms a plume that is modeled by the equation $y = \sqrt{x}$. Use this information to answer the following questions.

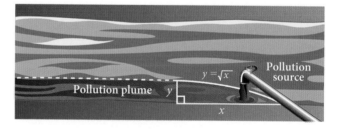

a. How wide is the plume 25 meters down river from the source of the pollution?

b. How wide is the plume 100 meters down river from the source of the pollution?

c. How far down river from the source of the pollution does the plume reach halfway across the river?

d. How far down the river from the source of the pollution does the plume reach the other side of the river?

Learning Objectives Assessment

The following problems can be used to help assess if you have successfully met the learning objectives for this section.

59. Solve: $\sqrt{3x - 2} - 3 = 1$.

   a. $\varnothing$ **b.** $\dfrac{4}{3}$ **c.** -2 **d.** 6

60. Solve: $\sqrt{2x + 5} + 1 = 0$.

   a. -2 **b.** $\varnothing$ **c.** -3 **d.** $-\dfrac{9}{2}$

61. Solve: $\sqrt[3]{2x - 3} = -4$.

   a. $\dfrac{19}{2}$ **b.** $\varnothing$ **c.** $\dfrac{67}{2}$ **d.** $-\dfrac{61}{2}$

62. Solve: $(2x + 1)^{1/2} = 9$.

   a. 1 **b.** 40 **c.** $\varnothing$ **d.** 4

Getting Ready for the Next Section

Simplify.

63. $\sqrt{25}$ **64.** $\sqrt{49}$ **65.** $\sqrt{12}$ **66.** $\sqrt{50}$

67. $(-1)^{15}$ **68.** $(-1)^{20}$ **69.** $(-1)^{50}$ **70.** $(-1)^{5}$

Solve.

71. $3x = 12$ **72.** $4 = 8y$ **73.** $4x - 3 = 5$ **74.** $7 = 2y - 1$

Perform the indicated operation.

75. $(3 + 4x) + (7 - 6x)$ **76.** $(2 - 5x) + (-1 + 7x)$

77. $(7 + 3x) - (5 + 6x)$ **78.** $(5 - 2x) - (9 - 4x)$

79. $(3 - 4x)(2 + 5x)$ **80.** $(8 + x)(7 - 3x)$

81. $2x(4 - 6x)$ **82.** $3x(7 + 2x)$

83. $(2 + 3x)^2$ **84.** $(3 + 5x)^2$

85. $(2 - 3x)(2 + 3x)$ **86.** $(4 - 5x)(4 + 5x)$

Complex Numbers

Learning Objectives

In this section, we will learn how to:

1. Write the square root of a negative number as a pure imaginary number.

2. Simplify powers of i.

3. Add and subtract complex numbers.

4. Multiply and divide complex numbers.

Introduction

The equation $x^2 = -1$ has no real number solutions because the square of a real number is always nonnegative. We have been unable to work with square roots of negative numbers like $\sqrt{-25}$ and $\sqrt{-16}$ for the same reason. Complex numbers allow us to expand our work with radicals to include square roots of negative numbers and to solve equations like $x^2 = -9$ and $x^2 = -64$.

Imaginary Numbers

Our work with complex numbers begins with the following definition.

> **(dēf) DEFINITION** *the number i*
>
> The **number i** is such that $i = \sqrt{-1}$ (which is the same as saying $i^2 = -1$).

The number i, called the **imaginary unit**, is not a real number. It is what we call a **pure imaginary number**. In general, a pure imaginary number is the product of a real number, b, and the imaginary unit i.

> **(dēf) DEFINITION** *pure imaginary number*
>
> A **pure imaginary number** is any number that can be expressed in the form bi, where b is a real number and i is the imaginary unit.

Note The first use of the term "imaginary number" was by René Descartes in his work *La Géométrie*, published in 1637. A century later, Leonhard Euler introduced the use of i to represent $\sqrt{-1}$ in 1777.

Some examples of pure imaginary numbers are

$$3i \qquad -\frac{2}{7}i \qquad \sqrt{5}\,i$$

Because of the way we have defined i, we can use it to simplify square roots of negative numbers.

> **⎰△≠∑⎱ *Square Roots of Negative Numbers***
>
> If b is a positive number, then $\sqrt{-b}$ can always be written as the pure imaginary number $i\sqrt{b}$. That is,
>
> $$\sqrt{-b} = i\sqrt{b} \qquad \text{if } b \text{ is a positive number}$$

To justify our rule, we simply square the quantity $i\sqrt{b}$ to obtain $-b$. This is what it looks like when we do so:

$$(i\sqrt{b})^2 = i^2 \cdot (\sqrt{b})^2$$
$$= -1 \cdot b$$
$$= -b$$

Here are some examples that illustrate the use of our new rule.

VIDEO EXAMPLES

SECTION 7.7

EXAMPLES   Write each square root as a pure imaginary number.

1. $\sqrt{-25} = i\sqrt{25} = i \cdot 5 = 5i$

2. $\sqrt{-49} = i\sqrt{49} = i \cdot 7 = 7i$

3. $\sqrt{-12} = i\sqrt{12} = i \cdot 2\sqrt{3} = 2i\sqrt{3}$

4. $\sqrt{-17} = i\sqrt{17}$

In Examples 3 and 4, we wrote i before the radical simply to avoid confusion. If we were to write the answer to 3 as $2\sqrt{3}i$, some people would think the i was under the radical sign, but it is not.

Powers of i

If we assume all the properties of exponents hold when the base is i, we can write any power of i as i, -1, $-i$, or 1. Using the fact that $i^2 = -1$, we have

$$i^1 = i$$
$$i^2 = -1$$
$$i^3 = i^2 \cdot i = -1(i) = -i$$
$$i^4 = i^2 \cdot i^2 = -1(-1) = 1$$

Because $i^4 = 1$, i^5 will simplify to i, and we will begin repeating the sequence i, -1, $-i$, 1 as we simplify higher powers of i: Any power of i simplifies to i, -1, $-i$, or 1. The easiest way to simplify higher powers of i is to write them in terms of i^4. For instance, to simplify i^{21}, we would write it as

$$(i^4)^5 \cdot i \qquad \text{because} \qquad 4 \cdot 5 + 1 = 21$$

Then, because $i^4 = 1$, we have

$$(1)^5 \cdot i = 1 \cdot i = i$$

EXAMPLES   Simplify each power of i.

5. $i^{30} = (i^4)^7 \cdot i^2 = (1)^7(-1) = -1$

6. $i^{11} = (i^4)^2 \cdot i^3 = (1)^2 \cdot (-i) = -i$

7. $i^{40} = (i^4)^{10} = (1)^{10} = 1$

Complex Numbers

Next, we consider numbers created by adding a real number and a pure imaginary number. Called **complex numbers**, these numbers are essential in many different fields, including digital signal and image processing, quantum mechanics, electrical engineering, and biology.

> **DEFINITION** *complex number*
>
> A *complex number* is any number that can be put in the form
>
> $$a + bi$$
>
> where a and b are real numbers and $i = \sqrt{-1}$. The form $a + bi$ is called *standard form* for complex numbers. The number a is called the **real part** of the complex number. The number b is called the **imaginary part** of the complex number.

Every real number is a complex number, whose imaginary part is 0. For example, 8 can be written as $8 + 0i$. Likewise, every pure imaginary number is a complex number whose real part is 0. For example, $2i$ can be written as $0 + 2i$. The remaining complex numbers, which cannot be classified as either real numbers or pure imaginary numbers, are called **compound numbers**. The diagram below shows all three subsets of the complex numbers, along with examples of the type of numbers that fall into those subsets.

Real Numbers	Compound Numbers	Pure Imaginary Numbers
When $a \neq 0$ and $b = 0$ Examples include: $-10, 0, 1, \sqrt{3}, \frac{5}{8}, \pi$	When neither a nor b is 0 Examples include: $5 + 4i, \frac{1}{3} + 4i, \sqrt{5} - i,$ $-6 + i\sqrt{5}$	When $a = 0$ and $b \neq 0$ Examples include: $-4i, i\sqrt{3}, -5i\sqrt{7}, \frac{3}{4}i$

©2009 James Robert Metz

Note See Section 1.2 for a review of the subsets of real numbers.

Note: The definition for compound numbers is from Jim Metz of Kapiolani Community College in Hawaii. Some textbooks use the phrase *imaginary numbers* to represent both the compound numbers and the pure imaginary numbers. In those books, the pure imaginary numbers are a subset of the imaginary numbers. We like the definition from Mr. Metz because it keeps the three subsets from overlapping.

Equality for Complex Numbers

Two complex numbers are equal if and only if their real parts are equal and their imaginary parts are equal. That is, for real numbers $a, b, c,$ and d,

$$a + bi = c + di \quad \text{if and only if} \quad a = c \quad \text{and} \quad b = d$$

EXAMPLE 8 Find x and y if $3x + 4i = 12 - 8yi$.

SOLUTION Because the two complex numbers are equal, their real parts are equal and their imaginary parts are equal:

$$3x = 12 \quad \text{and} \quad 4 = -8y$$

$$x = 4 \qquad \qquad y = -\frac{1}{2}$$

EXAMPLE 9   Find x and y if $(4x - 3) + 7i = 5 + (2y - 1)i$.

SOLUTION   The real parts are $4x - 3$ and 5. The imaginary parts are 7 and $2y - 1$:

$$4x - 3 = 5 \quad \text{and} \quad 7 = 2y - 1$$
$$4x = 8 \qquad\qquad 8 = 2y$$
$$x = 2 \qquad\qquad y = 4$$

Addition and Subtraction of Complex Numbers

To add two complex numbers, add their real parts and their imaginary parts. That is, if a, b, c, and d are real numbers, then

$$(a + bi) + (c + di) = (a + c) + (b + d)i$$

If we assume that the commutative, associative, and distributive properties hold for the number i, then the definition of addition is simply an extension of these properties, allowing us to add complex numbers by combining similar terms.

We define subtraction in a similar manner. If a, b, c, and d are real numbers, then

$$(a + bi) - (c + di) = (a - c) + (b - d)i$$

Notice that we can achieve this same result by simply distributing the negative sign and then combining similar terms.

EXAMPLES   Add or subtract as indicated.

10. $(3 + 4i) + (7 - 6i) = (3 + 7) + (4 - 6)i$

$$= 10 - 2i$$

11. $(7 + 3i) - (5 + 6i) = 7 + 3i - 5 - 6i$

$$= (7 - 5) + (3 - 6)i$$
$$= 2 - 3i$$

12. $(5 - 2i) - (9 - 4i) = 5 - 2i - 9 + 4i$

$$= (5 - 9) + (-2 + 4)i$$
$$= -4 + 2i$$

Multiplication of Complex Numbers

Because complex numbers have the same form as binomials, we find the product of two complex numbers the same way we find the product of two binomials.

EXAMPLE 13   Multiply $(3 - 4i)(2 + 5i)$.

SOLUTION   Multiplying each term in the second complex number by each term in the first, we have

$$\overset{\text{F}}{} \quad \overset{\text{O}}{} \quad \overset{\text{I}}{} \quad \overset{\text{L}}{}$$
$$(3 - 4i)(2 + 5i) = 3 \cdot 2 + 3 \cdot 5i - 2 \cdot 4i - 4i(5i)$$
$$= 6 + 15i - 8i - 20i^2$$

Combining similar terms and using the fact that $i^2 = -1$, we can simplify as follows:

$$6 + 15i - 8i - 20i^2 = 6 + 7i - 20(-1)$$
$$= 6 + 7i + 20$$
$$= 26 + 7i$$

The product of the complex numbers $3 - 4i$ and $2 + 5i$ is the complex number $26 + 7i$. ■

EXAMPLE 14   Multiply $2i(4 - 6i)$.

SOLUTION   Applying the distributive property gives us

$$2i(4 - 6i) = 2i \cdot 4 - 2i \cdot 6i$$
$$= 8i - 12i^2$$
$$= 8i - 12(-1)$$
$$= 12 + 8i$$

■

EXAMPLE 15   Expand $(3 + 5i)^2$.

SOLUTION   We treat this like the square of a binomial. Remember, $(a + b)^2 = a^2 + 2ab + b^2$:

$$(3 + 5i)^2 = 3^2 + 2(3)(5i) + (5i)^2$$
$$= 9 + 30i + 25i^2$$
$$= 9 + 30i - 25$$
$$= -16 + 30i$$

■

EXAMPLE 16   Multiply $(2 - 3i)(2 + 3i)$.

SOLUTION   This product has the form $(a - b)(a + b)$, which we know results in the difference of two squares, $a^2 - b^2$:

$$(2 - 3i)(2 + 3i) = 2^2 - (3i)^2$$
$$= 4 - 9i^2$$
$$= 4 + 9$$
$$= 13$$

■

The product of the two complex numbers $2 - 3i$ and $2 + 3i$ is the real number 13. The two complex numbers $2 - 3i$ and $2 + 3i$ are called *complex conjugates*. The fact that their product is a real number is very useful.

> (dĕf) **DEFINITION** *complex conjugates*
>
> The complex numbers $a + bi$ and $a - bi$ are called ***complex conjugates***. One important property they have is that their product is the real number $a^2 + b^2$. Here's why :
>
> $$\begin{aligned}(a + bi)(a - bi) &= a^2 - (bi)^2 \\ &= a^2 - b^2i^2 \\ &= a^2 - b^2(-1) \\ &= a^2 + b^2\end{aligned}$$

Division With Complex Numbers

The fact that the product of two complex conjugates is a real number is the key to division with complex numbers.

EXAMPLE 17   Divide $\dfrac{2 + i}{3 - 2i}$.

SOLUTION We want a complex number in standard form that is equivalent to the quotient $\frac{2+i}{3-2i}$. We need to eliminate i from the denominator. Multiplying the numerator and denominator by $3 + 2i$ will give us what we want:

$$\begin{aligned}\frac{2 + i}{3 - 2i} &= \frac{2 + i}{3 - 2i} \cdot \frac{(3 + 2i)}{(3 + 2i)} \\ &= \frac{6 + 4i + 3i + 2i^2}{9 - 4i^2} \\ &= \frac{6 + 7i - 2}{9 + 4} \\ &= \frac{4 + 7i}{13} \\ &= \frac{4}{13} + \frac{7}{13}i\end{aligned}$$

Dividing the complex number $2 + i$ by $3 - 2i$ gives the complex number $\frac{4}{13} + \frac{7}{13}i$.

EXAMPLE 18   Divide $\dfrac{7 - 4i}{2i}$.

SOLUTION The conjugate of the denominator is $-2i$. Multiplying numerator and denominator by this number, we have

$$\begin{aligned}\frac{7 - 4i}{2i} &= \frac{7 - 4i}{2i} \cdot \frac{-2i}{-2i} \\ &= \frac{-14i + 8i^2}{-4i^2} \\ &= \frac{-14i - 8}{4} \\ &= -\frac{8}{4} - \frac{14i}{4} \\ &= -2 - \frac{7}{2}i\end{aligned}$$

Fractal Geometry

A *fractal* is an infinitely complex pattern that is self-similar across different scales. In other words, a geometric pattern that looks the same under any level of magnification. An example of a fractal, called the Julia Set, is shown in Figure 1.

" Julia set (ice)"/Licensed under Public Domain via Wikimedia Commons

FIGURE 1

Fractals originated in 1918 in a paper published by the French mathematician Gaston Julia, but the subject of fractal geometry did not take off until 1975 with the advent of computers.

Most fractals are generated by taking complex numbers and repeating a simple set of operations on them. For instance, the Mandelbrot Set shown in Figure 2 is created by taking a complex number, squaring it, adding the original number, and then repeating this process many, many times.

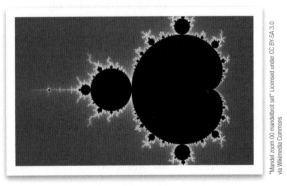

"Mandel zoom 00 mandelbrot set" Licensed under CC BY-SA 3.0 via Wikimedia Commons

FIGURE 2

Besides their artistic qualities, fractals have applications to biology, geology, economics, music, architecture, fiber optics, image compression, and chaos theory, among others.

Getting Ready for Class

After reading through the preceding section, respond in your own words and in complete sentences.

A. What is the number i?

B. What is a complex number?

C. What kind of number will always result when we multiply complex conjugates?

D. Explain how to divide complex numbers.

Problem Set 7.7

Write the following as pure imaginary numbers in terms of i, and simplify as much as possible.

1. $\sqrt{-36}$ **2.** $\sqrt{-49}$ **3.** $-\sqrt{-25}$ **4.** $-\sqrt{-81}$

5. $\sqrt{-72}$ **6.** $\sqrt{-48}$ **7.** $-\sqrt{-12}$ **8.** $-\sqrt{-75}$

Simplify each power of i.

9. i^{28} **10.** i^{31} **11.** i^{26} **12.** i^{37}

13. i^{75} **14.** i^{42}

Find x and y so each of the following equations is true.

15. $2x + 3yi = 6 - 3i$ **16.** $4x - 2yi = 4 + 8i$

17. $2 - 5i = -x + 10yi$ **18.** $4 + 7i = 6x - 14yi$

19. $2x + 10i = -16 - 2yi$ **20.** $4x - 5i = -2 + 3yi$

21. $(2x - 4) - 3i = 10 - 6yi$ **22.** $(4x - 3) - 2i = 8 + yi$

23. $(7x - 1) + 4i = 2 + (5y + 2)i$ **24.** $(5x + 2) - 7i = 4 + (2y + 1)i$

Add or subtract as indicated.

25. $(2 + 3i) + (3 + 6i)$ **26.** $(4 + i) + (3 + 2i)$

27. $(3 - 5i) + (2 + 4i)$ **28.** $(7 + 2i) + (3 - 4i)$

29. $(5 + 2i) - (3 + 6i)$ **30.** $(6 + 7i) - (4 + i)$

31. $(3 - 5i) - (2 + i)$ **32.** $(7 - 3i) - (4 + 10i)$

33. $[(3 + 2i) - (6 + i)] + (5 + i)$ **34.** $[(4 - 5i) - (2 + i)] + (2 + 5i)$

35. $[(7 - i) - (2 + 4i)] - (6 + 2i)$ **36.** $[(3 - i) - (4 + 7i)] - (3 - 4i)$

37. $(3 + 2i) - [(3 - 4i) - (6 + 2i)]$ **38.** $(7 - 4i) - [(-2 + i) - (3 + 7i)]$

39. $(4 - 9i) + [(2 - 7i) - (4 + 8i)]$ **40.** $(10 - 2i) - [(2 + i) - (3 - i)]$

Find the following products.

41. $3i(4 + 5i)$ **42.** $2i(3 + 4i)$ **43.** $6i(4 - 3i)$

44. $11i(2 - i)$ **45.** $(3 + 2i)(4 + i)$ **46.** $(2 - 4i)(3 + i)$

47. $(4 + 9i)(3 - i)$ **48.** $(5 - 2i)(1 + i)$ **49.** $(-1 + 2i)(6 - 5i)$

50. $(7 + 3i)(-3 - i)$ **51.** $(2 - i)^3$ **52.** $(2 + i)^3$

53. $(2 + 5i)^2$ **54.** $(3 + 2i)^2$ **55.** $(1 - i)^2$

56. $(1 + i)^2$ **57.** $(3 - 4i)^2$ **58.** $(6 - 5i)^2$

59. $(2 + i)(2 - i)$ **60.** $(3 + i)(3 - i)$ **61.** $(6 - 2i)(6 + 2i)$

62. $(5 + 4i)(5 - 4i)$ **63.** $(2 + 3i)(2 - 3i)$ **64.** $(2 - 7i)(2 + 7i)$

65. $(10 + 8i)(10 - 8i)$ **66.** $(11 - 7i)(11 + 7i)$

Simplify.

67. $(2 - 5i)^2 + (3 + i)^2$

68. $(4 + 3i)^2 + (5 - i)^2$

69. $(4 + i)(4 - i) - (1 + 2i)^2$

70. $(1 + 5i)(1 - 5i) - (3 - 2i)^2$

Find the following quotients. Write all answers in standard form for complex numbers.

71. $\dfrac{2 - 3i}{i}$

72. $\dfrac{3 + 4i}{-i}$

73. $\dfrac{5 + 2i}{-3i}$

74. $\dfrac{4 - 3i}{2i}$

75. $\dfrac{4}{2 - 3i}$

76. $\dfrac{3}{4 - 5i}$

77. $\dfrac{6}{-3 + 2i}$

78. $\dfrac{-1}{-2 - 5i}$

79. $\dfrac{2 + 3i}{2 - 3i}$

80. $\dfrac{4 - 7i}{4 + 7i}$

81. $\dfrac{5 + 4i}{3 + 6i}$

82. $\dfrac{2 + i}{5 - 6i}$

Applying the Concepts

83. Electric Circuits Complex numbers may be applied to electrical circuits. Electrical engineers use the fact that resistance R to electrical flow of the electrical current I and the voltage V are related by the formula $V = RI$. (Voltage is measured in volts, resistance in ohms, and current in amperes.) Find the resistance to electrical flow in a circuit that has a voltage $V = (80 + 20i)$ volts and current $I = (-6 + 2i)$ amps.

84. Electric Circuits Refer to the information about electrical circuits in Problem 83, and find the current in a circuit that has a resistance of $(4 + 10i)$ ohms and a voltage of $(5 - 7i)$ volts.

Learning Objectives Assessment

The following problems can be used to help assess if you have successfully met the learning objectives for this section.

85. Write $\sqrt{-44}$ as a pure imaginary number.

    a. $-2\sqrt{11}$ **b.** $2\sqrt{11}i$ **c.** $2i\sqrt{11}$ **d.** $11i\sqrt{2}$

86. Simplify: i^{61}.

    a. i **b.** $-i$ **c.** 1 **d.** -1

87. Subtract: $(5 - 3i) - (-7 - 4i)$.

    a. $12 + i$ **b.** $12 - 7i$ **c.** $-2 - 7i$ **d.** $-2 + i$

88. Mulitply: $(2 + i)(8 - 3i)$.

    a. $16 - 3i$ **b.** $13 + 2i$ **c.** 19 **d.** $19 + 2i$

Maintaining Your Skills

Solve each equation.

89. $\dfrac{t}{3} - \dfrac{1}{2} = -1$

90. $\dfrac{x}{x-2} + \dfrac{2}{3} = \dfrac{2}{x-2}$

91. $2 + \dfrac{5}{y} = \dfrac{3}{y^2}$

92. $1 - \dfrac{1}{y} = \dfrac{12}{y^2}$

Solve each application problem.

93. The sum of a number and its reciprocal is $\dfrac{41}{20}$. Find the number.

94. It takes an inlet pipe 8 hours to fill a tank. The drain can empty the tank in 6 hours. If the tank is full and both the inlet pipe and drain are open, how long will it take to drain the tank?

Chapter 7 Summary

EXAMPLES

Square Roots [7.1]

1. The number 49 has two square roots, 7 and -7. They are written like this:

$$\sqrt{49} = 7 \qquad -\sqrt{49} = -7$$

Every positive real number x has two square roots. The **positive square root** (or **principal square root**) of x is written $\sqrt{x}$, and the **negative square root** of x is written $-\sqrt{x}$. Both the positive and the negative square roots of x are numbers we square to get x; that is,

$$\left.\begin{array}{l}(\sqrt{x})^2 = x \\ \text{and} \quad (-\sqrt{x})^2 = x\end{array}\right\} \text{ for } x \geq 0$$

Higher Roots [7.1]

2. $\sqrt[3]{8} = 2$

$\sqrt[3]{-27} = -3$

In the expression $\sqrt[n]{a}$, n is the **index**, a is the **radicand**, and $\sqrt{}$ is the **radical sign**. The expression $\sqrt[n]{a}$ is such that

$$(\sqrt[n]{a})^n = a \qquad a \geq 0 \text{ when } n \text{ is even}$$

Property of Radicals [7.1]

3. $(\sqrt[3]{-13})^3 = -13$

If a is a nonnegative real number whenever n is even, then

1. $(\sqrt[n]{a})^n = a$

Root Functions [7.1]

4. The function $f(x) = \sqrt{x - 6}$ has domain $\{x \mid x \geq 6\}$.

For any positive integer n, the function given by $f(x) = \sqrt[n]{x}$ is called the **nth root function**.

The domain of $f(x) = \sqrt[n]{x}$ is

$$\{x \mid x \geq 0\} \qquad \text{if } n \text{ is even}$$
$$\{x \mid x \text{ is any real number}\} \qquad \text{if } n \text{ is odd}$$

Rational Exponents [7.2]

5. $25^{1/2} = \sqrt{25} = 5$

$8^{2/3} = (\sqrt[3]{8})^2 = 2^2 = 4$

$9^{3/2} = (\sqrt{9})^3 = 3^3 = 27$

Rational exponents are used to indicate roots. The relationship between rational exponents and roots is as follows:

$$a^{1/n} = \sqrt[n]{a} \qquad \text{and} \qquad a^{m/n} = (a^{1/n})^m = (a^m)^{1/n}$$

$$a \geq 0 \text{ when } n \text{ is even}$$

Properties of Radicals [7.3]

6. $\sqrt{4 \cdot 5} = \sqrt{4}\,\sqrt{5} = 2\sqrt{5}$

$\sqrt{\dfrac{7}{9}} = \dfrac{\sqrt{7}}{\sqrt{9}} = \dfrac{\sqrt{7}}{3}$

$\sqrt{(x-5)^2} = |x-5|$

If a and b are nonnegative real numbers whenever n is even, then

2. $\sqrt[n]{ab} = \sqrt[n]{a}\,\sqrt[n]{b}$

3. $\sqrt[n]{\dfrac{a}{b}} = \dfrac{\sqrt[n]{a}}{\sqrt[n]{b}}$ $(b \neq 0)$

If a is any real number, then

4. $\sqrt[n]{a^n} = |a|$ if n is even

$\sqrt[n]{a^n} = a$ if n is odd

Simplified Form for Radicals [7.3]

7. $\sqrt{\dfrac{12}{25}} = \dfrac{\sqrt{12}}{\sqrt{25}}$

$= \dfrac{\sqrt{4}\sqrt{3}}{5}$

$= \dfrac{2\sqrt{3}}{5}$

A radical expression is said to be in **simplified form**

1.   If there is no factor of the radicand that can be written as a power greater than or equal to the index;

2.   If there are no fractions under the radical sign; and

3.   If there are no radicals in the denominator.

Addition and Subtraction of Radical Expressions [7.4]

8. $5\sqrt{3} - 7\sqrt{3} = (5-7)\sqrt{3}$

$= -2\sqrt{3}$

$\sqrt{20} + \sqrt{45} = 2\sqrt{5} + 3\sqrt{5}$

$= (2+3)\sqrt{5}$

$= 5\sqrt{5}$

We add and subtract radical expressions by using the distributive property to combine similar radicals. Similar radicals are radicals with the same index and the same radicand.

Multiplication of Radical Expressions [7.5]

9. $(\sqrt{x} + 2)(\sqrt{x} + 3)$

$= \sqrt{x}\,\sqrt{x} + 3\sqrt{x} + 2\sqrt{x} + 2 \cdot 3$

$= x + 5\sqrt{x} + 6$

We multiply radical expressions in the same way that we multiply polynomials. We can use the distributive property and the FOIL method.

Rationalizing the Denominator [7.5]

10. $\dfrac{3}{\sqrt{2}} = \dfrac{3}{\sqrt{2}} \cdot \dfrac{\sqrt{2}}{\sqrt{2}} = \dfrac{3\sqrt{2}}{2}$

$\dfrac{3}{\sqrt{5} - \sqrt{3}} = \dfrac{3}{\sqrt{5} - \sqrt{3}} \cdot \dfrac{\sqrt{5} + \sqrt{3}}{\sqrt{5} + \sqrt{3}}$

$= \dfrac{3\sqrt{5} + 3\sqrt{3}}{5 - 3}$

$= \dfrac{3\sqrt{5} + 3\sqrt{3}}{2}$

$\dfrac{2}{\sqrt[3]{25}} = \dfrac{2}{\sqrt[3]{25}} \cdot \dfrac{\sqrt[3]{5}}{\sqrt[3]{5}} = \dfrac{2\sqrt[3]{5}}{5}$

When a fraction contains a square root in the denominator, we rationalize the denominator by multiplying numerator and denominator by

1.   The square root itself if there is only one term in the denominator, or

2.   The conjugate of the denominator if there are two terms in the denominator.

3.   When the denominator involves a cube root, we multiply by a radical that will produce a perfect cube under the cube root sign in the denominator.

Squaring Property of Equality [7.6]

11. $\sqrt{2x+1} = 3$
$(\sqrt{2x+1})^2 = 3^2$
$2x + 1 = 9$
$x = 4$

We may square both sides of an equation any time it is convenient to do so, as long as we check all resulting solutions in the original equation.

Complex Numbers [7.7]

12. $3 + 4i$ is a complex number.

Addition
$(3 + 4i) + (2 - 5i) = 5 - i$

Multiplication
$(3 + 4i)(2 - 5i)$
$= 6 - 15i + 8i - 20i^2$
$= 6 - 7i + 20$
$= 26 - 7i$

Division
$$\frac{2}{3 + 4i} = \frac{2}{3 + 4i} \cdot \frac{3 - 4i}{3 - 4i}$$
$$= \frac{6 - 8i}{9 + 16}$$
$$= \frac{6}{25} - \frac{8}{25}i$$

A **complex number** is any number that can be put in the form

$$a + bi$$

where a and b are real numbers and $i = \sqrt{-1}$. The **real part** of the complex number is a, and b is the **imaginary part**.

If $a, b, c,$ and d are real numbers, then we have the following definitions associated with complex numbers:

1. Equality

$$a + bi = c + di \quad \text{if and only if} \quad a = c \text{ and } b = d$$

2. Addition and subtraction

$$(a + bi) + (c + di) = (a + c) + (b + d)i$$
$$(a + bi) - (c + di) = (a - c) + (b - d)i$$

3. Multiplication

$$(a + bi)(c + di) = (ac - bd) + (ad + bc)i$$

4. Division is similar to rationalizing the denominator.

Chapter 7 Test

Assume all variable bases are positive integers throughout this test.

Simplify each of the following. [7.1]

1. $-\sqrt{81}$ **2.** $\sqrt[3]{-125}$ **3.** $(\sqrt{7})^2$ **4.** $(\sqrt[3]{-15})^3$

5. $\sqrt{49x^8}$ **6.** $\sqrt[5]{32x^{10}y^{20}}$

Graph. [7.1]

7. $f(x) = \sqrt{x-2}$ **8.** $f(x) = \sqrt[3]{x} + 3$

Find the domain of each function. [7.1]

9. $f(x) = \sqrt{9-x}$ **10.** $f(x) = \sqrt[3]{x+7}$

Simplify. [7.2]

11. $27^{-2/3}$ **12.** $\left(\dfrac{25}{49}\right)^{-1/2}$ **13.** $a^{3/4} \cdot a^{-1/3}$

14. $(x^{3/5})^{5/6}$ **15.** $\dfrac{x^{2/3}y^{-3}}{x^{3/4}y^{1/2}}$ **16.** $\dfrac{(36a^8b^4)^{1/2}}{(27a^9b^6)^{1/3}}$

17. Write $\sqrt[3]{2a}$ using a rational exponent. [7.2]

18. Use rational exponents to simplify $\sqrt{36x^6y^{18}}$. [7.2]

Write in simplified form. [7.3]

19. $\sqrt{125x^3y^5}$ **20.** $\sqrt[3]{40x^7y^8}$ **21.** $\sqrt{\dfrac{2}{9}}$ **22.** $\sqrt{\dfrac{12a^4b^3}{25c^6}}$

Combine. [7.4]

23. $3\sqrt{12} - 4\sqrt{27}$ **24.** $\sqrt[3]{24a^3b^3} - 5a\sqrt[3]{3b^3}$

Multiply. [7.5]

25. $(\sqrt{x} + 7)(\sqrt{x} - 4)$ **26.** $(3\sqrt{2} - \sqrt{3})^2$

Rationalize the denominator. [7.5]

27. $\sqrt{\dfrac{5}{6}}$ **28.** $\sqrt[3]{\dfrac{3x}{4y}}$ **29.** $\dfrac{5}{\sqrt{3}-1}$ **30.** $\dfrac{\sqrt{x}-\sqrt{2}}{\sqrt{x}+\sqrt{2}}$

Solve for x. [7.6]

31. $\sqrt{3x+1} = x - 3$ **32.** $\sqrt[3]{2x+7} = -1$

33. $\sqrt{x+3} = \sqrt{x+4} - 1$

34. Solve for x and y so that the following equation is true [7.7]:

$$(2x + 5) - 4i = 6 - (y - 3)i$$

Perform the indicated operations. [7.7]

35. $(3 + 2i) - [(7 - i) - (4 + 3i)]$ **36.** $(2 - 3i)(4 + 3i)$

37. $(5 - 4i)^2$ **38.** $\dfrac{2 - 3i}{2 + 3i}$

39. Simplify i^{38}. [7.7]

Quadratic Equations and Functions

Chapter Outline

Fir0002/Flagstaffotos
http://commons.wikimedia.org/wiki/Commons:GNU_Free_Documentation_License,_version_1.2

If you have been to the circus or the county fair recently, you may have witnessed one of the more spectacular acts, the human cannonball. The human cannonball shown in the photograph will reach a height of 70 feet, and travel a distance of 160 feet, before landing in a safety net. In this chapter, we use this information to derive the function

$$f(x) = -\frac{7}{640}(x - 80)^2 + 70 \quad \text{for } 0 \le x \le 160$$

which describes the path flown by this particular cannonball. The table and graph below were constructed from this equation. The function $f(x)$ is called a *quadratic function*, and its graph is a parabola.

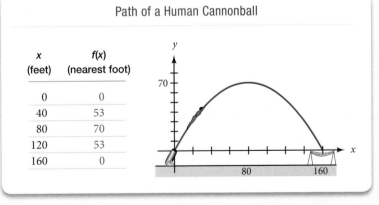

Path of a Human Cannonball

x (feet)	f(x) (nearest foot)
0	0
40	53
80	70
120	53
160	0

All objects that are projected into the air, whether they are basketballs, bullets, arrows, or coins, follow parabolic paths like the one shown in the graph. Studying the material in this chapter will give you a more mathematical hold on the world around you.

Success Skills

Never mistake activity for achievement.

— John Wooden, legendary UCLA basketball coach

You may think that the John Wooden quote above has to do with being productive and efficient, or using your time wisely, but it is really about being honest with yourself. I have had students come to me after failing a test saying, "I can't understand why I got such a low grade after I put so much time in studying." One student even had help from a tutor and felt she understood everything that we covered. After asking her a few questions, it became clear that she spent all her time studying with a tutor and the tutor was doing most of the work. The tutor can work all the homework problems, but the student cannot. She has mistaken activity for achievement.

Can you think of situations in your life when you are mistaking activity for achievement?

How would you describe someone who is mistaking activity for achievement in the way they study for their math class?

Which of the following best describes the idea behind the John Wooden quote?

- ▸ Always be efficient.
- ▸ Don't kid yourself.
- ▸ Take responsibility for your own success.
- ▸ Study with purpose.

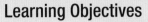

The Square Root Property and Completing the Square

8.1

Learning Objectives

In this section, we will learn how to:

1. Use the Square Root Property for Equations to solve a quadratic equation.

2. Solve a quadratic equation by completing the square.

3. Solve applied problems involving quadratic equations.

Introduction

If a baseball is dropped from a balcony that is 32 feet high, then the time it takes for the baseball to strike the ground can be found by solving the equation

$$-16t^2 + 32 = 0$$

If, instead, the baseball is thrown upward from the balcony with an initial velocity of 48 feet per second, then we would need to solve the equation

$$-16t^2 + 48t + 32 = 0$$

to find how long it would take for the baseball to hit the ground. Both of these equations are examples of a **quadratic equation**. Quadratic equations are one of the main topics we will study in this chapter. Here is the formal definition.

> **Note** For a quadratic equation written in standard form, the first term ax^2 is called the *quadratic term*, the second term bx is the *linear term*, and the last term c is called the *constant term*.

> (def **DEFINITION** *quadratic equation*
>
> Any equation that can be written in the form
> $$ax^2 + bx + c = 0$$
> where a, b, and c are constants and a is not 0 ($a \neq 0$), is called a **quadratic equation**. The form $ax^2 + bx + c = 0$ is called **standard form** for quadratic equations.

In Chapter 5 we learned how to solve certain quadratic equations using the Zero-Factor Property and factoring. In this section, we will develop two new methods of solving quadratic equations. The second of these methods is called **completing the square**. Completing the square on a quadratic equation allows us to obtain solutions, regardless of whether the equation can be factored. But before we solve equations by completing the square, we need to learn how to solve equations by taking square roots of both sides.

Square Root Property for Equations

Consider the equation

$$x^2 = 16$$

We could solve it by writing it in standard form, factoring the left side, and proceeding as we did in Chapter 5. We can shorten our work considerably, however, if we simply notice that x must be either the positive square root of 16 or the negative square root of 16. That is,

$$\text{If} \qquad x^2 = 16$$
$$\text{Then} \qquad x = \sqrt{16} \quad \text{or} \qquad x = -\sqrt{16}$$
$$x = 4 \qquad\qquad x = -4$$

We can generalize this result as follows.

> **[Δ≠Σ] PROPERTY** *Square Root Property for Equations*
>
> If $a^2 = b$, where b is a real number, then $a = \sqrt{b}$ or $a = -\sqrt{b}$.

Notation The expression $a = \sqrt{b}$ or $a = -\sqrt{b}$ can be written in shorthand form as $a = \pm\sqrt{b}$. The symbol $\pm$ is read "plus or minus."

VIDEO EXAMPLES

SECTION 8.1

EXAMPLE 1 Solve: $-16t^2 + 32 = 0$.

SOLUTION To use the Square Root Property for Equations, we must first isolate the quantity that is squared.

$$-16t^2 + 32 = 0$$
$$-16t^2 = -32$$
$$t^2 = 2$$

Now, applying the property, we have

$$t = \pm\sqrt{2}$$

The solution set is $\{-\sqrt{2}, \sqrt{2}\}$.

From the introduction to this section, we now know it would take $\sqrt{2} \approx 1.4$ seconds for the baseball to strike the ground if dropped from a balcony 32 feet high.

We can apply the Square Root Property for Equations to more complicated quadratic equations.

EXAMPLE 2 Solve: $(2x - 3)^2 = 25$.

SOLUTION

$$(2x - 3)^2 = 25$$

$2x - 3 = \pm\sqrt{25}$ Square Root Property for Equations

$2x - 3 = \pm 5$ $\sqrt{25} = 5$

$2x = 3 \pm 5$ Add 3 to both sides

$x = \dfrac{3 \pm 5}{2}$ Divide both sides by 2

The last equation can be written as two separate statements:

$$x = \frac{3 + 5}{2} \quad \text{or} \quad x = \frac{3 - 5}{2}$$
$$= \frac{8}{2} \qquad\qquad = \frac{-2}{2}$$
$$= 4 \qquad\qquad = -1$$

The solution set is $\{-1, 4\}$.

Notice that we could have solved the equation in Example 2 by expanding the left side, writing the resulting equation in standard form, and then factoring. The problem would look like this:

$$(2x - 3)^2 = 25 \qquad \text{Original equation}$$
$$4x^2 - 12x + 9 = 25 \qquad \text{Expand the left side}$$
$$4x^2 - 12x - 16 = 0 \qquad \text{Add } -25 \text{ to each side}$$
$$4(x^2 - 3x - 4) = 0 \qquad \text{Begin factoring}$$
$$4(x - 4)(x + 1) = 0 \qquad \text{Factor completely}$$
$$x - 4 = 0 \quad \text{or} \quad x + 1 = 0 \qquad \text{Set variable factors equal to 0}$$
$$x = 4 \qquad\qquad x = -1$$

As you can see, solving the equation by factoring leads to the same two solutions.

EXAMPLE 3   Solve for x: $(3x - 1)^2 = -12$

SOLUTION

$$(3x - 1)^2 = -12$$
$$3x - 1 = \pm\sqrt{-12} \qquad \text{Square Root Property for Equations}$$
$$3x - 1 = \pm 2i\sqrt{3} \qquad \sqrt{-12} = 2i\sqrt{3}$$
$$3x = 1 \pm 2i\sqrt{3} \qquad \text{Add 1 to both sides}$$
$$x = \frac{1}{3} \pm \frac{2\sqrt{3}}{3} i \qquad \text{Divide both sides by 3}$$

The solution set is $\left\{ \dfrac{1}{3} + \dfrac{2\sqrt{3}}{3} i, \dfrac{1}{3} - \dfrac{2\sqrt{3}}{3} i \right\}$. ∎

Note We cannot solve the equation in Example 3 by factoring. If we expand the left side and write the resulting equation in standard form, we have

$$(3x - 1)^2 = -12$$
$$9x^2 - 6x + 1 = -12$$
$$9x^2 - 6x + 13 = 0$$

which is not factorable.

EXAMPLE 4   Solve: $x^2 + 6x + 9 = 12$.

SOLUTION   We can solve this equation using the Square Root Property for Equations if we first write the left side as a perfect square.

$$x^2 + 6x + 9 = 12 \qquad \text{Original equation}$$
$$(x + 3)^2 = 12 \qquad \text{Write } x^2 + 6x + 9 \text{ as } (x + 3)^2$$
$$x + 3 = \pm\sqrt{12} \qquad \text{Square Root Property for Equations}$$
$$x + 3 = \pm 2\sqrt{3} \qquad \text{Simplify the radical}$$
$$x = -3 \pm 2\sqrt{3} \qquad \text{Add } -3 \text{ to each side}$$

We have two irrational solutions: $-3 + 2\sqrt{3}$ and $-3 - 2\sqrt{3}$. What is important about this problem, however, is the fact that the equation was easy to solve because the left side was a perfect square trinomial. ∎

Method of Completing the Square

The method of completing the square is simply a way of transforming any quadratic equation into an equation of the form found in the preceding three examples.

The key to understanding the method of completing the square lies in recognizing the relationship between the last two terms of any perfect square trinomial whose leading coefficient is 1.

Consider the formula for the square of a binomial

$$(a + b)^2 = a^2 + 2ab + b^2$$

Replacing a with x, we have

$$(x + b)^2 = x^2 + 2xb + b^2$$

Notice that the coefficient of x in the trinomial, $2b$, is twice the second term in the binomial, and the third term in the trinomial, b^2, is the square of the second term in the binomial. In other words, to be a perfect square trinomial, the third term must be the square of half the coefficient of the second term. So, if we are given

$$x^2 + ax$$

then

$$x^2 + ax + \left(\frac{a}{2}\right)^2$$

will be a perfect square trinomial whose factored form is $\left(x + \frac{a}{2}\right)^2$.

We can use these observations to build our own perfect square trinomials and, in doing so, solve some quadratic equations. The key is to add the term $\left(\frac{1}{2}a\right)^2$ in order to "complete" the perfect square trinomial.

▨ **EXAMPLE 5**   Solve $x^2 - 6x + 5 = 0$ by completing the square.

SOLUTION   We begin by adding -5 to both sides of the equation. We want just $x^2 - 6x$ on the left side so that we can add on our own third term to get a perfect square trinomial:

$$x^2 - 6x + 5 = 0$$
$$x^2 - 6x = -5 \qquad \text{Add } -5 \text{ to both sides}$$

The coefficient of x is -6, so we must add $(-6/2)^2 = (-3)^2 = 9$ to both sides in order to complete the perfect square trinomial.

$$x^2 - 6x + 9 = -5 + 9 \qquad \text{Add 9 to both sides}$$
$$(x - 3)^2 = 4 \qquad \text{Factor the left side, simplify the right side}$$

The resulting equation can now be solved using the Square Root Property for Equations.

$$x - 3 = \pm 2$$
$$x = 3 \pm 2 \qquad \text{Add 3 to both sides}$$
$$x = 3 + 2 \quad \text{or} \quad x = 3 - 2$$
$$x = 5 x = 1$$

The two solutions are 5 and 1.   ▨

> *Note*   The equation in Example 5 can be solved quickly by factoring:
>
> $$x^2 - 6x + 5 = 0$$
> $$(x - 5)(x - 1) = 0$$
> $$x - 5 = 0 \quad \text{or} \quad x - 1 = 0$$
> $$x = 5 x = 1$$
>
> The reason we didn't solve it by factoring is we want to practice completing the square on some simple equations.

▨ **EXAMPLE 6**   Solve by completing the square: $x^2 + 5x - 2 = 0$

SOLUTION   We must begin by adding 2 to both sides. (The left side of the equation, as it is, is not a perfect square, because it does not have the correct constant term. We will simply "move" that term to the other side and use our own constant term.)

$$x^2 + 5x = 2 \qquad \text{Add 2 to each side}$$

We complete the square by adding the square of half the coefficient of the linear term to both sides:

$$x^2 + 5x + \frac{25}{4} = 2 + \frac{25}{4}$$ Half of 5 is $\frac{5}{2}$, the square of which is $\frac{25}{4}$

$$\left(x + \frac{5}{2}\right)^2 = \frac{33}{4}$$ $2 + \frac{25}{4} = \frac{8}{4} + \frac{25}{4} = \frac{33}{4}$

$$x + \frac{5}{2} = \pm\sqrt{\frac{33}{4}}$$ Square Root Property for Equations

$$x + \frac{5}{2} = \pm\frac{\sqrt{33}}{2}$$ Simplify the radical

$$x = -\frac{5}{2} \pm \frac{\sqrt{33}}{2}$$ Add $-\frac{5}{2}$ to both sides

$$x = \frac{-5 \pm \sqrt{33}}{2}$$

The solution set is $\left\{ \dfrac{-5 + \sqrt{33}}{2}, \dfrac{-5 - \sqrt{33}}{2} \right\}$.

We can use a calculator to get decimal approximations to these solutions. If $\sqrt{33} \approx 5.74$, then

$$\frac{-5 + 5.74}{2} = 0.37 \quad \text{and} \quad \frac{-5 - 5.74}{2} = -5.37$$ ◼

The method we have developed for completing the perfect square trinomial is only valid when the coefficient of the quadratic term is 1. If this is not the case, then we must first modify the equation to obtain a leading coefficient of 1. The next two examples illustrate how this is done.

EXAMPLE 7   Solve: $-16t^2 + 48t + 32 = 0$

SOLUTION   Before we can complete the square, we must have a leading coefficient of 1. We begin by dividing both sides by -16, and then proceeding as before.

$$-16t^2 + 48t + 32 = 0$$

$$t^2 - 3t - 2 = 0$$ Divide both sides by -16

$$t^2 - 3t = 2$$ Add 2 to both sides

$$t^2 - 3t + \frac{9}{4} = 2 + \frac{9}{4}$$ Add $\left(-\frac{3}{2}\right)^2 = \frac{9}{4}$ to both sides

$$\left(t - \frac{3}{2}\right)^2 = \frac{17}{4}$$ Factor the left side, simplify the right side

$$t - \frac{3}{2} = \pm\sqrt{\frac{17}{4}}$$ Square Root Property for Equations

$$t - \frac{3}{2} = \pm\frac{\sqrt{17}}{2}$$ Simplify the radical

$$t = \frac{3}{2} \pm \frac{\sqrt{17}}{2}$$ Add $\frac{3}{2}$ to both sides

$$t = \frac{3 \pm \sqrt{17}}{2}$$

The solutions are $t = \dfrac{3 + \sqrt{17}}{2}$ and $t = \dfrac{3 - \sqrt{17}}{2}$.   ◼

As a result of our work in Example 7, we can now see it would take

$$\frac{3 + \sqrt{17}}{2} \approx 3.6 \text{ seconds}$$

for the baseball described in the introduction to this section to strike the ground if it were thrown upward at a speed of 48 feet per second.

EXAMPLE 8 Solve for x: $3x^2 - 8x + 7 = 0$.

SOLUTION

$$3x^2 - 8x + 7 = 0$$

$$3x^2 - 8x = -7 \qquad \text{Add } -7 \text{ to both sides}$$

We cannot complete the square on the left side because the leading coefficient is not 1. We take an extra step and divide both sides by 3

$$\frac{3x^2}{3} - \frac{8x}{3} = -\frac{7}{3}$$

$$x^2 - \frac{8}{3}x = -\frac{7}{3}$$

Half of $\frac{8}{3}$ is $\frac{4}{3}$, the square of which is $\frac{16}{9}$:

$$x^2 - \frac{8}{3}x + \frac{16}{9} = -\frac{7}{3} + \frac{16}{9} \qquad \text{Add } \frac{16}{9} \text{ to both sides}$$

$$\left(x - \frac{4}{3}\right)^2 = -\frac{5}{9} \qquad \text{Factor the left side, simplify right side}$$

$$x - \frac{4}{3} = \pm\sqrt{-\frac{5}{9}} \qquad \text{Square Root Property for Equations}$$

$$x - \frac{4}{3} = \pm\frac{i\sqrt{5}}{3} \qquad \sqrt{-\frac{5}{9}} = \frac{\sqrt{-5}}{3} = \frac{i\sqrt{5}}{3}$$

$$x = \frac{4}{3} \pm \frac{\sqrt{5}}{3}i \qquad \text{Add } \frac{4}{3} \text{ to both sides}$$

The solution set is $\left\{\frac{4}{3} + \frac{\sqrt{5}}{3}i, \frac{4}{3} - \frac{\sqrt{5}}{3}i\right\}$.

HOW TO *Solve a Quadratic Equation by Completing the Square*

To summarize the method used in the preceding two examples, we list the following steps:

Step 1: Write the equation in the form $ax^2 + bx = c$.

Step 2: If the leading coefficient is not 1, divide both sides by the coefficient so that the resulting equation has a leading coefficient of 1. That is, if $a \neq 1$, then divide both sides by a.

Step 3: Add the square of half the coefficient of the linear term to both sides of the equation.

Step 4: Write the left side of the equation as the square of a binomial, and simplify the right side if possible.

Step 5: Apply the Square Root Property for Equations, and solve as usual.

Applications

We conclude this section with some real-life applications of quadratic equations that can be solved using our two new methods.

EXAMPLE 9 The vertical rise of the Forest Double chair lift at the Northstar at Tahoe Ski Resort is 1,170 feet and the length of the chair lift as 5,750 feet. To the nearest foot, find the horizontal distance covered by a person riding this lift.

SOLUTION Figure 1 is a model of the Forest Double chair lift. A rider gets on the lift at point A and exits at point B. The length of the lift is AB.

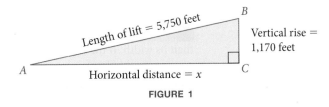

FIGURE 1

To find the horizontal distance covered by a person riding the chair lift, we use the Pythagorean theorem.

$$5,750^2 = x^2 + 1,170^2 \qquad \text{Pythagorean theorem}$$

$$33,062,500 = x^2 + 1,368,900 \qquad \text{Simplify squares}$$

$$x^2 = 33,062,500 - 1,368,900 \qquad \text{Solve for } x^2$$

$$x^2 = 31,693,600 \qquad \text{Simplify the right side}$$

$$x = \sqrt{31,693,600} \qquad \text{Square Root Property for Equations}$$

$$\approx 5,630 \text{ feet} \qquad \text{to the nearest foot}$$

Note When we use the Square Root Property for Equations in Example 9, we do not need to consider the negative square root because x represents a distance and must be positive.

A rider getting on the lift at point A and riding to point B will cover a horizontal distance of approximately 5,630 feet. ◼

EXAMPLE 10 Two boats leave from an island port at the same time. One travels due north at a speed of twelve miles per hour, and the other travels due west at a speed of 16 miles per hour. How long until the distance between the two boats is 50 miles?

SOLUTION If we let t represent the time, then the distance traveled by the boat going north is $12t$ and the distance traveled by the boat going west is $16t$. Figure 2 shows a diagram for the problem. We see that the distances traveled by the two boats form the legs of a right triangle. The hypotenuse of the triangle will be the distance between the boats, which is 50 miles.

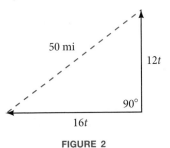

FIGURE 2

By the Pythagorean theorem, we have

$$(16t)^2 + (12t)^2 = 50^2 \qquad \text{Pythagorean theorem}$$
$$256t^2 + 144t^2 = 2500 \qquad \text{Simplify squares}$$
$$400t^2 = 2500 \qquad \text{Combine similar terms}$$
$$t^2 = \frac{25}{4} \qquad \text{Divide both sides by 400}$$
$$t = \sqrt{\frac{25}{4}} \qquad \text{Square Root Property for Equations (} t \text{ must be positive)}$$
$$t = \frac{5}{2} \qquad \text{Simplify the radical}$$

The two boats will be 50 miles apart after 2.5 hours.

EXAMPLE 11   A rectangular cement patio has a length that is 6 feet longer than its width. If the area of the patio is 120 square feet, how wide is the patio?

SOLUTION   We let x represent the width of the patio. The length will then be $x + 6$. Figure 3 illustrates the situation.

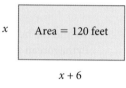

$x + 6$

FIGURE 3

Since the area is given by the product of the length and the width, we have

$$x(x + 6) = 120$$
$$x^2 + 6x = 120$$

This quadratic equation cannot be solved by factoring (try it), so we use the method of completing the square.

$$x^2 + 6x + 9 = 120 + 9 \qquad \text{Add } \left(\frac{6}{2}\right)^2 = 9 \text{ to both sides}$$
$$(x + 3)^2 = 129 \qquad \text{Factor and simplify}$$
$$x + 3 = \pm\sqrt{129} \qquad \text{Square Root Property for Equations}$$
$$x = -3 \pm \sqrt{129}$$

Because x must be a positive number, the only valid solution is

$$x = -3 + \sqrt{129} \approx 8.4 \text{ feet for the width}$$

Getting Ready for Class

After reading through the preceding section, respond in your own words and in complete sentences.

A. What kind of equation do we solve using the Square Root Property for Equations?

B. What kind of equation do we solve using the method of completing the square?

C. Explain in words how you would complete the square on $x^2 - 16x = 4$.

D. What are the first two steps in solving $2x^2 + 7x - 1 = 0$ by completing the square?

Solve using the Square Root Property for Equations.

1. $x^2 = 25$ **2.** $x^2 = 16$ **3.** $a^2 = -9$ **4.** $a^2 = -49$

5. $y^2 = \dfrac{3}{4}$ **6.** $y^2 = \dfrac{5}{9}$ **7.** $x^2 + 12 = 0$ **8.** $x^2 + 8 = 0$

9. $4a^2 - 45 = 0$ **10.** $9a^2 - 20 = 0$ **11.** $3x^2 + 28 = 0$ **12.** $5x^2 + 18 = 0$

13. $(2y - 1)^2 = 25$ **14.** $(3y + 7)^2 = 1$

15. $(2a + 3)^2 = -9$ **16.** $(3a - 5)^2 = -49$

17. $(5x + 2)^2 = -8$ **18.** $(6x - 7)^2 = -75$

19. $x^2 + 8x + 16 = 27$ **20.** $x^2 - 12x + 36 = 8$

21. $4a^2 - 12a + 9 = -4$ **22.** $9a^2 - 12a + 4 = -9$

Copy each of the following, and fill in the blanks so the left side of each is a perfect square trinomial. That is, complete the square.

23. $x^2 + 12x + \underline{\quad} = (x + \underline{\quad})^2$ **24.** $x^2 + 6x + \underline{\quad} = (x + \underline{\quad})^2$

25. $x^2 - 4x + \underline{\quad} = (x - \underline{\quad})^2$ **26.** $x^2 - 2x + \underline{\quad} = (x - \underline{\quad})^2$

27. $a^2 - 10a + \underline{\quad} = (a - \underline{\quad})^2$ **28.** $a^2 - 8a + \underline{\quad} = (a - \underline{\quad})^2$

29. $x^2 + 5x + \underline{\quad} = (x + \underline{\quad})^2$ **30.** $x^2 + 3x + \underline{\quad} = (x + \underline{\quad})^2$

31. $y^2 - 7y + \underline{\quad} = (y - \underline{\quad})^2$ **32.** $y^2 - y + \underline{\quad} = (y - \underline{\quad})^2$

33. $x^2 + \dfrac{1}{2}x + \underline{\quad} = (x + \underline{\quad})^2$ **34.** $x^2 - \dfrac{3}{4}x + \underline{\quad} = (x - \underline{\quad})^2$

35. $x^2 + \dfrac{2}{3}x + \underline{\quad} = (x + \underline{\quad})^2$ **36.** $x^2 - \dfrac{4}{5}x + \underline{\quad} = (x - \underline{\quad})^2$

Solve each of the following quadratic equations by completing the square.

37. $x^2 + 4x = 12$ **38.** $x^2 - 2x = 8$ **39.** $x^2 + 12x = -27$

40. $x^2 - 6x = 16$ **41.** $a^2 - 2a + 5 = 0$ **42.** $a^2 + 10a + 22 = 0$

43. $y^2 - 8y + 1 = 0$ **44.** $y^2 + 6y + 19 = 0$ **45.** $x^2 - 5x - 3 = 0$

46. $x^2 - 5x - 2 = 0$ **47.** $2x^2 - 4x - 8 = 0$ **48.** $3x^2 - 9x - 12 = 0$

49. $3t^2 - 8t + 1 = 0$ **50.** $5t^2 + 12t - 1 = 0$ **51.** $4x^2 - 3x + 5 = 0$

52. $7x^2 - 5x + 2 = 0$ **53.** $3x^2 + 4x - 1 = 0$ **54.** $2x^2 + 6x - 1 = 0$

55. $2x^2 - 10x = 11$ **56.** $25x^2 - 20x = 1$ **57.** $4x^2 - 10x + 11 = 0$

58. $4x^2 - 6x + 9 = 0$

59. For the equation $x^2 = -9$
     a. Can it be solved by factoring?   **b.** Solve it.

60. For the equation $x^2 - 10x + 18 = 0$
     a. Can it be solved by factoring?   **b.** Solve it.

61. Solve the equation $x^2 - 6x = 0$
     a. by factoring   **b.** by completing the square

62. Solve the equation $x^2 + ax = 0$
     a. by factoring   **b.** by completing the square

63. Solve the equation $x^2 + 2x = 35$
     a. by factoring   **b.** by completing the square

64. Solve the equation $8x^2 - 10x - 25 = 0$

    a. by factoring   **b.** by completing the square

65. Is $x = -3 + \sqrt{2}$ a solution to $x^2 - 6x = 7$?

66. Is $x = 2 - \sqrt{5}$ a solution to $x^2 - 4x = 1$?

67. Solve each equation.

    a. $5x - 7 = 0$ **b.** $5x - 7 = 8$ **c.** $(5x - 7)^2 = 8$

    d. $\sqrt{5x - 7} = 8$ **e.** $\dfrac{5}{2} - \dfrac{7}{2x} = \dfrac{4}{x}$

68. Solve each equation.

    a. $5x + 11 = 0$ **b.** $5x + 11 = 9$ **c.** $(5x + 11)^2 = 9$

    d. $\sqrt{5x + 11} = 9$ **e.** $\dfrac{5}{3} - \dfrac{11}{3x} = \dfrac{3}{x}$

69. Paying Attention to Instructions Work each problem according to the instructions given.

    a. Factor: $(2x - 3)^2 - 16$.

    b. Simplify: $(2x - 3)^2 - 16$.

    c. Solve: $(2x - 3)^2 - 16 = 0$.

    d. Solve: $(2x - 3)^2 = -16$.

70. Paying Attention to Instructions Work each problem according to the instructions given.

    a. Evaluate: $(3x + 4)^2 + 9$ if $x = -2$.

    b. Simplify: $(3x + 4)^2 + 9$.

    c. Solve: $(3x + 4)^2 + 9 = 0$.

    d. Solve: $(3x + 4)^2 = 9$.

Applying the Concepts

71. Geometry If the length of a side of a square is 1 inch, find the length of a diagonal of the square.

72. Geometry If the length of the shorter sides of a $45° - 45° - 90°$ triangle is x, find the length of the hypotenuse, in terms of x (Figure 4).

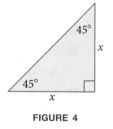

FIGURE 4

73. Chair Lift The Bear Paw Double chair lift at the Northstar at Tahoe Ski Resort is 790 feet long and has a vertical rise of 120 feet. Find the horizontal distance covered by a person riding this lift. Round your answer to the nearest foot.

74. Fermat's Last Theorem As mentioned in a previous chapter, the postage stamp shows Fermat's last theorem, which states that if n is an integer greater than 2, then there are no positive integers x, y, and z that will make the formula $x^n + y^n = z^n$ true. Use the formula $x^n + y^n = z^n$ to

a. find z if $n = 2$, $x = 6$, and $y = 8$.   **b.** find y if $n = 2$, $x = 5$, and $z = 13$.

75. Interest Rate Suppose a deposit of $3,000 in a savings account that paid an annual interest rate r (compounded yearly) is worth $3,456 after 2 years. Using the formula $A = P(1 + r)^t$, we have

$$3{,}456 = 3{,}000(1 + r)^2$$

Solve for r to find the annual interest rate.

76. Length of an Escalator An escalator in a department store is made to carry people a horizontal distance of 30 feet and a vertical distance of 20 feet between floors. How long is the escalator? (See Figure 5.)

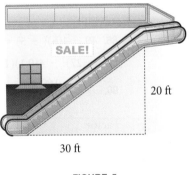

SALE!

20 ft

30 ft

FIGURE 5

7 ft

FIGURE 6

77. Right Triangle A 25-foot ladder is leaning against a building. The base of the ladder is 7 feet from the side of the building (Figure 6). How high does the ladder reach along the side of the building?

78. Right Triangle Noreen wants to place a 13-foot ramp against the side of her house so the top of the ramp rests on a ledge that is 4 feet above the ground. How far will the base of the ramp be from the house?

79. Distance Two cyclists leave from an intersection at the same time. One travels due north at a speed of 15 miles per hour, and the other travels due east at a speed of 20 miles per hour. How long until the distance between the two cyclists is 70 miles?

80. Distance Two airplanes leave from an airport at the same time. One travels due south at a speed of 480 miles per hour, and the other travels due west at a speed of 360 miles per hour. How long until the distance between the two airplanes is 2700 miles?

81. Rectangle The length of a rectangle is 3 feet more than twice the width. If the area is 25 square feet, find the width of the rectangle.

82. Poster A rectangular poster is 12 inches longer than it is wide. If the area of the poster is 650 square inches, find the dimensions of the poster.

Learning Objectives Assessment

The following problems can be used to help assess if you have successfully met the learning objectives for this section.

83. Solve: $2x^2 - 9 = 0$.

    a. $\pm\dfrac{3\sqrt{2}}{2}$ **b.** $\pm\dfrac{3\sqrt{2}}{2}i$ **c.** $\dfrac{3}{2}$ **d.** $\pm\dfrac{3}{2}$

84. In solving $2x^2 + 6x = 5$ by completing the square, what term should be added to both sides?

    a. 36   **b.** 9   **c.** $\dfrac{9}{4}$ **d.** $\dfrac{3}{2}$

85. Find the length of a diagonal of a rectangle that is 4 meters wide and 5 meters long.

    a. 3 m   **b.** 9 m   **c.** 41 m   **d.** $\sqrt{41}$ m

Getting Ready for the Next Section

Simplify.

86. $49 - 4(6)(-5)$ **87.** $49 - 4(6)(2)$

88. $(-27)^2 - 4(0.1)(1{,}700)$ **89.** $25 - 4(4)(-10)$

90. $-7 + \dfrac{169}{12}$ **91.** $-7 - \dfrac{169}{12}$

Factor.

92. $27t^3 - 8$ **93.** $125t^3 + 1$

The Quadratic Formula 8.2

Learning Objectives

In this section, we will learn how to:

1. Solve a quadratic equation using the quadratic formula.

2. Solve applied problems involving quadratic equations.

Introduction

As we mentioned in the previous section, the method of completing the square can be used to solve any quadratic equation. We will now take this a step further and solve *every* quadratic equation.

The idea of finding general solutions to different types of equations is threaded throughout the history of mathematics. The general solution for the quadratic equation, which we are about to show you, was first published in Europe in 1145 in a book written by Abraham bar Hiyya Ha-Nasi. It is presented here as a theorem.

⎰Δ≠Σ THEOREM   *The Quadratic Theorem*

For any quadratic equation in the form $ax^2 + bx + c = 0$, $a \neq 0$, the two solutions are

$$x = \frac{-b + \sqrt{b^2 - 4ac}}{2a} \qquad \text{and} \qquad x = \frac{-b - \sqrt{b^2 - 4ac}}{2a}$$

Proof We will prove the quadratic theorem by completing the square on $ax^2 + bx + c = 0$:

$$ax^2 + bx + c = 0$$

$$ax^2 + bx = -c \qquad \text{Add } -c \text{ to both sides}$$

$$x^2 + \frac{b}{a}x = -\frac{c}{a} \qquad \text{Divide both sides by } a$$

To complete the square on the left side, we add the square of $\frac{1}{2}$ of $\frac{b}{a}$ to both sides $\left(\frac{1}{2} \text{ of } \frac{b}{a} \text{ is } \frac{b}{2a}\right)$.

$$x^2 + \frac{b}{a}x + \left(\frac{b}{2a}\right)^2 = -\frac{c}{a} + \left(\frac{b}{2a}\right)^2$$

We now simplify the right side as a separate step. We combine the two terms by writing each with the least common denominator $4a^2$:

$$-\frac{c}{a} + \left(\frac{b}{2a}\right)^2 = -\frac{c}{a} + \frac{b^2}{4a^2} = \frac{4a}{4a}\left(\frac{-c}{a}\right) + \frac{b^2}{4a^2} = \frac{-4ac + b^2}{4a^2}$$

It is convenient to write this last expression as

$$\frac{b^2 - 4ac}{4a^2}$$

Continuing with the proof, we have

$$x^2 + \frac{b}{a}x + \left(\frac{b}{2a}\right)^2 = \frac{b^2 - 4ac}{4a^2}$$

$$\left(x + \frac{b}{2a}\right)^2 = \frac{b^2 - 4ac}{4a^2} \qquad \text{Write left side as a binomial square}$$

$$x + \frac{b}{2a} = \pm\sqrt{\frac{b^2 - 4ac}{4a^2}} \qquad \text{Square Root Property for Equations}$$

577

$$x + \frac{b}{2a} = \pm \frac{\sqrt{b^2 - 4ac}}{2a} \qquad \text{Simplify the radical}$$

$$x = -\frac{b}{2a} \pm \frac{\sqrt{b^2 - 4ac}}{2a} \qquad \text{Add } -\frac{b}{2a} \text{ to both sides}$$

$$= \frac{-b \pm \sqrt{b^2 - 4ac}}{2a}$$

Our proof is now complete. What we have is this: if our equation is in the form $ax^2 + bx + c = 0$ (standard form), where $a \neq 0$, the two solutions are always given by the formula

$$x = \frac{-b \pm \sqrt{b^2 - 4ac}}{2a}$$

This formula is known as the **quadratic formula**. If we substitute the coefficients a, b, and c of any quadratic equation in standard form into the formula, we need only perform some basic arithmetic to arrive at the solution set.

VIDEO EXAMPLES

SECTION 8.2

EXAMPLE 1   Solve $x^2 - 5x - 6 = 0$ by using the quadratic formula.

SOLUTION To use the quadratic formula, we must make sure the equation is in standard form; identify a, b, and c; substitute them into the formula; and work out the arithmetic.

For the equation $x^2 - 5x - 6 = 0$, $a = 1$, $b = -5$, and $c = -6$:

$$x = \frac{-b \pm \sqrt{b^2 - 4ac}}{2a}$$

$$= \frac{-(-5) \pm \sqrt{(-5)^2 - 4(1)(-6)}}{2(1)}$$

$$= \frac{5 \pm \sqrt{49}}{2}$$

$$= \frac{5 \pm 7}{2}$$

$$x = \frac{5 + 7}{2} \quad \text{or} \quad x = \frac{5 - 7}{2}$$

$$x = \frac{12}{2} \qquad\qquad x = -\frac{2}{2}$$

$$x = 6 \qquad\qquad x = -1$$

The two solutions are 6 and -1.

Note: Whenever the solutions to our quadratic equations turn out to be rational numbers, as in Example 1, it means the original equation could have been solved by factoring. (We didn't solve the equation in Example 1 by factoring because we were trying to get some practice with the quadratic formula.)

EXAMPLE 2   Solve for x: $2x^2 = -4x + 3$.

SOLUTION Before we can identify a, b, and c, we must write the equation in standard form. To do so, we add $4x$ and -3 to each side of the equation:

$$2x^2 = -4x + 3$$

$$2x^2 + 4x - 3 = 0 \qquad \text{Add } 4x \text{ and } -3 \text{ to each side}$$

Now that the equation is in standard form, we see that $a = 2$, $b = 4$, and $c = -3$. Using the quadratic formula we have:

$$x = \frac{-b \pm \sqrt{b^2 - 4ac}}{2a}$$

$$= \frac{-4 \pm \sqrt{4^2 - 4(2)(-3)}}{2(2)}$$

$$= \frac{-4 \pm \sqrt{40}}{4}$$

$$= \frac{-4 \pm 2\sqrt{10}}{4}$$

We can reduce the final expression in the preceding equation to lowest terms by factoring 2 from the numerator and denominator and then dividing it out:

$$x = \frac{\cancel{2}(-2 \pm \sqrt{10})}{\cancel{2} \cdot 2}$$

$$= \frac{-2 \pm \sqrt{10}}{2}$$

Our two solutions are $\dfrac{-2 + \sqrt{10}}{2}$ and $\dfrac{-2 - \sqrt{10}}{2}$

EXAMPLE 3 Solve: $x^2 - 6x = -7$.

SOLUTION We begin by writing the equation in standard form:

$$x^2 - 6x = -7$$

$$x^2 - 6x + 7 = 0 \qquad \text{Add 7 to each side}$$

Using $a = 1$, $b = -6$, and $c = 7$ in the quadratic formula

$$x = \frac{-b \pm \sqrt{b^2 - 4ac}}{2a}$$

we have:

$$x = \frac{-(-6) \pm \sqrt{(-6)^2 - 4(1)(7)}}{2(1)}$$

$$= \frac{6 \pm \sqrt{36 - 28}}{2}$$

$$= \frac{6 \pm \sqrt{8}}{2}$$

$$= \frac{6 \pm 2\sqrt{2}}{2}$$

The two terms in the numerator have a 2 in common. We reduce to lowest terms by factoring the 2 from the numerator and then dividing numerator and denominator by 2:

$$x = \frac{\cancel{2}(3 \pm \sqrt{2})}{\cancel{2}}$$

$$= 3 \pm \sqrt{2}$$

The two solutions are $3 + \sqrt{2}$ and $3 - \sqrt{2}$.

EXAMPLE 4 Solve for x: $\dfrac{1}{10}x^2 - \dfrac{1}{5}x = -\dfrac{1}{2}$.

SOLUTION It will be easier to apply the quadratic formula if we clear the equation of fractions. Multiplying both sides of the equation by the LCD 10 and then writing it in standard form gives us

$$10\left(\frac{1}{10}x^2 - \frac{1}{5}x\right) = \left(-\frac{1}{2}\right)10 \qquad \text{Multiply both sides by 10}$$

$$x^2 - 2x = -5 \qquad \text{Simplify}$$

$$x^2 - 2x + 5 = 0 \qquad \text{Add 5 to both sides}$$

Applying the quadratic formula with $a = 1$, $b = -2$, and $c = 5$, we have:

$$x = \frac{-(-2) \pm \sqrt{(-2)^2 - 4(1)(5)}}{2(1)}$$

$$= \frac{2 \pm \sqrt{-16}}{2}$$

$$= \frac{2 \pm 4i}{2}$$

$$= 1 \pm 2i$$

The two solutions are $1 + 2i$ and $1 - 2i$.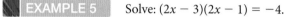

EXAMPLE 5   Solve: $(2x - 3)(2x - 1) = -4$.

SOLUTION   We multiply the binomials on the left side and then add 4 to each side to write the equation in standard form. From there we identify a, b, and c and apply the quadratic formula:

$$(2x - 3)(2x - 1) = -4$$

$$4x^2 - 8x + 3 = -4 \qquad \text{Multiply binomials on left side}$$

$$4x^2 - 8x + 7 = 0 \qquad \text{Add 4 to each side}$$

Placing $a = 4$, $b = -8$, and $c = 7$ in the quadratic formula we have:

$$x = \frac{-(-8) \pm \sqrt{(-8)^2 - 4(4)(7)}}{2(4)}$$

$$= \frac{8 \pm \sqrt{64 - 112}}{8}$$

$$= \frac{8 \pm \sqrt{-48}}{8}$$

$$= \frac{8 \pm 4i\sqrt{3}}{8} \qquad \sqrt{-48} = i\sqrt{48} = i\sqrt{16}\sqrt{3} = 4i\sqrt{3}$$

$$= \frac{8}{8} \pm \frac{4i\sqrt{3}}{8}$$

$$= 1 \pm \frac{\sqrt{3}}{2}i$$

Although the equation in our next example is not a quadratic equation, we solve it by using both factoring and the quadratic formula.

EXAMPLE 6   Solve: $27t^3 - 8 = 0$.

SOLUTION   It would be a mistake to add 8 to each side of this equation and then take the cube root of each side because we would lose two of our solutions. Instead, we factor the left side, and then set the factors equal to 0:

$$27t^3 - 8 = 0 \qquad \text{Equation in standard form}$$

$$(3t - 2)(9t^2 + 6t + 4) = 0 \qquad \text{Factor as the difference of two cubes.}$$

$$3t - 2 = 0 \quad \text{or} \quad 9t^2 + 6t + 4 = 0 \qquad \text{Set each factor equal to 0}$$

The first equation leads to a solution of $t = \frac{2}{3}$. The second equation does not factor, so we use the quadratic formula with $a = 9$, $b = 6$, and $c = 4$:

$$t = \frac{-6 \pm \sqrt{6^2 - 4(9)(4)}}{2(9)}$$

$$= \frac{-6 \pm \sqrt{36 - 144}}{18}$$

$$= \frac{-6 \pm \sqrt{-108}}{18}$$

$$= \frac{-6 \pm 6i\sqrt{3}}{18} \qquad \sqrt{-108} = i\sqrt{36 \cdot 3} = 6i\sqrt{3}$$

$$= -\frac{6}{18} \pm \frac{6i\sqrt{3}}{18}$$

$$= -\frac{1}{3} \pm \frac{\sqrt{3}}{3}i$$

The three solutions to our original equation are

$$\frac{2}{3}, \qquad -\frac{1}{3} + \frac{\sqrt{3}}{3}i, \quad \text{and} \quad -\frac{1}{3} - \frac{\sqrt{3}}{3}i$$

Applications

We conclude this section with some applied problems that can be solved using the quadratic formula.

EXAMPLE 7 One leg of a right triangle is 3 centimeters longer than the other leg. If the hypotenuse is 9 centimeters, find the lengths of the two legs. Approximate your answers to the nearest tenth of a centimeter.

SOLUTION We let x represent the length of the shorter leg. Then the longer leg will have length $x + 3$. Figure 1 shows the triangle.

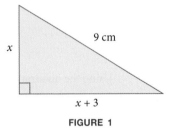

FIGURE 1

From the Pythagorean theorem, we have

$$x^2 + (x + 3)^2 = 9^2 \qquad \text{Pythagorean theorem}$$

$$x^2 + x^2 + 6x + 9 = 81 \qquad \text{Expand } (x + 3)^2$$

$$2x^2 + 6x - 72 = 0 \qquad \text{Standard form}$$

$$x^2 + 3x - 36 = 0 \qquad \text{Divide both sides by 2}$$

This equation cannot be factored, so we use the quadratic formula to solve for x.

$$x = \frac{-3 \pm \sqrt{3^2 - 4(1)(-36)}}{2(1)} \qquad \text{Quadratic formula}$$

$$= \frac{-3 \pm \sqrt{153}}{2} \qquad \text{Simplify}$$

$$= \frac{-3 \pm 3\sqrt{17}}{2} \qquad \sqrt{153} = \sqrt{9}\sqrt{17} = 3\sqrt{17}$$

Using a calculator to approximate $\sqrt{17} \approx 4.1$, the two solutions are $x = 4.7$ or $x = -7.7$. Because x is a length it must be positive. Discarding the negative solution, we find the lengths of the legs are 4.7 centimeters and $4.7 + 3 = 7.7$ centimeters.

EXAMPLE 8   A photographer wants to make a matte for an 8 x 10 inch frame that is the same width on all sides, leaving an open area of 50 square inches. Find the width of the matte. Approximate your answer to the nearest tenth of an inch.

SOLUTION   Figure 2 shows a diagram of the problem, where x represents the width of the matte.

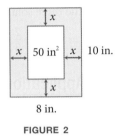

FIGURE 2

The inner rectangle will have a width of $8 - 2x$ and a height of $10 - 2x$. Since the area of this rectangle is 50, we have

$$(8 - 2x)(10 - 2x) = 50 \qquad \text{Area equals 50}$$

$$80 - 36x + 4x^2 = 50 \qquad \text{FOIL}$$

$$4x^2 - 36x + 30 = 0 \qquad \text{Standard form}$$

$$2x^2 - 18x + 15 = 0 \qquad \text{Divide both sides by 2}$$

Using the quadratic formula, we have

$$x = \frac{18 \pm \sqrt{(-18)^2 - 4(2)(15)}}{2(2)} \qquad \text{Quadratic formula}$$

$$= \frac{18 \pm \sqrt{204}}{4} \qquad \text{Simplify}$$

$$= \frac{18 \pm 2\sqrt{51}}{4} \qquad \sqrt{204} = \sqrt{4}\sqrt{51} = 2\sqrt{51}$$

$$= \frac{9 \pm \sqrt{51}}{2} \qquad \text{Reduce}$$

Approximating to the nearest tenth of an inch, we obtain $x = 8.1$ or $x = 0.9$. Although both of these values are positive, the width of the matte cannot be 8.1 inches because that exceeds the width of the original frame. Therefore, the photographer should make the matte 0.9 inches wide.

Getting Ready for Class

After reading through the preceding section, respond in your own words and in complete sentences.

A. State the quadratic formula.

B. Explain what the quadratic formula represents.

C. Under what circumstances should the quadratic formula be applied?

D. When would the quadratic formula result in complex solutions?

Solve each equation using the quadratic formula.

1. $a^2 - 4a + 1 = 0$ **2.** $a^2 + 4a + 1 = 0$ **3.** $2x^2 - x - 5 = 0$

4. $3x^2 + 4x - 1 = 0$ **5.** $12y^2 - 7y = 10$ **6.** $18y^2 + 8 = 51y$

7. $2x + 3 = -2x^2$ **8.** $2x - 3 = 3x^2$ **9.** $4x^2 - 28x + 49 = 0$

10. $9x^2 + 24x + 16 = 0$ **11.** $0.01x^2 + 0.06x - 0.08 = 0$

12. $0.02x^2 - 0.03x + 0.05 = 0$ **13.** $\dfrac{1}{6}x^2 - \dfrac{1}{2}x + \dfrac{1}{3} = 0$

14. $\dfrac{1}{6}x^2 + \dfrac{1}{2}x + \dfrac{1}{3} = 0$ **15.** $\dfrac{x^2}{2} + 1 = \dfrac{2x}{3}$ **16.** $\dfrac{x^2}{2} + \dfrac{2}{3} = -\dfrac{2x}{3}$

17. $\dfrac{2t^2}{3} - t = -\dfrac{1}{6}$ **18.** $\dfrac{t^2}{3} - \dfrac{t}{2} = -\dfrac{3}{2}$ **19.** $\dfrac{1}{2}r^2 = \dfrac{1}{6}r - \dfrac{2}{3}$

20. $\dfrac{1}{4}r^2 = \dfrac{2}{5}r + \dfrac{1}{10}$ **21.** $(x - 3)(x - 5) = 1$ **22.** $(x - 3)(x + 1) = -6$

Multiply both sides of each equation by its LCD. Then solve the resulting equation.

23. $\dfrac{1}{x + 1} - \dfrac{1}{x} = \dfrac{1}{2}$ **24.** $\dfrac{1}{x + 1} + \dfrac{1}{x} = \dfrac{1}{3}$ **25.** $\dfrac{1}{y - 1} + \dfrac{1}{y + 1} = 1$

26. $\dfrac{2}{y + 2} + \dfrac{3}{y - 2} = 1$ **27.** $\dfrac{1}{x + 2} + \dfrac{1}{x + 3} = 1$ **28.** $\dfrac{1}{x + 3} + \dfrac{1}{x + 4} = 1$

29. $\dfrac{6}{r^2 - 1} - \dfrac{1}{2} = \dfrac{1}{r + 1}$ **30.** $2 + \dfrac{5}{r - 1} = \dfrac{12}{(r - 1)^2}$

Solve each equation. In each case you will have three solutions.

31. $x^3 - 8 = 0$ **32.** $x^3 - 27 = 0$ **33.** $8a^3 + 27 = 0$

34. $27a^3 + 8 = 0$ **35.** $125t^3 - 1 = 0$ **36.** $64t^3 + 1 = 0$

Each of the following equations has three solutions. Look for the greatest common factor; then use the quadratic formula to find all solutions.

37. $2x^3 + 2x^2 + 3x = 0$ **38.** $6x^3 - 4x^2 + 6x = 0$ **39.** $3y^4 = 6y^3 - 6y^2$

40. $4y^4 = 16y^3 - 20y^2$ **41.** $6t^5 + 4t^4 = -2t^3$ **42.** $8t^5 + 2t^4 = -10t^3$

43. Which two of the expressions below are equivalent?

     a. $\dfrac{6 + 2\sqrt{3}}{4}$ **b.** $\dfrac{3 + \sqrt{3}}{2}$ **c.** $6 + \dfrac{\sqrt{3}}{2}$

44. Which two of the expressions below are equivalent?

     a. $\dfrac{8 - 4\sqrt{2}}{4}$ **b.** $2 - 4\sqrt{3}$ **c.** $2 - \sqrt{2}$

45. Solve $3x^2 - 5x = 0$

     a. by factoring   **b.** by the quadratic formula

46. Solve $3x^2 + 23x - 70 = 0$

     a. by factoring   **b.** by the quadratic formula

47. Can the equation $x^2 - 4x + 7 = 0$ be solved by factoring? Solve it.

48. Can the equation $x^2 = 5$ be solved by factoring? Solve it.

49. Is $x = -1 + i$ a solution to $x^2 + 2x = -2$?

50. Is $x = 2 + 2i$ a solution to $(x - 2)^2 = -4$?

Solve each equation using an appropriate method (factoring, Square Root Property for Equations, completing the square, or the quadratic formula).

51. $x^2 + 5x + 6 = 0$ **52.** $x^2 + 5x - 6 = 0$ **53.** $2y^2 + 10y = 0$

54. $30x^2 + 40x = 0$ **55.** $4a^2 - 27 = 0$ **56.** $6a^2 + 30 = 0$

57. $y^2 = 5y$ **58.** $50x^2 = 20x$ **59.** $2x^2 + 5x = 6$

60. $3x^2 + 13 = 12x$ **61.** $100x^2 - 200x + 100 = 0$

62. $100x^2 - 600x + 900 = 0$ **63.** $(x + 3)^2 + (x - 8)(x - 1) = 16$

64. $(x - 4)^2 + (x + 2)(x + 1) = 9$ **65.** $\dfrac{x^2}{3} - \dfrac{5x}{6} = \dfrac{1}{2}$

66. $\dfrac{x^2}{6} + \dfrac{5}{6} = -\dfrac{x}{3}$ **67.** $(19y - 31)^2 - 121 = 0$

68. $\left(\dfrac{1}{23}y - \dfrac{1}{13}\right)^2 + \dfrac{1}{16} = 0$

69. Solve each equation using an appropriate method.

    a. $(2x + 3)(2x - 3) = 0$

    b. $(2x + 3)(2x - 3) = 7$

    c. $(2x + 3)^2 = 7$

    d. $2x + 3 = 7x^2$

70. Solve each equation using an appropriate method.

    a. $(3x + 2)(3x - 4) = 0$

    b. $(3x + 2)(3x - 4) = 1$

    c. $(3x + 2)(3x - 4) = -6x$

    d. $3x + 2 = -6x^2$

Applying the Concepts

71. Right Triangle One leg of a right triangle is 2 meters shorter than the other leg. If the hypotenuse is 12 meters, find the lengths of the two legs. Approximate your answers to the nearest tenth of a meter.

72. Geometry A triangle has a height that is 4 feet longer than its base. If the area of the triangle is 18 square feet, find the length of the base and the height. Approximate your answers to the nearest tenth of a foot.

73. Rectangle The length of a rectangle is 5 centimeters less than 4 times its width. If the rectangle has an area of 60 square centimeters, find the dimensions of the rectangle. Approximate your answers to the nearest tenth of a centimeter.

74. Gravel Path A rectangular plot of ground that measures 40 feet by 60 feet is to be used for a garden surrounded by a gravel path (Figure 3). The path will be the same width on all sides. If the area inside the gravel path needs to be 2,000 square feet, find the width of the path. Round your answer to the nearest tenth of a foot.

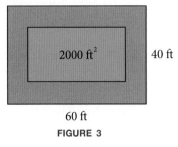

2000 ft^2 40 ft

60 ft

FIGURE 3

75. Area In the following diagram, $ABCD$ is a rectangle with diagonal AC. Find its area.

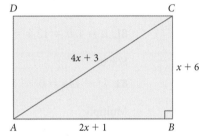

D C

$4x + 3$

$x + 6$

A $2x + 1$ B

76. Area and Perimeter A total of 160 yards of fencing is to be used to enclose part of a lot that borders on a river. This situation is shown in the following diagram.

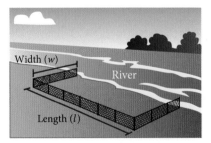

Width (w)

River

Length (l)

a. Write an equation that gives the relationship between the length and width and the 160 yards of fencing.

b. The formula for the area that is enclosed by the fencing and the river is $A = lw$. Solve the equation in part a for l, and then use the result to write the area in terms of w only.

c. Make a table that gives at least five possible values of w and associated area A.

d. From the pattern in your table shown in part c, what is the largest area that can be enclosed by the 160 yards of fencing? (Try some other table values if necessary.)

Learning Objectives Assessment

The following problems can be used to help assess if you have successfully met the learning objectives for this section.

77. Use the quadratic formula to solve: $2x^2 + x = 5$.

   a. $\dfrac{-1 \pm \sqrt{41}}{4}$ **b.** $\dfrac{1 \pm \sqrt{11}}{4}$ **c.** $-\dfrac{1}{4} \pm \dfrac{\sqrt{39}}{4}i$ **d.** $\dfrac{1}{2} \pm \dfrac{3}{2}i$

78. Right Triangle A right triangle has one leg that is 1 inch longer than the shorter leg, and the hypotenuse is 4 inches longer than the shorter leg. Find the length of the shortest leg.

   a. 11.9 inches   **b.** 3 inches   **c.** 8.9 inches   **d.** 7.9 inches

Getting Ready for the Next Section

Find the value of $b^2 - 4ac$ when

79. $a = 1, b = -3, c = -40$ **80.** $a = 2, b = 3, c = 4$

81. $a = 4, b = 12, c = 9$ **82.** $a = -3, b = 8, c = -1$

Solve.

83. $k^2 - 144 = 0$ **84.** $36 - 20k = 0$

Multiply.

85. $(x - 3)(x + 2)$ **86.** $(t - 5)(t + 3)$

87. $(x - 3)(x - 3)$ **88.** $(t - 5)(t + 5)$

Learning Objectives

In this section, we will learn how to:

1. Find the value of the discriminant.

2. Use the discriminant to determine the type of solutions a quadratic equation has.

3. Find a quadratic equation by reversing the factoring method.

4. Find a quadratic equation by reversing the Square Root Property for Equations method.

Introduction

In this section, we will do two things. First, we will define the discriminant and use it to find the kind of solutions a quadratic equation has without solving the equation. Second, we will reverse the Zero-Factor Property or Square Root Property for Equations to build equations from their solutions.

The Discriminant

The quadratic formula

$$x = \frac{-b \pm \sqrt{b^2 - 4ac}}{2a}$$

gives the solutions to any quadratic equation in standard form. There are times, when working with quadratic equations, that it is important only to know what kind of solutions the equation has.

> (dĕf) **DEFINITION** *discriminant*
>
> The expression under the radical in the quadratic formula is called the *discriminant*:
>
> $$\text{Discriminant} = D = b^2 - 4ac$$

The discriminant indicates the number and type of solutions to a quadratic equation, when the original equation has integer coefficients. For example, if we were to use the quadratic formula to solve the equation $2x^2 + 2x + 3 = 0$, we would find the discriminant to be

$$b^2 - 4ac = 2^2 - 4(2)(3) = -20$$

Because the discriminant appears under a square root symbol, we have the square root of a negative number in the quadratic formula. Our solutions would therefore be complex numbers. Similarly, if the discriminant were 0, the quadratic formula would yield

$$x = \frac{-b \pm \sqrt{0}}{2a} = \frac{-b \pm 0}{2a} = \frac{-b}{2a}$$

and the equation would have one rational solution, the number $\frac{-b}{2a}$.

The following table gives the relationship between the discriminant and the type of solutions to the equation.

For the equation $ax^2 + bx + c = 0$ where a, b, and c are integers and $a \neq 0$:

If the Discriminant b^2-4ac is	Then the Equation Will Have
Negative	Two non-real complex solutions containing i
Zero	One rational solution
A positive number that is also a perfect square	Two rational solutions
A positive number that is not a perfect square	Two irrational solutions

In the first case, when a quadratic equation has non-real complex solutions, they will always be complex conjugates. In the second and third cases, when the discriminant is 0 or a positive perfect square, the solutions are rational numbers. The quadratic equations in these two cases are the ones that can be factored.

VIDEO EXAMPLES

SECTION 8.3

EXAMPLES   For each equation, give the number and kind of solutions.

1. $x^2 - 3x - 40 = 0$

SOLUTION   Using $a = 1$, $b = -3$, and $c = -40$ in $b^2 - 4ac$, we have

$$(-3)^2 - 4(1)(-40) = 9 + 160 = 169.$$

The discriminant is a perfect square. The equation therefore has two rational solutions.

2. $2x^2 - 3x + 4 = 0$

SOLUTION   Using $a = 2$, $b = -3$, and $c = 4$, we have

$$b^2 - 4ac = (-3)^2 - 4(2)(4) = 9 - 32 = -23$$

The discriminant is negative, implying the equation has two complex solutions that contain i.

3. $4x^2 - 12x + 9 = 0$

SOLUTION   Using $a = 4$, $b = -12$, and $c = 9$, the discriminant is

$$b^2 - 4ac = (-12)^2 - 4(4)(9) = 144 - 144 = 0$$

Because the discriminant is 0, the equation will have one rational solution.

4. $x^2 + 6x = 8$

SOLUTION   We must first put the equation in standard form by adding -8 to each side. If we do so, the resulting equation is

$$x^2 + 6x - 8 = 0$$

Now we identify a, b, and c as 1, 6, and -8, respectively:

$$b^2 - 4ac = 6^2 - 4(1)(-8) = 36 + 32 = 68$$

The discriminant is a positive number, but not a perfect square. The equation will therefore have two irrational solutions.

EXAMPLE 5   Find an appropriate k so that the equation $4x^2 - kx = -9$ has exactly one rational solution.

SOLUTION   We begin by writing the equation in standard form:

$$4x^2 - kx + 9 = 0$$

Using $a = 4$, $b = -k$, and $c = 9$, we have

$$b^2 - 4ac = (-k)^2 - 4(4)(9)$$

$$= k^2 - 144$$

An equation has exactly one rational solution when the discriminant is 0. We set the discriminant equal to 0 and solve:

$$k^2 - 144 = 0$$
$$k^2 = 144$$
$$k = \pm 12$$

Choosing k to be 12 or -12 will result in an equation with one rational solution.

Building Equations From Their Solutions

Suppose we are given the two solutions to a quadratic equation, and we would like to work backwards and find what the original equation was. We can do this by reversing the steps of the factoring method or Square Root Property for Equations. Before continuing to the next examples, you may want to refresh your memory on these two methods by reviewing the examples from Sections 5.8 and 8.1.

EXAMPLE 6   Find a quadratic equation that has solutions $x = 3$ and $x = -2$. Write your answer in standard form.

SOLUTION   We can always use the factoring method in reverse to find an equation given the solutions. First, let's write our solutions as equations with 0 on the right side:

If	$x = 3$	First solution
then	$x - 3 = 0$	Add -3 to each side
and if	$x = -2$	Second solution
then	$x + 2 = 0$	Add 2 to each side

The quantities $x - 3$ and $x + 2$ must have been factors in our original equation. Because both $x - 3$ and $x + 2$ are 0, their product must be 0 also. We can therefore write

| $(x - 3)(x + 2) = 0$ | Zero-factor property |
| $x^2 - x - 6 = 0$ | Multiply out the left side |

In standard form, $x^2 - x - 6 = 0$ is a quadratic equation having $x = 3$ and $x = -2$ as solutions.

You may be wondering why the previous example asked us to find "a" quadratic equation instead of find "the" quadratic equation with the given solutions. The reason is that many equations have 3 and -2 as solutions. For example, any constant multiple of $x^2 - x - 6 = 0$, such as $5x^2 - 5x - 30 = 0$, also has 3 and -2 as solutions.

EXAMPLE 7   Find a quadratic equation with solutions $x = -\dfrac{2}{3}$ and $x = \dfrac{4}{5}$. Write your answer in standard form.

SOLUTION   The solution $x = -\frac{2}{3}$ can be rewritten as $3x + 2 = 0$ as follows:

$x = -\dfrac{2}{3}$	The first solution
$3x = -2$	Multiply each side by 3
$3x + 2 = 0$	Add 2 to each side

Similarly, the solution $x = \frac{4}{5}$ can be rewritten as $5x - 4 = 0$:

$$x = \frac{4}{5} \qquad \text{The second solution}$$

$$5x = 4 \qquad \text{Multiply each side by 5}$$

$$5x - 4 = 0 \qquad \text{Add } -4 \text{ to each side}$$

Because both $3x + 2$ and $5x - 4$ are 0, their product is 0 also, giving us the equation we are looking for:

$$(3x + 2)(5x - 4) = 0 \qquad \text{Zero-factor property}$$

$$15x^2 - 2x - 8 = 0 \qquad \text{Multiplication} \qquad ▨$$

If the two solutions are identical except for a $\pm$ sign, then another way to find an equation is to reverse the steps of the Square Root Property for Equations method. We illustrate how this is done in the next two examples.

 **EXAMPLE 8**   Find a quadratic equation in standard form with solutions $x = -\sqrt{5}$ and $x = \sqrt{5}$.

SOLUTION   We begin by writing the solutions using a $\pm$ symbol, then squaring both sides to eliminate the radical.

$$x = \pm\sqrt{5}$$

$$x^2 = (\pm\sqrt{5})^2 \qquad \text{Square both sides}$$

$$x^2 = 5 \qquad \text{Simplify}$$

$$x^2 - 5 = 0 \qquad \text{Standard form} \qquad ▨$$

Note: The $\pm$ symbol is eliminated when squaring both sides because the square of a positive number and the square of a negative number are both positive.

▨ **EXAMPLE 9**   Find a quadratic equation with solutions $x = 2 + 3i$ and $x = 2 - 3i$.

SOLUTION   In this case, we isolate the $\pm$ term before squaring both sides in order to eliminate the imaginary unit i.

$$x = 2 \pm 3i$$

$$x - 2 = \pm 3i \qquad \text{Subtract 2 from both sides}$$

$$(x - 2)^2 = (\pm 3i)^2 \qquad \text{Square both sides}$$

$$(x - 2)^2 = -9 \qquad (\pm 3i)^2 = 9i^2 = 9(-1) = -9$$

$$x^2 - 4x + 4 = -9 \qquad \text{FOIL}$$

$$x^2 - 4x + 13 = 0 \qquad \text{Standard form} \qquad ▨$$

USING TECHNOLOGY   *Graphing Calculators*

Solving Equations

Now that we have explored the relationship between equations and their solutions, we can look at how a graphing calculator can be used in the solution process. To begin, let's solve the equation $x^2 = x + 2$ using techniques from algebra: writing it in standard form, factoring, and then setting each factor equal to 0.

$x^2 - x - 2 = 0$	Standard form
$(x - 2)(x + 1) = 0$	Factor
$x - 2 = 0$ or $x + 1 = 0$	Set each factor equal to 0
$x = 2$ or $x = -1$	Solve

Our original equation, $x^2 = x + 2$, has two solutions: $x = 2$ and $x = -1$. To solve the equation using a graphing calculator, we need to associate it with an equation (or equations) in two variables. One way to do this is to associate the left side with the equation $y = x^2$ and the right side of the equation with $y = x + 2$. To do so, we set up the functions list in our calculator this way:

$$Y_1 = X^2$$
$$Y_2 = X + 2$$

Window: X from -5 to 5, Y from -5 to 5

Graphing these functions in this window will produce a graph similar to the one shown in Figure 1.

If we use the Trace feature to find the coordinates of the points of intesection, we find that the two curves intersect at $(-1, 1)$ and $(2, 4)$. We note that the x-coordinates of these two points match the solutions to the equation $x^2 = x + 2$, which we found using algebraic techniques. This makes sense because if two

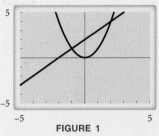

FIGURE 1

graphs intersect at a point (x, y), then the coordinates of that point satisfy both equations. If a point (x, y) satisfies both $y = x^2$ and $y = x + 2$, then for that particular point, $x^2 = x + 2$. From this, we conclude that the x-coordinates of the points of intersection are solutions to our original equation. Here is a summary of what we have discovered:

Conclusion 1   If the graphs of two functions $y = f(x)$ and $y = g(x)$ intersect in the coordinate plane, then the x-coordinates of the points of intersection are solutions to the equation $f(x) = g(x)$.

USING TECHNOLOGY   *Graphing Calculators Continued*

A second method of solving our original equation $x^2 = x + 2$ graphically requires the use of one function instead of two. To begin, we write the equation in standard form as $x^2 - x - 2 = 0$. Next, we graph the function $y = x^2 - x - 2$. The x-intercepts of the graph are the points with y-coordinates of 0. They therefore satisfy the equation $0 = x^2 - x - 2$, which is equivalent to our original equation. The graph in Figure 2 shows $Y_1 = X^2 - X - 2$ in a window with X from -5 to 5 and Y from -5 to 5.

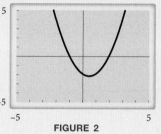

FIGURE 2

Using the Trace feature, we find that the x-intercepts of the graph are $x = -1$ and $x = 2$, which match the solutions to our original equation $x^2 = x + 2$. We can summarize the relationship between solutions to an equation and the intercepts of its associated graph this way:

Conclusion 2   If $y = f(x)$ is a function, then any x-intercept on the graph of $y = f(x)$ is a solution to the equation $f(x) = 0$.

Getting Ready for Class

After reading through the preceding section, respond in your own words and in complete sentences.

A. What is the discriminant?

B. What kind of solutions do we get to a quadratic equation when the discriminant is negative?

C. When will a quadratic equation have two rational solutions?

D. Describe two different methods you could use to find a quadratic equation having solutions $x = 4$ and $x = -4$.

Use the discriminant to find the number and kind of solutions for each of the following equations.

1. $x^2 - 6x + 5 = 0$ **2.** $x^2 - x - 12 = 0$

3. $4x^2 - 4x = -1$ **4.** $9x^2 + 12x = -4$

5. $x^2 + x - 1 = 0$ **6.** $x^2 - 2x + 3 = 0$

7. $2y^2 = 3y + 1$ **8.** $3y^2 = 4y - 2$

9. $x^2 - 9 = 0$ **10.** $4x^2 - 81 = 0$

11. $5a^2 - 4a = 5$ **12.** $3a = 4a^2 - 5$

Determine k so that each of the following has exactly one rational solution.

13. $x^2 - kx + 25 = 0$ **14.** $x^2 + kx + 25 = 0$

15. $x^2 = kx - 36$ **16.** $x^2 = kx - 49$

17. $4x^2 - 12x + k = 0$ **18.** $9x^2 + 30x + k = 0$

19. $kx^2 - 40x = 25$ **20.** $kx^2 - 2x = -1$

21. $3x^2 - kx + 2 = 0$ **22.** $5x^2 + kx + 1 = 0$

For each of the following problems, find a quadratic equation that has the given solutions. Write your answer in standard form.

23. $x = 5, x = 2$ **24.** $x = -5, x = -2$

25. $t = -3, t = 6$ **26.** $t = -4, t = 2$

27. $y = 2, y = -2$ **28.** $y = 1, y = -1$

29. $x = \dfrac{1}{2}, x = 3$ **30.** $x = \dfrac{1}{3}, x = 5$

31. $t = -\dfrac{3}{4}, t = 3$ **32.** $t = -\dfrac{4}{5}, t = 2$

33. $x = 3, x = -3$ **34.** $x = 5, x = -5$

35. $a = -\dfrac{1}{2}, a = \dfrac{3}{5}$ **36.** $a = -\dfrac{1}{3}, a = \dfrac{4}{7}$

37. $x = -\dfrac{2}{3}, x = \dfrac{2}{3}$ **38.** $x = -\dfrac{4}{5}, x = \dfrac{4}{5}$

39. $x = \sqrt{7}, x = -\sqrt{7}$ **40.** $x = -\sqrt{3}, x = \sqrt{3}$

41. $x = 5i, x = -5i$ **42.** $x = -2i, x = 2i$

43. $y = 3 + \sqrt{11}, y = 3 - \sqrt{11}$ **44.** $y = 4 - \sqrt{5}, y = 4 + \sqrt{5}$

45. $t = -6 - \sqrt{2}, t = -6 + \sqrt{2}$ **46.** $t = -5 + \sqrt{6}, t = -5 - \sqrt{6}$

47. $x = 1 + i, x = 1 - i$ **48.** $x = 2 + 3i, x = 2 - 3i$

49. $x = -2 - 3i, x = -2 + 3i$ **50.** $x = -1 + i, x = -1 - i$

51. $x = 7 + i\sqrt{3}, x = 7 - i\sqrt{3}$ **52.** $x = -8 - i\sqrt{7}, x = -8 + i\sqrt{7}$

53. $x = -3 - 5i\sqrt{2}, x = -3 + 5i\sqrt{2}$ **54.** $x = 4 + 2i\sqrt{5}, x = 4 - 2i\sqrt{5}$

55. Find a quadratic equation with solutions $x = 1 \pm \sqrt{10}$ using the following method:
  a. Zero-Factor Property
  b. squaring both sides

56. Find a quadratic equation with solutions $x = -8 \pm \sqrt{3}$ using the following method:
  a. Zero-Factor Property
  b. squaring both sides

57. Find a quadratic equation with solutions $x = -2 \pm i\sqrt{2}$ using the following method:
  a. Zero-Factor Property
  b. squaring both sides

58. Find a quadratic equation with solutions $x = 5 \pm 2i\sqrt{3}$ using the following method:
  a. Zero-Factor Property
  b. squaring both sides

Learning Objectives Assessment

The following problems can be used to help assess if you have successfully met the learning objectives for this section.

59. Find the value of the discriminant for the quadratic equation: $2x^2 + 5x = 3$.
  a. 1   **b.** 24   **c.** 7   **d.** 49

60. If a quadratic equation has discriminant $D = 20$, then it will have
  a. two irrational solutions   **b.** two rational solutions
  c. two complex solutions   **d.** one rational solution

61. Find a quadratic equation with solutions $x = \frac{1}{3}$ and $x = 2$.
  a. $3x^2 - 7x + 2 = 0$ **b.** $3x^2 + 7x + 2 = 0$
  c. $x^2 - 3x + 2 = 0$ **d.** $x^2 + 3x - 2 = 0$

62. Find a quadratic equation with solutions $x = 3 + i$ and $x = 3 - i$.
  a. $x^2 - 10 = 0$ **b.** $x^2 + 6x + 10 = 0$
  c. $x^2 - 6x + 8 = 0$ **d.** $x^2 - 6x + 10 = 0$

Getting Ready for the Next Section

Simplify.

63. $(x + 3)^2 - 2(x + 3) - 8$

64. $(x - 2)^2 - 3(x - 2) - 10$

65. $(2a - 3)^2 - 9(2a - 3) + 20$

66. $(3a - 2)^2 + 2(3a - 2) - 3$

67. $2(4a + 2)^2 - 3(4a + 2) - 20$

68. $6(2a + 4)^2 - (2a + 4) - 2$

Solve.

69. $x^2 = \dfrac{1}{4}$

70. $x^2 = -2$

71. $x^3 = \dfrac{1}{8}$

72. $x^3 = -27$

73. $\sqrt{x} = -3$

74. $\sqrt{x} = 2$

75. $\sqrt[3]{x} = -4$

76. $\sqrt[3]{x} = \dfrac{1}{3}$

77. $x + 3 = 4$

78. $x + 3 = -2$

79. $y^2 - 2y - 8 = 0$

80. $y^2 + y - 6 = 0$

81. $4y^2 + 7y - 2 = 0$

82. $6x^2 - 13x - 5 = 0$

SPOTLIGHT ON SUCCESS   *Student Instructor Aaron*

Sometimes you have to take a step back in order to get a running start forward.
—Anonymous

As a high school senior I was encouraged to go to college immediately after graduating. I earned good grades in high school and I knew that I would have a pretty good group of schools to pick from. Even though I felt like "more school" was not quite what I wanted, the counselors had so much faith and had done this process so many times that it was almost too easy to get the applications out. I sent out applications to schools I knew I could get into and a "dream school."

One night in my email inbox there was a letter of acceptance from my dream school. There was just one problem with getting into this school. It was going to be difficult and I still had senioritis. Going into my first quarter of college was as exciting and difficult as I knew it would be. But after my first quarter I could see that this was not the time for me to be here. I was interested in the subject matter but I could not find my motivating purpose like I had in high school. Instead of dropping out completely, I decided a community college would be a good way for me to stay on track. Without necessarily knowing my direction, I could take the general education classes and get those out of the way while figuring out exactly what and where I felt a good place for me to be.

Now I know what I want to go to school for and the next time I walk onto a four year campus it will be on my terms with my reasons for being there driving me to succeed. I encourage everyone to continue school after high school, even if you have no clue as to what you want to study. There are always stepping stones, like community colleges, that can help you get a clearer picture of what you want to strive for.

Quadratic Form

8.4

Learning Objectives

In this section, we will learn how to:

1. Solve equations that are quadratic in form using substitution.

2. Solve formulas that are quadratic in form.

Introduction

We are now in a position to put our knowledge of quadratic equations to work to solve a variety of equations. In this section, we will solve equations that are not quadratic, but have a structure similar to a quadratic equation. We call this structure *quadratic form*.

> **(def) DEFINITION** *quadratic form*
>
> An equation is *quadratic in form* if it can be written in the standard form
> $$a \blacksquare^2 + b \blacksquare + c = 0$$
> where $\blacksquare$ is some variable expression and $a \neq 0$.

Here are some examples of equations that are quadratic in form:

$$4x^4 + 7x^2 - 2 = 0 \qquad \blacksquare = x^2$$
$$x^{2/3} - x^{1/3} - 6 = 0 \qquad \blacksquare = x^{1/3}$$
$$(x + 3)^{-2} - 2(x + 3)^{-1} - 8 = 0 \qquad \blacksquare = (x + 3)^{-1}$$

We can solve equations that are quadratic in form using the method of substitution. We assign a new variable to represent the variable quantity $\blacksquare$ from the definition, thus turning the equation into a quadratic. Then we can use any of the methods from the previous two sections to solve the quadratic. Here is an outline of the process.

> **HOW TO** *Solve an Equation Quadratic in Form Using Substitution*
>
> **Step 1:** Write the equation in standard quadratic form.
> **Step 2:** Assign a new variable to represent the variable quantity from the middle term.
> **Step 3:** Substitute using the new variable.
> **Step 4:** Solve the resulting quadratic equation for the new variable.
> **Step 5:** Substitute back for the original variable.
> **Step 6:** Solve for the original variable.
> **Step 7:** Check your solutions.

Here are some examples of how this is done:

■ **EXAMPLE 1**   Solve: $(x + 3)^2 - 2(x + 3) - 8 = 0$.

SOLUTION   This equation is quadratic in terms of the variable expression $x + 3$. If we let $y = x + 3$, then replacing $x + 3$ with y we have

$$(x + 3)^2 - 2(x + 3) - 8 = 0$$

$$y^2 - 2y - 8 = 0 \qquad \text{Substitute}$$

This is now a quadratic equation in standard form. We can solve this equation by factoring.

$$y^2 - 2y - 8 = 0$$

$$(y - 4)(y + 2) = 0 \qquad \text{Factor}$$

$$y - 4 = 0 \quad \text{or} \quad y + 2 = 0 \quad \text{Zero-factor property}$$

$$y = 4 \qquad\qquad y = -2$$

Because our original equation was written in terms of the variable x, we want our solutions in terms of x also. Replacing y with $x + 3$ and then solving for x, we have

$$x + 3 = 4 \quad \text{or} \quad x + 3 = -2$$

$$x = 1 \quad \text{or} \qquad x = -5$$

The solutions to our original equation are 1 and -5.   ■

Notice, however, that the original equation in Example 1 was quadratic to begin with. In this case, there is another method that works just as well. Let's solve Example 1 again, but this time, let's begin by expanding $(x + 3)^2$ and $2(x + 3)$.

$$(x + 3)^2 - 2(x + 3) - 8 = 0$$

$$x^2 + 6x + 9 - 2x - 6 - 8 = 0 \qquad \text{Multiply}$$

$$x^2 + 4x - 5 = 0 \qquad \text{Combine similar terms}$$

$$(x - 1)(x + 5) = 0 \qquad \text{Factor}$$

$$x - 1 = 0 \quad \text{or} \quad x + 5 = 0 \quad \text{Zero-factor property}$$

$$x = 1 \qquad\qquad x = -5$$

As you can see, either method produces the same result.

■ **EXAMPLE 2**   Solve: $4x^4 + 7x^2 = 2$.

SOLUTION   If we write the equation as

$$4(x^2)^2 + 7x^2 - 2 = 0$$

we can see that this equation is quadratic in terms of the expression x^2. We can use the substitution $y = x^2$. Replacing x^2 with y and then solving the resulting equation, we have

$$4(x^2)^2 + 7x^2 - 2 = 0$$

$$4y^2 + 7y - 2 = 0 \qquad \text{Substitute}$$

$$(4y - 1)(y + 2) = 0 \qquad \text{Factor}$$

$$4y - 1 = 0 \quad \text{or} \quad y + 2 = 0 \quad \text{Zero-factor property}$$

$$y = \frac{1}{4} \qquad\qquad y = -2$$

Now we replace y with x^2 to solve for x:

$$x^2 = \frac{1}{4} \qquad \text{or} \quad x^2 = -2$$

$$x = \pm\sqrt{\frac{1}{4}} \qquad \text{or} \quad x = \pm\sqrt{-2} \qquad \text{Square Root Property for Equations}$$

$$x = \pm\frac{1}{2} \qquad \text{or} \quad x = \pm i\sqrt{2}$$

The solution set is $\left\{ \frac{1}{2}, -\frac{1}{2}, i\sqrt{2}, -i\sqrt{2} \right\}$.   ◼

■ EXAMPLE 3   Solve: $x + \sqrt{x} - 6 = 0$.

SOLUTION   To see that this equation is quadratic in form, we have to notice that $(\sqrt{x})^2 = x$. That is, the equation can be rewritten as

$$(\sqrt{x})^2 + \sqrt{x} - 6 = 0$$

In this case, we let $y = \sqrt{x}$. Replacing $\sqrt{x}$ with y and solving as usual, we have

$$(\sqrt{x})^2 + \sqrt{x} - 6 = 0$$

$$y^2 + y - 6 = 0 \qquad\qquad \text{Substitute}$$

$$(y + 3)(y - 2) = 0$$

$$y + 3 = 0 \qquad \text{or} \qquad y - 2 = 0$$

$$y = -3 \qquad\qquad y = 2$$

To find x, we replace y with $\sqrt{x}$ and solve:

$$\sqrt{x} = -3 \quad \text{or} \quad \sqrt{x} = 2$$

$$x = 9 \quad \text{or} \quad x = 4 \qquad \text{Square both sides of each equation}$$

Because we squared both sides of each equation, we have the possibility of obtaining extraneous solutions. We have to check both solutions in our original equation.

When	$x = 9$	When	$x = 4$
the equation	$x + \sqrt{x} - 6 = 0$	the equation	$x + \sqrt{x} - 6 = 0$
becomes	$9 + \sqrt{9} - 6 \stackrel{?}{=} 0$	becomes	$4 + \sqrt{4} - 6 \stackrel{?}{=} 0$
	$9 + 3 - 6 \stackrel{?}{=} 0$		$4 + 2 - 6 \stackrel{?}{=} 0$
	$6 \neq 0$		$0 = 0$
	This means 9 is extraneous		This means 4 is a solution

The only solution to the equation $x + \sqrt{x} - 6 = 0$ is $x = 4$.   ◼

We should note here that the two possible solutions, 9 and 4, to the equation in Example 3 can be obtained by another method. Instead of substituting for $\sqrt{x}$, we can isolate it on one side of the equation and then square both sides to clear the equation of radicals.

$$x + \sqrt{x} - 6 = 0$$

$$\sqrt{x} = -x + 6 \qquad \text{Isolate } \sqrt{x}$$

$$x = x^2 - 12x + 36 \qquad \text{Square both sides}$$

$$0 = x^2 - 13x + 36 \qquad \text{Add } -x \text{ to both sides}$$

$$0 = (x - 4)(x - 9) \qquad \text{Factor}$$

$$x - 4 = 0 \quad \text{or} \quad x - 9 = 0$$

$$x = 4 \qquad\qquad x = 9$$

We obtain the same two possible solutions. Because we squared both sides of the equation to find them, we would have to check each one in the original equation. As was the case in Example 3, only $x = 4$ is a solution; $x = 9$ is extraneous.

EXAMPLE 4   Solve: $x^{2/3} - x^{1/3} - 6 = 0$.

SOLUTION   Because $(x^{1/3})^2 = x^{2/3}$, this equation is quadratic in terms of the expression $x^{1/3}$. We substitute $y = x^{1/3}$ and then solve by factoring.

$$(x^{1/3})^2 - x^{1/3} - 6 = 0$$

$$y^2 - y - 6 = 0 \qquad \text{Substitute}$$

$$(y - 3)(y + 2) = 0 \qquad \text{Factor}$$

$$y - 3 = 0 \quad \text{or} \quad y + 2 = 0$$

$$y = 3 \qquad\qquad y = -2$$

Now we substitute back in terms of x and solve for x.

$$x^{1/3} = 3 \qquad \text{or} \qquad x^{1/3} = -2$$

$$\sqrt[3]{x} = 3 \qquad\qquad \sqrt[3]{x} = -2 \qquad \text{Definition of rational exponent}$$

$$x = 3^3 \qquad\qquad x = (-2)^3 \qquad \text{Cube both sides}$$

$$x = 27 \qquad\qquad x = -8$$

The solution set is $\{-8, 27\}$. If you check, you will see that both solutions satisfy the original equation.

EXAMPLE 5   Solve: $6(2x - 1)^{-2} - 7(2x - 1)^{-1} + 2 = 0$.

SOLUTION   We notice that $(2x - 1)^{-2} = \left[(2x - 1)^{-1} \right]^2$, so this equation is quadratic in the expression $(2x - 1)^{-1}$. Substituting $y = (2x - 1)^{-1}$ gives us

$$6\left[(2x - 1)^{-1} \right]^2 - 7(2x - 1)^{-1} + 2 = 0$$

$$6y^2 - 7y + 2 = 0 \qquad \text{Substitute}$$

$$(3y - 2)(2y - 1) = 0 \qquad \text{Factor}$$

$$3y - 2 = 0 \quad \text{or} \quad 2y - 1 = 0$$

$$3y = 2 \qquad\qquad 2y = 1$$

$$y = \frac{2}{3} \qquad\qquad y = \frac{1}{2}$$

Now we replace y with $(2x - 1)^{-1}$ and solve for x, keeping in mind an exponent of -1 indicates the reciprocal.

$$(2x - 1)^{-1} = \frac{2}{3} \quad \text{or} \quad (2x - 1)^{-1} = \frac{1}{2}$$

$$\frac{1}{2x - 1} = \frac{2}{3} \qquad\qquad \frac{1}{2x - 1} = \frac{1}{2} \qquad\qquad \text{Reciprocal}$$

$$3 = 4x - 2 \qquad\qquad 2 = 2x - 1 \qquad\qquad \text{Cross-multiplication}$$

$$5 = 4x \qquad\qquad\qquad 3 = 2x$$

$$\frac{5}{4} = x \qquad\qquad\qquad \frac{3}{2} = x$$

Because of the negative exponent, the original equation would be undefined if $x = \frac{1}{2}$. This value does not appear as one of our potential solutions, so the solution set is $\left\{ \frac{5}{4}, \frac{3}{2} \right\}$. ◼

Another way to solve the equation in Example 5 is to write the equation as the rational equation

$$\frac{6}{(2x - 1)^2} - \frac{7}{2x - 1} + 2 = 0$$

and then solve it using the methods we introduced in Section 6.5.

◼ EXAMPLE 6   Solve: $x^4 - 8x^2 + 14 = 0$.

SOLUTION   Writing x^4 as $(x^2)^2$, we see that this equation is quadratic in the expression x^2. We can substitute $y = x^2$ to obtain

$$(x^2)^2 - 8x^2 + 14 = 0$$

$$y^2 - 8y + 14 = 0$$

This equation does not factor, so we use the quadratic formula to solve for y.

$$y = \frac{-(-8) \pm \sqrt{(-8)^2 - 4(1)(14)}}{2(1)}$$

$$= \frac{8 \pm \sqrt{8}}{2}$$

$$= \frac{8 \pm 2\sqrt{2}}{2}$$

$$= 4 \pm \sqrt{2}$$

Replacing y with x^2, we can then solve for x by taking a cube root of both sides.

$$x^2 = 4 \pm \sqrt{2}$$

$$x = \pm\sqrt{4 \pm \sqrt{2}}$$

We will not attempt to simplify further. There are four solutions.

$$\sqrt{4 + \sqrt{2}}, \sqrt{4 - \sqrt{2}}, -\sqrt{4 + \sqrt{2}}, -\sqrt{4 - \sqrt{2}}$$

◼

Quadratic Formulas

Recall that a formula is an equation that contains more than one variable. Our last example involves a formula that is quadratic in terms of one of the variables.

EXAMPLE 7   If an object is tossed into the air with an upward velocity of 12 feet per second from the top of a building h feet high, the time it takes for the object to hit the ground below is given by the formula

$$16t^2 - 12t - h = 0$$

Solve this formula for t.

SOLUTION   The formula is in standard form and is quadratic in t. The coefficients a, b, and c that we need to apply to the quadratic formula are $a = 16$, $b = -12$, and $c = -h$. Substituting these quantities into the quadratic formula, we have

$$t = \frac{12 \pm \sqrt{144 - 4(16)(-h)}}{2(16)}$$

$$= \frac{12 \pm \sqrt{144 + 64h}}{32}$$

We can factor the perfect square 16 from the two terms under the radical and simplify our radical somewhat:

$$t = \frac{12 \pm \sqrt{16(9 + 4h)}}{32}$$

$$= \frac{12 \pm 4\sqrt{9 + 4h}}{32}$$

Now we can reduce to lowest terms by factoring a 4 from the numerator and denominator:

$$t = \frac{\cancel{4}(3 \pm \sqrt{9 + 4h})}{\cancel{4} \cdot 8}$$

$$= \frac{3 \pm \sqrt{9 + 4h}}{8}$$

If we were given a value of h, we would find that one of the solutions to this last formula would be a negative number. Because time is always measured in positive units, we wouldn't use that solution.

USING TECHNOLOGY *Graphing Calculators*

More About Example 1

As we mentioned before, algebraic expressions entered into a graphing calculator do not have to be simplified to be evaluated. This fact also applies to equations. We can graph the equation $y = (x + 3)^2 - 2(x + 3) - 8$ to assist us in solving the equation in Example 1. The graph is shown in Figure 1. Using the Zoom and Trace features at the x-intercepts gives us $x = 1$ and $x = -5$ as the solutions to the equation $0 = (x + 3)^2 - 2(x + 3) - 8$.

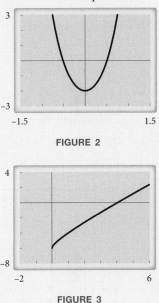

FIGURE 1

More About Example 2

Figure 2 shows the graph of $y = 4x^4 + 7x^2 - 2$. As we expect, the x-intercepts give the real number solutions to the equation $0 = 4x^4 + 7x^2 - 2$. The complex solutions do not appear on the graph.

FIGURE 2

More About Example 3

In solving the equation in Example 3, we found that one of the possible solutions was an extraneous solution. If we solve the equation $x + \sqrt{x} - 6 = 0$ by graphing the function $y = x + \sqrt{x} - 6$, we find that the extraneous solution, 9, is not an x-intercept. Figure 3 shows that the only solution to the equation occurs at the x-intercept 4.

FIGURE 3

Getting Ready for Class

After reading through the preceding section, respond in your own words and in complete sentences.

A. What does it mean for an equation to be quadratic in form?

B. What are all the circumstances in solving equations (that we have studied) in which it is necessary to check for extraneous solutions?

C. How would you start to solve the equation $x + \sqrt{x} - 6 = 0$?

D. What substitution would you use the solve $x^{1/2} + x^{1/4} = 12$?

Problem Set 8.4

Solve each equation.

1. $(x - 3)^2 + 3(x - 3) + 2 = 0$ **2.** $(x + 4)^2 - (x + 4) - 6 = 0$

3. $(2a - 3)^2 - 9(2a - 3) = -20$ **4.** $(3a - 2)^2 + 2(3a - 2) = 3$

5. $x^4 - 6x^2 - 27 = 0$ **6.** $x^4 + 2x^2 - 8 = 0$

7. $x^4 + 9x^2 = -20$ **8.** $x^4 - 11x^2 = -30$

9. $6t^4 = -t^2 + 5$ **10.** $3t^4 = -2t^2 + 8$

11. $9x^4 - 49 = 0$ **12.** $25x^4 - 9 = 0$

13. $8x^6 + 7x^3 - 1 = 0$ **14.** $27x^6 - 26x^3 - 1 = 0$

15. $x^6 - 28x^3 + 27 = 0$ **16.** $x^6 - 9x^3 + 8 = 0$

17. $t^8 + 81 = 82t^4$ **18.** $16t^8 + 1 = 17t^4$

19. $x^{-2} - 2x^{-1} - 15 = 0$ **20.** $8x^{-2} - 6x^{-1} + 1 = 0$

21. $x^{-4} - 14x^{-2} + 45 = 0$ **22.** $x^{-4} + 7x^{-2} + 12 = 0$

23. $2(x + 4)^{-2} + 5(x + 4)^{-1} - 12 = 0$ **24.** $3(x - 5)^{-2} + 14(x - 5)^{-1} - 5 = 0$

25. $2(4a + 2)^{-2} = 3(4a + 2)^{-1} + 20$ **26.** $6(2a + 4)^{-2} = (2a + 4)^{-1} + 2$

Solve each of the following equations. Remember, if you square both sides of an equation in the process of solving it, you have to check all solutions in the original equation.

27. $x - 7\sqrt{x} + 10 = 0$ **28.** $x - 6\sqrt{x} + 8 = 0$

29. $t - 2\sqrt{t} - 15 = 0$ **30.** $t - 3\sqrt{t} - 10 = 0$

31. $(a - 2) - 11\sqrt{a - 2} + 30 = 0$ **32.** $(a - 3) - 9\sqrt{a - 3} + 20 = 0$

33. $(2x + 1) - 8\sqrt{2x + 1} + 15 = 0$ **34.** $(2x - 3) - 7\sqrt{2x - 3} + 12 = 0$

35. $6x + 11x^{1/2} = 35$ **36.** $2x + x^{1/2} = 15$

37. $20x^{2/3} - 3 = 11x^{1/3}$ **38.** $3x^{2/3} + 4x^{1/3} = 4$

39. $4x^{4/3} - 37x^{2/3} + 9 = 0$ **40.** $9x^{4/3} - 13x^{2/3} + 4 = 0$

41. $27x^3 + 19x^{3/2} - 8 = 0$ **42.** $1{,}000x^3 + 117x^{3/2} - 1 = 0$

43. $12a^{-1} - 8a^{-1/2} + 1 = 0$ **44.** $4a^{-1} - 11a^{-1/2} - 3 = 0$

Use substitution with the quadratic formula to solve each of the following equations.

45. $x^4 - 8x^2 + 1 = 0$ **46.** $x^4 - 4x^2 + 1 = 0$

47. $x^4 + 10x^2 + 22 = 0$ **48.** $2x^4 - 4x^2 - 8 = 0$

49. $x - 2\sqrt{x} - 1 = 0$ **50.** $x + 4\sqrt{x} - 1 = 0$

51. Solve the formula $16t^2 - vt - h = 0$ for t.

52. Solve the formula $16t^2 + vt + h = 0$ for t.

53. Solve the formula $kx^2 + 8x + 4 = 0$ for x.

54. Solve the formula $k^2x^2 + kx + 4 = 0$ for x.

55. Solve $x^2 + 2xy + y^2 = 0$ for x by using the quadratic formula with $a = 1$, $b = 2y$, and $c = y^2$.

56. Solve $x^2 - 2xy + y^2 = 0$ for x by using the quadratic formula, with $a = 1$, $b = -2y$, $c = y^2$.

Applying the Concepts

For Problems 57 and 58, t is in seconds.

57. Falling Object   An object is tossed into the air with an upward velocity of 8 feet per second from the top of a building h feet high. The time it takes for the object to hit the ground below is given by the formula $16t^2 - 8t - h = 0$. Solve this formula for t.

58. Falling Object   An object is tossed into the air with an upward velocity of 6 feet per second from the top of a building h feet high. The time it takes for the object to hit the ground below is given by the formula $16t^2 - 6t - h = 0$. Solve this formula for t.

Learning Objectives Assessment

The following problems can be used to help assess if you have successfully met the learning objectives for this section.

59. Solve: $4x^4 + 13x^2 - 12 = 0$.

    a. $\pm 2i, \pm \dfrac{\sqrt{3}}{2}$ **b.** $16, \dfrac{9}{16}$ **c.** $-4, \dfrac{3}{4}$ **d.** $4, -\dfrac{3}{4}$

60. Solve the formula: $-\dfrac{1}{2}gt^2 + vt + h = 0$ for t.

    a. $\pm \dfrac{\sqrt{2gh}}{g}$ **b.** $\sqrt{2h}$ **c.** $\dfrac{v \pm \sqrt{v^2 + 2gh}}{g}$ **d.** $\dfrac{v \pm 2\sqrt{v^2 - 2gh}}{g}$

Getting Ready for the Next Section

Evaluate each function for $x = -2$, $x = -1$, $x = 0$, $x = 1$, and $x = 2$.

61. $f(x) = x^2$ **62.** $f(x) = -x^2$ **63.** $f(x) = 2x^2$

64. $f(x) = \dfrac{1}{2}x^2$ **65.** $f(x) = -\dfrac{1}{4}x^2$ **66.** $f(x) = -4x^2$

67. $f(x) = (x + 2)^2$ **68.** $f(x) = (x - 2)^2$ **69.** $f(x) = x^2 + 2$

70. $f(x) = x^2 - 2$

Quadratic Functions and Transformations

Learning Objectives

In this section, we will learn how to:

1. Graph the basic quadratic function.

2. Identify transformations of a quadratic function.

3. Use transformations to graph a quadratic function.

4. Determine the vertex of a quadratic function written in vertex form.

Introduction

In Example 5 of Section 3.5, we considered the function $h(t) = 32t - 16t^2$, which gave the height of a softball thrown into the air after t seconds for $0 \leq t \leq 2$. In this section, we will explore further these types of functions and their graphs.

Up to now in this chapter, we have been studying quadratic equations. If we take a quadratic equation in standard form and use it instead as the formula for a function, the result is called a **quadratic function**. Here is a formal definition:

> **(dĕf) DEFINITION** *quadratic function*
>
> Any function that can be written in the form
>
> $$f(x) = ax^2 + bx + c$$
>
> where a, b, and c are constants with $a \neq 0$, is called a **quadratic function**. We refer to this form as **standard form**.

The Basic Quadratic Function

The simplest of all quadratic functions results when we let $a = 1$ and $b = c = 0$. We will refer to $f(x) = x^2$ as the **basic quadratic function**. Table 1 gives some ordered pairs for this function, and the corresponding graph is shown in Figure 1. This graph is an example of a **parabola**.

x	$f(x) = x^2$
-3	9
-2	4
-1	1
0	0
1	1
2	4
3	9

TABLE 1

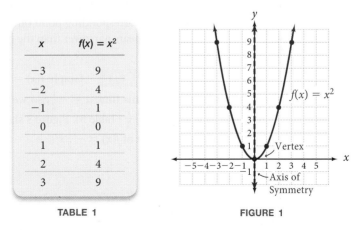

FIGURE 1

From the graph we can see that the domain of this function is the set of all real numbers and the range is $\{y \mid y \geq 0\}$. The point $(0, 0)$ where the parabola changes direction (the lowest point on the parabola) is called the **vertex**.

We also observe that the graph is symmetric about the y-axis, meaning the right half of the graph is a mirror image of the left half. For this reason, the vertical line $x = 0$ passing through the vertex is called the ***axis of symmetry***.

All quadratic functions have a graph that is a parabola and a domain that is the set of all real numbers. However, the shape, direction, and position of the parabola can vary as we will see in the following segment.

Transformations

Let's consider quadratic functions of the form $f(x) = ax^2$. In this case, all we are doing is taking each y-value (output) from the basic quadratic function $y = x^2$ and multiplying it by a factor of a. As a result, we can change the shape and direction of the basic parabola. We illustrate how this is done in the next two examples.

VIDEO EXAMPLES

SECTION 8.5

▓ EXAMPLE 1   Graph: $f(x) = 2x^2$.

SOLUTION Because $a = 2$, we take each y-coordinate from the basic parabola $y = x^2$ and double it. Table 2 shows how this is done for several values of x. The graph of the basic parabola $y = x^2$ and the graph of $f(x) = 2x^2$ are shown together in Figure 2.

x	$y = x^2$	$f(x) = 2x^2$
-3	9	$2 \cdot 9 = 18$
-2	4	$2 \cdot 4 = 8$
-1	1	$2 \cdot 1 = 2$
0	0	$2 \cdot 0 = 0$
1	1	$2 \cdot 1 = 2$
2	4	$2 \cdot 4 = 8$
3	9	$2 \cdot 9 = 18$

TABLE 2

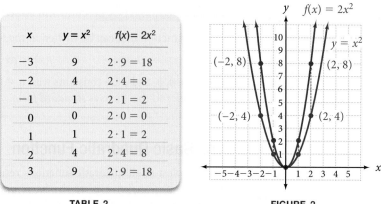

FIGURE 2

Notice that each point on the graph of $f(x) = 2x^2$ is twice the distance from the x-axis as the corresponding point on the basic quadratic. As a result, the graph of $f(x) = 2x^2$ appears narrower than the basic parabola.   ▓

▓ EXAMPLE 2   Graph: $f(x) = -\frac{1}{2}x^2$.

SOLUTION This time $a = -\frac{1}{2}$, so we multiply each y-coordinate from the basic parabola by $-\frac{1}{2}$ (Table 3). Figure 3 shows the graph of the basic quadratic $y = x^2$ and the graph of $f(x) = -\frac{1}{2}x^2$.

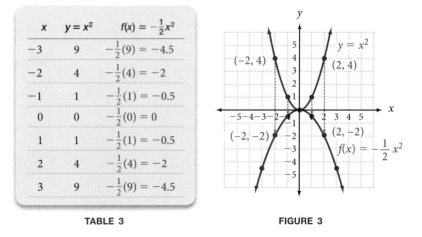

x	$y = x^2$	$f(x) = -\frac{1}{2}x^2$
-3	9	$-\frac{1}{2}(9) = -4.5$
-2	4	$-\frac{1}{2}(4) = -2$
-1	1	$-\frac{1}{2}(1) = -0.5$
0	0	$-\frac{1}{2}(0) = 0$
1	1	$-\frac{1}{2}(1) = -0.5$
2	4	$-\frac{1}{2}(4) = -2$
3	9	$-\frac{1}{2}(9) = -4.5$

TABLE 3   FIGURE 3

Observe that each point on the graph of $f(x) = -\frac{1}{2}x^2$ is half the distance from the x-axis as the corresponding point on the basic quadratic. Also, because of the negative sign, each point is now on the opposite side of the x-axis. As a result, the graph of $f(x) = -\frac{1}{2}x^2$ appears wider than the basic parabola and opens downward instead of upward. The vertex, though still at $(0, 0)$, is now the highest point on the graph. ◼

Here is a formal definition that summarizes our observations from the previous two examples.

> **(dĕf DEFINITION**   *vertical scaling and reflection*
>
> If $f(x) = ax^2$ and a is any nonzero real number, then for
>
> $|a| \neq 1$ The basic parabola will be *scaled vertically* by a factor of a.
>
> If $|a| > 1$, there is a *vertical expansion* and the graph of $y = ax^2$ will appear narrower than the basic parabola.
>
> If $|a| < 1$, there is a *vertical contraction* and the graph of $y = ax^2$ will appear wider than the basic parabola.
>
> $a < 0$ The basic parabola will be *reflected about the x-axis*. The graph of $y = ax^2$ will open downward instead of upward, and the vertex will be the highest point on the graph.

Next, let's consider quadratic functions of the form $f(x) = (x - h)^2 + k$. As the following examples illustrate, the terms h and k can change the position of the basic parabola on a rectangular coordinate system.

◼ **EXAMPLE 3**   Graph $f(x) = (x + 2)^2$ and $g(x) = (x - 2)^2$.

SOLUTION   Table 4 shows the results of calculating values of x^2, $(x + 2)^2$, and $(x - 2)^2$ for several values of x. By plotting points, we obtain the three graphs shown in Figure 4.

x	$y = x^2$	$f(x) = (x + 2)^2$	$g(x) = (x - 2)^2$
-5	25	9	49
-4	16	4	36
-3	9	1	25
-2	4	0	16
-1	1	1	9
0	0	4	4
1	1	9	1
2	4	16	0
3	9	25	1
4	16	36	4
5	25	49	9

TABLE 4

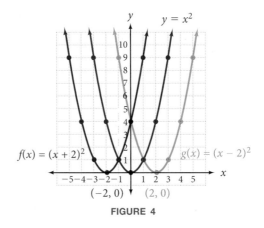

FIGURE 4

Looking at Table 4, notice that in order for the outputs of $f(x) = (x + 2)^2$ to match those of $y = x^2$, we must use x-values that are 2 units less, and with $g(x) = (x - 2)^2$ we must use x-values that are 2 units greater. From Figure 4, we can see that the graph of $f(x) = (x + 2)^2$ is the same shape as $y = x^2$, only shifted 2 units to the left. Likewise, the graph of $g(x) = (x - 2)^2$ is identical to the basic parabola, except shifted 2 units to the right.

EXAMPLE 4   Graph $f(x) = x^2 + 2$ and $g(x) = x^2 - 2$.

SOLUTION   In this case, we can obtain values of $f(x)$ and $g(x)$ by adding 2 or subtracting 2 from each y-coordinate of the basic quadratic $y = x^2$. Table 5 shows these calculations for several values of x. The graphs of $y = x^2$, $f(x) = x^2 + 2$, and $g(x) = x^2 - 2$ are shown in Figure 5.

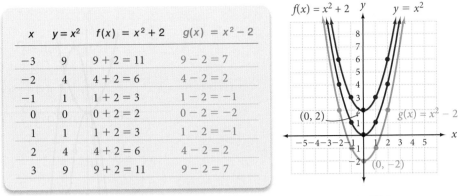

x	$y = x^2$	$f(x) = x^2 + 2$	$g(x) = x^2 - 2$
-3	9	$9 + 2 = 11$	$9 - 2 = 7$
-2	4	$4 + 2 = 6$	$4 - 2 = 2$
-1	1	$1 + 2 = 3$	$1 - 2 = -1$
0	0	$0 + 2 = 2$	$0 - 2 = -2$
1	1	$1 + 2 = 3$	$1 - 2 = -1$
2	4	$4 + 2 = 6$	$4 - 2 = 2$
3	9	$9 + 2 = 11$	$9 - 2 = 7$

TABLE 5   FIGURE 5

As you can see in Figure 5, the graph of $f(x) = x^2 + 2$ is identical to the basic parabola $y = x^2$, only shifted upward 2 units. Likewise, the graph of $g(x) = x^2 - 2$ is identical to the basic parabola, but shifted 2 units downward.

We can summarize our observations from Examples 3 and 4 as follows:

> **(dĕf DEFINITION**   *horizontal and vertical translation*
>
> If $f(x) = (x - h)^2 + k$, and h and k are real numbers, then h represents a **horizontal translation** and k represents a **vertical translation**. Specifically, if h and k are positive numbers, then
>
> | The graph of | is the graph of $y = x^2$ translated |
> |--------------|---------------------------------------|
> | $f(x) = (x - h)^2$ | h units to the right |
> | $f(x) = (x + h)^2$ | h units to the left |
> | $f(x) = x^2 + k$ | k units upward |
> | $f(x) = x^2 - k$ | k units downward |

It is important to note that of the three types of transformations (vertical scaling, vertical reflection, and translations), only the translations can affect the position of the vertex of the parabola.

In our next example, we tie these concepts together in graphing a quadratic function involving all three types of transformation.

■ EXAMPLE 5   Graph $f(x) = 5 - 3(x - 1)^2$, and state the vertex and range.

SOLUTION Writing the function in the form $f(x) = -3(x - 1)^2 + 5$, we see that there is a vertical expansion by a factor of 3 and reflection about the x-axis, a horizontal translation of 1 unit to the right, and a vertical translation of 5 units upward.

The process is shown in Figure 6. We begin by graphing the basic parabola $y = x^2$. Next, we multiply each y-coordinate from the basic parabola by -3 to obtain the graph of $y = -3x^2$. Finally, we shift each point on this new graph 1 unit to the right and 5 units upward.

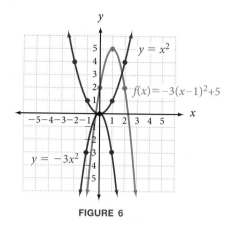

FIGURE 6

From the graph, we can see that the vertex is $(1, 5)$ and the range of the function is $\{y \mid y \le 5\}$, $(-\infty, 5]$ using interval notation.

Vertex Form

Now that we have a good understanding of transformations, we can give an alternate definition for a quadratic function.

Looking back at Example 5, the vertex ended up at the point $(1, 5)$ because we had to translate the parabola 1 unit to the right and 5 units upward.

$$f(x) = -3 \, (x - 1)^2 + 5 \quad \text{Vertex} = (1, 5)$$

$$\uparrow \qquad \uparrow$$

Shift 1   Shift 5
unit right   units upward

Notice how the coordinates of the vertex appear in the equation of the function. We can generalize this result as follows.

> **(def) DEFINITION** *vertex form*
>
> Any quadratic function that can be written in the *vertex form* as
>
> $$f(x) = a(x - h)^2 + k$$
>
> where a, h, and k are constants with $a \ne 0$. The vertex of the parabola will be the point (h, k).

EXAMPLE 6   If $f(x) = \frac{2}{3}(x + 4)^2 - 7$, state the vertex, domain, range, and axis of symmetry.

SOLUTION   Writing the function in vertex form as

$$f(x) = \frac{2}{3}(x - (-4))^2 - 7$$

we have $h = -4$ and $k = -7$. Therefore, the vertex is $(-4, -7)$. The domain for any quadratic function is all real numbers. Because the parabola has not been reflected about the x-axis, it opens upward and the vertex will be the lowest point on the graph. Thus, the range is $\{y \mid y \ge -7\}$. Finally, the axis of symmetry is the vertical line passing through the vertex, so it must have the equation $x = -4$.

Getting Ready for Class

After reading through the preceding section, respond in your own words and in complete sentences.

A. What is a parabola?

B. What are the two forms for writing the equation of a quadratic function?

C. How do you know by looking at the equation if the graph of a quadratic function opens upward or downward?

D. Explain how to tell the difference between the horizontal translation and the vertical translation in the equation of a parabola.

Problem Set 8.5

Complete each table using the given function.

1. $f(x) = 3x^2$

x	y
-3	
-2	
-1	
0	
1	
2	
3	

2. $f(x) = \frac{1}{3}x^2$

x	y
-3	
-2	
-1	
0	
1	
2	
3	

3. $f(x) = -\frac{3}{4}x^2$

x	y
-3	
-2	
-1	
0	
1	
2	
3	

4. $f(x) = -\frac{5}{2}x^2$

x	y
-3	
-2	
-1	
0	
1	
2	
3	

5. $f(x) = (x - 3)^2$

x	y
-3	
-2	
-1	
0	
1	
2	
3	

6. $f(x) = (x + 3)^2$

x	y
-3	
-2	
-1	
0	
1	
2	
3	

7. $f(x) = x^2 + 3$

8. $f(x) = x^2 - 3$

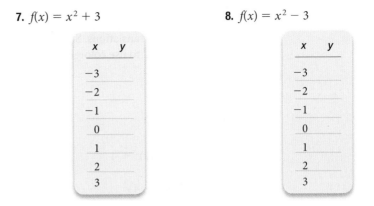

x	y
−3	
−2	
−1	
0	
1	
2	
3	

x	y
−3	
−2	
−1	
0	
1	
2	
3	

For each quadratic function, identify any vertical scaling or vertical reflection. Also state if the graph opens upward or downward, and whether the graph would be narrower or wider than the basic parabola $y = x^2$.

9. $f(x) = 4x^2$

10. $f(x) = \dfrac{1}{4}x^2$

11. $f(x) = -\dfrac{2}{3}x^2$

12. $f(x) = -\dfrac{5}{3}x^2$

For each quadratic function, identify any horizontal or vertical translations.

13. $f(x) = x^2 + 4$

14. $f(x) = (x + 4)^2$

15. $f(x) = (x - 1)^2$

16. $f(x) = x^2 - 1$

17. $f(x) = (x - 2)^2 + 5$

18. $f(x) = (x + 5)^2 - 2$

19. $f(x) = 3 - (x + 6)^2$

20. $f(x) = -3 - (x - 4)^2$

For each quadratic function, identify any transformations.

21. $f(x) = 3(x + 5)^2 - 2$

22. $f(x) = \dfrac{1}{2}(x - 1)^2 - 3$

23. $f(x) = 1 - 4(x - 3)^2$

24. $f(x) = 6 - (x + 4)^2$

Graph each of the following. Use one coordinate system for each problem.

25. a. $y = \dfrac{1}{2}x^2$ **b.** $y = \dfrac{1}{2}x^2 - 2$ **c.** $y = \dfrac{1}{2}x^2 + 2$

26. a. $y = 2x^2$ **b.** $y = 2x^2 - 8$ **c.** $y = 2x^2 + 1$

27. a. $y = -2x^2$ **b.** $y = -2(x - 3)^2$ **c.** $y = -2(x + 3)^2$

28. a. $y = -\dfrac{1}{2}x^2$ **b.** $y = -\dfrac{1}{2}(x + 4)^2$ **c.** $y = -\dfrac{1}{2}(x - 4)^2$

Use transformations to graph each of the following quadratic functions. Begin with the basic parabola $y = x^2$, then graph any vertical scaling and/or reflection, and finally graph any translations. Show all graphs on a single coordinate system (you may find it helpful to use colored pencils).

29. $f(x) = 3x^2$

30. $f(x) = \frac{1}{3}x^2$

31. $f(x) = -\frac{1}{4}x^2$

32. $f(x) = -4x^2$

33. $f(x) = (x - 2)^2 + 4$

34. $f(x) = (x + 1)^2 - 5$

35. $f(x) = -(x - 4)^2$

36. $f(x) = \frac{1}{4}(x - 3)^2$

37. $f(x) = 2x^2 + 6$

38. $f(x) = -3x^2 - 2$

39. $f(x) = \frac{1}{2}(x - 1)^2 - 3$

40. $f(x) = -4(x + 3)^2 + 1$

41. $f(x) = 2 - \frac{3}{2}(x + 4)^2$

42. $f(x) = 4 - \frac{3}{4}(x - 1)^2$

State the vertex for each of the following quadratic functions, and determine if the vertex is the highest or lowest point on the graph. Then state the range of each function.

43. $y = 2(x - 1)^2 + 3$

44. $y = 2(x + 1)^2 - 3$

45. $f(x) = -(x + 2)^2 + 4$

46. $f(x) = -(x - 3)^2 + 1$

47. $g(x) = \frac{1}{2}(x - 2)^2 - 4$

48. $g(x) = \frac{1}{3}(x - 3)^2 - 3$

49. $f(x) = -2(x - 4)^2 - 1$

50. $f(x) = -4(x - 1)^2 + 4$

Find the equation of the quadratic function, $f(x)$, in vertex form, whose graph is shown.

51.

52.

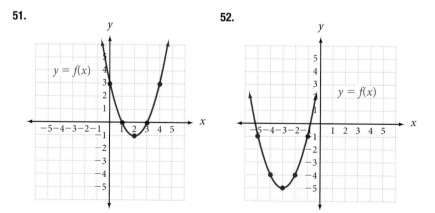

53.

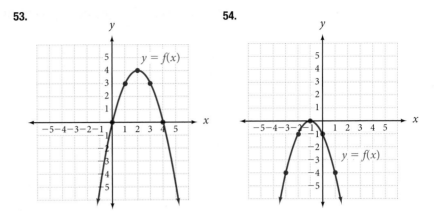

54.

Learning Objectives Assessment

The following problems can be used to help assess if you have successfully met the learning objectives for this section.

55. Which point lies on the graph of the basic quadratic function?

    a. $(4, 2)$ **b.** $(4, -2)$ **c.** $(-2, 4)$ **d.** $(-2, -4)$

56. In the quadratic function $f(x) = 1 - 5(x + 3)^2$, the value 5 represents a

    a. horizontal translation   **b.** vertical translation

    c. vertical contraction   **d.** vertical expansion

57. Which of the following could be the graph of $f(x) = -\frac{1}{2}(x + 1)^2$?

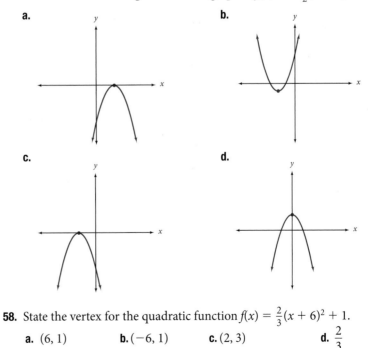

a.   **b.**

c.   **d.**

58. State the vertex for the quadratic function $f(x) = \frac{2}{3}(x + 6)^2 + 1$.

    a. $(6, 1)$ **b.** $(-6, 1)$ **c.** $(2, 3)$ **d.** $\frac{2}{3}$

Getting Ready for the Next Section

59. Evaluate $y = 3x^2 - 6x + 1$ for $x = 1$.

60. Evaluate $y = -2x^2 + 6x - 5$ for $x = \frac{3}{2}$.

61. Let $P(x) = -0.1x^2 + 27x - 500$ and find $P(135)$.

62. Let $P(x) = -0.1x^2 + 12x - 400$ and find $P(600)$.

Solve.

63. $0 = a(80)^2 + 70$ **64.** $0 = a(80)^2 + 90$

65. $x^2 - 6x + 5 = 0$ **66.** $x^2 - 3x - 4 = 0$

67. $-x^2 - 2x + 3 = 0$ **68.** $-x^2 + 4x + 12 = 0$

69. $2x^2 - 6x + 5 = 0$ **70.** $x^2 - 4x + 5 = 0$

Fill in the blanks to complete the square.

71. $x^2 - 6x + \square = (x - \square)^2$ **72.** $x^2 - 10x + \square = (x - \square)^2$

73. $y^2 + 2y + \square = (y + \square)^2$ **74.** $y^2 - 12y + \square = (x - \square)^2$

Learning Objectives

In this section, we will learn how to:

1. Find the vertex of a quadratic function written in standard form.

2. Find the x-intercepts and y-intercept for a quadratic function.

3. Graph a quadratic function using the vertex and intercepts.

4. Convert a quadratic function from standard form into vertex form by completing the square.

Introduction

In this section, we will continue to explore the behavior of quadratic functions and their graphs.

First, let's summarize what we learned in the previous section.

$\sqrt{\Delta \neq \Sigma}$ *Quadratic Functions in Vertex Form*

The graph of $f(x) = a(x - h)^2 + k$, $a \neq 0$, will be a parabola with vertex (h, k).

If $a > 0$, the parabola will open upward and the vertex will be the lowest point on the graph.

If $a < 0$, the parabola will open downward and the vertex will be the highest point on the graph.

If the quadratic function is in standard form, however, we do not yet have a way to identify the vertex easily. We will consider how to do this next.

Finding the Vertex in Standard Form

The vertex for the graph of $f(x) = ax^2 + bx + c$ will always occur when

$$x = -\frac{b}{2a}$$

To see this, we must transform the right side of $f(x) = ax^2 + bx + c$ into an expression that contains x in just one of its terms. This is accomplished by completing the square on the first two terms. Here is what it looks like:

$$f(x) = ax^2 + bx + c$$
$$= a\left(x^2 + \frac{b}{a}x\right) + c$$
$$= a\left[x^2 + \frac{b}{a}x + \left(\frac{b}{2a}\right)^2 - \left(\frac{b}{2a}\right)^2\right] + c$$
$$= a\left[x^2 + \frac{b}{a}x + \left(\frac{b}{2a}\right)^2\right] + c - a\left(\frac{b}{2a}\right)^2$$
$$= a\left(x + \frac{b}{2a}\right)^2 + \frac{4ac}{4a} - \frac{b^2}{4a}$$
$$= a\left(x - \left(-\frac{b}{2a}\right)\right)^2 + \frac{4ac - b^2}{4a}$$

This last line is now expressed in vertex form with

$$h = -\frac{b}{2a} \qquad \text{and} \qquad k = \frac{4ac - b^2}{4a}$$

This leads us to the following result, which we will refer to as the ***vertex formula***.

$[\Delta \neq \Sigma]$ **PROPERTY**   *vertex formula*

The graph of the quadratic function $f(x) = ax^2 + bx + c,\ a \neq 0$, will be a parabola with

$$\text{vertex} = \left(-\frac{b}{2a},\ \frac{4ac - b^2}{4a}\right)$$

VIDEO EXAMPLES

SECTION 8.6

EXAMPLE 1   Find the vertex of $f(x) = 2x^2 - 12x + 23$.

SOLUTION   To find the coordinates of the vertex, we calculate

$$x = -\frac{b}{2a} \qquad\qquad y = \frac{4ac - b^2}{4a}$$

$$= -\frac{(-12)}{2(2)} \qquad\qquad = \frac{4(2)(23) - (-12)^2}{4(2)}$$

$$= \frac{12}{4} \qquad\qquad\qquad = \frac{40}{8}$$

$$= 3 \qquad\qquad\qquad = 5$$

The vertex is the point $(3, 5)$.

Another way to find the y-coordinate of the vertex, which is often easier, is to simply evaluate the function at the x-coordinate of the vertex. Using this method, we have

$$\text{If} \qquad x = 3$$
$$\text{then} \qquad f(3) = 2(3)^2 - 12(3) + 23$$
$$= 18 - 36 + 23$$
$$= 5$$

As you can see, we obtain the same value with either method.

Based on our observations in the previous example, we offer the following alternative formula for finding the vertex from standard form.

$[\Delta \neq \Sigma]$ **ALTERNATE PROPERTY**   *vertex formula*

The vertex of the quadratic function $f(x) = ax^2 + bx + c,\ a \neq 0$, is given by

$$\left(-\frac{b}{2a},\ f\left(-\frac{b}{2a}\right)\right)$$

Intercepts

Recall that the x-intercepts (if they exist) are the points where a graph intersects the x-axis, and the y-intercept is the point where the graph intersects the y-axis.

EXAMPLE 2   Find any intercepts for $f(x) = -3(x - 1)^2 + 5$.

SOLUTION   To find the x-intercepts, we let $y = 0$ and solve for x. Because the function is in vertex form, this is most easily done using the Square Root Property for Equations.

When $\quad y = 0$

we have $\qquad 0 = -3(x - 1)^2 + 5$ $\qquad$ Replace $f(x)$ with 0

$$-5 = -3(x - 1)^2$$

$$\frac{5}{3} = (x - 1)^2$$

$$\pm\sqrt{\frac{5}{3}} = x - 1 \qquad\qquad \text{Square Root Property}$$

$$\pm\frac{\sqrt{15}}{3} = x - 1 \qquad\qquad \sqrt{\frac{5}{3}} = \frac{\sqrt{5}}{\sqrt{3}} \cdot \frac{\sqrt{3}}{\sqrt{3}} = \frac{\sqrt{15}}{3}$$

$$1 \pm \frac{\sqrt{15}}{3} = x$$

Using a calculator, we can approximate these values as $x \approx -0.3$ and $x \approx 2.3$.

To find the y-intercept, we let $x = 0$ and solve for y. This can be done by evaluating the function for $x = 0$.

$$f(0) = -3(0 - 1)^2 + 5$$

$$= -3(1) + 5$$

$$= 2$$

The graph will cross the y-axis at the point $(0, 2)$.

We actually graphed this function in the previous section. Looking at Figure 6 in Section 8.5, you can see that the final graph is consistent with these values. ▩

▩ **EXAMPLE 3** $\qquad$ Find any intercepts for $f(x) = 3x^2 - 6x + 1$.

SOLUTION To find the x-intercepts, we replace y with 0 and solve for x. The resulting equation does not factor, so we use the quadratic formula.

When $\quad y = 0$

we have $\quad 0 = 3x^2 - 6x + 1$

$$x = \frac{-(-6) \pm \sqrt{(-6)^2 - 4(3)(1)}}{2(3)}$$

$$= \frac{6 \pm \sqrt{24}}{6}$$

$$= \frac{6 \pm 2\sqrt{6}}{6}$$

$$= \frac{3 \pm \sqrt{6}}{3}$$

Approximating with a calculator, we obtain $x \approx 0.2$ and $x \approx 1.8$.

The y-intercept is found by evaluating the function for $x = 0$.

$$f(0) = 3(0)^2 - 6(0) + 1$$

$$= 1 \qquad\qquad\qquad ▩$$

Notice in Example 3 that the y-intercept was equal to the constant term in the equation of the function. If the quadratic is in standard form,

$$f(x) = ax^2 + bx + c$$

then for $x = 0$ we have

$$f(0) = a(0)^2 + b(0) + c$$

$$= c$$

So the y-intercept will always be the point $(0, c)$. Here is a summary of the process for finding the intercepts of a quadratic function.

> ### HOW TO   *Find Intercepts of a Quadratic Function*
>
> If the quadratic function is in vertex form, $f(x) = a(x - h)^2 + k$, then
>
> **Step 1:** Find the x-intercepts by solving the equation $a(x - h)^2 + k = 0$ using the Square Root Property for Equations.
> **Step 2:** Evaluate $f(0)$ to find the y-intercept.
>
> If the quadratic function is in standard form, $f(x) = ax^2 + bx + c$, then
>
> **Step 1:** Find the x-intercepts by solving the equation $ax^2 + bx + c = 0$ by factoring or using the quadratic formula.
> **Step 2:** The y-intercept is c.

Graphing Quadratic Functions

We can use the vertex along with the x- and y-intercepts to sketch the graph of any quadratic function.

EXAMPLE 4   Graph: $f(x) = -x^2 - 2x + 3$.

SOLUTION   The y-intercept is 3. To find the x-intercepts, we solve

$$-x^2 - 2x + 3 = 0$$
$$x^2 + 2x - 3 = 0 \qquad \text{Multiply each side by } -1$$
$$(x + 3)(x - 1) = 0$$
$$x = -3 \quad \text{or} \quad x = 1$$

The x-coordinate of the vertex is given by

$$x = -\frac{b}{2a} = -\frac{(-2)}{2(-1)} = -1$$

To find the y-coordinate of the vertex, we evaluate the function at $x = -1$ to get

$$f(-1) = -(-1)^2 - 2(-1) + 3 = -1 + 2 + 3 = 4$$

Our parabola has x-intercepts at -3 and 1, a y-intercept at 3, and a vertex at $(-1, 4)$. Figure 1 shows the graph.

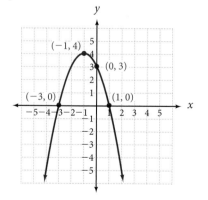

FIGURE 1

▨ **EXAMPLE 5**   Graph: $f(x) = 3x^2 - 6x + 1$.

SOLUTION We already found the intercepts for this graph back in Example 3. The *y*-intercept is the point (0, 1) and the *x*-intercepts are

$$\frac{3 - \sqrt{6}}{3} \approx 0.2 \quad \text{and} \quad \frac{3 + \sqrt{6}}{3} \approx 1.8$$

All we need now is to find the vertex. Using the formula that gives us the *x*-coordinate of the vertex, we have

$$x = -\frac{b}{2a} = -\frac{(-6)}{2(3)} = 1$$

Evaluating the function at $x = 1$ gives us the *y*-coordinate of the vertex.

$$f(1) = 3 \cdot 1^2 - 6 \cdot 1 + 1 = -2$$

Plotting the intercepts along with the vertex (1, −2), we have the graph shown in Figure 2.

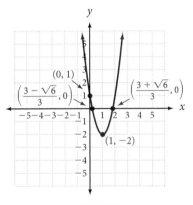

FIGURE 2

▨

▨ **EXAMPLE 6**   Graph: $f(x) = 2(x + 1)^2 - 3$.

SOLUTION To find the *x*-intercepts, we solve

$$2(x + 1)^2 - 3 = 0$$

$$2(x + 1)^2 = 3$$

$$(x + 1)^2 = \frac{3}{2}$$

$$x + 1 = \pm\sqrt{\frac{3}{2}} \qquad \text{Square Root Property for Equations}$$

$$x + 1 = \pm\frac{\sqrt{6}}{2} \qquad \sqrt{\frac{3}{2}} = \frac{\sqrt{3}}{\sqrt{2}} \cdot \frac{\sqrt{2}}{\sqrt{2}} = \frac{\sqrt{6}}{2}$$

$$x = -1 \pm\frac{\sqrt{6}}{2}$$

Using a calculator, we can approximate these values, obtaining $x \approx -2.2$ and $x \approx 0.2$.

For the y-intercept, we evaluate the function at $x = 0$.

$$f(0) = 2(0 + 1)^2 - 3$$

$$= 2(1) - 3$$

$$= -1$$

The vertex is the point $(-1, -3)$. Plotting these points, we have the graph shown in Figure 3.

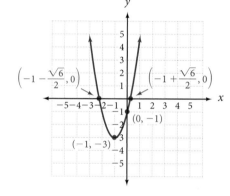

FIGURE 3

EXAMPLE 7   Graph: $f(x) = -2x^2 + 6x - 5$.

SOLUTION   The y-intercept is -5. To find the x-intercepts, we solve

$$-2x^2 + 6x - 5 = 0$$

The left side of this equation does not factor. The discriminant is $b^2 - 4ac = 36 - 4(-2)(-5) = -4$, which indicates that the solutions are complex numbers. This means that our original equation does not have x-intercepts. The graph does not cross the x-axis.

Let's find the vertex. Using our formula for the x-coordinate of the vertex, we have

$$x = -\frac{b}{2a} = -\frac{6}{2(-2)} = \frac{3}{2}$$

To find the y-coordinate, we evaluate $f(x)$ for $x = \frac{3}{2}$:

$$f\left(\frac{3}{2}\right) = -2\left(\frac{3}{2}\right)^2 + 6\left(\frac{3}{2}\right) - 5$$

$$= \frac{-18}{4} + \frac{18}{2} - 5$$

$$= \frac{-18 + 36 - 20}{4}$$

$$= -\frac{1}{2}$$

The vertex is $\left(\frac{3}{2}, -\frac{1}{2}\right)$. Because this is only two points so far, we must find at least one more. Let's try $x = 3$:

When $x = 3$

$$f(3) = -2(3)^2 + 6(3) - 5$$

$$= -18 + 18 - 5$$

$$= -5$$

An additional point on the graph is $(3, -5)$. Figure 4 shows the graph.

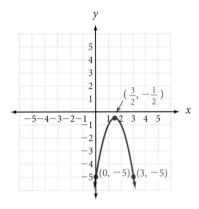

FIGURE 4

Converting Between Forms

We conclude this section by considering how to convert between standard form and vertex form.

To convert vertex form into standard form is a simple matter of squaring the binomial and then simplifying.

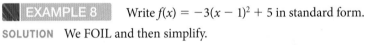 **EXAMPLE 8**   Write $f(x) = -3(x - 1)^2 + 5$ in standard form.

SOLUTION   We FOIL and then simplify.

$$f(x) = -3(x - 1)^2 + 5$$

$$= -3(x^2 - 2x + 1) + 5 \qquad \text{FOIL}$$

$$= -3x^2 + 6x - 3 + 5 \qquad \text{Distribute}$$

$$= -3x^2 + 6x + 2 \qquad \text{Standard form}$$

Now that the quadratic function is in standard form, we can see that the y-intercept is $y = 2$, which agrees with the value we obtained in Example 2.

To convert from standard form into vertex form, we complete the square on the first two terms. Because we are not solving an equation, this process is a little different from the one we followed in Section 8.1. Our last example illustrates how it is done.

EXAMPLE 9   Write $f(x) = -2x^2 + 6x - 5$ in vertex form.

SOLUTION   This is the same function we looked at in Example 7. We cannot complete the square unless the coefficient of x^2 is equal to 1. Instead of dividing by -2, we factor -2 from the first two terms.

$$f(x) = -2x^2 + 6x - 5$$

$$= -2(x^2 - 3x) - 5$$

To create a perfect square trinomial, we must add $\left(-\frac{3}{2}\right)^2 = \frac{9}{4}$. However, to maintain an equivalent expression, we add and subtract $\frac{9}{4}$ so the net result is the same as adding 0 to the right side.

$$f(x) = -2\left(x^2 - 3x + \frac{9}{4} - \frac{9}{4}\right) - 5 \qquad \text{Add and subtract } \tfrac{9}{4}$$

$$= -2\left(x^2 - 3x + \frac{9}{4}\right) - 2\left(-\frac{9}{4}\right) - 5 \qquad \text{Distribute } -2$$

$$= -2\left(x - \frac{3}{2}\right)^2 + \frac{9}{2} - 5 \qquad \text{Factor}$$

$$= -2\left(x - \frac{3}{2}\right)^2 - \frac{1}{2} \qquad \text{Simplify}$$

We now have the function in vertex form, so the vertex will be $\left(\frac{3}{2}, -\frac{1}{2}\right)$, which confirms our work in Example 7. ∎

One advantage of converting a quadratic function from standard form into vertex form is that it allows us to graph the function using transformations instead of plotting points.

   For example, we can now see that the graph of the function in Example 7 will be a basic parabola that is vertically expanded by 2, reflected about the x-axis, translated horizontally $\frac{3}{2}$ units to the right, and translated vertically $\frac{1}{2}$ unit downward.

Getting Ready for Class

After reading through the preceding section, respond in your own words and in complete sentences.

A. Explain how to find the vertex of a quadratic function in standard form.

B. In which form is the constant term equal to the y-intercept?

C. What is the best method to use in finding the x-intercepts when the quadratic function is in standard form? In vertex form?

D. How is the process of completing the square in this section different from the one we described in Section 8.1?

For each of the following quadratic functions, find the coordinates of the vertex and indicate whether the vertex is the highest point on the graph or the lowest point on the graph. Then state the range.

1. $f(x) = x^2 - 6x + 5$ **2.** $f(x) = -x^2 + 6x - 5$ **3.** $f(x) = -x^2 + 2x + 8$

4. $f(x) = x^2 - 2x - 8$ **5.** $f(x) = 12 + 4x - x^2$ **6.** $f(x) = -12 - 4x + x^2$

7. $f(x) = -x^2 - 8x$ **8.** $f(x) = x^2 + 8x$ **9.** $f(x) = x^2 - 4$

10. $f(x) = 6 - x^2$ **11.** $f(x) = 3x^2 - 6x + 14$ **12.** $f(x) = 2x^2 + 12x + 13$

13. $f(x) = -4x^2 - 16x - 17$ **14.** $f(x) = -3x^2 - 30x - 73$

15. $f(x) = 4x^2 - 4x + 19$ **16.** $f(x) = 4x^2 + 12x - 1$

For each of the following quadratic functions, give the x-intercepts (if they exist), the y-intercept, and the coordinates of the vertex. Then use these points to sketch the graph. Plot additional points as necessary.

17. $y = x^2 + 2x - 3$ **18.** $y = x^2 - 2x - 3$ **19.** $y = -x^2 - 4x + 5$

20. $y = x^2 + 4x - 5$ **21.** $y = x^2 - 1$ **22.** $y = x^2 - 4$

23. $y = -x^2 + 9$ **24.** $y = -x^2 + 1$ **25.** $f(x) = (x + 1)^2 - 4$

26. $f(x) = (x - 2)^2 - 9$ **27.** $y = 2x^2 - 4x - 6$ **28.** $y = 2x^2 + 4x - 6$

29. $y = x^2 - 2x - 4$ **30.** $y = x^2 - 2x - 2$ **31.** $f(x) = 2 - (x - 3)^2$

32. $f(x) = 3 - (x + 4)^2$ **33.** $y = x^2 - 4x - 4$ **34.** $y = x^2 - 2x + 3$

35. $y = -x^2 + 2x - 5$ **36.** $y = -x^2 + 4x - 2$ **37.** $f(x) = x^2 + 1$

38. $f(x) = x^2 + 4$ **39.** $y = -x^2 - 3$ **40.** $y = -x^2 - 2$

41. $g(x) = 3x^2 + 4x + 1$ **42.** $g(x) = 2x^2 + 4x + 3$

43. $f(x) = 2(x - 1)^2 - 6$ **44.** $f(x) = 3(x + 2)^2 + 3$

45. $f(x) = -3(x + 3)^2 - 1$ **46.** $f(x) = -4(x - 4)^2 + 20$

47. $f(x) = x^2 + 3x$ **48.** $f(x) = x^2 - x$ **49.** $f(x) = 8x - 2x^2$

50. $f(x) = -6x - 3x^2$ **51.** $f(x) = 6x^2 - 7x - 3$ **52.** $f(x) = 8x^2 - 10x + 3$

53. $f(x) = -2x^2 - x + 5$ **54.** $f(x) = 3x^2 - x + 2$ **55.** $f(x) = 3 + 3x^2 - 2x$

56. $f(x) = x + 1 - 4x^2$

Write each of the following quadratic functions in vertex form by completing the square. Then state the vertex.

57. $f(x) = x^2 + 4x$ **58.** $f(x) = x^2 - 6x$

59. $f(x) = -x^2 + 2x$ **60.** $f(x) = -x^2 - 2x$

61. $f(x) = 2x^2 - 8x + 13$ **62.** $f(x) = 3x^2 + 18x + 23$

63. $f(x) = -3x^2 - 3x - 2$ **64.** $f(x) = -2x^2 + 6x + 1$

65. $f(x) = 4x^2 + 8x + 4$ **66.** $f(x) = -5x^2 - 20x - 20$

67. Find a general formula for the x-intercepts of $f(x) = a(x - h)^2 + k$ by replacing $f(x)$ with 0 and solving for x.

68. Find a general formula for the y-intercept of $f(x) = a(x - h)^2 + k$ by evaluating $f(0)$.

Applying the Concepts

69. Use $h(t) = 32t - 16t^2$ to complete the table. Then sketch the graph of the function by plotting points.

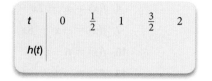

t	0	$\frac{1}{2}$	1	$\frac{3}{2}$	2
$h(t)$					

70. Use $h(t) = 48t - 16t^2$ to complete the table. Then sketch the graph of the function by plotting points.

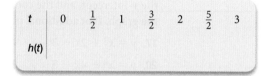

t	0	$\frac{1}{2}$	1	$\frac{3}{2}$	2	$\frac{5}{2}$	3
$h(t)$							

71. Softball Toss Chaudra is tossing a softball into the air with an underhand motion. The function $h(t) = -16t^2 + 64t$ gives the height of the softball t seconds after it is thrown, where $h(t)$ is measured in feet.

a. Use this function to complete the table.

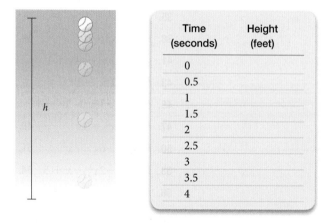

Time (seconds)	Height (feet)
0	
0.5	
1	
1.5	
2	
2.5	
3	
3.5	
4	

b. Sketch the graph of the functions by plotting points.

72. Model Rocket A small rocket is projected straight up into the air with a velocity of 128 feet per second. The function $h(t) = -16t^2 + 128t$ gives the height of the rocket t seconds after it is launched, where $h(t)$ is measured in feet.

a. Use this function to complete the table.

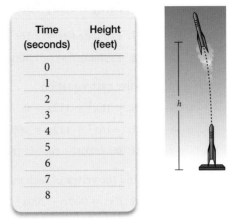

Time (seconds)	Height (feet)
0	
1	
2	
3	
4	
5	
6	
7	
8	

b. Sketch the graph of the function by plotting points.

Learning Objectives Assessment

The following problems can be used to help assess if you have successfully met the learning objectives for this section.

73. Find the vertex for $f(x) = 2x^2 + 4x^2 - 1$.

   a. $(0, -1)$ **b.** $(1, 4)$ **c.** $(-1, -3)$ **d.** $(-2, -1)$

74. Find the y-intercept of $f(x) = -(x + 3)^2 + 4$.

   a. -3 **b.** 4   **c.** -5 **d.** 3

75. Which graph could be the correct graph of $f(x) = -x^2 + 4x - 5$?

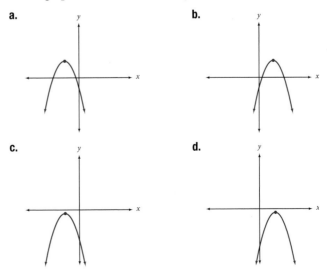

a.   **b.**

c.   **d.**

76. Convert $f(x) = 2x^2 + 20x + 38$ into vertex form by completing the square.

    a. $f(x) = 2(x + 10)^2 - 68$ **b.** $f(x) = 2(x - 10)^2 + 68$

    c. $f(x) = 2(x - 5)^2 + 12$ **d.** $f(x) = 2(x + 5)^2 - 12$

Getting Ready for the Next Section

77. Evaluate $f(t) = -16t^2 + 112t$ for $t = 1$ and $t = 3$.

78. Evaluate $f(p) = 1{,}300p - 100p^2$ for $p = 2$ and $p = 5$.

Solve each equation.

79. $-16x^2 + 112x = 0$ **80.** $1{,}300p - 100p^2 = 0$

81. $-16x^2 + 112x = 160$ **82.** $1{,}300p - 100p^2 = 3{,}600$

83. $96 + 80t - 16t^2 = 160$ **84.** $32 + 48t - 16t^2 = 52$

Find the vertex of each parabola.

85. $f(x) = 900x - 300x^2$ **86.** $f(x) = -16x^2 + 96x$

Applications of Quadratic Functions

Learning Objectives

In this section, we will learn how to:

1. Solve applied problems involving projectile motion.
2. Solve applied problems involving profit and revenue.
3. Solve optimization problems.
4. Find the equation of a quadratic function from the graph.

Introduction

An air-powered Stomp Rocket can be propelled over 200 feet using a blast of air. The harder you stomp on the Launch Pad, the farther the rocket flies.

If the rocket is launched straight up into the air with a velocity of 112 feet per second, then the quadratic function

$$h(t) = -16t^2 + 112t$$

gives the height h of the rocket t seconds after it is launched. We can use this formula to find the height of the rocket 3.5 seconds after launch by substituting $t = 3.5$:

$$h(t) = -16(3.5)^2 + 112(3.5) = 196$$

At 3.5 seconds, the rocket reaches a height of 196 feet.

Quadratic functions are good models for many types of real-life situations. We will see how quadratic functions can be used to solve a number of applications in this section.

Projectile Motion

The stomp rocket we just described is an example of **projectile motion**, which refers to the motion of an object that is somehow projected into the air. If gravity is the only force acting on the object, then the following formula can be used to model its height over time.

[△≠∑ PROPERTY *projectile motion formula*

If an object is projected vertically into the air with an initial speed v, in feet per second, from an initial height s, in feet, then the height of the object after t seconds is given by the quadratic function

$$h(t) = -16t^2 + vt + s$$

where h is measured in feet.

VIDEO EXAMPLES

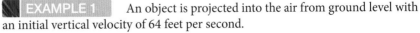

SECTION 8.7

EXAMPLE 1   An object is projected into the air from ground level with an initial vertical velocity of 64 feet per second.

a. Find the height of the object after 1.5 seconds.

b. At what times will the object be at a height of 48 feet?

SOLUTION   Substituting $v = 64$ and $s = 0$ into the projectile motion formula, we have

$$h(t) = -16t^2 + 64t$$

which is a quadratic function.

631

a. Substituting $t = 1.5$, we have

$$h(1.5) = -16(1.5)^2 + 64(1.5)$$
$$= 60$$

The object will be 60 feet high after 1.5 seconds.

b. We let $h(t) = 48$ and solve for t:

When	$h(t) = 48$
the function	$h(t) = -16t^2 + 64t$
becomes	$48 = -16t^2 + 64t$

We write it in standard form and solve by factoring:

$$16t^2 - 64t + 48 = 0$$
$$t^2 - 4t + 3 = 0 \qquad \text{Divide each side by 16.}$$
$$(t - 1)(t - 3) = 0 \qquad \text{Factor.}$$
$$t - 1 = 0 \quad \text{or} \quad t - 3 = 0$$
$$t = 1 \quad \text{or} \qquad t = 3$$

The object will be 48 feet above the ground after 1 second and again after 3 seconds. That is, it passes 48 feet going up and also coming back down.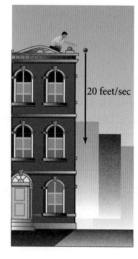

20 feet/sec

EXAMPLE 2   If an object is thrown downward with an initial velocity of 20 feet per second, the distance $s(t)$, in feet, it travels in t seconds is given by the function $s(t) = 20t + 16t^2$. How long does it take the object to fall 40 feet?

SOLUTION   We let $s(t) = 40$, and solve for t:

When	$s(t) = 40$
the function	$s(t) = 20t + 16t^2$
becomes	$40 = 20t + 16t^2$
or	$16t^2 + 20t - 40 = 0$
	$4t^2 + 5t - 10 = 0 \qquad \text{Divide by 4}$

Using the quadratic formula, we have

$$t = \frac{-5 \pm \sqrt{25 - 4(4)(-10)}}{2(4)}$$

$$= \frac{-5 \pm \sqrt{185}}{8}$$

$$= \frac{-5 + \sqrt{185}}{8} \quad \text{or} \quad t = \frac{-5 - \sqrt{185}}{8}$$

The second solution is impossible because it is a negative number and time t must be positive. It takes

$$t = \frac{-5 + \sqrt{185}}{8} \quad \text{or approximately} \quad \frac{-5 + 13.60}{8} \approx 1.08 \text{ seconds}$$

for the object to fall 40 feet.

USING TECHNOLOGY *Graphing Calculators*

More About Example 2

We can solve the problem discussed in Example 2 by graphing the function $Y_1 = 20X + 16X^2$ in a window with X from 0 to 2 (because X is taking the place of t and we know t is a positive quantity) and Y from 0 to 50 (because we are looking for X when Y_1 is 40). Graphing Y_1 gives a graph similar to the graph in Figure 1. Using the Zoom and Trace features at $Y_1 = 40$ gives us X = 1.08 to the nearest hundredth, matching the results we obtained by solving the original equation algebraically.

FIGURE 1

EXAMPLE 3 A water balloon is thrown upward with an initial velocity of 40 feet per second from a balcony that is 12 feet off the ground. When will the water balloon strike the ground?

SOLUTION Substituting $v = 40$ and $s = 12$ into the projectile motion formula, we have

$$h(t) = -16t^2 + 40t + 12$$

When the water balloon strikes the ground, its height will be zero. We substitute $h(t) = 0$ and solve for t.

When	$h(t) = 0$	
the function	$h(t) = -16t^2 + 40t + 12$	
becomes	$0 = -16t^2 + 40t + 12$	
	$0 = 4t^2 - 10t - 3$	Divide both sides by -4

The right side does not factor, so we can use the quadratic formula to solve for t.

$$t = \frac{-(-10) \pm \sqrt{(-10)^2 - 4(4)(-3)}}{2(4)}$$

$$= \frac{10 \pm \sqrt{148}}{8}$$

$$= \frac{10 \pm 2\sqrt{37}}{8} \qquad \sqrt{148} = \sqrt{4 \cdot 37} = 2\sqrt{37}$$

$$= \frac{5 \pm \sqrt{37}}{4} \qquad \text{Reduce}$$

Using a calculator to approximate both values, we obtain $t \approx -0.27$ and $t \approx 2.77$.

The negative value of t is not possible because the time must be positive. Thus, the water balloon will strike the ground after

$$t = \frac{5 + \sqrt{37}}{4} \approx 2.77 \text{ seconds}$$

Revenue, Cost, and Profit

In Chapter 5, we introduced the concepts of profit, revenue, and cost. If you recall, the relationship between the three quantities is known as the profit equation:

$$\text{Profit} = \text{Revenue} - \text{Cost}$$

$$P(x) = R(x) - C(x)$$

The revenue obtained from selling x items is the product of the number of items sold and the price per item, p. That is,

$$\text{Revenue} = (\text{Number of items sold})(\text{Price of each item})$$

$$R = xp$$

■ EXAMPLE 4   A manufacturer of small calculators knows that the number of calculators she can sell each week is related to the price of the calculators by the equation $x = 1,300 - 100p$, where x is the number of calculators and p is the price per calculator. What price should she charge for each calculator if she wants the weekly revenue to be $4,000?

SOLUTION   The formula for total revenue is $R = xp$. Because we want R in terms of p, we substitute $1,300 - 100p$ for x in the equation $R = xp$:

$$\text{If} \qquad R = xp$$
$$\text{and} \qquad x = 1,300 - 100p$$
$$\text{then} \qquad R = (1,300 - 100p)p$$

We want to find p when R is 4,000. Substituting for R in the formula gives us

$$4,000 = (1,300 - 100p)p$$
$$4,000 = 1,300p - 100p^2$$

This is a quadratic equation. To write it in standard form, we add $100p^2$ and $-1,300p$ to each side, giving us

$$100p^2 - 1,300p + 4,000 = 0$$
$$p^2 - 13p + 40 = 0 \qquad \text{Divide each side by 100}$$
$$(p - 5)(p - 8) = 0$$
$$p - 5 = 0 \quad \text{or} \quad p - 8 = 0$$
$$p = 5 \quad \text{or} \qquad p = 8$$

If she sells the calculators for $5 each or for $8 each, she will have a weekly revenue of $4,000.   ■

■ EXAMPLE 5   A company produces and sells copies of an accounting program for home computers. The total weekly cost (in dollars) to produce x copies of the program is $C(x) = 8x + 500$, and the weekly revenue for selling all x copies of the program is $R(x) = 35x - 0.1x^2$. How many programs must be sold each week for the weekly profit to be $1,200?

SOLUTION Substituting the given expressions for $R(x)$ and $C(x)$ in the equation $P(x) = R(x) - C(x)$, we have a quadratic function that represents the weekly profit $P(x)$:

$$P(x) = R(x) - C(x)$$
$$= 35x - 0.1x^2 - (8x + 500)$$
$$= 35x - 0.1x^2 - 8x - 500$$
$$= -500 + 27x - 0.1x^2$$

Setting this expression equal to 1,200, we have a quadratic equation to solve that gives us the number of programs x that need to be sold each week to bring in a profit of $1,200:

$$1,200 = -500 + 27x - 0.1x^2$$

We can write this equation in standard form by adding the opposite of each term on the right side of the equation to both sides of the equation. Doing so produces the following equation:

$$0.1x^2 - 27x + 1,700 = 0$$

Applying the quadratic formula to this equation with $a = 0.1$, $b = -27$, and $c = 1,700$, we have

$$x = \frac{27 \pm \sqrt{(-27)^2 - 4(0.1)(1,700)}}{2(0.1)}$$
$$= \frac{27 \pm \sqrt{729 - 680}}{0.2}$$
$$= \frac{27 \pm \sqrt{49}}{0.2}$$
$$= \frac{27 \pm 7}{0.2}$$

Writing this last expression as two separate expressions, we have our two solutions:

$$x = \frac{27 + 7}{0.2} \quad \text{or} \quad x = \frac{27 - 7}{0.2}$$
$$= \frac{34}{0.2} \qquad\qquad = \frac{20}{0.2}$$
$$= 170 \qquad\qquad\quad = 100$$

The weekly profit will be $1,200 if the company produces and sells 100 programs or 170 programs. ◾

What is interesting about this last example is that it has rational solutions, meaning it could have been solved by factoring. But looking back at the equation, factoring does not seem like a reasonable method of solution because the coefficients are either very large or very small. So, there are times when using the quadratic formula is a faster method of solution, even though the equation you are solving is factorable.

More About Example 5

To visualize the functions in Example 5, we set up our calculator this way:

$$Y_1 = 35X - .1X^2 \qquad \text{Revenue function}$$

$$Y_2 = 8X + 500 \qquad \text{Cost function}$$

$$Y_3 = Y_1 - Y_2 \qquad \text{Profit function}$$

Window: X from 0 to 350, Y from 0 to 3,500

Graphing these functions produces graphs similar to the ones shown in Figure 2. The lowest graph is the graph of the profit function. Using the Zoom and Trace features on the lowest graph at $Y_3 = 1{,}200$ produces two corresponding values of X, 170 and 100, which match the results in Example 5.

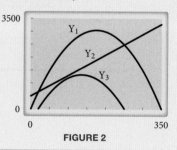

FIGURE 2

Optimization

The term **optimization** refers to the process of finding an optimal solution to some problem. In many cases, the optimal solution will be the maximum value or minimum value for some quantity. For example, a business would be interested in finding the maximum profit or minimum cost.

If the quantity in question is being modeled by a quadratic function, then the maximum or minimum value of the function (the optimal value) will correspond to the vertex.

EXAMPLE 6   For the water balloon described in Example 3, find the maximum height reached by the water balloon, and determine when it reaches this height.

SOLUTION   From Example 3, the function giving the height of the water balloon after t seconds is

$$h(t) = -16t^2 + 40t + 12$$

The graph of $h(t)$ is a parabola that opens downward, so the highest point on the graph is the vertex. Therefore, the vertex represents the point where the water balloon is at its maximum height.

Because the quadratic function is in standard form, we can use the vertex formula to find the coordinates of the vertex.

$$t = -\frac{b}{2a} = -\frac{40}{2(-16)} = \frac{40}{32} = 1.25$$

Now we evaluate the function at 1.25 to find the second coordinate.

$$h(1.25) = -16(1.25)^2 + 40(1.25) + 12$$

$$= 37$$

The vertex is the point (1.25, 37). This means the water balloon will reach a maximum height of 37 feet after 1.25 seconds.

EXAMPLE 7   A company selling copies of an accounting program for home computers finds that it will make a weekly profit of P dollars from selling x copies of the program, according to the equation

$$P(x) = -0.1x^2 + 27x - 500$$

How many copies of the program should it sell to make the largest possible profit, and what is the largest possible profit?

SOLUTION   Because the coefficient of x^2 is negative, we know the graph of this parabola will be concave down, meaning that the vertex is the highest point of the curve. We find the vertex by first finding its x-coordinate:

$$x = -\frac{b}{2a} = -\frac{27}{2(-0.1)} = \frac{27}{0.2} = 135$$

This represents the number of programs the company needs to sell each week to make a maximum profit. To find the maximum profit, we substitute 135 for x in the original equation. (A calculator is helpful for these kinds of calculations.)

$$P(135) = -0.1(135)^2 + 27(135) - 500$$

$$= -0.1(18,225) + 3,645 - 500$$

$$= -1,822.5 + 3,645 - 500$$

$$= 1,322.5$$

The maximum weekly profit is $1,322.50 and is obtained by selling 135 programs a week.

EXAMPLE 8   An art supply store finds that they can sell x sketch pads each week at p dollars each, according to the equation $x = 900 - 300p$. Graph the revenue equation $R = xp$. Then use the graph to find the price p that will bring in the maximum revenue. Finally, find the maximum revenue.

SOLUTION   As it stands, the revenue equation contains three variables. Because we are asked to find the value of p that gives us the maximum value of R, we rewrite the equation using just the variables R and p. Because $x = 900 - 300p$, we have

$$R = xp = (900 - 300p)p$$

The graph of this equation is shown in Figure 3. The graph appears in the first quadrant only, because R and p are both positive quantities.

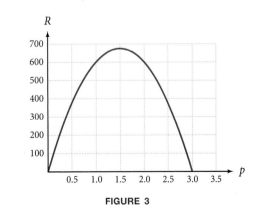

FIGURE 3

From the graph, we see that the maximum value of R occurs when $p = \$1.50$. We can calculate the maximum value of R from the equation:

When $\qquad\qquad\qquad\qquad\qquad p = 1.5$

the equation $\qquad\qquad\qquad\qquad R = (900 - 300p)p$

becomes $\qquad\qquad\qquad\qquad R = (900 - 300 \cdot 1.5)1.5$

$$= (900 - 450)1.5$$

$$= 450 \cdot 1.5$$

$$= 675$$

The maximum revenue is $\$675$. It is obtained by setting the price of each sketch pad at $p = \$1.50$.

Finding the Equation from the Graph

Before we conclude this section, we will look at a couple of examples of *curve-fitting*, which refers to finding the equation of a graph that meets certain criteria. In our case, we will find the equation of a quadratic function (parabola) given some information about the graph.

EXAMPLE 9   At a past Washington County Fair in Oregon, David Smith, Jr., The Bullet, was shot from a cannon. As a human cannonball, he reached a height of 70 feet before landing in a net 160 feet from the cannon. Sketch the graph of his path, and then find the equation of the graph.

SOLUTION   We assume that the path taken by the human cannonball is a parabola. If the origin of the coordinate system is at the opening of the cannon, then the net that catches him will be at 160 on the x-axis. Figure 4 shows the graph.

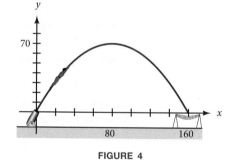

FIGURE 4

Because the curve is a parabola, we know the equation will have the form

$$y = a(x - h)^2 + k$$

Because the vertex of the parabola is at $(80, 70)$, we can fill in two of the three constants in our equation, giving us

$$y = a(x - 80)^2 + 70$$

To find a, we note that the landing point will be $(160, 0)$. Substituting the coordinates of this point into the equation, we solve for a:

$$0 = a(160 - 80)^2 + 70$$

$$0 = a(80)^2 + 70$$

$$0 = 6{,}400a + 70$$

$$a = -\frac{70}{6{,}400} = -\frac{7}{640}$$

The equation that describes the path of the human cannonball is

$$y = -\frac{7}{640}(x - 80)^2 + 70 \quad \text{for} \quad 0 \leq x \leq 160$$

USING TECHNOLOGY   *Graphing Calculators*

Graph the equation found in Example 8 on a graphing calculator using the window shown here.

$$\text{Window: } \text{X from 0 to 180, increment 20}$$
$$\text{Y from 0 to 80, increment 10}$$

On the TI-84, an increment of 20 for X means Xscl $=20$.

EXAMPLE 10   Find the equation of a quadratic function in standard form whose graph passes through the points $(-1, 4)$, $(1, -2)$, and $(2, 1)$.

SOLUTION   The standard form for a quadratic function is $y = ax^2 + bx + c$. The three given points will lie on the graph of the function if all three ordered pairs satisfy this equation.

By substituting each pair of values of x and y, we obtain the following:

Using $\qquad\qquad\quad x = -1$ and $y = 4$

we have $\qquad\qquad 4 = a(-1)^2 + b(-1) + c$

$$\qquad\qquad\qquad 4 = a - b + c$$

Using $\qquad\qquad\quad x = 1$ and $y = -2$

we have $\qquad\quad -2 = a(1)^2 + b(1) + c$

$$\qquad\qquad\quad -2 = a + b + c$$

Using $\qquad\qquad\quad x = 2$ and $y = 1$

we have $\qquad\qquad 1 = a(2)^2 + b(2) + c$

$$\qquad\qquad\qquad 1 = 4a + 2b + c$$

We must find values of a, b, and c that make all three of these equations true. In other words, we must solve the system of equations

$$a - b + c = \ \ 4 \qquad (1)$$

$$a + b + c = -2 \qquad (2)$$

$$4a + 2b + c = \ \ 1 \qquad (3)$$

Multiplying equation (1) by -1 and adding the result to equation (2) gives us

$$2b = -6$$
$$b = -3$$

If we multiply equation (2) by -1 and add the result to equation (3), we have

$$3a + b = 3$$

Substituting $b = -3$ into this last equation and solving for a gives us $a = 2$. Using $a = 2$ and $b = -3$ in equation (1), (2), or (3) and solving for c results in $c = -1$.

Therefore, the equation of the quadratic function whose graph passes through the three points is

$$y = 2x^2 - 3x - 1$$

Figure 5 shows the three given points and the graph of this function. As you can see, the parabola passes through all of the points.

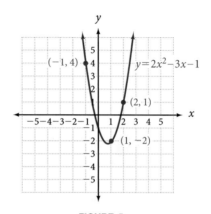

FIGURE 5

Getting Ready for Class

After reading through the preceding section, respond in your own words and in complete sentences.

A. What does projectile motion refer to?

B. State the formulas for revenue and profit.

C. Explain what optimization is.

D. What will be the optimal value for a quadratic function?

1. Find R if $p = 1.5$ and $R = (900 - 300p)p$

2. Find R if $p = 2.5$ and $R = (900 - 300p)p$

3. Find P if $P = -0.1x^2 + 27x + 1,700$ and
   a. $x = 100$ b. $x = 170$

4. Find P if $P = -0.1x^2 + 27x + 1,820$ and
   a. $x = 130$ b. $x = 140$

5. Find h if $h = 16 + 32t - 16t^2$ and
   a. $t = \dfrac{1}{4}$ b. $t = \dfrac{7}{4}$

6. Find h if $h = 64t - 16t^2$ and
   a. $t = 1$ b. $t = 3$

7. **Saint Louis Arch** The shape of the famous "Gateway to the West" arch in Saint Louis can be modeled by a parabola. The equation for one such parabola is:

$$y = -\frac{1}{150}x^2 + \frac{21}{5}x$$

   a. Sketch the graph of the arch's equation on a coordinate axis.

   b. Approximately how far do you have to walk to get from one side of the arch to the other?

8. **Interpreting Graphs** The graph below shows the different paths taken by the human cannonball when his velocity out of the cannon is 50 miles/hour, and his cannon is inclined at varying angles.

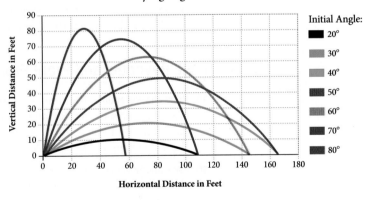

Initial Velocity: 50 miles per hour
Angle: 20°, 30°, 40°, 50°, 60°, 70°, 80°

   a. If his landing net is placed 104 feet from the cannon, at what angle should the cannon be inclined so that he lands in the net?

   b. Approximately where do you think he would land if the cannon was inclined at 45°?

   c. If the cannon was inclined at 45°, approximately what height do you think he would attain?

   d. Do you think there is another angle for which he would travel the same distance he travels at 80°? Give an estimate of that angle.

   e. The fact that every landing point can come from two different paths makes us think that the equations that give us the landing points must be what type of equations?

Projectile Motion Problems

9. **Height of a Bullet**   A bullet is fired into the air with an initial upward velocity of 80 feet per second from the top of a building 96 feet high. The function that gives the height of the bullet at any time t is $h(t) = 96 + 80t - 16t^2$. At what times will the bullet be 192 feet in the air?

10. **Height of an Arrow**   An arrow is shot into the air with an upward velocity of 48 feet per second from a hill 32 feet high. The function that gives the height of the arrow at any time t is $h(t) = 32 + 48t - 16t^2$. Find the times at which the arrow will be 64 feet above the ground.

11. **Velocity and Height** If an object is thrown straight up into the air with an initial velocity of 32 feet per second, then its height h (in feet) above the ground at any time t (in seconds) is given by the function $h(t) = 32t - 16t^2$. Find the times at which the object is on the ground by letting $h(t) = 0$ in the equation and solving for t.

12. **Falling Object** An object is thrown downward with an initial velocity of 5 feet per second. The relationship between the distance s it travels and time t is given by $s = 5t + 16t^2$. How long does it take the object to fall 74 feet?

13. **Coin Toss** A coin is tossed upward with an initial velocity of 32 feet per second from a height of 16 feet above the ground. The function giving the object's height h at any time t is $h(t) = 16 + 32t - 16t^2$. Does the object ever reach a height of 32 feet?

14. **Slingshot** A pebble is launched straight upward using a slingshot. If the pebble is released with an initial velocity of 120 feet per second from a height of 6 feet, how long will it take for the pebble to strike the ground?

Cost, Revenue, and Profit Problems

15. **Price and Revenue**   The relationship between the number of calculators x a company sells per day and the price of each calculator p is given by the equation $x = 1{,}700 - 100p$. At what price should the calculators be sold if the daily revenue is to be $7,000?

16. **Price and Revenue**   The relationship between the number of pencil sharpeners x a company can sell each week and the price of each sharpener p is given by the equation $x = 1{,}800 - 100p$. At what price should the sharpeners be sold if the weekly revenue is to be $7,200?

17. **Revenue** A company manufactures and sells DVDs. The revenue obtained by selling x DVDs is given by the function

$$R(x) = 11.5x - 0.05x^2$$

Find the number of DVDs they must sell to receive $650 in revenue.

18. **Revenue** A software company sells licenses to its office management suite. The revenue obtained by selling x licenses per month is given by the function

$$R(x) = 48x - 0.06x^2$$

Find the number of licenses they must sell each month to receive $7,200 in revenue.

19. **Profit** The total cost (in dollars) for a company to manufacture and sell x items per week is $C = 60x + 300$, whereas the revenue brought in by selling all x items is $R = 100x - 0.5x^2$. How many items must be sold to obtain a weekly profit of $300?

20. **Profit** Suppose a company manufactures and sells x picture frames each month with a total cost of $C = 1{,}200 + 3.5x$ dollars. If the revenue obtained by selling x frames is $R = 9x - 0.002x^2$, find the number of frames it must sell each month if its monthly profit is to be $2,300.

Optimization Problems

21. **Maximum Profit** A company finds that it can make a profit of P dollars each month by selling x patterns, according to the formula $P(x) = -0.002x^2 + 3.5x - 800$. How many patterns must it sell each month to have a maximum profit? What is the maximum profit?

22. **Maximum Profit** A company selling picture frames finds that it can make a profit of P dollars each month by selling x frames, according to the formula $P(x) = -0.002x^2 + 5.5x - 1{,}200$. How many frames must it sell each month to have a maximum profit? What is the maximum profit?

23. **Maximum Height** Chaudra is tossing a softball into the air with an underhand motion. The distance of the ball above her hand at any time is given by the function

$$h(t) = 32t - 16t^2 \quad \text{for} \quad 0 \le t \le 2$$

where $h(t)$ is the height of the ball (in feet) and t is the time (in seconds). Find the times at which the ball is in her hand, and the maximum height of the ball.

24. **Maximum Area** Justin wants to fence three sides of a rectangular exercise yard for his dog. The fourth side of the exercise yard will be a side of the house. He has 80 feet of fencing available. Find the dimensions of the exercise yard that will enclose the maximum area.

25. **Maximum Revenue** A company that manufactures typewriter ribbons knows that the number of ribbons x it can sell each week is related to the price p of each ribbon by the equation $x = 1{,}200 - 100p$. Graph the revenue equation $R = xp$. Then use the graph to find the price p that will bring in the maximum revenue. Finally, find the maximum revenue.

26. **Maximum Revenue** A company that manufactures diskettes for home computers finds that it can sell x diskettes each day at p dollars per diskette, according to the equation $x = 800 - 100p$. Graph the revenue equation $R = xp$. Then use the graph to find the price p that will bring in the maximum revenue. Finally, find the maximum revenue.

27. **Maximum Revenue** The relationship between the number of calculators x a company sells each day and the price p of each calculator is given by the equation $x = 1{,}700 - 100p$. Graph the revenue equation $R = xp$, and use the graph to find the price p that will bring in the maximum revenue. Then find the maximum revenue.

28. Maximum Revenue The relationship between the number x of pencil sharpeners a company sells each week and the price p of each sharpener is given by the equation $x = 1{,}800 - 100p$. Graph the revenue equation $R = xp$, and use the graph to find the price p that will bring in the maximum revenue. Then find the maximum revenue.

29. Union Dues A labor union has 10,000 members. For every $10 increase in union dues, membership is decreased by 200 people. If the current dues are $100, what should be the new dues (to the nearest multiple of $10) so income from dues is greatest, and what is that income? *Hint:* Because Income = (membership)(dues), we can let x = the number of $10 increases in dues, and then this will give us income of $y = (10{,}000 - 200x)(100 + 10x)$.

30. Bookstore Receipts The owner of a used book store charges $2 for quality paperbacks and usually sells 40 per day. For every 10-cent increase in the price of these paperbacks, he thinks that he will sell two fewer per day. What is the price he should charge (to the nearest 10 cents) for these books to maximize his income, and what would be that income? *Hint:* Let x = the number of 10-cent increases in price.

31. Jiffy-Lube The owner of a quick oil-change business charges $20 per oil change and has 40 customers per day. If each increase of $2 results in 2 fewer daily customers, what price should the owner charge (to the nearest $2) for an oil change if the income from this business is to be as great as possible?

32. Computer Sales A computer manufacturer charges $2,200 for its basic model and sells 1,500 computers per month at this price. For every $200 increase in price, it is believed that 75 fewer computers will be sold. What price should the company place on its basic model of computer (to the nearest $100) to have the greatest income?

Curve Fitting Problems

33. Human Cannonball A human cannonball is shot from a cannon at the county fair. He reaches a height of 60 feet before landing in a net 180 feet from the cannon. Sketch the graph of his path, and then find the equation of the graph.

34. Human Cannonball A human cannonball is shot from a cannon at the state fair. He reaches a height of 80 feet before landing in a net 120 feet from the cannon. Sketch the graph of his path, and then find the equation of the graph.

35. Gateway Arch The Gateway Arch in St. Louis, Missouri, has a shape that can be approximated using a parabola. The height and width of the arch are both 630 feet. Find the equation for this parabola.

36. Stone Bridge The Konitsa Bridge in Epirus, Greece, contains an arch whose shape can be approximated using a parabola. If the arch is 40 meters long and 20 meters high, find the equation for this parabola.

37. Parabola Find the equation of the quadratic function, in standard form, passing through the points $(-4, -5)$, $(-1, -8)$, and $(2, 7)$.

38. Parabola Find the equation of the quadratic function, in standard form, passing through the points $(-2, -1)$, $(-1, 2)$, and $(1, -4)$.

iStockPhoto.com/
© PanosKarapanagiotis

Learning Objectives Assessment

The following problems can be used to help assess if you have successfully met the learning objectives for this section.

39. Baseball A baseball is thrown upward with an initial velocity of 44 feet per second and released from a height of 6 feet. When will the baseball strike the ground?

    a. 2.5 sec   **b.** 2.9 sec   **c.** 3.1 sec   **d.** 2.7 sec

40. Revenue The relationship between the number of coffee makers x a company sells each week and the price of each coffee maker p is given by the equation $x = 800 - 40p$. At what price should the coffee makers be sold if the weekly revenue is to be $3,000?

    a. $4, $12   **b.** $9, $15   **c.** $10   **d.** $5, $15

41. Maximum Height Find the maximum height of the baseball described in Problem 39.

    a. 36.25 ft   **b.** 30.25 ft   **c.** 41.5 ft   **d.** 44 ft

42. Human Cannonball A human cannonball is shot from a cannon at a county fair. He reaches a height of 50 feet before landing in a net 110 feet from the cannon. Find the equation of his path.

    a. $y = -\dfrac{4}{121}(x + 55)^2 + 50$ **b.** $y = -\dfrac{2}{121}(x - 55)^2 + 50$

    c. $y = -\dfrac{5}{11}(x - 110)^2 + 50$ **d.** $y = -\dfrac{4}{11}(x + 110)^2 + 50$

Getting Ready for the Next Section

Solve.

43. $x^2 - 2x - 8 = 0$ **44.** $x^2 - x - 12 = 0$

45. $6x^2 - x = 2$ **46.** $3x^2 - 5x = 2$

47. $x^2 - 6x + 9 = 0$ **48.** $x^2 + 8x + 16 = 0$

Learning Objectives

In this section, we will learn how to:

1. Solve a quadratic inequality.

2. Solve a rational inequality.

Quadratic Inequalities

Quadratic inequalities in one variable are inequalities of the form

$$ax^2 + bx + c < 0 \qquad ax^2 + bx + c > 0$$
$$ax^2 + bx + c \leq 0 \qquad ax^2 + bx + c \geq 0$$

where a, b, and c are constants, with $a \neq 0$. The technique we will use to solve inequalities of this type involves graphing. Suppose, for example, we want to find the solution set for the inequality $x^2 - x - 6 > 0$. We begin by factoring the left side to obtain

$$(x - 3)(x + 2) > 0$$

We have two real numbers $x - 3$ and $x + 2$ whose product $(x - 3)(x + 2)$ is greater than zero. That is, their product is positive. The only way the product can be positive is either if both factors, $(x - 3)$ and $(x + 2)$, are positive or if they are both negative. To help visualize where $x - 3$ is positive and where it is negative, we draw a real number line and label it accordingly:

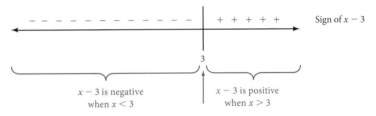

Here is a similar diagram showing where the factor $x + 2$ is positive and where it is negative:

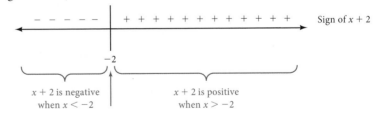

Drawing the two number lines together and eliminating the unnecessary numbers, we have

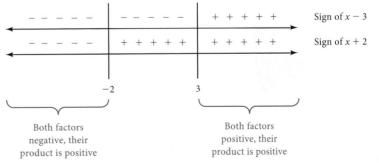

We can see from the preceding diagram that the graph of the solution to $x^2 - x - 6 > 0$ is

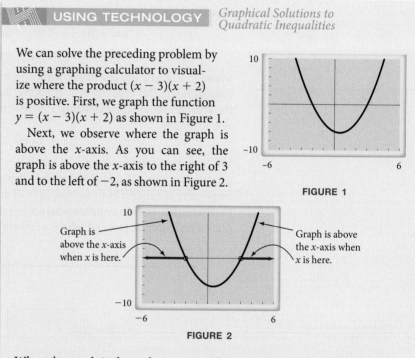

$$x < -2 \quad \text{or} \quad x > 3$$

In set-builder notation, we can write the solution set as $\{x \mid x < -2 \text{ or } x > 3\}$. Using interval notation, it would be $(-\infty, -2) \cup (3, \infty)$.

USING TECHNOLOGY   *Graphical Solutions to Quadratic Inequalities*

We can solve the preceding problem by using a graphing calculator to visualize where the product $(x - 3)(x + 2)$ is positive. First, we graph the function $y = (x - 3)(x + 2)$ as shown in Figure 1.

Next, we observe where the graph is above the x-axis. As you can see, the graph is above the x-axis to the right of 3 and to the left of -2, as shown in Figure 2.

FIGURE 1

Graph is above the x-axis when x is here.

Graph is above the x-axis when x is here.

FIGURE 2

When the graph is above the x-axis, we have points whose y-coordinates are positive. Because these y-coordinates are the same as the expression $(x - 3)(x + 2)$, the values of x for which the graph of $y = (x - 3)(x + 2)$ is above the x-axis are the values of x for which the inequality $(x - 3)(x + 2) > 0$ is true. Our solution set is therefore

$$x < -2 \quad \text{or} \quad x > 3$$

VIDEO EXAMPLES

SECTION 8.8

EXAMPLE 1   Solve: $x^2 - 2x - 8 \leq 0$.

ALGEBRAIC SOLUTION   We begin by factoring:

$$x^2 - 2x - 8 \leq 0$$
$$(x - 4)(x + 2) \leq 0$$

The product $(x - 4)(x + 2)$ is negative or zero. Either one of the factors must equal 0, or the factors must have opposite signs. The product will equal zero if

$$x - 4 = 0 \qquad \text{or} \qquad x + 2 = 0$$
$$x = 4 \qquad\qquad\qquad x = -2$$

We draw a diagram showing where each factor is positive and where each factor is negative:

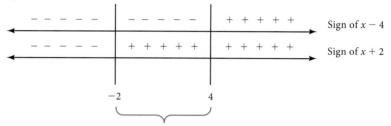

The product will equal zero if $x = -2$ or $x = 4$. From the diagram, we can see that the factors will have opposite signs if x is between -2 and 4. Therefore, the graph of the solution set is

We can write the solution set using set-builder notation or interval notation as

$$\{x \mid -2 \le x \le 4\} \text{ or } [-2, 4]$$

GRAPHICAL SOLUTION To solve this inequality with a graphing calculator, we graph the function $y = (x - 4)(x + 2)$ and observe where the graph is below the x-axis. These points have negative y-coordinates, which means that the product $(x - 4)(x + 2)$ is negative for these points. Figure 3 shows the graph of $y = (x - 4)(x + 2)$, along with the region on the x-axis where the graph contains points with negative y-coordinates.

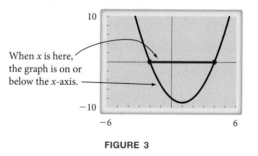

FIGURE 3

As you can see, the graph is below the x-axis when x is between -2 and 4. Because our original inequality includes the possibility that $(x - 4)(x + 2)$ is 0, we include the endpoints, -2 and 4, with our solution set.

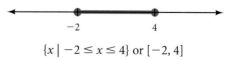

$$\{x \mid -2 \le x \le 4\} \text{ or } [-2, 4]$$

EXAMPLE 2 Solve: $6x^2 - x \geq 2$.

ALGEBRAIC SOLUTION

$$6x^2 - x \geq 2$$

$$6x^2 - x - 2 \geq 0 \qquad \text{Standard form}$$

$$(3x - 2)(2x + 1) \geq 0$$

The product is positive or zero, so the factors must equal zero or agree in sign. For the product to be zero, either

$$3x - 2 = 0 \qquad \text{or} \qquad 2x + 1 = 0$$

$$x = \frac{2}{3} \qquad\qquad x = -\frac{1}{2}$$

We draw a diagram to see where the factors will agree in sign.

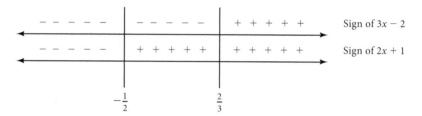

Because the factors agree in sign to the left of $-\frac{1}{2}$ and to the right of $\frac{2}{3}$, the graph of the solution set is

We can write the solution set as

$$\{x \mid x \leq -\frac{1}{2} \text{ or } x \geq \frac{2}{3}\} \quad \text{or} \quad \left(-\infty, -\frac{1}{2}\right] \cup \left[\frac{2}{3}, \infty\right)$$

GRAPHICAL SOLUTION To solve this inequality with a graphing calculator, we graph the function $y = (3x - 2)(2x + 1)$ and observe where the graph is above the x-axis. These are the points that have positive y-coordinates, which means that the product $(3x - 2)(2x + 1)$ is positive for these points. Figure 4 shows the graph of $y = (3x - 2)(2x + 1)$, along with the regions on the x-axis where the graph is on or above the x-axis.

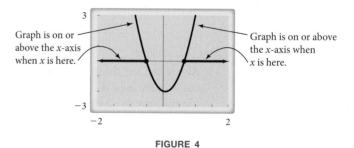

FIGURE 4

To find the points where the graph crosses the x-axis, we need to use either the Trace and Zoom features to zoom in on each point, or the calculator function that finds the intercepts automatically (on the TI-84 this is the root/zero function under the CALC key). Whichever method we use, we will obtain the following result:

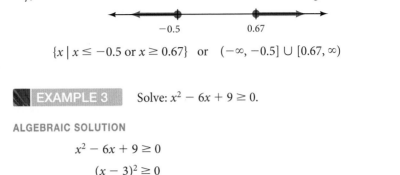

$$\{x \mid x \leq -0.5 \text{ or } x \geq 0.67\} \quad \text{or} \quad (-\infty, -0.5] \cup [0.67, \infty)$$

EXAMPLE 3 Solve: $x^2 - 6x + 9 \geq 0$.

ALGEBRAIC SOLUTION

$$x^2 - 6x + 9 \geq 0$$

$$(x - 3)^2 \geq 0$$

This is a special case in which both factors are the same. Because $(x - 3)^2$ is always positive or zero, the solution set is all real numbers. That is, any real number that is used in place of x in the original inequality will produce a true statement.

GRAPHICAL SOLUTION The graph of $y = (x - 3)^2$ is shown in Figure 5.

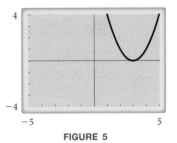

FIGURE 5

Notice that it touches the x-axis at 3 and is above the x-axis everywhere else. This means that every point on the graph has a y-coordinate greater than or equal to 0, no matter what the value of x. The conclusion that we draw from the graph is that the inequality $(x - 3)^2 \geq 0$ is true for all values of x.

Rational Inequalities

Our last two examples involve inequalities that contain rational expressions.

EXAMPLE 4 Solve: $\dfrac{x - 4}{x + 1} \leq 0$.

SOLUTION The inequality indicates that the quotient of $(x - 4)$ and $(x + 1)$ is negative or 0 (less than or equal to 0). We can use the same reasoning we used to solve the first three examples, because quotients are positive or negative under the same conditions that products are positive or negative. The quotient will equal zero if the numerator is zero, which is true if $x = 4$. If $x = -1$, the denominator would be zero, making the expression undefined.

Here is the diagram that shows where each factor is positive and where each factor is negative:

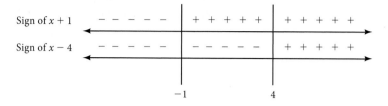

Sign of $x + 1$

Sign of $x - 4$

-1 4

Between -1 and 4, the quotient is negative because the factors have opposite signs. The solution set and its graph are shown here:

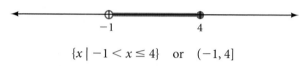

-1 4

$$\{x \mid -1 < x \leq 4\} \quad \text{or} \quad (-1, 4]$$

Notice that the left endpoint is open—that is, it is not included in the solution set—because $x = -1$ would make the denominator in the original inequality 0. It is important to check all endpoints of solution sets to inequalities that involve rational expressions.

EXAMPLE 5   Solve: $\dfrac{3}{x - 2} - \dfrac{2}{x - 3} > 0.$

SOLUTION   We begin by adding the two rational expressions on the left side. The common denominator is $(x - 2)(x - 3)$:

$$\frac{3}{x - 2} \cdot \frac{(x - 3)}{(x - 3)} - \frac{2}{x - 3} \cdot \frac{(x - 2)}{(x - 2)} > 0$$

$$\frac{3x - 9 - 2x + 4}{(x - 2)(x - 3)} > 0$$

$$\frac{x - 5}{(x - 2)(x - 3)} > 0$$

This time the quotient involves three factors. Here is the diagram that shows the signs of the three factors:

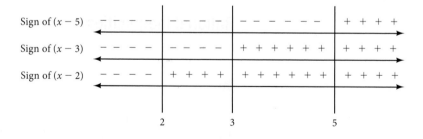

Sign of $(x - 5)$

Sign of $(x - 3)$

Sign of $(x - 2)$

2 3 5

The original inequality indicates that the quotient is positive. For this to happen, either all three factors must be positive, or exactly two factors must be negative. Looking back at the diagram, we see the regions that satisfy these conditions are between 2 and 3 or to the right of 5. Here is our solution set:

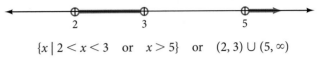

$$\{x \mid 2 < x < 3 \quad \text{or} \quad x > 5\} \quad \text{or} \quad (2, 3) \cup (5, \infty)$$

Getting Ready for Class

After reading through the preceding section, respond in your own words and in complete sentences.

A. What is the first step in solving a quadratic inequality?

B. How do you show that the endpoint of a line segment is not part of the graph of a quadratic inequality?

C. How would you use the graph of $y = ax^2 + bx + c$ to help you solve $ax^2 + bx + c < 0$?

D. How would the solution set to Example 5 be different if the inequality sign was $\geq$ instead of $>$?

Problem Set 8.8

Solve each of the following inequalities, and graph the solution set.

1. $x^2 + x - 6 > 0$ **2.** $x^2 + x - 6 < 0$ **3.** $x^2 - x - 12 \leq 0$

4. $x^2 - x - 12 \geq 0$ **5.** $x^2 + 5x \geq -6$ **6.** $x^2 - 5x > 6$

7. $6x^2 < 5x - 1$ **8.** $4x^2 \geq -5x + 6$ **9.** $x^2 - 9 < 0$

10. $x^2 - 16 \geq 0$ **11.** $4x^2 - 9 \geq 0$ **12.** $9x^2 - 4 < 0$

13. $2x^2 - x - 3 < 0$ **14.** $3x^2 + x - 10 \geq 0$ **15.** $x^2 - 4x + 4 \geq 0$

16. $x^2 - 4x + 4 < 0$ **17.** $x^2 - 10x + 25 < 0$ **18.** $x^2 - 10x + 25 > 0$

19. $(x - 2)(x - 3)(x - 4) > 0$ **20.** $(x - 2)(x - 3)(x - 4) < 0$

21. $(x + 1)(x + 2)(x + 3) \leq 0$ **22.** $(x + 1)(x + 2)(x + 3) \geq 0$

23. $\dfrac{x - 1}{x + 4} \leq 0$ **24.** $\dfrac{x + 4}{x - 1} \leq 0$

25. $\dfrac{3x}{x + 6} - \dfrac{8}{x + 6} < 0$ **26.** $\dfrac{5x}{x + 1} - \dfrac{3}{x + 1} < 0$

27. $\dfrac{4}{x - 6} + 1 > 0$ **28.** $\dfrac{2}{x - 3} + 1 \geq 0$

29. $\dfrac{x - 2}{(x + 3)(x - 4)} < 0$ **30.** $\dfrac{x - 1}{(x + 2)(x - 5)} < 0$

31. $\dfrac{2}{x - 4} - \dfrac{1}{x - 3} > 0$ **32.** $\dfrac{4}{x + 3} - \dfrac{3}{x + 2} > 0$

33. $\dfrac{x + 7}{2x + 12} + \dfrac{6}{x^2 - 36} \leq 0$ **34.** $\dfrac{x + 1}{2x - 2} - \dfrac{2}{x^2 - 1} \leq 0$

35. The graph of $y = x^2 - 4$ is shown in Figure 6. Use the graph to write the solution set for each of the following:

     a. $x^2 - 4 < 0$ **b.** $x^2 - 4 > 0$ **c.** $x^2 - 4 = 0$

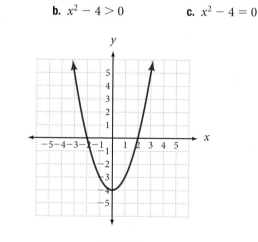

FIGURE 6

36. The graph of $y = 4 - x^2$ is shown in Figure 7. Use the graph to write the solution set for each of the following:

     a. $4 - x^2 < 0$ **b.** $4 - x^2 > 0$ **c.** $4 - x^2 = 0$

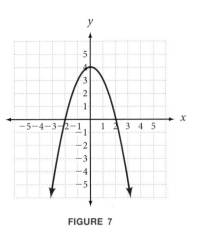

FIGURE 7

37. The graph of $y = x^2 - 3x - 10$ is shown in Figure 8. Use the graph to write the solution set for each of the following:

    a. $x^2 - 3x - 10 < 0$ **b.** $x^2 - 3x - 10 > 0$ **c.** $x^2 - 3x - 10 = 0$

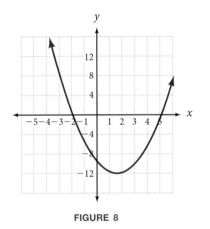

FIGURE 8

38. The graph of $y = x^2 + x - 12$ is shown in Figure 9. Use the graph to write the solution set for each of the following:

    a. $x^2 + x - 12 < 0$ **b.** $x^2 + x - 12 > 0$ **c.** $x^2 + x - 12 = 0$

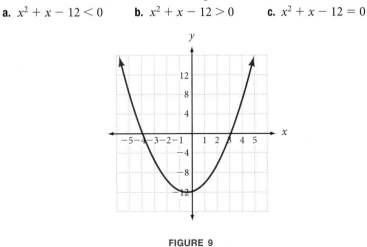

FIGURE 9

39. The graph of $y = x^3 - 3x^2 - x + 3$ is shown in Figure 10. Use the graph to write the solution set for each of the following:

a. $x^3 - 3x^2 - x + 3 < 0$ **b.** $x^3 - 3x^2 - x + 3 > 0$

c. $x^3 - 3x^2 - x + 3 = 0$

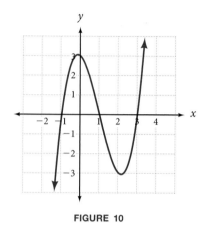

FIGURE 10

40. The graph of $y = x^3 + 4x^2 - 4x - 16$ is shown in Figure 11. Use the graph to write the solution set for each of the following:

a. $x^3 + 4x^2 - 4x - 16 < 0$ **b.** $x^3 + 4x^2 - 4x - 16 > 0$

c. $x^3 + 4x^2 - 4x - 16 = 0$

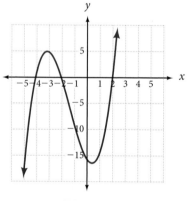

FIGURE 11

Applying the Concepts

41. Dimensions of a Rectangle The length of a rectangle is 3 inches more than twice the width. If the area is to be at least 44 square inches, what are the possibilities for the width?

42. Dimensions of a Rectangle The length of a rectangle is 5 inches less than three times the width. If the area is to be less than 12 square inches, what are the possibilities for the width?

43. **Revenue** A manufacturer of portable radios knows that the weekly revenue produced by selling x radios is given by the equation $R = 1{,}300p - 100p^2$, where p is the price of each radio (in dollars). What price should be charged for each radio if the weekly revenue is to be at least $4,000?

44. **Revenue** A manufacturer of small calculators knows that the weekly revenue produced by selling x calculators is given by the equation $R = 1{,}700p - 100p^2$, where p is the price of each calculator (in dollars). What price should be charged for each calculator if the revenue is to be at least $7,000 each week?

Learning Objectives Assessment

The following problems can be used to help assess if you have successfully met the learning objectives for this section.

45. Solve: $2x^2 - 3x - 20 < 0$.

   a. $x < -\dfrac{5}{2}$ or $x > 4$

   b. $x \le -\dfrac{5}{2}$ or $x \ge 4$

   c. $-\dfrac{5}{2} \le x \le 4$

   d. $-\dfrac{5}{2} < x < 4$

46. Solve: $\dfrac{x-2}{x+3} \ge 0$.

   a. $-3 \le x \le 2$

   b. $-3 < x \le 2$

   c. $x < -3$ or $x \ge 2$

   d. $x \le -3$ or $x \ge 2$

Maintaining Your Skills

Use a calculator to evaluate, give answers to 4 decimal places

47. $\dfrac{50{,}000}{32{,}000}$

48. $\dfrac{2.4362}{1.9758} - 1$

49. $\dfrac{1}{2}\left(\dfrac{4.5926}{1.3876} - 2\right)$

50. $1 + \dfrac{0.06}{12}$

Solve each equation

51. $2\sqrt{3t - 1} = 2$

52. $\sqrt{4t + 5} + 7 = 3$

53. $\sqrt{x + 3} = x - 3$

54. $\sqrt{x + 3} = \sqrt{x} - 3$

Graph each equation

55. $y = \sqrt{x - 1}$

56. $y = \sqrt[3]{x} - 1$

Chapter 8 Summary

The Square Root Property for Equations [8.1]

1. If $(x - 3)^2 = 25$
then $x - 3 = \pm 5$
$x = 3 \pm 5$
$x = 8$ or $x = -2$

If $a^2 = b$, where b is a real number, then
$$a = \sqrt{b} \qquad \text{or} \qquad a = -\sqrt{b}$$
which can be written as $a = \pm\sqrt{b}$.

To Solve a Quadratic Equation by Completing the Square [8.1]

2. Solve: $x^2 - 6x - 6 = 0$
$x^2 - 6x = 6$
$x^2 - 6x + 9 = 6 + 9$
$(x - 3)^2 = 15$
$x - 3 = \pm\sqrt{15}$
$x = 3 \pm \sqrt{15}$

Step 1: Write the equation in the form $ax^2 + bx = c$.
Step 2: If $a \neq 1$, divide through by the constant a so the coefficient of x^2 is 1.
Step 3: Complete the square on the left side by adding the square of $\frac{1}{2}$ the coefficient of x to both sides.
Step 4: Write the left side of the equation as the square of a binomial. Simplify the right side if possible.
Step 5: Apply the square root property for equations, and solve as usual.

The Quadratic Theorem [8.2]

3. If $2x^2 + 3x - 4 = 0$, then

$$x = \frac{-3 \pm \sqrt{9 - 4(2)(-4)}}{2(2)}$$

$$= \frac{-3 \pm \sqrt{41}}{4}$$

For any quadratic equation in the form $ax^2 + bx + c = 0$, $a \neq 0$, the two solutions are
$$x = \frac{-b \pm \sqrt{b^2 - 4ac}}{2a}$$
This last equation is known as the **quadratic formula**.

The Discriminant [8.3]

4. The discriminant for
$x^2 + 6x + 9 = 0$
is $D = 36 - 4(1)(9) = 0$, which means the equation has one rational solution.

The expression $b^2 - 4ac$ that appears under the radical sign in the quadratic formula is known as the **discriminant**.

We can classify the solutions to $ax^2 + bx + c = 0$ as follows:

The solutions are	When the discriminant is
Two complex numbers containing i	Negative
One rational number	Zero
Two rational numbers	A positive perfect square
Two irrational numbers	A positive number, but not a perfect square

Equations Quadratic in Form [8.4]

5. The equation $x^4 - x^2 - 12 = 0$ is quadratic in x^2. Letting $y = x^2$ we have
$$y^2 - y - 12 = 0$$
$$(y - 4)(y + 3) = 0$$
$$y = 4 \quad \text{or} \quad y = -3$$

Resubstituting x^2 for y, we have
$$x^2 = 4 \quad \text{or} \quad x^2 = -3$$
$$x = \pm 2 \quad \text{or} \quad x = \pm i\sqrt{3}$$

There are a variety of equations whose form is quadratic. We solve most of them by making a substitution so the equation becomes quadratic, and then solving the equation by factoring or the quadratic formula. For example,

The equation	*is quadratic in*
$(2x - 3)^2 + 5(2x - 3) - 6 = 0$	$2x - 3$
$4x^4 - 7x^2 - 2 = 0$	x^2
$2x - 7\sqrt{x} + 3 = 0$	$\sqrt{x}$

Transformations [8.5]

6. The graph of
$f(x) = -2(x + 1)^2 + 3$ is the graph of $y = x^2$, but with a vertical expansion by a factor of 2, an x-axis reflection, a horizontal translation of 1 unit left, and a vertical translation of 3 units upward.

The graph of $f(x) = ax^2$, $a \neq 0$, will be a
    vertical expansion of $y = x^2$ if $|a| > 1$
    vertical contraction of $y = x^2$ if $|a| < 1$
    x-axis reflection of $y = x^2$ if $a < 0$

The graph of $f(x) = (x - h)^2$ is the graph of $y = x^2$ translated h units to the right.

The graph of $f(x) = (x + h)^2$ is the graph of $y = x^2$ translated h units to the left.

The graph of $f(x) = x^2 + k$ is the graph of $y = x^2$ translated k units upward.

The graph of $f(x) = x^2 - k$ is the graph of $y = x^2$ translated k units downward.

Vertex Form [8.6]

7. The vertex of the function
$f(x) = -2(x + 1)^2 + 3$ is the point $(-1, 3)$, which is the highest point on the graph.

The graph of the function
$$f(x) = a(x - h)^2 + k, \, a \neq 0$$
is a parabola with vertex (h, k).

If $a > 0$, the parabola opens upward and the vertex is the lowest point on the graph.

If $a < 0$, the parabola opens downward and the vertex is the highest point on the graph.

Graphing Quadratic Functions in Standard Form [8.6]

8. The graph of $y = x^2 - 4$ will be a parabola. It will cross the x-axis at 2 and -2, and the vertex will be $(0, -4)$.

The graph of any function of the form
$$f(x) = ax^2 + bx + c \qquad a \neq 0$$
is a **parabola**. The graph opens upward if $a > 0$ and opens downward if $a < 0$. The **vertex** is the point
$$\left(-\frac{b}{2a}, f\left(-\frac{b}{2a}\right)\right)$$

Projectile Motion [8.7]

9. If a baseball is thrown upward with an initial velocity of 48 feet per second and released from a height of 4 feet, then its height after t seconds is given by

$$h(t) = -16t^2 + 48t + 4$$

The vertex represents the highest point reached by the baseball, which is 40 feet after 1.5 seconds.

If an object is projected vertically into the air with an initial speed v, in feet per second, from an initial height s, in feet, then the height of the object after t seconds is given by the quadratic function

$$h(t) = -16t^2 + vt + s$$

where h is measured in feet.

Profit, Cost, and Revenue [8.7]

10. An art supply store can sell x paint sets each week at p dollars each according to the equation $x = 700 - 20p$. The revenue is

$$\begin{aligned} R &= xp \\ &= (700 - 20p)p \end{aligned}$$

For the revenue to be $6,000, we solve

$$\begin{aligned} 700p - 20p^2 &= 6,000 \\ 20p^2 - 700p + 6,000 &= 0 \\ p^2 - 35p + 300 &= 0 \\ (p - 15)(p - 20) &= 0 \\ p = 15 \quad &\text{or} \quad p = 20 \end{aligned}$$

The relationship between profit, cost, and revenue is given by

$$\text{Profit} = \text{Revenue} - \text{Cost}$$

$$P(x) = R(x) - C(x)$$

If x items are sold and p is the selling price of each item, then

$$R = xp$$

Quadratic Inequalities [8.8]

11. Solve: $x^2 - 2x - 8 > 0$.

We factor and draw the sign diagram:

$$(x - 4)(x + 2) > 0$$

$- - - -$	$- - - -$	$+ + + +$	$(x - 4)$
$- - - -$	$+ + + +$	$+ + + +$	$(x + 2)$
	-2	4	

The solution is $x < -2$ or $x > 4$.

We solve quadratic inequalities by manipulating the inequality to get 0 on the right side and then factoring the left side. We then make a diagram that indicates where the factors are positive and where they are negative. From this sign diagram and the original inequality we graph the appropriate solution set.

Chapter 8 Test

Solve each equation. [8.1, 8.2]

1. $(2x + 4)^2 = 25$

2. $(2x - 6)^2 = -8$

3. $y^2 - 10y + 25 = -4$

4. $(y + 1)(y - 3) = -6$

5. $8t^3 - 125 = 0$

6. $\dfrac{1}{a + 2} - \dfrac{1}{3} = \dfrac{1}{a}$

7. Solve the formula $64(1 + r)^2 = A$ for r. [8.1]

8. Solve $x^2 - 4x = -2$ by completing the square. [8.1]

9. Find k so that $kx^2 = 12x - 4$ has one rational solution. [8.3]

10. Use the discriminant to identify the number and kind of solutions to $2x^2 - 5x = 7$. [8.3]

Find equations that have the given solutions. [8.3]

11. $x = 5, x = -\dfrac{2}{3}$

12. $x = 2i, x = -2i$

Solve each equation. [8.4]

13. $4x^4 - 7x^2 - 2 = 0$

14. $(2t + 1)^2 - 5(2t + 1) + 6 = 0$

15. $2t - 7\sqrt{t} + 3 = 0$

16. Projectile Motion An object is tossed into the air with an upward velocity of 14 feet per second from the top of a building h feet high. The time it takes for the object to hit the ground below is given by the formula $16t^2 - 14t - h = 0$. Solve this formula for t. [8.4]

Sketch the graph of each of the following functions. Find any intercepts and give the coordinates of the vertex in each case. [8.5, 8.6]

17. $f(x) = 3 - (x + 2)^2$

18. $f(x) = 2(x - 1)^2 - 3$

19. $f(x) = x^2 - 2x - 3$

20. $f(x) = -x^2 + 2x + 8$

21. Projectile Motion An object projected upward with an initial velocity of 32 feet per second will rise and fall according to the equation $s(t) = 32t - 16t^2$, where s is its distance above the ground at time t. At what times will the object be 12 feet above the ground? [8.7]

20. Revenue The total weekly cost for a company to make x ceramic coffee cups is given by the formula $C(x) = 2x + 100$. If the weekly revenue from selling all x cups is $R(x) = 25x - 0.2x^2$, how many cups must it sell a week to make a profit of $200 a week? [8.7]

23. Profit Find the maximum weekly profit for a company with weekly costs of $C = 5x + 100$ and weekly revenue of $R = 25x - 0.1x^2$. [8.7]

Graph each of the following inequalities. [8.8]

24. $x^2 - x - 6 \le 0$

25. $2x^2 + 5x > 3$

Exponential and Logarithmic Functions

Chapter Outline

iStockphoto.com © ZoneCreative

If you have had any problems with or had testing done on your thyroid gland, then you may have come in contact with radioactive iodine-131. Like all radio-active elements, iodine-131 decays naturally. The half-life of iodine-131 is 8 days, which means that every 8 days a sample of iodine-131 will decrease to half of its original amount. The following table and graph show what happens to a 1,600-microgram sample of iodine-131 over time.

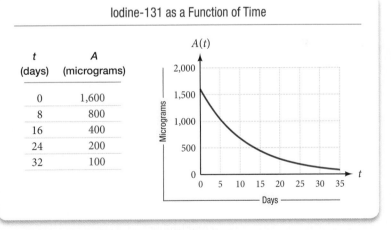

Iodine-131 as a Function of Time

t (days)	A (micrograms)
0	1,600
8	800
16	400
24	200
32	100

The function represented by the information in the table and graph is

$$A(t) = 1{,}600 \cdot \frac{1}{2}^{\,t/8}$$

It is one of the types of functions we will study in this chapter.

Success Skills

"Once you replace negative thoughts with positive ones, you'll start having positive results."
— Willie Nelson

Do you complain to your classmates about your teacher? If you do, it could be getting in the way of your success in the class.

I have students that tell me that they like the way I teach and that they are enjoying my class. I have other students, in the same class, that complain to each other about me. They say I don't explain things well enough. Are the complaining students giving themselves a reason for not doing well in the class? I think so. They are shifting the responsibility for their success from themselves to me. It's not their fault they are not doing well, it's mine. When these students are alone, trying to do homework, they start thinking about how unfair everything is and they lose their motivation to study. Without intending to, they have set themselves up to fail by making their complaints more important than their progress in the class.

What happens when you stop complaining? You put yourself back in charge of your success. When there is no one to blame if things don't go well, you are more likely to do well. I have had students tell me that, once they stopped complaining about a class, the teacher became a better teacher and they started to actually enjoy going to class.

If you find yourself complaining to your friends about a class or a teacher, make a decision to stop. When other people start complaining to each other about the class or the teacher, walk away; don't participate in the complaining session. Try it for a day, or a week, or for the rest of the term. It may be difficult to do at first, but I'm sure you will like the results, and if you don't, you can always go back to complaining.

Exponential Functions

Learning Objectives

In this section, we will learn how to:

1. Evaluate an exponential function.

2. Sketch the graph of an exponential function.

3. Solve problems involving compound interest.

4. Graph the natural exponential function.

Introduction

To obtain an intuitive idea of how exponential functions behave, we can consider the heights attained by a bouncing ball. When a ball used in the game of racquetball is dropped from any height, the first bounce will reach a height that is $\frac{2}{3}$ of the original height. The second bounce will reach $\frac{2}{3}$ of the height of the first bounce, and so on, as shown in Figure 1.

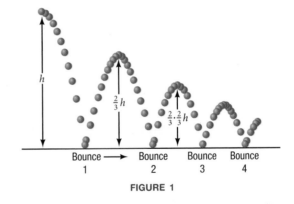

FIGURE 1

If the ball is initially dropped from a height of 1 meter, then during the first bounce it will reach a height of $\frac{2}{3}$ meter. The height of the second bounce will reach $\frac{2}{3}$ of the height reached on the first bounce. The maximum height of any bounce is $\frac{2}{3}$ of the height of the previous bounce.

$$\text{Initial height: } h = 1$$

$$\text{Bounce 1:} \quad h = \frac{2}{3}(1) = \frac{2}{3}$$

$$\text{Bounce 2:} \quad h = \frac{2}{3}\left(\frac{2}{3}\right) = \left(\frac{2}{3}\right)^2$$

$$\text{Bounce 3:} \quad h = \frac{2}{3}\left(\frac{2}{3}\right)^2 = \left(\frac{2}{3}\right)^3$$

$$\text{Bounce 4:} \quad h = \frac{2}{3}\left(\frac{2}{3}\right)^3 = \left(\frac{2}{3}\right)^4$$

$$\cdot \qquad \cdot$$
$$\cdot \qquad \cdot$$
$$\cdot \qquad \cdot$$

$$\text{Bounce } n: \quad h = \frac{2}{3}\left(\frac{2}{3}\right)^{n-1} = \left(\frac{2}{3}\right)^n$$

This last equation is exponential in form. In fact, this relationship can be expressed in function notation as

$$h(n) = \left(\frac{2}{3}\right)^n$$

Because the variable is an exponent, we call this an **exponential function**. We classify all exponential functions together with the following definition.

> **(dĕf DEFINITION**   *exponential function*
>
> An **exponential function** is any function that can be written in the form
>
> $$f(x) = b^x$$
>
> where b is a positive real number other than 1. The constant b is called the **base**.

Each of the following is an exponential function:

$$f(x) = 2^x \qquad y = 3^x \qquad f(x) = \left(\frac{1}{4}\right)^x$$

Notice the fundamental difference between an exponential function and a function involving a polynomial, such as a quadratic function.

$$\text{Exponential Function} \qquad \text{Quadratic function}$$
$$f(x) = 2^x \qquad\qquad f(x) = x^2$$

With a quadratic function, we have a constant exponent on a variable base. With the exponential function, the base is a constant and the exponent is a variable.

Evaluating Exponential Functions

We can evaluate an exponential function by substituting values for x and then evaluating the exponent.

EXAMPLE 1   If $f(x) = 2^x$, then

$$f(0) = 2^0 = 1$$

$$f(3) = 2^3 = 8$$

$$f(-2) = 2^{-2} = \frac{1}{2^2} = \frac{1}{4}$$

$$f\left(\frac{1}{2}\right) = 2^{1/2} = \sqrt{2} \approx 1.4$$

$$f\left(-\frac{3}{4}\right) = 2^{-3/4} = \frac{1}{2^{3/4}} = \frac{1}{\sqrt[4]{2^3}} \approx 0.6$$

$$f(\sqrt{5}) = 2^{\sqrt{5}} \approx 2^{2.24} \approx 4.7$$

With the last three values we used a calculator to obtain an approximation.

As you can see in Example 1, we can use any real number as the input for an exponential function. All exponential functions have a domain that is the set of all real numbers.

In the introduction to this chapter, we indicated that the half-life of iodine-131 is 8 days, which means that every 8 days a sample of iodine-131 will decrease to half of its original amount. If we start with A_0 micrograms of iodine-131, then after t days the sample will contain

$$A(t) = A_0 \cdot \left(\frac{1}{2}\right)^{t/8}$$

micrograms of iodine-131. This is an exponential function with a base of $\frac{1}{2}$.

EXAMPLE 2   A patient is administered a 1,200-microgram dose of iodine-131. How much iodine-131 will be in the patient's system after 10 days, and after 16 days?

SOLUTION The initial amount of iodine-131 is $A_0 = 1,200$, so the function that gives the amount left in the patient's system after t days is

$$A(t) = 1,200 \cdot \left(\frac{1}{2}\right)^{t/8}$$

After 10 days, the amount left in the patient's system is

$$A(10) = 1,200 \cdot \left(\frac{1}{2}\right)^{10/8} = 1,200 \cdot \left(\frac{1}{2}\right)^{1.25} \approx 504.5 \text{ micrograms}$$

After 16 days, the amount left in the patient's system is

$$A(16) = 1,200 \cdot \left(\frac{1}{2}\right)^{16/8} = 1,200 \cdot \left(\frac{1}{2}\right)^{2} = 300 \text{ micrograms}$$

Graphs of Exponential Functions

We now turn our attention to the graphs of exponential functions.

EXAMPLE 3   Sketch the graph of the exponential function $f(x) = 2^x$.

SOLUTION First we make a table of ordered pairs by choosing convenient values of x (see Example 1 for reference). Graphing the ordered pairs given in the table and connecting them with a smooth curve, we have the graph of $y = 2^x$ shown in Figure 2.

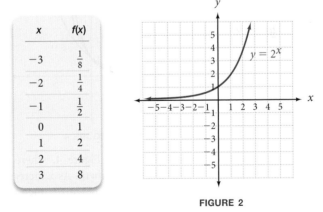

x	$f(x)$
-3	$\frac{1}{8}$
-2	$\frac{1}{4}$
-1	$\frac{1}{2}$
0	1
1	2
2	4
3	8

FIGURE 2

As you can see from Figure 2, the range of the function is $\{y \mid y > 0\}$. Notice that the graph does not cross the x-axis. It *approaches* the x-axis — in fact, we can get it as close to the x-axis as we want without it actually intersecting the x-axis. For the graph of $y = 2^x$ to intersect the x-axis, we would have to find a value of x that would make $2^x = 0$. Because no such value of x exists, the graph of $y = 2^x$ cannot intersect the x-axis. The graph has a horizontal asymptote at the x-axis, which we can express as the line $y = 0$.

Although the graph does not have an x-intercept, it does have a y-intercept of 1.

EXAMPLE 4   Sketch the graph of $f(x) = \left(\dfrac{1}{3}\right)^x$.

SOLUTION   The table beside Figure 3 gives some ordered pairs that satisfy the equation. Using the ordered pairs from the table, we have the graph shown in Figure 3.

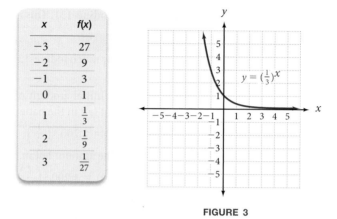

x	f(x)
−3	27
−2	9
−1	3
0	1
1	$\dfrac{1}{3}$
2	$\dfrac{1}{9}$
3	$\dfrac{1}{27}$

FIGURE 3

Once again, the range of the function is $\{y \mid y > 0\}$ and the graph has a horizontal asymptote at $y = 0$. The y-intercept is 1. There is no x-intercept.

Figures 4 and 5 show some families of exponential curves to help you become more familiar with them on an intuitive level.

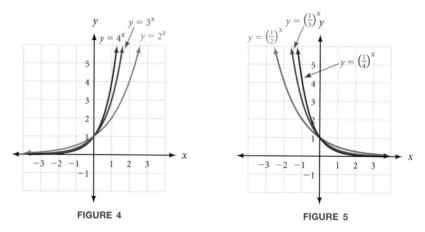

FIGURE 4   **FIGURE 5**

The graphs in Figure 4, where the base b is larger than 1, are examples of *exponential growth*. As x increases, y increases very rapidly. If the base b is less than 1, such as in Figure 5, then we refer to the behavior as *exponential decay*. As x increases, y decreases towards 0.

Here is a summary of the behavior of the graph for an exponential function.

> **[△≠Σ] RULE** *Graphs of Exponential Functions*
>
> The graph of the $f(x) = b^x$, for $b > 0$ and $b \neq 1$, has the following:
>
> $$\text{domain} = \text{all real numbers}$$
> $$\text{range} = \{y \mid y > 0\}$$
> $$y\text{-intercept} = (0, 1)$$
> $$\text{horizontal asymptote at } y = 0$$
>
> If $b > 1$ the graph represents exponential growth, and if $b < 1$ the graph represents exponential decay.

EXAMPLE 5 Sketch the graph of $f(x) = 3 - 2^x$, and then state the range of the function.

SOLUTION We can use transformations to sketch the graph. Figure 2 in Example 3 shows the graph of $y = 2^x$. To obtain the graph of $f(x) = 3 - 2^x$, we do the following:

1. reflect the graph of $y = 2^x$ about the x-axis

2. translate the result upward 3 units.

Figure 6 shows all the steps and the final graph.

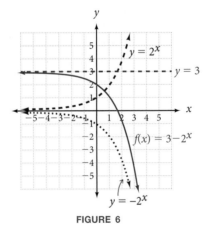

FIGURE 6

Notice that, as a result of the vertical translation, the horizontal asymptote is the line $y = 3$. The range is $\{y \mid y < 3\}$. ∎

Compound Interest

Among the many applications of exponential functions are the applications having to do with interest-bearing accounts. Suppose you invest P dollars (this is referred to as the ***principal***) in an account that earns an annual interest rate r, and that your interest is calculated n times each year. Then with each calculation the rate applied will be $\frac{r}{n}$.

So, after the first interest period, your new balance will be

$$\text{Balance} = \text{Principal} + \text{Interest}$$

$$= P + \left(\frac{r}{n}\right)P$$

$$= \left(1 + \frac{r}{n}\right)P$$

In other words, at each calculation your previous balance is multiplied by a factor of

$$\left(1 + \frac{r}{n}\right)$$

If you leave your money in the account for t years and do not make any withdrawals, then we have the following.

> ### $\lceil\Delta\neq\Sigma\rceil$ *Compound Interest*
>
> If P dollars are deposited in an account with annual interest rate r, compounded n times per year, then the amount of money in the account after t years is given by the formula
>
> $$A(t) = P\left(1 + \frac{r}{n}\right)^{nt}$$

 EXAMPLE 6   Suppose you deposit $500 in an account with an annual interest rate of 2.4%. Find the amount of money in your account after 5 years if the interest is compounded:

a. semiannually

b. quarterly

c. monthly

SOLUTION   First, we note that $P = 500$, $r = 0.024$, and $t = 5$. Substituting these values into the compound interest formula gives us

$$A(5) = 500\left(1 + \frac{0.024}{n}\right)^{n \cdot 5}$$

All we need to do is determine the correct value of n and then perform the calculation.

a. Semiannually means twice each year, so $n = 2$:

$$A(5) = 500\left(1 + \frac{0.024}{2}\right)^{2(5)} = 500(1.012)^{10} \approx \$563.35$$

b. Quarterly means four times each year, so now $n = 4$.

$$A(5) = 500\left(1 + \frac{0.024}{4}\right)^{4(5)} = 500(1.006)^{20} \approx \$563.55$$

c. There are 12 months in a year, so we use $n = 12$.

$$A(5) = 500\left(1 + \frac{0.024}{12}\right)^{12(5)} = 500(1.002)^{60} \approx \$563.68$$ ∎

As you can see in Example 6, we earn more money by compounding interest more frequently. This is because our interest begins earning its own interest earlier in the year.

What would happen if we let the number of compounding periods become larger and larger, so that we compounded the interest every week, then every day, then every hour, then every second, and so on? If we take this as far as it can go, we end up compounding the interest every moment. When this happens, we have an account with interest that is compounded continuously.

To see how this affects our compound interest formula, we let $x = \frac{n}{r}$, giving us

$$A(t) = P\left(1 + \frac{r}{n}\right)^{nt}$$

$$= P\left(1 + \frac{1}{x}\right)^{xrt}$$

$$= P\left[\left(1 + \frac{1}{x}\right)^{x}\right]^{rt}$$

Table 1 shows values of the expression

$$\left(1 + \frac{1}{x}\right)^{x}$$

as n, and thus x, increases.

x	$\left(1 + \frac{1}{x}\right)^{x}$
1	2
10	2.59374
100	2.70481
1,000	2.71692
10,000	2.71815
100,000	2.71827

TABLE 1

Notice the values in the right column seem to be approaching some specific number. This number is denoted as e. The number e is a number like π. It is irrational and occurs in many formulas that describe the world around us. Like π, it can be approximated with a decimal number. Whereas π is approximately 3.1416, e is approximately 2.7183.

Replacing the expression $\left(1 + \frac{1}{x}\right)^{x}$ with e in our compound interest formula leads us to the following.

 Continuously Compounded Interest

If P dollars are deposited in an account with annual interest rate r, compounded continuously, then the amount of money in the account after t years is given by the formula

$$A(t) = Pe^{rt}$$

EXAMPLE 7   Suppose you deposit $500 in an account with an annual interest rate of 2.4% compounded continuously. Find an equation that gives the amount of money in the account after t years. Then find the amount of money in the account after 5 years.

SOLUTION   Because the interest is compounded continuously, we use the formula $A(t) = Pe^{rt}$. Substituting $P = 500$ and $r = 0.024$ into this formula, we have

$$A(t) = 500e^{0.024t}$$

After 5 years, this account will contain

$$A(5) = 500e^{0.024 \cdot 5} = 500e^{0.12} \approx \$563.75$$

to the nearest cent.

The Natural Exponential Function

Because e is a positive number, we can use it as the base of an exponential function, giving us $f(x) = e^x$. This is called the **natural exponential** function.

The table below shows some values of the natural exponential function. The graph is shown in Figure 7. Notice that this graph is another example of exponential growth.

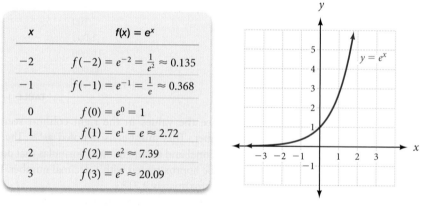

x	$f(x) = e^x$
-2	$f(-2) = e^{-2} = \frac{1}{e^2} \approx 0.135$
-1	$f(-1) = e^{-1} = \frac{1}{e} \approx 0.368$
0	$f(0) = e^0 = 1$
1	$f(1) = e^1 = e \approx 2.72$
2	$f(2) = e^2 \approx 7.39$
3	$f(3) = e^3 \approx 20.09$

FIGURE 7

Getting Ready for Class

After reading through the preceding section, respond in your own words and in complete sentences.

A. What is an exponential function?

B. In an exponential function, explain why the base b cannot equal 1. (What kind of function would you get if the base was equal to 1?)

C. What characteristics do the graphs of $y = 2^x$ and $y = \left(\frac{1}{2}\right)^x$ have in common?

D. What is meant by continuously compounded interest?

Let $f(x) = 3^x$ and $g(x) = \left(\dfrac{1}{2}\right)^x$, and evaluate each of the following.

1. $g(0)$ **2.** $f(0)$ **3.** $g(-1)$ **4.** $g(-4)$

5. $f(-3)$ **6.** $f(-1)$ **7.** $f(2) + g(-2)$ **8.** $f(2) - g(-2)$

Let $f(x) = 4^x$ and $g(x) = \left(\dfrac{1}{3}\right)^x$. Evaluate each of the following.

9. $f(-1) + g(1)$ **10.** $f(2) + g(-2)$ **11.** $\dfrac{f(-2)}{g(1)}$ **12.** $f(3) - f(2)$

Let $f(x) = 2^x$. Use a calculator to approximate each of the following. Round to the nearest hundredth.

13. $f\left(\dfrac{1}{3}\right)$ **14.** $f(\sqrt{3})$ **15.** $f(-\sqrt{7})$ **16.** $f\left(-\dfrac{3}{2}\right)$

Let $f(x) = e^x$. Use a calculator to approximate each of the following. Round to the nearest hundredth. (If your calculator does not have an e^x key, use $e \approx 2.7183$.)

17. $f(3)$ **18.** $f(-3)$ **19.** $f(-0.5)$ **20.** $f(1.5)$

21. $f\left(\dfrac{1}{5}\right)$ **22.** $f\left(\dfrac{5}{3}\right)$ **23.** $f(\pi)$ **24.** $f(e)$

Graph each of the following functions.

25. $y = 4^x$ **26.** $y = 3^x$ **27.** $y = \left(\dfrac{1}{2}\right)^x$ **28.** $y = \left(\dfrac{1}{4}\right)^x$

29. $y = \left(\dfrac{1}{3}\right)^{-x}$ **30.** $y = \left(\dfrac{1}{2}\right)^{-x}$ **31.** $y = e^x$ **32.** $y = e^{-x}$

Use transformations to graph each function. Then state the horizontal asymptote and the range.

33. $y = 2^{x+1}$ **34.** $y = 2^{x-3}$ **35.** $y = 3^x + 2$ **36.** $y = 3^x - 1$

37. $y = -2^x$ **38.** $y = -e^x$ **39.** $y = 2e^x$ **40.** $y = \dfrac{1}{2} \cdot 3^x$

41. $y = 4 - 2^x$ **42.** $y = -1 - 3^x$ **43.** $y = e^x + 2$ **44.** $y = -e^x + 2$

Graph each of the following functions on the same coordinate system for positive values of x only.

45. $y = 2x, y = x^2, y = 2^x$ **46.** $y = 3x, y = x^3, y = 3^x$

47. On a graphing calculator, graph the family of curves $y = b^x$, $b = 2, 4, 6, 8$.

48. On a graphing calculator, graph the family of curves $y = b^x$, $b = \dfrac{1}{2}, \dfrac{1}{4}, \dfrac{1}{6}, \dfrac{1}{8}$.

Applying the Concepts

49. Bouncing Ball Suppose the ball mentioned in the introduction to this section is dropped from a height of 6 feet above the ground. Find an exponential equation that gives the height h the ball will attain during the nth bounce. How high will it bounce on the fifth bounce?

50. **Bouncing Ball** A golf ball is manufactured so that if it is dropped from A feet above the ground onto a hard surface, the maximum height of each bounce will be one half of the height of the previous bounce. Find an exponential equation that gives the height h the ball will attain during the nth bounce. If the ball is dropped from 10 feet above the ground onto a hard surface, how high will it bounce on the eighth bounce?

51. **Cost of Freon** Automobiles built before 1993 use Freon in their air conditioners. The federal government now prohibits the manufacture of Freon. Because the supply of Freon is decreasing, the price per pound is increasing exponentially. Current estimates put the formula for the price per pound of Freon at $p(t) = 17.6(1.25)^t$, where t is the number of years since 2000. Find the price of Freon in 2000 and 2010. How much will Freon cost in the year 2025?

52. **Airline Travel** The number of airline passengers in 1990 was 466 million. The number of passengers traveling by airplane each year has increased exponentially according to the model, $P(t) = 466 \cdot 1.035^t$, where t is the number of years since 1990 (U.S. Census Bureau).

    a. How many passengers traveled in 2010?

    b. How many passengers will travel in 2025?

53. **Bacteria Growth** Suppose it takes 12 hours for a certain strain of bacteria to reproduce by dividing in half. If 50 bacteria are present to begin with, then the total number present after x days will be $f(x) = 50 \cdot 4^x$. Find the total number present after 1 day, 2 days, and 3 days.

54. **Bacteria Growth** Suppose it takes 1 day for a certain strain of bacteria to reproduce by dividing in half. If 100 bacteria are present to begin with, then the total number present after x days will be $f(x) = 100 \cdot 2^x$. Find the total number present after 1 day, 2 days, 3 days, and 4 days. How many days must elapse before over 100,000 bacteria are present?

55. **Cost Increase** The cost of a can of Coca Cola in 1960 was \$0.10. The exponential function that models the cost of a Coca Cola by year is given below, where t is the number of years since 1960.

$$C(t) = 0.10e^{0.0576t}$$

    a. What was the expected cost of a can of Coca Cola in 2000?

    b. What was the expected cost of a can of Coca Cola in 2015?

    c. What is the expected cost of a can of Coca Cola in 2050?

56. **Bacteria Decay** You are conducting a biology experiment and begin with 5,000,000 cells, but some of those cells are dying each minute. The rate of death of the cells is modeled by the function $A(t) = A_0 \cdot e^{-0.598t}$, where A_0 is the original number of cells, t is time in minutes, and A is the number of cells remaining after t minutes.

    a. How may cells remain after 5 minutes?

    b. How many cells remain after 10 minutes?

    c. How many cells remain after 20 minutes?

57. Compound Interest Suppose you deposit $1,200 in an account with an annual interest rate of 6% compounded quarterly.

  a. Find an equation that gives the amount of money in the account after t years.

  b. Find the amount of money in the account after 8 years.

  c. If the interest were compounded continuously, how much money would the account contain after 8 years?

58. Compound Interest Suppose you deposit $500 in an account with an annual interest rate of 8% compounded monthly.

  a. Find an equation that gives the amount of money in the account after t years.

  b. Find the amount of money in the account after 5 years.

  c. If the interest were compounded continuously, how much money would the account contain after 5 years?

59. Compound Interest Valerie invests $10,000 in a CD earning 1.5% annual interest. If the CD matures in 5 years, what will be its value if interest is compounded:

  a. quarterly

  b. monthly

  c. continuously

60. Compound Interest Daniel purchases a U.S. Savings Bond for $100 that earns 0.5% annual interest. How much will the bond be worth in 30 years if the interest is compounded:

  a. semi-annually

  b. monthly

  c. continuously

61. Compound Interest Mark has $2,500 he wants to invest in an interest earning account for 4 years. One bank offers 1.4% annual interest, compounded quarterly. Another bank offers 1.35% annual interest, compounded continuously. Which is the better deal?

62. Compound Interest Ashley deposits $800 in a savings account that earns 2% annual interest, compounded continuously. At the end of two years whe withdraws $500. What is the balance in her account after an additional 2 years pass?

63. Value of a Painting A painting is purchased as an investment for $150. If the painting's value doubles every 3 years, then its value is given by the function

$$V(t) = 150 \cdot 2^{t/3} \text{ for } t \geq 0$$

where t is the number of years since it was purchased, and $V(t)$ is its value (in dollars) at that time. Graph this function.

64. Value of a Painting A painting is purchased as an investment for $125. If the painting's value doubles every 5 years, then its value is given by the function

$$V(t) = 125 \cdot 2^{t/5} \text{ for } t \geq 0$$

where t is the number of years since it was purchased, and $V(t)$ is its value (in dollars) at that time. Graph this function.

65. **Bankruptcy Model** The model for the number of bankruptcies filed under the Bankruptcy Reform Act is $B(t) = 0.798 \cdot 1.164^t$, where t is the number of years since 1994 and B is the number of bankruptcies filed in terms of millions. (*Source*: Administrative Office of the U.S. Courts, Statistical Tables for the Federal Judiciary)

     a. What is the expected number of bankruptcy filings in 2030?

     b. Graph this function for $0 \le t \le 40$.

66. **Health Care** In 1990, $699 billion were spent on health care expenditures. The amount of money, E, in billions spent on health care expenditures can be estimated using the function $E(t) = 78.16(1.11)^t$, where t is time in years since 1970.   (*Source*: U.S. Census Bureau)

     a. What are the expected health care expenditures in 2020, 2025, and 2030?

     b. Graph this function for $0 \le t \le 60$.

Declining-Balance Depreciation The declining-balance method of depreciation is an accounting method businesses use to deduct most of the cost of new equipment during the first few years of purchase. Unlike other methods, the declining-balance formula does not consider salvage value.

67. **Value of a Crane** The function

$$V(t) = 450{,}000\,(1 - 0.30)^t,$$

     where V is value and t is time in years, can be used to find the value of a crane for the first 6 years of use.

     a. What is the value of the crane after 3 years and 6 months?

     b. State the domain of this function.

     c. Sketch the graph of this function.

     d. State the range of this function.

     e. After how many years will the crane be worth only $85,000?

68. **Value of a Printing Press** The function $V(t) = 375{,}000(1 - 0.25)^t$, where V is value and t is time in years, can be used to find the value of a printing press during the first 7 years of use.

     a. What is the value of the printing press after 4 years and 9 months?

     b. State the domain of this function.

     c. Sketch the graph of this function.

     d. State the range of this function.

     e. After how many years will the printing press be worth only $65,000?

69. **Getting Close to e** Use a calculator to complete the following table.

x	$(1 + x)^{1/x}$
1	
0.5	
0.1	
0.01	
0.001	
0.0001	
0.00001	

What number does the expression $(1 + x)^{1/x}$ seem to approach as x gets closer and closer to zero?

70. Getting Close to e Use a calculator to complete the following table.

x	$\left(1 + \frac{1}{x}\right)^x$
1	
10	
50	
100	
500	
1,000	
10,000	
1,000,000	

What number does the expression $\left(1 + \frac{1}{x}\right)^x$ seem to approach as x gets larger and larger?

Learning Objectives Assessment

The following problems can be used to help assess if you have successfully met the learning objectives for this section.

71. Evaluate $f(-2)$ if $f(x) = \left(\frac{1}{2}\right)^x$.

    a. -1 **b.** $\dfrac{1}{4}$ **c.** 4   **d.** $-\dfrac{1}{4}$

72. Which function has the graph shown in Figure 8?

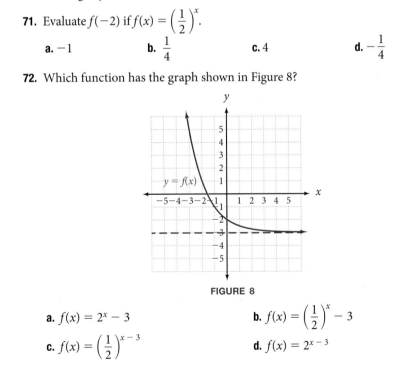

FIGURE 8

    a. $f(x) = 2^x - 3$ **b.** $f(x) = \left(\frac{1}{2}\right)^x - 3$

    c. $f(x) = \left(\frac{1}{2}\right)^{x-3}$ **d.** $f(x) = 2^{x-3}$

73. If $600 is invested in an account earning 3.5% annual interest compounded monthly, how much is in the account after 8 years?

    a. $650.98   **b.** $793.88   **c.** $792.91   **d.** $793.55

74. Which point is on the graph of $f(x) = e^x$?

    a. $(0, 2.7183)$ **b.** $(3, 8.1548)$

    c. $(-1, -2.7183)$ **d.** $(-2, 0.1353)$

Getting Ready for the Next Section

Solve each equation for y.

75. $x = 2y - 3$ **76.** $x = \dfrac{y + 7}{5}$

77. $x = y^2 - 3$ **78.** $x = (y + 4)^3$

79. $x = \dfrac{y - 4}{y - 2}$ **80.** $x = \dfrac{y + 5}{y - 3}$

81. $x = \sqrt{y - 3}$ **82.** $x = \sqrt{y} + 5$

The Inverse of a Function

Learning Objectives

In this section, we will learn how to:

1. Find the Inverse of a function as a set of ordered pairs.

2. Graph the inverse of a function.

3. Find the equation of an inverse function.

4. Use the horizontal line test to identify a one-to-one function.

5. Use composition to identify functions that are inverses.

Introduction

The following diagram (Figure 1) shows the route Justin takes to school. He leaves his home and drives 3 miles east, and then turns left and drives 2 miles north. When he leaves school to drive home, he drives the same two segments, but in the reverse order and the opposite direction; that is, he drives 2 miles south, turns right, and drives 3 miles west. When he arrives home from school, he is right where he started. His route home "undoes" his route to school, leaving him where he began.

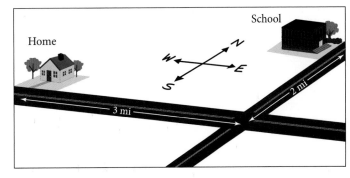

FIGURE 1

The relationship between a function and its inverse function is similar to the relationship between Justin's route from home to school and his route from school to home. The purpose of an inverse function is to undo, or reverse, whatever action the function performed. In other words, if a function contains the ordered pair (a, b), then the inverse should reverse this and contain the ordered pair (b, a). Since the function "turns a into b," the inverse should "turn b back into a."

This leads us to the following definition.

(def) DEFINITION *inverse of a relation*

The *inverse of a relation* R is obtained by interchanging the components of each ordered pair contained in R.

VIDEO EXAMPLES

SECTION 9.2

EXAMPLE 1 Find the inverse of the function given by $f = \{(1, 4), (2, 5), (3, 6), (4, 7)\}$.

SOLUTION The inverse of f is obtained by reversing the order of the coordinates in each ordered pair in f. The inverse of f is the relation given by

$$g = \{(4, 1), (5, 2), (6, 3), (7, 4)\}$$

Looking at Example 1, it is obvious that the domain of f is now the range of g, and the range of f is now the domain of g. With inverses, the domain and range simply switch roles.

Suppose a function f is defined with an equation instead of a list of ordered pairs. We can obtain the equation of the inverse of f by interchanging the role of x and y in the equation for f.

EXAMPLE 2   If the function f is defined by $f(x) = 2x - 3$, find the equation that represents the inverse of f.

SOLUTION   Because the inverse of f is obtained by interchanging the components of all the ordered pairs belonging to f, every value of x takes on the role of a y-value, and every value of y assumes the role of an x-value. We simply exchange x and y in the equation $y = 2x - 3$ to get the formula for the inverse of f:

$$x = 2y - 3$$

We now solve this equation for y in terms of x:

$$x + 3 = 2y$$
$$\frac{x + 3}{2} = y$$
$$y = \frac{x + 3}{2}$$

The last line gives the equation that defines the inverse of f.

Let's compare the graphs of f and its inverse from Example 2. (See Figure 2.)

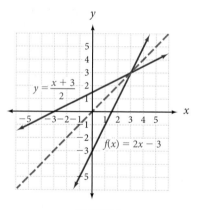

FIGURE 2

The graphs of f and its inverse have symmetry about the line $y = x$. This is a reasonable result since the one function was obtained from the other by interchanging x and y in the equation. The ordered pairs (a, b) and (b, a) always have symmetry about the line $y = x$. This gives us the following property.

PROPERTY   *graph of an inverse relation*

The graph of the inverse of a relation can be found by reflecting the graph of the original relation about the line $y = x$.

EXAMPLE 3   Graph the function $y = x^2 - 2$ and its inverse. Give the equation for the inverse.

SOLUTION   We can obtain the graph of the inverse of $y = x^2 - 2$ by graphing $y = x^2 - 2$ by the usual methods, and then reflecting the graph about the line $y = x$.

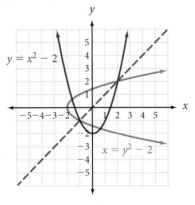

FIGURE 3

The equation that corresponds to the inverse of $y = x^2 - 2$ is obtained by interchanging x and y to get $x = y^2 - 2$.

We can solve the equation $x = y^2 - 2$ for y in terms of x as follows:

$$x = y^2 - 2$$
$$x + 2 = y^2$$
$$y = \pm\sqrt{x + 2}$$

Comparing the graphs in Figures 2 and 3, we observe that the inverse of a function is not always a function. In Example 2, both f and its inverse have graphs that are nonvertical straight lines and therefore both represent functions. In Example 3, the inverse of function f is not a function, since a vertical line crosses it in more than one place.

One-to-One Functions

We can distinguish between those functions with inverses that are also functions and those functions with inverses that are not functions with the following definition.

> **(dĕf** **DEFINITION**   *one-to-one functions*
>
> A function is a *one-to-one function* if every element in the range comes from exactly one element in the domain.

This definition indicates that a one-to-one function will yield a set of ordered pairs in which no two different ordered pairs have the same second coordinates. For example, the function

$$f = \{(2, 3), (-1, 3), (5, 8)\}$$

is not one-to-one because the element 3 in the range comes from both 2 and -1 in the domain.

On the other hand, the function

$$g = \{(5, 7), (3, -1), (4, 2)\}$$

is a one-to-one function because every element in the range comes from only one element in the domain.

Horizontal Line Test

If we have the graph of a function, we can determine if the function is one-to-one with the following test. If a horizontal line crosses the graph of a function in more than one place, then the function is not a one-to-one function because the points at which the horizontal line crosses the graph will be points with the same y-coordinates, but different x-coordinates. Therefore, the function will have an element in the range (the y-coordinate) that comes from more than one element in the domain (the x-coordinates).

Of the functions we have covered previously, all the (non-horizontal) linear functions and exponential functions are one-to-one functions because no horizontal lines can be found that will cross their graphs in more than one place.

Functions Whose Inverses Are Also Functions

Because one-to-one functions do not repeat second coordinates, when we reverse the components of the ordered pairs in a one-to-one function, we obtain a relation in which no two ordered pairs have the same first coordinate — by definition, this relation must be a function. In other words, every one-to-one function has an inverse that is itself a function. Because of this, we can use function notation to represent that inverse.

> **⌈△≠∑⌋** *Inverse Function Notation*
>
> If $f(x)$ is a one-to-one function, then the inverse of f is also a function and can be denoted by $f^{-1}(x)$.

Note   The notation f^{-1} does not represent the reciprocal of f. That is, the -1 in this notation is not an exponent. The notation f^{-1} is defined as representing the inverse function for a one-to-one function.

To illustrate, in Example 2 we found that the inverse of $f(x) = 2x - 3$ was the function $y = \frac{x + 3}{2}$. We can write this inverse function with inverse function notation as

$$f^{-1}(x) = \frac{x + 3}{2}$$

On the other hand, the inverse of the function in Example 2 is not itself a function, so we do not use the notation $f^{-1}(x)$ to represent it.

◼ EXAMPLE 4   Find the inverse of $g(x) = \dfrac{x - 4}{x - 2}$.

SOLUTION   To find the inverse for g, we begin by replacing $g(x)$ with y to obtain

$$y = \frac{x - 4}{x - 2} \qquad \text{The original function}$$

To find an equation for the inverse, we exchange x and y.

$$x = \frac{y - 4}{y - 2} \qquad \text{The inverse of the original function}$$

To solve for y, we first multiply each side by $y - 2$ to obtain

$$x(y - 2) = y - 4$$

$$xy - 2x = y - 4 \qquad \text{Distributive property}$$

$$xy - y = 2x - 4 \qquad \text{Collect all terms containing } y \text{ on the left side}$$

$$y(x - 1) = 2x - 4 \qquad \text{Factor } y \text{ from each term on the left side}$$

$$y = \frac{2x - 4}{x - 1} \qquad \text{Divide each side by } x - 1$$

Because our original function is one-to-one, as verified by the graph in Figure 4, its inverse is also a function. Therefore, we can use inverse function notation to write

$$g^{-1}(x) = \frac{2x - 4}{x - 1}$$

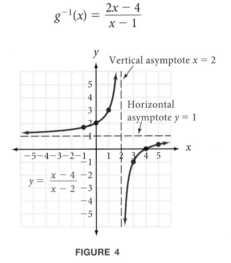

FIGURE 4

Graphing the Inverse Function

EXAMPLE 5   Graph the function $y = 2^x$ and its inverse $x = 2^y$.

SOLUTION   We graphed $y = 2^x$ in the preceding section. We simply reflect its graph about the line $y = x$ to obtain the graph of its inverse $x = 2^y$.

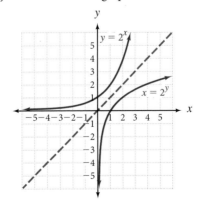

FIGURE 5

As you can see from the graph, $x = 2^y$ is a function. We do not have the mathematical tools to solve this equation for y, however. Therefore, we are unable to use the inverse function notation to represent this function. In the next section, we will give a definition that solves this problem. For now, we simply leave the equation as $x = 2^y$.

Inverses and Composition

One of the most powerful applications of an inverse function occurs when we perform function composition with a function and its inverse. Consider the following example.

EXAMPLE 6   Given $f(x) = 2x - 3$ and $f^{-1}(x) = \dfrac{x + 3}{2}$, find $(f^{-1} \circ f)(x)$ and $(f \circ f^{-1})(x)$.

SOLUTION

$$(f^{-1} \circ f)(x) = f^{-1}[f(x)]$$

$$= f^{-1}(2x - 3)$$

$$= \frac{2x - 3 + 3}{2}$$

$$= \frac{2x}{2}$$

$$= x$$

$$(f \circ f^{-1})(x) = f[f^{-1}(x)]$$

$$= f\left(\frac{x + 3}{2}\right)$$

$$= 2\left(\frac{x + 3}{2}\right) - 3$$

$$= x + 3 - 3$$

$$= x$$

Notice that in both cases the result is x, which was our input into both composition functions.

Because a function and its inverse "undo" each other, when we use them together under function composition the second function reverses whatever the first function did, returning us to our original value.

PROPERTY   *Composition of Inverses*

For any one-to-one function $f(x)$ and its inverse function $f^{-1}(x)$,

$$(f \circ f^{-1})(x) = x \qquad \text{for all } x \text{ in the domain of } f^{-1}(x)$$

and

$$(f^{-1} \circ f)(x) = x \qquad \text{for all } x \text{ in the domain of } f(x)$$

We can use this property to identify functions that are inverses of each other. Given functions $f(x)$ and $g(x)$, if $(f \circ g)(x) = x$ and $(g \circ f)(x) = x$, then f and g *must* be inverses. That is, $g(x) = f^{-1}(x)$ and $f(x) = g^{-1}(x)$.

EXAMPLE 7 Show that $f(x) = \dfrac{x}{x+2}$ and $g(x) = \dfrac{2x}{1-x}$ are inverses.

SOLUTION We must show $(f \circ g)(x) = x$ and $(g \circ f)(x) = x$.

$$(f \circ g)(x) = f[g(x)]$$

$$= f\left(\frac{2x}{1-x}\right)$$

$$= \frac{\dfrac{2x}{1-x}}{\dfrac{2x}{1-x} + 2}$$

$$= \frac{2x}{2x + 2(1-x)} \qquad \text{Multiply numerator and denominator by } 1-x$$

$$= \frac{2x}{2x + 2 - 2x}$$

$$= \frac{2x}{2}$$

$$= x$$

Also,

$$(g \circ f)(x) = g[f(x)]$$

$$= g\left(\frac{x}{x+2}\right)$$

$$= \frac{2\left(\dfrac{x}{x+2}\right)}{1 - \dfrac{x}{x+2}}$$

$$= \frac{2x}{x+2-x} \qquad \text{Multiply numerator and denominator by } x+2$$

$$= \frac{2x}{2}$$

$$= x$$

Because both compositions simplify to x, the functions must be inverses.

Functions, Relations, and Inverses—A Summary

Here is a summary of some of the things we know about functions, relations, and their inverses:

1. Every function is a relation, but not every relation is a function.

2. Every function has an inverse, but only one-to-one functions have inverses that are also functions.

3. The domain of a function is the range of its inverse, and the range of a function is the domain of its inverse.

4. If $f(x)$ is a one-to-one function, then we can use the notation $f^{-1}(x)$ to represent its inverse function.

5. The graph of a function and its inverse have symmetry about the line $y = x$.

6. If (a, b) belongs to the function f, then the point (b, a) belongs to its inverse.

7. For a function $f(x)$ and its inverse function $f^{-1}(x)$, $(f \circ f^{-1})(x) = x$ and $(f^{-1} \circ f)(x) = x$.

Getting Ready for Class

After reading through the preceding section, respond in your own words and in complete sentences.

A. What is the inverse of a function?

B. What is the relationship between the graph of a function and the graph of its inverse?

C. Explain why only one-to-one functions have inverses that are also functions.

D. Describe the vertical line test, and explain the difference between the vertical line test and the horizontal line test.

Find the inverse of each function, and then determine whether the inverse itself is a function or not.

1. $\{(1, 0), (2, 1), (3, 2), (4, 3)\}$ **2.** $\{(-2, 3), (-1, 2), (0, 1), (1, -1)\}$

3. $\{(-4, 3), (-2, -1), (1, 3), (3, -2)\}$ **4.** $\{(5, 1), (5, -1), (6, 2), (6, -2)\}$

5. $\{(-3, 4), (0, 4), (3, 4)\}$ **6.** $\{(-3, -3), (0, 0), (3, 3)\}$

For each of the following relations, sketch the graph of the relation and its inverse, and write an equation for the inverse.

7. $y = 2x - 1$ **8.** $y = 3x + 1$ **9.** $y = x^2 - 3$

10. $y = x^2 + 1$ **11.** $y = x^2 - 2x - 3$ **12.** $y = x^2 + 2x - 3$

13. $y = 3^x$ **14.** $y = \left(\dfrac{1}{2}\right)^x$ **15.** $y = 4$

16. $y = -2$ **17.** $y = \dfrac{1}{2}x^3$ **18.** $y = x^3 - 2$

19. $y = \dfrac{1}{2}x + 2$ **20.** $y = \dfrac{1}{3}x - 1$ **21.** $y = \sqrt{x + 2}$

22. $y = \sqrt{x} + 2$

Determine if the following functions are one-to-one.

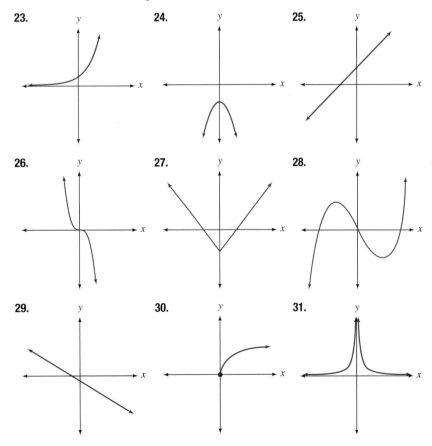

23.   **24.**   **25.**

26.   **27.**   **28.**

29.   **30.**   **31.**

32.

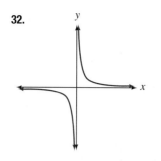

Could the following tables of values represent ordered pairs from one-to-one functions? Explain your answer.

33.

x	y
−2	5
−1	4
0	3
1	4
2	5

34.

x	y
1.5	0.1
2.0	0.2
2.5	0.3
3.0	0.4
3.5	0.5

35.

x	y
0	63
10	35
20	16
30	7
40	4

36.

x	y
0	1
2.5	−1
5	1
7.5	−1
10	1

For each of the following one-to-one functions, find the equation of the inverse. Write the inverse using the notation $f^{-1}(x)$.

37. $f(x) = 3x - 1$ **38.** $f(x) = 2x - 5$ **39.** $f(x) = x^3$

40. $f(x) = x^3 - 2$ **41.** $f(x) = \dfrac{x - 3}{x - 1}$ **42.** $f(x) = \dfrac{x - 2}{x - 3}$

43. $f(x) = \dfrac{x - 3}{4}$ **44.** $f(x) = \dfrac{x + 7}{2}$ **45.** $f(x) = \dfrac{1}{2}x - 3$

46. $f(x) = \dfrac{1}{3}x + 1$ **47.** $f(x) = \dfrac{2}{3}x - 3$ **48.** $f(x) = -\dfrac{1}{2}x + 4$

49. $f(x) = x^3 - 4$ **50.** $f(x) = -3x^3 + 2$ **51.** $f(x) = \dfrac{4x - 3}{2x + 1}$

52. $f(x) = \dfrac{3x - 5}{4x + 3}$ **53.** $f(x) = \dfrac{2x + 1}{3x + 1}$ **54.** $f(x) = \dfrac{3x + 2}{5x + 1}$

55. If $f(x) = 3x - 2$, then $f^{-1}(x) = \dfrac{x + 2}{3}$. Use these two functions to find

     a. $f(2)$ **b.** $f^{-1}(2)$ **c.** $f[f^{-1}(2)]$

     d. $f^{-1}[f(2)]$ **e.** $(f \circ f^{-1})(x)$ **f.** $(f^{-1} \circ f)(x)$

56. If $f(x) = \dfrac{1}{2}x + 5$, then $f^{-1}(x) = 2x - 10$. Use these two functions to find

a. $f(-4)$

b. $f^{-1}(-4)$

c. $f[f^{-1}(-4)]$

d. $f^{-1}[f(-4)]$

e. $(f \circ f^{-1})(x)$

f. $(f^{-1} \circ f)(x)$

57. Let $f(x) = \dfrac{1}{x}$, and find $f^{-1}(x)$.

58. Let $f(x) = \dfrac{a}{x}$, and find $f^{-1}(x)$. (a is a real number constant.)

Sketch the graph of the inverse of each function.

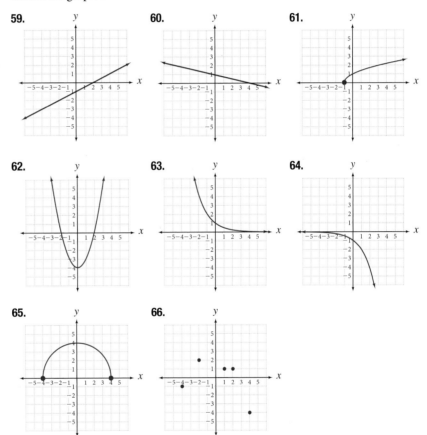

59.

60.

61.

62.

63.

64.

65.

66.

Use function composition to show that each pair of functions are inverses. That is, show $(f \circ g)(x) = x$ and $(g \circ f)(x) = x$.

67. $f(x) = 3x + 4$, $g(x) = \dfrac{x - 4}{3}$

68. $f(x) = \dfrac{1}{3}x + 2$, $g(x) = 3x - 6$

69. $f(x) = \sqrt{x + 4}$, $g(x) = x^2 - 4$
$(x \geq 0)$

70. $f(x) = x^3 + 5$, $g(x) = \sqrt[3]{x - 5}$

71. $f(x) = \dfrac{1}{x + 1}$, $g(x) = \dfrac{1 - x}{x}$

72. $f(x) = \dfrac{x + 2}{x - 3}$, $g(x) = \dfrac{3x + 2}{x - 1}$

Applying the Concepts

Inverse Functions in Words Inverses may also be found by *inverse reasoning*. For example, to find the inverse of $f(x) = 3x + 2$, first list, in order, the operations done to variable x:

Step 1: Multiply by 3.

Step 2: Add 2.

Then, to find the inverse, simply apply the inverse operations, in reverse order, to the variable x. That is:

Step 3:   Subtract 2.

Step 4:   Divide by 3.

The inverse function then becomes $f^{-1}(x) = \frac{x-2}{3}$.

73. Use this method of "inverse reasoning" to find the inverse of the function $f(x) = \frac{x}{7} - 2$.

74. Inverse Functions in Words Use *inverse reasoning* to find the following inverses:

a. $f(x) = 2x + 7$

b. $f(x) = \sqrt{x} - 9$

c. $f(x) = x^3 - 4$

d. $f(x) = \sqrt{x^3 - 4}$

75. Reading Tables Evaluate each of the following functions using the functions defined by Tables 1 and 2.

a. $f[g(-3)]$

b. $g[f(-6)]$

c. $g[f(2)]$

d. $f[g(3)]$

e. $f[g(-2)]$

f. $g[f(3)]$

What can you conclude about the relationship between functions f and g?

TABLE 1	
x	$f(x)$
-6	3
2	-3
3	-2
6	4

TABLE 2	
x	$g(x)$
-3	2
-2	3
3	-6
4	6

76. Reading Tables Use the functions defined in Tables 1 and 2 in Problem 75 to answer the following questions.

a. What are the domain and range of f?

b. What are the domain and range of g?

c. How are the domain and range of f related to the domain and range of g?

d. Is f a one-to-one function?

e. Is g a one-to-one function?

77. Social Security A function that models the billions of dollars of Social Security payment (as shown in the chart) per year is $s(t) = 16t + 249.4$, where t is time in years since 1990 (U.S. Census Bureau).

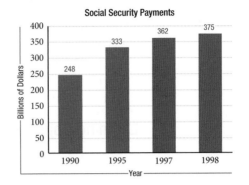

Social Security Payments

a. Use the model to estimate the amount of Social Security payments to be paid in 2020.
b. Write the inverse of the function.
c. Using the inverse function, estimate the year in which payments will reach $1 trillion.

78. Families The function for the percentage of one-parent families (as shown in the following chart) is $f(x) = 0.417x + 24$, when x is the time in years since 1990.   (*Source*: U.S. Census Bureau)

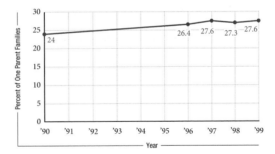

a. Use the function to predict the percentage of families with one parent in the year 2020.
b. Determine the inverse of the function, and estimate the year in which approximately 50% of the families are one-parent families.

79. Speed The fastest type of plane, a rocket plane, can travel at a speed of 4,520 miles per hour. The function $f(m) = \frac{22m}{15}$ converts miles per hour, m, to feet per second.   (*Source*: World Book Encyclopedia)

a. Use the function to convert the speed of the rocket plane to feet per second.
b. Write the inverse of the function.
c. Using the inverse function, convert 2 feet per second to miles per hour.

80. **Speed** A Lockheed SR-71A airplane set a world record (as reported by Air Force Armament Museum in 1996) with an absolute speed record of 2,193.167 miles per hour. The function $s(h) = 0.4468424h$ converts miles per hour, h, to meters per second, s.

    a. What is the absolute speed of the Lockheed SR-71A in meters per second?

    b. What is the inverse of this function?

    c. Using the inverse function, determine the speed of an airplane in miles per hour that flies 150 meters per second.

Learning Objectives Assessment

The following problems can be used to help assess if you have successfully met the learning objectives for this section.

81. If $(2, -5)$ is an element of function f, which ordered pair is an element of the inverse of f?

    a. $(-2, 5)$ **b.** $(-5, 2)$ **c.** $(-2, -5)$ **d.** $(2, 5)$

82. Graph the inverse of the function shown in the figure below.

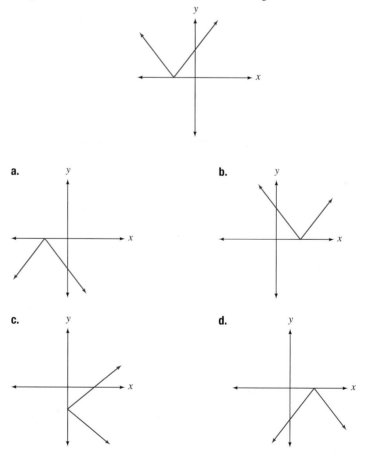

83. If $f(x) = 4x + 12$, find $f^{-1}(x)$.

    a. $f^{-1}(x) = \dfrac{1}{4}x - 3$ **b.** $f^{-1}(x) = \dfrac{1}{4}x - 12$

    c. $f^{-1}(x) = \dfrac{1}{4x + 12}$ **d.** $f^{-1}(x) = 4y + 12$

84. Which of the following is a one-to-one function?

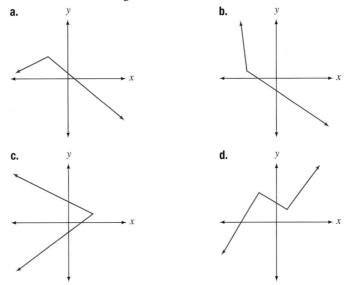

85. Which of the following would help prove that f and g are inverse functions?

    a. $g(f(x)) = x$ **b.** $f(g(x)) = 1$

    c. $f(x) = -g(x)$ **d.** $g(x) = \dfrac{1}{f(x)}$

Getting Ready for the Next Section

Simplify.

86. 3^{-2} **87.** 2^3

Solve.

88. $2 = 3x$ **89.** $3 = 5x$ **90.** $4 = x^3$ **91.** $12 = x^2$

Fill in the boxes to make each statement true.

92. $8 = 2^{\square}$ **93.** $27 = 3^{\square}$

94. $10{,}000 = 10^{\square}$ **95.** $1{,}000 = 10^{\square}$

96. $81 = 3^{\square}$ **97.** $81 = 9^{\square}$

98. $6 = 6^{\square}$ **99.** $1 = 5^{\square}$

The price of success is hard work, dedication to the job at hand, and the determination that whether we win or lose, we have applied the best of ourselves to the task at hand.
                                          —*Vince Lombardi*

My earliest memory of math is from elementary school, having to continuously take the division and times tables test because I couldn't finish it in the allotted time. I have never been naturally gifted at math, but I enjoy a challenge. I wouldn't allow my setbacks to stop me from succeeding in my classes. As a high school freshman I struggled in my geometry class, and as a senior, I thrived in my AP Calculus class. The difference in those short four years was the effort I put in, as well as having great teachers, who provided the necessary knowledge and support along the way.

However, to continue to thrive in math, I had to realize that my success wasn't solely in the hands of the teacher. I saw students fail classes taught by some of the best teachers on campus. It all stemmed from the personal goals each student held. There were multiple times when I told myself, "You can do this. You can learn the concepts." I'm not sure I would have moved past prealgebra without those personal words of encouragement.

Having confidence in yourself is necessary to be successful in all aspects of life. It's easy to give up on something that doesn't come naturally, but the reward of achieving something you thought to be impossible is worth the hardship.

Learning Objectives

In this section, we will learn how to:

1. Write an exponential expression in logarithmic form.

2. Write a logarithmic expression in exponential form.

3. Evaluate a logarithm.

4. Graph a logarithmic function.

Introduction

In March 2011, a major earthquake occurred off the coast of Japan. The sudden displacement of water in the Pacific Ocean caused a large tsunami, resulting in massive destruction and loss of life. The USGS (United States Geological Survey) reported the strength of the quake by indicating that it measured 9.0 on the Richter scale. For comparison, Table 1 gives the Richter magnitude of a number of other earthquakes.

Although the size of the numbers in the table do not seem to be very different, the intensity of the earthquakes they measure can be very different. For example, the 2004 earthquake off the coast of Northern Sumatra was 10 times stronger than the 1985 earthquake in Mexico City. The reason behind this is that the Richter scale is a ***logarithmic scale***.

TABLE 1 Earthquakes

Year	Earthquake	Richter Magnitude
1985	Mexico City	8.1
1994	Northridge	6.6
2004	Northern Sumatra	9.1
2008	Eastern Sichuan, China	7.9
2010	Haiti	7.0
2011	Honshu, Japan	9.0

In this section, we start our work with logarithms, which will give you an understanding of the Richter scale. Let's begin.

In Section 9.1 we introduced exponential functions

$$f(x) = b^x \qquad b > 0, b \neq 1$$

From our work in that section, we know that the graph of an exponential function for any base will pass the horizontal line test. This means that exponential function are one-to-one and have an inverse that is also a function.

To find the inverse, we exchange x and y:

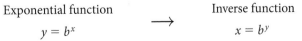

Exponential function   Inverse function

$$y = b^x \qquad \longrightarrow \qquad x = b^y$$

To isolate the y in the equation on the right, we must define a new notation.

> (dēf) **DEFINITION** *logarithm*
>
> The expression $y = \log_b x$ is read "y is the logarithm to the base b of x" and is equivalent to the expression
>
> $$x = b^y \qquad b > 0, b \neq 1$$
>
> In words, we say "$\log_b x$ is the exponent we raise b to in order to get x."

Notation   When an expression is in the form $x = b^y$, it is said to be in ***exponential form***. On the other hand, if an expression is in the form $y = \log_b x$, it is said to be in ***logarithmic form***.

Here are some equivalent statements written in both forms.

Exponential Form		Logarithmic Form
$8 = 2^3$	$\Leftrightarrow$	$\log_2 8 = 3$
$25 = 5^2$	$\Leftrightarrow$	$\log_5 25 = 2$
$0.1 = 10^{-1}$	$\Leftrightarrow$	$\log_{10} 0.1 = -1$
$\frac{1}{8} = 2^{-3}$	$\Leftrightarrow$	$\log_2 \frac{1}{8} = -3$
$r = z^s$	$\Leftrightarrow$	$\log_z r = s$

Evaluating Logarithms

One of the most important things to remember about logarithms is that a logarithm always represents an exponent. To evaluate a logarithm means to find the exponent that it represents. We illustrate this with the following examples.

VIDEO EXAMPLES

SECTION 9.3

Note Remember that a rational exponent can be used to represent a root. We used this fact in Examples 5 and 6:

$$5 = \sqrt{25} = 25^{1/2}$$
$$\frac{1}{4} = \frac{1}{\sqrt[3]{64}} = 64^{-1/3}$$

⬛ **EXAMPLE 1**   Evaluate: $\log_3 9$.

SOLUTION   We must find the exponent of 3 that will result in 9.

$$3^? = 9$$

Since the correct exponent is 2, we have $\log_3 9 = 2$.   ⬛

⬛ **EXAMPLES**

	To find	We ask	And get		
2.	$\log_2 \frac{1}{8}$	$2^? = \frac{1}{8}$	$? = -3$	so	$\log_2 \frac{1}{8} = -3$
3.	$\log_7 7$	$7^? = 7$	$? = 1$	so	$\log_7 7 = 1$
4.	$\log_9 1$	$9^? = 1$	$? = 0$	so	$\log_9 1 = 0$
5.	$\log_{25} 5$	$25^? = 5$	$? = \frac{1}{2}$	so	$\log_{25} 5 = \frac{1}{2}$
6.	$\log_{64} \frac{1}{4}$	$64^? = \frac{1}{4}$	$? = -\frac{1}{3}$	so	$\log_{64} \frac{1}{4} = -\frac{1}{3}$ ⬛

EXAMPLE 7   Find: $\log_8 16$.

SOLUTION   We must find an exponent y such that

$$8^y = 16$$

Because $8 = 2^3$ and $16 = 2^4$, we have

$$16 = (\sqrt[3]{8})^4 = 8^{4/3}$$

Therefore $\log_8 16 = \dfrac{4}{3}$.

Logarithmic Functions

Earlier in this section we found that the inverse of the exponential function $f(x) = b^x$ is given by the equation $x = b^y$. Using the logarithmic form of this equation now allows us to make the following definition.

---

dëf   DEFINITION   *logarithmic function*

A *logarithmic function* is any function that can be written in the form

$$f(x) = \log_b x$$

where b is a positive real number other than 1.

We also state the following property, which summarizes the relationship between exponential functions and logarithmic functions.

Δ≠Σ   PROPERTY

For a given base b ($b > 0$, $b \neq 1$), the exponential function with base b and the logarithmic function with base b are inverses. That is,

$$\text{if} \quad f(x) = b^x \quad \text{then} \quad f^{-1}(x) = \log_b x$$

The exponential function $y = b^x$ has a domain of all real numbers and a range of all positive real numbers. Because they are inverses, the roles will be reversed for the logarithmic function. That is, the function $y = \log_b x$ has a domain of all positive real numbers and a range of all real numbers. It is important to keep in mind that the input to a basic logarithmic function must be a positive number.

Graphing Logarithmic Functions

One way to graph a logarithmic function is to use the graph of an exponential function and the fact that the graphs of inverse functions have symmetry about the line $y = x$. Here's an example to illustrate.

EXAMPLE 8   Graph: $f(x) = \log_2 x$.

SOLUTION   The logarithmic function $y = \log_2 x$ is the inverse of the exponential function $y = 2^x$. The graph of $y = 2^x$ was given in Figure 2 of Section 9.1. We simply reflect the graph of $y = 2^x$ about the line $y = x$ to get the graph of $f(x) = \log_2 x$. (See Figure 1.)

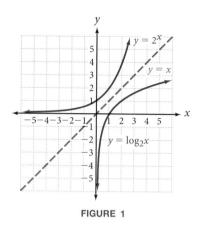

FIGURE 1

It is apparent from the graph that $y = \log_2 x$ is a function, because no vertical line will cross its graph in more than one place. Note also that the graph of $y = \log_2 x$ appears to the right of the y-axis, indicating that the domain consists of positive values of x only. There is an x-intercept at the point $(1, 0)$.

Finally, we can see in Figure 1 that the graph of the logarithmic function approaches the y-axis but never intersects it. This means the logarithmic function has a vertical asymptote at $x = 0$. ◼

Here is a summary of the behavior of the graph of a logarithmic function.

> **⌈∆≠∑⌉ *Graphs of Logarithmic Functions***
>
> The graph of $f(x) = \log_b x$, for $b > 0$ and $b \neq 1$, has the following:
>
> > domain $= \{x \mid x > 0\}$
> > range $=$ all real numbers
> > x-intercept $= (1, 0)$
> > vertical asymptote at $x = 0$

Another way to graph a logarithmic function that is more direct (not requiring we graph an exponential function first) is to simply write the equation for the function in exponential form and then make a table of ordered pairs. We illustrate this procedure in our next example.

◼ **EXAMPLE 9**   Graph: $f(x) = \log_{1/2} x$.

SOLUTION   Writing the equation $y = \log_{1/2} x$ in exponential form, we have

$$x = \left(\frac{1}{2}\right)^y$$

We can now choose values of y and evaluate the exponent to get the corresponding values of x.

y	$x = \left(\frac{1}{2}\right)^{y}$	x	Solutions
-2	$x = \left(\frac{1}{2}\right)^{-2}$	4	$(4, -2)$
-1	$x = \left(\frac{1}{2}\right)^{-1}$	2	$(2, -1)$
0	$x = \left(\frac{1}{2}\right)^{0}$	1	$(1, 0)$
1	$x = \left(\frac{1}{2}\right)^{1}$	$\frac{1}{2}$	$\left(\frac{1}{2}, 1\right)$
2	$x = \left(\frac{1}{2}\right)^{2}$	$\frac{1}{4}$	$\left(\frac{1}{4}, 2\right)$

Plotting these points gives us the graph shown in Figure 2.

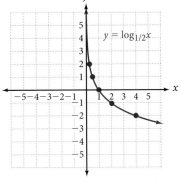

FIGURE 2

This graph represents the inverse of the exponential decay function $y = \left(\frac{1}{2}\right)^{x}$.

EXAMPLE 10   Find the domain of $f(x) = \log_4 (3 - x)$.

SOLUTION   Because a logarithmic function is only defined for a positive value, we need the expression $3 - x$ to be positive. We can express this as an inequality and solve for x.

$$3 - x > 0 \qquad \text{3 − x must be positive}$$

$$-x > -3 \qquad \text{Subtract 3 from both sides}$$

$$x < 3 \qquad \text{Multiply both sides by −1}$$

The domain is $\{x \mid x < 3\}$.

EXAMPLE 11   Sketch the graph of $f(x) = \log_2 (x + 4)$ and state the vertical asymptote.

SOLUTION   We can use transformations to sketch the graph. The graph of $y = \log_2 x$ is shown in Figure 1. To obtain the graph of $f(x) = \log_2 (x + 4)$, we must translate this graph 4 units to the left. Figure 3 shows how this is done.

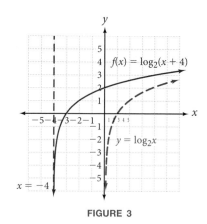

FIGURE 3

As you can see, the vertical asymptote translates 4 units to the left as well, and has the equation $x = -4$.

Application

As we mentioned in the introduction to this section, one application of logarithms is in measuring the magnitude of an earthquake. If an earthquake has a shock wave T times greater than the smallest shock wave that can be measured on a seismograph, then the magnitude M of the earthquake, as measured on the Richter scale, is given by the formula

$$M = \log_{10} T$$

(When we talk about the size of a shock wave, we are talking about its amplitude. The amplitude of a wave is half the difference between its highest point and its lowest point.)

To illustrate the discussion, an earthquake that produces a shock wave that is 10,000 times greater than the smallest shock wave measurable on a seismograph will have a magnitude M on the Richter scale of

$$M = \log_{10} 10{,}000 = 4$$

EXAMPLE 12   If an earthquake has a magnitude of $M = 5$ on the Richter scale, what can you say about the size of its shock wave?

SOLUTION   To answer this question, we put $M = 5$ into the formula $M = \log_{10} T$ to obtain

$$5 = \log_{10} T$$

Writing this expression in exponential form, we have

$$T = 10^5 = 100{,}000$$

We can say that an earthquake that measures 5 on the Richter scale has a shock wave 100,000 times greater than the smallest shock wave measurable on a seismograph.

From Example 12 and the discussion that preceded it, we find that an earthquake of magnitude 5 has a shock wave that is 10 times greater than an earthquake of magnitude 4, because 100,000 is 10 times 10,000.

Getting Ready for Class

After reading through the preceding section, respond in your own words and in complete sentences.

A. What is a logarithm?

B. What is the relationship between $y = 2^x$ and $y = \log_2 x$? How are their graphs related?

C. Will the graph of $y = \log_b x$ ever appear in the second or third quadrants? Explain why or why not.

D. Explain why $\log_2 0 = x$ has no solution for x.

Problem Set 9.3

Write each of the following expressions in logarithmic form.

1. $2^4 = 16$ **2.** $3^2 = 9$ **3.** $125 = 5^3$ **4.** $16 = 4^2$

5. $0.01 = 10^{-2}$ **6.** $0.001 = 10^{-3}$ **7.** $2^{-5} = \dfrac{1}{32}$ **8.** $4^{-2} = \dfrac{1}{16}$

9. $\left(\dfrac{1}{2}\right)^{-3} = 8$ **10.** $\left(\dfrac{1}{3}\right)^{-2} = 9$ **11.** $27 = 3^3$ **12.** $81 = 3^4$

Write each of the following expressions in exponential form.

13. $\log_{10} 100 = 2$ **14.** $\log_2 8 = 3$ **15.** $\log_2 64 = 6$

16. $\log_2 32 = 5$ **17.** $\log_8 1 = 0$ **18.** $\log_9 9 = 1$

19. $\log_{10} 0.001 = -3$ **20.** $\log_{10} 0.0001 = -4$ **21.** $\log_6 36 = 2$

22. $\log_7 49 = 2$ **23.** $\log_5 \dfrac{1}{25} = -2$ **24.** $\log_3 \dfrac{1}{81} = -4$

Evaluate each of the following.

25. $\log_2 16$ **26.** $\log_3 27$ **27.** $\log_{10} 1{,}000$ **28.** $\log_{10} 10{,}000$

29. $\log_3 3$ **30.** $\log_4 4$ **31.** $\log_5 1$ **32.** $\log_{10} 1$

33. $\log_8 \dfrac{1}{8}$ **34.** $\log_3 \dfrac{1}{3}$ **35.** $\log_4 \dfrac{1}{16}$ **36.** $\log_2 \dfrac{1}{32}$

37. $\log_{10} 0.01$ **38.** $\log_{10} 0.0001$ **39.** $\log_{16} 4$ **40.** $\log_{81} 9$

41. $\log_{64} 4$ **42.** $\log_{16} 2$ **43.** $\log_{25} 125$ **44.** $\log_9 27$

45. $\log_4 8$ **46.** $\log_{100} 1000$ **47.** $\log_{32} 16$ **48.** $\log_{64} 16$

Sketch the graph of each of the following logarithmic functions.

49. $y = \log_3 x$ **50.** $y = \log_{1/2} x$ **51.** $y = \log_{1/3} x$ **52.** $y = \log_4 x$

53. $y = \log_5 x$ **54.** $y = \log_{1/5} x$ **55.** $y = \log_{10} x$ **56.** $y = \log_{1/4} x$

Fore each function $y = f(x)$, state the inverse function $y = f^{-1}(x)$.

57. $f(x) = \log_4 x$ **58.** $f(x) = \log_{1/5} x$ **59.** $f(x) = \left(\dfrac{1}{8}\right)^x$ **60.** $f(x) = 10^x$

Find the domain of each logarithmic function.

61. $f(x) = \log_9 x$ **62.** $f(x) = \log_{1/9} x$

63. $f(x) = \log_5 (x + 6)$ **64.** $f(x) = \log_2 (x - 5)$

65. $f(x) = \log_3 (1 - x)$ **66.** $f(x) = \log_{1/3} (4 - x)$

67. $f(x) = \log_{1/2} (2x + 3)$ **68.** $f(x) = \log_{10} (3x - 2)$

Use transformations to sketch the graph of each function. Then state the vertical asymptote.

69. $f(x) = \log_2 (x - 3)$ **70.** $f(x) = \log_3 (x + 2)$

71. $f(x) = 3 + \log_2 x$ **72.** $f(x) = -2 + \log_3 x$

73. $f(x) = -2 \log_4 x$ **74.** $f(x) = -\dfrac{1}{2} \log_5 x$

Each of the following graphs has an equation of the form $y = b^x$ or $y = \log_b x$. Find the equation for each graph.

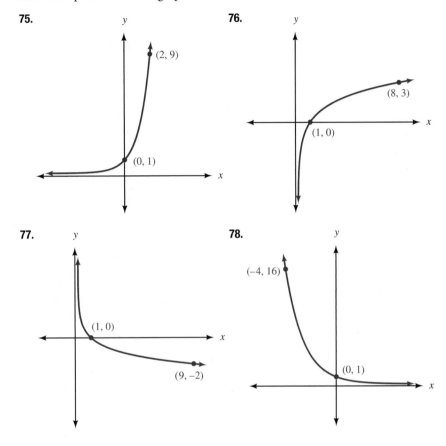

75.

y

$(2, 9)$

$(0, 1)$

x

76.

y

$(8, 3)$

$(1, 0)$

x

77.

y

$(1, 0)$

$(9, -2)$

x

78.

y

$(-4, 16)$

$(0, 1)$

x

Applying the Concepts

79. Metric System The metric system uses logical and systematic prefixes for multiplication. For instance, to multiply a unit by 100, the prefix "hecto" is applied, so a hectometer is equal to 100 meters. For each of the prefixes in the following table find the logarithm, base 10, of the multiplying factor.

Prefix	Multiplying Factor	$\log_{10}$ (Multiplying Factor)
Nano	0.000 000 001	
Micro	0.000 001	
Deci	0.1	
Giga	1,000,000,000	
Peta	1,000,000,000,000,000	

80. **Domain and Range** Use the graphs of $y = 2^x$ and $y = \log_2 x$ shown in Figure 1 of this section to find the domain and range for each function. Explain how the domain and range found for $y = 2^x$ relate to the domain and range found for $y = \log_2 x$.

81. **Magnitude of an Earthquake** Find the magnitude M of an earthquake with a shock wave that measures T = 100 on a seismograph.

82. **Magnitude of an Earthquake** Find the magnitude M of an earthquake with a shock wave that measures T = 100,000 on a seismograph.

83. **Shock Wave** If an earthquake has a magnitude of 8 on the Richter scale, how many times greater is its shock wave than the smallest shock wave measurable on a seismograph?

84. **Shock Wave** If the 1999 Colombia earthquake had a magnitude of 6 on the Richter scale, how many times greater was its shock wave than the smallest shock wave measurable on a seismograph?

Earthquake The table below categorizes earthquake by the magnitude and identifies the average annual occurrence.

Earthquakes		
Descriptor	Magnitude	Average Annual Occurrence
Great	≥8.0	1
Major	7–7.9	18
Strong	6–6.9	120
Moderate	5–5.9	800
Light	4–4.9	6,200
Minor	3–3.9	49,000
Very Minor	2–2.9	1,000 per day
Very Minor	1–1.9	8,000 per day

Source: USGS National Earthquake Information.

85. What is the average number of earthquakes that occur per year when the number of times the associated shockwave is greater than the smallest measurable shockwave, T, is 1,000,000?

86. What is the average number of earthquakes that occur per year when $T = 1,000,000$ or greater?

Learning Objectives Assessment

The following problems can be used to help assess if you have successfully met the learning objectives for this section.

87. Write $5^x = 100$ in logarithmic form.

    a. $\log_5 100 = x$ **b.** $\log_{100} 5 = x$ **c.** $\log_x 5 = 100$ **d.** $\log_5 x = 100$

88. Write $\log_{1/3} 81 = -4$ in exponential form.

    a. $(-4)^{1/3} = 81$ **b.** $\frac{1}{3}(81) = -4$ **c.** $81^{1/3} = -4$ **d.** $\left(\frac{1}{3}\right)^{-4} = 81$

89. Evaluate: $\log_6 \dfrac{1}{36}$.

    a. -2 **b.** $-\dfrac{1}{2}$ **c.** 2 **d.** $\dfrac{1}{2}$

90. Graph: $f(x) = \log_6 x$.

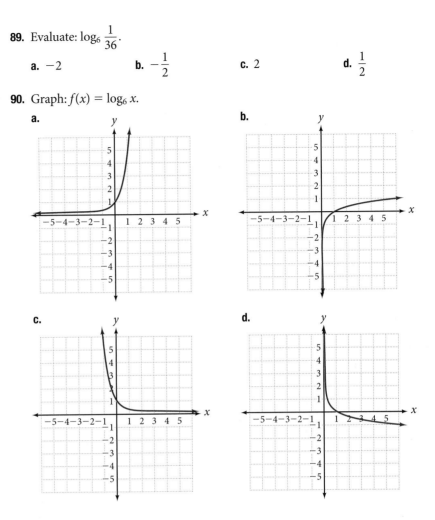

  a.   **b.**

  c.   **d.**

Getting Ready for the Next Section

Simplify.

91. $8^{2/3}$

92. $27^{2/3}$

Solve.

93. $(x + 2)(x) = 2^3$

94. $(x + 3)(x) = 2^2$

95. $\dfrac{x - 2}{x + 1} = 9$

96. $\dfrac{x + 1}{x - 4} = 25$

Write in exponential form.

97. $\log_2 [(x + 2)(x)] = 3$

98. $\log_4 [x(x - 6)] = 2$

99. $\log_3 \left(\dfrac{x - 2}{x + 1} \right) = 4$

100. $\log_3 \left(\dfrac{x - 1}{x - 4} \right) = 2$

Properties of Logarithms

Learning Objectives

In this section, we will learn how to:

1. Use properties of logarithms to evaluate logarithms.

2. Use properties of logarithms to expand a logarithm.

3. Use properties of logarithms to consolidate logarithmic expressions.

Introduction

If we search for a definition of the word **decibel**, we find the following: A unit used to express relative difference in power or intensity, usually between two acoustic or electric signals, equal to ten times the common logarithm of the ratio of the two levels.

Decibels	Comparable to
10	A light whisper
20	Quiet conversation
30	Normal conversation
40	Light traffic
50	Typewriter, loud conversation
60	Noisy office
70	Normal traffic, quiet train
80	Rock music, subway
90	Heavy traffic, thunder
100	Jet plane at takeoff

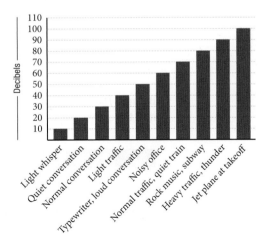

The precise definition for a **decibel** is

$$D = 10 \log_{10}\left(\frac{I}{I_0}\right)$$

where I is the intensity of the sound being measured, and I_0 is the intensity of the least audible sound. (Sound intensity is related to the amplitude of the sound wave that models the sound and is given in units of watts per meter^2.) In this section, we will see that the preceding formula can also be written as

$$D = 10(\log_{10} I - \log_{10} I_0)$$

The rules we use to rewrite expressions containing logarithms are called the *properties of logarithms*. There are seven of them.

Seven Properties of Logarithms

From our work in the previous section, we know that

$$\text{because}\quad b^0 = 1, \text{we have}\quad \log_b 1 = 0$$

and

$$\text{because}\quad b^1 = b, \text{we have}\quad \log_b b = 1$$

This leads us to our first two properties of logarithms. All seven properties assume b is a positive number other than 1.

$[\Delta \neq \Sigma]$ **PROPERTY**   *Property 1*

$$\log_b 1 = 0$$

In words: The logarithm of 1 is equal to 0.

$[\Delta \neq \Sigma]$ **PROPERTY**   *Property 2*

$$\log_b b = 1$$

In words: The logarithm of the base is equal to 1.

If you recall from Section 9.2, if $y = f(x)$ is a one-to-one function, then it has an inverse function $f^{-1}(x)$ and

$$(f \circ f^{-1})(x) = f(f^{-1}(x)) = x$$

and

$$(f^{-1} \circ f)(x) = f^{-1}(f(x)) = x$$

Because $f(x) = b^x$ is one-to-one and $f^{-1}(x) = \log_b x$, substituting these functions into the above properties of inverses gives us the following.

$[\Delta \neq \Sigma]$ **PROPERTY**   *Property 3*

$$b^{\log_b x} = x \quad (x > 0)$$

In words: A base raised to a logarithm of the same base "undo" each other.

$[\Delta \neq \Sigma]$ **PROPERTY**   *Property 4*

$$\log_b(b^x) = x$$

In words: The logarithm of an exponential function of the same base "undo" each other.

VIDEO EXAMPLES

SECTION 9.4

EXAMPLES

1. $\log_{\sqrt{2}} 1 = 0$ Property 1

2. $\log_5 5 = 1$ Property 2

3. $\log_{10} 10^4 = 4$ Property 4

4. $17^{\log_{17} 2} = 2$ Property 3

EXAMPLE 5   Simplify: $\log_4 (\log_5 5)$.

SOLUTION   Because $\log_5 5 = 1$ (see Example 2), we have

$$\log_4 (\log_5 5) = \log_4 (1) \quad \text{Property 2}$$
$$= 0 \quad \text{Property 1}$$

EXAMPLE 6   Simplify: $\sqrt{6} \log_3 3^{\sqrt{2x}}$.

SOLUTION   By Property 4, we have

$$\sqrt{6} \log_3 3^{\sqrt{2x}} = \sqrt{6}(\sqrt{2x})$$
$$= \sqrt{12x}$$
$$= 2\sqrt{3x}$$

For our last three properties, x and y are both positive real numbers and r is any real number.

Note The reason logarithms were invented in the early 1600s was because of these properties, which allow products, quotients, and powers to be performed using addition, subtraction, and multiplication.

PROPERTY   *Property 5*

$$\log_b(xy) = \log_b x + \log_b y$$

In words: The logarithm of a *product* is the *sum* of the logarithms.

PROPERTY   *Property 6*

$$\log_b\left(\frac{x}{y}\right) = \log_b x - \log_b y$$

In words: The logarithm of a *quotient* is the *difference* of the logarithms.

PROPERTY   *Property 7*

$$\log_b x^r = r \log_b x$$

In words: The logarithm of a number raised to a *power* is the *product* of the power and the logarithm of the number.

Proof of Property 5   To prove Property 5, we simply apply Property 3 and our knowledge of exponents:

$$b^{\log_b xy} = xy = (b^{\log_b x})(b^{\log_b y}) = b^{\log_b x + \log_b y}$$

Because the first and last expressions are equal and the bases are the same, the exponents $\log_b xy$ and $\log_b x + \log_b y$ must be equal. Therefore,

$$\log_b xy = \log_b x + \log_b y$$

The proofs of Properties 6 and 7 proceed in much the same manner, so we will omit them here. The examples that follow show how these last three properties can be used.

EXAMPLE 7   Simplify: $\log_5 \left(25 \cdot 5^{\sqrt{3}}\right)$.

SOLUTION   We must use Property 5 first, which will then allow us to use Property 4.

$$\log_5 \left(25 \cdot 5^{\sqrt{3}}\right) = \log_5 25 + \log_5 5^{\sqrt{3}} \qquad \text{Property 5}$$

$$= \log_5 25 + \sqrt{3} \qquad \text{Property 4}$$

$$= 2 + \sqrt{3} \qquad \log_5 25 = 2$$

EXAMPLE 8   Expand, using the properties of logarithms: $\log_5 \dfrac{3xy}{z}$

SOLUTION   Applying Property 6, we can write the quotient of $3xy$ and z in terms of a difference:

$$\log_5 \frac{3xy}{z} = \log_5 3xy - \log_5 z$$

Now we can apply Property 5 to the product $3xy$, writing it in terms of addition:

$$\log_5 \frac{3xy}{z} = \log_5 3 + \log_5 x + \log_5 y - \log_5 z$$

EXAMPLE 9   Expand, using the properties of logarithms:

$$\log_2 \frac{x^4}{\sqrt{y} \cdot z^3}$$

SOLUTION   We write $\sqrt{y}$ as $y^{1/2}$ and apply the properties:

$$\log_2 \frac{x^4}{\sqrt{y} \cdot z^3} = \log_2 \frac{x^4}{y^{1/2} z^3} \qquad \sqrt{y} = y^{1/2}$$

$$= \log_2 x^4 - \log_2(y^{1/2} \cdot z^3) \qquad \text{Property 6}$$

$$= \log_2 x^4 - (\log_2 y^{1/2} + \log_2 z^3) \qquad \text{Property 5}$$

$$= \log_2 x^4 - \log_2 y^{1/2} - \log_2 z^3 \qquad \begin{array}{l}\text{Remove parentheses}\\ \text{and distribute } -1\end{array}$$

$$= 4 \log_2 x - \frac{1}{2} \log_2 y - 3 \log_2 z \qquad \text{Property 7}$$

EXAMPLE 10   Expand, using the properties of logarithms:

$$\log_4 \frac{(x + 2)^3}{4x}$$

SOLUTION   We begin with Property 6 to write the quotient as a difference.

$$\log_4 \frac{(x + 2)^3}{4x} = \log_4 (x + 2)^3 - \log_4 (4x) \qquad \text{Property 6}$$

$$= 3 \log_4 (x + 2) - \log_4 (4x) \qquad \text{Property 7}$$

$$= 3 \log_4 (x + 2) - [\log_4 4 + \log_4 x] \qquad \text{Property 5}$$

$$= 3 \log_4 (x + 2) - [1 + \log_4 x] \qquad \text{Property 2}$$

$$= 3 \log_4 (x + 2) - 1 - \log_4 x \qquad \text{Distribute the } -1$$

⚠ COMMON MISTAKES

In Example 10, it is very tempting for students to want to further expand $\log_4 (x + 2)$ as $\log_4 x + \log_4 2$. However, these two expressions are not equal! There is no property allowing us to expand the logarithm of a sum or difference. Because this is such a common mistake, we offer the following caution.

$$\log_b (x + y) \neq \log_b x + \log_b y$$

$$\log_b (x - y) \neq \log_b x - \log_b y$$

Next, we consider some examples where we consolidate an expression containing multiple logarithms into a single logarithm. It is important to understand that in order to consolidate logarithms using Property 5 or Property 6, the logarithms must have the same base and have a coefficient of 1.

EXAMPLE 11   Evaluate: $\log_6 30 - \log_6 5$.

SOLUTION   This problem is very difficult if we leave the logarithms separate, because we do not know the exponents they represent. That is,

$$6^? = 30 \quad \text{and} \quad 6^? = 5$$

do not have simple answers. However, we can use Property 6 to consolidate the expression into a single logarithm.

$$\log_6 30 - \log_6 5 = \log_6 \left(\frac{30}{5} \right) \qquad \text{Property 6}$$

$$= \log_6 6 \qquad \frac{30}{5} = 6$$

$$= 1 \qquad \text{Property 2}$$

EXAMPLE 12   Write as a single logarithm:

$$2 \log_{10} a + 3 \log_{10} b - \frac{1}{3} \log_{10} c$$

SOLUTION   Notice all three logarithms have the same base. We begin by applying Property 7 to write each coefficient as an exponent.

$$
\begin{aligned}
2 \log_{10} a + 3 \log_{10} b - \frac{1}{3} \log_{10} c &= \log_{10} a^2 + \log_{10} b^3 - \log_{10} c^{1/3} &\quad \text{Property 7} \\
&= \log_{10} (a^2 \cdot b^3) - \log_{10} c^{1/3} &\quad \text{Property 5} \\
&= \log_{10} \frac{a^2 b^3}{c^{1/3}} &\quad \text{Property 6} \\
&= \log_{10} \frac{a^2 b^3}{\sqrt[3]{c}} &\quad c^{1/3} = \sqrt[3]{c}
\end{aligned}
$$

Getting Ready for Class

After reading through the preceding section, respond in your own words and in complete sentences.

A. Explain why the following statement is false: "The logarithm of a product is the product of the logarithms."

B. Explain the difference between $\log_b m + \log_b n$ and $\log_b(m + n)$. Are they equivalent?

C. Is $\log_b (a \cdot b^x) = ax$? Explain why or why not.

D. What conditions must be met in order to combine logarithms using Property 5 or Property 6?

Evaluate each expression by using the properties of logarithms.

1. $\log_{17} 1$ **2.** $\log_{17} 17$ **3.** $4 \log_9 9$ **4.** $9 \log_4 1$

5. $8^{\log_8 3}$ **6.** $5^{\log_5 10}$ **7.** $\log_2 2^{\sqrt{2}}$ **8.** $\log_3 3^{\sqrt{6}}$

9. $3 \log_7 7^4$ **10.** $\sqrt{5} \log_{11} 11^{\sqrt{5}}$ **11.** $6^{2 \log_6 9}$ **12.** $4^{3 \log_4 3}$

13. $\log_9 81^2$ **14.** $\log_{10} 0.1^5$ **15.** $\log_3 (\log_2 8)$ **16.** $\log_5 (\log_{32} 2)$

17. $\log_{1/2} (\log_3 81)$ **18.** $\log_9 (\log_8 2)$ **19.** $\log_3 (\log_6 6)$ **20.** $\log_5 (\log_3 3)$

21. $\log_4 [\log_2(\log_2 16)]$ **22.** $\log_4 [\log_3(\log_2 8)]$

Use the properties of logarithms given in this section to expand each expression as much as possible.

23. $\log_3 4x$ **24.** $\log_2 5x$ **25.** $\log_6 \dfrac{5}{x}$ **26.** $\log_3 \dfrac{x}{5}$

27. $\log_2 y^5$ **28.** $\log_7 y^3$ **29.** $\log_9 \sqrt[3]{z}$ **30.** $\log_8 \sqrt{z}$

31. $\log_6 x^2 y^4$ **32.** $\log_{10} x^2 y^4$ **33.** $\log_5 (\sqrt{x} \cdot y^4)$ **34.** $\log_8 \sqrt[3]{xy^6}$

35. $\log_b \dfrac{xy}{z}$ **36.** $\log_b \dfrac{3x}{y}$ **37.** $\log_{10} \dfrac{4}{xy}$ **38.** $\log_{10} \dfrac{5}{4y}$

39. $\log_{10} \dfrac{x^2 y}{\sqrt{z}}$ **40.** $\log_{10} \dfrac{\sqrt{x} \cdot y}{z^3}$ **41.** $\log_{10} \dfrac{x^3 \sqrt{y}}{z^4}$ **42.** $\log_{10} \dfrac{x^4 \sqrt[3]{y}}{\sqrt{z}}$

43. $\log_b \sqrt[3]{\dfrac{x^2 y}{z^4}}$ **44.** $\log_b \sqrt[4]{\dfrac{x^4 y^3}{z^5}}$ **45.** $\log_3 \sqrt[3]{\dfrac{x^2 y}{z^6}}$ **46.** $\log_8 \sqrt[4]{\dfrac{x^5 y^6}{z^3}}$

47. $\log_a \dfrac{4x^5}{9a^2}$ **48.** $\log_b \dfrac{16b^2}{25y^3}$ **49.** $\log_4 x^2(x + 2)$ **50.** $\log_5 y(y - 3)^3$

51. $\log_b (5b^7)$ **52.** $\log_b (3b^4)$ **53.** $\log_8 (8x^9)$ **54.** $\log_{10} (10y^2)$

55. $\log_2 8(x - 1)^5$ **56.** $\log_3 9(x + 4)^8$ **57.** $\log_6 \dfrac{x^2 z^3}{\sqrt{x + z}}$ **58.** $\log_7 \dfrac{\sqrt{x - y}}{xy^4}$

59. $\log_9 \sqrt{\dfrac{x + 3}{x - 3}}$ **60.** $\log_{11} \sqrt[3]{\dfrac{x + y}{x - y}}$

Evaluate each expression by first writing the expression as a single logarithm.

61. $\log_6 3 + \log_6 12$ **62.** $\log_{10} 25 + \log_{10} 40$ **63.** $\log_5 50 - \log_5 2$

64. $\log_2 48 - \log_2 3$ **65.** $\log_4 100 - 2 \log_4 5$ **66.** $2 \log_2 6 - \log_2 9$

Write each expression as a single logarithm.

67. $\log_b x + \log_b z$ **68.** $\log_b x - \log_b z$

69. $2 \log_3 x - 3 \log_3 y$ **70.** $4 \log_2 x + 5 \log_2 y$

71. $\dfrac{1}{2} \log_{10} x + \dfrac{1}{3} \log_{10} y$ **72.** $\dfrac{1}{3} \log_{10} x - \dfrac{1}{4} \log_{10} y$

73. $3 \log_2 x + \dfrac{1}{2} \log_2 y - \log_2 z$ **74.** $2 \log_3 x + 3 \log_3 y - \log_3 z$

75. $\dfrac{1}{2} \log_2 x - 3 \log_2 y - 4 \log_2 z$ **76.** $3 \log_{10} x - \log_{10} y - \log_{10} z$

77. $\frac{3}{2} \log_{10} x - \frac{3}{4} \log_{10} y - \frac{4}{5} \log_{10} z$ **78.** $3 \log_{10} x - \frac{4}{3} \log_{10} y - 5 \log_{10} z$

79. $\frac{1}{2} \log_5 x + \frac{2}{3} \log_5 y - 4 \log_5 z$ **80.** $\frac{1}{4} \log_7 x + 5 \log_7 y - \frac{1}{3} \log_7 z$

81. $2\log_b x + 3\log_b (x - 10)$ **82.** $3\log_b y + 2\log_b (y + 8)$

83. $4\log_6 x + 5\log_6 z - 2\log_6 (y + z)$ **84.** $9\log_7 y + \log_7 z - 3\log_7 (x - y)$

85. $\log_3(x^2 - 16) - 2\log_3(x + 4)$ **86.** $\log_4(x^2 - x - 6) - \log_4(x^2 - 9)$

Applying the Concepts

87. Decibel Formula Use the properties of logarithms to rewrite the decibel formula $D = 10 \log_{10}\left(\frac{I}{I_0}\right)$ as

$$D = 10(\log_{10} I - \log_{10} I_0).$$

88. Decibel Formula In the decibel formula $D = 10 \log_{10}\left(\frac{I}{I_0}\right)$, the threshold of hearing, I_0, is

$$I_0 = 10^{-12} \text{ watts/meter}^2$$

Substitute 10^{-12} for I_0 in the decibel formula, then show that it simplifies to

$$D = 10(\log_{10} I + 12)$$

89. Finding Logarithms If $\log_{10} 8 = 0.903$ and $\log_{10} 5 = 0.699$, find the following without using a calculator.

    a. $\log_{10} 40$ **b.** $\log_{10} 320$ **c.** $\log_{10} 1,600$

90. Matching Match each expression in the first column with an equivalent expression in the second column:

    a. $\log_2(ab)$ **i.** b

    b. $\log_2\left(\frac{a}{b}\right)$ **ii.** 2

    c. $\log_5 a^b$ **iii.** $\log_2 a + \log_2 b$

    d. $\log_a b^a$ **iv.** $\log_2 a - \log_2 b$

    e. $\log_a a^b$ **v.** $a \log_a b$

    f. $\log_3 9$ **vi.** $b \log_5 a$

91. Henderson–Hasselbalch Formula Doctors use the Henderson–Hasselbalch formula to calculate the pH of a person's blood. pH is a measure of the acidity and/or the alkalinity of a solution. This formula is represented as

$$pH = 6.1 + \log_{10}\left(\frac{x}{y}\right)$$

where x is the base concentration and y is the acidic concentration. Rewrite the Henderson–Hasselbalch formula so that the logarithm of a quotient is not involved.

92. Food Processing The formula $M = 0.21(\log_{10} a - \log_{10} b)$ is used in the food processing industry to find the number of minutes M of heat processing a certain food should undergo at 250°F to reduce the probability of survival of *Clostridium botulinum* spores. The letter a represents the number of spores per can before heating, and b represents the number of spores per can after heating. Find M if $a = 1$ and $b = 10^{-12}$. Then find M using the same values for a and b in the formula $M = 0.21 \log_{10} \frac{a}{b}$.

93. Acoustic Powers The formula $N = \log_{10} \frac{P_1}{P_2}$ is used in radio electronics to find the ratio of the acoustic powers of two electric circuits in terms of their electric powers. Find N if P_1 is 100 and P_2 is 1. Then use the same two values of P_1 and P_2 to find N in the formula $N = \log_{10} P_1 - \log_{10} P_2$.

Learning Objectives Assessment

The following problems can be used to help assess if you have successfully met the learning objectives for this section.

94. Evaluate: $\log_2 72 - \log_2 9$.

    a. 3   **b.** 8   **c.** 126   **d.** 63

95. Use properties of logarithms to expand $\log_{10} \frac{xy}{z^3}$.

    a. $\log_{10} x + \log_{10} y - 3\log_{10} z$ **b.** $\log_{10} x - \log_{10} y + 3\log_{10} z$

    c. $3\log_{10} xy - \log_{10} z$ **d.** $x + y - 3z$

96. Write $3\log_4 x - 5\log_4 y$ as a single logarithm.

    a. $\log_4 \left(\frac{x}{y}\right)^{15}$ **b.** $\log_4 (x^3 y^5)$ **c.** $\frac{\log_4 x^3}{\log_4 y^5}$ **d.** $\log_4 \frac{x^3}{y^5}$

Getting Ready for the Next Section

Simplify.

    97. 5^0 **98.** 4^1 **99.** $\log_3 3$ **100.** $\log_5 5$

    101. $\log_b b^4$ **102.** $\log_a a^k$

Common Logarithms, Natural Logarithms, and Change of Base

Learning Objectives

In this section, we will learn how to:

1. Find a common logarithm.

2. Find a natural logarithm.

3. Simplify expressions involving common or natural logarithms.

4. Change a logarithm from one base to another.

Introduction

Acid rain was first discovered in the 1960s by Gene Likens and his research team who studied the damage caused by acid rain to Hubbard Brook in New Hampshire. Acid rain is rain with a pH of 5.6 and below. As you will see as you work your way through this section, pH is defined in terms of common logarithms—one of the topics we present in this section. So, when you are finished with this section, you will have a more detailed knowledge of pH and acid rain.

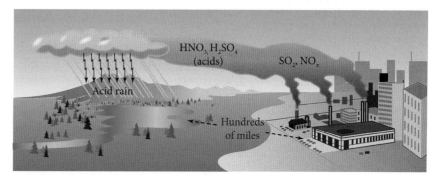

Two kinds of logarithms occur more frequently than other logarithms. Logarithms with a base of 10 are very common because our number system is a base-10 number system. For this reason, we call base-10 logarithms *common logarithms*.

> ### (def) DEFINITION *common logarithms*
>
> A *common logarithm* is a logarithm with a base of 10. Because common logarithms are used so frequently, it is customary, in order to save time, to omit notating the base. That is,
>
> $$\log_{10} x = \log x$$
>
> When the base is not shown, it is assumed to be 10.

Common Logarithms

Common logarithms of powers of 10 are simple to evaluate. We need only recognize that $\log 10 = \log_{10} 10$ and apply Property 4 of logarithms.

$$\log 1{,}000 \quad = \log_{10} 10^3 \quad = \quad 3$$
$$\log 100 \quad = \log_{10} 10^2 \quad = \quad 2$$
$$\log 10 \quad = \log_{10} 10^1 \quad = \quad 1$$
$$\log 1 \quad = \log_{10} 10^0 \quad = \quad 0$$
$$\log 0.1 \quad = \log_{10} 10^{-1} \quad = \quad -1$$
$$\log 0.01 \quad = \log_{10} 10^{-2} \quad = \quad -2$$
$$\log 0.001 \quad = \log_{10} 10^{-3} \quad = \quad -3$$

To find common logarithms of numbers that are not powers of 10, we use a calculator with a $\boxed{\text{LOG}}$ key.

VIDEO EXAMPLES

SECTION 9.5

EXAMPLE 1   Use a calculator to approximate log 2,760.

SOLUTION
$$\log 2{,}760 \approx 3.4409$$

To work this problem on a scientific calculator, we simply enter the number 2,760 and press the key labeled $\boxed{\text{LOG}}$. On a graphing calculator we press the $\boxed{\text{LOG}}$ key first, then 2,760.

The 3 in the answer is called the ***characteristic***, and the decimal part of the logarithm is called the ***mantissa***.

EXAMPLES

2. $\log 0.0391 \approx -1.4078$

3. $\log 0.00523 \approx -2.2815$

4. $\log 9.99 \approx 0.9996$

EXAMPLE 5   Write $\log x - 2\log (x + 1)$ as a single logarithm.

SOLUTION   Because the base of both logarithms is 10, we can use the properties of logarithms to consolidate the expression.

$$\log x - 2\log (x + 1) = \log x - \log (x + 1)^2 \qquad \text{Property 7}$$

$$= \log \frac{x}{(x + 1)^2} \qquad \text{Property 6}$$

EXAMPLE 6   Expand: $\log 10x^5$.

SOLUTION

$$\log 10x^5 = \log 10 + \log x^5 \qquad \text{Property 5}$$

$$= 1 + \log x^5 \qquad \log 10 = \log_{10} 10 = 1$$

$$= 1 + 5\log x \qquad \text{Property 7}$$

In Section 9.3, we found that the magnitude M of an earthquake that produces a shock wave T times larger than the smallest shock wave that can be measured on a seismograph is given by the formula

$$M = \log_{10} T$$

We can rewrite this formula using our shorthand notation for common logarithms as

$$M = \log T$$

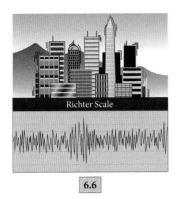

Richter Scale

6.6

EXAMPLE 7 The San Fernando earthquake of 1971 measured 6.6 on the Richter scale. The shockwave for the San Francisco earthquake of 1906 was estimated to be 50 times larger than the shockwave for the San Fernando earthquake. Based on this information, find the magnitude of the 1906 San Francisco earthquake.

SOLUTION If we assume T represents the relative size of the shockwave for the 1971 earthquake, then the size of the shockwave for the 1906 earthquake would be $50T$. Therefore, the magnitude of the San Francisco earthquake was

$$M = \log(50T) \qquad \text{Formula}$$
$$= \log 50 + \log T \qquad \text{Property 5}$$
$$= \log 50 + 6.6 \qquad \log T = 6.6$$
$$\approx 1.7 + 6.6 \qquad \log 50 \approx 1.7$$
$$= 8.3$$

In chemistry, the pH of a solution is the measure of the acidity of the solution. The definition for pH involves common logarithms. Here it is:

$$\text{pH} = -\log[\text{H}^+]$$

where $[\text{H}^+]$ is the concentration of the hydrogen ion in moles per liter. The range for pH is from 0 to 14. Pure water, a neutral solution, has a pH of 7. An acidic solution, such as vinegar, will have a pH less than 7, and an alkaline solution, such as ammonia, has a pH above 7.

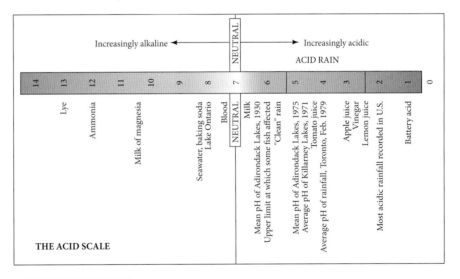

EXAMPLE 8 The concentration of the hydrogen ion in a sample of acid rain known to kill fish is 3.2×10^{-5} mole per liter. Find the pH of this acid rain to the nearest tenth.

SOLUTION Substituting 3.2×10^{-5} for $[\text{H}^+]$ in the formula $\text{pH} = -\log[\text{H}^+]$, we have

$$\text{pH} = -\log[3.2 \times 10^{-5}] \qquad \text{Substitution}$$
$$\approx -(-4.5) \qquad \text{Evaluate the logarithm}$$
$$\approx 4.5 \qquad \text{Simplify}$$

Natural Logarithms

Note The notation ln x is short for "logarithm naturale of x," which is the language that was used in the 17th century.

(d̆éf̆) **DEFINITION**   *natural logarithms*

A *natural logarithm* is a logarithm with a base of e. The natural logarithm of x is denoted by ln x. That is,

$$\ln x = \log_e x$$

We can assume that all our properties of exponents and logarithms hold for expressions with a base of e, because e is a real number. Here are some examples intended to make you more familiar with the number e and natural logarithms.

EXAMPLE 9   Simplify each of the following expressions.

a. $e^0 = 1$

b. $e^1 = e$

c. $\ln e = 1$ Property 2

d. $\ln 1 = 0$ Property 1

e. $\ln e^3 = 3$ Property 4

f. $\ln e^{-4} = -4$ Property 4

g. $\ln e^t = t$ Property 4

EXAMPLE 10   Use the properties of logarithms to expand the expression $\ln Ae^{5t}$.

SOLUTION   Because the properties of logarithms hold for natural logarithms, we have

$$\ln Ae^{5t} = \ln A + \ln e^{5t} \qquad \text{Property 5}$$

$$= \ln A + 5t \qquad \text{Property 4}$$

EXAMPLE 11   If $\ln 2 = 0.6931$ and $\ln 3 = 1.0986$, find

a. $\ln 6$ **b.** $\ln 0.5$ **c.** $\ln 8$

SOLUTION

a. Because $6 = 2 \cdot 3$, we have

$$\ln 6 = \ln 2 \cdot 3$$

$$= \ln 2 + \ln 3$$

$$= 0.6931 + 1.0986$$

$$= 1.7917$$

b. Writing 0.5 as $\frac{1}{2}$ and applying Property 6 for logarithms gives us

$$\ln 0.5 = \ln \frac{1}{2}$$
$$= \ln 1 - \ln 2$$
$$= 0 - 0.6931$$
$$= -0.6931$$

c. Writing 8 as 2^3 and applying Property 7 for logarithms, we have

$$\ln 8 = \ln 2^3$$
$$= 3 \ln 2$$
$$= 3(0.6931)$$
$$= 2.0793$$

Most calculators will only compute common logarithms and natural logarithms. If we need to approximate a logarithm with a base other than 10 or e, say $\log_8 24$, then we can use the gollowing formula to change one base to another.

Change of Base Formula

If a and b are both positive numbers other than 1, and if $x > 0$, then

$$\log_a x = \frac{\log_b x}{\log_b a}$$

Base a Base b

The logarithm on the left side has a base of a, and both logarithms on the right side have a base of b. This allows us to change from base a to any other base b that is a positive number other than 1. Here is a proof.

Proof We begin by writing the identity (Property 3 for logarithms)

$$a^{\log_a x} = x$$

Taking the logarithm base b of both sides and writing the exponent $\log_a x$ as a coefficient, we have

$$\log_b (a^{\log_a x}) = \log_b (x)$$
$$\log_a x \cdot \log_b a = \log_b x$$

Dividing both sides by $\log_b a$, we have the desired result:

$$\frac{\log_a x \cdot \log_b a}{\log_b a} = \frac{\log_b x}{\log_b a}$$
$$\log_a x = \frac{\log_b x}{\log_b a}$$

Our last example illustrates the use of this property.

EXAMPLE 12 Find: $\log_8 24$.

SOLUTION Because we do not have base-8 logarithms on our calculators, we can change this expression to an equivalent expression that contains only base-10 logarithms:

$$\log_8 24 = \frac{\log 24}{\log 8} \qquad \text{Change of base formula}$$

Don't be confused. We did not just drop the base, we changed to base 10. We could have written the last line like this:

$$\log_8 24 = \frac{\log_{10} 24}{\log_{10} 8}$$

From our calculators, we write

$$\log_8 24 \approx \frac{1.3802}{0.9031}$$

$$\approx 1.5283$$

Getting Ready for Class

After reading through the preceding section, respond in your own words and in complete sentences.

A. What is a common logarithm?

B. What is a natural logarithm?

C. What is the change of base formula used for?

D. Explain how you would approximate $\log_2 5$.

Evaluate the following logarithms.

1. $\log 1$ **2.** $\ln 1$ **3.** $\ln e$ **4.** $\log 10$

5. $\log 10{,}000$ **6.** $\log 0.001$ **7.** $\ln e^5$ **8.** $\ln e^{-3}$

9. $\log \sqrt{1000}$ **10.** $\log \sqrt[3]{10{,}000}$ **11.** $\ln \dfrac{1}{e^3}$ **12.** $\ln \sqrt{e}$

Approximate the following logarithms. Round your answers to four decimal places.

13. $\log 378$ **14.** $\log 37.8$ **15.** $\ln 345$ **16.** $\ln 10$

17. $\log 0.4260$ **18.** $\log 0.00971$ **19.** $\ln 0.345$ **20.** $\ln 0.0345$

Simplify each of the following expressions.

21. $\ln e^x$ **22.** $\ln e^y$ **23.** $\log 10^x$ **24.** $\log 10^z$

25. $10^{\log 3x}$ **26.** $e^{\ln 2y}$ **27.** $e^{4\ln x}$ **28.** $10^{5\log y}$

Use the properties of logarithms to expand each of the following expressions.

29. $\ln 10e^{3t}$ **30.** $\ln 10e^{4t}$ **31.** $\ln Ae^{-2t}$

32. $\ln Ae^{-3t}$ **33.** $\log [100(1.01)^{3t}]$ **34.** $\log \left[\dfrac{1}{10}\, (1.5)^{t+2} \right]$

35. $\ln (Pe^{rt})$ **36.** $\ln \left(\dfrac{1}{2}\, e^{-kt} \right)$ **37.** $-\log (4.2 \times 10^{-3})$

38. $-\log (5.7 \times 10^{-10})$

Write each expression as a single logarithm.

39. $\log x + \log (x - 2)$ **40.** $\log (x + 3) - \log (x - 1)$

41. $\ln (x + 1) - \ln (x + 4)$ **42.** $\ln x + \ln (x - 3)$

43. $2\log x - 5\log y$ **44.** $3\ln x + 4\ln z$

If $\ln 2 = 0.6931$, $\ln 3 = 1.0986$, and $\ln 5 = 1.6094$, find each of the following.

45. $\ln 15$ **46.** $\ln 10$ **47.** $\ln \dfrac{1}{3}$ **48.** $\ln \dfrac{1}{5}$

49. $\ln 9$ **50.** $\ln 25$ **51.** $\ln 16$ **52.** $\ln 81$

53. Graph: $f(x) = \log x$.   **54.** Graph: $f(x) = \ln x$.

Use the change-of-base property and a calculator to find a decimal approximation to each of the following logarithms. Round your answers to four decimal places.

55. $\log_8 16$ **56.** $\log_9 27$ **57.** $\log_{16} 8$ **58.** $\log_{27} 9$

59. $\log_7 15$ **60.** $\log_3 12$ **61.** $\log_{15} 7$ **62.** $\log_{12} 3$

63. $\log_8 240$ **64.** $\log_6 180$ **65.** $\log_4 321$ **66.** $\log_5 462$

Applying the Concepts

67. **Atomic Bomb Tests** The formula for determining the magnitude, M, of an earthquake on the Richter scale is $M = \log_{10} T$, where T is the number of times the shockwave is greater than the smallest measurable shockwave. The Bikini Atoll in the Pacific Ocean was used as a location for atomic bomb tests by the United States government in the 1950s. One such test resulted in an earthquake. If the 1906 San Francisco earthquake of estimated magnitude 8.3 had a shockwave that was approximately 2,000 times greater, find the magnitude of the earthquake caused by this test.

68. **Atomic Bomb Tests** Today's nuclear weapons are 1,000 times more powerful than the atomic bombs tested in the Bikini Atoll mentioned in Problem 67. Determine the Richter scale measurement of a nuclear test today.

69. **Earthquake** The chart below is a partial listing of earthquakes that were recorded in Canada during one year. Complete the chart by computing the magnitude on the Richter Scale, M.

Location	Date	Magnitude M	Shockwave T
Moresby Island	Jan. 23		1.00×10^4
Vancouver Island	Apr. 30		1.99×10^5
Quebec City	June 29		1.58×10^3
Mould Bay	Nov. 13		1.58×10^5
St. Lawrence	Dec. 14		5.01×10^3

Source: National Resources Canada, National Earthquake Hazards Program.

70. **Memory** A class of students take a test on the mathematics concept of solving quadratic equations. That class agrees to take a similar form of the test each month for the next 6 months to test their memory of the topic since instruction. The function of the average score earned each month on the test is $m(x) = 75 - 5 \ln(x + 1)$, where x represents time in months. Complete the table to indicate the average score earned by the class at each month.

Time, x	Score, m
0	
1	
2	
3	
4	
5	
6	

Use the formula $pH = -\log[H^+]$, where $[H^+]$ is the concentration of the hydrogen ion in moles per liter, to solve problems 71-72.

71. **pH** Find the pH of orange juice if the concentration of the hydrogen ion in the juice is $[H^+] = 6.50 \times 10^{-4}$.

72. **pH** Find the pH of milk if the concentration of the hydrogen ions in milk is $[H^+] = 1.88 \times 10^{-6}$.

Learning Objectives Assessment

The following problems can be used to help assess if you have successfully met the learning objectives for this section.

73. Find: log 0.001.

    a. -6.9 **b.** 3   **c.** -2 **d.** -3

74. Approximate: ln 100.

    a. 271.8   **b.** 2.7   **c.** 2   **d.** 4.6

75. Use the properties of logarithms to expand $\ln 50e^{0.6t}$.

    a. $\ln 50 + 0.6t$ **b.** $50 + 0.6t$ **c.** $30t$ **d.** $0.6t(\ln 50)$

76. Approximate: $\log_{13} 31$.

    a. 1.3388   **b.** 0.7469   **c.** 34.5322   **d.** 0.1147

Getting Ready for the Next Section

Solve.

77. $5(2x + 1) = 12$ **78.** $4(3x - 2) = 21$

Use a calculator to evaluate, give answers to 4 decimal places.

79. $\dfrac{100,000}{32,000}$ **80.** $\dfrac{1.4982}{6.5681} + 3$

81. $\dfrac{1}{2}\left(\dfrac{-0.6931}{1.4289} + 3\right)$ **82.** $1 + \dfrac{0.04}{52}$

Use the power rule to rewrite the following logarithms.

83. $\log 1.05^t$ **84.** $\log 1.033^t$

Use identities to simplify.

85. $\ln e^{0.05t}$ **86.** $\ln e^{-0.000121t}$

Use a calculator to find each of the following. Write your answer in scientific notation with the first number in each answer rounded to the nearest tenth.

87. $10^{-5.6}$ **88.** $10^{-4.1}$

Divide and round to the nearest whole number.

89. $\dfrac{2.00 \times 10^8}{3.96 \times 10^6}$ **90.** $\dfrac{3.25 \times 10^{12}}{1.72 \times 10^{10}}$

Exponential and Logarithmic Equations

Learning Objectives

In this section, we will learn how to:

1. Solve an exponential equation.
2. Solve a logarithmic equation.
3. Solve applications involving exponential and logarithmic equations.

Introduction

For items involved in exponential growth, the time it takes for a quantity to double is called the **doubling time**. For example, if you invest $5,000 in an account that pays 5% annual interest, compounded quarterly, you may want to know how long it will take for your money to double in value. You can find this doubling time if you can solve the equation

$$10{,}000 = 5{,}000\,(1.0125)^{4t}$$

You will see, as you progress through this section, logarithms are the key to solving equations of this type.

Exponential Equations

The equation mentioned above is an example of an **exponential equation**. An exponential equation is simply an equation that contains one or more exponential functions.

One way to solve an exponential equation is to isolate the exponential function, and then write the equation in logarithmic form. The following example shows how this is done.

VIDEO EXAMPLES

SECTION 9.6

EXAMPLE 1   Solve: $5^x = 12$.

SOLUTION   The exponential function, 5^x, is already isolated on the left side. We solve for x by writing the equation in logarithmic form.

$$5^x = 12 \qquad \text{Original equation}$$

$$x = \log_5 12 \qquad \text{Logarithmic form}$$

The solution is the number $\log_5 12$. If we want to approximate this value, we can use the change of base formula.

$$\log_5 12 = \frac{\log 12}{\log 5}$$

$$\approx 1.5440$$

A second method for solving exponential equations involves taking a logarithm of both sides by applying the following property.

⎰Δ≠Σ   PROPERTY

If x and y are positive real numbers and $x = y$, then for any base b ($b > 0$, $b \neq 1$):

$$\log_b x = \log_b y$$

Here is how the previous example would look if we solved the equation using this second method.

EXAMPLE 2   Use the second method for solving exponential equations to solve $5^x = 12$.

SOLUTION   Because 5^x and 12 are equal, we have

$$5^x = 12$$

$$\log 5^x = \log 12 \qquad \text{Apply the property}$$

$$x \log 5 = \log 12 \qquad \text{Property 7}$$

$$x = \frac{\log 12}{\log 5} \qquad \text{Divide both sides by log 5}$$

$$\approx 1.5440 \qquad \text{Approximate}$$

As you can see, we obtain the same result.

HOW TO   *Solve an Exponential Equation*

Step 1: Isolate the exponential function.

Method 1

Step 2: Write the equation in logarithmic form.

Step 3: Isolate the variable.

Step 4: If necessary, use the change of base formula to approximate the logarithm.

Method 2

Step 2: Take a logarithm of both sides.

Step 3: Apply Property 7 for logarithms (or Property 4 if the bases are the same).

Step 4: Isolate the variable.

EXAMPLE 3   Solve: $25^{2x+1} = 15$.

SOLUTION   The exponential function, 25^{2x+1}, is already isolated.

Method 1:

$$25^{2x+1} = 15$$

$$2x + 1 = \log_{25} 15 \qquad \text{Logarithmic form}$$

$$2x = \log_{25} 15 - 1 \qquad \text{Subtract 1 from both sides}$$

$$x = \frac{1}{2}(\log_{25} 15 - 1) \qquad \text{Multiply both sides by } \frac{1}{2}$$

$$x = \frac{1}{2}\left(\frac{\log 15}{\log 25} - 1\right) \qquad \text{Change of base formula}$$

$$\approx \frac{1}{2}(0.8413 - 1)$$

$$\approx -0.079$$

Method 2:

$$25^{2x+1} = 15$$

$$\log 25^{2x+1} = \log 15 \qquad \text{Take the log of both sides}$$

$$(2x + 1)\log 25 = \log 15 \qquad \text{Property 7}$$

$$2x + 1 = \frac{\log 15}{\log 25} \qquad \text{Divide by log 25}$$

$$2x = \frac{\log 15}{\log 25} - 1 \qquad \text{Add } -1 \text{ to both sides}$$

$$x = \frac{1}{2}\left(\frac{\log 15}{\log 25} - 1\right) \qquad \text{Multiply both sides by } \frac{1}{2}$$

$$\approx \frac{1}{2}(0.8413 - 1)$$

$$\approx -0.079$$

USING TECHNOLOGY   *Graphing Calculators*

We can evaluate many logarithmic expressions on a graphing calculator by using the fact that logarithmic functions and exponential functions are inverses.

EXAMPLE 4   Evaluate the logarithmic expression $\log_3 7$ from the graph of an exponential function.

SOLUTION   First, we let $\log_3 7 = x$. Next, we write this expression in exponential form as $3^x = 7$. We can solve this equation graphically by finding the intersection of the graphs $Y_1 = 3^x$ and $Y_2 = 7$, as shown in Figure 1.

Using the calculator, we find the two graphs intersect at $(1.77, 7)$. Therefore, $\log_3 7 = 1.77$ to the nearest hundredth. We can check our work by evaluating the expression $3^{1.77}$ on our calculator with the key strokes

$$3 \;\boxed{\wedge}\; 1.77 \;\boxed{\text{ENTER}}$$

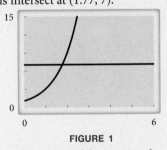

FIGURE 1

The result is 6.99 to the nearest hundredth, which seems reasonable since 1.77 is accurate to the nearest hundredth. To get a result closer to 7, we would need to find the intersection of the two graphs more accurately.

Applications of Exponential Equations

If you invest P dollars in an account with an annual interest rate r that is compounded n times a year, then t years later the amount of money in that account will be

$$A = P\left(1 + \frac{r}{n}\right)^{nt}$$

EXAMPLE 5 How long does it take for $5,000 to double if it is deposited in an account that yields 5% interest compounded once a year?

SOLUTION Substituting $P = 5,000$, $r = 0.05$, $n = 1$, and $A = 10,000$ into our formula, we have

$$10,000 = 5,000(1 + 0.05)^t$$

$$10,000 = 5,000(1.05)^t$$

$$2 = (1.05)^t \qquad\qquad \text{Divide by 5,000}$$

This is an exponential equation. We solve by taking the logarithm of both sides (Method 2):

$$\log 2 = \log(1.05)^t$$

$$= t \log 1.05$$

Dividing both sides by $\log 1.05$, we have

$$t = \frac{\log 2}{\log 1.05}$$

$$\approx 14.2$$

It takes a little over 14 years for $5,000 to double if it earns 5% interest per year, compounded once a year.

EXAMPLE 6 Suppose that the population in a small city is 32,000 in the beginning of 2010 and that the city council assumes that the population size t years later can be estimated by the equation

$$P = 32,000e^{0.05t}$$

Approximately when will the city have a population of 50,000?

SOLUTION We substitute 50,000 for P in the equation and solve for t:

$$50,000 = 32,000e^{0.05t}$$

$$1.5625 = e^{0.05t} \qquad\qquad \text{Divide both sides by 32,000}$$

To solve this equation for t, we can take the natural logarithm of each side:

$$\ln 1.5625 = \ln e^{0.05t}$$

$$= 0.05t \qquad\qquad \text{Property 4}$$

$$t = \frac{\ln 1.5625}{0.05} \qquad\qquad \text{Divide each side by 0.05}$$

$$\approx 8.93 \text{ years}$$

Note In Example 6, because the base of the logarithm and the base of the exponential function are both e, we can use Property 4 for logarithms to simplify the right side of the equation instead of Property 7.

We can estimate that the population will reach 50,000 toward the end of 2018.

Logarithmic Equations

Now we will turn our attention to solving *logarithmic equations*, which are equations containing one or more logarithmic functions.

> **HOW TO** *Solve Logarithmic Equations*
>
> **Step 1:** Isolate the logarithmic function. If necessary, combine logarithms using Property 5 or Property 6.
>
> **Step 2:** Write the equation in exponential form.
>
> **Step 3:** Isolate the variable.
>
> **Step 4:** Check for extraneous solutions.

EXAMPLE 7 Solve: $\log_3 x = -2$.

SOLUTION In exponential form, the equation looks like this:

$$x = 3^{-2}$$

or

$$x = \frac{1}{9}$$

The solution is $\frac{1}{9}$.

EXAMPLE 8 Solve: $\log_x 4 = 3$.

SOLUTION Again, we use the definition of logarithms to write the expression in exponential form:

$$4 = x^3$$

Taking the cube root of both sides, we have

$$\sqrt[3]{4} = \sqrt[3]{x^3}$$

$$x = \sqrt[3]{4}$$

The solution set is $\{\sqrt[3]{4}\}$.

EXAMPLE 9 Solve: $\log x = 3.8774$.

SOLUTION

$$\log x = 3.8774$$

$$x = 10^{3.8774} \qquad \text{Exponential form}$$

$$\approx 7{,}540 \qquad \text{Approximate}$$

In the specific circumstance where the logarithm is a common logarithm, the solution 7,540 is called the *antilogarithm*, or just *antilog*, of 3.8774. That is, 7,540 is the number whose logarithm is 3.8774.

If more than one logarithmic function appears in the equation, we may be able to use a property to combine the logarithms as long as they have the same base.

EXAMPLE 10 Solve: $\log_2(x + 2) + \log_2 x = 3$.

SOLUTION Applying Property 5 to the left side of the equation allows us to write it as a single logarithm:

$$\log_2(x + 2) + \log_2 x = 3$$
$$\log_2[(x + 2)(x)] = 3$$

The last line can be written in exponential form using the definition of logarithms:

$$(x + 2)(x) = 2^3$$

Solve as usual:

$$x^2 + 2x = 8$$
$$x^2 + 2x - 8 = 0$$
$$(x + 4)(x - 2) = 0$$
$$x + 4 = 0 \quad \text{or} \quad x - 2 = 0$$
$$x = -4 \qquad\qquad x = 2$$

In a previous section, we noted the fact that x in the expression $y = \log_b x$ cannot be a negative number. Because substitution of $x = -4$ into the original equation gives

$$\log_2(-2) + \log_2(-4) = 3$$

which contains logarithms of negative numbers, we cannot use -4 as a solution. It is an extraneous solution. The solution set is $\{2\}$. ∎

Notice that we obtained an extraneous solution in Example 10. Extraneous solutions can occur whenever we use Properties 5, 6, or 7 for logarithms to help solve a logarithmic equation. As a result, it is a good idea to check for extraneous solutions when solving any logarithmic equation.

EXAMPLE 11 Solve: $\ln(3x + 7) = \ln(2x + 5)$.

SOLUTION First we get both logarithms on the same side of the equation, and then combine them using Property 6.

$$\ln(3x + 7) = \ln(2x + 5)$$

$$\ln(3x + 7) - \ln(2x + 5) = 0 \qquad\qquad \text{Subtract } \ln(2x + 5) \text{ from both sides}$$

$$\ln \frac{3x + 7}{2x + 5} = 0 \qquad\qquad \text{Property 6}$$

$$\log_e \frac{3x + 7}{2x + 5} = 0 \qquad\qquad \ln \text{ means } \log_e$$

$$\frac{3x + 7}{2x + 5} = e^0 \qquad\qquad \text{Exponential form}$$

$$\frac{3x + 7}{2x + 5} = 1 \qquad\qquad e^0 = 1$$

$$3x + 7 = 2x + 5 \qquad\qquad \text{Multiply both sides by } 2x + 5$$

$$x + 7 = 5 \qquad\qquad \text{Subtract } 2x \text{ from both sides}$$

$$x = -2$$

> *Note* We do not discard $x = -2$ as a solution simply because it is negative. As you can see, both logarithms end up being evaluated at a positive number (1) which is in their domain.

Now we check to see if $x = -2$ is an extraneous solution. Replacing x with -2 in the original equation gives us

$$\ln(3(-2) + 7) \overset{?}{=} \ln(2(-2) + 5)$$

$$\ln(-6 + 7) \overset{?}{=} \ln(-4 + 5)$$

$$\ln 1 = \ln 1$$

$$0 = 0 \qquad \text{A true statement}$$

Therefore, $x = -2$ is a solution to the equation.

Applications of Logarithmic Equations

In the previous section we introduced the formula for pH of a solution:

$$\text{pH} = -\log[\text{H}^+]$$

where $[\text{H}^+]$ is the concentration of the hydrogen ion in moles per liter.

EXAMPLE 12 Normal rainwater has a pH of 5.6. What is the concentration of the hydrogen ion in normal rainwater?

SOLUTION Substituting 5.6 for pH in the formula $\text{pH} = -\log[\text{H}^+]$, we have

$$5.6 = -\log[\text{H}^+] \qquad \text{Substitution}$$

$$\log[\text{H}^+] = -5.6 \qquad \text{Isolate the logarithm}$$

$$[\text{H}^+] = 10^{-5.6} \qquad \text{Write in exponential form}$$

$$\approx 2.5 \times 10^{-6} \text{ moles per liter} \qquad \text{Answer in scientific notation}$$

Getting Ready for Class

After reading through the preceding section, respond in your own words and in complete sentences.

A. What is an exponential equation?

B. How do logarithms help you solve exponential equations?

C. Describe the process for solving a logarithmic equation.

D. What can cause extraneous solutions when solving a logarithmic equation?

Problem Set 9.6

Solve each exponential equation. Use a calculator to write the answer in decimal form accurate to four decimal places.

1. $3^x = 5$ **2.** $4^x = 3$ **3.** $5^x = 3$ **4.** $3^x = 4$

5. $5^{-x} = 12$ **6.** $7^{-x} = 8$ **7.** $12^{-x} = 5$ **8.** $8^{-x} = 7$

9. $8^{x+1} = 4$ **10.** $9^{x+1} = 3$ **11.** $4^{x-1} = 4$ **12.** $3^{x-1} = 9$

13. $3^{2x+1} = 2$ **14.** $2^{2x+1} = 3$ **15.** $3^{1-2x} = 2$ **16.** $2^{1-2x} = 3$

17. $15^{3x-4} = 10$ **18.** $10^{3x-4} = 15$ **19.** $6^{5-2x} = 4$ **20.** $9^{7-3x} = 5$

21. $3^{-4x} = 81$ **22.** $2^{5x} = \dfrac{1}{16}$ **23.** $5^{3x-2} = 15$ **24.** $7^{4x+3} = 200$

25. $100e^{3t} = 250$ **26.** $150e^{0.065t} = 400$

27. $1200\left(1 + \dfrac{0.072}{4}\right)^{4t} = 25000$ **28.** $2700\left(1 + \dfrac{0.086}{12}\right)^{12t} = 10000$

29. $50e^{-0.0742t} = 32$ **30.** $19e^{-0.000243t} = 12$

Solve each logarithmic equation.

31. $\log_3 x = 2$ **32.** $\log_4 x = 3$ **33.** $\log_5 x = -3$ **34.** $\log_2 x = -4$

35. $\log_x 4 = 2$ **36.** $\log_x 16 = 4$ **37.** $\log_x 5 = 3$ **38.** $\log_x 8 = 2$

39. $\log_x 36 = 2$ **40.** $\log_5 x = -2$ **41.** $\log_8 x = -2$ **42.** $\log_x \dfrac{1}{25} = 2$

43. $\log x = 1$ **44.** $\log x = -1$ **45.** $\log x = -2$

46. $\log x = 4$ **47.** $\ln x = -1$ **48.** $\ln x = 4$

49. $\log x = 10$ **50.** $\log x = 20$ **51.** $\log x = -20$

52. $\log x = -10$ **53.** $\log x = \log_2 8$ **54.** $\log x = \log_3 9$

Solve each logarithmic equation. Use a calculator to write the answer in decimal form.

55. $\log x = 2.8802$ **56.** $\log x = 4.8802$ **57.** $\log x = -2.1198$

58. $\log x = -3.1198$ **59.** $\ln x = 3.1553$ **60.** $\ln x = 5.5911$

61. $\ln x = -5.3497$ **62.** $\ln x = -1.5670$

Solve each of the following logarithmic equations. Be sure to check for any extraneous solutions.

63. $\log_2 x + \log_2 3 = 1$ **64.** $\log_3 x + \log_3 3 = 1$

65. $\log_3 x - \log_3 2 = 2$ **66.** $\log_3 x + \log_3 2 = 2$

67. $\log_3 x + \log_3(x - 2) = 1$ **68.** $\log_6 x + \log_6(x - 1) = 1$

69. $\log_3(x + 3) - \log_3(x - 1) = 1$ **70.** $\log_4(x - 2) - \log_4(x + 1) = 1$

71. $\log_2 x + \log_2(x - 2) = 3$ **72.** $\log_4 x + \log_4(x + 6) = 2$

73. $\log_8 x + \log_8(x - 3) = \dfrac{2}{3}$ **74.** $\log_{27} x + \log_{27}(x + 8) = \dfrac{2}{3}$

75. $\log_3(x + 2) - \log_3 x = 1$ **76.** $\log_2(x + 3) - \log_2(x - 3) = 2$

77. $\log_2(x + 1) + \log_2(x + 2) = 1$

78. $\log_3 x + \log_3(x + 6) = 3$

79. $\log_9 \sqrt{x} + \log_9 \sqrt{2x + 3} = \dfrac{1}{2}$

80. $\log_8 \sqrt{x} + \log_8 \sqrt{5x + 2} = \dfrac{2}{3}$

81. $4 \log_3 x - \log_3 x^2 = 6$

82. $9 \log_4 x - \log_4 x^3 = 12$

83. $\log_5 \sqrt{x} + \log_5 \sqrt{6x + 5} = 1$

84. $\log_2 \sqrt{x} + \log_2 \sqrt{6x + 5} = 1$

85. $\log x = 2 \log 5$

86. $\log x = -\log 4$

87. $\ln x = -3 \ln 2$

88. $\ln x = 5 \ln 3$

Applying the Concepts

89. Compound Interest How long will it take for $500 to double if it is invested at 6% annual interest compounded 2 times a year?

90. Compound Interest How long will it take for $500 to double if it is invested at 6% annual interest compounded 12 times a year?

91. Compound Interest How long will it take for $1,000 to triple if it is invested at 12% annual interest compounded 6 times a year?

92. Compound Interest How long will it take for $1,000 to become $4,000 if it is invested at 12% annual interest compounded 6 times a year?

93. Doubling Time How long does it take for an amount of money P to double itself if it is invested at 8% interest compounded 4 times a year?

94. Tripling Time How long does it take for an amount of money P to triple itself if it is invested at 8% interest compounded 4 times a year?

95. Tripling Time If a $25 investment is worth $75 today, how long ago must that $25 have been invested at 6% interest compounded twice a year?

96. Doubling Time If a $25 investment is worth $50 today, how long ago must that $25 have been invested at 6% interest compounded twice a year?

Recall from Section 9.1 that if P dollars are invested in an account with annual interest rate r, compounded continuously, then the amount of money in the account after t years is given by the formula

$$A(t) = Pe^{rt}$$

97. Continuously Compounded Interest Repeat Problem 89 if the interest is compounded continuously.

98. Continuously Compounded Interest Repeat Problem 92 if the interest is compounded continuously.

99. Continuously Compounded Interest How long will it take $500 to triple if it is invested at 6% annual interest, compounded continuously?

100. Continuously Compounded Interest How long will it take $500 to triple if it is invested at 12% annual interest, compounded continuously?

101. Continuously Compounded Interest How long will it take for $1,000 to be worth $2,500 at 8% interest, compounded continuously?

102. Continuously Compounded Interest How long will it take for $1,000 to be worth $5,000 at 8% interest, compounded continuously?

103. **Exponential Decay** Twinkies on the shelf of a convenience store lose their fresh tastiness over time. We say that the taste quality is 1 when the Twinkies are first put on the shelf at the store, and that the quality of tastiness declines according to the function $Q(t) = 0.85^t$ (t in days). Determine when the taste quality will be one half of its original value.

104. **Henderson–Hasselbalch Formula** Doctors use the Henderson–Hasselbalch formula to calculate the pH of a person's blood. pH is a measure of the acidity and/or the alkalinity of a solution. This formula is represented as

$$pH = 6.1 + \log_{10}\left(\frac{x}{y}\right)$$

where x is the base concentration and y is the acidic concentration. If most people have a blood pH of 7.4, use the Henderson–Hasselbalch formula to find the ratio of $\frac{x}{y}$ for an average person.

105. **University Enrollment** The percentage of students enrolled in a university who are between the ages of 25 and 34 can be modeled by the formula $s = 5\ln x$, where s is the percentage of students and x is the number of years since 1989. Predict the year in which approximately 20% of students will enroll in a university who are between the ages of 25 and 34.

106. **Earthquake** On January 6, 2001, an earthquake with a magnitude of 7.7 on the Richter Scale hit southern India (*National Earthquake Information Center*). By what factor was this earthquake's shockwave greater than the smallest measurable shockwave?

The Richter Scale Find the relative size T of the shock wave of earthquakes with the following magnitudes, as measured on the Richter scale.

107. 5.5   108. 6.6   109. 8.3   110. 8.7

111. **pH** Find the concentration of hydrogen ions in a glass of wine if the pH is 4.75.

112. **pH** Find the concentration of hydrogen ions in a bottle of vinegar if the pH is 5.75.

113. **Exponential Growth** Suppose that the population in a small city is 32,000 at the beginning of 2005 and that the city council assumes that the population size t years later can be estimated by the equation

$$P(t) = 32{,}000e^{0.05t}$$

Approximately when will the city have a population of 80,000?

114. **Exponential Growth** Suppose the population of a city is given by the equation

$$P(t) = 100{,}000e^{0.05t}$$

where t is the number of years from the present time. How large is the population now? (*Now* corresponds to a certain value of t. Once you realize what that value of t is, the problem becomes very simple.)

115. **Airline Travel** The number of airline passengers in 1990 was 466 million. The number of passengers traveling by airplane each year has increased exponentially according to the model, $P(t) = 466 \cdot 1.035^t$, where t is the number of years since 1990 (U.S. Census Bureau). In what year is it predicted that 1.9 billion passengers will travel by airline?

116. **Bankruptcy Model** In 1997, there were a total of 1,316,999 bankruptcies filed under the Bankruptcy Reform Act. The model for the number of bankruptcies filed is $B(t) = 0.798 \cdot 1.164^t$, where t is the number of years since 1994 and B is the number of bankruptcies filed in terms of millions (Administrative Office of the U.S. Courts, *Statistical Tables for the Federal Judiciary*). In what year is it predicted that 50 million bankruptcies will be filed?

117. **Health Care** In 1990, $699 billion was spent on health care expenditures. The amount of money, E, in billions spent on health care expenditures can be estimated using the function $E(t) = 78.16(1.11)^t$, where t is time in years since 1970 (*U.S. Census Bureau*). In what year is it estimated that $40 trillion will be spent on health care expenditures?

118. **Value of a Car** As a car ages, its value decreases. The value of a particular car with an original purchase price of $25,600 is modeled by the function $c(t) = 25,600(1 - 0.22)^t$, where c is the value at time t (Kelly Blue Book). How old is the car when its value is $10,000?

119. **Carbon Dating** Scientists use Carbon-14 dating to find the age of fossils and other artifacts. The amount of Carbon-14 in an organism will yield information concerning its age. A formula used in Carbon-14 dating is $A(t) = A_0 \cdot 2^{-t/5600}$, where A_0 is the amount of carbon originally in the organism, t is time in years, and A is the amount of carbon remaining after t years. Determine the number of years since an organism died if it originally contained 1,000 gram of Carbon-14 and it currently contains 600 gram of Carbon-14.

120. **Online Banking Use** The number of households using online banking services has increased from 754,000 in 1995 to 12,980,000 in 2000. The formula $H(t) = 0.76e^{0.55t}$ models the number of households, H, in millions when time is t years since 1995 according to the Home Banking Report. According to this model, in what year did 50,000,000 households use online banking services?

Depreciation The annual rate of depreciation r on a car that is purchased for P dollars and is worth W dollars t years later can be found from the formula.

$$\log(1 - r) = \frac{1}{t} \log \frac{W}{P}$$

121. Find the annual rate of depreciation on a car that is purchased for $9,000 and sold 5 years later for $4,500.

122. Find the annual rate of depreciation on a car that is purchased for $9,000 and sold 4 years later for $3,000.

Two cars depreciate in value according to the following depreciation tables. In each case, find the annual rate of depreciation.

123.

Age in Years	Value in Dollars
New	7,550
5	5,750

124.

Age in Years	Value in Dollars
New	7,550
3	5,750

Learning Objectives Assessment

The following problems can be used to help assess if you have successfully met the learning objectives for this section.

125. Solve: $3^{2x-9} = 30$.

    a. 6.05   **b.** 9.5   **c.** 2.89   **d.** 5

126. Solve: $\log_5 x + \log_5 (x + 4) = 1$.

    a. $\{-2 \pm \sqrt{5}\}$ **b.** $\left\{\dfrac{1}{2}\right\}$ **c.** $\{1\}$ **d.** $\{-5, 1\}$

127. Cost Increase The cost of a can of Coca Cola in 1960 was \$0.10. The function that models the cost of a Coca Cola by year is $C(t) = 0.10e^{0.0576t}$, where t is the number of years since 1960. In what year was it expected that a can of Coca Cola would cost \$1.00?

    a. 1985   **b.** 1977   **c.** 1993   **d.** 2000

Maintaining Your Skills

The following problems review material we covered in Section 8.5.

Find the vertex for each of the following parabolas, and then indicate if it is the highest or lowest point on the graph.

128. $y = 2x^2 + 8x - 15$ **129.** $y = 3x^2 - 9x - 10$

130. $y = 12x - 4x^2$ **131.** $y = 18x - 6x^2$

132. Maximum Height An object is projected into the air with an initial upward velocity of 64 feet per second. Its height h at any time t is given by the formula $h = 64t - 16t^2$. Find the time at which the object reaches its maximum height. Then, find the maximum height.

133. Maximum Height An object is projected into the air with an initial upward velocity of 64 feet per second from the top of a building 40 feet high. If the height h of the object t seconds after it is projected into the air is $h = 40 + 64t - 16t^2$, find the time at which the object reaches its maximum height. Then, find the maximum height it attains.

Chapter 9 Summary

Exponential Functions [9.1]

1. For the exponential function
$f(x) = 2^x$,
$$f(0) = 2^0 = 1$$
$$f(1) = 2^1 = 2$$
$$f(2) = 2^2 = 4$$
$$f(3) = 2^3 = 8$$

Any function of the form
$$f(x) = b^x$$
where $b > 0$ and $b \neq 1$, is an **exponential function**. The domain is the set of all real numbers and the range is $\{y \mid y > 0\}$. The graph has a horzontal asymptote at $y = 0$.

Compound Interest [9.1]

2. If \$500 is invested at 3% annual interest compounded quarterly, then after 2 years the balance in the account will be
$$A(4) = 500\left(1 + \frac{0.03}{4}\right)^{4(2)}$$
$$\approx \$530.80$$

If P dollars are deposited in an account with annual interest rate r, compounded n times per year, then the amount of money in the account after t years is given by
$$A(t) = P\left(1 + \frac{r}{n}\right)^{nt}$$
If the interest is compounded continuously, then the formula used is
$$A(t) = Pe^{rt}$$

One-to-One Functions [9.2]

3. The function $f(x) = x^2$ is not one-to-one because 9, which is in the range, comes from both 3 and -3 in the domain.

A function is a **one-to-one function** if every element in the range comes from exactly one element in the domain.

Inverse Functions [9.2]

4. The inverse of $f(x) = 2x - 3$ is
$$f^{-1}(x) = \frac{x + 3}{2}$$

The **inverse** of a function is obtained by reversing the order of the coordinates of the ordered pairs belonging to the function. Only one-to-one functions have inverses that are also functions.

Definition of Logarithms [9.3]

5. The definition allows us to write expressions like
$$y = \log_3 27$$
equivalently in exponential form as
$$3^y = 27$$
which makes it apparent that y is 3.

If b is a positive number not equal to 1, then the expression
$$y = \log_b x$$
is equivalent to $x = b^y$; that is, in the expression $y = \log_b x$, y is the exponent to which we raise b in order to get x. Expressions written in the form $y = \log_b x$ are said to be in **logarithmic form**. Expressions like $x = b^y$ are in **exponential form**.

Logarithmic Functions [9.3]

6. The function $y = \log_2 x$ is the inverse of the exponential function $y = 2^x$.

Any function of the form

$$f(x) = \log_b x$$

where $b > 0$ and $b \neq 1$, is a **logarithmic function**. The domain is $\{x \mid x > 0\}$ and the range is the set of all real numbers. The graph has a vertical asymptote at $x = 0$.

   The logarithmic function with base b and the exponential function with base b are inverse functions.

Properties of Logarithms [9.4]

7. We can rewrite the expression

$$\log_{10} \frac{45^6}{273}$$

using the properties of logarithms, as

$$6 \log_{10} 45 - \log_{10} 273$$

If x, y, and b are positive real numbers, $b \neq 1$, and r is any real number, then:

1. $\log_b 1 = 0$

2. $\log_b b = 1$

3. $b^{\log_b x} = x$

4. $\log_b b^x = x$

5. $\log_b (xy) = \log_b x + \log_b y$

6. $\log_b \left(\dfrac{x}{y} \right) = \log_b x - \log_b y$

7. $\log_b x^r = r \log_b x$

Common Logarithms [9.5]

8. $\log_{10} 10{,}000 = \log 10{,}000$
$= \log 10^4$
$= 4$

Common logarithms are logarithms with a base of 10. To save time in writing, we omit the base when working with common logarithms; that is,

$$\log x = \log_{10} x$$

Natural Logarithms [9.5]

9. $\ln e = 1$
$\ln 1 = 0$

Natural logarithms, written *ln x*, are logarithms with a base of e, where the number e is an irrational number (like the number π). A decimal approximation for e is 2.7183. All the properties of exponents and logarithms hold when the base is e.

$$\ln x = \log_e x$$

Change of Base [9.5]

10. $\log_6 475 = \dfrac{\log 475}{\log 6}$

$\approx \dfrac{2.6767}{0.7782}$

≈ 3.44

If x, a, and b are positive real numbers, $a \neq 1$ and $b \neq 1$, then

$$\log_a x = \frac{\log_b x}{\log_b a}$$

Exponential Equations [9.6]

11.
$$10e^{2x} = 50$$
$$e^{2x} = 5$$
$$\ln{(e^{2x})} = \ln{(5)}$$
$$2x = \ln 5$$
$$x = \frac{1}{2}\ln 5 \approx 0.8047$$

To solve an exponential equation, we isolate the exponential function. Then we either write the equation in logarithmic form or take a logarithm of both sides.

Logarithmic Equations [9.6]

12. $\log_2 x + \log_2 (x - 2) = 3$
$$\log_2 x(x - 2) = 3$$
$$x^2 - 2x = 2^3$$
$$x^2 - 2x - 8 = 0$$
$$(x - 4)(x + 2) = 0$$
$$x = 4, -2$$
$x = -2$ is extraneous, so the only solution is $x = 4$.

To solve a logarithmic equation, we first isolate the logarithmic function, and then write the equation in exponential form. If it is necessary to use properties of logarithms, then we should check for extraneous solutions.

> **⚠ COMMON MISTAKE**
>
> The most common mistakes that occur with logarithms come from trying to apply the properties of logarithms to situations in which they don't apply. For example, a very common mistake looks like this:
>
> $$\frac{\log 3}{\log 2} = \log 3 - \log 2 \qquad \text{Mistake}$$
>
> This is not a property of logarithms. To write the equation $\log 3 - \log 2$, we would have to start with
>
> $$\log \frac{3}{2} \qquad NOT \qquad \frac{\log 3}{\log 2}$$
>
> There is a difference.

Chapter 9 Test

Graph each exponential function. [9.1]

1. $f(x) = 2^x$

2. $g(x) = 3^{-x}$

3. Graph the relation $y = x^2 - 2$, and then graph the inverse of the relation. [9.2]

4. Graph the function $f(x) = 2x - 3$ and its inverse. Then find the equation for $f^{-1}(x)$. [9.2]

Graph each of the following. [9.3]

5. $y = \log_2 x$

6. $y = \log_{1/2} x$

Evaluate each of the following. [9.3, 9.5]

7. $\log_7 \dfrac{1}{49}$

8. $\log_8 4$

9. $\log 23,400$

10. $\ln 0.0462$

Use the properties of logarithms to expand each expression. [9.4]

11. $\log_2 \dfrac{8x^2}{y}$

12. $\log \dfrac{\sqrt{x}}{(y^4)\sqrt[5]{z}}$

Write each expression as a single logarithm. [9.4]

13. $2\log_3 x - \dfrac{1}{2}\log_3 y$

14. $\dfrac{1}{3}\log x - \log y - 2\log z$

Use the change of base formula to find each logarithm. [9.5]

15. $\log_9 100$

16. $\log_{100} 0.9$

Use a calculator to find x. [9.6]

17. $\log x = 4.8476$

18. $\log x = -2.6478$

Solve for x. [9.6]

19. $\log_4 x = 3$

20. $\log_x 5 = 2$

21. $5 = 3^x$

22. $4^{2x-1} = 8$

23. $\log_5 x - \log_5 3 = 1$

24. $\log_2 x + \log_2 (x - 7) = 3$

25. pH Find the pH of a solution in which $[H^+] = 6.6 \times 10^{-7}$. [9.5]

26. Compound Interest If $400 is deposited in an account that earns 10% annual interest compounded twice a year, how much money will be in the account after 5 years? [9.1]

27. Depreciation If a car depreciates in value 20% per year for the first 5 years after it is purchased for P_0 dollars, then its value after t years will be $V(t) = P_0(1 - r)^t$ for $0 \le t \le 5$. To the nearest dollar, find the value of a car 4 years after it is purchased for $18,000. [9.1]

28. Compound Interest How long will it take $600 to become $1,800 if the $600 is deposited in an account that earns 8% annual interest compounded 4 times a year? [9.6]

Sequences and Series

10

iStockphoto.com © Dangubic

S uppose you run up a balance of $1,000 on a credit card that charges 1.65% interest each month (i.e., an annual rate of 19.8%). If you stop using the card and make the minimum payment of $20 each month, how long will it take you to pay off the balance on the card? The answer can be found by using the formula

$$U_n = (1.0165)U_{n-1} - 20$$

where U_n stands for the current unpaid balance on the card, and U_{n-1} is the previous month's balance. The table and figure were created from this formula and a graphing calculator. As you can see from the table, the balance on the credit card decreases very little each month.

Monthly Credit Card Balances

Previous Balance U_{n-1}	Monthly Interest Rate	Payment Number n	Monthly Payment	New Balance U_n
$1,000.00	1.65%	1	$20	$996.00
$996.00	1.65%	2	$20	$992.94
$992.94	1.65%	3	$20	$989.32
$989.32	1.65%	4	$20	$985.64
$985.64	1.65%	5	$20	$981.90

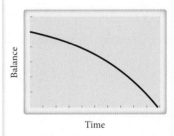

The formula for U_n is called a recursion formula, and the values of U_n it generates

1,000.00, 996.00, 992.94, 989.94, 989.32, 985.32

form a sequence. In this chapter, we will learn about different types of sequences and explore some of their applications.

iStockphoto.com © Rawpixel Ltd

Think about the most successful people you have met or heard about. What are the qualities they tend to have in common? One of these qualities usually involves making a resolute commitment. If you are not firmly committed to something, then you will tend to give less than your full effort. Consider this quote from Faust by Johann Wolfgang Von Goethe:

> Until one is committed, there is hesitancy, the chance to draw back, always ineffectiveness. Concerning all acts of initiative and creation, there is one elementary truth the ignorance of which kills countless ideas and splendid plans: that the moment one definitely commits oneself, then providence moves too. All sorts of things occur to help one that would never otherwise have occurred. A whole stream of events issues from the decision, raising in one's favor all manner of unforeseen incidents, meetings and material assistance which no man could have dreamed would have come his way. Whatever you can do or dream you can, begin it. Boldness has genius, power and magic in it.

Successful people do not give up easily. They forge ahead, even when confronted by difficulties or when the odds seem stacked against them.

Take a moment to reflect on your own life experiences. When have you been the most successful? Can you think of a time when providence has moved in your favor, perhaps unexpectedly?

Sequences

Learning Objectives

In this section, we will learn how to:

1. Use the nth term to find terms of a sequence.

2. Find terms of a sequence using a recursion formula.

3. Find a formula for the general term of a sequence.

Introduction

Informally, a sequence is simply a list of numbers that have been placed in a particular order. Many of the sequences in this chapter will be familiar to you on an intuitive level because you have worked with them for some time now. Here are some of those sequences:

The sequence of odd numbers

$$1, 3, 5, 7,\ldots$$

The sequence of even numbers

$$2, 4, 6, 8,\ldots$$

The sequence of squares

$$1^2, 2^2, 3^2, 4^2,\ldots = 1, 4, 9, 16,\ldots$$

The numbers in each of these sequences can be found from formulas that define functions. For example, the sequence of even numbers can be found from the function

$$f(x) = 2x$$

by finding $f(1), f(2), f(3), f(4)$, and so forth. This gives us justification for the formal definition of a sequence.

> (def DEFINITION *sequence*
>
> A *sequence* is a function whose domain is the set of positive integers $1, 2, 3, 4,\ldots$.

As you can see, sequences are simply functions with a specific domain. If we want to form a sequence from the function $f(x) = 3x + 5$, we simply find $f(1), f(2), f(3), f(4)$, and so on. Doing so gives us the sequence

$$8, 11, 14, 17,\ldots$$

because $f(1) = 3(1) + 5 = 8$, $f(2) = 3(2) + 5 = 11$, $f(3) = 3(3) + 5 = 14$, and $f(4) = 3(4) + 5 = 17$. Notice that the terms of the sequence are just the outputs from the function written in order.

Notation Because the domain for a sequence is always the set $\{1, 2, 3,\ldots\}$, we can simplify the notation we use to represent the terms of a sequence. Using the letter a instead of f, and subscripts instead of numbers enclosed by parentheses, we can represent the sequence from the previous discussion as follows:

$$a_n = 3n + 5$$

Instead of $f(1)$, we write a_1 for the *first term* of the sequence.
Instead of $f(2)$, we write a_2 for the *second term* of the sequence.
Instead of $f(3)$, we write a_3 for the *third term* of the sequence.
Instead of $f(4)$, we write a_4 for the *fourth term* of the sequence.
Instead of $f(n)$, we write a_n for the nth *term* of the sequence.

The nth term is also called the **general term** of the sequence. The general term is used to define the other terms of the sequence. That is, if we are given the formula for the general term a_n, we can find any other term in the sequence. The following examples illustrate.

VIDEO EXAMPLES

SECTION 10.1

EXAMPLE 1   Find the first four terms of the sequence whose general term is given by $a_n = 2n - 1$.

SOLUTION The subscript notation a_n works the same way function notation works. To find the first, second, third, and fourth terms of this sequence, we simply substitute 1, 2, 3, and 4 for n in the formula $2n - 1$:

If   the general term is $a_n = 2n - 1$

then   the first term is $a_1 = 2(1) - 1 = 1$

the second term is $a_2 = 2(2) - 1 = 3$

the third term is $a_3 = 2(3) - 1 = 5$

the fourth term is $a_4 = 2(4) - 1 = 7$

The first four terms of this sequence are the odd numbers 1, 3, 5, and 7. The whole sequence can be written as

$$1, 3, 5, 7,\ldots, 2n - 1,\ldots$$

Because each term in this sequence is greater than the preceding term, we say the sequence is an **increasing sequence**.

EXAMPLE 2   Write the first four terms of the sequence defined by

$$a_n = \frac{1}{n + 1}$$

SOLUTION Replacing n with 1, 2, 3, and 4, we have, respectively, the first four terms:

$$\text{First term} = a_1 = \frac{1}{1 + 1} = \frac{1}{2}$$

$$\text{Second term} = a_2 = \frac{1}{2 + 1} = \frac{1}{3}$$

$$\text{Third term} = a_3 = \frac{1}{3 + 1} = \frac{1}{4}$$

$$\text{Fourth term} = a_4 = \frac{1}{4 + 1} = \frac{1}{5}$$

The sequence defined by

$$a_n = \frac{1}{n+1}$$

can be written as

$$\frac{1}{2}, \frac{1}{3}, \frac{1}{4}, \frac{1}{5}, \dots, \frac{1}{n+1}, \dots$$

Because each term in the sequence is less than the term preceding it, the sequence is said to be a ***decreasing sequence***. ◼

EXAMPLE 3   Find the fifth and sixth terms of the sequence whose general term is given by

$$a_n = \frac{(-1)^n}{n^2}$$

SOLUTION   For the fifth term, we replace n with 5. For the sixth term, we replace n with 6:

$$\text{Fifth term} = a_5 = \frac{(-1)^5}{5^2} = \frac{-1}{25}$$

$$\text{Sixth term} = a_6 = \frac{(-1)^6}{6^2} = \frac{1}{36}$$ ◼

The sequence in Example 3 can be written as

$$-1, \frac{1}{4}, -\frac{1}{9}, \frac{1}{16}, \dots, \frac{(-1)^n}{n^2}, \dots$$

Because the terms alternate in sign — if one term is positive, then the next term is negative — we call this an ***alternating sequence***. The first three examples all illustrate how we work with a sequence in which we are given a formula for the general term.

USING TECHNOLOGY   *Finding Sequences on a Graphing Calculator*

Method 1: Using a Table

We can use the table function on a graphing calculator to view the terms of a sequence. To view the terms of the sequence $a_n = 3n + 5$, we set $Y_1 = 3X + 5$. Then we use the table setup feature on the calculator to set the table minimum to 1, and the table increment to 1 also. Here is the setup and result for a TI-84.

Table Setup	Y Variables Setup	Resulting Table	
		X	Y
Table minimum $=1$	$Y_1 = 3X + 5$		
Table increment $=1$		1	8
Independent variable: Auto		2	11
Dependent variable: Auto		3	14
		4	17
		5	20

To find any particular term of a sequence, we change the independent variable setting to Ask, and then input the number of the term of the sequence we want to find. For example, if we want term a_{100}, then we input 100 for the independent variable, and the table gives us the value of 305 for that term.

USING TECHNOLOGY   *Finding Sequences on a Graphing Calculator Continued*

Method 2: Using the Built-in seq(Command

Using this method, first find the seq(command. On a TI-84 it is found in the $\boxed{\text{LIST}}$ OPS menu. To find terms a_1 through a_7 for $a_n = 3n + 5$, we first bring up the seq(command on our calculator, then we input the following four items, in order, separated by commas: 3X+5, X, 1, 7. Then we close the parentheses. Our screen will look like this:

$$\text{seq}(3X+5, X, 1, 7)$$

Pressing $\boxed{\text{ENTER}}$ displays the first five terms of the sequence. Pressing the right arrow key repeatedly brings the remaining members of the sequence into view.

Method 3: Using the Built-in Seq Mode

Press the $\boxed{\text{MODE}}$ key on your TI-84 and then select Seq (it's next to Func Par and Pol). Go to the Y variables list and set $n\text{Min} = 1$ and $u(n) = 3n+5$. Then go to the $\boxed{\text{TBLSET}}$ key to set up your table like the one shown in Method 1. Pressing $\boxed{\text{TABLE}}$ will display the sequence you have defined.

Recursion Formulas

Let's go back to one of the first sequences we looked at in this section:

$$8, 11, 14, 17,\ldots$$

Each term in the sequence can be found by simply substituting positive integers for n in the formula $a_n = 3n + 5$. Another way to look at this sequence, however, is to notice that each term can be found by adding 3 to the preceding term; so, we could give all the terms of this sequence by simply saying

Start with 8, and then add 3 to each term to get the next term.

The same idea, expressed in symbols, looks like this:

$$a_1 = 8 \quad \text{and} \quad a_n = a_{n-1} + 3 \quad \text{for } n > 1$$

This formula is called a ***recursion formula*** because each term is written recursively, meaning defined in relation to the term or terms that precede it.

EXAMPLE 4   Write the first four terms of the sequence given recursively by

$$a_1 = 4 \quad \text{and} \quad a_n = 5a_{n-1} \quad \text{for } n > 1$$

SOLUTION   The formula tells us to start the sequence with the number 4, and then multiply each term by 5 to get the next term. Therefore,

$$a_1 = 4$$
$$a_2 = 5a_1 = 5(4) = 20$$
$$a_3 = 5a_2 = 5(20) = 100$$
$$a_4 = 5a_3 = 5(100) = 500$$

The sequence is 4, 20, 100, 500,….

Recursion Formulas on a Graphing Calculator

We can use a TI-84 graphing calculator to view the sequence defined recursively as

$$a_1 = 8, a_n = a_{n-1} + 3$$

First, put your TI-84 calculator in sequence mode by pressing the MODE key, and then selecting Seq (it's next to Func Par and Pol). Go to the Y variables list and set $n\text{Min} = 1$, $u(n) = u(n-1)+3$, and $u(n\text{Min}) = 8$. (The u is above the 7, and the n is on the X, T, θ, n and is automatically displayed if that key is pressed when the calculator is in the Seq mode.) Pressing TABLE will display the sequence you have defined.

Finding the General Term

In the first four examples, we found some terms of a sequence after being given the general term. In the next two examples, we will do the reverse. That is, given some terms of a sequence, we will find a formula for the general term.

EXAMPLE 5 Find a formula for the nth term of the sequence 2, 8, 18, 32,....

SOLUTION Solving a problem like this involves some guessing. Looking over the first four terms, we see each is twice a perfect square:

$$2 = 2(1)$$
$$8 = 2(4)$$
$$18 = 2(9)$$
$$32 = 2(16)$$

If we write each square with an exponent of 2, the formula for the nth term becomes obvious:

$$a_1 = 2 \ = 2(1)^2$$
$$a_2 = 8 \ = 2(2)^2$$
$$a_3 = 18 = 2(3)^2$$
$$a_4 = 32 = 2(4)^2$$
$$\vdots$$
$$a_n = 2(n)^2 = 2n^2$$

The general term of the sequence 2, 8, 18, 32,...is $a_n = 2n^2$.

EXAMPLE 6   Find the general term for the sequence $2, \dfrac{3}{8}, \dfrac{4}{27}, \dfrac{5}{64}, \ldots$

SOLUTION   The first term can be written as $\dfrac{2}{1}$. The denominators are all perfect cubes. The numerators are all 1 more than the base of the cubes in the denominators:

$$a_1 = \frac{2}{1} = \frac{1+1}{1^3}$$

$$a_2 = \frac{3}{8} = \frac{2+1}{2^3}$$

$$a_3 = \frac{4}{27} = \frac{3+1}{3^3}$$

$$a_4 = \frac{5}{64} = \frac{4+1}{4^3}$$

Observing this pattern, we recognize the general term to be

$$a_n = \frac{n+1}{n^3}$$

> *Note*   Finding the nth term of a sequence from the first few terms is not always automatic. That is, it sometimes takes awhile to recognize the pattern. Don't be afraid to guess at the formula for the general term. Many times, an incorrect guess leads to the correct formula.

Getting Ready for Class

After reading through the preceding section, respond in your own words and in complete sentences.

A. How are subscripts used to denote the terms of a sequence?

B. What is the relationship between the subscripts used to denote the terms of a sequence and function notation?

C. What is a decreasing sequence?

D. What is meant by a recursion formula for a sequence?

Write the first five terms of the sequences with the following general terms.

1. $a_n = 3n + 1$

2. $a_n = 2n + 3$

3. $a_n = 4n - 1$

4. $a_n = n + 4$

5. $a_n = n$

6. $a_n = -n$

7. $a_n = n^2 + 3$

8. $a_n = n^3 + 1$

9. $a_n = \dfrac{n}{n + 3}$

10. $a_n = \dfrac{n + 1}{n + 2}$

11. $a_n = \dfrac{1}{n^2}$

12. $a_n = \dfrac{1}{n^3}$

13. $a_n = 2^n$

14. $a_n = 3^{-n}$

15. $a_n = 1 + \dfrac{1}{n}$

16. $a_n = n - \dfrac{1}{n}$

17. $a_n = (-2)^n$

18. $a_n = (-3)^n$

19. $a_n = 4 + (-1)^n$

20. $a_n = 10 + (-2)^n$

21. $a_n = (-1)^{n+1} \cdot \dfrac{n}{2n - 1}$

22. $a_n = (-1)^n \cdot \dfrac{2n + 1}{2n - 1}$

23. $a_n = n^2 \cdot 2^{-n}$

24. $a_n = n^n$

25. If $a_n = n^2 + 2n$, find a_8.

26. If $a_n = 3n - n^2$, find a_{10}.

27. If $a_n = \dfrac{(-1)^n}{2n + 3}$, find a_{100}.

28. If $a_n = \dfrac{(-1)^{n+1}}{2^n - 3}$, find a_{12}.

Write the first five terms of the sequences defined by the following recursion formulas.

29. $a_1 = 3$ $\quad$ $a_n = -3a_{n-1}$ $\quad$ $n > 1$

30. $a_1 = 3$ $\quad$ $a_n = a_{n-1} - 3$ $\quad$ $n > 1$

31. $a_1 = 1$ $\quad$ $a_n = 2a_{n-1} + 3$ $\quad$ $n > 1$

32. $a_1 = 1$ $\quad$ $a_n = a_{n-1} + n$ $\quad$ $n > 1$

33. $a_1 = 2$ $\quad$ $a_n = 2a_{n-1} - 1$ $\quad$ $n > 1$

34. $a_1 = -4$ $\quad$ $a_n = -2a_{n-1}$ $\quad$ $n > 1$

35. $a_1 = 5$ $\quad$ $a_n = 3a_{n-1} - 4$ $\quad$ $n > 1$

36. $a_1 = -3$ $\quad$ $a_n = -2a_{n-1} + 5$ $\quad$ $n > 1$

37. $a_1 = 4$ $\quad$ $a_n = 2a_{n-1} - a_1$ $\quad$ $n > 1$

38. $a_1 = -3$ $\quad$ $a_n = -2a_{n-1} - n$ $\quad$ $n > 1$

Determine a formula for the general term for each of the following sequences.

39. 4, 8, 12, 16, 20,...

40. 7, 10, 13, 16,...

41. 1, 4, 9, 16,...

42. 3, 12, 27, 48,...

43. 4, 8, 16, 32,...

44. −2, 4, −8, 16,...

45. $\dfrac{1}{4}, \dfrac{1}{8}, \dfrac{1}{16}, \dfrac{1}{32},\ldots$

46. $\dfrac{1}{4}, \dfrac{2}{9}, \dfrac{3}{16}, \dfrac{4}{25},\ldots$

47. 5, 8, 11, 14,...

48. 7, 5, 3, 1,...

49. −2, −6, −10, −14,...

50. −2, 2, −2, 2,...

51. 1, −2, 4, −8,...

52. −1, 3, −9, 27,...

53. $\log_2 3, \log_3 4, \log_4 5, \log_5 6,\ldots$

54. $0, \dfrac{3}{5}, \dfrac{8}{10}, \dfrac{15}{17},\ldots$

Applying the Concepts

55. Salary Increase The entry level salary for a teacher is $28,000 with 4% increases after every year of service.

a. Write a sequence for this teacher's salary for the first 5 years.

b. Find the general term of the sequence in part *a*.

56. Holiday Account To save money for holiday presents, a person deposits $5 in a savings account on January 1, and then deposits an additional $5 every week thereafter until Christmas.

a. Write a sequence for the money in that savings account for the first 10 weeks of the year.

b. Write the general term of the sequence in part *a*.

c. If there are 50 weeks from January 1 to Christmas, how much money will be available for spending on Christmas presents?

57. Akaka Falls If a boulder fell from the top of Akaka Falls in Hawaii, the distance, in feet, the boulder would fall in each consecutive second would be modeled by a sequence whose general term is $a_n = 32n - 16$, where n represents the number of seconds.

a. Write a sequence for the first 5 seconds the boulder falls.

b. What is the total distance the boulder fell in 5 seconds?

c. If Akaka Falls is approximately 420 feet high, will the boulder hit the ground within 5 seconds?

58. Polygons The formula for the sum of the interior angles of a polygon with n sides is $a_n = 180°(n - 2)$.

a. Write a sequence to represent the sum of the interior angles of a polygon with 3, 4, 5, and 6 sides.

b. What would be the sum of the interior angles of a polygon with 20 sides?

c. What happens when $n = 2$ to indicate that a polygon cannot be formed with only two sides?

59. Pendulum A pendulum swings 10 feet left to right on its first swing. On each swing following the first, the pendulum swings $\frac{4}{5}$ of the previous swing.

    a. Write a sequence for the distance traveled by the pendulum on the first, second, and third swing.

    b. Write a general term for the sequence, where n represents the number of the swing.

    c. How far will the pendulum swing on its tenth swing? (Round to the nearest hundredth.)

Learning Objectives Assessment

The following problems can be used to help assess if you have successfully met the learning objectives for this section.

60. Find the fifth term of the sequence $a_n = (-1)^n(n^2 - 2)$.

    a. 23   **b.** -14 **c.** -23 **d.** 14

61. Find the fifth term of the sequence defined by the recursion formula $a_1 = 6, a_n = 2a_{n-1} + 1 \ (n > 1)$.

    a. 6   **b.** 11   **c.** 111   **d.** 223

62. Find the general term for the sequence 2, 5, 10, 17, 26,….

    a. $a_n = 3n - 1$ **b.** $a_n = n^2 + 1$

    c. $a_n = 4n - 2$ **d.** $a_n = 2n^2$

Getting Ready for the Next Section

Simplify.

63. $-2 + 6 + 4 + 22$ **64.** $9 - 27 + 81 - 243$

65. $-8 + 16 - 32 + 64$ **66.** $-4 + 8 - 16 + 32 - 64$

67. $(1 - 3) + (4 - 3) + (9 - 3) + (16 - 3)$

68. $(1 - 3) + (9 + 1) + (16 + 1) + (25 + 1) + (36 + 1)$

69. $-\dfrac{1}{3} + \dfrac{1}{9} - \dfrac{1}{27} + \dfrac{1}{81}$ **70.** $\dfrac{1}{2} + \dfrac{2}{3} + \dfrac{3}{4} + \dfrac{4}{5}$

71. $\dfrac{1}{3} + \dfrac{1}{2} + \dfrac{3}{5} + \dfrac{2}{3}$ **72.** $\dfrac{1}{16} + \dfrac{1}{32} + \dfrac{1}{64}$

Learning Objectives

In this section, we will learn how to:

1. Evaluate a series expressed in summation notation.
2. Write a series using summation notation.

Introduction

There is an interesting relationship between the sequence of odd numbers and the sequence of squares that is found by adding the terms in the sequence of odd numbers.

$$
\begin{aligned}
1 &= 1 \\
1 + 3 &= 4 \\
1 + 3 + 5 &= 9 \\
1 + 3 + 5 + 7 &= 16
\end{aligned}
$$

When we add the terms of a sequence, the result is called a series.

> **(def) DEFINITION** *series*
>
> The sum of a number of terms in a sequence is called a *series*.

A sequence can be finite or infinite, depending on whether the sequence ends with a particular term or continues indefinitely. For example,

$$1, 3, 5, 7, 9$$

is a finite sequence, but

$$1, 3, 5, \ldots$$

is an infinite sequence. Associated with each of the preceding sequences is a series found by adding the terms of the sequence:

$$1 + 3 + 5 + 7 + 9 \qquad \text{Finite series}$$

$$1 + 3 + 5 + \ldots \qquad \text{Infinite series}$$

Summation Notation

In this section, we will consider only finite series. We can introduce a new kind of notation here that is a compact way of indicating a finite series. The notation is called *summation notation*, or *sigma notation* because it is written using the Greek letter sigma. The expression

$$\sum_{i=1}^{4} (8i - 10)$$

Note The Greek letter sigma Σ is a capital S, which was chosen as a reminder that a series is a *sum*.

is an example of summation notation. This expression is used to indicate the sum of the first through the fourth terms of the sequence $a_i = 8i - 10$. That is,

$$
\begin{aligned}
\sum_{i=1}^{4} (8i - 10) &= a_1 + a_2 + a_3 + a_4 \\
&= (8 \cdot 1 - 10) + (8 \cdot 2 - 10) + (8 \cdot 3 - 10) + (8 \cdot 4 - 10) \\
&= -2 + 6 + 14 + 22 \\
&= 40
\end{aligned}
$$

The letter i as used here is called the **index of summation**, or just **index** for short. Here are some examples illustrating the use of summation notation.

EXAMPLE 1   Expand and simplify: $\sum\limits_{i=1}^{5}(i^2-1)$.

SOLUTION   We replace i in the expression i^2-1 with all consecutive integers from 1 up to 5, including 1 and 5:

$$\sum_{i=1}^{5}(i^2-1)=(1^2-1)+(2^2-1)+(3^2-1)+(4^2-1)+(5^2-1)$$
$$=0+3+8+15+24$$
$$=50$$

EXAMPLE 2   Expand and simplify: $\sum\limits_{i=3}^{6}(-2)^i$.

SOLUTION   We replace i in the expression $(-2)^i$ with the consecutive integers beginning at 3 and ending at 6:

$$\sum_{i=3}^{6}(-2)^i=(-2)^3+(-2)^4+(-2)^5+(-2)^6$$
$$=-8+16+(-32)+64$$
$$=40$$

USING TECHNOLOGY   *Evaluating a Series on a Graphing Calculator*

A TI-84 graphing calculator has a built-in sum(command that, when used with the seq(command, allows us to add the terms of a series. Let's repeat Example 1 using our graphing calculator. First, we go to $\boxed{\text{LIST}}$ and select MATH. The fifth option in that list is sum(, which we select. Then we go to $\boxed{\text{LIST}}$ again and select OPS. From that list we select seq(. Next we enter X^2−1, X, 1, 5, and then we close both sets of parentheses. Our screen shows the following:

sum(seq(X^2−1, X, 1, 5))   which will give us $\sum\limits_{i=1}^{5}(i^2-1)$

When we press $\boxed{\text{ENTER}}$ the calculator displays 50, which is the same result we obtained in Example 1.

EXAMPLE 3   Expand: $\sum\limits_{i=2}^{5}(x^i-3)$.

SOLUTION   We must be careful not to confuse the letter x with i. The index i is the quantity we replace by the consecutive integers from 2 to 5, not x:

$$\sum_{i=2}^{5}(x^i-3)=(x^2-3)+(x^3-3)+(x^4-3)+(x^5-3).$$

Because we were only asked to expand, but not simplify, the series, we leave our answer in this form.

In the first three examples, we were given an expression with summation notation and asked to expand it. The next examples in this section illustrate how we can write a series as an expression involving summation notation.

EXAMPLE 4 Write with summation notation: $1 + 3 + 5 + 7 + 9$.

SOLUTION A formula that gives us the terms of this sum is

$$a_i = 2i - 1$$

where i ranges from 1 up to and including 5. Notice we are using the subscript i in exactly the same way we used the subscript n in the previous section — to indicate the general term. Writing the sum

$$1 + 3 + 5 + 7 + 9$$

with summation notation looks like this:

$$\sum_{i=1}^{5} (2i - 1)$$

EXAMPLE 5 Write with summation notation: $3 + 12 + 27 + 48$.

SOLUTION We need a formula, in terms of i, that will give each term in the sum. Writing the sum as

$$3 \cdot 1^2 + 3 \cdot 2^2 + 3 \cdot 3^2 + 3 \cdot 4^2$$

we see the formula

$$a_i = 3 \cdot i^2$$

where i ranges from 1 up to and including 4. Using this formula and summation notation, we can represent the sum

$$3 + 12 + 27 + 48$$

as

$$\sum_{i=1}^{4} 3i^2$$

EXAMPLE 6 Write with summation notation:

$$\frac{x+3}{x^3} + \frac{x+4}{x^4} + \frac{x+5}{x^5} + \frac{x+6}{x^6}$$

SOLUTION A formula that gives each of these terms is

$$a_i = \frac{x+i}{x^i}$$

where i assumes all integer values between 3 and 6, including 3 and 6. The sum can be written as

$$\sum_{i=3}^{6} \frac{x+i}{x^i}$$

Getting Ready for Class

After reading through the preceding section, respond in your own words and in complete sentences.

A. What is the difference between a sequence and a series?

B. What does the symbol Σ stand for?

C. Explain what the 1 and 4 indicate in the series $\sum\limits_{i=1}^{4}(2i + 3)$.

D. Determine for what values of n the series $\sum\limits_{i=1}^{n}(-1)^i$ will be equal to 0. Explain your answer.

Expand and simplify each of the following.

1. $\displaystyle\sum_{i=1}^{4}(2i+4)$ 2. $\displaystyle\sum_{i=1}^{5}(2i+4)$ 3. $\displaystyle\sum_{i=2}^{3}(i^2-1)$ 4. $\displaystyle\sum_{i=3}^{6}(i^2+1)$

5. $\displaystyle\sum_{i=1}^{4}(i^2-3)$ 6. $\displaystyle\sum_{i=2}^{6}(2i^2+1)$ 7. $\displaystyle\sum_{i=1}^{4}\frac{i}{1+i}$ 8. $\displaystyle\sum_{i=1}^{3}\frac{i^2}{2i-1}$

9. $\displaystyle\sum_{i=1}^{4}(-3)^i$ 10. $\displaystyle\sum_{i=1}^{4}\left(-\frac{1}{3}\right)^i$ 11. $\displaystyle\sum_{i=3}^{6}(-2)^i$ 12. $\displaystyle\sum_{i=4}^{6}\left(-\frac{1}{2}\right)^i$

13. $\displaystyle\sum_{i=2}^{6}(-2)^i$ 14. $\displaystyle\sum_{i=2}^{5}(-3)^i$ 15. $\displaystyle\sum_{i=1}^{5}\left(-\frac{1}{2}\right)^i$ 16. $\displaystyle\sum_{i=3}^{6}\left(-\frac{1}{3}\right)^i$

17. $\displaystyle\sum_{i=2}^{5}\frac{i-1}{i+1}$ 18. $\displaystyle\sum_{i=2}^{4}\frac{i^2-1}{i^2+1}$

Expand the following. Do not simplify.

19. $\displaystyle\sum_{i=1}^{5}(x+i)$ 20. $\displaystyle\sum_{i=2}^{7}(x+1)^i$ 21. $\displaystyle\sum_{i=1}^{4}(x-2)^i$ 22. $\displaystyle\sum_{i=2}^{5}\left(x+\frac{1}{i}\right)^2$

23. $\displaystyle\sum_{i=1}^{5}\frac{x+i}{x-1}$ 24. $\displaystyle\sum_{i=1}^{6}\frac{x-3i}{x+3i}$ 25. $\displaystyle\sum_{i=3}^{8}(x+i)^i$ 26. $\displaystyle\sum_{i=1}^{5}(x+i)^{i+1}$

27. $\displaystyle\sum_{i=3}^{6}(x-2i)^{i+3}$ 28. $\displaystyle\sum_{i=5}^{8}\left(\frac{x-i}{x+i}\right)^{2i}$

Write each of the following sums with summation notation. Do not calculate the sum. Use i as the index of summation. (*Note*: Other answers are possible.)

29. $2+4+8+16$

30. $3+5+7+9+11$

31. $4+8+16+32+64$

32. $3+8+15+24$

33. $5+9+13+17+21$

34. $3-6+12-24+48$

35. $-4+8-16+32$

36. $15+24+35+48+63$

37. $\dfrac{3}{4}+\dfrac{4}{5}+\dfrac{5}{6}+\dfrac{6}{7}+\dfrac{7}{8}$

38. $\dfrac{1}{2}+\dfrac{2}{3}+\dfrac{3}{4}+\dfrac{4}{5}$

39. $\dfrac{1}{3}+\dfrac{2}{5}+\dfrac{3}{7}+\dfrac{4}{9}$

40. $\dfrac{3}{1}+\dfrac{5}{3}+\dfrac{7}{5}+\dfrac{9}{7}$

41. $(x-2)^6+(x-2)^7+(x-2)^8+(x-2)^9$

42. $(x+1)^3+(x+2)^4+(x+3)^5+(x+4)^6+(x+5)^7$

43. $\left(1+\dfrac{1}{x}\right)^2+\left(1+\dfrac{2}{x}\right)^3+\left(1+\dfrac{3}{x}\right)^4+\left(1+\dfrac{4}{x}\right)^5$

44. $\dfrac{x-1}{x+2}+\dfrac{x-2}{x+4}+\dfrac{x-3}{x+6}+\dfrac{x-4}{x+8}+\dfrac{x-5}{x+10}$

45. $\dfrac{x}{x+3}+\dfrac{x}{x+4}+\dfrac{x}{x+5}$ 46. $\dfrac{x-3}{x^3}+\dfrac{x-4}{x^4}+\dfrac{x-5}{x^5}+\dfrac{x-6}{x^6}$

47. $x^2(x+2)+x^3(x+3)+x^4(x+4)$ 48. $x(x+2)^2+x(x+3)^3+x(x+4)^4$

49. **Repeating Decimals** Any repeating, nonterminating decimal may be viewed as a series. For instance, $\frac{2}{3} = 0.6 + 0.06 + 0.006 + 0.0006 + \cdots$. Write the following fractions as series.

    a. $\dfrac{1}{3}$ **b.** $\dfrac{2}{9}$ **c.** $\dfrac{3}{11}$

50. **Repeating Decimals** Refer to the previous exercise, and express the following repeating decimals as fractions.

    a. 0.55555...   **b.** 1.33333...   **c.** 0.29292929...

Applying the Concepts

51. **Skydiving** A skydiver jumps from a plane and falls 16 feet the first second, 48 feet the second second, and 80 feet the third second. If he continues to fall in the same manner, how far will he fall the seventh second? What is the distance he falls in 7 seconds?

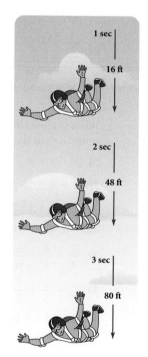

52. **Bacterial Growth** After 1 day, a colony of 50 bacteria reproduces to become 200 bacteria. After 2 days, they reproduce to become 800 bacteria. If they continue to reproduce at this rate, how many bacteria will be present after 4 days?

53. **Akaka Falls** In Section 10.1, when a boulder fell from the top of Akaka Falls in Hawaii, the sequence generated during the first 5 seconds the boulder fell was 16, 48, 80, 112, and 144.

   a. Write a finite series that represents the sum of this sequence.

   b. The general term of this sequence was given as $a_n = 32n - 16$. Write the series produced in part **a** in summation notation.

54. **Pendulum** A pendulum swings 12 feet left to right on its first swing, and on each swing following the first, swings $\frac{3}{4}$ of the previous swing. The distance the pendulum traveled in 5 seconds can be expressed in summation notation

$$\sum_{i=1}^{5} 12\left(\frac{3}{4}\right)^{i-1}$$

   Expand the summation notation and simplify. Round your final answer to the nearest tenth.

Learning Objectives Assessment

The following problems can be used to help assess if you have successfully met the learning objectives for this section.

55. Expand and simplify: $\displaystyle\sum_{i=3}^{5} (i^2 - 2i)$.

   a. 15   **b.** 26   **c.** 360   **d.** 3, 8, 15

56. Write $\frac{1}{2} + \frac{1}{4} + \frac{1}{8} + \frac{1}{16} + \frac{1}{32}$ using summation notation.

   a. $\displaystyle\sum_{i=1}^{16} \frac{1}{2i}$ **b.** $\displaystyle\sum_{i=2}^{32} \frac{1}{i}$

   c. $\displaystyle\sum_{i=1}^{5} \frac{1}{2^i}$ **d.** $\displaystyle\sum_{i=\frac{1}{2}}^{\frac{1}{32}} \frac{1}{2^i}$

Getting Ready for the Next Section

Simplify.

57. $2 + 9(8)$

58. $\frac{1}{2} + 9\left(\frac{1}{2}\right)$

59. $\frac{10}{2}\left(\frac{1}{2} + 5\right)$

60. $\frac{10}{2}(2 + 74)$

61. $3 + (n - 1)2$

62. $7 + (n - 1)3$

Solve each system of equations.

63. $x + 2y = 7$
    $x + 7y = 17$

64. $x + 3y = 14$
    $x + 9y = 32$

SPOTLIGHT ON SUCCESS   *Student Instructor Andrew*

*Somewhere, something incredible
is waiting to be known."*
—*Carl Sagan*

Don't give up even if you think you'll never understand math! I often felt discouraged by math as a kid. It seemed tedious and full of memorization and routine. I failed algebra the first time I took it and neither my teachers nor my parents proved very helpful. After scraping my way through high school, I worked one summer in a machine shop for a sheet metal company, which gave me a glimpse into what my future would be if I didn't go further with my education. That fall I went to community college where I met a professor who helped change the way I felt about math, and in turn, helped change the course of my life. Professor Carol Paxton was the first person to show me that I could succeed in math class. She had a way of demystifying the complex and making it tangible even for me, a failed math student. She taught me trigonometry and calculus, two classes I never would have dreamed I could take, let alone pass. I later became her TA for those classes and decided to change my major to mathematics. While some people have innate mathematical talent, that was never me. Professor Paxton helped me realize that with hard work, there is beauty and joy to be found in mathematics. In my current role, try to be that life-changing person for others who feel as I once did about math.

Arithmetic Sequences

Learning Objectives

In this section, we will learn how to:

1. Identify an arithmetic sequence.

2. Find the common difference for an arithmetic sequence.

3. Find a formula for the general term of an arithmetic sequence.

4. Find the nth partial sum for an arithmetic sequence.

Introduction

In this and the following section, we will investigate two major types of sequences—arithmetic sequences and geometric sequences.

> **(dĕf) DEFINITION** *arithmetic sequence*
>
> An *arithmetic sequence* is a sequence of numbers in which each term is obtained from the preceding term by adding the same amount each time. An arithmetic sequence is also called an *arithmetic progression.*

The sequence

$$2, 6, 10, 14,\ldots$$

is an example of an arithmetic sequence, because each term is obtained from the preceding term by adding 4 each time. The amount we add each time — in this case, 4 — is called the **common difference**, because it can be obtained by subtracting any two consecutive terms. The common difference is denoted by d, and given by $d = a_n - a_{n-1}$ for any $n > 1$.

VIDEO EXAMPLES

SECTION 10.3

▨ **EXAMPLE 1** Give the common difference d for the arithmetic sequence 4, 10, 16, 22,…

SOLUTION Because each term can be obtained from the preceding term by adding 6, the common difference is 6. That is, $d = 6$. ▨

▨ **EXAMPLE 2** Give the common difference for 100, 93, 86, 79,….

SOLUTION The common difference in this case is $d = -7$, since adding -7 to any term always produces the next consecutive term. ▨

▨ **EXAMPLE 3** Give the common difference for $\frac{1}{2}, 1, \frac{3}{2}, 2,$….

SOLUTION The common difference is $d = \frac{1}{2}$. ▨

The General Term

The general term a_n of an arithmetic progression can always be written in terms of the first term a_1 and the common difference d. Consider the sequence from Example 1:

$$4, 10, 16, 22,\ldots$$

We can write each number in terms of the first term 4 and the common difference 6:

$$4, \qquad 4 + (1 \cdot 6), \qquad 4 + (2 \cdot 6), \qquad 4 + (3 \cdot 6),\dots$$
$$a_1, \qquad\qquad a_2, \qquad\qquad a_3, \qquad\qquad a_4,\dots$$

Observing the relationship between the subscript on the terms in the second line and the coefficients of the 6's in the first line, we write the general term for the sequence as

$$a_n = 4 + (n - 1)6$$

We generalize this result to obtain the general term of any arithmetic sequence.

$\lceil \Delta \neq \Sigma$ *Arithmetic Sequences*

The *general term* of an arithmetic progression with first term a_1 and common difference d is given by

$$a_n = a_1 + (n - 1)d$$

EXAMPLE 4   Find the general term for the sequence

$$7, 10, 13, 16,\dots$$

SOLUTION The first term is $a_1 = 7$, and the common difference is $d = 3$. Substituting these numbers into the formula given earlier, we have

$$a_n = 7 + (n - 1)3$$

which we can simplify, if we choose, to

$$a_n = 7 + 3n - 3$$
$$= 3n + 4$$

∎

EXAMPLE 5   Find the general term of the arithmetic progression whose third term a_3 is 7 and whose eighth term a_8 is 17.

SOLUTION According to the formula for the general term, the third term can be written as $a_3 = a_1 + 2d$, and the eighth term can be written as $a_8 = a_1 + 7d$. Because these terms are also equal to 7 and 17, respectively, we can write

$$a_3 = a_1 + 2d = 7$$
$$a_8 = a_1 + 7d = 17$$

To find a_1 and d, we simply solve the system:

$$a_1 + 2d = 7$$
$$a_1 + 7d = 17$$

We add the opposite of the top equation to the bottom equation. The result is

$$5d = 10$$
$$d = 2$$

To find a_1, we simply substitute 2 for d in either of the original equations and get

$$a_1 = 3$$

The general term for this progression is

$$a_n = 3 + (n - 1)2$$

which we can simplify to

$$a_n = 2n + 1$$

Looking at the simplified formulas in Examples 4 and 5, notice that we obtained a linear function in both cases. This is true in general. An arithmetic sequence can always be expressed as a linear function with slope given by d.

Arithmetic Series

The sum of the first n terms of an arithmetic sequence is denoted by S_n. The following theorem gives the formula for finding S_n, which is sometimes called the **nth partial sum**.

> $\lceil\Delta\neq\Sigma$ **THEOREM** *10.1*
>
> The sum of the first n terms of an arithmetic sequence whose first term is a_1 and whose nth term is a_n is given by
>
> $$S_n = \sum_{i=1}^{n}(a_1 + (i - 1)d) = \frac{n}{2}(a_1 + a_n)$$

Proof We can write S_n in expanded form as

$$S_n = a_1 + [a_1 + d] + [a_1 + 2d] + \cdots + [a_1 + (n - 1)d]$$

We can arrive at this same series by starting with the last term a_n and subtracting d each time. Writing S_n this way, we have

$$S_n = a_n + [a_n - d] + [a_n - 2d] + \cdots + [a_n - (n - 1)d]$$

If we add the preceding two expressions term by term, we have

$$2S_n = (a_1 + a_n) + (a_1 + a_n) + (a_1 + a_n) + \cdots + (a_1 + a_n)$$

$$2S_n = n(a_1 + a_n)$$

$$S_n = \frac{n}{2}(a_1 + a_n)$$

EXAMPLE 6   Find the sum of the first 10 terms of the arithmetic progression 2, 10, 18, 26,....

SOLUTION The first term is 2, and the common difference is 8. The tenth term is

$$a_{10} = 2 + 8(10 - 1)$$

$$= 2 + 72$$

$$= 74$$

Substituting $n = 10$, $a_1 = 2$, and $a_{10} = 74$ into the formula

$$S_n = \frac{n}{2}(a_1 + a_n)$$

we have

$$S_{10} = \frac{10}{2}(2 + 74)$$
$$= 5(76)$$
$$= 380$$

The sum of the first 10 terms is 380.

Getting Ready for Class

After reading through the preceding section, respond in your own words and in complete sentences.

A. Explain how to determine if a sequence is arithmetic.

B. What is a common difference?

C. Suppose the value of a_5 is given. What other possible pieces of information could be given to have enough information to obtain the first 10 terms of the sequence?

D. Explain the formula $a_n = a_1 + (n - 1)d$ in words so that someone who wanted to find the nth term of an arithmetic sequence could do so from your description.

Determine which of the following sequences are arithmetic progressions. For those that are arithmetic progressions, identify the common difference d.

1. $1, 2, 3, 4,\ldots$ **2.** $4, 6, 8, 10,\ldots$ **3.** $1, 2, 4, 7,\ldots$ **4.** $1, 2, 4, 8,\ldots$

5. $50, 45, 40,\ldots$ **6.** $1, \dfrac{1}{2}, \dfrac{1}{4}, \dfrac{1}{8},\ldots$ **7.** $1, 4, 9, 16,\ldots$ **8.** $5, 7, 9, 11,\ldots$

9. $\dfrac{1}{3}, 1, \dfrac{5}{3}, \dfrac{7}{3},\ldots$ **10.** $5, 11, 17,\ldots$

Each of the following problems refers to arithmetic sequences.

11. If $a_1 = 3$ and $d = 4$, find a_n and a_{24}.

12. If $a_1 = 5$ and $d = 10$, find a_n and a_{100}.

13. If $a_1 = 6$ and $d = -2$, find a_{10} and S_{10}.

14. If $a_1 = 7$ and $d = -1$, find a_{24} and S_{24}.

15. If $a_6 = 17$ and $a_{12} = 29$, find a_1, the common difference d, and then find a_{30}.

16. If $a_5 = 23$ and $a_{10} = 48$, find a_1, the common difference d, and then find a_{40}.

17. If the third term is 16 and the eighth term is 26, find the first term, the common difference, and then find a_{20} and S_{20}.

18. If the third term is 16 and the eighth term is 51, find the first term, the common difference, and then find a_{50} and S_{50}.

19. If $a_1 = 3$ and $d = 4$, find a_{20} and S_{20}.

20. If $a_1 = 40$ and $d = -5$, find a_{25} and S_{25}.

21. If $a_4 = 14$ and $a_{10} = 32$, find a_{40} and S_{40}.

22. If $a_7 = 0$ and $a_{11} = -\dfrac{8}{3}$, find a_{61} and S_{61}.

23. If $a_6 = -17$ and $S_6 = -12$, find a_1 and d.

24. If $a_{10} = 12$ and $S_{10} = 40$, find a_1 and d.

25. Find a_{85} for the sequence $14, 11, 8, 5,\ldots$

26. Find S_{100} for the sequence $-32, -25, -18, -11,\ldots$

27. If $S_{20} = 80$ and $a_1 = -4$, find d and a_{39}.

28. If $S_{24} = 60$ and $a_1 = 4$, find d and a_{116}.

29. Find the sum of the first 100 terms of the sequence $5, 9, 13, 17,\ldots.$

30. Find the sum of the first 50 terms of the sequence $8, 11, 14, 17,\ldots.$

31. Find a_{35} for the sequence $12, 7, 2, -3,\ldots.$

32. Find a_{45} for the sequence $25, 20, 15, 10,\ldots.$

33. Find the tenth term and the sum of the first 10 terms of the sequence $\dfrac{1}{2}, 1, \dfrac{3}{2}, 2,\ldots.$

34. Find the 15th term and the sum of the first 15 terms of the sequence $-\dfrac{1}{3}, 0, \dfrac{1}{3}, \dfrac{2}{3},\ldots.$

Applying the Concepts

Straight-Line Depreciation Recall from a previous section that straight-line depreciation is an accounting method used to help spread the cost of new equipment over a number of years. The value at any time during the life of the machine can be found with a linear equation in two variables. For income tax purposes, however, it is the value at the end of the year that is most important, and for this reason sequences can be used.

35. **Value of a Copy Machine** A large copy machine sells for $18,000 when it is new. Its value decreases $3,300 each year after that. We can use an arithmetic sequence to find the value of the machine at the end of each year. If we let a_0 represent the value when it is purchased, then a_1 is the value after 1 year, a_2 is the value after 2 years, and so on.

   a. Write the first 5 terms of the sequence.

   b. What is the common difference?

   c. Construct a line graph for the first 5 terms of the sequence.

   d. Use the line graph to estimate the value of the copy machine 2.5 years after it is purchased.

   e. Write the sequence from part a using a recursive formula.

36. **Value of a Forklift** An electric forklift sells for $125,000 when new. Each year after that, it decreases $16,500 in value.

   a. Write an arithmetic sequence that gives the value of the forklift at the end of each of the first 5 years after it is purchased.

   b. What is the common difference for this sequence?

   c. Construct a line graph for this sequence.

   d. Use the line graph to estimate the value of the forklift 3.5 years after it is purchased.

   e. Write the sequence from part a using a recursive formula.

37. **Distance** A rocket travels vertically 1,500 feet in its first second of flight, and then about 40 feet less each succeeding second. Use these estimates to answer the following questions.

   a. Write a sequence of the vertical distance traveled by a rocket in each of its first 6 seconds.

   b. Is the sequence in part a an arithmetic sequence? Explain why or why not.

   c. What is the general term of the sequence in part a?

38. **Depreciation** Suppose an automobile sells for N dollars new, and then depreciates 40% each year.

   a. Write a sequence for the value of this automobile (in terms of N) for each year.

   b. What is the general term of the sequence in part a?

   c. Is the sequence in part a an arithmetic sequence? Explain why it is or is not.

39. Triangular Numbers The first four triangular numbers are 1, 3, 6, 10, . . ., and are illustrated in the following diagram.

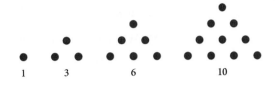

1   3   6   10

  a. Write a sequence of the first 15 triangular numbers.

  b. Write the recursive general term for the sequence of triangular numbers.

  c. Is the sequence of triangular numbers an arithmetic sequence? Explain why it is or is not.

40. Arithmetic Means Three (or more) arithmetic means between two numbers may be found by forming an arithmetic sequence using the original two numbers and the arithmetic means. For example, three arithmetic means between 10 and 34 may be found by examining the sequence 10, a, b, c, 34. For the sequence to be arithmetic, the common difference must be 6; therefore, $a = 16$, $b = 22$, and $c = 28$. Use this idea to answer the following questions.

  a. Find four arithmetic means between 10 and 35.

  b. Find three arithmetic means between 2 and 62.

  c. Find five arithmetic means between 4 and 28.

41. Paratroopers At the Ft. Campbell Army Base, soldiers in the 101st Airborne Division are trained to be paratroopers. A paratrooper's free fall drop per second can be modeled by an arithmetic sequence whose first term is 16 feet and whose common difference is 32 feet.

  a. Write the general term for this arithmetic progression.

  b. How far would a paratrooper fall during the tenth second of free fall?

  c. Using the sum formula for an arithmetic sequence, what would be the total distance a paratrooper fell during 10 seconds of free fall?

42. College Fund When Jack's first grandchild, Kayla, was born, he decided to establish a college fund for her. On Kayla's first birthday, Jack deposited $1,000 into an account and decided that each year on her birthday he would deposit an amount $500 more than the previous year, through her 18th birthday.

  a. Write a sequence to represent the amount Jack deposited on her 1st through 5th birthdays. Does this sequence represent an arithmetic progression? If so, what is the common difference?

  b. How much will Jack deposit on Kayla's 18th birthday?

  c. What would be the total amount (excluding any interest earned) Jack has saved for Kayla's college fund by her 18th birthday?

Learning Objectives Assessment

The following problems can be used to help assess if you have successfully met the learning objectives for this section.

43. Which of the following is an arithmetic sequence?

    a. $7, 4, 1, -2,...$

    b. $2, 4, 8, 16,...$

    c. $\dfrac{1}{2}, \dfrac{1}{4}, \dfrac{1}{6}, \dfrac{1}{8},...$

    d. $1, 1.1, 1.11, 1.111,...$

44. What is the common difference for the sequence $90, 80, 70, 60,...$?

    a. 10 **b.** $\dfrac{15}{2}$ **c.** 30 **d.** -10

45. Find the tenth term for the arithmetic sequence $15, 21, 27,....$

    a. 81 **b.** 75 **c.** 60 **d.** 69

46. Find the sum of the first 20 terms of the arithmetic sequence with $a_1 = 5$ and $d = 2.5$.

    a. 57.5 **b.** 600 **c.** 575 **d.** 150

Getting Ready for the Next Section

Simplify.

47. $\dfrac{1}{8}\left(\dfrac{1}{2}\right)$

48. $\dfrac{1}{4}\left(\dfrac{1}{2}\right)$

49. $\dfrac{3\sqrt{3}}{3}$

50. $\dfrac{3}{\sqrt{3}}$

51. $2 \cdot 2^{n-1}$

52. $3 \cdot 3^{n-1}$

53. $\dfrac{ar^6}{ar^3}$

54. $\dfrac{ar^7}{ar^4}$

55. $\dfrac{\dfrac{1}{5}}{1 - \dfrac{1}{2}}$

56. $\dfrac{\dfrac{9}{10}}{1 - \dfrac{1}{10}}$

57. $\dfrac{3[(-2)^8 - 1]}{-2 - 1}$

58. $\dfrac{4\left[\left(\dfrac{1}{2}\right)^6 - 1\right]}{\dfrac{1}{2} - 1}$

Geometric Sequences   10.4

Learning Objectives

In this section, we will learn how to:

1. Identify a geometric sequence.

2. Find the common ratio for a geometric sequence.

3. Find the general term for a geometric sequence.

4. Find the sum for a finite or infinite geometric series.

Introduction

This section is concerned with the second major classification of sequences, called geometric sequences. The problems in this section are similar to the problems in the preceding section.

> **(def) DEFINITION** *geometric sequence*
>
> A sequence of numbers in which each term is obtained from the previous term by multiplying by the same amount each time is called a **geometric sequence**. Geometric sequences are also called **geometric progressions**.

The sequence

$$3, 6, 12, 24,\ldots$$

is an example of a geometric progression. Each term is obtained from the previous term by multiplying by 2. The amount by which we multiply each time — in this case, 2 — is called the **common ratio**. The common ratio, denoted by r, is found by taking the ratio of any two consecutive terms $r = \dfrac{a_n}{a_{n-1}}$ for any $n > 1$.

VIDEO EXAMPLES

SECTION 10.4

> **EXAMPLE 1**   Find the common ratio for the geometric progression.
>
> $$\frac{1}{2}, \frac{1}{4}, \frac{1}{8}, \frac{1}{16},\ldots$$

SOLUTION Because each term can be obtained from the term before it by multiplying by $\frac{1}{2}$, the common ratio is $\frac{1}{2}$. That is, $r = \frac{1}{2}$. ∎

> **EXAMPLE 2**   Find the common ratio for $\sqrt{3}, 3, 3\sqrt{3}, 9,\ldots$

SOLUTION   If we take the ratio of the third term to the second term, we have

$$\frac{3\sqrt{3}}{3} = \sqrt{3}$$

The common ratio is $r = \sqrt{3}$. ∎

> **⌈Δ≠Σ⌉   Geometric Sequences**
>
> The **general term** a_n of a geometric sequence with first term a_1 and common ratio r is given by
>
> $$a_n = a_1 r^{n-1}$$

To see how we arrive at this formula, consider the following geometric progression whose common ratio is 3:

$$2, 6, 18, 54, \ldots$$

We can write each number in the sequence in terms of the first term 2 and the common ratio 3:

$$2 \cdot 3^0, \quad 2 \cdot 3^1, \quad 2 \cdot 3^2, \quad 2 \cdot 3^3, \ldots$$
$$a_1, \quad\quad a_2, \quad\quad a_3, \quad\quad a_4, \ldots$$

Observing the relationship between the two preceding lines, we find we can write the general term of this progression as

$$a_n = 2 \cdot 3^{n-1}$$

Because the first term can be designated by a_1 and the common ratio by r, the formula

$$a_n = 2 \cdot 3^{n-1}$$

coincides with the formula

$$a_n = a_1 r^{n-1}$$

EXAMPLE 3   Find the general term for the geometric progression

$$5, 10, 20, \ldots$$

SOLUTION   The first term is $a_1 = 5$, and the common ratio is $r = 2$. Using these values in the formula

$$a_n = a_1 r^{n-1}$$

we have

$$a_n = 5 \cdot 2^{n-1}$$

EXAMPLE 4   Find the tenth term of the sequence $3, \dfrac{3}{2}, \dfrac{3}{4}, \dfrac{3}{8}, \ldots$.

SOLUTION   The sequence is a geometric progression with first term $a_1 = 3$ and common ratio $r = \frac{1}{2}$. The tenth term is

$$a_{10} = 3\left(\frac{1}{2}\right)^9 = \frac{3}{512}$$

EXAMPLE 5   Find the general term for the geometric progression whose fourth term is 16 and whose seventh term is 128.

SOLUTION   The fourth term can be written as $a_4 = a_1 r^3$, and the seventh term can be written as $a_7 = a_1 r^6$.

$$a_4 = a_1 r^3 = 16$$
$$a_7 = a_1 r^6 = 128$$

We can solve for r by using the ratio $\frac{a_7}{a_4}$.

$$\frac{a_7}{a_4} = \frac{a_1 r^6}{a_1 r^3} = \frac{128}{16}$$
$$r^3 = 8$$
$$r = 2$$

The common ratio is 2. To find the first term, we substitute $r = 2$ into either of the original two equations. The result is

$$a_1 = 2$$

The general term for this progression is

$$a_n = 2 \cdot 2^{n-1}$$

which we can simplify by adding exponents, because the bases are equal:

$$a_n = 2^n$$

Finite Geometric Series

As was the case in the preceding section, the sum of the first n terms of a geometric progression is denoted by S_n, which is called the ***nth partial sum*** of the progression.

[△≠Σ] THEOREM   *10.2*

The sum of the first n terms of a geometric progression with first term a_1 and common ratio r is given by the formula

$$S_n = \sum_{i=1}^{n} a_1 r^{i-1} = \frac{a_1(r^n - 1)}{r - 1}$$

Proof   We can write the sum of the first n terms in expanded form:

$$S_n = a_1 + a_1 r + a_1 r^2 + \cdots + a_1 r^{n-1} \qquad (1)$$

Then multiplying both sides by r, we have

$$r S_n = a_1 r + a_1 r^2 + a_1 r^3 + \cdots + a_1 r^n \qquad (2)$$

If we subtract the left side of equation (1) from the left side of equation (2) and do the same for the right sides, we end up with

$$r S_n - S_n = a_1 r^n - a_1$$

We factor S_n from both terms on the left side and a_1 from both terms on the right side of this equation:

$$S_n(r - 1) = a_1(r^n - 1)$$

Dividing both sides by $r - 1$ gives the desired result:

$$S_n = \frac{a_1(r^n - 1)}{r - 1}$$

EXAMPLE 6   Find the sum of the first 10 terms of the geometric progression 5, 15, 45, 135,....

SOLUTION   The first term is $a_1 = 5$, and the common ratio is $r = 3$. Substituting these values into the formula for S_{10}, we have the sum of the first 10 terms of the sequence:

$$S_{10} = \frac{5(3^{10} - 1)}{3 - 1}$$

$$= \frac{5(3^{10} - 1)}{2}$$

The answer can be left in this form. A calculator will give the result as 147,620.

Infinite Geometric Series

Suppose the common ratio for a geometric sequence is a number whose absolute value is less than 1—for instance, $\frac{1}{2}$. The sum of the first n terms is given by the formula

$$S_n = \frac{a_1\left[\left(\frac{1}{2}\right)^n - 1\right]}{\frac{1}{2} - 1}$$

As n becomes larger and larger, the term $\left(\frac{1}{2}\right)^n$ will become closer and closer to 0. That is, for $n = 10$, 20, and 30, we have the following approximations:

$$\left(\frac{1}{2}\right)^{10} \approx 0.001$$

$$\left(\frac{1}{2}\right)^{20} \approx 0.000001$$

$$\left(\frac{1}{2}\right)^{30} \approx 0.000000001$$

so that for large values of n, there is little difference between the expression

$$\frac{a_1(r^n - 1)}{r - 1}$$

and the expression

$$\frac{a_1(0 - 1)}{r - 1} = \frac{-a_1}{r - 1} = \frac{a_1}{1 - r} \qquad \text{if} \qquad |r| < 1$$

In fact, the sum of the terms of a geometric sequence in which $|r| < 1$ actually becomes the expression

$$\frac{a_1}{1 - r}$$

as n approaches infinity. To summarize, we have the following:

> **⎡Δ≠Σ⎤**   *The Sum of an Infinite Geometric Sequence*
>
> If a geometric sequence has first term a_1 and common ratio r such that $|r| < 1$, then the following is called an *infinite geometric series*:
>
> $$S = \sum_{i=0}^{\infty} a_1 r^i = a_1 + a_1 r + a_1 r^2 + a_1 r^3 + \cdots$$
>
> Its sum is given by the formula
>
> $$S = \frac{a_1}{1 - r}$$

EXAMPLE 7   Find the sum of the infinite geometric series

$$\frac{1}{5} + \frac{1}{10} + \frac{1}{20} + \frac{1}{40} + \cdots$$

SOLUTION   The first term is $a_1 = \frac{1}{5}$, and the common ratio is $r = \frac{1}{2}$, which has an absolute value less than 1. Therefore, the sum of this series is

$$S = \frac{a_1}{1 - r} = \frac{\frac{1}{5}}{1 - \frac{1}{2}} = \frac{\frac{1}{5}}{\frac{1}{2}} = \frac{2}{5}$$

EXAMPLE 8 Show that 0.999... is equal to 1.

SOLUTION We begin by writing 0.999... as an infinite geometric series:

$$0.999... = 0.9 + 0.09 + 0.009 + 0.0009 + \cdots$$

$$= \frac{9}{10} + \frac{9}{100} + \frac{9}{1,000} + \frac{9}{10,000} + \cdots$$

$$= \frac{9}{10} + \frac{9}{10}\left(\frac{1}{10}\right) + \frac{9}{10}\left(\frac{1}{10}\right)^2 + \frac{9}{10}\left(\frac{1}{10}\right)^3 + \cdots$$

As the last line indicates, we have an infinite geometric series with $a_1 = \frac{9}{10}$ and $r = \frac{1}{10}$. The sum of this series is given by

$$S = \frac{a_1}{1-r} = \frac{\frac{9}{10}}{1 - \frac{1}{10}} = \frac{\frac{9}{10}}{\frac{9}{10}} = 1$$

Getting Ready for Class

After reading through the preceding section, respond in your own words and in complete sentences.

A. What is a common ratio?

B. Explain the formula $a_n = a_1 r^{n-1}$ in words so that someone who wanted to find the nth term of a geometric sequence could do so from your description.

C. When is the sum of an infinite geometric series a finite number?

D. Explain how a repeating decimal can be represented as an infinite geometric series.

Problem Set 10.4

Identify those sequences that are geometric progressions. For those that are geometric, give the common ratio r.

1. 1, 5, 25, 125,...

2. 6, 12, 24, 48,...

3. $\frac{1}{2}, \frac{1}{6}, \frac{1}{18}, \frac{1}{54}, \ldots$

4. 5, 10, 15, 20,...

5. 4, 9, 16, 25,...

6. $-1, \frac{1}{3}, -\frac{1}{9}, \frac{1}{27}, \ldots$

7. $-2, 4, -8, 16,\ldots$

8. 1, 8, 27, 64,...

9. 4, 6, 8, 10,...

10. 1, -3, 9, -27,...

Each of the following problems gives some information about a specific geometric progression.

11. If $a_1 = 4$ and $r = 3$, find a_n.

12. If $a_1 = 5$ and $r = 2$, find a_n.

13. If $a_1 = -2$ and $r = -\frac{1}{2}$, find a_6.

14. If $a_1 = 25$ and $r = -\frac{1}{5}$, find a_6.

15. If $a_1 = 3$ and $r = -1$, find a_{20}.

16. If $a_1 = -3$ and $r = -1$, find a_{20}.

17. If $a_1 = 10$ and $r = 2$, find S_{10}.

18. If $a_1 = 8$ and $r = 3$, find S_5.

19. If $a_1 = 1$ and $r = -1$, find S_{20}.

20. If $a_1 = 1$ and $r = -1$, find S_{21}.

21. Find a_8 for $\frac{1}{5}, \frac{1}{10}, \frac{1}{20}, \ldots$

22. Find a_8 for $\frac{1}{2}, \frac{1}{10}, \frac{1}{50}, \ldots$

23. Find S_5 for $-\frac{1}{2}, -\frac{1}{4}, -\frac{1}{8}, \ldots$

24. Find S_6 for $-\frac{1}{2}, 1, -2, \ldots$

25. Find a_{10} and S_{10} for $\sqrt{2}, 2, 2\sqrt{2}, \ldots$

26. Find a_8 and S_8 for $\sqrt{3}, 3, 3\sqrt{3}, \ldots$

27. Find a_6 and S_6 for 100, 10, 1,...

28. Find a_6 and S_6 for 100, -10, 1,...

29. If $a_4 = 40$ and $a_6 = 160$, find r.

30. If $a_5 = \frac{1}{8}$ and $a_8 = \frac{1}{64}$, find r.

31. Given the sequence $-3, 6, -12, 24,\ldots$, find a_8 and S_8.

32. Given the sequence 4, 2, 1, $\frac{1}{2}, \ldots$, find a_9 and S_9.

33. Given $a_7 = 13$ and $a_{10} = 104$, find r.

34. Given $a_5 = -12$ and $a_8 = 324$, find r.

Find the sum of each infinite geometric series.

35. $\frac{1}{2} + \frac{1}{4} + \frac{1}{8} + \cdots$

36. $\frac{1}{3} + \frac{1}{9} + \frac{1}{27} + \cdots$

37. $4 + 2 + 1 + \cdots$

38. $8 + 4 + 2 + \cdots$

39. $2 + 1 + \frac{1}{2} + \cdots$

40. $3 + 1 + \frac{1}{3} + \cdots$

41. $\frac{4}{3} - \frac{2}{3} + \frac{1}{3} + \cdots$

42. $6 - 4 + \frac{8}{3} + \cdots$

43. $\frac{2}{5} + \frac{4}{25} + \frac{8}{125} + \cdots$

44. $\frac{3}{4} + \frac{9}{16} + \frac{27}{64} + \cdots$

45. $\frac{3}{4} + \frac{1}{4} + \frac{1}{12} + \cdots$

46. $\frac{5}{3} + \frac{1}{3} + \frac{1}{15} + \cdots$

47. Show that 0.444... is the same as $\dfrac{4}{9}$.

48. Show that 0.333...is the same as $\dfrac{1}{3}$.

49. Show that 0.272727...is the same as $\dfrac{3}{11}$.

50. Show that 0.545454...is the same as $\dfrac{6}{11}$.

Applying the Concepts

Declining-Balance Depreciation The declining-balance method of depreciation is an accounting method businesses use to deduct most of the cost of new equipment during the first few years of purchase. The value at any time during the life of the machine can be found with an exponential equation in two variables. For income tax purposes, however, it is the value at the end of the year that is most important, and for this reason sequences can be used.

51. Value of a Crane A construction crane sells for $450,000 if purchased new. After that, the value decreases by 30% each year. We can use a geometric sequence to find the value of the crane at the end of each year. If we let a_0 represent the value when it is purchased, then a_1 is the value after 1 year, a_2 is the value after 2 years, and so on.

  a. Write the first five terms of the sequence.

  b. What is the common ratio?

  c. Construct a line graph for the first five terms of the sequence.

  d. Use the line graph to estimate the value of the crane 3.5 years after it is purchased.

  e. Write the sequence from part *a* using a recursive formula.

52. Value of a Printing Press A large printing press sells for $375,000 when it is new. After that, its value decreases 25% each year.

  a. Write a geometric sequence that gives the value of the press at the end of each of the first 5 years after it is purchased.

  b. What is the common ratio for this sequence?

  c. Construct a line graph for this sequence.

  d. Use the line graph to estimate the value of the printing press 1.5 years after it is purchased.

  e. Write the sequence from part *a* using a recursive formula.

53. Adding Terms Given the geometric series

$$\frac{1}{3} + \frac{1}{9} + \frac{1}{27} + \cdots$$

  a. Find the sum of all the terms.

  b. Find the sum of the first six terms.

  c. Find the sum of all but the first six terms.

54. Bouncing Ball A ball is dropped from a height of 20 feet. Each time it bounces it returns to $\frac{7}{8}$ of the height it fell from. If the ball is allowed to bounce an infinite number of times, find the total vertical distance that the ball travels.

55. Stacking Paper Assume that a thin sheet of paper is 0.002 inch thick. The paper is torn in half, and the two halves placed together.

a. How thick is the pile of torn paper?

b. The pile of paper is torn in half again, and then the two halves placed together and torn in half again. The paper is large enough so this process may be performed a total of 5 times. How thick is the pile of torn paper?

c. Refer to the tearing and piling process described in part *b* Assuming that somehow the original paper is large enough, how thick is the pile of torn paper if 25 tears are made?

56. Pendulum A pendulum swings 15 feet left to right on its first swing. On each swing following the first, the pendulum swings $\frac{4}{5}$ of the previous swing.

a. Write the general term for this geometric sequence.

b. If the pendulum is allowed to swing an infinite number of times, what is the total distance the pendulum will travel?

57. Salary Increases After completing her MBA degree, an accounting firm offered Jane a job with a starting salary of $60,000, with a guaranteed annual raise of 7% of her previous year's salary.

a. Write a finite sequence to represent Jane's first 5 years of income with this company. (Round each calculation to the nearest dollar.)

b. Write the general term for this geometric sequence.

c. Find the sum of Jane's income for the first 10 years with the company. Round to the nearest dollar.

Learning Objectives Assessment

The following problems can be used to help assess if you have successfully met the learning objectives for this section.

58. Which of the following is a geometric sequence?

a. 7, 4, 1, −2,...   **b.** 2, 4, 8, 16,...

c. $\frac{1}{2}, \frac{1}{4}, \frac{1}{6}, \frac{1}{8},...$ **d.** 1, 1.1, 1.11, 1.111,...

59. What is the common ratio for the sequence 5, 0.5, 0.05, 0.005,...?

a. 0.1   **b.** 10   **c.** −10   **d.** −0.1

60. Find a_6 for the geometric sequence 64, 96, 144,....

a. 729   **b.** 224   **c.** 486   **d.** 324

61. Find the sum of the infinite geometric series 64, −16, 4, −1,....

a. $\frac{64}{5}$ **b.** $\frac{256}{3}$ **c.** 51   **d.** $\frac{256}{5}$

Getting Ready for the Next Section

Simplify.

62. $(x + y)^0$

63. $(x + y)^1$

Expand and multiply.

64. $(x + y)^2$

65. $(x + y)^3$

Simplify.

66. $\dfrac{6 \cdot 5 \cdot 4 \cdot 3 \cdot 2 \cdot 1}{(2 \cdot 1)(4 \cdot 3 \cdot 2 \cdot 1)}$

67. $\dfrac{7 \cdot 6 \cdot 5 \cdot 4 \cdot 3 \cdot 2 \cdot 1}{(5 \cdot 4 \cdot 3 \cdot 2 \cdot 1)(2 \cdot 1)}$

SPOTLIGHT ON SUCCESS   *Student Instructor Nathan*

Keep steadily before you the fact that all true success depends at last upon yourself.
—Theodore T. Hunger

Math has always come fairly easily for me and is the academic subject I have enjoyed most. I knew I wanted to attend Cal Poly San Luis Obispo for its high job placement and prestige, but I had no idea what I wanted to study. I decided to major in Mathematics because it is so universal but not so specialized or concentrated that I would get stuck in a field that I did not enjoy. I felt that if I kept studying math and its related fields, I would set myself up to be successful later in life, as math is the foundation for engineering, physics, and other science related fields. I have not looked back on my decision. I know it will be a degree that I am proud to have achieved.

I appreciate the consistency that math offers in its problems and in its solutions. I like that math can be simplified into smaller easier-to-understand parts, and its answers are almost always definite. It provides challenges that I enjoy solving, like completing a puzzle piece by piece. In the end, I am able to enjoy the success I have put together for myself.

Learning Objectives

In this section, we will learn how to:

1. Expand the power of a binomial using Pascal's Triangle.

2. Evaluate a factorial.

3. Use the Binomial Theorem to find a given term in a binomial expansion.

Introduction

The purpose of this section is to write the expansion of expressions of the form $(x + y)^n$, where n is any positive integer. If n is small, it is not too difficult to find the expansion using multiplication.

VIDEO EXAMPLES

SECTION 10.5

> **EXAMPLE 1**   Expand: $(x + y)^3$.
>
> **SOLUTION**   We write the expression as a product and then multiply.
>
> $$(x + y)^3 = \underbrace{(x + y)(x + y)}(x + y)$$
>
> $$= (x^2 + 2xy + y^2)(x + y) \qquad \text{FOIL}$$
>
> $$= x^3 + x^2y + 2x^2y + 2xy^2 + xy^2 + y^3 \quad \text{Multiply}$$
>
> $$= x^3 + 3x^2y + 3xy^2 + y^3 \qquad \text{Combine like terms}$$

As n becomes larger, however, this process becomes very tedious and time consuming. For example, imagine multiplying ten factors of $(x + y)$ in order to expand $(x + y)^{10}$.

Pascal's Triangle

It turns out there is a convenient formula that will allow us to expand a binomial power much more efficiently. To write the formula, we must generalize the information in the following chart:

$$(x + y)^0 = \qquad\qquad\qquad\qquad 1$$
$$(x + y)^1 = \qquad\qquad\qquad\quad x \;+\; y$$
$$(x + y)^2 = \qquad\qquad\quad x^2 \;+\; 2xy \;+\; y^2$$
$$(x + y)^3 = \qquad\qquad x^3 \;+\; 3x^2y \;+\; 3xy^2 \;+\; y^3$$
$$(x + y)^4 = \qquad x^4 \;+\; 4x^3y \;+\; 6x^2y^2 \;+\; 4xy^3 \;+\; y^4$$
$$(x + y)^5 = x^5 \;+\; 5x^4y \;+\; 10x^3y^2 \;+\; 10x^2y^3 \;+\; 5xy^4 \;+\; y^5$$

Note   The polynomials to the right have been found by expanding the binomials on the left — we just haven't shown the work.

There are a number of similarities to notice among the polynomials on the right. Here is a list:

1. The number of terms in each polynomial is one greater than the exponent on the binomial at the left.

2. In each polynomial, the sequence of exponents on the variable x decreases to 0 from the exponent on the binomial at the left. (The exponent 0 is not shown, since $x^0 = 1$.)

3. In each polynomial, the exponents on the variable y increase from 0 to the exponent on the binomial at the left. (Because $y^0 = 1$, it is not shown in the first term.)

4. The sum of the exponents on the variables in any single term is equal to the exponent on the binomial at the left.

5. The coefficients of the first and last terms are both equal to 1.

The pattern in the coefficients of the polynomials on the right can best be seen by writing the right side again without the variables. It looks like this:

row 0						1					
row 1					1		1				
row 2				1		2		1			
row 3			1		3		3		1		
row 4		1		4		6		4		1	
row 5	1		5		10		10		5		1

This triangle-shaped array of coefficients is called **Pascal's Triangle**. Each row begins and ends with the number 1. Each internal entry in the triangular array is obtained by adding the two numbers above it. If we were to continue Pascal's Triangle, the next two rows would be

```
row 5        1      5     10     10      5      1
              \ + /  \ + /  \ + /  \ + /  \ + /
row 6      1      6     15     20     15      6      1
            \ + /  \ + /  \ + /  \ + /  \ + /  \ + /
row 7    1      7     21     35     35     21      7      1
```

The coefficients for the terms in the expansion of $(x + y)^n$ are given in the nth row of Pascal's Triangle, where we count rows beginning with 0.

Pascal's Triangle is named after Blaise Pascal, a seventeenth century mathematician and philosopher. However, we now know the triangle had been discovered as early as the eleventh century by both the Persians and the Chinese. Figure 1 shows a version of the triangle by Yang Hui from 1261.

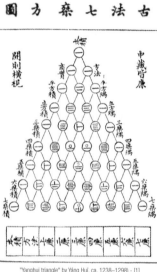

"Yanghui triangle" by Yáng Huī, ca. 1238–1298) - [1]
Licensed under Public Domain via Wikimedia Commons

FIGURE 1

EXAMPLE 2   Expand: $(3x + 2y)^4$.

SOLUTION   First, we find the expansion of $(a + b)^4$. There are 5 terms in the expansion. The powers of a will decrease from 4 and the powers of b will increase from 0. The coefficients can be found in the fourth row of Pascal's Triangle, which are 1, 4, 6, 4, 1. This gives us

$$(a + b)^4 = 1a^4b^0 + 4a^3b^1 + 6a^2b^2 + 4a^1b^3 + 1a^0b^4$$

Now, we simplify $a^0 = b^0 = 1$ and then substitute $a = 3x$ and $b = 2y$.

$$(3x + 2y)^4 = 1(3x)^4 \cdot (1) + 4(3x)^3(2y) + 6(3x)^2(2y)^2 + 4(3x)(2y)^3 + (1) \cdot 1(2y)^4$$

To simplify, we first evaluate each exponent, and the perform all of the multiplications within each term.

$$(3x + 2y)^4 = 1(81x^4)(1) + 4(27x^3)(2y) + 6(9x^2)(4y^2) + 4(3x)(8y^3) + 1(1)(16y^4)$$
$$= 81x^4 + 216x^3y + 216x^2y^2 + 96xy^3 + 16y^4$$

Factorial Notation

There is an alternative method of finding these coefficients that does not involve Pascal's Triangle. The alternative method involves **factorial notation**.

> (def) **DEFINITION**   *n!*
>
> The expression *n!* is read "*n* factorial" and is the product of all the consecutive integers from *n* down to 1. For example,
>
> $$1! = 1$$
> $$2! = 2 \cdot 1 = 2$$
> $$3! = 3 \cdot 2 \cdot 1 = 6$$
> $$4! = 4 \cdot 3 \cdot 2 \cdot 1 = 24$$
> $$5! = 5 \cdot 4 \cdot 3 \cdot 2 \cdot 1 = 120$$
>
> As a special case, we define $0! = 1$.

We use factorial notation to define binomial coefficients. The following formula comes from probability theory, and is used to count the number of combinations of *r* items selected from *n* items.

> (def) **DEFINITION**   *binomial coefficient*
>
> The expression $_nC_r$, sometimes written as $\binom{n}{r}$, is called a ***binomial coefficient*** and is defined by
>
> $$_nC_r = \binom{n}{r} = \frac{n!}{r!(n-r)!}$$

EXAMPLE 3   Calculate the following binomial coefficients:
$$_7C_5, \binom{6}{2}, {_3C_0}$$

SOLUTION   We simply apply the definition for binomial coefficients:

$$_7C_5 = \frac{7!}{5!(7-5)!}$$
$$= \frac{7!}{5! \cdot 2!}$$
$$= \frac{7 \cdot 6 \cdot 5 \cdot 4 \cdot 3 \cdot 2 \cdot 1}{(5 \cdot 4 \cdot 3 \cdot 2 \cdot 1)(2 \cdot 1)}$$
$$= \frac{42}{2}$$
$$= 21$$

$$\binom{6}{2} = \frac{6!}{2!(6-2)!}$$
$$= \frac{6!}{2! \cdot 4!}$$
$$= \frac{6 \cdot 5 \cdot 4 \cdot 3 \cdot 2 \cdot 1}{(2 \cdot 1)(4 \cdot 3 \cdot 2 \cdot 1)}$$
$$= \frac{30}{2}$$
$$= 15$$

$$_3C_0 = \frac{3!}{0!(3-0)!}$$

$$= \frac{3!}{0! \cdot 3!}$$

$$= \frac{3 \cdot 2 \cdot 1}{(1)(3 \cdot 2 \cdot 1)}$$

$$= 1$$

USING TECHNOLOGY *Binomial Coefficients on a Calculator*

Because the binomial coefficient formula is also used for statistics, many scientific and graphing calculators have this formula built-in.

For instance, on a TI-84 graphing calculator we can find $_7C_5$ using the $_nC_r$ command. First input a 7, then press MATH followed by the left arrow key. Select the $_nC_r$ command and then input a 5. Our screen will look like this:

$$7 \ _nC_r \ 5$$

Pressing ENTER gives us a result of 21, which agrees with our answer in Example 3.

If we were to calculate all the binomial coefficients in the following array, we would find they match exactly with the numbers in Pascal's Triangle. That is why they are called binomial coefficients — because they are the coefficients of the expansion of $(x + y)^n$. The value $_nC_r = \binom{n}{r}$ will be the number in Pascal's Triangle located in position r of row n, where rows are counted beginning with 0, and position is counted in a given row from the left and also beginning with 0 (See Figure 2).

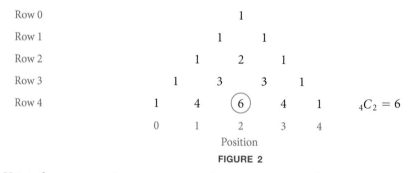

Row 0					1				
Row 1				1		1			
Row 2			1		2		1		
Row 3		1		3		3		1	
Row 4	1		4		6		4		1	$_4C_2 = 6$

	0	1	2	3	4
			Position		

FIGURE 2

Using the new notation to represent the entries in Pascal's Triangle, we can summarize everything we have noticed about the expansion of binomial powers of the form $(x + y)^n$.

EXAMPLE 4 Expand $(x - 2)^3$.

SOLUTION Applying the binomial coefficient formula, we have

$$(x - 2)^3 = \binom{3}{0} x^3(-2)^0 + \binom{3}{1} x^2(-2)^1 + \binom{3}{2} x^1(-2)^2 + \binom{3}{3} x^0(-2)^3$$

The coefficients

$$\binom{3}{0}, \binom{3}{1}, \binom{3}{2}, \text{ and } \binom{3}{3}$$

can be found in the third row of Pascal's Triangle. They are 1, 3, 3, and 1:

$$(x - 2)^3 = 1x^3(-2)^0 + 3x^2(-2)^1 + 3x^1(-2)^2 + 1x^0(-2)^3$$

$$= x^3 - 6x^2 + 12x - 8$$

EXAMPLE 5 Write the first three terms in the expansion of $(x + 5)^9$.

SOLUTION The coefficients of the first three terms are

$$_9C_0, \ _9C_1, \text{ and } _9C_2$$

which we calculate as follows:

$$_9C_0 = \frac{9!}{0! \cdot 9!} = \frac{9 \cdot 8 \cdot 7 \cdot 6 \cdot 5 \cdot 4 \cdot 3 \cdot 2 \cdot 1}{(1)(9 \cdot 8 \cdot 7 \cdot 6 \cdot 5 \cdot 4 \cdot 3 \cdot 2 \cdot 1)} = \frac{1}{1} = 1$$

$$_9C_1 = \frac{9!}{1! \cdot 8!} = \frac{9 \cdot 8 \cdot 7 \cdot 6 \cdot 5 \cdot 4 \cdot 3 \cdot 2 \cdot 1}{(1)(8 \cdot 7 \cdot 6 \cdot 5 \cdot 4 \cdot 3 \cdot 2 \cdot 1)} = \frac{9}{1} = 9$$

$$_9C_2 = \frac{9!}{2! \cdot 7!} = \frac{9 \cdot 8 \cdot 7 \cdot 6 \cdot 5 \cdot 4 \cdot 3 \cdot 2 \cdot 1}{(2 \cdot 1)(7 \cdot 6 \cdot 5 \cdot 4 \cdot 3 \cdot 2 \cdot 1)} = \frac{72}{2} = 36$$

We write the first three terms:

$$(x + 5)^9 = 1 \cdot x^9 + 9 \cdot x^8(5) + 36x^7(5)^2 + \cdots$$
$$= x^9 + 45x^8 + 900x^7 + \cdots$$

The Binomial Theorem

If we look at each term in the expansion of $(x + y)^n$ as a term in a sequence, a_1, a_2, a_3, . . . , we can write

$$a_1 = \binom{n}{0} x^n y^0 \quad = _nC_0 x^n y^0$$

$$a_2 = \binom{n}{1} x^{n-1} y^1 \quad = _nC_1 x^{n-1} y^1$$

$$a_3 = \binom{n}{2} x^{n-2} y^2 \quad = _nC_2 x^{n-2} y^2$$

$$a_4 = \binom{n}{3} x^{n-3} y^3 \quad = _nC_3 x^{n-3} y^3 \quad \text{and so on}$$

To write the formula for the general term, we simply notice that the number used for the binomial coefficient and appearing in the exponents of the variables is always 1 less than the term number. This observation allows us to write the following:

> **⟨Δ≠Σ⟩** *The General Term of a Binomial Expansion*
>
> The kth term in the expansion of $(x + y)^n$ is
>
> $$a_k = \binom{n}{k - 1} x^{n - (k-1)} y^{k-1}$$

Then the expansion of $(x + y)^n$ can be treated as a finite series

$$(x + y)^n = a_1 + a_2 + a_3 + \ldots + a_{n+1}$$

Using summation notation, we can express this series as

$$(x + y)^n = \sum_{k=1}^{n+1} a_k$$

To simplify the summation notation and general term formula, we let $r = k - 1$.

Then r will take on the values $0, 1, 2, \ldots, n$. This leads us to the following theorem:

 Binomial Theorem

If n is a positive integer, then

$$(x + y)^n = \sum_{r=0}^{n} \binom{n}{r} x^{n-r} y^r = \sum_{r=0}^{n} {}_nC_r\, x^{n-r} y^r$$

EXAMPLE 6   Find the fifth term in the expansion of $(2x - 3y)^{12}$.

SOLUTION   Applying the general term formula with $k = 5$, we have

$$a_5 = \binom{12}{5-1}(2x)^{12-(5-1)}(-3y)^{5-1}$$

$$= \binom{12}{4}(2x)^8(-3y)^4$$

$$= \frac{12!}{4! \cdot 8!}(2x)^8(-3y)^4$$

Notice that once we have one of the exponents, the other exponent and the denominator of the coefficient are determined: The two exponents add to 12 and match the numbers in the denominator of the coefficient.

Making the calculations from the preceding formula, we have

$$a_5 = 495(256x^8)(81y^4)$$

$$= 10{,}264{,}320x^8y^4$$

 Getting Ready for Class

After reading through the preceding section, respond in your own words and in complete sentences.

A. What is Pascal's Triangle?

B. Why is $\binom{n}{0} = 1$ for any natural number?

C. State the Binomial Theorem.

D. When is the Binomial Theorem more efficient than using Pascal's Triangle to expand a binomial raised to a whole-number exponent?

Use Pascal's Triangle to expand each of the following.

1. $(x + 2)^4$ **2.** $(x - 2)^5$ **3.** $(x + y)^6$ **4.** $(x - 1)^6$

5. $(2x + 1)^5$ **6.** $(2x - 1)^4$ **7.** $(x - 2y)^5$ **8.** $(2x + y)^5$

9. $(3x - 2)^4$ **10.** $(2x - 3)^4$ **11.** $(4x - 3y)^3$ **12.** $(3x - 4y)^3$

13. $(x^2 + 2)^4$ **14.** $(x^2 - 3)^3$ **15.** $(x^2 + y^2)^3$ **16.** $(x^2 - 3y)^4$

17. $(2x + 3y)^4$ **18.** $(2x - 1)^5$ **19.** $\left(\frac{x}{2} + \frac{y}{3}\right)^3$ **20.** $\left(\frac{x}{3} - \frac{y}{2}\right)^4$

21. $\left(\frac{x}{2} - 4\right)^3$ **22.** $\left(\frac{x}{3} + 6\right)^3$ **23.** $\left(\frac{x}{3} + \frac{y}{2}\right)^4$ **24.** $\left(\frac{x}{2} - \frac{y}{3}\right)^4$

Evaluate the factorial.

25. $6!$ **26.** $9!$ **27.** $10!$ **28.** $12!$

Find each of the following binomial coefficients.

29. $\binom{10}{0}$ **30.** $\binom{10}{10}$ **31.** $_8C_1$ **32.** $_8C_7$

33. $\binom{15}{11}$ **34.** $\binom{11}{4}$ **35.** $_{20}C_7$ **36.** $_{30}C_{25}$

Use the Binomial Theorem to write the first four terms in the expansion of the following.

37. $(x + 2)^9$ **38.** $(x - 2)^9$ **39.** $(x - y)^{10}$ **40.** $(x + 2y)^{10}$

41. $(x + 3)^{25}$ **42.** $(x - 1)^{40}$ **43.** $(x - 2)^{60}$ **44.** $\left(x + \frac{1}{2}\right)^{30}$

45. $(x - y)^{18}$ **46.** $(x - 2y)^{65}$

Use the Binomial Theorem to write the first three terms in the expansion of each of the following.

47. $(x + 1)^{15}$ **48.** $(x - 1)^{15}$ **49.** $(x - y)^{12}$ **50.** $(x + y)^{12}$

51. $(x + 2)^{20}$ **52.** $(x - 2)^{20}$

Use the Binomial Theorem to write the first two terms in the expansion of each of the following.

53. $(x + 2)^{100}$ **54.** $(x - 2)^{50}$ **55.** $(x + y)^{50}$ **56.** $(x - y)^{100}$

Use the Binomial Theorem for Problems 57-68.

57. Find the ninth term in the expansion of $(2x + 3y)^{12}$.

58. Find the sixth term in the expansion of $(2x + 3y)^{12}$.

59. Find the fifth term of $(x - 2)^{10}$.

60. Find the fifth term of $(2x - 1)^{10}$.

61. Find the sixth term in the expansion of $(x - 2)^{12}$.

62. Find the ninth term in the expansion of $(7x - 1)^{10}$.

63. Find the third term in the expansion of $(x - 3y)^{25}$.

64. Find the 24th term in the expansion of $(2x - y)^{26}$.

65. Write the formula for the 12th term of $(2x + 5y)^{20}$. Do not simplify.

66. Write the formula for the eighth term of $(2x + 5y)^{20}$. Do not simplify.

67. Write the first three terms of the expansion of $(x^2y - 3)^{10}$.

68. Write the first three terms of the expansion of $\left(x - \dfrac{1}{x}\right)^{50}$.

Applying the Concepts

69. Probability The third term in the expansion of $\left(\frac{1}{2} + \frac{1}{2}\right)^7$ will give the probability that in a family with 7 children, 5 will be boys and 2 will be girls. Find the third term.

70. Probability The fourth term in the expansion of $\left(\frac{1}{2} + \frac{1}{2}\right)^8$ will give the probability that in a family with 8 children, 3 will be boys and 5 will be girls. Find the fourth term.

Learning Objectives Assessment

The following problems can be used to help assess if you have successfully met the learning objectives for this section.

71. Use Pascal's Triangle to expand $(x - 2y)^4$.

    a. $x^4 + 4x^3y + 6x^2y^2 + 4xy^3 + y^4$

    b. $x^4 - 8x^3y + 24x^2y^2 - 32xy^3 + 16y^4$

    c. $x^4 + 16y^4$

    d. $x^4 - 8x^3y - 12x^2y^2 - 8xy^3 - 2y^4$

72. Find $\dbinom{10}{6}$.

    a. $\dfrac{5}{3}$ **b.** 5,040   **c.** 210   **d.** 151,200

73. Find the fourth term in the expansion of $(3x - 2y)^7$.

    a. $-22,680x^4y^3$ **b.** $35x^4y^3$

    c. $15,120x^3y^4$ **d.** $-210x^3y^4$

Maintaining Your Skills

Solve each equation. Write your answers to the nearest hundredth.

74. $5^x = 7$ **75.** $10^x = 15$ **76.** $8^{2x-1} = 16$ **77.** $9^{3x-1} = 27$

78. Compound Interest How long will it take \$400 to double if it is invested in an account with an annual interest rate of 10% compounded four times a year?

79. Compound Interest How long will it take \$200 to become \$800 if it is invested in an account with an annual interest rate of 8% compounded four times a year?

Find each of the following to the nearest hundredth.

80. $\log_4 20$ **81.** $\log_7 21$ **82.** $\ln 576$ **83.** $\ln 5{,}760$

84. Solve the formula $A = 10e^{5t}$ for t.   **85.** Solve the formula $A = Pe^{-5t}$ for t.

Chapter 10 Summary

Sequences [10.1]

1. In the sequence $a_n = 2n - 1$,
$$a_1 = 2(1) - 1 = 1$$
$$a_2 = 2(2) - 1 = 3$$
$$a_3 = 2(3) - 1 = 5$$
resulting in 1, 3, 5,...

A **sequence** is a function whose domain is the set of positive integers. The terms of a sequence are denoted by

$$a_1, a_2, a_3, \ldots, a_n, \ldots$$

where a_1 (read "a sub 1") is the first term, a_2 the second term, and a_n the nth or **general term**.

Summation Notation [10.2]

2. $\displaystyle\sum_{i=3}^{6} (-2)^i$

$$= (-2)^3 + (-2)^4 + (-2)^5 + (-2)^6$$
$$= -8 + 16 + (-32) + 64$$
$$= 40$$

The notation

$$\sum_{i=1}^{n} a_i = a_1 + a_2 + a_3 + \cdots + a_n$$

is called **summation notation** or **sigma notation**. The letter i as used here is called the **index of summation** or just **index**.

Arithmetic Sequences [10.3]

3. For the sequence 3, 7, 11, 15,...,
$a_1 = 3$ and $d = 4$. The general term is
$$a_n = 3 + (n - 1)4$$
$$= 4n - 1$$
Using this formula to find the tenth term, we have
$$a_{10} = 4(10) - 1 = 39$$
The sum of the first 10 terms is
$$S_{10} = \frac{10}{2}(3 + 39) = 210$$

An **arithmetic sequence** is a sequence in which each term comes from the preceding term by adding a constant amount each time. If the first term of an arithmetic sequence is a_1 and the amount we add each time (called the **common difference**) is d, then the nth term of the progression is given by

$$a_n = a_1 + (n - 1)d$$

The sum of the first n terms of an arithmetic sequence is

$$S_n = \frac{n}{2}(a_1 + a_n)$$

S_n is called the **nth partial sum**.

Geometric Sequences [10.4]

4. For the geometric progression
3, 6, 12, 24,..., $a_1 = 3$ and $r = 2$.
The general term is
$$a_n = 3 \cdot 2^{n-1}$$
The sum of the first 10 terms is
$$S_{10} = \frac{3(2^{10} - 1)}{2 - 1} = 3{,}069$$

A **geometric sequence** is a sequence of numbers in which each term comes from the previous term by multiplying by a constant amount each time. The constant by which we multiply each term to get the next term is called the **common ratio**. If the first term of a geometric sequence is a_1 and the common ratio is r, then the formula that gives the general term a_n is

$$a_n = a_1 r^{n-1}$$

The sum of the first n terms of a geometric sequence is given by the formula

$$S_n = \frac{a_1(r^n - 1)}{r - 1}$$

The Sum of an Infinite Geometric Series [10.4]

5. The sum of the series

$$\frac{1}{3} + \frac{1}{6} + \frac{1}{12} + \cdots$$

is

$$S = \frac{\frac{1}{3}}{1 - \frac{1}{2}} = \frac{\frac{1}{3}}{\frac{1}{2}} = \frac{2}{3}$$

If a geometric sequence has first term a_1 and common ratio r such that $|r| < 1$, then the following is called an *infinite geometric series*:

$$S = \sum_{i=0}^{\infty} a_1 r^i = a_1 + a_1 r + a_1 r^2 + a_1 r^3 + \cdots$$

Its sum is given by the formula

$$S = \frac{a_1}{1 - r}$$

Factorials [10.5]

The notation $n!$ is called **n factorial** and is defined to be the product of each consecutive integer from n down to 1. That is,

$$0! = 1 \qquad \text{(By definition)}$$
$$1! = 1$$
$$2! = 2 \cdot 1$$
$$3! = 3 \cdot 2 \cdot 1$$
$$4! = 4 \cdot 3 \cdot 2 \cdot 1$$

and so on.

Binomial Coefficients [10.5]

6.
$$\binom{7}{3} = \frac{7!}{3!(7-3)!}$$
$$= \frac{7!}{3! \cdot 4!}$$
$$= \frac{7 \cdot 6 \cdot 5 \cdot 4 \cdot 3 \cdot 2 \cdot 1}{(3 \cdot 2 \cdot 1)(4 \cdot 3 \cdot 2 \cdot 1)}$$
$$= 35$$

The notation $\binom{n}{r}$, or $_nC_r$, is called a *binomial coefficient* and is defined by

$$_nC_r = \binom{n}{r} = \frac{n!}{r!(n-r)!}$$

Binomial coefficients can be found by using the formula above or by *Pascal's Triangle*, which is

$$
\begin{array}{ccccccccccc}
 & & & & & 1 & & & & & \\
 & & & & 1 & & 1 & & & & \\
 & & & 1 & & 2 & & 1 & & & \\
 & & 1 & & 3 & & 3 & & 1 & & \\
 & 1 & & 4 & & 6 & & 4 & & 1 & \\
1 & & 5 & & 10 & & 10 & & 5 & & 1 \\
\end{array}
$$

and so on.

Binomial Theorem [10.5]

7. $(x + 2)^4$
$$= x^4 + 4x^3 \cdot 2 + 6x^2 \cdot 2^2 + 4x \cdot 2^3 + 2^4$$
$$= x^4 + 8x^3 + 24x^2 + 32x + 16$$

If n is a positive integer, then the formula for expanding $(x + y)^n$ is given by

$$(x + y)^n = \sum_{r=0}^{n} \binom{n}{r} x^{n-r} y^r = \sum_{r=0}^{n} {_nC_r}\, x^{n-r} y^r$$

Write the first five terms of the sequences with the following general terms or recursive formulas. [10.1]

1. $a_n = 3n - 5$

2. $a_1 = 3, a_n = a_{n-1} + 4, n > 1$

3. $a_n = n^2 + 1$

4. $a_n = 2n^3$

5. $a_n = \dfrac{n+1}{n^2}$

6. $a_1 = 4, a_n = -2a_{n-1}, n > 1$

Give the general term for each sequence. [10.1]

7. 6, 10, 14, 18,...

8. 1, 2, 4, 8,...

9. $\dfrac{1}{2}, \dfrac{1}{4}, \dfrac{1}{8}, \dfrac{1}{16},...$

10. $-3, 9, -27, 81,...$

11. Expand and simplify each of the following. [10.2]

a. $\displaystyle\sum_{i=1}^{5}(5i + 3)$

b. $\displaystyle\sum_{i=3}^{5}(2^i - 1)$

c. $\displaystyle\sum_{i=2}^{6}(i^2 + 2i)$

12. Find the first term of an arithmetic progression if $a_5 = 11$ and $a_9 = 19$. [10.3]

13. Find the second term of a geometric progression for which $a_3 = 18$ and $a_5 = 162$. [10.4]

Find the sum of the first 10 terms of the following arithmetic progressions. [10.3]

14. 5, 11, 17,...

15. 25, 20, 15,...

16. Write a formula for the sum of the first 50 terms of the geometric progression 3, 6, 12,.... [10.4]

17. Find the sum of $\dfrac{1}{2} + \dfrac{1}{6} + \dfrac{1}{18} + \dfrac{1}{54} + \cdots$. [10.4]

Use Pascal's Triangle to expand each of the following. [10.5]

18. $(x - 3)^4$

19. $(2x - 1)^5$

Find each binomial coefficient. [10.5]

20. $\dbinom{12}{9}$

21. $_{15}C_5$

22. Find the first 3 terms in the expansion of $(x - 1)^{20}$. [10.5]

23. Find the sixth term in $(2x - 3y)^8$. [10.5]

Conic Sections

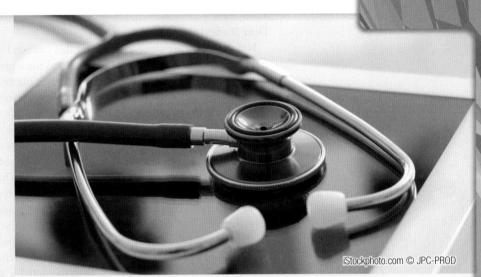

iStockphoto.com © JPC-PROD

One of the curves we will study in this chapter has interesting reflective properties. Figure 1(A) shows how you can draw one of these curves (an ellipse) using thumbtacks, string, pencil, and paper. Elliptical surfaces will reflect sound waves that originate at one focus through the other focus. This property of ellipses allows doctors to treat patients with kidney stones using a procedure called lithotripsy. A lithotripter is an elliptical device that creates sound waves that crush the kidney stone into small pieces, without surgery. The sound wave originates at one focus of the lithotripter. The energy from it reflects off the surface of the lithotripter and converges at the other focus, where the kidney stone is positioned. Below (Figure 1(B)) is a cross-section of a lithotripter, with a patient positioned so the kidney stone is at the other focus.

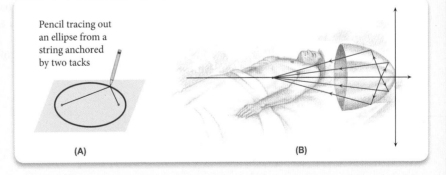

Pencil tracing out an ellipse from a string anchored by two tacks

(A)　　　　　　　　(B)

By studying the conic sections in this chapter, you will be better equipped to understand some of the more technical equipment that exists in the world outside of class.

Success Skills

Dear Student,

Now that you are close to finishing this course, I want to pass on a couple of things that have helped me a great deal with my career. I'll introduce each one with a quote:

Do something for the person you will be 5 years from now.

I have always made sure that I arranged my life so that I was doing something for the person I would be 5 years later. For example, when I was 20 years old, I was in college. I imagined that the person I would be as a 25-year-old, would want to have a college degree, so I made sure I stayed in school. That's all there is to this. It is not a hard, rigid philosophy. It is a soft, behind the scenes, foundation. It does not include ideas such as "Five years from now I'm going to graduate at the top of my class from the best college in the country." Instead, you think, "five years from now I will have a college degree, or I will still be in school working towards it."

This philosophy led to a community college teaching job, writing textbooks, doing videos with the textbooks, then to MathTV and the book you are reading right now. Along the way there were many other options and directions that I didn't take, but all the choices I made were due to keeping the person I would be in 5 years in mind.

It's easier to ride a horse in the direction it is going.

I started my college career thinking that I would become a dentist. I enrolled in all the courses that were required for dental school. When I completed the courses, I applied to a number of dental schools, but wasn't accepted. I kept going to school, and applied again the next year, again, without success. My life was not going in the direction of dental school, even though I had worked hard to put it in that direction. So I did a little inventory of the classes I had taken and the grades I earned, and realized that I was doing well in mathematics. My life was actually going in that direction so I decided to see where that would take me. It was a good decision.

It is a good idea to work hard toward your goals, but it is also a good idea to take inventory every now and then to be sure you are headed in the direction that is best for you.

I wish you good luck with the rest of your college years, and with whatever you decide you want to do as a career.

Pat McKeague

Learning Objectives

In this section, we will learn how to:

1. Find the distance between two points in the plane.

2. Find the equation of a circle given the center and radius.

3. Determine the center and radius of a circle from its equation.

4. Sketch the graph of a circle.

Introduction

Conic sections include ellipses, circles, hyperbolas, and parabolas. They are called conic sections because each can be found by slicing a cone with a plane as shown in Figure 1.

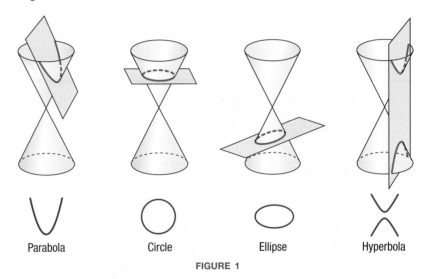

Parabola   Circle   Ellipse   Hyperbola

FIGURE 1

The Distance Formula

We begin our work with conic sections by studying circles. Before we find the general equation of a circle, we must first derive what is known as the ***distance formula***. Suppose (x_1, y_1) and (x_2, y_2) are any two points in the first quadrant. (Actually, we could choose the two points to be anywhere on the coordinate plane. It is just more convenient to have them in the first quadrant.) We can name the points P_1 and P_2, respectively, and draw the diagram shown in Figure 2.

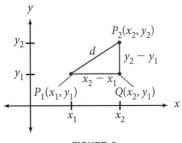

FIGURE 2

Notice the coordinates of point Q. The x-coordinate is x_2 because Q is directly below point P_2. The y-coordinate of Q is y_1 because Q is directly across from point P_1. It is evident from the diagram that the length of P_2Q is $y_2 - y_1$ and the length of P_1Q is $x_2 - x_1$. Using the Pythagorean theorem, we have

$$(P_1P_2)^2 = (P_1Q)^2 + (P_2Q)^2$$

or

$$d^2 = (x_2 - x_1)^2 + (y_2 - y_1)^2$$

Taking the square root of both sides, we have

$$d = \sqrt{(x_2 - x_1)^2 + (y_2 - y_1)^2}$$

We know this is the positive square root, because d is the distance from P_1 to P_2 and must therefore be positive. This formula is called the **distance formula**.

$[\Delta \neq \Sigma]$ **FORMULA**   *Distance Formula*

If (x_1, y_1) and (x_2, y_2) are any two points in the coordinate plane, then the distance between them is given by

$$d = \sqrt{(x_2 - x_1)^2 + (y_2 - y_1)^2}$$

VIDEO EXAMPLES

SECTION 11.1

EXAMPLE 1   Find the distance between $(3, 5)$ and $(2, -1)$.

SOLUTION   If we let $(3, 5)$ be (x_1, y_1) and $(2, -1)$ be (x_2, y_2) and apply the distance formula, we have

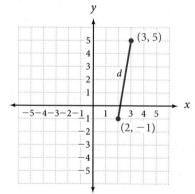

$$d = \sqrt{(2 - 3)^2 + (-1 - 5)^2}$$

$$= \sqrt{(-1)^2 + (-6)^2}$$

$$= \sqrt{1 + 36}$$

$$= \sqrt{37}$$

FIGURE 3

EXAMPLE 2   Find x if the distance from $(x, 5)$ to $(3, 4)$ is $\sqrt{2}$.

SOLUTION   Using the distance formula, we have

$\sqrt{2} = \sqrt{(x - 3)^2 + (5 - 4)^2}$	Distance formula
$2 = (x - 3)^2 + 1^2$	Square each side
$2 = x^2 - 6x + 9 + 1$	Expand $(x - 3)^2$
$0 = x^2 - 6x + 8$	Simplify
$0 = (x - 4)(x - 2)$	Factor
$x = 4$ or $x = 2$	Set factors equal to 0

The two solutions are 4 and 2, which indicates that two points, $(4, 5)$ and $(2, 5)$, are $\sqrt{2}$ units from $(3, 4)$.

Circles

Because of their perfect symmetry, circles have been used for thousands of years in many disciplines, including art, science, and religion. The photograph on the left is of Stonehenge, a 4,500-year-old site in England. The arrangement of the stones is based on a circular plan that is thought to have both religious and astronomical significance. More recently, the design shown in the photo on the right began appearing in agricultural fields in England in the 1990s. Whoever made these designs chose the circle as their basic shape.

We can model circles very easily in algebra by using equations that are based on the distance formula.

© Masterfile

© Masterfile

> **⌈∆≠∑   THEOREM**   *Circle Theorem*
>
> The *equation of the circle* with center at (h, k) and radius r is given by
> $$(x - h)^2 + (y - k)^2 = r^2$$

Proof   By definition, all points on the circle are a distance r from the center (h, k). If we let (x, y) represent any point on the circle, then (x, y) is r units from (h, k). Applying the distance formula, we have

$$r = \sqrt{(x - h)^2 + (y - k)^2}$$

Squaring both sides of this equation gives the equation of the circle:

$$(x - h)^2 + (y - k)^2 = r^2$$

We can use the circle theorem to find the equation of a circle, given its center and radius, or to find its center and radius, given the equation.

EXAMPLE 3   Find the equation of the circle with center at $(-3, 2)$ having a radius of 5.

SOLUTION   We have $(h, k) = (-3, 2)$ and $r = 5$. Applying our theorem for the equation of a circle yields

$$[x - (-3)]^2 + (y - 2)^2 = 5^2$$
$$(x + 3)^2 + (y - 2)^2 = 25$$

EXAMPLE 4   Give the equation of the circle with radius 3 whose center is at the origin.

SOLUTION   The coordinates of the center are $(0, 0)$, and the radius is 3. The equation must be

$$(x - 0)^2 + (y - 0)^2 = 3^2$$
$$x^2 + y^2 = 9$$

We can see from Example 4 that the equation of any circle with its center at the origin and radius r will be

$$x^2 + y^2 = r^2$$

EXAMPLE 5   Find the center and radius, and sketch the graph of the circle whose equation is

$$(x - 1)^2 + (y + 3)^2 = 4$$

SOLUTION   Writing the equation in the form

$$(x - h)^2 + (y - k)^2 = r^2$$

we have

$$(x - 1)^2 + [y - (-3)]^2 = 2^2$$

The center is at $(1, -3)$, and the radius is 2. The graph is shown in Figure 4.

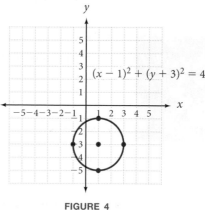

FIGURE 4

 EXAMPLE 6   Sketch the graph of $x^2 + y^2 = 9$.

SOLUTION   Because the equation can be written in the form

$$(x - 0)^2 + (y - 0)^2 = 3^2$$

it must have its center at $(0, 0)$ and a radius of 3. The graph is shown in Figure 5.

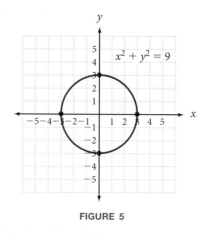

FIGURE 5

 EXAMPLE 7   Sketch the graph of $x^2 + y^2 + 6x - 4y - 12 = 0$.

SOLUTION   To sketch the graph, we must find the center and radius of our circle. We can do so easily if the equation is in standard form. That is, if it has the form

$$(x - h)^2 + (y - k)^2 = r^2$$

To put our equation in standard form, we start by using the addition property of equality to group all the constant terms together on the right side of the equation.

In this case, we add 12 to each side of the equation. We do this because we are going to add our own constants later to complete the square.

$$x^2 + y^2 + 6x - 4y = 12$$

Next, we group all the terms containing x together and all terms containing y together, and we leave some space at the end of each group for the numbers we will add when we complete the square on each group.

$$x^2 + 6x \quad + y^2 - 4y \quad = 12$$

To complete the square on x, we add 9 to each side of the equation. To complete the square on y, we add 4 to each side of the equation.

$$x^2 + 6x + 9 + y^2 - 4y + 4 = 12 + 9 + 4$$

The first three terms on the left side can be written as $(x + 3)^2$. Likewise, the last three terms on the left side can be factored as $(y - 2)^2$. The right side simplifies to 25.

$$(x + 3)^2 + (y - 2)^2 = 25$$

Writing 25 as 5^2, we have our equation in standard form.

$$(x + 3)^2 + (y - 2)^2 = 5^2$$

From this last line, it is apparent that the center is at $(-3, 2)$ and the radius is 5. Using this information, we create the graph shown in Figure 6.

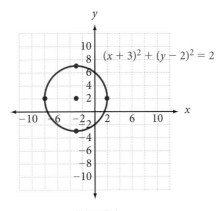

FIGURE 6

Getting Ready for Class

After reading through the preceding section, respond in your own words and in complete sentences.

A. Describe the distance formula in words, as if you were explaining to someone how they should go about finding the distance between two points.

B. What is the mathematical definition of a circle?

C. How are the distance formula and the equation of a circle related?

D. When graphing a circle from its equation, why is completing the square sometimes useful?

Problem Set 11.1

Find the distance between the following points.

1. $(3, 7)$ and $(6, 3)$ **2.** $(4, 7)$ and $(8, 1)$

3. $(0, 9)$ and $(5, 0)$ **4.** $(-3, 0)$ and $(0, 4)$

5. $(3, -5)$ and $(-2, 1)$ **6.** $(-8, 9)$ and $(-3, -2)$

7. $(-1, -2)$ and $(-10, 5)$ **8.** $(-3, -8)$ and $(-1, 6)$

9. Find x so the distance between $(x, 2)$ and $(1, 5)$ is $\sqrt{13}$.

10. Find x so the distance between $(-2, 3)$ and $(x, 1)$ is 3.

11. Find x so the distance between $(x, 5)$ and $(3, 9)$ is 5.

12. Find y so the distance between $(-4, y)$ and $(2, 1)$ is 8.

13. Find x so the distance between $(x, 4)$ and $(2x + 1, 6)$ is 6.

14. Find y so the distance between $(3, y)$ and $(7, 3y - 1)$ is 6.

Write the equation of the circle with the given center and radius.

15. Center $(3, -2)$; $r = 3$ **16.** Center $(-2, 4)$; $r = 1$

17. Center $(-5, -1)$; $r = \sqrt{5}$ **18.** Center $(-7, -6)$; $r = \sqrt{3}$

19. Center $(0, -5)$; $r = 1$ **20.** Center $(0, -1)$; $r = 7$

21. Center $(0, 0)$; $r = 2$ **22.** Center $(0, 0)$; $r = 5$

Give the center and radius, and sketch the graph of each of the following circles.

23. $x^2 + y^2 = 4$ **24.** $x^2 + y^2 = 16$

25. $(x - 1)^2 + (y - 3)^2 = 25$ **26.** $(x - 4)^2 + (y - 1)^2 = 36$

27. $(x + 2)^2 + (y - 4)^2 = 8$ **28.** $(x - 3)^2 + (y + 1)^2 = 12$

29. $(x + 2)^2 + (y - 4)^2 = 17$ **30.** $x^2 + (y + 2)^2 = 11$

31. $x^2 + y^2 + 2x - 4y = 4$ **32.** $x^2 + y^2 - 4x + 2y = 11$

33. $x^2 + y^2 - 6y = 7$ **34.** $x^2 + y^2 - 4y = 5$

35. $x^2 + y^2 + 2x = 1$ **36.** $x^2 + y^2 + 10x = 0$

37. $x^2 + y^2 - 4x - 6y = -4$ **38.** $x^2 + y^2 - 4x + 2y = 4$

39. $x^2 + y^2 + 2x + y = \dfrac{11}{4}$ **40.** $x^2 + y^2 - 6x - y = -\dfrac{1}{4}$

41. $4x^2 + 4y^2 - 4x + 8y = 11$ **42.** $36x^2 + 36y^2 - 24x - 12y = 31$

Each of the following circles passes through the origin. In each case, find the equation.

43.

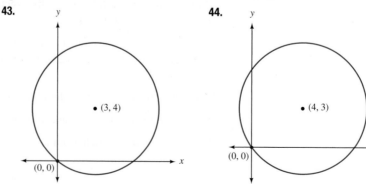

44.

45. Find the equations of circles *A*, *B*, and *C* in the following diagram. The three points are the centers of the three circles.

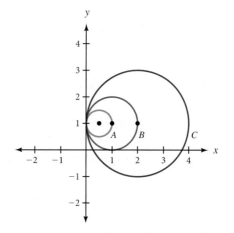

46. Each of the following circles passes through the origin. The centers are as shown. Find the equation of each circle.

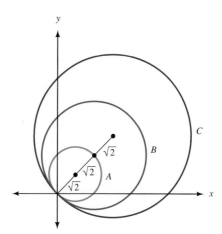

47. Find the equation of the circle with center at the origin that contains the point (3, 4).

48. Find the equation of the circle with center at the origin that contains the point (−5, 12).

49. Find the equation of the circle with center at the origin and *x*-intercepts 3 and −3.

50. Find the equation of the circle with *y*-intercepts 4 and −4 and center at the origin.

51. A circle with center at (−1, 3) passes through the point (4, 3). Find the equation.

52. A circle with center at (2, 5) passes through the point (−1, 4). Find the equation.

53. Find the equation of the circle with center at (−2, 5), which passes through the point (1, −3).

54. Find the equation of the circle with center at $(4, -1)$, which passes through the point $(6, -5)$.

55. Find the equation of the circle with center on the y-axis and y-intercepts at -2 and 6.

56. Find the equation of the circle with center on the x-axis and x-intercepts at -8 and 2.

57. Find the circumference and area of the circle $x^2 + (y - 3)^2 = 18$. Leave your answer in terms of π.

58. Find the circumference and area of the circle $(x + 2)^2 + (y + 6)^2 = 12$. Leave your answer in terms of π.

59. Find the circumference and area of the circle $x^2 + y^2 + 4x + 2y = 20$. Leave your answer in terms of π.

60. Find the circumference and area of the circle $x^2 + y^2 - 6x + 2y = 6$. Leave your answer in terms of π.

Applying the Concepts

61. **Search Area** A 3-year-old child has wandered away from home. The police have decided to search a circular area with a radius of 6 blocks. The child turns up at his grandmother's house, 5 blocks East and 3 blocks North of home. Was he found within the search area?

62. **Placing a Bubble Fountain** A circular garden pond with a diameter of 12 feet is to have a bubble fountain. The water from the bubble fountain falls in a circular pattern with a radius of 1.5 feet. If the center of the bubble fountain is placed 4 feet West and 3 feet North of the center of the pond, will all the water from the fountain fall inside the pond? What is the farthest distance from the center of the pond that water from the fountain will fall?

63. **Ferris Wheel** A giant Ferris wheel has a diameter of 240 feet and sits 12 feet above the ground. As shown in the diagram below, the wheel is 500 feet from the entrance to the park. The xy-coordinate system containing the wheel has its origin on the ground at the center of the entrance. Write an equation that models the shape of the wheel.

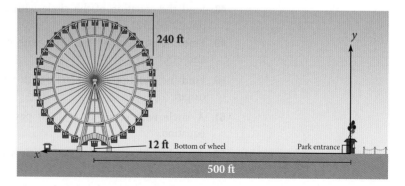

64. Magic Rings A magician is holding two rings
that seem to lie in the same plane and intersect
in two points. Each ring is 10 inches in diameter.

  a. Find the equation of each ring if a coordinate
system is placed with its origin at the center of the
first ring and the *X*-axis contains the center of the
second ring.

  b. Find the equation of each ring if a coordinate system
is placed with its origin at the center of the second
ring and the *x*-axis contains the center of the first
ring.

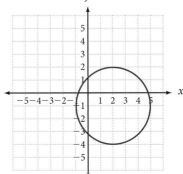

Learning Objectives Assessment

The following problems can be used to help assess if you have successfully met
the learning objectives for this section.

65. Find the distance between the points $(5, -1)$ and $(2, 4)$.

  a. $3\sqrt{2}$ **b.** 4 **c.** $\sqrt{58}$ **d.** $\sqrt{34}$

66. Find the equation of the circle with center $(1, -2)$ and radius 4.

  a. $x^2 - 2y^2 = 16$ **b.** $(x+2)^2 - (y-1)^2 = 4$

  c. $(x-1)^2 + (y+2)^2 = 16$ **d.** $(x+1)^2 + (y-2)^2 = 4$

67. Find the center of the circle $x^2 + y^2 + 4y = 5$.

  a. $(0, -2)$ **b.** $(0, 2)$ **c.** $(0, 0)$ **d.** $(2, 0)$

68. Graph the circle $(x-2)^2 + (y+1)^2 = 3$.

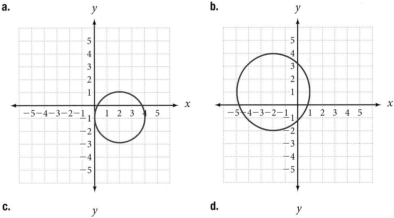

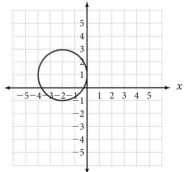

Getting Ready for the Next Section

Solve for p.

69. $4p = 12$

70. $4p = -8$

71. $4p = -1$

72. $4p = \dfrac{1}{2}$

Find the vertex for each quadratic function.

73. $y = 3(x - 4)^2 + 1$

74. $y = -2(x + 3)^2 - 5$

Sketch the graph of each function.

75. $y = 2x^2$

76. $y = -x^2$

77. $y = 2(x - 1)^2 - 3$

78. $y = -(x + 1)^2 + 2$

Learning Objectives

In this section, we will learn how to:

1. Graph a parabola.

2. Find the vertex for a parabola.

3. Find the focus for a parabola.

Introduction

In Chapter 8 we introduced the parabola as the graph of a quadratic function. In this section we will revisit the parabola, but from the context of a conic section.

As with all the conic sections, the parabola has a reflective property that makes it very useful in many modern devices, such as satellite television dishes. Satellite dishes are built in the shape of parabolas, because a parabola will reflect the incoming signal waves through a common point, called the *focus* (Figure 1).

FIGURE 1

Graphing Parabolas

If we position the parabola on a coordinate system with its vertex at the origin, and let p represent the coordinate of the focus, then we have the following.

$[\Delta \neq \Sigma]$ *Parabolas*

The graph of the equation

$$x^2 = 4py$$

will be a *parabola* with vertex at the origin and *focus* $(0, p)$. The *axis of symmetry* is $x = 0$ (the y-axis). If $p > 0$, the parabola opens upward, and if $p < 0$, the parabola opens downward.

The graph of the equation

$$y^2 = 4px$$

will be a *parabola* with vertex at the origin and *focus* $(p, 0)$. The *axis of symmetry* is $y = 0$ (the x-axis). If $p > 0$, the parabola opens to the right, and if $p < 0$, the parabola opens to the left.

VIDEO EXAMPLES

SECTION 11.2

EXAMPLE 1   Graph $x^2 = 8y$ and state the focus.

SOLUTION   The vertex is the origin. To find additional points on the graph, we can isolate the variable y and then substitute values for x.

$$8y = x^2$$

$$y = \frac{x^2}{8}$$

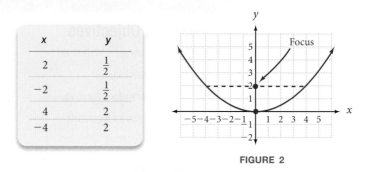

x	y
2	$\frac{1}{2}$
−2	$\frac{1}{2}$
4	2
−4	2

FIGURE 2

The graph is shown in Figure 2. To find the focus, we notice that the coefficent of y is equal to $4p$. Therefore

$$4p = 8$$

$$p = 2$$

The focus is the point $(0, 2)$ (see Figure 2). The focus will always lie on the axis of symmetry.

The length of the dashed line segment shown in Figure 2 is called the *focal width*. It tells us how wide the parabola is at the focus. Notice that the length of this segment is 8 units, which coincides with the coefficient of y in the equation from Example 1. This leads us to the following definition.

> (def **DEFINITION**   *focal width*
>
> For any parabola, the *focal width* is the distance across the parabola, measured through the focus and perpendicular to the axis of symmetry. Its value is given by
>
> $$\text{focal width} = |4p|$$

The focal width gives us a convenient way of graphing a parabola more easily, as the following example illustrates.

EXAMPLE 2   Graph $y^2 = -3x$ and state the focus.

SOLUTION   The vertex is $(0, 0)$. Because the squared variable is y, the axis of symmetry is the x-axis. To find p, we have

$$4p = -3$$

$$p = -\frac{3}{4}$$

The focus is the point $\left(-\frac{3}{4}, 0\right)$. The focal width is $|-3| = 3$. The graph is shown in Figure 3. Notice how we have used the focal width to locate two other points on the graph.

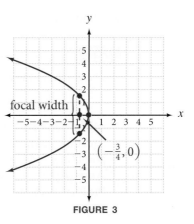

FIGURE 3

Translated Parabolas

From our work in Chapter 8, we know the graph of the equation $y = \frac{1}{4}(x-1)^2 + 2$ is a parabola with vertex $(1, 2)$. If we isolate the squared binomial, we have

$$\frac{1}{4}(x-1)^2 + 2 = y$$

$$\frac{1}{4}(x-1)^2 = y - 2$$

$$(x-1)^2 = 4(y-2)$$

This is now expressed in the form $x^2 = 4py$, except that the vertex is no longer located at the origin because the terms subtracted from x and y represent translations. As you can see in Figure 4, the graph of $(x-1)^2 = 4(y-2)$ is identical to the graph of $x^2 = 4y$, except that it has been translated horizontally 1 unit to the right and vertically 2 units upward.

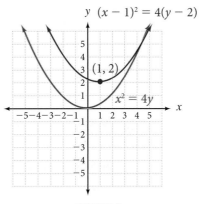

FIGURE 4

We summarize the information above with the following.

> **⟨Δ≠Σ⟩ RULE**   *Parabolas with Vertex at (h, k)*
>
> The graphs of the equations
>
> $$(x - h)^2 = 4p(y - k) \quad \text{and} \quad (y - k)^2 = 4p(x - h)$$
>
> will be *parabolas with their vertex at (h, k)*. The graph of the first equation will have an axis of symmetry $x = h$ and the focus will be the point $(h, p + k)$. The graph of the second equation will have an axis of symmetry $y = k$ and the focus will be the point $(p + h, k)$. In either case, the focal width is $|4p|$.

█ EXAMPLE 3   Graph: $(y + 3)^2 = -3(x - 2)$.

SOLUTION   If we write the equation in the form

$$(y - (-3))^2 = -3(x - 2)$$

then we see that $h = 2$ and $k = -3$. The vertex will be the point $(2, -3)$. Also, $4p = -3$, so $p = -\frac{3}{4}$. The graph will be identical to the parabola shown in Figure 3, except that there is a horizontal translation 2 units to the right and a vertical translation of 3 units downward. The graph is shown in Figure 5.

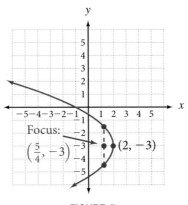

FIGURE 5

The axis of symmetry is $y = -3$ and the focus is the point

$$\left(-\tfrac{3}{4} + 2, -3\right) = \left(\tfrac{5}{4}, -3\right)$$

The focal width is $|-3| = 3$, the same as it was in Example 2.   █

EXAMPLE 4 Graph the parabola $y^2 - x - 4y - 2 = 0$.

SOLUTION To identify the vertex and focus, we must complete the square on y. To begin, we rearrange the terms so that those containing y are together, and any other terms are on the other side of the equal sign. Doing so gives us the following equation:

$$y^2 - 4y = x + 2$$

To complete the square on y, we add 4 to each side of the equation. Then we can factor each side to obtain standard form.

$$y^2 - 4y + 4 = x + 2 + 4 \qquad \text{Add 4 to both sides}$$

$$(y - 2)^2 = x + 6 \qquad \text{Factor the left side}$$

$$(y - 2)^2 = 1(x + 6) \qquad \text{Factor the right side}$$

The graph is a parabola with vertex $(-6, 2)$. The axis of symmetry is $y = 2$. To find the focus, we have

$$4p = 1$$

$$p = \frac{1}{4}$$

The focus is the point $\left(\frac{1}{4} + (-6), 2\right) = \left(-\frac{23}{4}, 2\right)$ and the focal width is 1. The graph is shown in Figure 6.

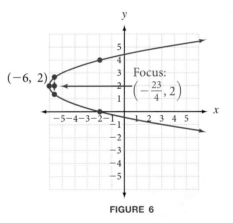

FIGURE 6

We can find extra points on the graph as necessary by choosing values for one variable and then solving for the other. For example, if $x = -2$, then

$$(y - 2)^2 = 1(-2 + 6)$$

$$(y - 2)^2 = 4$$

$$y - 2 = \pm 2$$

$$y = 2 \pm 2$$

$$y = 4 \text{ or } 0$$

The points $(-2, 4)$ and $(-2, 0)$ also lie on the graph (see Figure 6).

Application

As we mentioned in the introduction to this section, parabolas have many practical uses in the real world because of their reflective property.

EXAMPLE 5   A satellite television company uses a parabolic dish to collect the signal from their satellite. The dish is 10 inches deep and measures 30 inches across. The receiver for the dish will be located at the focus. How far from the base of the dish should the receiver be placed?

SOLUTION   We assume the vertex is positioned at the origin and use the equation $x^2 = 4py$. Based on the dimensions of the dish, the point $(15, 10)$ must lie on the graph of the parabola (see Figure 7).

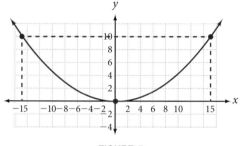

FIGURE 7

Substituting $x = 15$ and $y = 10$, we can solve for p.

$$15^2 = 4p(10)$$

$$215 = 40p$$

$$5.625 = p$$

The focus is the point $(0, 5.625)$. This means the receiver should be placed 5.625 inches from the base of the dish.

Getting Ready for Class

After reading through the preceding section, respond in your own words and in complete sentences.

A. How can you tell the difference between the equation of a parabola with a horizontal axis of symmetry and the equation of a parabola with a vertical axis of symmetry?

B. How do we find the coordinates of the focus for a parabola?

C. What is the focal width for a parabola?

D. Explain how to use the focal width to help sketch the graph of a parabola.

Graph each of the following parabolas. State the coordinates of the focus.

1. $x^2 = 4y$ **2.** $x^2 = -6y$ **3.** $x^2 = y$

4. $x^2 = -y$ **5.** $x^2 = -\dfrac{1}{2}y$ **6.** $x^2 = \dfrac{1}{4}y$

7. $y^2 = -2x$ **8.** $y^2 = 8x$ **9.** $y^2 = -x$

10. $y^2 = x$ **11.** $y^2 = \dfrac{1}{3}x$ **12.** $y^2 = -\dfrac{1}{2}x$

Graph the parabola. In each case, label the coordinates of the vertex and focus.

13. $(x + 1)^2 = 2(y + 3)$ **14.** $(x + 2)^2 = -10(y + 4)$

15. $(x - 4)^2 = -(y - 2)$ **16.** $(x - 3)^2 = (y - 5)$

17. $(x + 2)^2 = -\dfrac{1}{3}(y - 1)$ **18.** $(x - 2)^2 = \dfrac{1}{4}(y + 3)$

19. $(y - 1)^2 = 3(x - 2)$ **20.** $(y + 2)^2 = 4(x + 5)$

21. $(y + 3)^2 = x - 3$ **22.** $(y + 1)^2 = -(x - 4)$

23. $(y - 2)^2 = -\dfrac{1}{3}(x - 5)$ **24.** $(y - 4)^2 = -\dfrac{1}{2}(x + 3)$

25. $x^2 - 4x - y + 1 = 0$ **26.** $x^2 - 4x + y - 3 = 0$

27. $x^2 + 2x + 8y - 23 = 0$ **28.** $x^2 + 4x - 10y - 6 = 0$

29. $y^2 - x + 4y + 7 = 0$ **30.** $y^2 + x - 6y + 5 = 0$

31. $y^2 + 6x - 2y + 7 = 0$ **32.** $y^2 - 9x + 4y + 22 = 0$

Applying the Concepts

33. Telescope A parabolic mirror for a telescope measures 10 inches across and 1 inch deep. If a lens is to be located at the focus, how far from the base of the mirror should the lens be placed?

34. Headlamp The reflector for a car headlamp is in the shape of a parabolic dish measuring 16 centimeters across and 10 centimeters deep. If the bulb is located at the focus of the parabola, how far from the base of the reflector should the bulb be placed?

35. Satellite Dish The parabolic dish for a satellite television company is 24 inches across and 12 inches deep. If the signal receiver is located at the focus of the parabola, how far should it be placed from the base of the dish?

36. Telescope The primary mirror for the telescope at the Keck Observatory is a parabolic dish that is 10 meters across and has a focal length of 17.5 meters (which means the focus is located 17.5 meters from the base of the mirror). What is the depth of the mirror (see Figure 8)?

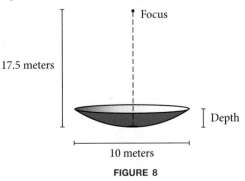

FIGURE 8

Learning Objectives Assessment

The following problems can be used to help assess if you have successfully met the learning objectives for this section.

37. Graph: $y^2 = -7x$.

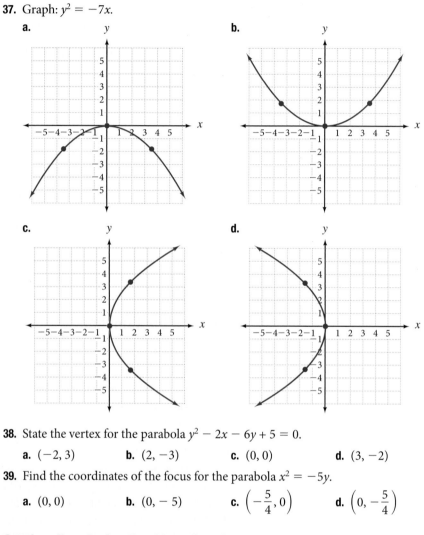

a.

b.

c.

d.

38. State the vertex for the parabola $y^2 - 2x - 6y + 5 = 0$.

    a. $(-2, 3)$ **b.** $(2, -3)$ **c.** $(0, 0)$ **d.** $(3, -2)$

39. Find the coordinates of the focus for the parabola $x^2 = -5y$.

    a. $(0, 0)$ **b.** $(0, -5)$ **c.** $\left(-\dfrac{5}{4}, 0\right)$ **d.** $\left(0, -\dfrac{5}{4}\right)$

Getting Ready for the Next Section

Solve.

40. $y^2 = 9$ **41.** $x^2 = 25$ **42.** $-y^2 = 4$

43. $-x^2 = 16$ **44.** $-x^2 = 9$ **45.** $y^2 = 100$

46. Divide $4x^2 + 9y^2$ by 36.   **47.** Divide $25x^2 + 4y^2$ by 100.

Find the x-intercepts and the y-intercepts

48. $3x - 4y = 12$ **49.** $y = 3x^2 + 5x - 2$

50. If $\dfrac{x^2}{25} + \dfrac{y^2}{9} = 1$, find y when x is 3.   **51.** If $\dfrac{x^2}{25} + \dfrac{y^2}{9} = 1$, find y when x is -4.

Learning Objectives

In this section, we will learn how to:

1. Graph an ellipse.

2. Graph a hyperbola.

3. Find the center for an ellipse or hyperbola.

4. Find the vertices of an ellipse or hyperbola.

Introduction

The photograph below shows Halley's comet as it passed close to earth in 1986. Like the planets in our solar system, it orbits the sun in an elliptical path. While it takes the earth 1 year to complete one orbit around the sun, it takes Halley's comet 76 years. The first known sighting of Halley's comet was in 239 B.C. Its most famous appearance occurred in 1066 A.D., when it was seen at the Battle of Hastings.

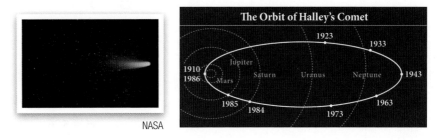

NASA

This section is concerned with the last of the two conic sections, which are ellipses and hyperbolas. To begin, we will consider only those graphs that are centered about the origin.

Ellipses

Suppose we want to graph the equation

$$\frac{x^2}{25} + \frac{y^2}{9} = 1$$

We can find the y-intercepts by letting $x = 0$, and we can find the x-intercepts by letting $y = 0$:

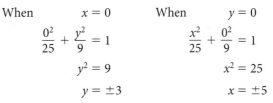

When $x = 0$ When $y = 0$

$$\frac{0^2}{25} + \frac{y^2}{9} = 1 \qquad \frac{x^2}{25} + \frac{0^2}{9} = 1$$

$$y^2 = 9 \qquad\qquad x^2 = 25$$

$$y = \pm 3 \qquad\qquad x = \pm 5$$

The graph crosses the y-axis at $(0, 3)$ and $(0, -3)$ and the x-axis at $(5, 0)$ and $(-5, 0)$. Graphing these points and then connecting them with a smooth curve gives the graph shown in Figure 1.

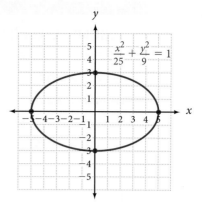

FIGURE 1

We can find other ordered pairs on the graph by substituting in values for x (or y) and then solving for y (or x). For example, if we let $x = 3$, then

$$\frac{3^2}{25} + \frac{y^2}{9} = 1$$

$$\frac{9}{25} + \frac{y^2}{9} = 1$$

$$0.36 + \frac{y^2}{9} = 1$$

$$\frac{y^2}{9} = 0.64 \qquad \text{Add } -0.36 \text{ to each side}$$

$$y^2 = 5.76 \qquad \text{Multiply each side by 9}$$

$$y = \pm 2.4 \qquad \text{Square root of each side}$$

This would give us the two ordered pairs $(3, -2.4)$ and $(3, 2.4)$.

A graph of the type shown in Figure 1 is called an **ellipse**. If we were to find some other ordered pairs that satisfy our original equation, we would find that their graphs lie on the ellipse. Also, the coordinates of any point on the ellipse will satisfy the equation. We can generalize these results as follows.

Note The vertices are always the two points on the ellipse that are furthest from each other.

⎡Δ≠Σ⎤ *The Ellipse*

The graph of any equation of the form

$$\frac{x^2}{a^2} + \frac{y^2}{b^2} = 1 \qquad \text{Standard form}$$

will be an **ellipse** centered at the origin. The ellipse will cross the x-axis at $(a, 0)$ and $(-a, 0)$. It will cross the y-axis at $(0, b)$ and $(0, -b)$. When a and b are equal, the ellipse will be a circle.

If $a^2 > b^2$, then the x-intercepts, $(\pm a, 0)$, are called the **vertices** and the y-intercepts, $(0, \pm b)$, are called the **co-vertices**. If $b^2 > a^2$, then the vertices are the points $(0, \pm b)$ and the co-vertices are the points $(\pm a, 0)$.

The most convenient way to graph an ellipse is to locate the intercepts (vertices and co-vertices).

EXAMPLE 1 Sketch the graph of $4x^2 + 9y^2 = 36$.

SOLUTION To write the equation in the form

$$\frac{x^2}{a^2} + \frac{y^2}{b^2} = 1$$

we must divide both sides by 36:

$$\frac{4x^2}{36} + \frac{9y^2}{36} = \frac{36}{36}$$

$$\frac{x^2}{9} + \frac{y^2}{4} = 1$$

The graph crosses the x-axis at $(3, 0)$, $(-3, 0)$ and the y-axis at $(0, 2)$, $(0, -2)$. (See Figure 2.) The vertices are the points $(\pm 3, 0)$, and the co-vertices are the points $(0, \pm 2)$.

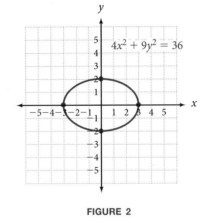

FIGURE 2

Hyperbolas

The photo below shows Europa, one of Jupiter's moons, as it was photographed by the Galileo space probe in the late 1990s. To speed up the trip from Earth to Jupiter—nearly a billion miles—Galileo made use of the ***slingshot effect***. This involves flying a hyperbolic path very close to a planet, so that gravity can be used to gain velocity as the space probe hooks around the planet (Figure 3).

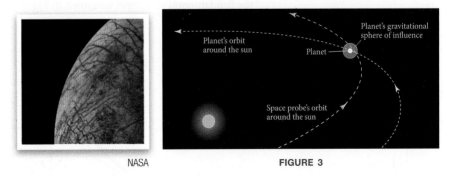

NASA

FIGURE 3

Consider the equation

$$\frac{x^2}{9} - \frac{y^2}{4} = 1$$

If we were to find a number of ordered pairs that are solutions to the equation and connect their graphs with a smooth curve, we would have Figure 4.

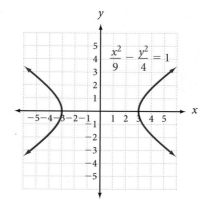

FIGURE 4

This graph is an example of a **hyperbola**. Notice that the graph has x-intercepts at $(3, 0)$ and $(-3, 0)$. These intercepts are the vertices for the hyperbola. The graph has no y-intercepts and hence does not cross the y-axis. We can show this by substituting $x = 0$ into the equation

$$\frac{0^2}{9} - \frac{y^2}{4} = 1 \qquad \text{Substitute 0 for } x$$

$$-\frac{y^2}{4} = 1 \qquad \text{Simplify left side}$$

$$y^2 = -4 \qquad \text{Multiply each side by } -4$$

for which there is no real solution.

   We want to produce reasonable sketches of hyperbolas without having to build extensive tables. We can produce the graphs we are after by using what is called the *fundamental rectangle*. The hyperbola is the only conic section whose graph has a pair of asymptotes, and the position of these asymptotes is determined by the fundamental rectangle.

   The shape of the fundamental rectangle is based on the square roots of the denominators of the two variable terms. For the hyperbola shown in Figure 4, the sides of the fundamental rectangle will pass through the points $(\pm 3, 0)$ and $(0, \pm 2)$. The asymptotes will be the lines passing through the opposite corners of the fundamental rectangle, as shown in Figure 5.

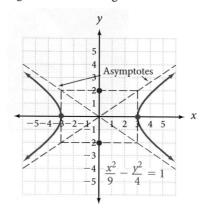

FIGURE 5

Notice that both asymptotes pass through the origin, and that their slopes are $\frac{2}{3}$ and $-\frac{2}{3}$. Thus, the equations of the asymptotes are

$$y = \frac{2}{3}x \quad \text{and} \quad y = -\frac{2}{3}x$$

The further we get from the origin, the closer the hyperbola is to these lines.

EXAMPLE 2   Graph the equation $\frac{y^2}{9} - \frac{x^2}{16} = 1$.

SOLUTION   In this case the y-intercepts are 3 and -3, and there are no x-intercepts. The vertices are the points $(0, \pm 3)$. The sides of the fundamental rectangle used to draw the asymptotes must pass through 3 and -3 on the y-axis, and 4 and -4 on the x-axis. Figure 6 shows the fundamental rectangle, the asymptotes, and the hyperbola.

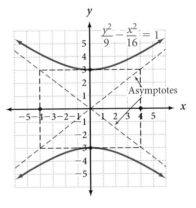

FIGURE 6

Here is a summary of what we have for hyperbolas.

⟦Δ≠Σ⟧ *Hyberbolas Centered at the Origin*

The graph of the equation

$$\frac{x^2}{a^2} - \frac{y^2}{b^2} = 1$$

will be a *hyperbola centered at the origin.* The graph will have x-intercepts (*vertices*) at $(\pm a, 0)$.

The graph of the equation

$$\frac{y^2}{b^2} - \frac{x^2}{a^2} = 1$$

will be a *hyperbola centered at the origin.* The graph will have y-intercepts (*vertices*) at $(0, \pm b)$.

As an aid in sketching either of these equations, the asymptotes can be found by drawing lines through opposite corners of the fundamental rectangle whose sides pass through $(\pm a, 0)$ and $(0, \pm b)$. The asymptotes are given by the lines

$$y = \frac{b}{a}x \quad \text{and} \quad y = -\frac{b}{a}x$$

Ellipses and Hyperbolas not Centered at the Origin

Consider the equation

$$\frac{(x-4)^2}{9} + \frac{(y-1)^2}{4} = 1$$

The terms subtracted from x and y in the grouping symbols represent translations. The graph of this equation will be identical to the ellipse

$$\frac{x^2}{9} + \frac{y^2}{4} = 1$$

whose graph is shown in Figure 2, except translated 4 units to the right and translated vertically upward 1 unit. This means the center of the ellipse will be shifted from the origin to the point (4, 1). Figure 7 shows the resulting graph.

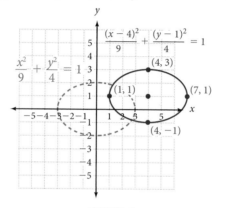

FIGURE 7

Note that the horizontal distance from the center to the vertices is 3—the square root of the denominator of the $(x-4)^2$ term. Likewise, the vertical distance from the center to the co-vertices is 2—the square root of the denominator of the $(y-1)^2$ term.

We summarize the information above with the following:

⌈Δ≠Σ⌉   *An Ellipse with Center at (h, k)*

The graph of the equation

$$\frac{(x-h)^2}{a^2} + \frac{(y-k)^2}{b^2} = 1$$

will be an *ellipse with center at (h, k).* The vertices and co-vertices of the ellipse will be at the points $(h + a, k)$, $(h - a, k)$, $(h, k + b)$, and $(h, k - b)$.

EXAMPLE 3   Graph the ellipse: $x^2 + 9y^2 + 4x - 54y + 76 = 0$

SOLUTION   To identify the coordinates of the center, we must complete the square on x and also on y. To begin, we rearrange the terms so that those containing x are together, those containing y are together, and the constant term is on the other side of the equal sign. Doing so gives us the following equation:

$$x^2 + 4x \qquad + 9y^2 - 54y \qquad = -76$$

Before we can complete the square on y, we must factor 9 from each term containing y:

$$x^2 + 4x \qquad + 9(y^2 - 6y \qquad) = -76$$

To complete the square on x, we add 4 to each side of the equation. To complete the square on y, we add 9 inside the parentheses. This increases the left side of the equation by 81 since each term within the parentheses is multiplied by 9. Therefore, we must add 81 to the right side of the equation also.

$$x^2 + 4x + 4 + 9(y^2 - 6y + 9) = -76 + 4 + 81$$

$$(x + 2)^2 + 9(y - 3)^2 = 9$$

To obtain standard form, we divide each term on both sides by 9:

$$\frac{(x + 2)^2}{9} + \frac{9(y - 3)^2}{9} = \frac{9}{9}$$

$$\frac{(x + 2)^2}{9} + \frac{(y - 3)^2}{1} = 1$$

The graph is an ellipse with center at $(-2, 3)$, as shown in Figure 8. The vertices are the points located 3 units to the left and right of the center at $(-5, 3)$, and $(1, 3)$. The co-vertices are located 2 units above and below the center at $(-2, 4)$ and $(-2, 2)$.

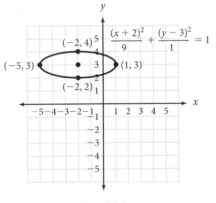

FIGURE 8

The ideas associated with graphing hyperbolas whose centers are not at the origin parallel the ideas just presented about graphing ellipses whose centers have been moved off the origin.

$\lceil \Delta \neq \Sigma \rceil$ *Hyperbolas with Centers at (h, k)*

The graphs of the equations

$$\frac{(x - h)^2}{a^2} - \frac{(y - k)^2}{b^2} = 1 \quad \text{and} \quad \frac{(y - k)^2}{b^2} - \frac{(x - h)^2}{a^2} = 1$$

will be *hyperbolas with their centers at (h, k)*. The vertices of the graph of the first equation will be at the points $(h + a, k)$ and $(h - a, k)$, and the vertices for the graph of the second equation will be at $(h, k + b)$ and $(h, k - b)$.
In either case, the asymptotes can be found by connecting opposite corners of the fundamental rectangle that contains the four points $(h + a, k)$, $(h - a, k)$, $(h, k + b)$, and $(h, k - b)$.

EXAMPLE 4   Graph the hyperbola: $4x^2 - y^2 + 4y - 20 = 0$

SOLUTION   To identify the coordinates of the center of the hyperbola, we need to complete the square on y. (Because there is no linear term in x, we do not need to complete the square on x. The x-coordinate of the center will be $x = 0$.)

$$4x^2 - y^2 + 4y - 20 = 0$$

$$4x^2 - y^2 + 4y = 20 \qquad \text{Add 20 to each side}$$

$$4x^2 - 1(y^2 - 4y) = 20 \qquad \text{Factor } -1 \text{ from each term containing } y$$

To complete the square on y, we add 4 to the terms inside the parentheses. Doing so adds -4 to the left side of the equation because everything inside the parentheses is multiplied by -1. To keep from changing the equation we must add -4 to the right side also.

$$4x^2 - 1(y^2 - 4y + 4) = 20 - 4 \qquad \text{Add } -4 \text{ to each side}$$

$$4x^2 - 1(y - 2)^2 = 16 \qquad y^2 - 4y + 4 = (y - 2)^2$$

$$\frac{4x^2}{16} - \frac{(y - 2)^2}{16} = \frac{16}{16} \qquad \text{Divide each side by 16}$$

$$\frac{x^2}{4} - \frac{(y - 2)^2}{16} = 1 \qquad \text{Simplify each term}$$

This is the equation of a hyperbola with center at $(0, 2)$. The graph opens to the right and left as shown in Figure 9. The vertices are located 2 units to the left and right of the center at $(-2, 2)$ and $(2, 2)$.

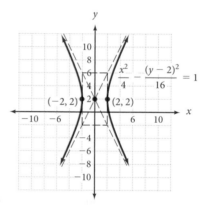

FIGURE 9

Getting Ready for Class

After reading through the preceding section, respond in your own words and in complete sentences.

A. How do we find the x-intercepts of a graph from the equation?

B. What is an ellipse?

C. How can you tell by looking at an equation if its graph will be an ellipse or a hyperbola?

D. How do we know if the graph of a hyperbola will open to the left and right, or open upward and downward?

Graph each of the following ellipses. Be sure to label both the x- and y-intercepts.

1. $\dfrac{x^2}{9} + \dfrac{y^2}{16} = 1$ **2.** $\dfrac{x^2}{25} + \dfrac{y^2}{4} = 1$ **3.** $\dfrac{x^2}{16} + \dfrac{y^2}{9} = 1$

4. $\dfrac{x^2}{4} + \dfrac{y^2}{25} = 1$ **5.** $\dfrac{x^2}{3} + \dfrac{y^2}{4} = 1$ **6.** $\dfrac{x^2}{4} + \dfrac{y^2}{3} = 1$

7. $4x^2 + 25y^2 = 100$ **8.** $4x^2 + 9y^2 = 36$ **9.** $x^2 + 8y^2 = 16$

10. $12x^2 + y^2 = 36$

Graph each of the following hyperbolas. Show the intercepts and the asymptotes in each case.

11. $\dfrac{x^2}{9} - \dfrac{y^2}{16} = 1$ **12.** $\dfrac{x^2}{25} - \dfrac{y^2}{4} = 1$ **13.** $\dfrac{x^2}{16} - \dfrac{y^2}{9} = 1$

14. $\dfrac{x^2}{4} - \dfrac{y^2}{25} = 1$ **15.** $\dfrac{y^2}{9} - \dfrac{x^2}{16} = 1$ **16.** $\dfrac{y^2}{25} - \dfrac{x^2}{4} = 1$

17. $\dfrac{y^2}{36} - \dfrac{x^2}{4} = 1$ **18.** $\dfrac{y^2}{4} - \dfrac{x^2}{36} = 1$ **19.** $x^2 - 4y^2 = 4$

20. $y^2 - 4x^2 = 4$ **21.** $16y^2 - 9x^2 = 144$ **22.** $4y^2 - 25x^2 = 100$

Find the x- and y-intercepts, if they exist, for each of the following. Do not graph.

23. $0.4x^2 + 0.9y^2 = 3.6$ **24.** $1.6x^2 + 0.9y^2 = 14.4$ **25.** $\dfrac{x^2}{0.04} - \dfrac{y^2}{0.09} = 1$

26. $\dfrac{y^2}{0.16} - \dfrac{x^2}{0.25} = 1$ **27.** $\dfrac{25x^2}{9} + \dfrac{25y^2}{4} = 1$ **28.** $\dfrac{16x^2}{9} + \dfrac{16y^2}{25} = 1$

Graph each of the following ellipses. In each case, label the coordinates of the center and the vertices.

29. $\dfrac{(x-4)^2}{4} + \dfrac{(y-2)^2}{9} = 1$ **30.** $\dfrac{(x-2)^2}{4} + \dfrac{(y-4)^2}{9} = 1$

31. $4x^2 + y^2 - 4y - 12 = 0$ **32.** $4x^2 + y^2 - 24x - 4y + 36 = 0$

33. $x^2 + 9y^2 + 4x - 54y + 76 = 0$ **34.** $4x^2 + y^2 - 16x + 2y + 13 = 0$

Graph each of the following hyperbolas. In each case, label the coordinates of the center and the vertices and show the asymptotes.

35. $\dfrac{(x-2)^2}{16} - \dfrac{y^2}{4} = 1$ **36.** $\dfrac{(y-2)^2}{16} - \dfrac{x^2}{4} = 1$

37. $9y^2 - x^2 - 4x + 54y + 68 = 0$ **38.** $4x^2 - y^2 - 24x + 4y + 28 = 0$

39. $4y^2 - 9x^2 - 16y + 72x - 164 = 0$ **40.** $4x^2 - y^2 - 16x - 2y + 11 = 0$

41. Find x when $y = 4$ in the equation $\dfrac{x^2}{25} + \dfrac{y^2}{16} = 1$.

42. Find x when $y = 3$ in the equation $\dfrac{x^2}{25} + \dfrac{y^2}{16} = 1$.

43. Find y when $x = -3$ in the equation $\dfrac{x^2}{9} + \dfrac{y^2}{16} = 1$.

44. Find y when $x = -2$ in the equation $\dfrac{x^2}{9} + \dfrac{y^2}{16} = 1$.

45. The line segment connecting the vertices of an ellipse is called the *major axis* of the ellipse. Give the length of the major axis of the ellipse you graphed in Problem 3.

46. The line segment connecting the co-vertices of an ellipse is called the *minor axis* of the ellipse. Give the length of the minor axis of the ellipse you graphed in Problem 3.

Applying the Concepts

Some of the problems that follow use the major and minor axes mentioned in Problems 45 and 46. The diagram below shows the minor axis and the major axis for an ellipse.

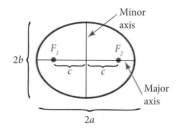

For an ellipse in which $a^2 > b^2$, the length of the major axis is $2a$ and the length of the minor axis is $2b$ (these are the same a and b that appear in the general equation of an ellipse). Each of the points F_1 and F_2 shown above on the major axis is a *focus* of the ellipse. If the distance from the center of the ellipse to each focus is c, then it is always true that $a^2 = b^2 + c^2$. You will need this information for some of the problems that follow.

47. The Colosseum The Colosseum in Rome seated 50,000 spectators around a central elliptical arena. The base of the Colosseum measured 615 feet long and 510 feet wide. Write an equation for the elliptical shape of the Colosseum. Assume the ellipse is centered at the origin and that the major axis is horizontal.

48. Archway A new theme park is planning an archway at its main entrance. The arch is to be in the form of a semi-ellipse (half of an ellipse) with the major axis as the span. If the span is to be 40 feet and the height at the center is to be 10 feet, what is the equation of the ellipse? How far left and right of center could a 6-foot man walk upright under the arch? Assume the ellipse is centered at the origin and that the major axis is horizontal.

49. The Ellipse President's Park, located between the White House and the Washington Monument in Washington, DC, is also called The Ellipse. The park is enclosed by an elliptical path with major axis 458 meters and minor axis 390 meters. What is the equation for the path around The Ellipse? Assume the ellipse is centered at the origin and that the major axis is horizontal.

50. Garden Trellis John is planning to build an arched trellis for the entrance to his botanical garden. If the arch is to be in the shape of the upper half of an ellipse that is 6 feet wide at the base and 9 feet high, what is the equation for the ellipse? Assume the ellipse is centered at the origin and that the major axis is vertical.

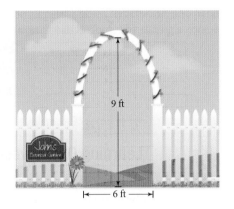

51. Elliptical Pool Table A children's science museum plans to build an elliptical pool table to demonstrate that a ball rolled from a particular point (focus) will always go into a hole located at another particular point (the other focus). The focus needs to be 1 foot from the vertex of the ellipse. If the table is to be 8 feet long, how wide should it be? *Hint:* The distance from the center to each focus point is represented by c and is found by using the equation $a^2 = b^2 + c^2$.

52. Entering the Zoo A zoo is planning a new entrance. Visitors are to be "funneled" into the zoo between two tall brick fences. The bases of the fences will be in the shape of a hyperbola. The narrowest passage East and West between the fences will be 24 feet. The total North–South distance of the fences is to be 50 feet. Write an equation for the hyperbolic shape of the fences if the center of the hyperbola is to be placed at the origin of the coordinate system and the vertices are on the x-axis.

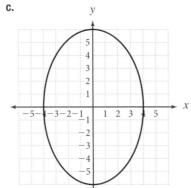

Learning Objectives Assessment

The following problems can be used to help assess if you have successfully met the learning objectives for this section.

53. Graph the ellipse $\dfrac{x^2}{4} + \dfrac{y^2}{6} = 1$.

a.

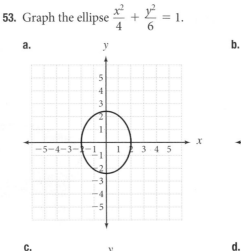

b.

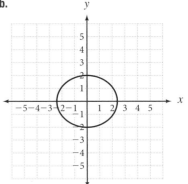

c.

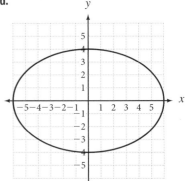

d.

54. Graph the hyperbola $\dfrac{x^2}{9} - \dfrac{y^2}{4} = 1$.

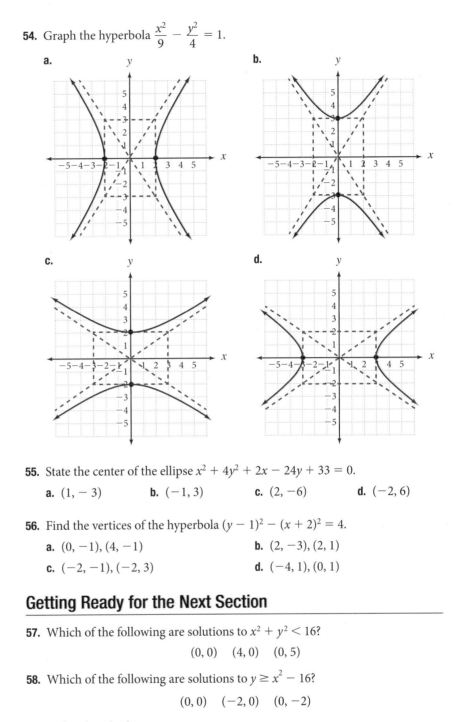

a.

b.

c.

d.

55. State the center of the ellipse $x^2 + 4y^2 + 2x - 24y + 33 = 0$.

    a. $(1, -3)$ **b.** $(-1, 3)$ **c.** $(2, -6)$ **d.** $(-2, 6)$

56. Find the vertices of the hyperbola $(y - 1)^2 - (x + 2)^2 = 4$.

    a. $(0, -1), (4, -1)$ **b.** $(2, -3), (2, 1)$

    c. $(-2, -1), (-2, 3)$ **d.** $(-4, 1), (0, 1)$

Getting Ready for the Next Section

57. Which of the following are solutions to $x^2 + y^2 < 16$?

$$(0, 0) \quad (4, 0) \quad (0, 5)$$

58. Which of the following are solutions to $y \geq x^2 - 16$?

$$(0, 0) \quad (-2, 0) \quad (0, -2)$$

Expand and Multiply.

59. $(2y + 4)^2$ **60.** $(y + 3)^2$

61. Solve $x - 2y = 4$ for x.   **62.** Solve $2x + 3y = 6$ for y.

Simplify.

63. $x^2 - 2(x^2 - 3)$

64. $x^2 + (x^2 - 4)$

Factor.

65. $5y^2 + 16y + 12$

66. $3x^2 + 17x - 28$

Solve.

67. $y^2 = 4$

68. $x^2 = 25$

69. $-x^2 + 6 = 2$

70. $5y^2 + 16y + 12 = 0$

Second-Degree Inequalities and Nonlinear Systems

Learning Objectives

In this section, we will learn how to:

1. Graph the solution set for a second-degree inequality.

2. Solve a system of nonlinear equations using the substitution method.

3. Solve a system of nonlinear equations using the addition method.

4. Graph the solution set for a system of nonlinear inequalities.

Introduction

In Section 3.4, we graphed linear inequalities by first graphing the boundary and then choosing a test point not on the boundary to determine the region used for the solution set. The problems in this section are very similar. We will use the same general methods for graphing the inequalities in this section that we used in Section 3.4.

Second-Degree Inequalities

EXAMPLE 1 Graph $x^2 + y^2 < 16$.

SOLUTION The boundary is $x^2 + y^2 = 16$, which is a circle with center at the origin and a radius of 4. Because the inequality sign is $<$, the boundary is not included in the solution set and must therefore be represented with a broken line. The graph of the boundary is shown in Figure 1.

The solution set for $x^2 + y^2 < 16$ is either the region inside the circle or the region outside the circle. To see which region represents the solution set, we choose a convenient point not on the boundary and test it in the original inequality. The origin $(0, 0)$ is a convenient point. Because the origin satisfies the inequality $x^2 + y^2 < 16$, all points in the same region will also satisfy the inequality. Therefore, we shade the region inside the circle. The graph of the solution set is shown in Figure 2.

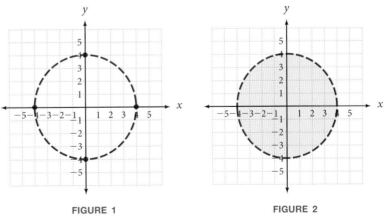

FIGURE 1   FIGURE 2

EXAMPLE 2   Graph the inequality $y \le x^2 - 2$.

SOLUTION   The parabola $y = x^2 - 2$ is the boundary and is included in the solution set. Using $(0, 0)$ as the test point, we see that $0 \le 0^2 - 2$ is a false statement, which means that the region containing $(0, 0)$ is not in the solution set. Therefore, we shade the region to the outside of the parabola, as shown in Figure 3.

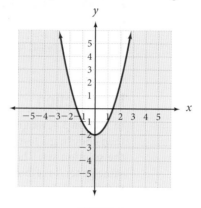

FIGURE 3

EXAMPLE 3   Graph $4y^2 - 9x^2 < 36$.

SOLUTION   The boundary is the hyperbola $4y^2 - 9x^2 = 36$, which we can write in standard form as

$$\frac{y^2}{9} - \frac{x^2}{4} = 1$$

It is not included in the solution set. Testing $(0, 0)$ in the original inequality yields a true statement, which means that the region containing the origin is the solution set. (See Figure 4.)

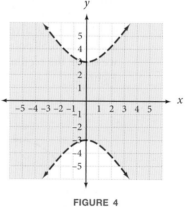

FIGURE 4

Nonlinear Systems of Equations

In Chapter 4 we learned how to solve systems of linear equations in two and three variables. Now we will consider systems of equations that are not linear. We will rely once again on the addition method and substitution method that was introduced back in Section 4.1.

EXAMPLE 4   Solve the system.

$$x^2 + y^2 = 4$$
$$x - 2y = 4$$

SOLUTION In this case, the substitution method is the most convenient. Solving the second equation for x in terms of y, we have

$$x - 2y = 4$$

$$x = 2y + 4$$

We now substitute $2y + 4$ for x in the first equation in our original system and proceed to solve for y:

$$(2y + 4)^2 + y^2 = 4$$

$$4y^2 + 16y + 16 + y^2 = 4 \qquad \text{Expand } (2y + 4)^2$$

$$5y^2 + 16y + 16 = 4 \qquad \text{Simplify left side}$$

$$5y^2 + 16y + 12 = 0 \qquad \text{Add } -4 \text{ to each side}$$

$$(5y + 6)(y + 2) = 0 \qquad \text{Factor}$$

$$5y + 6 = 0 \quad \text{or} \quad y + 2 = 0 \qquad \text{Set factors equal to 0}$$

$$y = -\frac{6}{5} \quad \text{or} \qquad y = -2 \qquad \text{Solve}$$

These are the y-coordinates of the two solutions to the system. Substituting $y = -\frac{6}{5}$ into $x - 2y = 4$ and solving for x gives us $x = \frac{8}{5}$. Using $y = -2$ in the same equation yields $x = 0$. The two solutions to our system are $\left(\frac{8}{5}, -\frac{6}{5}\right)$ and $(0, -2)$. Although graphing the system is not necessary, it does help us visualize the situation. (See Figure 5.)

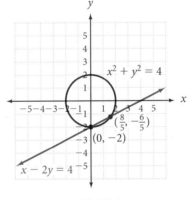

FIGURE 5

EXAMPLE 5 Solve the system.

$$16x^2 - 4y^2 = 64$$

$$x^2 + y^2 = 9$$

SOLUTION Because each equation is of the second degree in both x and y, it is easier to solve this system by eliminating one of the variables by addition. To eliminate y, we multiply the bottom equation by 4 and add the result to the top equation:

$$\begin{array}{rl} 16x^2 - 4y^2 = & 64 \\ 4x^2 + 4y^2 = & 36 \\ \hline 20x^2 \qquad = & 100 \end{array}$$

$$x^2 = 5$$

$$x = \pm\sqrt{5}$$

The x-coordinates of the points of intersection are $\sqrt{5}$ and $-\sqrt{5}$. We substitute each back into the second equation in the original system and solve for y:

When
$$x = \sqrt{5}$$
$$(\sqrt{5})^2 + y^2 = 9$$
$$5 + y^2 = 9$$
$$y^2 = 4$$
$$y = \pm 2$$

When
$$x = -\sqrt{5}$$
$$(-\sqrt{5})^2 + y^2 = 9$$
$$5 + y^2 = 9$$
$$y^2 = 4$$
$$y = \pm 2$$

The four points of intersection are $(-\sqrt{5},\ 2)$, $(-\sqrt{5},\ -2)$, $(\sqrt{5},\ 2)$, and $(\sqrt{5},\ -2)$. Graphically the situation is as shown in Figure 6.

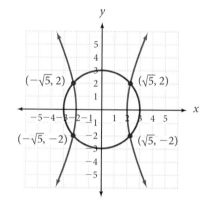

FIGURE 6

EXAMPLE 6   Solve the system.
$$x^2 - 2y = 2$$
$$y = x^2 - 3$$

SOLUTION   We can solve this system using the substitution method. Replacing y in the first equation with $x^2 - 3$ from the second equation, we have

$$x^2 - 2(x^2 - 3) = 2$$
$$-x^2 + 6 = 2$$
$$x^2 = 4$$
$$x = \pm 2$$

Using either $+2$ or -2 in the equation $y = x^2 - 3$ gives us $y = 1$. The system has two solutions: $(2, 1)$ and $(-2, 1)$.

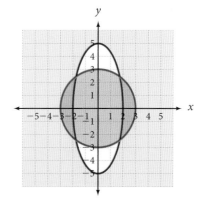 **EXAMPLE 7**   The sum of the squares of two numbers is 34. The difference of their squares is 16. Find the two numbers.

SOLUTION   Let x and y be the two numbers. The sum of their squares is $x^2 + y^2$, and the difference of their squares is $x^2 - y^2$. (We can assume here that x^2 is the larger number.) The system of equations that describes the situation is

$$x^2 + y^2 = 34$$
$$x^2 - y^2 = 16$$

We can eliminate y by simply adding the two equations. The result of doing so is

$$2x^2 = 50$$
$$x^2 = 25$$
$$x = \pm 5$$

Substituting $x = 5$ into either equation in the system gives $y = \pm 3$. Using $x = -5$ gives the same results, $y = \pm 3$. The four pairs of numbers that are solutions to the original problem are

$$(5, 3) \qquad (-5, 3) \qquad (5, -3) \qquad (-5, -3)$$

Systems of Nonlinear Inequalities

We now turn our attention to systems of inequalities. To solve a system of inequalities by graphing, we simply graph each inequality on the same set of axes. The solution set for the system is the region common to both graphs — the intersection of the individual solution sets.

EXAMPLE 8   Graph the solution set for the system

$$x^2 + y^2 \leq 9$$
$$\frac{x^2}{4} + \frac{y^2}{25} \geq 1$$

SOLUTION   The boundary for the top inequality is a circle with center at the origin and a radius of 3. The solution set includes the boundary and the region inside the boundary. The boundary for the second inequality is an ellipse. In this case, the solution set includes the boundary and the region outside the boundary. (See Figure 7.)

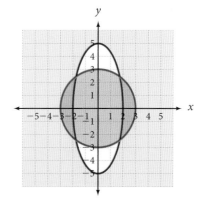

FIGURE 7

The solution set for the system is the intersection of the two individual solution sets, which consists of the two small regions inside the circle but outside the ellipse, and the portions of the boundaries that enclose these regions. Figure 8 shows the graph of the solution set, where we have kept the original boundaries as dashed curves for reference.

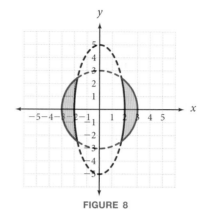

FIGURE 8

Getting Ready for Class

After reading through the preceding section, respond in your own words and in complete sentences.

A. What is the significance of a broken line when graphing inequalities?

B. Describe, in words, the set of points described by
$(x - 3)^2 + (y - 2)^2 < 9$.

C. When solving nonlinear systems of equations whose graphs are a line and a circle, how many possible solutions can you expect?

D. When solving nonlinear systems of equations whose graphs are both circles, how many possible solutions can you expect?

Graph each of the following second-degree inequalities.

1. $x^2 + y^2 \le 49$

2. $x^2 + y^2 < 49$

3. $(x - 2)^2 + (y + 3)^2 < 16$

4. $(x + 3)^2 + (y - 2)^2 \ge 25$

5. $y < x^2 - 6x + 7$

6. $y \ge x^2 + 2x - 8$

7. $\dfrac{x^2}{25} - \dfrac{y^2}{9} \ge 1$

8. $\dfrac{x^2}{25} - \dfrac{y^2}{9} \le 1$

9. $4x^2 + 25y^2 \le 100$

10. $25x^2 - 4y^2 > 100$

11. $\dfrac{(x + 2)^2}{25} + \dfrac{(y - 1)^2}{9} \le 1$

12. $\dfrac{(x - 1)^2}{16} - \dfrac{(y + 1)^2}{16} < 1$

13. $16x^2 - 9y^2 \ge 144$

14. $16y^2 - 9x^2 < 144$

15. $9x^2 + 4y^2 + 36x - 8y + 4 < 0$

16. $9x^2 - 4y^2 + 36x + 8y \ge 4$

17. $9y^2 - x^2 + 18y + 2x > 1$

18. $x^2 + y^2 - 6x - 4y \le 12$

Solve each of the following systems of equations.

19. $x^2 + y^2 = 9$
    $2x + y = 3$

20. $x^2 + y^2 = 9$
    $x + 2y = 3$

21. $x^2 + y^2 = 16$
    $x + 2y = 8$

22. $x^2 + y^2 = 16$
    $x - 2y = 8$

23. $x^2 + y^2 = 25$
    $x^2 - y^2 = 25$

24. $x^2 + y^2 = 4$
    $2x^2 - y^2 = 5$

25. $x^2 + y^2 = 9$
    $y = x^2 - 3$

26. $x^2 + y^2 = 4$
    $y = x^2 - 2$

27. $x^2 + y^2 = 16$
    $x = y^2 - 4$

28. $x^2 + y^2 = 1$
    $x = y^2 - 1$

29. $3x + 2y = 10$
    $y = x^2 - 5$

30. $4x + 2y = 10$
    $y = x^2 - 10$

31. $y = x^2 + 2x - 3$
    $y = -x + 1$

32. $y = -x^2 - 2x + 3$
    $y = x - 1$

33. $x = y^2 - 6y + 5$
    $x = y - 5$

34. $x = y^2 - 2y - 4$
    $x = y - 4$

35. $4x^2 - 9y^2 = 36$
    $4x^2 + 9y^2 = 36$

36. $4x^2 + 25y^2 = 100$
    $4x^2 - 25y^2 = 100$

37. $x - y = 4$
    $x^2 + y^2 = 16$

38. $x + y = 2$
    $x^2 - y^2 = 4$

39. $2x^2 - y = 1$
    $x^2 + y = 7$

40. $x^2 + y^2 = 52$
    $y = x + 2$

41. $y = x^2 - 3$
    $y = x^2 - 2x - 1$

42. $y = 8 - 2x^2$
    $y = x^2 - 1$

43. $4x^2 + 5y^2 = 40$
    $4x^2 - 5y^2 = 40$

44. $x + 2y^2 = -4$
    $3x - 4y^2 = -3$

Graph the solution sets to the following systems of nonlinear inequalities.

45. $x^2 + y^2 < 9$
$\quad\quad y \geq x^2 - 1$

46. $x^2 + y^2 \leq 16$
$\quad\quad y < x^2 + 2$

47. $\dfrac{x^2}{9} + \dfrac{y^2}{25} \leq 1$
$\quad\quad \dfrac{x^2}{4} - \dfrac{y^2}{9} > 1$

48. $\dfrac{x^2}{4} + \dfrac{y^2}{16} \geq 1$
$\quad\quad \dfrac{x^2}{9} - \dfrac{y^2}{25} < 1$

49. $4x^2 + 9y^2 \leq 36$
$\quad\quad y > x^2 + 2$

50. $9x^2 + 4y^2 \geq 36$
$\quad\quad y < x^2 + 1$

51. $x^2 + y^2 \leq 3$
$\quad\quad y^2 \leq 4x$

52. $x^2 + y^2 < 4$
$\quad\quad y^2 \geq 2x$

53. $\quad x + y \leq 3$
$\quad\quad y^2 + 4x > 0$

54. $\quad x - y > 3$
$\quad\quad x^2 + 2y \leq 0$

55. $x^2 - y^2 \leq 4$
$\quad\quad y \geq 0$

56. $y^2 - x^2 \geq 4$
$\quad\quad x \leq 0$

57. $x^2 - 4y \leq 0$
$\quad\quad y^2 - 4x \leq 0$

58. $x^2 - y^2 \geq 1$
$\quad\quad x + y^2 \leq 0$

59. $\quad x^2 + y^2 \leq 25$
$\quad\quad \dfrac{x^2}{9} - \dfrac{y^2}{16} > 1$

60. $\dfrac{x^2}{16} + \dfrac{y^2}{25} < 1$
$\quad\quad \dfrac{y^2}{4} - \dfrac{x^2}{1} \geq 1$

61. $\quad x + y \leq 2$
$\quad\quad y > x^2$

62. $\dfrac{x^2}{9} + \dfrac{y^2}{16} \leq 1$
$\quad\quad \dfrac{x^2}{16} + \dfrac{y^2}{9} > 1$

Applying the Concepts

63. Number Problem The sum of the squares of two numbers is 89. The difference of their squares is 39. Find the numbers.

64. Number Problem The difference of the squares of two numbers is 35. The sum of their squares is 37. Find the numbers.

65. Consider the equations for the three circles below. They are are

Circle A	Circle B	Circle C
$(x + 8)^2 + y^2 = 64$	$x^2 + y^2 = 64$	$(x - 8)^2 + y^2 = 64$

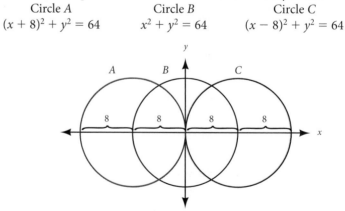

a. Find the points of intersection of circles A and B.

b. Find the points of intersection of circles B and C.

66. A magician is holding two rings that seem to lie in the same plane and intersect in two points. Each ring is 10 inches in diameter. If a coordinate system is placed with its origin at the center of the first ring and the x-axis contains the center of the second ring, then the equations are as follows:

First Ring	Second Ring
$x^2 + y^2 = 25$	$(x - 5)^2 + y^2 = 25$

Find the points of intersection of the two rings.

Learning Objectives Assessment

The following problems can be used to help assess if you have successfully met the learning objectives for this section.

67. Graph: $x^2 + y^2 > 16$.

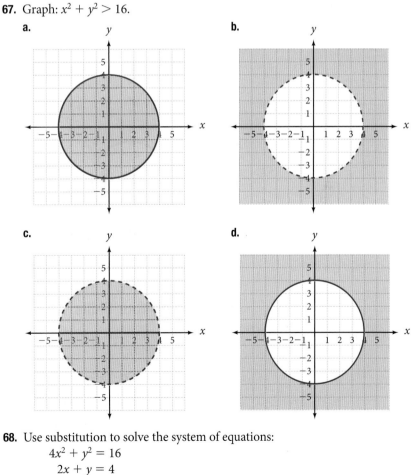

68. Use substitution to solve the system of equations:
$$4x^2 + y^2 = 16$$
$$2x + y = 4$$

a. $\{(-4, 0), (0, -2)\}$ **b.** $\{(-4, 0), (0, 2)\}$

c. $\{(2, 0), (0, -4)\}$ **d.** $\{(2, 0), (0, 4)\}$

69. Use the addition method to solve the system of equations:
$$4x^2 + y^2 = 16$$
$$x^2 - y^2 = 4$$

a. $\{(2, 0)\}$ **b.** $\{(0, 4)\}$

c. $\{(0, -4), (0, 4)\}$ **d.** $\{(-2, 0), (2, 0)\}$

70. Graph the solution set for the system of inequalities:
$$x^2 + y^2 \geq 9$$
$$x^2 - y^2 \leq 4$$

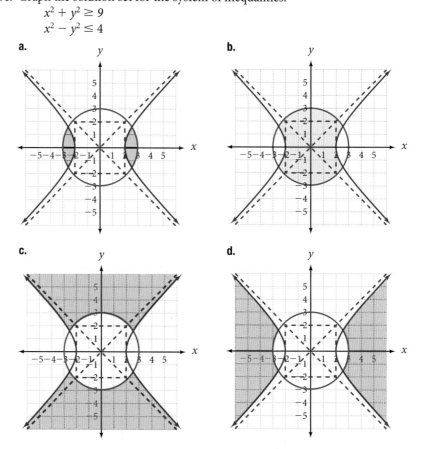

a. **b.** **c.** **d.**

Maintaining Your Skills

Expand and simplify.

71. $(x + 2)^4$ **72.** $(x - 2)^4$ **73.** $(2x + y)^3$ **74.** $(x - 2y)^3$

75. Find the first two terms in the expansion of $(x + 3)^{50}$.

76. Find the first two terms in the expansion of $(x - y)^{75}$.

Chapter 11 Summary

EXAMPLES

Distance Formula [11.1]

1. The distance between $(5, 2)$ and $(-1, 1)$ is
$$d = \sqrt{(5 + 1)^2 + (2 - 1)^2} = \sqrt{37}$$

The distance between the two points (x_1, y_1) and (x_2, y_2) is given by the formula
$$d = \sqrt{(x_2 - x_1)^2 + (y_2 - y_1)^2}$$

The Circle [11.1]

2. The graph of the circle $(x - 3)^2 + (y + 2)^2 = 25$ will have its center at $(3, -2)$ and the radius will be 5.

The graph of any equation of the form
$$(x - h)^2 + (y - k)^2 = r^2$$
will be a circle having its center at (h, k) and a radius of r.

The Parabola [11.2]

3. The parabola $y^2 = 12x$ has the vertex $(0, 0)$ and focus $(3, 0)$. The focal width is 12. The parabola will open to the right.

The graph of an equation that can be put into either of the forms
$$x^2 = 4py \qquad \text{or} \qquad y^2 = 4px$$
will be a parabola with vertex at $(0, 0)$ and focal width equal to $|4p|$. The focus for the first equation will be at $(0, p)$, and the focus for the second equation will be at $(p, 0)$.

The Ellipse [11.3]

4. The ellipse $\frac{x^2}{9} + \frac{y^2}{4} = 1$ will cross the x-axis at 3 and -3 and will cross the y-axis at 2 and -2.

Any equation that can be put in the form
$$\frac{x^2}{a^2} + \frac{y^2}{b^2} = 1$$
will have an ellipse for its graph. The x-intercepts will be at a and $-a$, and the y-intercepts will be at b and $-b$.

The Hyperbola [11.3]

5. The hyperbola $\frac{x^2}{4} - \frac{y^2}{9} = 1$ will cross the x-axis at 2 and -2. It will not cross the y-axis.

The graph of an equation that can be put in either of the forms
$$\frac{x^2}{a^2} - \frac{y^2}{b^2} = 1 \quad \text{or} \quad \frac{y^2}{b^2} - \frac{x^2}{a^2} = 1$$
will be a hyperbola. The x-intercepts, for the first equation, will be at a and $-a$. The y-intercepts, for the second equation, will be at b and $-b$. Two straight lines, called **asymptotes**, are associated with the graph of every hyperbola. Although the asymptotes are not part of the hyperbola, they are useful in sketching the graph.

Second-Degree Inequalities in Two Variables [11.4]

6. The graph of the inequality

$$x^2 + y^2 < 9$$

is all points inside the circle with center at the origin and radius 3. The circle itself is not part of the solution and therefore is shown with a broken curve.

We graph second-degree inequalities in two variables in much the same way that we graphed linear inequalities; that is, we begin by graphing the boundary, using a solid curve if the boundary is included in the solution (this happens when the inequality symbol is $\geq$ or $\leq$) or a broken curve if the boundary is not included in the solution (when the inequality symbol is $>$ or $<$). After we have graphed the boundary, we choose a test point that is not on the boundary and try it in the original inequality. A true statement indicates we are in the region of the solution. A false statement indicates we are not in the region of the solution. Then we shade the region that contains solutions.

Systems of Nonlinear Equations [11.4]

7. We can solve the system

$$x^2 + y^2 = 4$$
$$x = 2y + 4$$

by substituting $2y + 4$ from the second equation for x in the first equation:

$$(2y + 4)^2 + y^2 = 4$$
$$4y^2 + 16y + 16 + y^2 = 4$$
$$5y^2 + 16y + 12 = 0$$
$$(5y + 6)(y + 2) = 0$$
$$y = -\frac{6}{5} \quad \text{or} \quad y = -2$$

Substituting these values of y into the second equation in our system gives $x = \frac{8}{5}$ and $x = 0$. The solutions are $\left(\frac{8}{5}, -\frac{6}{5}\right)$ and $(0, -2)$.

A system of nonlinear equations is two or more equations, at least one of which is not linear, considered at the same time. The solution set for the system consists of all ordered pairs that satisfy all of the equations. In most cases we use the substitution method to solve these systems; however, the addition method can be used if like variables are raised to the same power in two equations. It is sometimes helpful to graph each equation in the system on the same set of axes to anticipate the number and approximate positions of the solutions.

1. Find x so that $(x, 2)$ is $2\sqrt{5}$ units from $(-1, 4)$. [11.1]

2. Give the equation of the circle with center at $(-2, 4)$ and radius 3. [11.1]

3. Give the equation of the circle with center at the origin that contains the point $(-3, -4)$. [11.1]

4. Find the center and radius of the circle $x^2 + y^2 - 10x + 6y = 5$. [11.1]

Graph each of the following. [11.1- 11.3]

5. $4x^2 - y^2 = 16$

6. $\dfrac{x^2}{25} + \dfrac{y^2}{4} = 1$

7. $x^2 + y^2 = 16$

8. $y^2 = 6x$

9. $9x^2 + 4y^2 - 72x - 16y + 124 = 0$

10. $(x - 1)^2 + (y + 3)^2 = 4$

11. $\dfrac{(y - 1)^2}{4} - \dfrac{(x - 3)^2}{1} = 1$

12. $x^2 + 4x + 4y + 8 = 0$

Graph the solution set to each inequality. [11.4]

13. $(x - 2)^2 + (y + 1)^2 \leq 9$

14. $9x^2 - 4y^2 < 36$

Solve the following systems. [11.4]

15. $x^2 + y^2 = 25$
    $2x + y = 5$

16. $x^2 + y^2 = 16$
    $y = x^2 - 4$

Graph the solution set to each system of inequalities. [11.4]

17. $\dfrac{x^2}{16} + \dfrac{y^2}{9} \leq 1$
    $y^2 - x^2 < 1$

18. $(x + 2)^2 + y^2 < 9$
    $y^2 + 4x \leq 0$

Chapter 1

PROBLEM SET 1.1

1-7.

9. $\frac{18}{24}$ **11.** $\frac{12}{24}$ **13.** $\frac{15}{24}$ **15.** $\frac{36}{60}$ **17.** $\frac{22}{60}$

19. $-\frac{50}{60}$ **21.** $-10, \frac{1}{10}, 10$ **23.** $-\frac{3}{4}, \frac{4}{3}, \frac{3}{4}$

25. $-\frac{11}{2}, \frac{2}{11}, \frac{11}{2}$ **27.** $3, -\frac{1}{3}, 3$ **29.** $\frac{2}{5}, -\frac{5}{2}, \frac{2}{5}$

31. $-x, \frac{1}{x}, |x|$ **33.** 2 **35.** $\frac{3}{4}$ **37.** π **39.** -4

41. -2 **43.** $-\frac{3}{4}$ **45.** $<$ **47.** $>$ **49.** $>$

51. $>$ **53.** $<$ **55.** $<$ **57.** $2 \cdot 7 \cdot 19$ **59.** $3 \cdot 37$

61. $3^2 \cdot 41$ **63.** $\frac{3}{7}$ **65.** $\frac{11}{21}$ **67.** $\frac{3}{5}$ **69.** $\frac{5}{9}$

71. $\frac{3}{4}$ **73.** 40 **75.** 4 in.; 1 in.^{2} **77.** 4.5 in.; 1.125 in.^{2}

79. 10.25 cm; 5 cm^2 **81.** $-15°$F

83. a. 260 applications **b.** False **c.** True

85. 93.5 square inches, 39 inches

87. 1,387 calories **89.** 654 more calories **91.** 20, -20

93. No, the absolute value of a number can never be negative.

95. 157 **97.** 121 or 161 **99.** True **101.** True

103. $\{x \mid x > 5\}$ **105.** $\{x \mid x \geq 5\}$ **107. a.** $\{1, 2\}$ **b.** $\{3, 4\}$

109. d **111.** d **113.** b **115.** c

PROBLEM SET 1.2

1. Commutative property of addition

3. Commutative property of multiplication

5. Additive inverse property

7. Commutative property of addition

9. Associative and commutative properties of multiplication

11. Commutative and associative properties of addition

13. Distributive property

15. Multiplicative inverse property

17. $6 + x$ **19.** $a + 8$ **21.** $15y$ **23.** x **25.** a

27. x **29.** $3x + 18$ **31.** $12x + 8$ **33.** $15a + 10b$

35. $\frac{4}{3}x + 2$ **37.** $2 + y$ **39.** $40t + 8$

41. $9x + 3y - 6z$ **43.** $3x + 7y$ **45.** $6x + 7y$

47. $3x + 1$ **49.** $2x - 1$ **51.** $x + 2$ **53.** $a - 3$

55. $x + 24$ **57.** $3x - 2y$ **59.** $3x + 8y$ **61.** $8x + 5y$

63. $15x + 10$ **65.** $8y + 32$ **67.** $15t + 9$

69. $28x + 11$ **71.** $\frac{7}{15}$ **73.** $\frac{29}{35}$ **75.** $\frac{35}{144}$

77. $\frac{949}{1260}$ **79.** 94 **81.** 15 **83.** $14a + 7$

85. $12x + 2$ **87.** $24a + 15$ **89.** $8x + 13$

91. $17x + 14y$ **93.** $17b + 9a$ **95.** $\{1, 2\}$

97. $\left\{ -6, -5.2, 0, 1, 2, 2.3, \frac{9}{2} \right\}$ **99.** $\{ -\sqrt{7}, -\pi, \sqrt{17} \}$

101. $\{0, 1, 2\}$

103. $12 + \frac{1}{4}(12) \overset{?}{=} 15$

$12 + 3 \overset{?}{=} 15$

$15 = 15$

105. 767 **107.** 1.1% **109.** About 85 times more likely

111. a. 2008/09 and 2010/11 **b.** $-\$99$ **c.** Answers will vary

113. c **115.** a

PROBLEM SET 1.3

1. 4 **3.** -4 **5.** -10 **7.** -4 **9.** $\frac{19}{12} = 1\frac{7}{12}$

11. $-\frac{32}{105}$ **13.** -8 **15.** -12 **17.** $-7x$ **19.** 13

21. -14 **23.** $6a$ **25.** -15 **27.** 15 **29.** -24

31. $-10x$ **33.** x **35.** y **37.** 39 **39.** 11

41. -5 **43.** 11 **45.** 7 **47.** -44 **49.** 2

51. $\frac{4}{3}$ **53.** 0 **55.** Undefined **57.** $-\frac{2}{3}$ **59.** 0

61. 6 **63.** 22 **65.** 3 **67.** 7 **69.** 3

71. $-8x + 6$ **73.** $-3a + 4$ **75.** $14x + 12$

77. $7m - 15$ **79.** $-2x + 9$ **81.** $7y + 10$

83. $-20x + 5$ **85.** $-11x + 10$ **87.** $4x + 13$

89. Undefined **91.** 0 **93.** $-\frac{2}{3}$ **95.** 32 **97.** 64

99. $-\frac{1}{18}$ **101.** $\frac{5}{3}$ **103.** 11 **105.** 12

107.

		Sum	Difference	Product	Quotient
a	b	$a + b$	$a - b$	ab	$\frac{a}{b}$
3	12	15	-9	36	$\frac{1}{4}$
-3	12	9	-15	-36	$-\frac{1}{4}$
3	-12	-9	15	-36	$-\frac{1}{4}$
-3	-12	-15	9	36	$\frac{1}{4}$

109.

x	$3(5x - 2)$	$15x - 6$	$15x - 2$
-2	-36	-36	-32
-1	-21	-21	-17
0	-6	-6	-2
1	9	9	13
2	24	24	28

111. a. 1 **b.** $\frac{3}{2}$ **c.** -1 **d.** 135 **113.** 1.5282

115. -0.0794 **117.** -0.0714 **119.** 3.4

121. 1.6 **123.** 1,200 **125.** 190 **127.** $-8, -2$

129. $-64°$F; $-54°$F **131.** -100 feet; -105 feet

133. 3.8 times

135. a. 0.068 stores/mi^2 **b.** 0.034 stores/mi^2 **c.** Georgia

137. d **139.** b

PROBLEM SET 1.4

1. 16 **3.** -16 **5.** -0.027 **7.** 32 **9.** $\frac{1}{8}$

11. $\frac{25}{36}$ **13.** $\frac{1}{10,000}$ **15.** $\frac{25}{36}$ **16.** $\frac{49}{64}$ **17.** $\frac{9}{49}$

19. x^9 **21.** 64 **23.** $-6a^6$ **25.** $16x^4$ **27.** $\frac{1}{9}$

29. $-\frac{1}{32}$ **31.** $\frac{16}{9}$ **33.** 17 **35.** $80x^3y^6$

37. $72x^3y^3$ **39.** $15xy$ **41.** $12x^7y^6$ **43.** $54a^6b^2c^4$

45. $36x^4y^6$ **47.** $2x^3$ **49.** $4x$ **51.** $-4xy^4$

53. $-2x^2y^2$ **55.** $\frac{1}{x^{10}}$ **57.** a^{10} **59.** $\frac{1}{x^6}$ **61.** x^{12}

63. x^{18} **65.** $\frac{1}{x^{22}}$ **67.** $\frac{a^3b^7}{4}$ **69.** $\frac{y^{38}}{x^{16}}$ **71.** $\frac{16y^{16}}{x^8}$

73. x^4y^6 **75.** 1 **77.** 1 **79. a.** 19 **b.** 27 **c.** 27

81. a. 16 **b.** 12 **c.** 18 **83. a.** 144 **b.** 74 **c.** 144

85. a. 23 **b.** 41 **c.** 65 **87.** 3.78×10^5

89. 4.9×10^3 **91.** 3.7×10^{-4} **93.** 4.95×10^{-3}

95. 5,340 **97.** 7,800,000 **99.** 0.00344 **101.** 0.49

103. 8×10^4 **105.** 2×10^9 **107.** 2.5×10^{-6}

109. 1.8×10^{-7} **111.** 50 **113.** 7.9×10^{-5}

115. 6.3×10^8 **117.** 1.003×10^{19} miles **119.** 2.478×10^{13} miles
121. About 2.478×10^{13} minutes or 70,000 years
123. a. 4.204×10^6 **b.** 7.995×10^6 **c.** 6.761×10^6
125. a. 8.0×10^{11} **b.** \$4,444
127.

Number of Bytes		
Unit	Exponential Form	Scientific Notation
Kilobyte	$2^{10} = 1,024$	1.024×10^3
Megabyte	$2^{20} \approx 1,049,000$	1.049×10^6
Gigabyte	$2^{30} \approx 1,074,000,000$	1.074×10^9
Terabyte	$2^{40} \approx 1,099,500,000,000$	1.100×10^{12}

129. a

CHAPTER 1 TEST
1. $\frac{9}{18}$ **2.** $\frac{15}{18}$ **3.** $<$ **4.** $<$ **5.** $>$ **6.** $<$
7. 16 **8.** -9 **9.** $-\frac{2}{7}, \frac{7}{2}$ **10.** $2^2 \cdot 3 \cdot 7^2$ **11.** $\frac{8}{13}$
12. $\frac{6}{17}$ **13.** $12x + 2$ **14.** $10x + 2$ **15.** $10x + 6y$
16. $3a + 10b + 7$ **17.** $2y + 3x + 9$ **18.** $10y - 12x$
19. $\frac{7}{6}$ **20.** $\frac{197}{204}$ **21.** $2x + 11$ **22.** $-2x + 2$
23. $-x + 3$ **24.** 15 **25.** 1 **26.** 11 **27.** $-\frac{1}{27x^{12}}$
28. $64x^6y^3$ **29.** $\frac{81}{25}$ **30.** $35x^5y^2$ **31.** $-\frac{18y^7}{x}$
32. $\frac{6a^3}{b^2}$ **33.** $\frac{3b^5}{4a^2}$ **34.** $27x^9y^3$ **35.** 1.253×10^7
36. 5.2×10^{-3} **37.** 0.00526 **38.** 490,000
39. 1.3×10^1 **40.** 1.7×10^8

Chapter 2

PROBLEM SET 2.1
1. 3 **3.** $-\frac{4}{3}$ **5.** 7,000 **7.** -3 **9.** No solution
11. All real numbers
13. a. $3x - 12$ **b.** $-x$ **c.** 4 **d.** $\frac{4}{3}$
15. a. $\frac{5}{9}$ **b.** 5 **c.** 2 **d.** No solution
17. $-\frac{9}{2}$ **19.** $-\frac{7}{640}$ **21.** $\frac{7}{10}$ **23.** 24 **25.** $\frac{46}{15}$
27. $\frac{10}{9}$ **29.** 11 **31.** -3 **33.** 5 **35.** 4
37. 6,000 **39.** $\frac{37}{18}$ **41.** $-\frac{2}{7}$ **43.** No solution
45. No solution **47.** All real numbers
49. 30,000 **51.** d **53.** a **55.** $\frac{5}{2}$ **57.** 1.25
59. $\frac{5}{2}$ **61.** $\frac{5}{4}$ **63.** 3

PROBLEM SET 2.2
1. -3 **3.** 0 **5.** $\frac{3}{2}$ **7.** 4 **9.** $\frac{8}{5}$ **11.** $-\frac{7}{640}$
13. $c = 2$ **15.** 3 **17.** 2,400 **19.** $\frac{25}{3}$ **21.** $\frac{14}{3}$
23. $r = \frac{d}{t}$ **25.** $t = \frac{d}{r+c}$ **27.** $\ell = \frac{A}{w}$ **29.** $t = \frac{I}{pr}$
31. $T = \frac{PV}{nR}$ **33.** $x = \frac{y-b}{m}$ **35.** $F = \frac{9}{5}C + 32$
37. $v = \frac{h - 16t^2}{t}$ **39.** $d = \frac{A-a}{n-1}$ **41.** $y = -\frac{2}{3}x + 2$
43. $y = \frac{3}{5}x + 3$ **45.** $y = \frac{1}{3}x + 2$ **47.** $x = \frac{5}{a-b}$
49. $h = \frac{S - \pi r^2}{2\pi r}$ **51.** $x = \frac{4}{3}y - 4$ **53.** $x = -\frac{10}{a-c}$
55. $y = \frac{1}{2}x + \frac{3}{2}$ **57.** $y = -2x - 5$ **59.** $y = -\frac{2}{3}x + 1$

61. $y = -\frac{1}{2}x + \frac{7}{2}$
63. a. $y = 4x - 1$ **b.** $y = -\frac{1}{2}x$ **c.** $y = -3$
65. $y = -\frac{1}{4}x + 2$ **67.** $y = \frac{3}{5}x - 3$
69. a. $-\frac{15}{4} = -3.75$ **b.** -7 **c.** $y = \frac{4}{5}x + 4$ **d.** $x = \frac{5}{4}y - 5$
71. $\frac{9}{5}$ tons **73.** 6 miles per hour
75. 42 miles per hour **77.** 6.8 feet per second **79.** 35
81. 13,330 KB **83.** 8
85. Shar: 128.4 beats per min, Sara: 140.4 beats per min
87. b **89.** c **91.** $2x - 3$ **93.** $x + y = 180$
95. 30 **97.** 8.5 **99.** 6,000

PROBLEM SET 2.3
1. 10 feet by 20 feet **3.** 7 feet **5.** 5 inches
7. 4 meters **9.** \$92.00 **11.** \$200.00 **13.** \$86.47
15. \$99.6 million **17.** $20°, 160°$
19. a. $20.4°, 69.6°$ **b.** $38.4°, 141.6°$
21. $27°, 72°, 81°$ **23.** $102°, 44°, 34°$
25. Base angles: $43°$; third angle: $94°$
27. \$6,000 at 8%; \$3,000 at 9%
29. \$5,000 at 12%; \$10,000 at 10%
31. \$4,000 at 8%; \$2,000 at 9%
33. 16 mph for Allegra, 20 mph for Eliana
35.

Speed (miles per hour)	Distance (miles)
20	10
30	15
40	20
50	25
60	30
70	35

37.

Time (hours)	Distance Upstream (miles)	Distance Downstream (miles)
1	6	14
2	12	28
3	18	42
4	24	56
5	30	70
6	36	84

39. 30 fathers, 45 sons **41.** \$54 **43.** 44 minutes

45.

x	2,000	3,000	4,000	5,000
y	160	240	320	400

47.

Hot Coffee Sales	
Year	Sales (billions of dollars)
2005	7
2006	7.5
2007	8
2008	8.6
2009	9.2

49.

w (ft)	l (ft)	A (ft²)
2	22	44
4	20	80
6	18	108
8	16	128
10	14	140
12	12	144

51.

Age (years)	Maximum Heart Rate (beats per minute)
18	202
19	201
20	200
21	199
22	198
23	197

53.

Resting Heart Rate (beats per minute)	Training Heart Rate (beats per minute)
60	144
62	144.8
64	145.6
68	147.2
70	148
72	148.8

55. b **57.** a
59. $<$ **61.** $>$ **63.** -5 **65.** 6

PROBLEM SET 2.4

1. $\{x \mid x < 6\}, (-\infty, 6)$
3. $\{x \mid x \geq -1\}, [-1, \infty)$

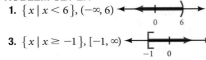

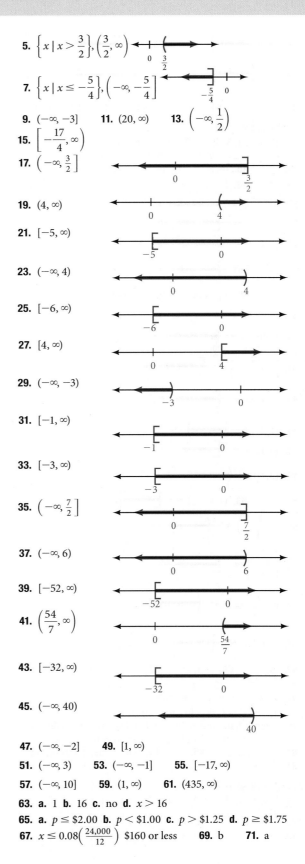

5. $\left\{x \mid x > \dfrac{3}{2}\right\}, \left(\dfrac{3}{2}, \infty\right)$

7. $\left\{x \mid x \leq -\dfrac{5}{4}\right\}, \left(-\infty, -\dfrac{5}{4}\right]$

9. $(-\infty, -3]$ **11.** $(20, \infty)$ **13.** $\left(-\infty, \dfrac{1}{2}\right)$

15. $\left[-\dfrac{17}{4}, \infty\right)$

17. $\left(-\infty, \dfrac{3}{2}\right]$

19. $(4, \infty)$

21. $[-5, \infty)$

23. $(-\infty, 4)$

25. $[-6, \infty)$

27. $[4, \infty)$

29. $(-\infty, -3)$

31. $[-1, \infty)$

33. $[-3, \infty)$

35. $\left(-\infty, \dfrac{7}{2}\right]$

37. $(-\infty, 6)$

39. $[-52, \infty)$

41. $\left(\dfrac{54}{7}, \infty\right)$

43. $[-32, \infty)$

45. $(-\infty, 40)$

47. $(-\infty, -2]$ **49.** $[1, \infty)$
51. $(-\infty, 3)$ **53.** $(-\infty, -1]$ **55.** $[-17, \infty)$
57. $(-\infty, 10]$ **59.** $(1, \infty)$ **61.** $(435, \infty)$
63. a. 1 **b.** 16 **c.** no **d.** $x > 16$
65. a. $p \leq \$2.00$ **b.** $p < \$1.00$ **c.** $p > \$1.25$ **d.** $p \geq \$1.75$
67. $x \leq 0.08\left(\dfrac{24{,}000}{12}\right)$ $\$160$ or less   **69.** b   **71.** a

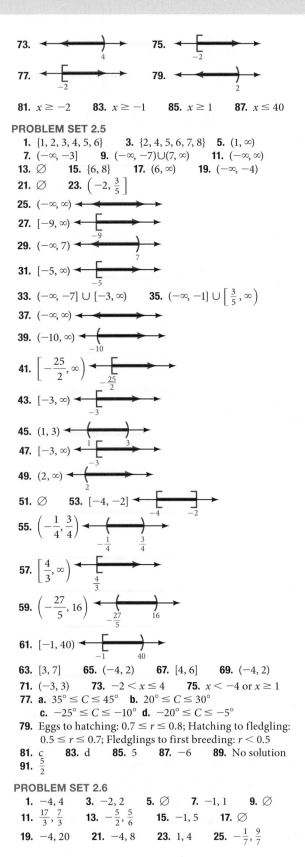

73. **75.**

77. **79.**

81. $x \geq -2$ **83.** $x \geq -1$ **85.** $x \geq 1$ **87.** $x \leq 40$

PROBLEM SET 2.5
1. $\{1, 2, 3, 4, 5, 6\}$ **3.** $\{2, 4, 5, 6, 7, 8\}$ **5.** $(1, \infty)$
7. $(-\infty, -3]$ **9.** $(-\infty, -7) \cup (7, \infty)$ **11.** $(-\infty, \infty)$
13. $\varnothing$ **15.** $\{6, 8\}$ **17.** $(6, \infty)$ **19.** $(-\infty, -4)$
21. $\varnothing$ **23.** $\left(-2, \frac{3}{5}\right]$

25. $(-\infty, \infty)$

27. $[-9, \infty)$

29. $(-\infty, 7)$

31. $[-5, \infty)$

33. $(-\infty, -7] \cup [-3, \infty)$ **35.** $(-\infty, -1] \cup \left[\frac{3}{5}, \infty\right)$

37. $(-\infty, \infty)$

39. $(-10, \infty)$

41. $\left[-\frac{25}{2}, \infty\right)$

43. $[-3, \infty)$

45. $(1, 3)$

47. $[-3, \infty)$

49. $(2, \infty)$

51. $\varnothing$ **53.** $[-4, -2]$

55. $\left(-\frac{1}{4}, \frac{3}{4}\right)$

57. $\left[\frac{4}{3}, \infty\right)$

59. $\left(-\frac{27}{5}, 16\right)$

61. $[-1, 40)$

63. $[3, 7]$ **65.** $(-4, 2)$ **67.** $[4, 6]$ **69.** $(-4, 2)$
71. $(-3, 3)$ **73.** $-2 < x \leq 4$ **75.** $x < -4$ or $x \geq 1$
77. a. $35° \leq C \leq 45°$ **b.** $20° \leq C \leq 30°$
c. $-25° \leq C \leq -10°$ **d.** $-20° \leq C \leq -5°$
79. Eggs to hatching: $0.7 \leq r \leq 0.8$; Hatching to fledgling:
$0.5 \leq r \leq 0.7$; Fledglings to first breeding: $r < 0.5$
81. c   **83.** d   **85.** 5   **87.** -6 **89.** No solution
91. $\frac{5}{2}$

PROBLEM SET 2.6
1. $-4, 4$ **3.** $-2, 2$ **5.** $\varnothing$ **7.** $-1, 1$ **9.** $\varnothing$
11. $\frac{17}{3}, \frac{7}{3}$ **13.** $-\frac{5}{2}, \frac{5}{6}$ **15.** $-1, 5$ **17.** $\varnothing$
19. $-4, 20$ **21.** $-4, 8$ **23.** $1, 4$ **25.** $-\frac{1}{7}, \frac{9}{7}$

27. $-3, 12$ **29.** -4 **31.** $\frac{2}{3}, -\frac{10}{3}$ **33.** $\varnothing$
35. $\frac{3}{2}, -1$ **37.** $-\frac{2}{3}$ **39.** $5, 25$ **41.** $-30, 26$
43. $-12, 28$ **45.** $-2, 0$ **47.** $-\frac{1}{2}, \frac{7}{6}$ **49.** $0, 15$
51. $-\frac{23}{7}, -\frac{11}{7}$ **53.** $-5, \frac{3}{5}$ **55.** $1, \frac{1}{9}$ **57.** $-\frac{1}{2}$
59. 0 **61.** $-\frac{1}{6}, -\frac{7}{4}$ **63.** All real numbers
65. All real numbers   **67.** $-\frac{3}{10}, \frac{3}{2}$ **69.** $-\frac{1}{10}, -\frac{3}{5}$
71. a. $\frac{5}{4} = 1.25$ **b.** $\frac{5}{4} = 1.25$ **c.** 2 **d.** $\frac{1}{2}, 2$ **e.** $\frac{1}{3}, 4$
73. 1987 and 1995   **75.** b   **77.** c   **79.** $1 < x$
81. $a \leq \frac{2}{3}$ **83.** $t \geq 3$

PROBLEM SET 2.7
1. $(-3, 3)$ **3.** $(-\infty, -2] \cup [2, \infty)$ **5.** $(-3, 3)$
7. $(-\infty, -7) \cup (7, \infty)$ **9.** $\varnothing$
11. All real numbers, $(-\infty, \infty)$ **13.** $(-4, 10)$
15. $(-\infty, -9] \cup [-1, \infty)$ **17.** $\varnothing$
19. All real numbers, $(-\infty, \infty)$
21. $\varnothing$
23. $-1 < x < 5$

25. $y \leq -5$ or $y \geq -1$

27. $k \leq -5$ or $k \geq 2$

29. $-1 < x < 7$

31. $a \leq -2$ or $a \geq 1$

33. $-6 < x < \frac{8}{3}$

35. $\varnothing$ **37.** $[-2, 8]$ **39.** $\left(-2, \frac{4}{3}\right)$
41. $(-\infty, -5] \cup [-3, \infty)$ **43.** $\left(-\infty, -\frac{7}{2}\right) \cup \left(-\frac{3}{2}, \infty\right)$
45. $\left[-1, \frac{11}{5}\right]$ **47.** $\left(\frac{5}{3}, 3\right)$
49. $x < 2$ or $x > 8$

51. $x \leq -3$ or $x \geq 12$

53. $x < 2$ or $x > 6$

55. $0.99 < x < 1.01$ **57.** $x \leq -\frac{3}{5}$ or $x \geq -\frac{2}{5}$
59. $\frac{5}{9} \leq x \leq \frac{7}{9}$ **61.** $x < -\frac{2}{3}$ or $x > 0$
63. $x \leq \frac{2}{3}$ or $x \geq 2$ **65.** $-\frac{1}{6} \leq x \leq \frac{3}{2}$
67. $-0.05 < x < 0.25$ **69.** $|x| \leq 4$ **71.** $|x - 5| \leq 1$

73. a. 3 **b.** $\left\{-2, \frac{4}{5}\right\}$ **c.** no **d.** $x < -2$ or $x > \frac{4}{5}$

75. $|x - 65| \leq 10$ **77.** d **79.** a **81.** $\frac{1}{9}$ **83.** $\frac{3x^2}{y^2}$

85. $\frac{x^7}{y^{12}}$ **87.** 5.4×10^4 **89.** 6,440 **91.** 1.2×10^4

CHAPTER 2 TEST

1. 28 **2.** -3 **3.** $-\frac{7}{4}$ **4.** 2 **5.** $w = \frac{P - 2l}{2}$

6. $B = \frac{2A - hb}{h}$ **7.** $y = \frac{5x - 10}{2}$ **8.** $y = 3x - 7$

9. 6 inches by 12 inches **10.** $55°, 125°$

11. $[-6, \infty)$

12. $(-\infty, 4)$

13. $(-\infty, 6)$

14. $[-52, \infty)$

15. $\{1, 2, 3, 4, 6\}$ **16.** $(-\infty, 4]$ **17.** $\{2\}$ **18.** $(-\infty, -1)$

19. All real numbers **20.** $(-\infty, -3)$ **21.** $\{-10, -2\}$

22. $\{2, 6\}$ **23.** $\varnothing$ **24.** $\left\{-\frac{3}{4}, -\frac{1}{6}\right\}$

25. $x < -1$ or $x > \frac{4}{3}$

26. $-\frac{2}{3} \leq x \leq 4$

27. All real numbers

28. $\varnothing$

Chapter 3

PROBLEM SET 3.1

1.

3. A. $(4, 1)$ **B.** $(-4, 3)$ **C.** $(-2, -5)$ **D.** $(2, -2)$
E. $(0, 5)$ **F.** $(-4, 0)$ **G.** $(1, 0)$

5. b **7.** b

9.

11.

13.

15.

17.

19. $(2, 0), (0, -3)$

21. $(0, 4), (2, 0)$

23. $(0, 0)$

25. $(0, -4), (2, 0)$

27. $(0, 1), (-2, 0)$

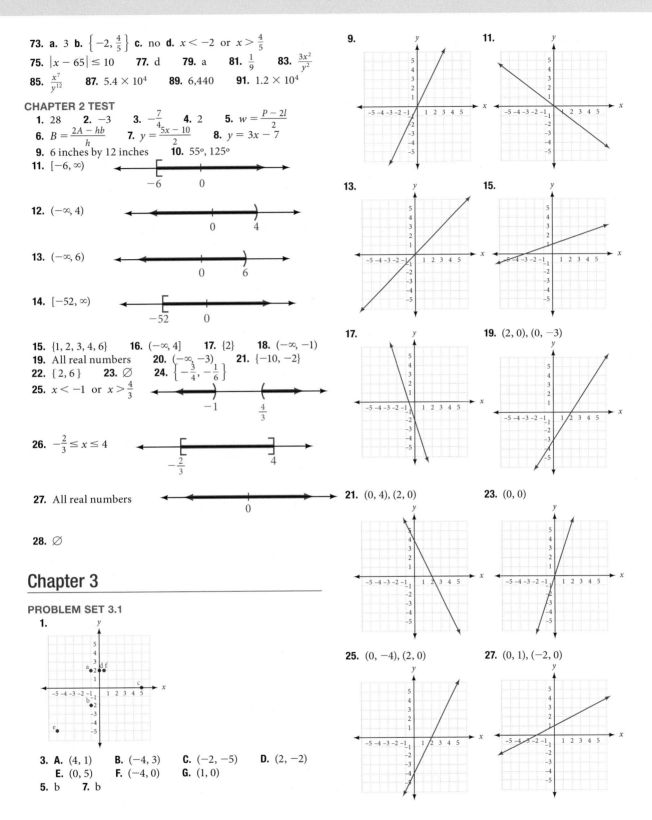

29. $(0, -3), (3, 0)$ **31.** $\left(0, \frac{5}{4}\right), \left(\frac{5}{2}, 0\right)$

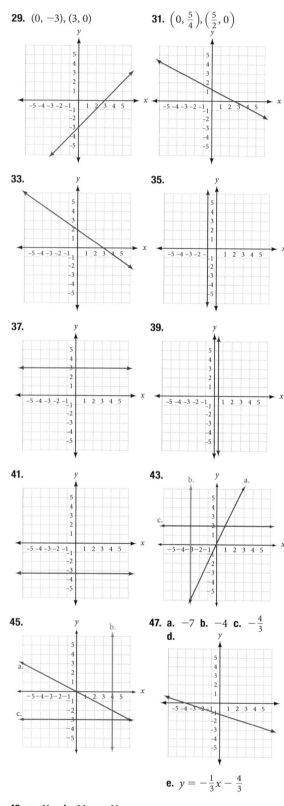

33.

35.

37.

39.

41.

43.

45.

47. a. -7 **b.** -4 **c.** $-\frac{4}{3}$
d.

e. $y = -\frac{1}{3}x - \frac{4}{3}$

49. a. Yes **b.** No **c.** Yes

51.

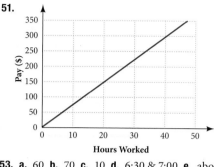

Hours Worked

53. a. 60 **b.** 70 **c.** 10 **d.** 6:30 & 7:00 **e.** about 22 minutes
55. a. $y = -19.7x + 701$ **b.** $543.40 **c.** 10 **d.** $504
57. a   **59.** b   **61.** $-\frac{6}{100}$ **63.** 1   **65.** $-\frac{4}{3}$
67. undefined   **69. a.** $\frac{3}{2}$ **b.** $-\frac{3}{2}$

PROBLEM SET 3.2

1. $\frac{3}{2}$ **3.** No slope   **5.** $\frac{2}{3}$ **7.** Slope $= \frac{3}{2}$
9. Slope $= -\frac{1}{2}$ **11.** Slope $= \frac{5}{3}$ **13.** Slope $= -3$
15. Slope $= -\frac{3}{2}$ **17.** Slope $= 0$ **19.** No slope
21. $a = 5$ **23.** $b = 2$

25. $m = -\frac{2}{3}$ **27.** $m = \frac{2}{3}$

x	y
0	2
3	0

x	y
0	-5
3	-3

29. $m = \frac{2}{3}$ **31.** Undefined   **33.** Parallel
35. Neither   **37.** Perpendicular   **39.** $\frac{1}{5}$ **41.** 0
43. -1 **45.** $-\frac{3}{2}$ **47. a.** Yes **b.** No   **49.** 17.5 mph
51. 120 ft/sec
53. a. 10 minutes **b.** 20 minutes **c.** 20°C per minute
   d. 10°C per minute **e.** 1st minute
55. 1,333 solar thermal collector shipments/year. Between
   1997 and 2006 the number of solar thermal collector
   shipments increased an average of 1,333 shipments per
   year.
57. a. .05 watts/lumen gained. For every additional lumen,
   the incandescent light bulb needs to use on average an
   extra .05 watts.
   b. .014 watts/lumen gained. For every additional lumen,
   the energy efficient light bulb needs to use on average an
   extra .014 watts.
   c. Energy efficient light bulb. Answers may vary.
59. d   **61.** a   **63.** -1 **65.** 2   **67.** $y = mx + b$
69. $y = -2x - 5$ **71.** 5

PROBLEM SET 3.3

1. $y = -4x - 3$ **3.** $y = -\frac{2}{3}x$ **5.** $y = -\frac{2}{3}x + \frac{1}{4}$
7. a. 3 **b.** $-\frac{1}{3}$ **9. a.** -3 **b.** $\frac{1}{3}$ **11. a.** $-\frac{2}{5}$ **b.** $\frac{5}{2}$
13. Slope $= 3$, y-intercept $= -2$, perpendicular slope $= -\frac{1}{3}$
15. Slope $= \frac{2}{3}$, y-intercept $= -4$, perpendicular slope $= -\frac{3}{2}$
17. Slope $= -\frac{4}{5}$, y-intercept $= 4$, perpendicular slope $= \frac{5}{4}$
19. Slope $= \frac{1}{2}$, y-intercept $= -4$, $y = \frac{1}{2}x - 4$

21. Slope $= -\frac{2}{3}$, y-intercept $= 3$, $y = -\frac{2}{3}x + 3$

23. $y = 2x - 1$ **25.** $y = -\frac{1}{2}x - 1$ **27.** $y = -3x + 1$

29. $y = \frac{2}{3}x + \frac{14}{3}$ **31.** $y = -\frac{1}{4}x - \frac{13}{4}$ **33.** $y = 6$

35. $3x + 5y = -1$ **37.** $x - 12y = -8$ **39.** $6x - 5y = 3$

41. $y = 2$ **43.** $3x = -4$ **45.** $(0, -4)$, $(2, 0)$; $y = 2x - 4$

47. $(-2, 0)$, $(0, 4)$; $y = 2x + 4$

49. a. x: $\frac{10}{3}$, y: -5 **b.** $(4, 1)$, answers may vary

    c. $y = \frac{3}{2}x - 5$ **d.** no

51. a. 2 **b.** $\frac{3}{2}$ **c.** -3 **d.**

    e. $y = 2x - 3$

53. a. Slope $= \frac{1}{2}$; x-intercept $= 0$; y-intercept $= 0$

    b. No slope; x-intercept $= 3$; y-intercept $=$ none

    c. Slope $= 0$; x-intercept $=$ none; y-intercept $= -2$

55. $3x - y = -7$ **57.** $5x + 2y = -26$ **59.** $x - 4y = -1$

61. $2x + 3y = 6$ **63. a.** Answers will vary **b.** $86°$

65. a. Answers will vary **b.** 15.4 cm

67. a. \$190,000 **b.** \$19 **c.** \$6.50

69. $y = 7.7x - 15{,}334.6$ **71.** c   **73.** d

75. $(0, 0)$, $(4, 0)$ **77.** $(0, 0)$, $(2, 0)$

PROBLEM SET 3.4

13.

15.

17. $y > -x + 4$ **19.** $y \le \frac{1}{2}x + 2$

21.

23.

25.

27.

29.

31.

33.

35.

37.

39.

1.

3.

5.

7.

9.

11.

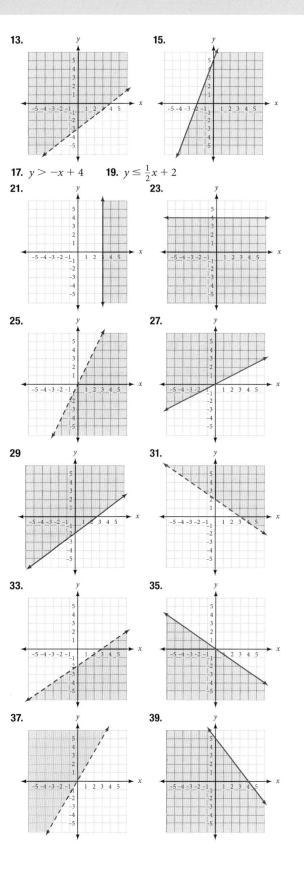

41.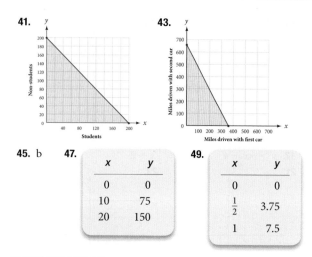

43.

c.

45. b   **47.**

x	y
0	0
10	75
20	150

49.

x	y
0	0
$\frac{1}{2}$	3.75
1	7.5

d. Domain = $\{x \mid 10 \le x \le 40\}$;
Range = $\{y \mid 85 \le y \le 340\}$

e. Minimum = \$85; Maximum = \$340

41. Domain = $\{2004, 2005, 2006, 2007, 2008, 2009, 2010\}$;
Range = $\{680, 730, 800, 900, 920, 990, 1030\}$

43. a. III **b.** I **c.** II **d.** IV   **45.** b and c   **47.** b

49. 113   **51.** −9   **53. a.** 6 **b.** 7.5   **55. a.** 27 **b.** 6

57. 1   **59.** −3   **61.** $-\frac{6}{5}$ **63.** $-\frac{35}{32}$

PROBLEM SET 3.5

1. Domain = $\{1, 3, 5, 7\}$; Range = $\{2, 4, 6, 8\}$; a function
3. Domain = $\{0, 1, 2, 3\}$; Range = $\{4, 5, 6\}$; a function
5. Domain = $\{a, b, c, d\}$; Range = $\{3, 4, 5\}$; a function
7. Domain = $\{a\}$; Range = $\{1, 2, 3, 4\}$; not a function
9. Yes   **11.** No   **13.** No   **15.** Yes   **17.** Yes
19. Yes
21. Domain = $\{x \mid -3 \le x \le 1\}$; Range = $\{y \mid -2 \le y \le 4\}$
23. Domain = $\{x \mid -5 \le x \le 3\}$, Range = $\{y \mid y = 3\}$
25. Domain = $\{x \mid -5 \le x \le 5\}$, Range = $\{y \mid 0 \le y \le 5\}$
27. Domain = $\{-4, -3, -1, 0, 2, 5\}$, Range = $\{-4, -1, 1, 3, 4\}$
29. Domain = All real numbers, Range = All real numbers,
A function
31. Domain = $\{4\}$, Range = All real numbers, Not a function
33. Domain = All real numbers, Range = $\{y \mid y \le 0\}$, A function
35. Domain = All real numbers, Range = $\{y \mid y \ge -1\}$, A function
37. Domain = $\{x \mid x \le 3\}$, Range = All real numbers; Not a function
39. a. $y = 8.5x$ for $10 \le x \le 40$
b.

PROBLEM SET 3.6

1. $(3, 8)$ **3.** $(-1, 5)$ **5.** $f(4) = 0$ **7.** $f(-1) = 9$
9. $f(10) = 0.1$ **11.** $f\left(-\frac{1}{5}\right) = -\frac{1}{10}$ **13.** −1
15. −11   **17.** 2   **19.** 4   **21.** $a^2 + 3a + 4$
23. $2a + 7$ **25.** 1   **27.** −9   **29.** 8
31. 0   **33.** $3a^2 - 4a + 1$ **35.** $3a^2 + 8a + 5$
37. $\frac{3}{10}$ **39.** $\frac{2}{5}$ **41.** Undefined   **43.** −1   **45.** −11
47. 99   **49.** 4   **51.** 0   **53.** 2   **55.** −2   **57.** 2
59. 1   **61.** 15   **63.** −1   **65.** 2   **67.** 0   **69.** −1
71. a. $a^2 - 7$ **b.** $a^2 - 6a + 5$ **c.** $x^2 - 2$ **d.** $x^2 + 4x$
e. $a^2 + 2ab + b^2 - 4$ **f.** $x^2 + 2xh + h^2 - 4$
g. $9x^2 - 4$ **h.** $3x^2 - 12$
73. 4   **75.** 1, 4, 9   **77.** $f(x) = -\frac{1}{3}x + 2$
79. $V(3) = 300$, the painting is worth \$300 in 3 years;
$V(6) = 600$, the painting is worth \$600 in 6 years.
81. a. True **b.** False **c.** True **d.** False **e.** True
83. a. 3.5 cm **b.** 4 lbs **c.**

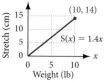

85. a. \$5,625 **b.** \$1,500 **c.** $\{t \mid 0 \le t \le 5\}$
d.

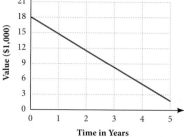

TABLE 7   Weekly Wages

Hours Worked	Function Rule	Gross Pay (\$)
x	y = 8.5x	y
10	y = 8.5(10)	85
20	y = 8.5(20)	170
30	y = 8.5(30)	255
40	y = 8.5(40)	340

e. $\{V(t) \mid 1,500 \le V(t) \le 18,000\}$ **f.** About 2.42 years
87. c   **89.** a   **91.** $-.1x^2 + 35x$ **93.** $4x^2 - 7x + 3$
95. $-.1x^2 + 27x - 500$ **97.** $2x^2 + 8x + 8$ **99.** 10

PROBLEM SET 3.7

1. 15 **3.** 98 **5.** $\frac{3}{2}$ **7.** 1 **9.** 40 **11.** 147

13. $6x + 2$ **15.** $-2x + 8$ **17.** $8x^2 + 14x - 15$

19. $\frac{2x + 5}{4x - 3}$ **21.** $4x - 7$ **23.** $3x^2 - 10x + 8$

25. $-2x + 3$ **27.** $3x^2 - 11x + 10$

29. $9x^3 - 48x^2 + 85x - 50$ **31.** $x - 2$

33. $\frac{1}{x - 2}$ **35.** $3x^2 - 7x + 3$ **37.** $6x^2 - 22x + 20$

39. a. 17 **b.** $x^2 + 7x - 1$ **c.** 17

41. a. 48 **b.** $15x^2 + 2x - 8$ **c.** 48 **43.** -3 **45.** 38

47. 81 **49.** 6 **51. a.** 81 **b.** 29 **c.** $(x + 4)^2$ **d.** $x^2 + 4$

53. a. -2 **b.** -1 **c.** $16x^2 + 4x - 2$ **d.** $4x^2 + 12x - 1$

55. $(f \circ g)(x) = 5\left[\frac{x + 4}{5}\right] - 4$
$$= x + 4 - 4$$
$$= x$$
$$(g \circ f)(x) = \frac{(5x - 4) + 4}{5}$$
$$= \frac{5x}{5}$$
$$= x$$

57. -6 **59.** -5 **61.** 12 **63.** $-\frac{1}{2}$ **65.** 0

67. -4 **69.** 5 **71.** 3 **73.** -3

75. 0 **77.** $\frac{1}{4}$ **79.** 4 **81.** 1 **83.** 3

85. a. $R(x) = 11.5x - 0.05x^2$ **b.** $C(x) = 2x + 200$
c. $P(x) = -0.05x^2 + 9.5x - 200$ **d.** $\overline{C}(x) = 2 + \frac{200}{x}$

87. a. $M(x) = 220 - x$ **b.** $M(24) = 196$ **c.** 142
d. 135 **e.** 128

89. c **91.** d **93.** 196 **95.** 4 **97.** 1.6 **99.** 3

101. 2,400

PROBLEM SET 3.8

1. Inverse, $K = 3$ **3.** Direct, $K = 2\pi$ **5.** Joint, $K = \frac{1}{2}$

7. Direct, $K = 0.5$ **9.** $z = K\sqrt{x}$ **11.** $F = \frac{Km^2}{d}$

13. $A = Kh(a + b)$ **15.** 30 **17.** -6 **19.** 40

21. $\frac{81}{5}$ **23.** 64 **25.** 108 **27.** 300 **29.** ± 2

31. 1600 **33.** ± 8 **35.** $\frac{50}{7}$ pounds

37. a. $T = 4P$ **b.**

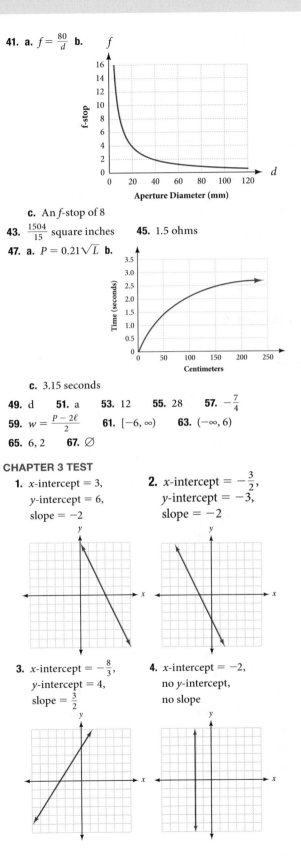

c. 70 pounds per square inch

39. 12 pounds per square inch

41. a. $f = \frac{80}{d}$ **b.**

c. An f-stop of 8

43. $\frac{1504}{15}$ square inches **45.** 1.5 ohms

47. a. $P = 0.21\sqrt{L}$ **b.**

c. 3.15 seconds

49. d **51.** a **53.** 12 **55.** 28 **57.** $-\frac{7}{4}$

59. $w = \frac{P - 2\ell}{2}$ **61.** $[-6, \infty)$ **63.** $(-\infty, 6)$

65. 6, 2 **67.** $\varnothing$

CHAPTER 3 TEST

1. x-intercept $= 3$,
y-intercept $= 6$,
slope $= -2$

2. x-intercept $= -\frac{3}{2}$,
y-intercept $= -3$,
slope $= -2$

3. x-intercept $= -\frac{8}{3}$,
y-intercept $= 4$,
slope $= \frac{3}{2}$

4. x-intercept $= -2$,
no y-intercept,
no slope

5. $y = 2x + 5$ **6.** $y = -\frac{3}{7}x + \frac{5}{7}$ **7.** $y = \frac{2}{5}x - 5$

8. $y = -\frac{1}{3}x - \frac{7}{3}$ **9.** $x = 4$

10. **11.**

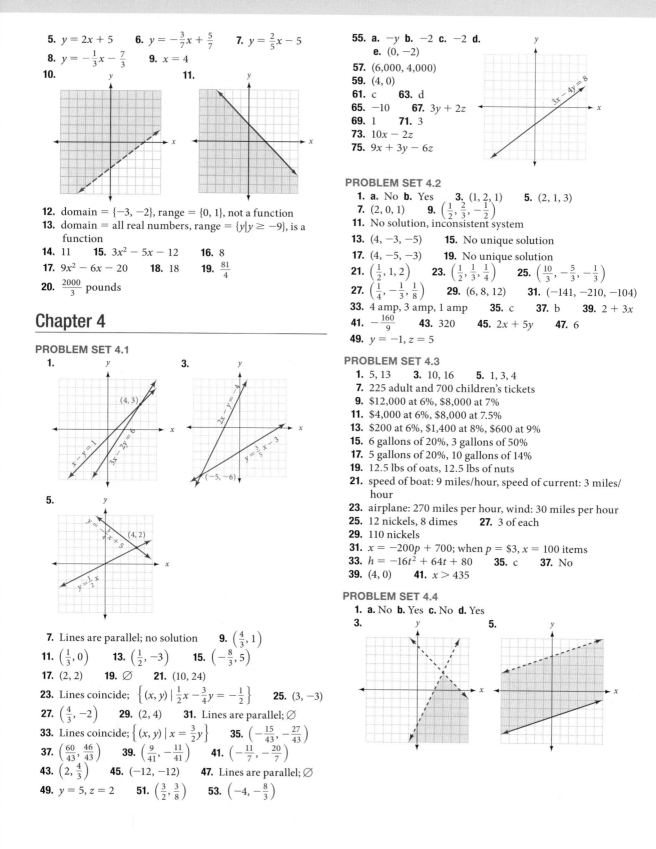

12. domain = $\{-3, -2\}$, range = $\{0, 1\}$, not a function

13. domain = all real numbers, range = $\{y | y \geq -9\}$, is a function

14. 11   **15.** $3x^2 - 5x - 12$ **16.** 8

17. $9x^2 - 6x - 20$ **18.** 18   **19.** $\frac{81}{4}$

20. $\frac{2000}{3}$ pounds

Chapter 4

PROBLEM SET 4.1

1. **3.** **5.**

7. Lines are parallel; no solution   **9.** $\left(\frac{4}{3}, 1\right)$

11. $\left(\frac{1}{3}, 0\right)$ **13.** $\left(\frac{1}{2}, -3\right)$ **15.** $\left(-\frac{8}{3}, 5\right)$

17. $(2, 2)$ **19.** $\varnothing$ **21.** $(10, 24)$

23. Lines coincide; $\left\{(x, y) \mid \frac{1}{2}x - \frac{3}{4}y = -\frac{1}{2}\right\}$ **25.** $(3, -3)$

27. $\left(\frac{4}{3}, -2\right)$ **29.** $(2, 4)$ **31.** Lines are parallel; $\varnothing$

33. Lines coincide; $\left\{(x, y) \mid x = \frac{3}{2}y\right\}$ **35.** $\left(-\frac{15}{43}, -\frac{27}{43}\right)$

37. $\left(\frac{60}{43}, \frac{46}{43}\right)$ **39.** $\left(\frac{9}{41}, -\frac{11}{41}\right)$ **41.** $\left(-\frac{11}{7}, -\frac{20}{7}\right)$

43. $\left(2, \frac{4}{3}\right)$ **45.** $(-12, -12)$ **47.** Lines are parallel; $\varnothing$

49. $y = 5, z = 2$ **51.** $\left(\frac{3}{2}, \frac{3}{8}\right)$ **53.** $\left(-4, -\frac{8}{3}\right)$

55. a. $-y$ **b.** -2 **c.** -2 **d.**
   e. $(0, -2)$

57. $(6,000, 4,000)$

59. $(4, 0)$

61. c   **63.** d

65. -10 **67.** $3y + 2z$

69. 1   **71.** 3

73. $10x - 2z$

75. $9x + 3y - 6z$

PROBLEM SET 4.2

1. a. No **b.** Yes   **3.** $(1, 2, 1)$ **5.** $(2, 1, 3)$

7. $(2, 0, 1)$ **9.** $\left(\frac{1}{2}, \frac{2}{3}, -\frac{1}{2}\right)$

11. No solution, inconsistent system

13. $(4, -3, -5)$ **15.** No unique solution

17. $(4, -5, -3)$ **19.** No unique solution

21. $\left(\frac{1}{2}, 1, 2\right)$ **23.** $\left(\frac{1}{2}, \frac{1}{3}, \frac{1}{4}\right)$ **25.** $\left(\frac{10}{3}, -\frac{5}{3}, -\frac{1}{3}\right)$

27. $\left(\frac{1}{4}, -\frac{1}{3}, \frac{1}{8}\right)$ **29.** $(6, 8, 12)$ **31.** $(-141, -210, -104)$

33. 4 amp, 3 amp, 1 amp   **35.** c   **37.** b   **39.** $2 + 3x$

41. $-\frac{160}{9}$ **43.** 320   **45.** $2x + 5y$ **47.** 6

49. $y = -1, z = 5$

PROBLEM SET 4.3

1. 5, 13   **3.** 10, 16   **5.** 1, 3, 4

7. 225 adult and 700 children's tickets

9. \$12,000 at 6%, \$8,000 at 7%

11. \$4,000 at 6%, \$8,000 at 7.5%

13. \$200 at 6%, \$1,400 at 8%, \$600 at 9%

15. 6 gallons of 20%, 3 gallons of 50%

17. 5 gallons of 20%, 10 gallons of 14%

19. 12.5 lbs of oats, 12.5 lbs of nuts

21. speed of boat: 9 miles/hour, speed of current: 3 miles/hour

23. airplane: 270 miles per hour, wind: 30 miles per hour

25. 12 nickels, 8 dimes   **27.** 3 of each

29. 110 nickels

31. $x = -200p + 700$; when $p = \$3, x = 100$ items

33. $h = -16t^2 + 64t + 80$ **35.** c   **37.** No

39. $(4, 0)$ **41.** $x > 435$

PROBLEM SET 4.4

1. a. No **b.** Yes **c.** No **d.** Yes

3. **5.**

7. **9.**

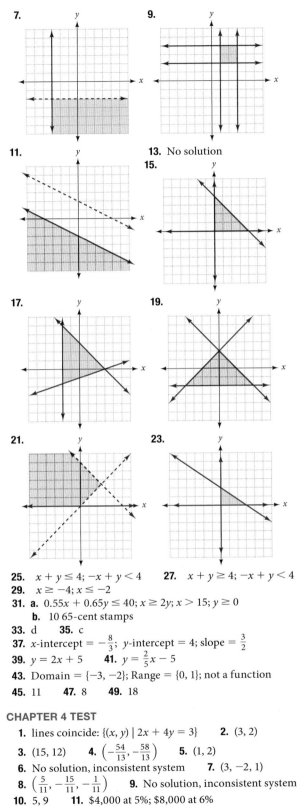

11. **13.** No solution

15.

17. **19.**

21. **23.**

25. $x + y \leq 4$; $-x + y < 4$ **27.** $x + y \geq 4$; $-x + y < 4$

29. $x \geq -4$; $x \leq -2$

31. a. $0.55x + 0.65y \leq 40$; $x \geq 2y$; $x > 15$; $y \geq 0$
   b. 10 65-cent stamps

33. d **35.** c

37. x-intercept $= -\frac{8}{3}$; y-intercept $= 4$; slope $= \frac{3}{2}$

39. $y = 2x + 5$ **41.** $y = \frac{2}{5}x - 5$

43. Domain $= \{-3, -2\}$; Range $= \{0, 1\}$; not a function

45. 11 **47.** 8 **49.** 18

CHAPTER 4 TEST

1. lines coincide: $\{(x, y) \mid 2x + 4y = 3\}$ **2.** (3, 2)

3. (15, 12) **4.** $\left(-\frac{54}{13}, -\frac{58}{13}\right)$ **5.** (1, 2)

6. No solution, inconsistent system **7.** (3, -2, 1)

8. $\left(\frac{5}{11}, -\frac{15}{11}, -\frac{1}{11}\right)$ **9.** No solution, inconsistent system

10. 5, 9 **11.** $4,000 at 5%; $8,000 at 6%

12. 340 adult, 410 children

13. 4 gallons of 30%, 12 gallons of 70%
14. boat: 8 mph; current: 2 mph
15. 11 nickels, 3 dimes, 1 quarter
16.

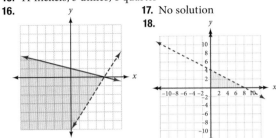

17. No solution
18.

Chapter 5

PROBLEM SET 5.1

1. Trinomial, degree 2, leading coefficient 5
3. Binomial, degree 1, leading coefficient 3
5. Trinomial, degree 5, leading coefficient 8
7. Polynomial, degree 3, leading coefficient 4
9. Monomial, degree 0, leading coefficient $-\frac{3}{4}$
11. Trinomial, degree 3, leading coefficient 6
13. $2x^2 + 7x - 15$ **15.** $12a^2 - 7ab - 10b^2$
17. $7x + 1$ **19.** $x^2 - 13x + 3$
21. $\frac{1}{4}x^2 - \frac{7}{12}x - \frac{1}{4}$ **23.** $-y^3 - y^2 - 4y + 7$
25. $2x^3 + x^2 - 3x - 17$ **27.** $\frac{1}{15}x^2 + \frac{1}{7}xy + \frac{5}{7}y^2$
29. $-3a^3 + 6a^2b - 5ab^2$ **31.** $3x^2 - 12xy$
33. $14a^2 - 2ab + 8b^2$ **35.** $-3x$ **37.** $17x^5 - 12$
39. $2 - x$ **41.** $10x - 5$ **43.** $9x - 35$
45. $9y - 4x$ **47.** $9a + 2$ **49.** -2
51. a. 208 **b.** 103 **53. a.** 51 **b.** -15
55. a. 110 **b.** -120 **57. a.** $-5,000$ **b.** 3,000
59. $3(43) + 2(23) + 54 = 229$ milligrams
61. 240 feet; 240 feet
63. $P(x) = -300 + 40x - 0.5x^2$; $P(60) = 300
65. $P(x) = -800 + 3.5x - 0.002x^2$; $P(1,000) = 700
67. $[6 + 0.05(x - 400)] = 6 + 0.05(592 - 400) = 15.60
69. a **71.** a **73.** $12a^2 - 7ab - 10b^2$
75. $3x^3 - 13x^2 + 8x + 12$ **77.** $-20x^3$ **79.** $15x^4$
81. a^8 **83.** 190

PROBLEM SET 5.2

1. $12x^3 - 10x^2 + 8x$ **3.** $-3a^5 + 18a^4 - 21a^2$
5. $2a^5b - 2a^3b^2 + 2a^2b^4$ **7.** $x^2 - 2x - 15$
9. $6x^4 - 19x^2 + 15$ **11.** $x^3 + 9x^2 + 23x + 15$
12. $x^3 - 7x^2 + 17x - 14$ **13.** $a^3 - b^3$ **15.** $8x^3 + y^3$
17. $2a^3 - a^2b - ab^2 - 3b^3$ **19.** $x^2 + x - 6$
21. $6a^2 + 13a + 6$ **23.** $20 - 2t - 6t^2$
25. $x^6 - 2x^3 - 15$ **27.** $20x^2 - 9xy - 18y^2$
29. $18t^2 - \frac{2}{9}$
31. a. $2x - 4$ **b.** $6x - 2$ **c.** $8x^2 - 2x - 3$
33. $25x^2 + 20xy + 4y^2$ **35.** $25 - 30t^3 + 9t^6$
37. $4a^2 - 9b^2$ **39.** $9r^4 - 49s^2$ **41.** $y^2 + 3y + \frac{9}{4}$
43. $a^2 - a + \frac{1}{4}$ **45.** $x^2 + \frac{1}{2}x + \frac{1}{16}$
47. $t^2 + \frac{2}{3}t + \frac{1}{9}$ **49.** $\frac{1}{9}x^2 - \frac{4}{25}$
51. $x^3 - 6x^2 + 12x - 8$ **53.** $x^3 - \frac{3}{2}x^2 + \frac{3}{4}x - \frac{1}{8}$

55. $3x^3 - 18x^2 + 33x - 18$ **57.** $a^2b^2 + b^2 + 8a^2 + 8$

59. $3x^2 + 12x + 14$ **61.** $24x$ **63.** $x^2 + 4x - 5$

65. $4a^2 - 30a + 56$ **67.** $32a^2 + 20a - 18$

69. a. $2^4 - 3^4 = -65$ **b.** $(2 - 3)^4 = 1$

    c. $(2^2 + 3^2)(2 + 3)(2 - 3) = -65$

71. $R(p) = 900p - 300p^2; R(x) = 3x - \frac{x^2}{300}; \$672$

73. $R(p) = 350p - 10p^2; R(x) = 35x - \frac{x^2}{10}; \$1{,}852.50$

75. $P(x) = -\frac{x^2}{10} + 30x - 500; P(60) = \$940$

77. $A = (10 + 2x)(40 + 2x) - 400; x = 5, A = 600 \text{ ft}^2;$
$x = 8, A = 1{,}056 \text{ ft}^2$

79. $V = (9 - 2x)(12 - 2x)x; x = 2, V = 80 \text{ in}^3$

81. $A = 100 + 400r + 600r^2 + 400r^3 + 100r^4$

83. $\$8.64$

85. 20 years ago: $\$1{,}101$, 30 years ago: $\$510$

87. b   **89.** c   **91.** 2   **93.** $-2x^2y^2$ **95.** 185.12

97. $4x^3 - 8x^2$ **99.** $4x^3 - 6x - 20$ **101.** $-3x + 9$

PROBLEM SET 5.3

1. $2x^2 - 4x + 3$ **3.** $-2x^2 - 3x + 4$

5. $2y^2 + \frac{5}{2} - \frac{3}{2y^2}$ **7.** $-\frac{5}{2}x + 4 + \frac{3}{x}$ **9.** $4ab^3 + 6a^2b$

11. $-xy + 2y^2 + 3xy^2$ **13.** $x - 7 + \frac{7}{x+2}$

15. $2x + 5 + \frac{2}{3x-4}$ **17.** $2x^2 - 5x + 1 + \frac{4}{x+1}$

19. $y^2 - 3y - 13$ **21.** $x - 3$

23. $3y^2 + 6y + 8 + \frac{37}{2y-4}$

25. $a^3 + 2a^2 + 4a + 6 + \frac{17}{a-2}$ **27.** $y^3 + 2y^2 + 4y + 8$

29. $x^2 - 2x + 1$ **31.** $x - 7 + \frac{20}{x+2}$ **33.** $3x - 1$

35. $x^2 + 4x + 11 + \frac{26}{x-2}$ **37.** $3x^2 + 8x + 26 + \frac{83}{x-3}$

38. $2x^2 - x - 1$ **39.** $2x^2 + 2x + 3$

41. $x^3 - 4x^2 + 18x - 72 + \frac{289}{x+4}$

43. $x^4 + x^2 - x - 3 - \frac{5}{x-2}$

45. $x + 2 + \frac{3}{x-1}$ **47.** $x^3 - x^2 + x - 1$

49. $x^2 + x + 1$ **51.** same

53. a.

x	1	5	10	15	20
$C(x)$	3.10	3.50	4.00	4.50	5.00

    b. $\overline{C}(x) = \frac{3}{x} + 0.10$

    c.

x	1	5	10	15	20
$\overline{C}(x)$	3.10	1.70	0.40	0.30	0.25

    d. It decreases.

55. a. $T(100) = \$11.95, T(400) = \$32.95, T(500) = \$39.95$

    b. $\frac{4.95}{m} + 0.07$

    c. $\overline{T}(100) = 0.1195, \overline{T}(400) = 0.0824, \overline{T}(500) = 0.0799$

57. c   **59.** b   **61.** $8a^2$ **63.** -48 **65.** $2a^3b$

67. $-3b^2$ **69.** $8a^3 - 8a^2 - 48a$ **71.** $5x + 5y + x^2 + xy$

PROBLEM SET 5.4

1. $5x^2(2x - 3)$ **3.** $9y^3(y^3 + 2)$ **5.** $3ab(3a - 2b)$

7. $7xy^2(3y^2 + x)$ **9.** $3(a^2 - 7a + 11)$

11. $4x(x^2 - 4x + 5)$ **13.** $10x^2y^2(x^2 + 2xy + 3y^2)$

---

15. $xy(-x + y - xy)$ **17.** $2xy^2z(2x^2 - 4xz + 3z^2)$

19. $5abc(4abc - 6b + 5ac)$ **21.** $(a - 2b)(5x - 3y)$

23. $3(x + y)^2(x^2 - 2y^2)$ **25.** $(x + 5)(2x^2 + 7x + 8)$

27. $(x + 1)(3y + 2a)$ **29.** $(xy + 1)(x + 3)$

31. $(x - 2)(3y^2 + 4)$ **33.** $(x - a)(x - b)$

35. $(b + 5)(a - 1)$ **37.** $(b^2 + 1)(a^4 - 5)$

39. $(x + 3)(x^2 + 4)$ **41.** $(x + 2)(x^2 + 25)$

43. $(2x + 3)(x^2 + 4)$ **45.** $(x + 3)(4x^2 + 9)$ **47.** 6

49. $P(1 + r) + P(1 + r)r$
$= (1 + r)(P + Pr)$
$= (1 + r)P(1 + r)$
$= P(1 + r)^2$

51. $\$28.50$ **53.** b   **55.** $3x^2(x^2 - 3xy - 6y^2)$

57. $(x - 3)(2x^2 - 4x - 3)$ **59.** $3x^2 + 5x - 2$

61. $3x^2 - 5x + 2$ **63.** $x^2 + 5x + 6$ **65.** $6y^2 + y - 35$

67. $20 - 19a + 3a^2$ **69.** $25 - 4x^2$

71.

Two Numbers a and b	Their Product ab	Their Sum $a + b$
1, −24	−24	−23
−1, 24	−24	23
2, −12	−24	−10
−2, 12	−24	10
3, −8	−24	−5
−3, 8	−24	5
4, −6	−24	−2
−4, 6	−24	2

73. 9, 4   **75.** 8, −5   **77.** −12, 4

PROBLEM SET 5.5

1. $(x + 3)(x + 4)$ **3.** $(x + 3)(x - 4)$

5. $(y + 3)(y - 2)$ **7.** $(2 - x)(8 + x)$

9. $(2 + x)(6 + x)$ **11.** $(4 + x)(4 - x)$

13. $(x + 2y)(x + y)$ **15.** $(a + 6b)(a - 3b)$

17. $(x - 8a)(x + 6a)$ **19.** $(x - 6b)^2$ **21.** $3(a - 2)(a - 5)$

23. $4x(x - 5)(x + 1)$ **25.** $3(x - 3y)(x + y)$

27. $2a^3(a^2 + 2ab + 2b^2)$ **29.** $10x^2y^2(x + 3y)(x - y)$

31. $(2x - 3)(x + 5)$ **33.** $(2x - 5)(x + 3)$

35. $(2x - 3)(x - 5)$ **37.** $(2x - 5)(x - 3)$

39. Prime   **41.** $(2 + 3a)(1 + 2a)$ **43.** $15(4y + 3)(y - 1)$

45. $x^2(3x - 2)(2x + 1)$ **47.** $10r(2r - 3)^2$

49. $(4x + y)(x - 3y)$ **51.** $(2x - 3a)(5x + 6a)$

53. $(3a + 4b)(6a - 7b)$ **55.** $200(1 + 2t)(3 - 2t)$

57. $y^2(3y - 2)(3y + 5)$ **59.** $2a^2(3 + 2a)(4 - 3a)$

61. $x^2y^2(8x + y)(x - 6y)$ **63.** $(5a^2 + 3)(4a^2 + 5)$

65. $2(1 + 3r^2)(1 - 5r^2)$ **67.** $(x + 2)(2x + 3)(x + 5)$

69. $(x + 1)(2x + 3)(x + 2)$ **71.** $(4x + 3)(x + 5)(x + 6)$

73. $(5x + 1)(2x - 7)(x + 4)$ **75.** $(4x + 3)(6x + 5)(x - 6)$

77. $(2x - 5)(7x - 2)(3x + 4)$

79. $(3x - 4)(5x + 6)(4x - 5)$ **81.** $49x^2 - 4y^2$

83. $a - 100$ **85.** $18x - 5$ **87.** $9x + 8$ **89.** $9x - 5$

91. $y = 3(3x - 1)(x + 4); y = 0$ when $x = \frac{1}{3}$ or $x = -4$,
$y = 168$ when $x = 3$

93. b   **95.** $\frac{2}{3}$ **97.** x^4 **99.** $9y^2$ **101.** $\frac{1}{3}$

103. x^4 **105.** $5y$ **107.** $10x$

PROBLEM SET 5.6

1. $(x - 3)^2$ **3.** $(a - 6)^2$ **5.** $(5 - t)^2$ **7.** $\left(\frac{1}{3}x + 3\right)^2$

9. $(2y^2 - 3)^2$ **11.** $(4a + 5b)^2$ **13.** $\left(\frac{1}{5} + \frac{1}{4}t^2\right)^2$

15. $\left(y + \frac{3}{2}\right)^2$ **17.** $\left(a - \frac{1}{2}\right)^2$ **19.** $\left(x - \frac{1}{4}\right)^2$

21. $\left(t + \frac{1}{3}\right)^2$ **23.** $4(2x - 3)^2$ **25.** $3a(5a + 1)^2$

27. $(x + 2 + 3)^2 = (x + 5)^2$ **29.** $(x + 3)(x - 3)$

31. $(7x + 8y)(7x - 8y)$ **33.** $\left(2a + \frac{1}{2}\right)\left(2a - \frac{1}{2}\right)$

35. $\left(x + \frac{3}{5}\right)\left(x - \frac{3}{5}\right)$ **37.** $(3x + 4y)(3x - 4y)$

39. $10(5 + t)(5 - t)$ **41.** $(x^2 + 9)(x + 3)(x - 3)$

43. $3(x^2 + 9)(x + 3)(x - 3)$ **45.** $(4a^2 + 9)(2a + 3)(2a - 3)$

47. $\left(\frac{1}{9} + \frac{y^2}{4}\right)\left(\frac{1}{3} + \frac{y}{2}\right)\left(\frac{1}{3} - \frac{y}{2}\right)$

49. $\left(\frac{x^2}{4} + \frac{4}{9}\right)\left(\frac{x}{2} + \frac{2}{3}\right)\left(\frac{x}{2} - \frac{2}{3}\right)$

51. $\left(a^2 + \frac{9}{16}\right)\left(a + \frac{3}{4}\right)\left(a - \frac{3}{4}\right)$

53. $(x - y)(x + y)(x^2 + xy + y^2)(x^2 - xy + y^2)$

55. $2a(a - 2)(a + 2)(a^2 + 2a + 4)(a^2 - 2a + 4)$

57. $(x + 1)(x - 5)$ **59.** $y(y + 8)$

61. $(x - 5 + y)(x - 5 - y)$ **63.** $(a + 4 + b)(a + 4 - b)$

65. $(x + y + a)(x + y - a)$ **67.** $(x + 3)(x + 2)(x - 2)$

69. $(x + 2)(x + 5)(x - 5)$ **71.** $(2x + 3)(x + 2)(x - 2)$

73. $(x + 3)(2x + 3)(2x - 3)$ **75.** $(2x - 15)(2x + 5)$

77. $(a - 3 - 4b)(a - 3 + 4b)$

79. $(a - 3 - 4b)(a - 3 + 4b)$ **81.** $(x + 4)(x - 3)^2$

83. $(x - y)(x^2 + xy + y^2)$ **85.** $(a + 2)(a^2 - 2a + 4)$

87. $(3 + x)(9 - 3x + x^2)$ **89.** $(y - 1)(y^2 + y + 1)$

91. $10(r - 5)(r^2 + 5r + 25)$

93. $(4 + 3a)(16 - 12a + 9a^2)$

95. $(2x - 3y)(4x^2 + 6xy + 9y^2)$

97. $\left(t + \frac{1}{3}\right)\left(t^2 - \frac{1}{3}t + \frac{1}{9}\right)$ **99.** $\left(3x - \frac{1}{3}\right)\left(9x^2 + x + \frac{1}{9}\right)$

101. $(4a + 5b)(16a^2 - 20ab + 25b^2)$

103. 30 and -30 **105.** 81   **107.** a   **109.** d

111. $2ab^3(b^2 + 4b + 1)$ **113.** $(2x - 3)(2x + a)$

115. $(3x - 2)(5a + 4)$ **117.** $(x + 2)(x - 2)$

119. $(A + 5)(A - 5)$ **121.** $(x - 3)^2$ **123.** $(x + 4y)^2$

125. $(3a - 4)(2a - 1)$ **127.** $(6x + 5)(2x - 7)$

129. $(x + 2)(x^2 - 2x + 4)$ **131.** $(2x - 3)(4x^2 + 6x + 9)$

PROBLEM SET 5.7

1. $(x + 9)(x - 9)$ **3.** $(x - 3)(x + 5)$

5. $(x + 2)(x + 3)^2$ **7.** $(x^2 + 2)(y^2 + 1)$

9. $2ab(a^2 + 3a + 1)$ **11.** Does not factor, prime

13. $3(2a + 5)(2a - 5)$ **15.** $(3x - 2y)^2$ **17.** $(5 - t)^2$

19. $4x(x^2 + 4y^2)$ **21.** $2y(y + 5)^2$

23. $a^4(a + 2b)(a^2 - 2ab + 4b^2)$

25. $(t + 3 + x)(t + 3 - x)$ **27.** $(x + 5)(x + 3)(x - 3)$

29. $5(a + b)^2$ **31.** Does not factor, prime

33. $3(x + 2y)(x + 3y)$ **35.** $\left(3a + \frac{1}{3}\right)^2$

37. $(x - 3)(x - 7)^2$ **39.** $(x + 8)(x - 8)$

41. $(2 - 5x)(4 + 3x)$ **43.** $a^5(7a + 3)(7a - 3)$

45. $\left(r + \frac{1}{5}\right)\left(r - \frac{1}{5}\right)$ **47.** Does not factor, prime

49. $100(x - 3)(x + 2)$ **51.** $a(5a + 3)(5a + 1)$

53. $(3x^2 + 1)(x^2 - 5)$ **55.** $3a^2b(2a - 1)(4a^2 + 2a + 1)$

57. $(4 - r)(16 + 4r + r^2)$ **59.** $5x^2(2x + 3)(2x - 3)$

61. $100(2t + 3)(2t - 3)$ **63.** $2x^3(4x - 5)(2x - 3)$

65. $(y + 1)(y - 1)(y^2 - y + 1)(y^2 + y + 1)$

67. $2(5 + a)(5 - a)$ **69.** $3x^2y^2(2x + 3y)^2$

71. $(x - 2 + y)(x - 2 - y)$ **73.** $\left(a - \frac{2}{3}b\right)^2$

75. $\left(x - \frac{2}{5}y\right)^2$ **77.** $\left(a - \frac{5}{6}b\right)^2$ **79.** $\left(x - \frac{4}{5}y\right)^2$

81. $(2x - 3)(x - 5)(x + 2)$ **83.** $(x - 4)^3(x - 3)$

85. $2(y - 3)(y^2 + 3y + 9)$

87. $2(a - 4b)(a^2 + 4ab + 16b^2)$

89. $2(x + 6y)(x^2 - 6xy + 36y^2)$ **91.** b

93. $2x^2 + 2x + 1$ **95.** $t^2 - 4t + 3$

97. $(x - 6)(x + 4)$ **99.** $x(2x + 1)(x - 3)$

101. $(x + 2)(x - 3)(x + 3)$ **103.** $(x + 2)(x^2 - 5)$

105. 6   **107.** $-\frac{1}{2}$

PROBLEM SET 5.8

1. $6, -1$ **3.** $\frac{1}{3}, -4$ **5.** $\frac{2}{3}, \frac{3}{2}$ **7.** $5, -5$

9. $-4, \frac{5}{2}$ **11.** $0, \frac{4}{3}$ **13.** $-\frac{1}{5}, \frac{1}{3}$ **15.** $-10, 0$

17. $-5, 1$ **19.** $1, 2$ **21.** $-2, 3$ **23.** $-2, \frac{1}{4}$

25. $-2, \frac{5}{3}$ **27.** $-1, 9$ **29.** $0, -3$ **31.** $-4, -2$

33. $\frac{1}{2}, 5$ **35.** $\frac{3}{2}, -6$ **37.** $9, 2$ **39.** $0, 2, 3$

41. $0, -3, 7$ **43.** $-3, -2, 2$ **45.** $-2, -5, 5$

47. $0, -\frac{4}{3}, \frac{4}{3}$ **49.** $-\frac{3}{2}, -2, 2$ **51.** $-3, -\frac{3}{2}, \frac{3}{2}$

53. a. $\frac{5}{8}$ **b.** $10x - 8$ **c.** $16x^2 - 34x + 15$ **d.** $\frac{3}{2}, \frac{5}{8}$

55. a. $\frac{25}{9}$ **b.** $-\frac{5}{3}, \frac{5}{3}$ **c.** $-3, 3$ **d.** $\frac{5}{3}$ **57.** $-\frac{3}{2}$

59. $-2, 8$ **61.** -3 **63.** $-7, 1$ **65.** $0, 5$

67. $-1, 8$ **69.** $0, 1$ **71.** $\frac{3}{2}$

73. 9 and 11, or -11 and -9 **75.** 8 and 6

77. 8 and 10   **79.** 7 and 9   **81.** $l = 14$ cm, $w = 9$ cm

83. $b = 16$ m, $h = 10$ m   **85.** $b = 9$ in., $h = 14$in.

87. $x = 5$ **89.** 6, 8, 10

91. 2 meters, 8 meters   **93.** 18 cm, 4 cm

95. 2 yards   **97.** 350 square feet   **99.** 12 feet

101. 0.5 inches   **103.** 1 and 2 seconds

105. 0 and $\frac{3}{2}$ seconds   **107.** 2 and 3 seconds

109. a   **111.** c   **113.** (1, 2)   **115.** (15, 12)

117.   **119.**

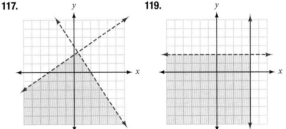

CHAPTER 5 TEST

1. $\frac{6}{5}x^3 - \frac{6}{5}x^2 - \frac{8}{5}x - \frac{6}{5}$ **2.** $-14x - 58$

3. $R(x) = 36x - 0.3x^2$ **4.** $P(x) = -0.3x^2 + 32x - 50$

5. \$600   **6.** \$450   **7.** \$150   **8.** $-5x^2 - 31x + 28$

9. $6x^3 + 14x^2 - 27x + 10$ **10.** $9a^8 - 42a^4 + 49$

11. $4x^2 - 9$ **12.** $3x^3 - 17x^2 - 28x$ **13.** $14x^2 - \frac{1}{14}$

14. $6x^2 + 3xy - 4y^2$ **15.** $x^2 - 4x - 2 + \frac{8}{2x - 1}$

16. $x^2 - 3x - 10$ **17.** $y^2 + y + 1 + \frac{17}{y - 1}$

18. $(x - 1)(x - 5)$ **19.** $3(5x^2 - 4)(x^2 + 3)$

20. $(3x + 2y)(3x - 2y)(9x^2 + 4y^2)$ **21.** $(6x - y)(a + 3b^2)$

22. $\left(y - \frac{1}{3}\right)\left(y^2 + \frac{1}{3}y + \frac{1}{9}\right)$ **23.** $3x^2y^4(x - 3y)(x + 8y)$

24. $(a - b - 6)(a - b + 6)$ **25.** $(2 - x)(2 + x)(4 + x^2)$

26. $-10, -\frac{1}{2}$ **27.** $0, 3$ **28.** $-6, 3$ **29.** $-5, -3, 3$
30. $-3, 5$ **31.** $-5, 4$ **32.** 3 **33.** 5 inches
34. 6 centimeters   **35.** 0 and 4 seconds

Chapter 6

PROBLEM SET 6.1
1. $5, 0$ **3.** $2, \frac{3}{4}$ **5.** $-\frac{6}{5}$ **7.** -2 **9.** 0 **11.** ± 2
13. $-2, 5$ **15.** $1, 2$ **17.** Defined for all real numbers
19. $\frac{x-4}{6}$ **21.** $(a^2+9)(a+3)$ **23.** $\frac{2y+3}{y+1}$
25. $\frac{x-2}{x-1}$ **27.** $\frac{x-3}{x+2}$ **29.** $\frac{x^2-x+1}{x-1}$ **31.** $\frac{-4a}{3}$
33. $\frac{b-1}{b+1}$ **35.** $\frac{7x-3}{7x+5}$ **37.** $\frac{4x+3}{4x-3}$ **39.** $\frac{x+5}{2x-7}$
41. $\frac{a^2-ab+b^2}{a-b}$ **43.** $\frac{2x-2}{x}$ **45.** $\frac{x+3}{y-4}$ **47.** $x+2$
49. $\frac{x^2+2x+4}{x+2}$ **51.** $\frac{4x^2+6x+9}{2x+3}$ **53.** 1
55. Not possible   **57.** -1 **59.** $-(y+6)$
61. $-\frac{3a+1}{3a-1}$ **63.** $-\frac{x-4}{x+2}$ **65.** 3 **67.** $x+a$
69. $x+2$ **71.** $a-3$ **73.** x^2-2x+4
75. a. $\frac{2}{3}$ **b.** $\frac{2}{x}$ **c.** $0, 1$ **d.** -2 **77.** c   **79.** d
81. $\frac{2}{3}$ **83.** $20x^2y^2$ **85.** $72x^4y^5$ **87.** $(x+2)(x-2)$
89. $x^2(x-y)$ **91.** $2(y+1)(y-1)$

PROBLEM SET 6.2
1. $\frac{1}{6}$ **3.** $\frac{9}{4}$ **5.** $\frac{1}{2}$ **7.** $\frac{15y}{x^2}$ **9.** $\frac{b}{a}$ **11.** $\frac{2y^5}{z^3}$
13. $\frac{x+3}{x+2}$ **15.** $y+1$ **17.** $\frac{3(x+4)}{x-2}$ **19.** $\frac{y^2}{xy+1}$
21. $\frac{x^2+9}{x^2-9}$ **23.** $\frac{1}{4}$ **25.** 1 **27.** $-\frac{(a-2)(a+2)}{a-5}$
29. $\frac{9t^2-6t+4}{4t^2-2t+1}$ **31.** $\frac{x+3}{x+4}$ **33.** $-\frac{a-b}{5}$
35. $\frac{5c-1}{3c-2}$ **37.** $\frac{5a-b}{9a^2+15ab+25b^2}$ **39.** 2
41. $x(x-1)(x+1)$ **43.** $\frac{(a+4b)(a-3b)}{(a-4b)(a+5b)}$ **45.** $-\frac{2y-1}{2y-3}$
47. $\frac{(y-2)(y+1)}{(y+2)(y-1)}$ **49.** $\frac{x-1}{x+1}$ **51.** $\frac{x-2}{x+3}$
53. $\frac{w(y-1)}{w-x}$ or $-\frac{w(y-1)}{x-w}$ **55.** $\frac{(m+2)(x+y)}{(2x+y)^2}$ **57.** $3x$
59. $2(x+5)$ **61.** $x-2$ **63.** $-(y-4)$ or $4-y$
65. $(a-5)(a+1)$
67. a. $\frac{5}{21}$ **b.** $\frac{5x+3}{25x^2+15x+9}$ **c.** $\frac{5x-3}{25x^2+15x+9}$ **d.** $\frac{5x+3}{5x-3}$
69. d   **71.** $\frac{2}{3}$ **73.** $\frac{47}{105}$ **75.** $x-7$
77. $(x+1)(x-1)$ **79.** $2(x+5)$
81. $(a-b)(a^2+ab+b^2)$

PROBLEM SET 6.3
1. $\frac{5}{4}$ **3.** $\frac{1}{3}$ **5.** $\frac{41}{24}$ **7.** $\frac{19}{144}$ **9.** $\frac{31}{24}$ **11.** 1
13. -1 **15.** $\frac{1}{x+y}$ **17.** $\frac{x-4}{x-2}$ **19.** 1 **21.** $\frac{5}{x+3}$
23. $\frac{x+3}{x-4}$ **25.** $x(x-3)$ **27.** $(x+4)(x-4)$
29. $(x+3)^2(x-4)$ **31.** $\frac{a^2+2a-3}{a^3}$ **33.** $\frac{1}{2}$
35. $\frac{1}{5}$ **37.** $\frac{7x-6}{x(x-3)}$ **39.** $\frac{5x+13}{(x+2)(x+3)}$ **41.** $\frac{x+3}{2(x+1)}$
43. $\frac{a-b}{a^2+ab+b^2}$ **45.** $\frac{2y-3}{4y^2+6y+9}$ **47.** $\frac{2(2x-3)}{(x-3)(x-2)}$
49. $\frac{1}{2t-7}$ **51.** $\frac{4}{(a-3)(a+1)}$
53. $\frac{-4x^2}{(2x+1)(2x-1)(4x^2+2x+1)}$
55. $\frac{2}{(2x+3)(4x+3)}$ **57.** $\frac{a}{(a+4)(a+5)}$
59. $\frac{x+1}{(x-2)(x+3)}$ **61.** $\frac{x-1}{(x+1)(x+2)}$

63. $\frac{1}{(x+2)(x+1)}$ **65.** $\frac{1}{(x+2)(x+3)}$ **67.** $\frac{4x+5}{2x+1}$
69. $\frac{22-5t}{4-t}$ **71.** $\frac{2x^2+3x-4}{2x+3}$ **73.** $\frac{2x-3}{2x}$ **75.** $\frac{1}{2}$
77. $\frac{3}{x+4}$ **79.** $\frac{2}{(x-4)(x-2)}$
81. a. $\frac{1}{16}$ **b.** $\frac{9}{4}$ **c.** $\frac{13}{24}$ **d.** $\frac{5x+15}{(x-3)^2}$ **e.** $\frac{x+3}{5}$ **f.** $\frac{x-2}{x-3}$
83. $\frac{51}{10} = 5.1$ **85.** $x + \frac{4}{x} = \frac{x^2+4}{x}$
87. $\frac{1}{x} + \frac{1}{x+1} = \frac{2x+1}{x(x+1)}$ **89.** a   **91.** a   **93.** $\frac{6}{5}$
95. $x+2$ **97.** $3-x$ **99.** $(x+2)(x-2)$

PROBLEM SET 6.4
1. $\frac{9}{8}$ **3.** $\frac{2}{15}$ **5.** $\frac{119}{20}$ **7.** $\frac{1}{x+1}$ **9.** $\frac{a+1}{a-1}$
11. $\frac{y-x}{y+x}$ **13.** $\frac{1}{(x+5)(x-2)}$ **15.** $\frac{1}{a^2-a+1}$
17. $\frac{x+3}{x+2}$ **19.** $\frac{a+3}{a-2}$ **21.** $\frac{x-3}{x}$ **23.** $\frac{x+4}{x+2}$
25. $\frac{x-3}{x+3}$ **27.** $\frac{a-1}{a+1}$ **29.** $-\frac{x}{3}$ **31.** $\frac{y^2+1}{2y}$
33. $\frac{-x^2+x-1}{x-1}$ **35.** $\frac{5}{3}$ **37.** $\frac{2x-1}{2x+3}$ **39.** $-\frac{1}{x(x+h)}$
41. $\frac{3c+4a-2b}{5}$ **43.** $\frac{(t-4)(t+1)}{(t+6)(t-3)}$ **45.** $\frac{(5b-1)(b+5)}{2(2b-11)}$
47. $-\frac{3}{2x+14}$ **49.** $2m-9$
51. a. As v approaches 0, the denominator approaches 1
     b. $v = \frac{fs}{h} - s$
53. a   **55.** $xy-2x$ **57.** $3x-18$ **59.** ab
61. $(y+5)(y-5)$ **63.** $x(a+b)$ **65.** 2

PROBLEM SET 6.5
1. $\frac{-35}{3}$ **3.** $-\frac{18}{5}$ **5.** $\frac{36}{11}$ **7.** 2 **9.** 5 **11.** 2
13. $-3, 4$ **15.** $1, -\frac{4}{3}$
17. Possible solution -1, which does not check; $\varnothing$
19. 5 **21.** $-\frac{1}{2}, \frac{5}{3}$ **23.** $\frac{2}{3}$ **25.** 18
27. Possible solution 4, which does not check; $\varnothing$
29. Possible solutions 3 and -4; only -4 checks; -4
31. -6 **33.** -5 **35.** $\frac{53}{17}$
37. Possible solutions 1 and 2; only 2 checks; 2
39. Possible solution 3, which does not check; $\varnothing$
41. $\frac{22}{3}$ **43. a.** $\frac{1}{3}$ **b.** 3 **c.** 9 **d.** 4 **e.** $\frac{1}{3}, 3$
45. a. $\frac{6}{(x-4)(x+3)}$ **b.** $\frac{x-3}{x-4}$ **c.** 5
47. $y = \frac{x-3}{x-1}$ **49.** $y = \frac{1-x}{3x-2}$ **51.** $\lambda = \frac{h}{P}$
53. $l = \frac{wa^2}{2R}$ **55.** $b = \frac{ay}{x}$ **57.** $b = a - \frac{a}{c}$
59. $x = \frac{3b}{a-b}$ **61.** $x = \frac{ab}{a-b}$ **63.** $y = \frac{3xz}{mz+6}$
65. $\frac{24}{5}$ feet   **67.** b   **69.** d   **71.** 2,358
73. 12.3   **75.** 3   **77.** $9, -1$ **79.** 60

PROBLEM SET 6.6
1. $\frac{1}{x} + \frac{1}{3x} = \frac{20}{3}$; $\frac{1}{5}$ and $\frac{3}{5}$ **3.** $x + \frac{1}{x} = \frac{10}{3}$; 3 or $\frac{1}{3}$
5. $\frac{1}{x} + \frac{1}{x+1} = \frac{7}{12}$; 3, 4   **7.** $\frac{7+x}{9+x} = \frac{5}{6}$; 3

9. a. b.

	d	r	t
Upstream	1.5	$5 - x$	$\dfrac{1.5}{5 - x}$
Downstream	3	$5 + x$	$\dfrac{3}{5 + x}$

  c. They are the same. $\dfrac{1.5}{5 - x} = \dfrac{3}{5 + x}$

  d. The speed of the current is $\dfrac{5}{3}$ mph.

11. $\dfrac{8}{x + 2} + \dfrac{8}{x - 2} = 3; 6$ mph

13. a. b.

	d	r	t
Train A	150	$x + 15$	$\dfrac{150}{x + 15}$
Train B	120	x	$\dfrac{120}{x}$

  c. They are the same. $\dfrac{150}{x + 15} = \dfrac{120}{x}$

  d. The speed of train A is 75 mph, and for train B it is 60 mph

15. 540 mph   **17.** 54 mph   **19.** 2.4 hours

21. 90 minutes   **23.** 16 hours   **25.** 15 hours

27. 5.25 minutes

29. $10 = \dfrac{1}{3}\left[\left(x + \dfrac{2}{3}x\right) + \dfrac{1}{3}\left(x + \dfrac{2}{3}x\right)\right]$ $x = \dfrac{27}{2}$

31. a. 30 grams **b.** 3.25 moles   **33.** d   **35.** a

37. 2   **39.** ± 3 **41.** $x + a$ **43.** $\dfrac{5}{2}$ **45.** 5, 6

47. $(2, 0)$ and $(0, -4)$

PROBLEM SET 6.7

1. $g(0) = -3, g(-3) = 0, g(3) = 3, g(-1) = -1,$
$g(1) =$ undefined

3. $h(0) = -3, h(-3) = 3, h(3) = 0, h(-1)$ is undefined,
$h(1) = -1$

5. $\{x \mid x \neq 1\}$ **7.** $\{x \mid x \neq 2\}$ **9.** $\{t \mid t \neq 4, t \neq -4\}$

11. a. 2 **b.** -4 **c.** Undefined **d.** 2   **13. a.** 4 **b.** 4

15. a. 5 **b.** 5   **17. a.** $x + a$ **b.** $2x + h$

19. a. $x + a$ **b.** $2x + h$ **21. a.** $x + a - 3$ **b.** $2x + h - 3$

23. a. $\dfrac{-4}{ax}$ **b.** $\dfrac{-1}{(x + 1)(a + 1)}$ **c.** $-\dfrac{a + x}{a^2 x^2}$

25. a. $-\dfrac{9}{5}, 5$ **b.** $\dfrac{9}{2}$ **c.** no solution

27

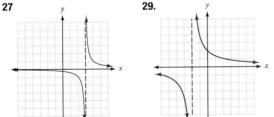

29.

31.

33.

35.

37.

39.

41.

Weeks	Weight (lb)
x	$W(x)$
0	200
1	194.3
4	184
12	173.3
24	168

43. b   **45.** d   **47.** $\dfrac{2}{3a}$ **49.** $(x - 3)(x + 2)$

51. 1   **53.** $\dfrac{3 - x}{x + 3}$ **55.** No solution

CHAPTER 6 TEST

1. $-5, 2$ **2.** Defined for all real numbers   **3.** $x + y$

4. $\dfrac{x - 1}{x + 1}$ **5.** $-\dfrac{1}{2}$ **6.** $-\dfrac{a + 2}{a - 2}$ **7.** $2(a + 4)$

8. $4(a + 3)$ **9.** $x + 3$ **10.** $\dfrac{38}{105}$ **11.** $\dfrac{7}{8}$

12. $\dfrac{1}{a - 3}$ **13.** $\dfrac{3(x - 1)}{x(x - 3)}$ **14.** $\dfrac{x}{(x + 4)(x + 5)}$

15. $\dfrac{x + 4}{(x + 1)(x + 2)}$ **16.** $\dfrac{3a + 8}{3a + 10}$ **17.** $\dfrac{x - 3}{x - 2}$ **18.** $-\dfrac{3}{5}$

19. No solution (3 does not check)   **20.** $\dfrac{3}{13}$ **21.** $-2, 3$

22. -7 **23.** 6 mph   **24.** 15 hours   **25.** $\{x \mid x \neq -2\}$

26. $\{t \mid t \neq 6, t \neq -6\}$ **27.** 4   **28.** $2x + h + 2$

29.

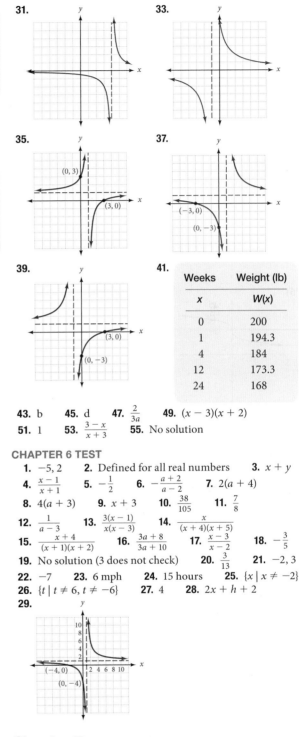

Chapter 7

PROBLEM SET 7.1

1. 12   **3.** Not a real number   **5.** -7 **7.** -3

9. 3   **11.** -2 **13.** 0.2   **15.** 0.2   **17.** $\dfrac{1}{6}$

19. $\dfrac{1}{2}$ **21.** 5   **23.** 2   **25.** 10   **27.** $7x$

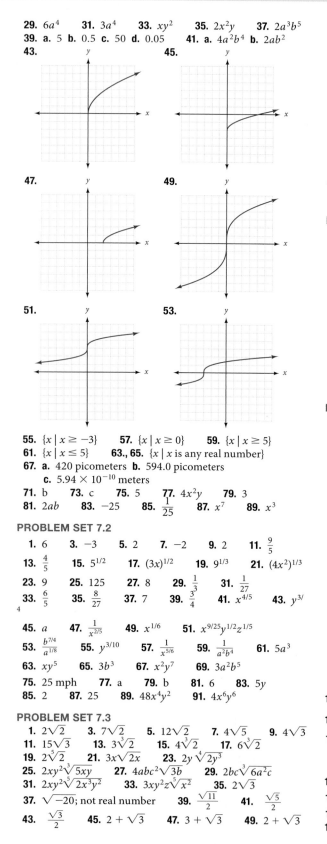

29. $6a^4$ **31.** $3a^4$ **33.** xy^2 **35.** $2x^2y$ **37.** $2a^3b^5$
39. a. 5 **b.** 0.5 **c.** 50 **d.** 0.05 **41. a.** $4a^2b^4$ **b.** $2ab^2$
43. **45.**

47. **49.**

51. **53.**

55. $\{x \mid x \geq -3\}$ **57.** $\{x \mid x \geq 0\}$ **59.** $\{x \mid x \geq 5\}$
61. $\{x \mid x \leq 5\}$ **63., 65.** $\{x \mid x \text{ is any real number}\}$
67. a. 420 picometers **b.** 594.0 picometers
c. 5.94×10^{-10} meters
71. b **73.** c **75.** 5 **77.** $4x^2y$ **79.** 3
81. $2ab$ **83.** -25 **85.** $\frac{1}{25}$ **87.** x^7 **89.** x^3

PROBLEM SET 7.2
1. 6 **3.** -3 **5.** 2 **7.** -2 **9.** 2 **11.** $\frac{9}{5}$
13. $\frac{4}{5}$ **15.** $5^{1/2}$ **17.** $(3x)^{1/2}$ **19.** $9^{1/3}$ **21.** $(4x^2)^{1/3}$
23. 9 **25.** 125 **27.** 8 **29.** $\frac{1}{3}$ **31.** $\frac{1}{27}$
33. $\frac{6}{5}$ **35.** $\frac{8}{27}$ **37.** 7 **39.** $\frac{3}{4}$ **41.** $x^{4/5}$ **43.** $y^{3/4}$
45. a **47.** $\frac{1}{x^{2/5}}$ **49.** $x^{1/6}$ **51.** $x^{9/25}y^{1/2}z^{1/5}$
53. $\frac{b^{7/4}}{a^{1/8}}$ **55.** $y^{3/10}$ **57.** $\frac{1}{x^{5/6}}$ **59.** $\frac{1}{a^2b^4}$ **61.** $5a^3$
63. xy^5 **65.** $3b^3$ **67.** x^2y^7 **69.** $3a^2b^5$
75. 25 mph **77.** a **79.** b **81.** 6 **83.** $5y$
85. 2 **87.** 25 **89.** $48x^4y^2$ **91.** $4x^6y^6$

PROBLEM SET 7.3
1. $2\sqrt{2}$ **3.** $7\sqrt{2}$ **5.** $12\sqrt{2}$ **7.** $4\sqrt{5}$ **9.** $4\sqrt{3}$
11. $15\sqrt{3}$ **13.** $3\sqrt[3]{2}$ **15.** $4\sqrt[3]{2}$ **17.** $6\sqrt[3]{2}$
19. $2\sqrt[5]{2}$ **21.** $3x\sqrt{2x}$ **23.** $2y\sqrt[4]{2y^3}$
25. $2xy^2\sqrt[3]{5xy}$ **27.** $4abc^2\sqrt[3]{3b}$ **29.** $2bc\sqrt[3]{6a^2c}$
31. $2xy^2\sqrt[5]{2x^3y^2}$ **33.** $3xy^2z\sqrt[5]{x^2}$ **35.** $2\sqrt{3}$
37. $\sqrt{-20}$; not real number **39.** $\frac{\sqrt{11}}{2}$ **41.** $\frac{\sqrt{5}}{2}$
43. $\frac{\sqrt{3}}{2}$ **45.** $2 + \sqrt{3}$ **47.** $3 + \sqrt{3}$ **49.** $2 + \sqrt{3}$

51. $\frac{-2 - 3\sqrt{3}}{6}$ **53.** $-2 - \sqrt{2}$ **55.** $\frac{\sqrt{7}}{5}$ **57.** $\frac{\sqrt{5x}}{6}$
59. $\frac{\sqrt[3]{3}}{4}$ **61.** $\frac{\sqrt[3]{2a}}{b}$ **63.** $\frac{\sqrt[3]{9}}{2}$ **65.** $\sqrt{5}$ **67.** $\sqrt[3]{3}$
69. $\frac{2x\sqrt{3}}{5}$ **71.** $\frac{x\sqrt{3x}}{2y^3}$ **73.** $\frac{b\sqrt[3]{15b}}{2a}$ **75.** $\frac{xy^2\sqrt[4]{9x^2y^2}}{2z^2}$
77. $5|x|$ **79.** $3|xy|\sqrt{3x}$ **81.** $|x - 5|$ **83.** $|2x + 3|$
85. $2|a(a + 2)|$ **87.** $2|x|\sqrt{x - 2}$
89. $\sqrt{9 + 16} \overset{?}{=} \sqrt{9} + \sqrt{16}$
$\sqrt{25} \overset{?}{=} 3 + 4$
$5 \neq 7$
91. $5\sqrt{13}$ feet **93.** $\sqrt{2}$
95. a. $\sqrt{2}:1 \approx 1.414:1$ **b.** $5:\sqrt{2}$ **c.** $5:4$ **97.** 1.618
103. b **105.** a **107.** $7x$ **109.** $27xy^2$ **111.** $\frac{5}{6}x$
113. $3\sqrt{2}$ **115.** $5|y|\sqrt{3xy}$ **117.** $2a\sqrt[3]{ab^2}$

PROBLEM SET 7.4
1. $7\sqrt{5}$ **3.** $-x\sqrt{7}$ **5.** $\sqrt[3]{10}$ **7.** $9\sqrt[5]{6}$ **9.** 0
11. $7\sqrt{2} + 3\sqrt{3}$ **13.** $2\sqrt{x} - 12\sqrt{y}$
15. $6\sqrt{3} + \sqrt[3]{3}$ **17.** $\sqrt{5}$ **19.** $-32\sqrt{2}$ **21.** $-3x\sqrt{2}$
23. $-2\sqrt[3]{2}$ **25.** $8x\sqrt[3]{xy^2}$ **27.** $3a^2b\sqrt{3ab}$
29. $11ab\sqrt[3]{3a^2b}$ **31.** $10xy\sqrt[3]{3y}$ **33.** $2\sqrt{3}$ **35.** $\frac{2\sqrt{5}}{3}$
37. $-\frac{\sqrt{x}}{6}$ **39.** $\frac{5\sqrt{2}}{6}$ **41.** $(4x + 15y)\sqrt{2}$
43. $(4x - 3)\sqrt[3]{2}$ **45.** $\sqrt{2} + \sqrt{3} \approx 3.146$; $\sqrt{5} \approx 2.236$
47. $\sqrt{8} + \sqrt{18} \approx 2.828 + 4.243 = 7.071$;
$\sqrt{50} \approx 7.071$; $\sqrt{26} \approx 5.099$
49. $8\sqrt{2x}$ **51.** 5 **53.** a **55.** 6
57. $4x^2 + 3xy - y^2$ **59.** $x^2 + 6x + 9$
61. $x^2 - 4$ **63.** $6\sqrt{2}$ **65.** 6 **67.** $9x$

PROBLEM SET 7.5
1. $3\sqrt{2}$ **3.** $10\sqrt{21}$ **5.** 720 **7.** 54 **9.** $\sqrt{6} - 9$
11. $24 + 6\sqrt[3]{4}$ **13.** $7 + 2\sqrt{6}$ **15.** $x + 2\sqrt{x} - 15$
17. $34 + 20\sqrt{3}$ **19.** $19 + 8\sqrt{3}$ **21.** $x - 6\sqrt{x} + 9$
23. $4a - 12\sqrt{ab} + 9b$ **25.** $x + 4\sqrt{x - 4}$
27. $x - 6\sqrt{x - 5} + 4$ **29.** 1 **31.** $a - 49$
33. $25 - x$ **35.** $x - 8$ **37.** $10 + 6\sqrt{3}$
39. $\sqrt{5}$ **41.** $2\sqrt{5}$ **43.** $\sqrt[3]{4}$ **45.** $\frac{\sqrt{2}}{6}$ **47.** $\frac{2x\sqrt{3}}{3}$
49. $11b\sqrt{2}$ **51.** $\frac{2\sqrt{3}}{3}$ **53.** $\frac{5\sqrt{6}}{6}$ **55.** $\frac{\sqrt{2}}{2}$
57. $\frac{\sqrt{5}}{5}$ **59.** $2\sqrt[3]{4}$ **61.** $\frac{2\sqrt[3]{3}}{3}$ **63.** $\frac{\sqrt[4]{24x^2}}{2x}$
65. $\frac{\sqrt[3]{8y^3}}{y}$ **67.** $\frac{\sqrt[3]{36xy^2}}{3y}$ **69.** $\frac{\sqrt[3]{6xy^2}}{3y}$ **71.** $\frac{3x\sqrt{15xy}}{5y}$
73. $\frac{5xy\sqrt{6xz}}{2z}$ **75. a.** $\frac{\sqrt{2}}{2}$ **b.** $\frac{\sqrt[4]{2}}{2}$ **c.** $\frac{\sqrt[4]{8}}{2}$
77. $\sqrt{2}$ **79.** $\frac{8\sqrt{5}}{15}$ **81.** $\frac{(x - 1)\sqrt{x}}{x}$ **83.** $\frac{3\sqrt{2}}{2}$
85. $\frac{5\sqrt{6}}{6}$ **87.** $\frac{8\sqrt[3]{25}}{5}$ **89.** $\frac{\sqrt{3} + 1}{2}$ **91.** $\frac{5 - \sqrt{5}}{4}$
93. $\frac{x + 3\sqrt{x}}{x - 9}$ **95.** $\frac{10 + 3\sqrt{5}}{11}$ **97.** $\frac{3\sqrt{x} + 3\sqrt{y}}{x - y}$
99. $2 + \sqrt{3}$ **101.** $\frac{11 - 4\sqrt{7}}{3}$
103. a. $2\sqrt{x}$ **b.** $x - 4$ **c.** $x + 4\sqrt{x} + 4$ **d.** $\frac{x + 4\sqrt{x} + 4}{x - 4}$
105. a. $\sqrt{6} + 2\sqrt{2}$ **b.** $2 + 2\sqrt{3}$ **c.** $1 + \sqrt{3}$ **d.** $\frac{-1 + \sqrt{3}}{2}$
107. $(\sqrt[3]{2} + \sqrt[3]{3})(\sqrt[3]{4} - \sqrt[3]{6} + \sqrt[3]{9}) =$
$\sqrt[3]{8} - \sqrt[3]{12} + \sqrt[3]{18} + \sqrt[3]{12} - \sqrt[3]{18} + \sqrt[3]{27}$
$= 2 + 3 = 5$
109. $10\sqrt{3}$ **111.** $x + 6\sqrt{x} + 9$ **113.** 75
115. $\frac{5\sqrt{2}}{4}$ second; $\frac{5}{2}$ second **117.** Answers will vary
119. Answers will vary **121.** a **123.** a
125. $t^2 + 10t + 25$ **127.** x **129.** 7 **131.** $-4, -3$

133. $-6, -3$ **135.** $-5, -2$ **137.** Yes   **139.** No

PROBLEM SET 7.6

1. 4   **3.** $\varnothing$ **5.** 5   **7.** $\varnothing$ **9.** $\frac{39}{2}$ **11.** $\varnothing$

13. 5   **15.** 3   **17.** $-\frac{32}{3}$ **19.** 3, 4

21. $-1, -2$ **22.** Possible solutions -1 and 6; only 6 checks

23. -1 **25.** $\varnothing$ **27.** 7   **29.** 0, 3   **31.** -4 **33.** 8

35. 0   **37.** 9   **39.** 0   **41.** 8

43. Possible solution 9, which does not check; $\varnothing$

45. a. 100 **b.** 40 **c.** $\varnothing$

     d. Possible solutions 5, 8; only 8 checks

47. a. 3 **b.** 9 **c.** 3 **d.** $\varnothing$ **e.** 4 **f.** $\varnothing$

     g. Possible solutions 1, 4; only 4 checks

49. 2   **51.** $\frac{7}{4}$ **53.** $-3, -1$ **55.** $h = 100 - 16t^2$

57. $\frac{392}{121} \approx 3.24$ feet

59. d   **61.** d   **63.** 5   **65.** $2\sqrt{3}$ **67.** -1 **69.** 1

71. 4   **73.** 2   **75.** $10 - 2x$ **77.** $2 - 3x$

79. $6 + 7x - 20x^2$ **81.** $8x - 12x^2$ **83.** $4 + 12x + 9x^2$

85. $4 - 9x^2$

PROBLEM SET 7.7

1. $6i$ **3.** $-5i$ **5.** $6i\sqrt{2}$ **7.** $-2i\sqrt{3}$ **9.** 1

11. -1 **13.** $-i$ **15.** $x = 3, y = -1$

17. $x = -2, y = -\frac{1}{2}$ **19.** $x = -8, y = -5$

21. $x = 7, y = \frac{1}{2}$ **23.** $x = \frac{3}{7}, y = \frac{2}{5}$ **25.** $5 + 9i$

27. $5 - i$ **29.** $2 - 4i$ **31.** $1 - 6i$ **33.** $2 + 2i$

35. $-1 - 7i$ **37.** $6 + 8i$ **39.** $2 - 24i$

41. $-15 + 12i$ **43.** $18 + 24i$ **45.** $10 + 11i$

47. $21 + 23i$ **49.** $4 + 17i$ **51.** $2 - 11i$

53. $-21 + 20i$ **55.** $-2i$ **57.** $-7 - 24i$

59. 5   **61.** 40   **63.** 13   **65.** 164

67. $-13 - 14i$ **69.** $20 - 4i$ **71.** $-3 - 2i$

73. $-\frac{2}{3} + \frac{5}{3}i$ **75.** $\frac{8}{13} + \frac{12}{13}i$ **77.** $-\frac{18}{13} - \frac{12}{13}i$

79. $-\frac{5}{13} + \frac{12}{13}i$ **81.** $\frac{13}{15} - \frac{2}{5}i$ **83.** $R = -11 - 7i$ ohms

85. c   **87.** a   **89.** $-\frac{3}{2}$ **91.** $-3, \frac{1}{2}$ **93.** $\frac{5}{4}$ or $\frac{4}{5}$

CHAPTER 7 TEST

1. -9 **2.** -5 **3.** 7   **4.** -15 **5.** $7x^4$

6. $2x^2y^4$

7.

8.

9. $\{x \mid x \le 9\}$ **10.** $\{x \mid x$ is any real number$\}$

11. $\frac{1}{9}$ **12.** $\frac{7}{5}$ **13.** $a^{5/12}$ **14.** $x^{1/2}$ **15.** $\frac{1}{x^{1/12}y^{7/2}}$

16. $2a$ **17.** $(2a)^{1/3}$ **18.** $6x^3y^9$ **19.** $5xy^2\sqrt{5xy}$

20. $2x^2y^2\sqrt[3]{5xy^2}$ **21.** $\frac{\sqrt{2}}{3}$ **22.** $\frac{2a^2b\sqrt{3b}}{5c^3}$

23. $-6\sqrt{3}$ **24.** $-3ab\sqrt[3]{3}$ **25.** $x + 3\sqrt{x} - 28$

26. $21 - 6\sqrt{6}$ **27.** $\frac{\sqrt{30}}{6}$ **28.** $\frac{\sqrt[3]{6xy^2}}{2y}$

29. $\frac{5 + 5\sqrt{3}}{2}$ **30.** $\frac{x - 2\sqrt{2x} + 2}{x - 2}$

31. 8 (1 does not check)   **32.** -4 **33.** -3

34. $x = \frac{1}{2}, y = 7$ **35.** $6i$ **36.** $17 - 6i$ **37.** $9 - 40i$

38. $-\frac{5}{13} - \frac{12}{13}i$ **39.** $i^{38} = (i^2)^{19} = (-1)^{19} = -1$

Chapter 8

PROBLEM SET 8.1

1. ± 5 **3.** $\pm 3i$ **5.** $\pm\frac{\sqrt{3}}{2}$ **7.** $\pm 2i\sqrt{3}$ **9.** $\pm\frac{3\sqrt{5}}{2}$

11. $\pm\frac{2\sqrt{21}}{3}i$ **13.** $-2, 3$ **15.** $-\frac{3}{2} \pm \frac{3}{2}i$

17. $-\frac{2}{5} \pm \frac{2\sqrt{2}}{5}i$ **19.** $-4 \pm 3\sqrt{3}$ **21.** $\frac{3}{2} \pm i$

23. $36, 6$ **25.** $4, 2$ **27.** $25, 5$ **29.** $\frac{25}{4}, \frac{5}{2}$

31. $\frac{49}{4}, \frac{7}{2}$ **33.** $\frac{1}{16}, \frac{1}{4}$ **35.** $\frac{1}{9}, \frac{1}{3}$ **37.** $-6, 2$

39. $-3, -9$ **41.** $1 \pm 2i$ **43.** $4 \pm \sqrt{15}$

45. $\frac{5 \pm \sqrt{37}}{2}$ **47.** $1 \pm \sqrt{5}$ **49.** $\frac{4 \pm \sqrt{13}}{3}$

51. $\frac{3}{8} \pm \frac{\sqrt{71}}{8}i$ **53.** $\frac{-2 \pm \sqrt{7}}{3}$ **55.** $\frac{5 \pm \sqrt{47}}{2}$

57. $\frac{5}{4} \pm \frac{\sqrt{19}}{4}i$ **59. a.** No **b.** $\pm 3i$

61. a. $0, 6$ **b.** $0, 6$ **63. a.** $-7, 5$ **b.** $-7, 5$ **65.** No

67. a. $\frac{7}{5}$ **b.** 3 **c.** $\frac{7 \pm 2\sqrt{2}}{5}$ **d.** $\frac{71}{5}$ **e.** 3

69. a. $(2x + 1)(2x - 7)$ **b.** $4x^2 - 12x - 7$ **c.** $-\frac{1}{2}, \frac{7}{2}$

     d. $\frac{3}{2} \pm 2i$

71. $\sqrt{2}$ inches   **73.** 781 feet

75. 7.3% to the nearest tenth   **77.** 24 ft   **79.** 2.8 hours

81. $\frac{-3 + \sqrt{209}}{4} \approx 2.9$ ft   **83.** a   **85.** d   **87.** 1

89. 185   **91.** $-\frac{253}{12}$ **93.** $(5t + 1)(25t^2 - 5t + 1)$

PROBLEM SET 8.2

1. $2 \pm \sqrt{3}$ **3.** $\frac{1 \pm \sqrt{41}}{4}$ **5.** $-\frac{2}{3}, \frac{5}{4}$

7. $-\frac{1}{2} \pm \frac{\sqrt{5}}{2}i$ **9.** $\frac{7}{2}$ **11.** $-3 \pm \sqrt{17}$ **13.** $1, 2$

15. $\frac{2}{3} \pm \frac{\sqrt{14}}{3}i$ **17.** $\frac{3 \pm \sqrt{5}}{4}$ **19.** $\frac{1}{6} \pm \frac{\sqrt{47}}{6}i$

21. $4 \pm \sqrt{2}$ **23.** $-\frac{1}{2} \pm \frac{\sqrt{7}}{2}i$ **25.** $1 \pm \sqrt{2}$

27. $\frac{-3 \pm \sqrt{5}}{2}$ **29.** $3, -5$ **31.** $2, -1 \pm i\sqrt{3}$

33. $-\frac{3}{2}, \frac{3}{4} \pm \frac{3\sqrt{3}}{4}i$ **35.** $\frac{1}{5}, -\frac{1}{10} \pm \frac{\sqrt{3}}{10}i$

37. $0, -\frac{1}{2} \pm \frac{\sqrt{5}}{2}i$ **39.** $0, 1 \pm i$ **41.** $0, -\frac{1}{3} \pm \frac{\sqrt{2}}{3}i$

43. a and b   **45. a.** $\frac{5}{3}, 0$ **b.** $\frac{5}{3}, 0$ **47.** No, $2 \pm i\sqrt{3}$

49. Yes   **51.** $-3, -2$ **53.** $0, -5$ **55.** $\pm\frac{3\sqrt{3}}{2}$

57. $0, 5$ **59.** $\frac{-5 \pm \sqrt{73}}{4}$ **61.** 1   **63.** $\frac{1}{2}, 1$

65. $-\frac{1}{2}, 3$ **67.** $\frac{20}{19}, \frac{42}{19}$

69. a. $\pm\frac{3}{2}$ **b.** ± 2 **c.** $\frac{-3 \pm \sqrt{7}}{2}$ **d.** $\frac{1 \pm \sqrt{22}}{7}$

71. 7.4 m, 9.4 m   **73.** 4.5 cm, 13 cm   **75.** $\frac{3120}{121}$

77. a   **79.** 169   **81.** 0   **83.** ± 12 **85.** $x^2 - x - 6$

87. $x^2 - 6x + 9$

PROBLEM SET 8.3

1. $D = 16$, two rational   **3.** $D = 0$, one rational

5. $D = 5$, two irrational   **7.** $D = 17$, two irrational

9. $D = 36$, two rational   **11.** $D = 116$, two irrational

13. ± 10 **15.** ± 12 **17.** 9   **19.** -16 **21.** $\pm 2\sqrt{6}$

23. $x^2 - 7x + 10 = 0$ **25.** $t^2 - 3t - 18 = 0$

27. $y^2 - 4 = 0$ **29.** $2x^2 - 7x + 3 = 0$

31. $4t^2 - 9t - 9 = 0$ **33.** $x^2 - 9 = 0$

35. $10a^2 - a - 3 = 0$ **37.** $9x^2 - 4 = 0$

39. $x^2 - 7 = 0$ **41.** $x^2 + 25 = 0$ **43.** $y^2 - 6y - 2 = 0$

45. $t^2 + 12t + 34 = 0$ **47.** $x^2 - 2x + 2 = 0$

49. $x^2 + 4x + 13 = 0$ **51.** $x^2 - 14x + 52 = 0$

53. $x^2 + 6x + 59 = 0$ **55.** $x^2 - 2x - 9 = 0$

57. $x^2 + 4x + 6 = 0$ **59.** d   **61.** a

63. $x^2 + 4x - 5$ **65.** $4a^2 - 30a + 56$

67. $32a^2 + 20a - 18$ **69.** $\pm\frac{1}{2}$ **71.** $\frac{1}{2}, -\frac{1}{4} \pm \frac{\sqrt{3}}{4}i$

73. No solution   **75.** -64 **77.** 1   **79.** $-2, 4$

81. $-2, \frac{1}{4}$

PROBLEM SET 8.4

1. $1, 2$ **3.** $\frac{7}{2}, 4$ **5.** $\pm 3, \pm i\sqrt{3}$ **7.** $\pm 2i, \pm i\sqrt{5}$

9. $\pm\frac{\sqrt{30}}{6}, \pm i$ **11.** $\pm\frac{\sqrt{21}}{3}, \pm\frac{\sqrt{21}}{3}i$

13. $-1, \frac{1}{2}, -\frac{1}{4} \pm \frac{\sqrt{3}}{4}i, \frac{1}{2} \pm \frac{\sqrt{3}}{2}i$

15. $3, -\frac{3}{2} \pm \frac{3\sqrt{3}}{2}i, 1, -\frac{1}{2} \pm \frac{\sqrt{3}}{2}i$

17. $\pm 1, \pm 3, \pm i, \pm 3i$ **19.** $-\frac{1}{3}, \frac{1}{5}$ **21.** $\pm\frac{1}{3}, \pm\frac{\sqrt{5}}{5}$

23. $-\frac{10}{3}, -\frac{17}{4}$ **25.** $-\frac{3}{5}, -\frac{7}{16}$ **27.** $4, 25$

29. only 25 checks   **31.** $27, 38$ **33.** $4, 12$

35. only $\frac{25}{9}$ checks   **37.** $-\frac{1}{125}, \frac{27}{64}$ **39.** $\pm\frac{1}{8}, \pm 27$

41. $\frac{4}{9}, -\frac{1}{2} \pm \frac{\sqrt{3}}{2}i$ **43.** $4, 36$ **45.** $\pm\sqrt{4 \pm \sqrt{15}}$

47. $\pm\sqrt{-5 \pm \sqrt{3}}$ **49.** only $3 + 2\sqrt{2}$ checks

51. $t = \frac{v \pm \sqrt{v^2 + 64h}}{32}$ **53.** $x = \frac{-4 \pm 2\sqrt{4 - k}}{k}$

55. $x = -y$ **57.** $t = \frac{1 \pm \sqrt{1 + h}}{4}$ **59.** a

61. $4, 1, 0, 1, 4$ **63.** $8, 2, 0, 2, 8$ **65.** $-1, -\frac{1}{4}, 0, -\frac{1}{4}, -1$

67. $0, 1, 4, 9, 16$ **69.** $6, 3, 2, 3, 6$

PROBLEM SET 8.5

1. $27, 12, 3, 0, 3, 12, 27$

3. $-\frac{27}{4}, -3, -\frac{3}{4}, 0, -\frac{3}{4}, -3, -\frac{27}{4}$

5. $36, 25, 16, 9, 4, 1, 0$ **7.** $12, 7, 4, 3, 4, 7, 12$

9. Vertical expansion by 4, upward, narrower

11. Vertical contraction by $\frac{2}{3}$, x-axis reflection, downward, wider

13. Vertical translation upward 4 units

15. Horizontal translation 1 unit right

17. Horizontal translation 2 units right, vertical translation upward 5 units

19. Vertical translation upward 3 units, horizontal translation 6 units left

21. Vertical expansion by 3, horizontal translation 5 units left, vertical translation downward 2 units

23. Vertical translation upward 1 unit, x-axis reflection, vertical expansion by 4, horizontal translation 3 units right

25.   **27.**

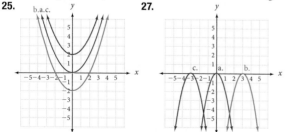

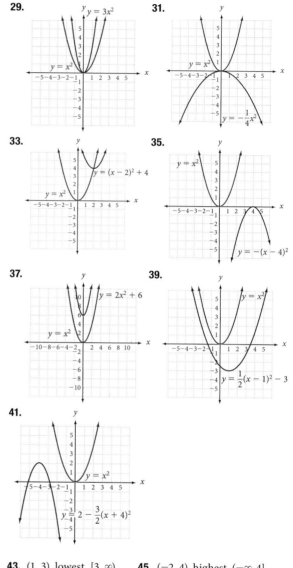

29.   **31.**

33.   **35.**

37.   **39.**

41.

43. $(1, 3)$, lowest, $[3, \infty)$ **45.** $(-2, 4)$, highest, $(-\infty, 4]$

47. $(2, -4)$, lowest, $[-4, \infty)$ **49.** $(4, -1)$, highest, $(-\infty, -1]$

51. $f(x) = (x - 2)^2 - 1$ **53.** $f(x) = -(x - 2)^2 + 4$

55. c   **57.** c   **59.** -2 **61.** $1,322.5$ **63.** $-\frac{7}{640}$

65. $1, 5$ **67.** $-3, 1$ **69.** $\frac{3}{2} \pm \frac{1}{2}i$ **71.** $9, 3$ **73.** $1, 1$

PROBLEM SET 8.6

1. $(3, -4)$, lowest, $[-4, \infty)$ **3.** $(1, 9)$, highest, $(-\infty, 9]$

5. $(2, 16)$, highest, $(-\infty, 16]$

7. $(-4, 16)$, highest, $(-\infty, 16]$ **9.** $(0, -4)$, lowest, $[-4, \infty)$

11. $(1, 11)$, lowest, $[11, \infty)$ **13.** $(-2, -1)$, highest, $(-\infty, -1]$

15. $\left(\frac{1}{2}, 18\right)$, lowest, $[18, \infty)$

17. x-intercepts $= -3, 1$; y-intercept $= -3$; vertex $= (-1, -4)$

19. x-intercepts $= -5, 1$; y-intercept $= 5$; vertex $= (-2, 9)$

21. x-intercepts $= -1, 1$; y-intercept $= -1$; vertex $= (0, -1)$

23. x-intercepts $= -3, 3$; y-intercept $= 9$;
vertex $= (0, 9)$
25. x-intercepts $= -3, 1$; y-intercept $= -3$;
vertex $= (-1, -4)$
27. x-intercepts $= -1, 3$; y-intercept $= -6$;
vertex $= (1, -8)$
29. x-intercepts $= 1 \pm \sqrt{5}$; y-intercept $= -4$;
vertex $= (1, -5)$
31. x-intercepts $= 3 \pm \sqrt{2}$; y-intercept $= -7$;
vertex $= (3, 2)$
33. x-intercepts $= 2 \pm 2\sqrt{2}$; y-intercept $= -4$;
vertex $= (2, -8)$
35. No x-intercepts; y-intercept $= -5$;
vertex $= (1, -4)$
37. No x-intercepts; y-intercept $= 1$;
vertex $= (0, 1)$
39. No x-intercepts; y-intercept $= -3$;
vertex $= (0, -3)$
41. x-intercepts $= -1, -\frac{1}{3}$; y-intercept $= 1$;
vertex $= \left(-\frac{2}{3}, -\frac{1}{3}\right)$
43. x-intercepts $= 1 \pm \sqrt{3}$; y-intercept $= -4$;
vertex $= (1, -6)$
45. No x-intercepts; y-intercept $= -28$;
vertex $= (-3, -1)$
47. x-intercepts $= -3, 0$; y-intercept $= 0$;
vertex $= \left(-\frac{3}{2}, -\frac{9}{4}\right)$
49. x-intercepts $= 0, 4$; y-intercept $= 0$; vertex $= (2, 8)$
51. x-intercepts $= -\frac{1}{3}, \frac{3}{2}$; y-intercept $= -3$;
vertex $= \left(\frac{7}{12}, -\frac{121}{24}\right)$
53. x-intercepts $= \frac{-1 \pm \sqrt{41}}{4}$; y-intercept $= 5$;
vertex $= \left(-\frac{1}{4}, \frac{41}{8}\right)$
55. No x-intercepts; y-intercept $= 3$; vertex $= \left(\frac{1}{3}, \frac{8}{3}\right)$
57. $f(x) = (x + 2)^2 - 4$, $(-2, -4)$
59. $f(x) = -(x - 1)^2 + 1$, $(1, 1)$
61. $f(x) = 2(x - 2)^2 + 5$, $(2, 5)$
63. $f(x) = -3\left(x + \frac{1}{2}\right)^2 - \frac{5}{4}$, $\left(-\frac{1}{2}, -\frac{5}{4}\right)$
65. $f(x) = 4(x + 1)^2$, $(-1, 0)$ **67.** $x = h \pm \sqrt{-\dfrac{k}{a}}$

69.

t	0	$\frac{1}{2}$	1	$\frac{3}{2}$	2
$h(t)$	0	12	16	12	0

$h(t)$

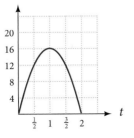

71.

Time (seconds)	Height (feet)
0	0
0.5	28
1	48
1.5	60
2	64
2.5	60
3	48
3.5	28
4	0

$h(t)$

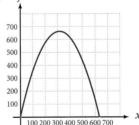

73. c   **75.** d   **77.** 96; 192   **79.** 0, 7   **81.** 2, 5
83. 1, 4   **85.** (1.5, 675)

PROBLEM SET 8.7
1. 675   **3. a.** 3,400 **b.** 3,400   **5. a.** 23 **b.** 23
7. a. y **b.** 630 ft

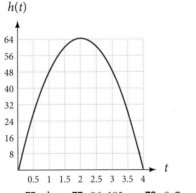

9. 2, 3 seconds   **11.** 0, 2 seconds
13. Yes, at $t = 1$ second   **15.** \$7 or \$10
17. 100 or 130 DVDs   **19.** 20 or 60 items
21. 875 patterns; maximum profit \$731.25
23. The ball is in her hand when $h(t) = 0$, which means
$t = 0$ or $t = 2$ seconds. Maximum height is $h(1) = 16$
feet.
25. Maximum $R = \$3,600$ when $p = \$6.00$
27. Maximum $R = \$7,225$ when $p = \$8.50$
29. \$300, \$1,800,000   **31.** \$30, \$900
33. $y = -\frac{1}{135}(x - 90)^2 + 60$
35. $y = -\frac{2}{315}(x - 315)^2 + 630$ **37.** $f(x) = x^2 + 4x - 5$
39. b   **41.** a   **43.** $-2, 4$ **45.** $-\frac{1}{2}, \frac{2}{3}$ **47.** 3

PROBLEM SET 8.8

1. $x < -3$ or $x > 2$ **3.** $-3 \leq x \leq 4$

5. $x \leq -3$ or $x \geq -2$ **7.** $\frac{1}{3} < x < \frac{1}{2}$

9. $-3 < x < 3$ **11.** $x \leq -\frac{3}{2}$ or $x \geq \frac{3}{2}$ 1

13. $-1 < x < \frac{3}{2}$ **15.** All real numbers   **17.** $\varnothing$

19. $2 < x < 3$ or $x > 4$ **21.** $x \leq -3$ or $-2 \leq x \leq -1$

23. $-4 < x \leq 1$ **25.** $-6 < x < \frac{8}{3}$ **27.** $x < 2$ or $x > 6$

29. $x < -3$ or $2 < x < 4$ **30.** $x < -2$ or $1 < x < 5$

31. $x > 4$ or $2 < x < 3$ **32.** $-3 < x < -2$ or $x > 1$

33. $5 \leq x < 6$

35. a. $-2 < x < 2$ **b.** $x < -2$ or $x > 2$ **c.** $x = -2$ or $x = 2$

37. a. $-2 < x < 5$ **b.** $x < -2$ or $x > 5$ **c.** $x = -2$ or $x = 5$

39. a. $x < -1$ or $1 < x < 3$ **b.** $-1 < x < 1$ or $x > 3$

   c. $x = -1$ or $x = 1$ or $x = 3$

41. $x \geq 4$; the width is at least 4 inches

43. $5 \leq p \leq 8$; charge at least \$5 but no more than \$8 for
   each radio

45. d   **47.** 1.5625   **49.** 0.6549   **51.** $\frac{2}{3}$

53. Possible solutions 1 and 6; only 6 checks; 6

55.

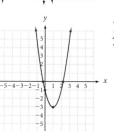

CHAPTER 8 TEST

1. $-\frac{9}{2}, \frac{1}{2}$ **2.** $3 \pm i\sqrt{2}$ **3.** $5 \pm 2i$ **4.** $1 \pm i\sqrt{2}$

5. $\frac{5}{2}, -\frac{5}{4} \pm \frac{5i\sqrt{3}}{4}$ **6.** $-1 \pm i\sqrt{5}$ **7.** $r = \pm\frac{\sqrt{A}}{8} - 1$

8. $2 \pm \sqrt{2}$ **9.** 9   **10.** $D = 81$; two rational solutions

11. $3x^2 - 13x - 10 = 0$ **12.** $x^2 + 4 = 0$

13. $\pm\sqrt{2}, \pm\frac{1}{2}i$ **14.** $\frac{1}{2}, 1$ **15.** $\frac{1}{4}, 9$

16. $t = \dfrac{7 + \sqrt{49 + 16h}}{16}$

17.

x-intercepts $= -2 \pm \sqrt{3}$;
y-intercept $= -1$;
vertex $= (-2, 3)$

18.

x-intercepts $= 1 \pm \dfrac{\sqrt{6}}{2}$;
y-intercept $= -1$
vertex $= (1, -3)$

19.

x-intercepts $= -1, 3$
y-intercept $= -3$
vertex $= (1, -4)$

20.

x-intercepts $= -2, 4$
y-intercept $= 8$
vertex $= (1, 9)$

21. $\frac{1}{2}$ or $\frac{3}{2}$ sec
22. 15 or 100 cups

23. profit $=$ \$900

24. $-2 \leq x \leq 3$

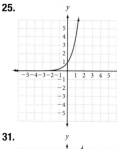

25. $x < -3$ or $x > \frac{1}{2}$

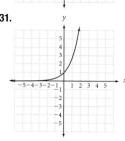

Chapter 9

PROBLEM SET 9.1

1. 1   **3.** 2   **5.** $\frac{1}{27}$ **7.** 13   **9.** $\frac{7}{12}$ **11.** $\frac{3}{16}$

13. 1.26   **15.** 0.16   **17.** 20.09   **19.** 0.61   **21.** 1.22

23. 23.14

25.

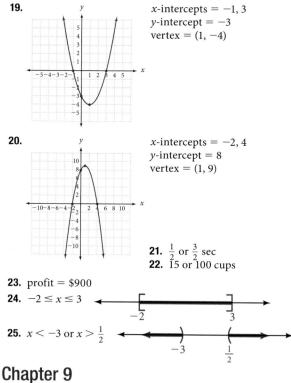

29.

31.

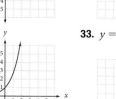

33. $y = 0, \{y \mid y > 0\}$

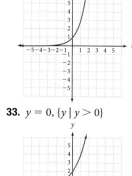

35. $y = 2, \{y \mid y > 2\}$ **37.** $y = 0, \{y \mid y < 0\}$

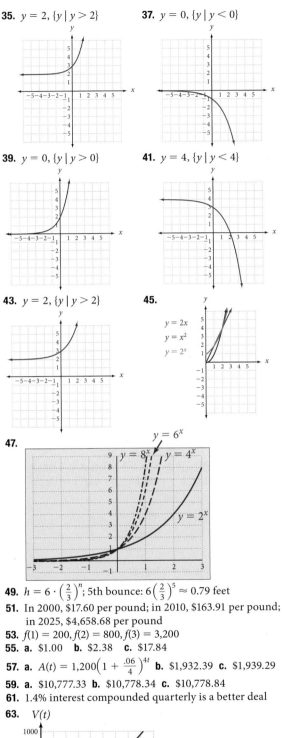

39. $y = 0, \{y \mid y > 0\}$ **41.** $y = 4, \{y \mid y < 4\}$

43. $y = 2, \{y \mid y > 2\}$ **45.**

47.

49. $h = 6 \cdot \left(\frac{2}{3}\right)^n$; 5th bounce: $6\left(\frac{2}{3}\right)^5 \approx 0.79$ feet

51. In 2000, \$17.60 per pound; in 2010, \$163.91 per pound; in 2025, \$4,658.68 per pound

53. $f(1) = 200, f(2) = 800, f(3) = 3{,}200$

55. a. \$1.00 **b.** \$2.38 **c.** \$17.84

57. a. $A(t) = 1{,}200\left(1 + \frac{.06}{4}\right)^{4t}$ **b.** \$1,932.39 **c.** \$1,939.29

59. a. \$10,777.33 **b.** \$10,778.34 **c.** \$10,778.84

61. 1.4% interest compounded quarterly is a better deal

63. $V(t)$

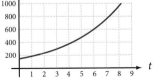

65. a. 188,934,028 **b.** $B(t)$

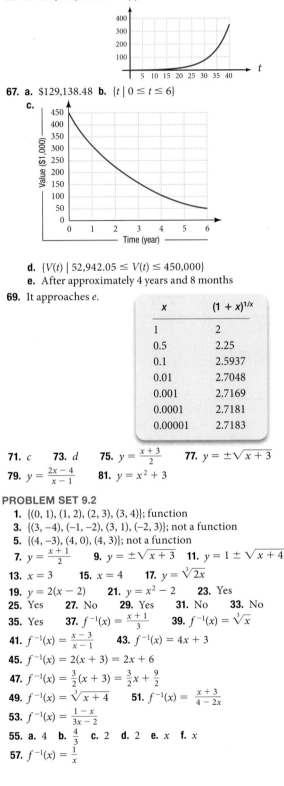

67. a. \$129,138.48 **b.** $\{t \mid 0 \le t \le 6\}$
c.

d. $\{V(t) \mid 52{,}942.05 \le V(t) \le 450{,}000\}$
e. After approximately 4 years and 8 months

69. It approaches e.

x	$(1 + x)^{1/x}$
1	2
0.5	2.25
0.1	2.5937
0.01	2.7048
0.001	2.7169
0.0001	2.7181
0.00001	2.7183

71. c **73.** d **75.** $y = \frac{x + 3}{2}$ **77.** $y = \pm\sqrt{x + 3}$

79. $y = \frac{2x - 4}{x - 1}$ **81.** $y = x^2 + 3$

PROBLEM SET 9.2

1. $\{(0, 1), (1, 2), (2, 3), (3, 4)\}$; function

3. $\{(3, -4), (-1, -2), (3, 1), (-2, 3)\}$; not a function

5. $\{(4, -3), (4, 0), (4, 3)\}$; not a function

7. $y = \frac{x + 1}{2}$ **9.** $y = \pm\sqrt{x + 3}$ **11.** $y = 1 \pm \sqrt{x + 4}$

13. $x = 3$ **15.** $x = 4$ **17.** $y = \sqrt[3]{2x}$

19. $y = 2(x - 2)$ **21.** $y = x^2 - 2$ **23.** Yes

25. Yes   **27.** No   **29.** Yes   **31.** No   **33.** No

35. Yes   **37.** $f^{-1}(x) = \frac{x + 1}{3}$ **39.** $f^{-1}(x) = \sqrt[3]{x}$

41. $f^{-1}(x) = \frac{x - 3}{x - 1}$ **43.** $f^{-1}(x) = 4x + 3$

45. $f^{-1}(x) = 2(x + 3) = 2x + 6$

47. $f^{-1}(x) = \frac{3}{2}(x + 3) = \frac{3}{2}x + \frac{9}{2}$

49. $f^{-1}(x) = \sqrt[3]{x + 4}$ **51.** $f^{-1}(x) = \frac{x + 3}{4 - 2x}$

53. $f^{-1}(x) = \frac{1 - x}{3x - 2}$

55. a. 4 **b.** $\frac{4}{3}$ **c.** 2 **d.** 2 **e.** x **f.** x

57. $f^{-1}(x) = \frac{1}{x}$

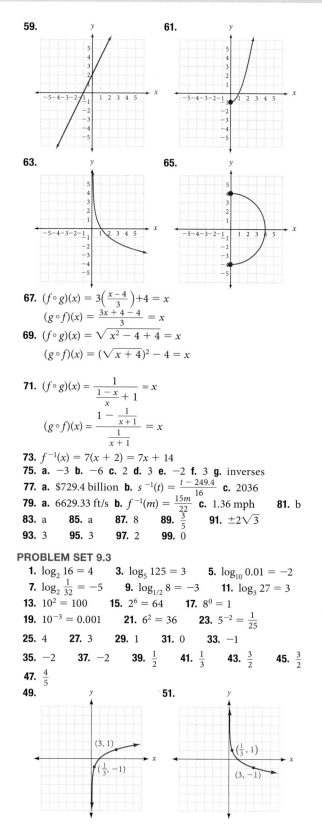

59.

61.

63.

65.

67. $(f \circ g)(x) = 3\left(\frac{x-4}{3}\right) + 4 = x$

$(g \circ f)(x) = \frac{3x + 4 - 4}{3} = x$

69. $(f \circ g)(x) = \sqrt{x^2 - 4 + 4} = x$

$(g \circ f)(x) = (\sqrt{x+4})^2 - 4 = x$

71. $(f \circ g)(x) = \dfrac{1}{\frac{1-x}{x} + 1} = x$

$(g \circ f)(x) = \dfrac{1 - \frac{1}{x+1}}{\frac{1}{x+1}} = x$

73. $f^{-1}(x) = 7(x + 2) = 7x + 14$

75. a. -3 **b.** -6 **c.** 2 **d.** 3 **e.** -2 **f.** 3 **g.** inverses

77. a. \$729.4 billion **b.** $s^{-1}(t) = \frac{t - 249.4}{16}$ **c.** 2036

79. a. 6629.33 ft/s **b.** $f^{-1}(m) = \frac{15m}{22}$ **c.** 1.36 mph   **81. b**

83. a   **85.** a   **87.** 8   **89.** $\frac{3}{5}$ **91.** $\pm 2\sqrt{3}$

93. 3   **95.** 3   **97.** 2   **99.** 0

PROBLEM SET 9.3

1. $\log_2 16 = 4$ **3.** $\log_5 125 = 3$ **5.** $\log_{10} 0.01 = -2$

7. $\log_2 \frac{1}{32} = -5$ **9.** $\log_{1/2} 8 = -3$ **11.** $\log_3 27 = 3$

13. $10^2 = 100$ **15.** $2^6 = 64$ **17.** $8^0 = 1$

19. $10^{-3} = 0.001$ **21.** $6^2 = 36$ **23.** $5^{-2} = \frac{1}{25}$

25. 4   **27.** 3   **29.** 1   **31.** 0   **33.** -1

35. -2 **37.** -2 **39.** $\frac{1}{2}$ **41.** $\frac{1}{3}$ **43.** $\frac{3}{2}$ **45.** $\frac{3}{2}$

47. $\frac{4}{5}$

49.

51.

53.

55.

57. $f^{-1}(x) = 4^x$ **59.** $f^{-1}(x) = \log_{1/8} x$

61. $\{x \mid x > 0\}$ **63.** $\{x \mid x > -6\}$ **65.** $\{x \mid x < 1\}$

67. $\left\{ x \mid x > -\frac{3}{2} \right\}$ **69.** $x = 3$ **71.** $x = 0$

73. $x = 0$ **75.** $y = 3^x$ **77.** $y = \log_{1/3} x$

79.

Prefix	Multiplying Factor	$\log_{10}$ (Multiplying Factor)
Nano	0.000 000 001	-9
Micro	0.000 001	-6
Deci	0.1	-1
Giga	1,000,000,000	9
Peta	1,000,000,000,000,000	15

81. 2   **83.** 10^8 times as large   **85.** 120   **87.** a

89. a   **91.** 4   **93.** $-4, 2$ **95.** $-\frac{11}{8}$

97. $2^3 = (x + 2)(x)$ **99.** $3^4 = \frac{x-2}{x+1}$

PROBLEM SET 9.4

1. 0   **3.** 4   **5.** 3   **7.** $\sqrt{2}$ **9.** 12   **11.** 81

13. 4   **15.** 1   **17.** -2 **19.** 0   **21.** $\frac{1}{2}$

23. $\log_3 4 + \log_3 x$ **25.** $\log_6 5 - \log_6 x$

27. $5 \log_2 y$ **29.** $\frac{1}{3} \log_9 z$ **31.** $2 \log_6 x + 4 \log_6 y$

33. $\frac{1}{2} \log_5 x + 4 \log_5 y$ **35.** $\log_b x + \log_b y - \log_b z$

37. $\log_{10} 4 - \log_{10} x - \log_{10} y$

39. $2 \log_{10} x + \log_{10} y - \frac{1}{2} \log_{10} z$

41. $3 \log_{10} x + \frac{1}{2} \log_{10} y - 4 \log_{10} z$

43. $\frac{2}{3} \log_b x + \frac{1}{3} \log_b y - \frac{4}{3} \log_b z$

45. $\frac{2}{3} \log_3 x + \frac{1}{3} \log_3 y - 2 \log_3 z$

47. $2\log_a 2 + 5\log_a x - 2\log_a 3 - 2$

49. $2\log_4 x + \log_4 (x + 2)$ **51.** $\log_b 5 + 7$

53. $1 + 9\log_8 x$ **55.** $3 + 5\log_2 (x - 1)$

57. $2\log_6 x + 3\log_6 z - \frac{1}{2}\log_6 (x + z)$

59. $\frac{1}{2}\log_9 (x + 3) - \frac{1}{2}\log_9 (x - 3)$ **61.** 2

63. 2   **65.** 1   **67.** $\log_b xz$ **69.** $\log_3 \frac{x^2}{y^3}$

71. $\log_{10} \sqrt{x}\sqrt[3]{y}$ **73.** $\log_2 \frac{x^3\sqrt{y}}{z}$ **75.** $\log_2 \frac{\sqrt{x}}{y^3 z^4}$

77. $\log_{10} \frac{x^{3/2}}{y^{3/4} z^{4/5}}$ **79.** $\log_5 \frac{\sqrt{x} \cdot \sqrt[3]{y^2}}{z^4}$

81. $\log_b x^2 (x - 10)^3$ **83.** $\log_6 \frac{x^4 z^5}{(y + z)^2}$ **85.** $\log_3 \frac{x-4}{x+4}$

89. a. 1.602   **b.** 2.505   **c.** 3.204

90. a. iii   **b.** iv   **c.** vi   **d.** v   **e.** i   **f.** ii

91. $pH = 6.1 + \log_{10} x - \log_{10} y$ **93.** 2   **95.** a

97. 1   **99.** 1   **101.** 4

PROBLEM SET 9.5

1. 0   **3.** 1   **5.** 4   **7.** 5   **9.** $\frac{3}{2}$ **11.** -3
13. 2.5775   **15.** 5.8435   **17.** -0.3706 **19.** -1.0642
21. x **23.** x **25.** $3x$ **27.** x^4 **29.** $\ln 10 + 3t$
31. $\ln A - 2t$ **33.** $2 + 3t\log 1.01$ **35.** $rt + \ln P$
37. $3 - \log 4.2$ **39.** $\log x(x-2)$ **41.** $\ln \frac{x+1}{x+4}$
43. $\log \frac{x^2}{y^5}$ **45.** 2.7080   **47.** -1.0986 **49.** 2.1972
51. 2.7724

53.

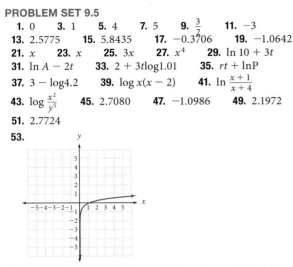

55. 1.3333   **57.** 0.7500   **59.** 1.3917   **61.** 0.7186
63. 2.6356   **65.** 4.1632   **67.** 5.0

69.

Location	Date	Magnitude M	Shockwave T
Moresby Island	Jan. 23	4.0	1.00×10^4
Vancouver Island	Apr. 30	5.3	1.99×10^5
Quebec City	June 29	3.2	1.58×10^3
Mould Bay	Nov. 13	5.2	1.58×10^5
St. Lawrence	Dec. 14	3.7	5.01×10^3

Source: National Resources Canada, National Earthquake Hazards Program

71. 3.19   **73.** d   **75.** a   **77.** $\frac{7}{10}$
79. 3.1250   **81.** 1.2575   **83.** $t\log 1.05$ **85.** $.05t$
87. 2.5×10^{-6} **89.** 51

PROBLEM SET 9.6

1. 1.4650   **3.** 0.6826   **5.** -1.5440 **7.** -0.6477
9. -0.3333 **11.** 2.000   **13.** -0.1845 **15.** 0.1845
17. 1.6168   **19.** 2.1131   **21.** -1.0000 **23.** 1.2275
25. 0.3054   **27.** 42.5528   **29.** 6.0147   **31.** 9
33. $\frac{1}{125}$ **35.** 2   **37.** $\sqrt[3]{5}$ **39.** 6   **41.** $\frac{1}{64}$
43. 10   **45.** $\frac{1}{100}$ **47.** $\frac{1}{e}$ **49.** 10^{10} **51.** 10^{-20}
53. 1,000   **55.** 758.93   **57.** 0.0075893   **59.** 23.460
61. 0.0047496   **63.** $\frac{2}{3}$ **65.** 18   **67.** 3   **69.** 3
71. 4   **73.** 4   **75.** 1   **77.** 0   **79.** $\frac{3}{2}$ **81.** 27
83. $\frac{5}{3}$ **85.** 25   **87.** $\frac{1}{8}$ **89.** 11.72 years
91. 9.25 years   **93.** 8.75 years   **95.** 18.58 years
97. 11.55 years   **99.** 18.31 years   **101.** 11.45 years
103. 4.27 days   **105.** 2043   **107.** 3.16×10^5
109. 2.00×10^8 **111.** 1.78×10^{-5} **113.** 2023
115. 2030   **117.** 2029
119. It has been approximately 4,127 years.   **121.** 12.9%
123. 5.3%   **125.** a   **127.** d   **129.** $\left(\frac{3}{2}, -\frac{67}{4}\right)$, lowest
131. $\left(\frac{3}{2}, \frac{27}{2}\right)$, highest   **133.** 2 seconds, 104 feet

CHAPTER 9 TEST

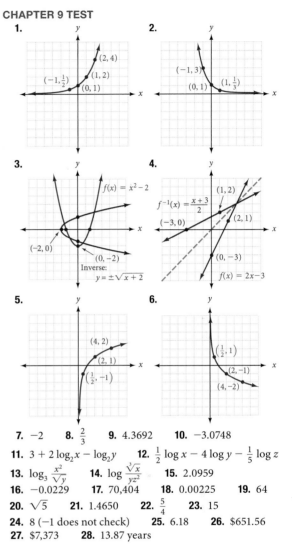

1. (graph with points $(-1, \frac{1}{2})$, $(0, 1)$, $(1, 2)$, $(2, 4)$)

2. (graph with points $(-1, 3)$, $(0, 1)$, $(1, \frac{1}{3})$)

3. (graph of $f(x) = x^2 - 2$ with points $(-2, 0)$, $(0, -2)$; Inverse: $y = \pm\sqrt{x+2}$)

4. (graph of $f(x) = 2x - 3$ and $f^{-1}(x) = \frac{x+3}{2}$ with points $(1, 2)$, $(2, 1)$, $(0, -3)$)

5. (graph with points $(\frac{1}{2}, -1)$, $(2, 1)$, $(4, 2)$)

6. (graph with points $(\frac{1}{2}, 1)$, $(2, -1)$, $(4, -2)$)

7. -2 **8.** $\frac{2}{3}$ **9.** 4.3692   **10.** -3.0748
11. $3 + 2\log_2 x - \log_2 y$ **12.** $\frac{1}{2}\log x - 4\log y - \frac{1}{5}\log z$
13. $\log_3 \frac{x^2}{\sqrt{y}}$ **14.** $\log \frac{\sqrt[3]{x}}{yz^2}$ **15.** 2.0959
16. -0.0229 **17.** 70,404   **18.** 0.00225   **19.** 64
20. $\sqrt{5}$ **21.** 1.4650   **22.** $\frac{5}{4}$ **23.** 15
24. 8 (-1 does not check)   **25.** 6.18   **26.** $651.56
27. $7,373   **28.** 13.87 years

Chapter 10

PROBLEM SET 10.1

1. 4, 7, 10, 13, 16   **3.** 3, 7, 11, 15, 19   **5.** 1, 2, 3, 4, 5
7. 4, 7, 12, 19, 28   **9.** $\frac{1}{4}, \frac{2}{5}, \frac{3}{6}, \frac{4}{7}, \frac{5}{8}$ **11.** 1, $\frac{1}{4}, \frac{1}{9}, \frac{1}{16}, \frac{1}{25}$
13. 2, 4, 8, 16, 32   **15.** 2, $\frac{3}{2}, \frac{4}{3}, \frac{5}{4}, \frac{6}{5}$
17. $-2, 4, -8, 16, -32$ **19.** 3, 5, 3, 5, 3
21. 1, $-\frac{2}{3}, \frac{3}{5}, -\frac{4}{7}, \frac{5}{9}$ **23.** $\frac{1}{2}, 1, \frac{9}{8}, 1, \frac{25}{32}$ **25.** 80
27. $\frac{1}{203}$ **29.** 3, $-9, 27, -81, 243$ **31.** 1, 5, 13, 29, 61
33. 2, 3, 5, 9, 17   **35.** 5, 11, 29, 83, 245   **37.** 4, 4, 4, 4, 4
39. $a_n = 4n$ **41.** $a_n = n^2$ **43.** $a_n = 2^{n+1}$
45. $a_n = \frac{1}{2^{n+1}}$ **47.** $a_n = 3n + 2$ **49.** $a_n = -4n + 2$
51. $a_n = (-2)^{n-1}$ **53.** $a_n = \log_{n+1}(n+2)$
55. **a.** $28,000, $29,120, $30,284.80, $31,496.19, $32,756.04
   b. $a_n = 28,000(1.04)^{n-1}$
57. **a.** 16 feet, 48 feet, 80 feet, 112 feet, 144 feet
   b. 400 feet   **c.** No

59. a. $10, 8, \frac{32}{5}$ **b.** $a_n = 10\left(\frac{4}{5}\right)^{n-1}$ **c.** 1.34 ft

61. c   **63.** 30   **65.** 40   **67.** 18   **69.** $-\frac{20}{81}$

71. $\frac{21}{10}$

PROBLEM SET 10.2

1. 36   **3.** 11   **5.** 18   **7.** $\frac{163}{60}$ **9.** 60   **11.** 40

13. 44   **15.** $-\frac{11}{32}$ **17.** $\frac{21}{10}$

19. $(x + 1) + (x + 2) + (x + 3) + (x + 4) + (x + 5)$

21. $(x - 2) + (x - 2)^2 + (x - 2)^3 + (x - 2)^4$

23. $\frac{x+1}{x-1} + \frac{x+2}{x-1} + \frac{x+3}{x-1} + \frac{x+4}{x-1} + \frac{x+5}{x-1}$

25. $(x + 3)^3 + (x + 4)^4 + (x + 5)^5 + (x + 6)^6 + (x + 7)^7 + (x + 8)^8$

27. $(x - 6)^6 + (x - 8)^7 + (x - 10)^8 + (x - 12)^9$

29. $\sum_{i=1}^{4} 2^i$ **31.** $\sum_{i=2}^{6} 2^i$ **33.** $\sum_{i=1}^{5} (4i + 1)$ **35.** $\sum_{i=2}^{5} -(-2)^i$

37. $\sum_{i=3}^{7} \frac{i}{i+1}$ **39.** $\sum_{i=1}^{4} \frac{i}{2i+1}$ **41.** $\sum_{i=6}^{9} (x - 2)^i$

43. $\sum_{i=1}^{4}\left(1 + \frac{i}{x}\right)^{i+1}$ **45.** $\sum_{i=3}^{5} \frac{x}{x+i}$ **47.** $\sum_{i=2}^{4} x^i(x + i)$

49. a. $0.3 + 0.03 + 0.003 + 0.0003 + \ldots$
   b. $0.2 + 0.02 + 0.002 + 0.0002 + \ldots$
   c. $0.27 + 0.0027 + 0.000027 + \ldots$

51. seventh second: 208 feet; total: 784 feet

53. a. $16 + 48 + 80 + 112 + 144$ **b.** $\sum_{i=1}^{5} (32i - 16)$

55. b   **57.** 74   **59.** $\frac{55}{2}$ **61.** $2n + 1$ **63.** $(3, 2)$

PROBLEM SET 10.3

1. Arithmetic, $d = 1$ **3.** Not arithmetic

5. Arithmetic, $d = -5$ **7.** Not arithmetic

9. Arithmetic, $d = \frac{2}{3}$ **11.** $a_n = 4n - 1; a_{24} = 95$

13. $a_{10} = -12; S_{10} = -30$ **15.** $a_1 = 7; d = 2; a_{30} = 65$

17. $a_1 = 12; d = 2; a_{20} = 50; S_{20} = 620$

19. $a_{20} = 79, S_{20} = 820$ **21.** $a_{40} = 122, S_{40} = 2540$

23. $a_1 = 13, d = -6$ **25.** $a_{85} = -238$

27. $d = \frac{16}{19}, a_{39} = 28$ **29.** 20,300   **31.** -158

33. $a_{10} = 5; S_{10} = \frac{55}{2}$

35. a. $18,000, $14,700, $11,400, $8,100, $4,800 **b.** $-$3,300
   c. $V(t)$ **d.** $9,750
   e. $a_0 = 18,000$,
      $a_n = a_{n-1} - 3,300$ for $n \geq 1$

37. a. $1,500, 1,460, 1,420, 1,380, 1,340, 1,300$
   b. It is arithmetic because the same amount is subtracted from each succeeding term.
   c. $a_n = 1,500 - (n - 1)40 = 1,540 - 40n$

39. a. $1, 3, 6, 10, 15, 21, 28, 36, 45, 55, 66, 78, 91, 105, 120$
   b. $a_1 = 1; a_n = n + a_{n-1}$ for $n \geq 2$
   c. No, it is not arithmetic because the same amount is not
   added to each term.

41. a. $a_n = 32n - 16$ **b.** $a_{10} = 304$ feet **c.** 1,600 feet

43. a   **45.** d   **47.** $\frac{1}{16}$ **49.** $\sqrt{3}$ **51.** 2^n

53. r^3 **55.** $\frac{2}{5}$ **57.** -255

PROBLEM SET 10.4

1. 5   **3.** $\frac{1}{3}$ **5.** Not geometric   **7.** -2

9. Not geometric   **11.** $a_n = 4 \cdot 3^{n-1}$

13. $a_6 = -2\left(-\frac{1}{2}\right)^5 = \frac{1}{16}$ **15.** $a_{20} = 3(-1)^{19} = -3$

17. 10,230   **19.** $S_{20} = \frac{1((-1)^{20} - 1)}{-1 - 1} = 0$

21. $a_8 = \frac{1}{5}\left(\frac{1}{2}\right)^7 = \frac{1}{640}$ **23.** $-\frac{31}{32}$ **25.** $32, 62 + 31\sqrt{2}$

27. $\frac{1}{1000}$, 111.111   **29.** $r = \pm 2$ **31.** $a_8 = 384, S_8 = 255$

33. $r = 2$ **35.** $S = \dfrac{\frac{1}{2}}{1 - \frac{1}{2}} = 1$ **37.** 8   **39.** 4

41. $\frac{8}{9}$ **43.** $S = \dfrac{\frac{2}{5}}{1 - \frac{2}{5}} = \frac{2}{3}$ **45.** $S = \dfrac{\frac{3}{4}}{1 - \frac{1}{3}} = \frac{9}{8}$

51. a. $450,000, $315,000, $220,500, $154,350, $108,045
   b. 0.7 **c.** $V(t)$

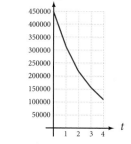

   d. $130,000 **e.** $a_0 = 450,000, a_n = 0.7a_{n-1}$

53. a. $\frac{1}{2}$ **b.** $\frac{364}{729}$ **c.** $\frac{1}{1,458}$

55. a. 0.004 inches **b.** 0.064 inches **c.** 67,108.864 inches

57. a. $60,000; $64,200; $68,694; $73,503; $78,648;
   b. $a_n = 60,000(1.07)^{n-1}$ **c.** $828,987

59. a   **61.** d   **63.** $x + y$ **65.** $x^3 + 3x^2y + 3xy^2 + y^3$

67. 21

PROBLEM SET 10.5

1. $x^4 + 8x^3 + 24x^2 + 32x + 16$

3. $x^6 + 6x^5y + 15x^4y^2 + 20x^3y^3 + 15x^2y^4 + 6xy^5 + y^6$

5. $32x^5 + 80x^4 + 80x^3 + 40x^2 + 10x + 1$

7. $x^5 - 10x^4y + 40x^3y^2 - 80x^2y^3 + 80xy^4 - 32y^5$

9. $81x^4 - 216x^3 + 216x^2 - 96x + 16$

11. $64x^3 - 144x^2y + 108xy^2 - 27y^3$

13. $x^8 + 8x^6 + 24x^4 + 32x^2 + 16$

15. $x^6 + 3x^4y^2 + 3x^2y^4 + y^6$

17. $16x^4 + 96x^3y + 216x^2y^2 + 216xy^3 + 81y^4$

19. $\frac{x^3}{8} + \frac{x^2y}{4} + \frac{xy^2}{6} + \frac{y^3}{27}$ **21.** $\frac{x^5}{8} - 3x^2 + 24x - 64$

23. $\frac{x^4}{81} + \frac{2x^3y}{27} + \frac{x^2y^2}{6} + \frac{xy^3}{6} + \frac{y^4}{16}$ **25.** 720

27. 3,628,800   **29.** 1   **31.** 8   **33.** 1,365

35. 77,520   **37.** $x^9 + 18x^8 + 144x^7 + 672x^6$

39. $x^{10} - 10x^9y + 45x^8y^2 - 120x^7y^3$

41. $x^{25} + 75x^{24} + 2,700x^{23} + 62,100x^{22}$

43. $x^{60} - 120x^{59} + 7,080x^{58} - 273,760x^{57}$

45. $x^{18} - 18x^{17}y + 153x^{16}y^2 - 816x^{15}y^3$

47. $x^{15} + 15x^{14} + 105x^{13}$ **49.** $x^{12} - 12x^{11}y + 66x^{10}y^2$

51. $x^{20} + 40x^{19} + 760x^{18}$ **53.** $x^{100} + 200x^{99}$

55. $x^{50} + 50x^{49}y$ **57.** $51,963,120x^4y^8$ **59.** $3,360x^6$

61. $-25,344x^7$ **63.** $2,700x^{23}y^2$

65. $\binom{20}{11}(2x)^9(5y)^{11}$

67. $x^{20}y^{10} - 30x^{18}y^9 + 405x^{16}y^8$ **69.** $\frac{21}{128}$ **71.** b

73. a   **75.** $x \approx 1.18$ **77.** $x = \frac{5}{6}$ **79.** ≈ 17.5 years

81. 1.56   **83.** 8.66   **85.** $t = -\frac{1}{5}\ln\left(\frac{A}{P}\right)$

CHAPTER 10 TEST

1. $-2, 1, 4, 7, 10$ **2.** $3, 7, 11, 15, 19$ **3.** $2, 5, 10, 17, 26$

4. $2, 16, 54, 128, 250$ **5.** $2, \frac{3}{4}, \frac{4}{9}, \frac{5}{16}, \frac{6}{25}$

6. $4, -8, 16, -32, 64$ **7.** $a_n = 4n + 2$

8. $a_n = 2^{n-1}$ **9.** $a_n = \frac{1}{2^n}$

10. $a_n = (-3)^n$ **11. a.** 90 **b.** 53 **c.** 130

12. 3   **13.** ± 6 **14.** 320   **15.** 25

16. $S_{50} = \frac{3(2^{50} - 1)}{2 - 1} = 3(2^{50} - 1)$ **17.** $\frac{3}{4}$

18. $x^4 - 12x^3 + 54x^2 - 108x + 81$

19. $32x^5 - 80x^4 + 80x^3 - 40x^2 + 10x - 1$

20. 220   **21.** 3003   **22.** $x^{20} - 20x^{19} + 190x^{18}$

23. $-108{,}864x^3y^5$

Chapter 11

PROBLEM SET 11.1

1. 5   **3.** $\sqrt{106}$ **5.** $\sqrt{61}$ **7.** $\sqrt{140}$ **9.** 3 or -1

11. 0 or 6   **13.** $x = -1 \pm 4\sqrt{2}$

15. $(x - 3)^2 + (y + 2)^2 = 9$ **17.** $(x + 5)^2 + (y + 1)^2 = 5$

19. $x^2 + (y + 5)^2 = 1$ **21.** $x^2 + y^2 = 4$

23. center $= (0, 0)$; radius $= 2$

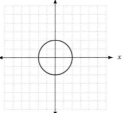

25. center $= (1, 3)$; radius $= 5$

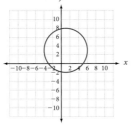

27. center $= (-2, 4)$; radius $= 2\sqrt{2}$

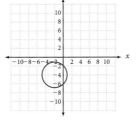

29. center $= (-2, 4)$; radius $= \sqrt{17}$

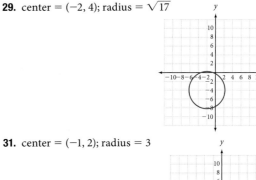

31. center $= (-1, 2)$; radius $= 3$

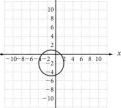

33. center $= (0, 3)$; radius $= 4$

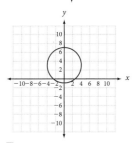

35. center $= (-1, 0)$; radius $= \sqrt{2}$

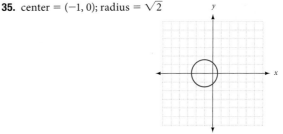

37. center $= (2, 3)$; radius $= 3$

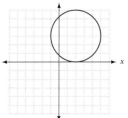

39. center $= \left(-1, -\frac{1}{2}\right)$; radius $= 2$

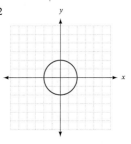

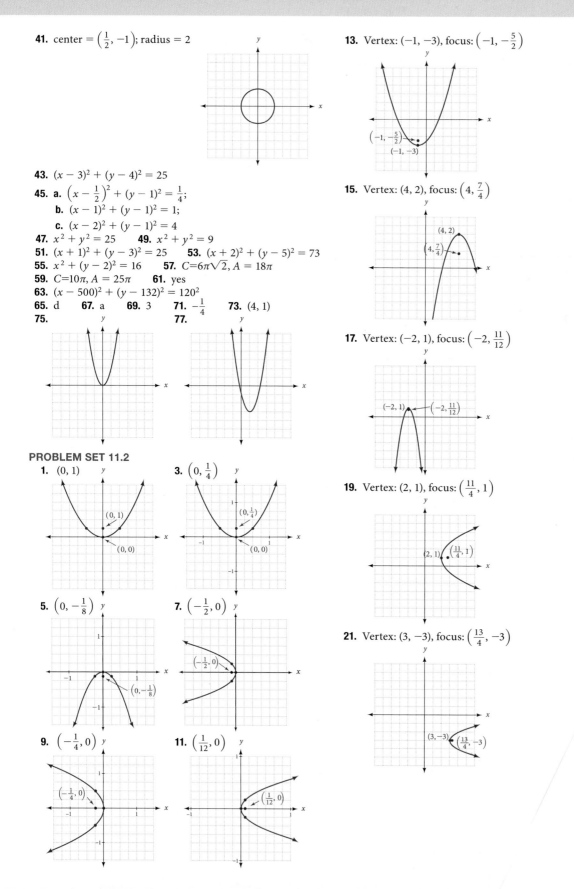

41. center $= \left(\frac{1}{2}, -1\right)$; radius $= 2$

43. $(x - 3)^2 + (y - 4)^2 = 25$

45. a. $\left(x - \frac{1}{2}\right)^2 + (y - 1)^2 = \frac{1}{4}$;
 b. $(x - 1)^2 + (y - 1)^2 = 1$;
 c. $(x - 2)^2 + (y - 1)^2 = 4$

47. $x^2 + y^2 = 25$ **49.** $x^2 + y^2 = 9$

51. $(x + 1)^2 + (y - 3)^2 = 25$ **53.** $(x + 2)^2 + (y - 5)^2 = 73$

55. $x^2 + (y - 2)^2 = 16$ **57.** $C = 6\pi\sqrt{2}$, $A = 18\pi$

59. $C = 10\pi$, $A = 25\pi$ **61.** yes

63. $(x - 500)^2 + (y - 132)^2 = 120^2$

65. d   **67.** a   **69.** 3   **71.** $-\frac{1}{4}$ **73.** $(4, 1)$

75.   **77.**

PROBLEM SET 11.2

1. $(0, 1)$ **3.** $\left(0, \frac{1}{4}\right)$

5. $\left(0, -\frac{1}{8}\right)$ **7.** $\left(-\frac{1}{2}, 0\right)$

9. $\left(-\frac{1}{4}, 0\right)$ **11.** $\left(\frac{1}{12}, 0\right)$

13. Vertex: $(-1, -3)$, focus: $\left(-1, -\frac{5}{2}\right)$

15. Vertex: $(4, 2)$, focus: $\left(4, \frac{7}{4}\right)$

17. Vertex: $(-2, 1)$, focus: $\left(-2, \frac{11}{12}\right)$

19. Vertex: $(2, 1)$, focus: $\left(\frac{11}{4}, 1\right)$

21. Vertex: $(3, -3)$, focus: $\left(\frac{13}{4}, -3\right)$

23. Vertex: $(5, 2)$, focus: $\left(\frac{59}{12}, 2\right)$

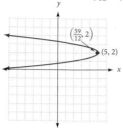

25. Vertex: $(2, -3)$, focus: $\left(2, -\frac{11}{4}\right)$

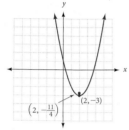

27. Vertex: $(-1, 3)$, focus: $(-1, 1)$

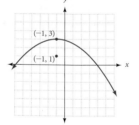

29. Vertex: $(3, -2)$, focus: $\left(\frac{13}{4}, -2\right)$

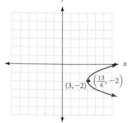

31. Vertex: $(-1, 1)$, focus: $\left(-\frac{5}{2}, 1\right)$

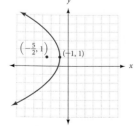

33. 6.25 inches   **34.** 1.6 cm   **35.** 3 inches

36. $\frac{5}{14}$ meters ≈ 0.357 meters   **37.** d   **38.** a

39. d   **40.** $y = \pm 3$ **41.** $x = \pm 5$ **42.** $y = \pm 2i$

43. $x = \pm 4i$ **44.** $x = \pm 3i$ **45.** $y = \pm 10$

47. $\frac{x^2}{4} + \frac{y^2}{25}$

48. x-intercept 4, y-intercept -3

49. x-intercepts $\frac{1}{3}$, -2; y-intercept -2 **50.** $\pm\frac{12}{5}$

51. $\pm\frac{9}{5}$

PROBLEM SET 11.3

1.

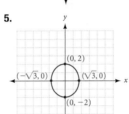

3.

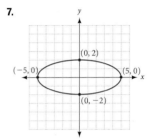

5.

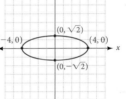

7.

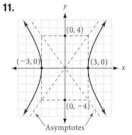

9.

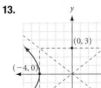

11.

13.

15.

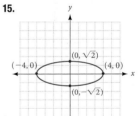

17.

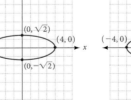

19.

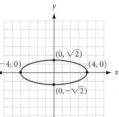

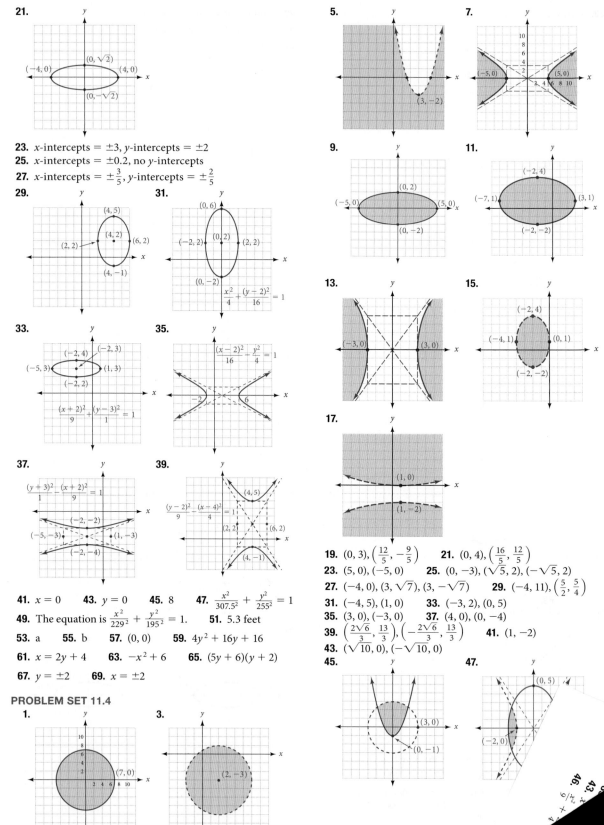

21.

23. x-intercepts $= \pm 3$, y-intercepts $= \pm 2$
25. x-intercepts $= \pm 0.2$, no y-intercepts
27. x-intercepts $= \pm \frac{3}{5}$, y-intercepts $= \pm \frac{2}{5}$

29.

31.

33.

35.

37.

39.

41. $x = 0$ **43.** $y = 0$ **45.** 8   **47.** $\dfrac{x^2}{307.5^2} + \dfrac{y^2}{255^2} = 1$

49. The equation is $\dfrac{x^2}{229^2} + \dfrac{y^2}{195^2} = 1$.   **51.** 5.3 feet

53. a   **55.** b   **57.** $(0, 0)$ **59.** $4y^2 + 16y + 16$

61. $x = 2y + 4$ **63.** $-x^2 + 6$ **65.** $(5y + 6)(y + 2)$

67. $y = \pm 2$ **69.** $x = \pm 2$

PROBLEM SET 11.4

1.

3.

5.

7.

9.

11.

13.

15.

17.

19. $(0, 3), \left(\dfrac{12}{5}, -\dfrac{9}{5} \right)$ **21.** $(0, 4), \left(\dfrac{16}{5}, \dfrac{12}{5} \right)$

23. $(5, 0), (-5, 0)$ **25.** $(0, -3), (\sqrt{5}, 2), (-\sqrt{5}, 2)$

27. $(-4, 0), (3, \sqrt{7}), (3, -\sqrt{7})$ **29.** $(-4, 11), \left(\dfrac{5}{2}, \dfrac{5}{4} \right)$

31. $(-4, 5), (1, 0)$ **33.** $(-3, 2), (0, 5)$

35. $(3, 0), (-3, 0)$ **37.** $(4, 0), (0, -4)$

39. $\left(\dfrac{2\sqrt{6}}{3}, \dfrac{13}{3} \right), \left(-\dfrac{2\sqrt{6}}{3}, \dfrac{13}{3} \right)$ **41.** $(1, -2)$

43. $(\sqrt{10}, 0), (-\sqrt{10}, 0)$

45.

47.

43. $\frac{x^2}{9} + 4$ **46.**

49.

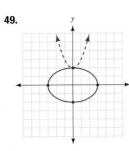

51.

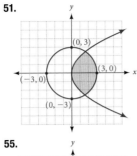

7.

8.

53.

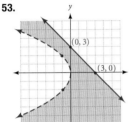

55.

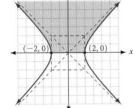

9.

10.

57.

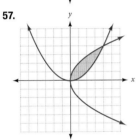

59.

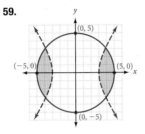

11.

12.

61.

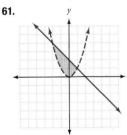

13.

14.

63. 8, 5 or −8, −5 or 8, −5 or −8, 5
65. a. $(-4, 4\sqrt{3})$ and $(-4, -4\sqrt{3})$
  b. $(4, 4\sqrt{3})$ and $(4, -4\sqrt{3})$

67. b   **69.** d   **71.** $x^4 + 8x^3 + 24x^2 + 32x + 16$

73. $8x^3 + 12x^2y + 6xy^2 + y^3$ **75.** $x^{50} + 150x^{49}$

15. $(0, 5), (4, -3)$ **16.** $(0, -4), (-\sqrt{7}, 3), (\sqrt{7}, 3)$

17.

18.

CHAPTER 11 TEST

1. −5, 3   **2.** $(x + 2)^2 + (y - 4)^2 = 9$ **3.** $x^2 + y^2 = 25$

4. $(5, -3), \sqrt{39}$

5.

6.

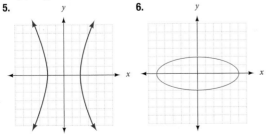

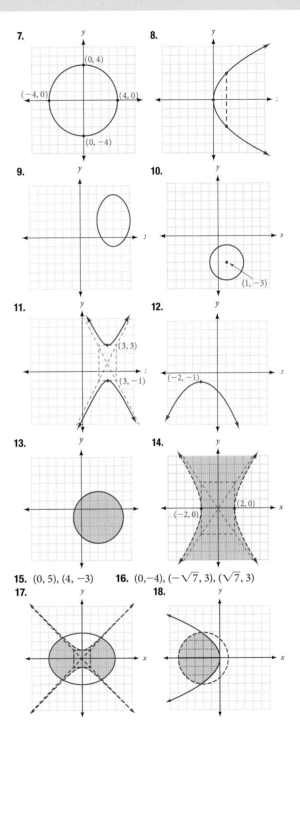

Index